D0218683

QUANTUM THEORY

QUANTUM THEORY

DAVID BOHM

Emeritus Professor of Theoretical Physics,

University of London

DOVER PUBLICATIONS, INC., New York

This Dover edition, first published in 1989, is an unabridged and unaltered republication of the work first published by Prentice-Hall, Inc., Englewood Cliffs, New Jersey, in 1951.

Manufactured in the United States of America
Dover Publications, Inc., 31 East 2nd Street, Mineola, N.Y. 11501

Library of Congress
Cataloging-in-Publication Data

Bohm, David.
Quantum theory / David Bohm.
p. cm.
Reprint. Originally published: New York : Prentice-Hall, 1951.
Bibliography: p.
Includes index.
ISBN 0-486-65969-0
1. Quantum theory. I. Title.
QC174.12.B632 1989
530.1′2—dc19 89-31187
CIP

PREFACE

The quantum theory is the result of long and successful efforts of physicists to account correctly for an extremely wide range of experimental results, which the previously existing classical theory could not even begin to explain. It is not generally realized, however, that the quantum theory represents a radical change, not only in the content of scientific knowledge, but also in the fundamental conceptual framework in terms of which such knowledge can be expressed. The true extent of this change of conceptual framework has perhaps been obscured by the contrast between the relatively pictorial and easily imagined terms in which classical theory has always been expressed, with the very abstract and mathematical form in which quantum theory obtained its original development. So strong is this contrast that an appreciable number of physicists were led to the conclusion that the quantum properties of matter imply a renunciation of the possibility of their being understood in the customary imaginative sense, and that instead, there remains only a self-consistent mathematical formalism which can, in some mysterious way, predict correctly the numerical results of actual experiments. Nevertheless, with the further development of the physical interpretation of the theory (primarily as a result of the work of Niels Bohr), it finally became possible to express the results of the quantum theory in terms of comparatively qualitative and imaginative concepts, which are, however, of a totally different nature from those appearing in the classical theory. To provide such a formulation of the quantum theory at a relatively elementary level is the central aim of this book.

The precise nature of the new quantum-theoretical concepts will be developed throughout the book, principally in Chapters 6, 7, 8, 22, and 23, but the most important conceptual changes can be briefly summarized here. First, the classical concept of a continuous and precisely defined trajectory is fundamentally altered by the introduction of a description of motion in terms of a series of indivisible transitions. Second, the rigid determinism of classical theory is replaced by the concept of causality as an approximate and statistical trend. Third, the classical assumption that elementary particles have an "intrinsic" nature which can never change is replaced by the assumption that they can act either like waves or like particles, depending on how they are treated by the surrounding environment. The application of these three new concepts results in the breakdown of an assumption which lies behind much of our customary language and way of thinking; namely, that the world can correctly be analyzed into distinct parts each having a separate existence, but working together according to exact causal laws to form the whole. Instead, quantum concepts imply that the world acts more like a single indivisible unit, in which even the "intrinsic" nature of each part (wave or particle) depends to some degree on its relationship to its surroundings. It is only at the microscopic (or quantum) level, however, that the indivisible unity of the various parts of the world produces significant effects, so that at the macroscopic (or classical) level, the parts act, to a very high degree of approximation, as if they did have a completely separate existence.

It has been the author's purpose throughout this book to present the main ideas of the quantum theory in non-mathematical terms. Experience shows, however, that some mathematics is needed in order to express these ideas in a more precisely defined form, and to indicate how typical problems in the quantum theory can be solved. The general plan adopted in this book has therefore been to supplement a basically qualitative and physical presentation of fundamental principles with a broad range of specific applications that are worked out in considerable mathematical detail.

In accordance with the general plan outlined above, unusual emphasis is placed (especially in Part I) on showing how the quantum theory can be developed in a natural way, starting from the previously existing classical theory and going step by step through the experimental facts and theoretical lines of reasoning which led to replacement of the classical theory by the quantum theory. In this way, one avoids the need for introducing the basic principles of quantum theory in terms of a complete set of abstract mathematical postulates, justified only by the fact that complex calculations based on these postulates happen to agree with experiment. Although the treatment adopted in this book is perhaps not as neat mathematically as the postulational approach, it has a threefold advantage. First, it shows more clearly why such a radically new kind of theory is needed. Second, it makes the physical meaning of the theory clearer. Third, it is less rigid in its conceptual structure, so that one can see more easily how small modifications in the theory can be made if complete agreement with experiment is not immediately obtained.

Although the qualitative and physical development of the quantum theory takes place mainly in Parts I and VI, a systematic effort is made throughout the whole book to explain the results of mathematical calculations in qualitative and physical terms. It is hoped, moreover, that the mathematics has been simplified sufficiently to allow the reader to follow the general line of reasoning without spending too much time on mathematical details. Finally, it should be stated that the relative de-emphasis on mathematics is not intended for the purpose of reducing the amount of thinking needed for a thorough grasp of the theory. Instead, it is hoped that the reader will thereby be stimulated to do more thinking, and thus to provide himself with a general point of view which serves to orient him for further reading and study in this fascinating field.

An appreciable part of the material in this book was suggested by remarks made by Professor J. R. Oppenheimer in a series of lectures on quantum theory delivered at the University of California at Berkeley, and by notes on part of these lectures taken by Professor B. Peters. A series of lectures by Niels Bohr, entitled "Atomic Theory and the Description of Nature" were of crucial importance in supplying the general philosophical basis needed for a rational understanding of quantum theory. Numerous discussions with students and faculty at Princeton University were very helpful in clarifying the presentation. Dr. A. Wightman, in particular, contributed significantly to the clarification of Chapter 22, which deals with the quantum theory of measurements. Members of the author's quantum theory class in 1947 and 1948 performed invaluable work, checking both the mathematics and the reasoning, while the manuscript was being written. Finally, the author wishes to express his gratitude to M. Weinstein, who read and criticized the manuscript, and who supplied many very useful suggestions, and to L. Schmid who edited the manuscript and read the proofs.

DAVID BOHM

CONTENTS

PART I

Physical Formulation of the Quantum Theory

PART II

Mathematical Formulation of the Quantum Theory

PART III

Applications to Simple Systems. Further Extensions of Quantum Theory Formulation

PART IV

Methods of Approximate Solution of Schrödinger's Equation

PART V

Theory of Scattering

PART VI

Quantum Theory of the Process of Measurement

QUANTUM THEORY

PART I

PHYSICAL FORMULATION OF THE QUANTUM THEORY

MODERN QUANTUM THEORY is unusual in two respects. First, it embodies a set of physical ideas that differ completely with much of our everyday experience, and also with most experiments in physics on a macroscopic scale. Second, the mathematical apparatus needed to apply this theory to even the simplest examples is much less familiar than that required in corresponding problems of classical physics. As a result, there has been a tendency to present the quantum theory as being inseparable from the mathematical problems that arise in its applications. This approach might be likened to introducing Newton's laws of motion to a student of elementary physics, as problems in the theory of differential equations. In this book, special emphasis is placed on developing the guiding physical principles that are useful not only when it is necessary to apply our ideas to a new problem, but also when we wish to forsee the general properties of the mathematical solutions without carrying out extensive calculations. The development of the special mathematical techniques that are necessary for obtaining quantitative results in complex problems should take place, for the most part, either in a mathematics course or in a special course concerned with the mathematics of quantum theory. It seems impossible, however, to develop quantum concepts extensively without Fourier analysis. It is, therefore, presupposed that the reader is moderately familiar with Fourier analysis.

In the first part of this book, an unusual amount of attention is given to the steps by which the quantum theory may be developed, starting with classical theory and with specific experiments that led to the replacement of classical theory by the quantum theory. The experiments are presented not in historical order, but rather in what may be called a logical order. An historical order would contain many confusing elements that would hide the inherent unity that the quantum theory possesses. In this book, the experimental and theoretical developments are presented in such a way as to emphasize this unity and to show that each new step is either based directly on experiment or else follows logically from the previous steps. In this manner, the quantum theory can be made to seem less like a strange and somewhat arbitrary prescription, justified only by the fact that the results of its abstruse mathematical calculations happen to agree with experiment.

As an integral part of our plan for developing the theory on a basis that is not too abstract for a beginner, a complete account of the relation between quantum theory and the previously existing classical theory is given. Wherever possible the meaning of the quantum theory is illustrated in simple physical terms. Moreover, the final chapter of Part I points out broad regions of everyday experience in which we continually use ways of thinking that are closer to quantum-theoretical than to classical concepts. In this chapter, we also discuss in detail some of the philosophical implications of the quantum theory, and show that these lead to a striking modification in our general view of the world, as compared with that suggested by classical theory.

The reader will notice that most of the problems are interspersed throughout the text. These problems should be read as part of the text, because the results obtained from them are often used directly in the development of ideas. It is usually possible to understand the significance of the results without solving the problems, but the reader is strongly urged to try to solve them. The main advantage of the interspersed problems is that they make the reader think more specifically about the subject previously discussed, thus facilitating his understanding of the subject.

Supplementary References

The following list of supplementary texts will prove very helpful to the reader and will be referred to throughout various parts of this book:

Bohr, N., *Atomic Theory and the Description of Nature.* London: Cambridge University Press, 1934.

Born, M., *Atomic Physics.* Glasgow: Blackie & Son, Ltd., 1945.

Born, M., *Mechanics of the Atom.* London: George Bell & Sons, Ltd., 1927.

Dirac, P. A. M., *The Principles of Quantum Mechanics.* Oxford: Clarendon Press, 1947.

Heisenberg, W., *The Physical Principles of the Quantum Theory.* Chicago: University of Chicago Press, 1930.

Kramers, H. A., *Die Grundlagen der Quantentheorie.* Leipzig: Akademische Verlagsgesellschaft, 1938.

Mott, N. F., *An Outline of Wave Mechanics.* London: Cambridge University Press, 1934.

Mott, N. F., and I. N. Sneddon, *Wave Mechanics and Its Applications.* Oxford: Clarendon Press, 1948.

Pauli, W., *Die Allgemeinen Prinzipen der Wellenmechanik.* Ann Arbor, Mich.: Edwards Bros., Inc., 1946. Reprinted from *Handbuch der Physik,* 2. Aufl., Band 24. 1. Teil.

Pauling, L., and E. Wilson, *Introduction to Quantum Mechanics.* New York: McGraw-Hill Book Company, Inc., 1935.

Richtmeyer, F. K., and E. H. Kennard, *Introduction to Modern Physics.* New York: McGraw-Hill Book Company, Inc., 1933.

Rojansky, V., *Introductory Quantum Mechanics*. New York: Prentice-Hall, Inc., 1938.

Ruark, A. E., and H. C. Urey, *Atoms, Molecules, and Quanta*. New York: McGraw-Hill Book Company, Inc., 1930.

Schiff, L., *Quantum Mechanics*. New York: McGraw-Hill Book Company, Inc., 1949.

CHAPTER 1

The Origin of the Quantum Theory

The Rayleigh-Jeans Law

1. Blackbody Radiation in Equilibrium. Historically, the quantum theory began with the attempt to account for the equilibrium distribution of electromagnetic radiation in a hollow cavity. We shall, therefore, begin with a brief description of the characteristics of this distribution of radiation. The radiant energy originates in the walls of the cavity, which continually emit waves of every possible frequency and direction, at a rate which increases very rapidly with the temperature. The amount of radiant energy in the cavity does not, however, continue to increase indefinitely with time, because the process of emission is opposed by the process of absorption that takes place at a rate proportional to the intensity of radiation already present in the cavity. In the state of thermodynamic equilibrium, the amount of energy $U(\nu)\,d\nu$, in the frequency range between ν and $\nu + d\nu$, will be determined by the condition that the rate at which the walls emit this frequency shall be balanced by the rate at which they absorb this frequency. It has been demonstrated both experimentally and theoretically,* that after equilibrium has been reached, $U(\nu)$ depends only on the temperature of the walls, and not on the material of which the walls are made nor on their structure.

To observe this radiation, we make a hole in the wall. If the hole is very small compared with the size of the cavity, it produces a negligible change in the distribution of radiant energy inside the cavity. The intensity of radiation per unit solid angle coming through the hole is then readily shown to be $I(\nu) = \frac{c}{4\pi} U(\nu)$, where c is the velocity of light.†

Measurements disclose that, at a particular temperature, the function $U(\nu)$ follows a curve resembling the solid curve of Fig. 1. At low frequencies the energy is proportional to ν^2, while at high frequencies it drops off exponentially. As the temperature is raised, the maximum is shifted in the direction of higher frequencies; this accounts for the change in the color of the radiation emitted by a body as it gets hotter.

By thermodynamic arguments† Wien showed that the distribution

* Richtmeyer and Kennard. (See list of references on p. 2.)

† See Richtmeyer and Kennard for a derivation of this formula and also for a more complete account of blackbody radiation. The term "blackbody" arose

must be of the form $U(\nu) = \nu^3 f(\nu/T)$. The function f, however, cannot be determined from thermodynamics alone. Wien obtained a fairly good, but not perfect, fit to the empirical curve with the formula

$$U(\nu)\,d\nu \sim \nu^3 e^{-h\nu/\kappa T}\,d\nu \qquad \text{(Wien's law)} \tag{1}$$

Here κ is Boltzmann's constant, and h is an experimentally determined constant (which later turned out to be the famous quantum of action).*

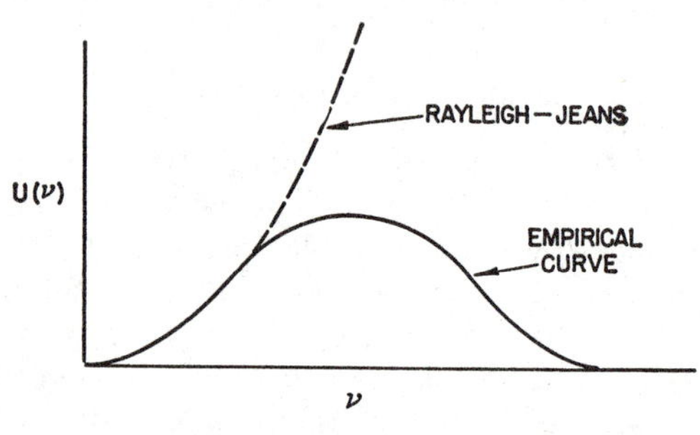

FIG. 1

Classical electrodynamics, on the other hand, leads to a perfectly definite and quite incorrect form for $U(\nu)$. This theoretical distribution, which will be derived in subsequent sections, is given by

$$U(\nu)\,d\nu \sim \kappa T \nu^2\,d\nu \qquad \text{(Rayleigh-Jeans law)} \tag{2}$$

Reference to Fig. 1 shows that the Rayleigh-Jeans law is in agreement with experiment at low frequencies, but gives too much radiation for high frequencies. In fact, if we attempt to integrate over all frequencies to find the total energy, the result diverges, and we are led to the absurd conclusion that the cavity contains an infinite amount of energy. Experimentally, the correct curve begins to deviate appreciably from the Rayleigh-Jeans law where $h\nu$ becomes of the order of κT. Hence, we must try to develop a theory that leads to the classical results for $h\nu < \kappa T$, but which deviates from classical theory at higher frequencies.

Before we proceed to discuss the way in which the classical theory must be modified, however, we shall find it instructive to examine in some detail the derivation of the Rayleigh-Jeans law. In the course of this deviation we shall not only gain insight into the ways in which classical physics fails, but we shall also be led to introduce certain classical physical concepts that are very helpful in the understanding of the quantum theory. In addition, the introduction of Fourier analysis to

---

because the radiation from a hole in such a cavity is identical with that coming from a perfectly black object.

* Wien did not actually introduce Planck's constant, h, but instead the constant h/κ.

deal with this classical problem will also constitute some preparation for its later use in the problems of quantum theory.

2. Electromagnetic Energy. According to classical electrodynamics, empty space containing electromagnetic radiation possesses energy. In fact, this radiant energy is responsible for the ability of a hollow cavity to absorb heat. In terms of the electric field, $\mathcal{E}(x, y, z, t)$, and the magnetic field, $\mathcal{H}(x, y, z, t)$, the energy can be shown to be*

$$E = \frac{1}{8\pi} \int (\mathcal{E}^2 + \mathcal{H}^2)\, d\tau \tag{3}$$

where $d\tau$ signifies integration over all the space available to the fields.

Our problem, then, is to determine the way in which this energy is distributed among the various frequencies present in the cavity when the walls are at a given temperature. The first step will be to use Fourier analysis for the fields and to express the energy as a sum of contributions from each frequency. In so doing, we shall see that the radiation field behaves, in every respect, like a collection of simple harmonic oscillators, the so-called "radiation oscillators." We shall then apply statistical mechanics to these oscillators and determine the mean energy of each oscillator when it is in equilibrium with the walls at the temperature T. Finally, we shall determine the number of oscillators in a given frequency range and, by multiplying this number by the mean energy of an oscillator, we shall obtain the equilibrium energy corresponding to this frequency, i.e., the Rayleigh-Jeans law.

3. Electromagnetic Potentials. We begin with a brief review of electrodynamics. The partial differential equations of the electromagnetic field, according to Maxwell, are given by

$$\nabla \times \mathcal{E} = -\frac{1}{c}\frac{\partial \mathcal{H}}{\partial t} \tag{4}$$

$$\nabla \cdot \mathcal{H} = 0 \tag{5}$$

$$\nabla \times \mathcal{H} = \frac{1}{c}\frac{\partial \mathcal{E}}{\partial t} + 4\pi \boldsymbol{j} \tag{6}$$

$$\nabla \cdot \mathcal{E} = 4\pi\rho \tag{7}$$

where $\boldsymbol{j}$ is the current density and ρ is the charge density. We can show from (4) and (5) that the most general electric and magnetic field can be expressed in terms of the vector and scalar potentials, $\boldsymbol{a}$ and ϕ, in the following way:

$$\mathcal{H} = \nabla \times \boldsymbol{a} \tag{8}$$

and

$$\mathcal{E} = -\frac{1}{c}\frac{\partial \boldsymbol{a}}{\partial t} - \nabla\phi \tag{9}$$

When $\mathcal{E}$ and $\mathcal{H}$ are expressed in this form, (4) and (5) are satisfied identically, and the equations for $\boldsymbol{a}$ and ϕ are then obtained by the substitution of relations (8) and (9) into (6) and (7).

Now, eqs. (8) and (9) do not define the potentials uniquely in terms

* Richtmeyer and Kennard, Chap. 2.

of the fields. If, for example, we add an arbitrary vector, $-\nabla\psi$, to the vector potential, the magnetic field is not changed because $\nabla \times \nabla\psi = 0$. If we simultaneously add the quantity $\frac{1}{c}\frac{\partial}{\partial t}\psi$ to the scalar potential, the electric field is also unchanged. Thus, we find that the electric and magnetic fields remain invariant under the following transformation of the potentials:*

$$\left.\begin{aligned} \boldsymbol{a}' &= \boldsymbol{a} - \nabla\psi \\ \phi' &= \phi + \frac{1}{c}\frac{\partial\psi}{\partial t} \end{aligned}\right\} \tag{10}$$

The above is called a "gauge transformation."

We can utilize the invariance of the fields to a gauge transformation for the purpose of simplifying the expressions for $\mathcal{E}$ and $\mathcal{H}$. A common choice is to make div $\boldsymbol{a} = 0$. To show that this is always possible, suppose that we start with an arbitrary set of potentials, $\boldsymbol{a}(x, y, z, t)$ and $\phi(x, y, z, t)$. We then make the gauge transformation of eq. (10) to a new set of potentials, A' and ϕ'. In order to obtain div $\boldsymbol{a}' = 0$, we must choose ψ such that

$$\text{div}\, \boldsymbol{a} - \nabla^2\psi = 0$$

But the above is just Poisson's equation defining ψ in terms of the specified function, div $\boldsymbol{a}$. Its solution can always be obtained and is, in fact, equal to

$$\psi = -\frac{1}{4\pi}\iiint \frac{\text{div}\, \boldsymbol{a}(x', y', z', t)\, dx'\, dy'\, dz'}{|\boldsymbol{r} - \boldsymbol{r}'|}$$

Thus, we prove that a gauge transformation that yields div $\boldsymbol{a}' = 0$ can always be carried out.

We now show that in *empty space* the choice div $\boldsymbol{a} = 0$ also leads to $\phi = 0$ and, therefore, to a considerable simplification in the representation of the electric field. To do this, we substitute eq. (9) into (7), setting $\rho = 0$ since, by hypothesis, there are no charges in empty space. The result is

$$\text{div}\, \mathcal{E} = -\frac{1}{c}\,\text{div}\,\frac{\partial \boldsymbol{a}}{\partial t} - \nabla^2\phi = 0$$

But since div $\boldsymbol{a} = 0$, we obtain

$$\nabla^2\phi = 0$$

This is, however, simply Laplace's equation. It is well known that the only solution of this equation that is regular over all of space is $\phi = 0$. (All other solutions imply the existence of charge at some points in space and, therefore, a failure of Laplace's equation at these points.) We

* $\mathcal{E}$ and $\mathcal{H}$ are the only physically significant quantities connected with the electromagnetic field.

should note, however, that the condition $\phi = 0$ follows only in empty space because, in the presence of charge, eq. (7) leads to $\nabla^2\phi = -4\pi\rho$, which is Poisson's equation. This equation has nonzero regular solutions, provided that ρ is not everywhere zero.

We conclude, then, that in empty space we obtain the following expressions for the fields:

$$\mathcal{H} = \nabla \times \boldsymbol{a} \tag{11}$$

$$\mathcal{E} = -\frac{1}{c}\frac{\partial \boldsymbol{a}}{\partial t} \tag{12}$$

subject to the condition that

$$\operatorname{div} \boldsymbol{a} = 0 \tag{13}$$

Finally, we obtain the partial differential equation defining $\boldsymbol{a}$ in empty space by inserting (11), (12), and (13) into (6), provided that we also assume that $\boldsymbol{j} = 0$, as is necessary in the absence of matter. We obtain

$$\nabla^2\boldsymbol{a} - \frac{1}{c^2}\frac{\partial^2 \boldsymbol{a}}{\partial t^2} = 0 \tag{14}$$

Equations (11), (12), (13), and (14), together with the boundary conditions, completely determine the electromagnetic fields in a cavity that contains no charges or currents.

4. Boundary Conditions. As pointed out in Sec. 1, it has been demonstrated both experimentally and theoretically* that the equilibrium distribution of energy density in a hollow cavity does not depend on the shape of the container or on the material in the walls. Hence, we are at liberty to choose the simplest possible boundary conditions consistent with equilibrium. We shall choose a set of boundary conditions that are somewhat artificial from an experimental point of view, but that greatly simplify the mathematical treatment.

Let us imagine a cube of side L with very thin walls of some material that is not an electrical conductor. We then imagine that this structure is repeated periodically through space in all directions, so that space is filled up with cubes of side L. Let us suppose, further, that the fields are the same at corresponding points of every cube.

We now assert that these boundary conditions will yield the same equilibrium radiation density as will any other boundary conditions at the walls.† To prove this, we need only ask why the equilibrium conditions are independent of the type of boundary. The answer is that, from

* The theoretical proof depends on the use of statistical mechanics. See, for example, R. C. Tolman, *The Principles of Statistical Mechanics*. Oxford, Clarendon Press, 1938.

† With these conditions, no walls are actually necessary, but the thermodynamic results are the same as for an arbitrary wall, including, for example, a perfect reflector or a perfect absorber.

the thermodynamic viewpoint, the wall merely serves to prevent the system from gaining or losing energy. Making the fields periodic must have the same effect because each cube can neither gain energy from the other cubes nor lose it to them; if this were not so, the system would cease to be periodic. Thus, we have a boundary condition that serves the essential function of keeping the energy in any individual cube constant. Although artificial, it must give the right answer, and it will make the calculations easier by simplifying the Fourier analysis of the fields.

5. Fourier Analysis. Now, $\mathbf{a}(x, y, z, t)$ may be any conceivable solution of Maxwell's equations, with the sole restriction, imposed by our boundary conditions, that it must be periodic in space with period L/n, where n is an integer.* It is a well-known mathematical theorem that an arbitrary periodic function,† $f(x, y, z, t)$, can be represented by means of a Fourier series in the following manner:

$$f(x, y, z, t) = \sum_{l,m,n} \left[a_{l,m,n}(t) \cos \frac{2\pi}{L} (lx+my+nz) + b_{l,m,n}(t) \sin \frac{2\pi}{L} (lx+my+nz) \right] \quad (15)$$

where l, m, n are integers running from $-\infty$ to ∞, including zero. Any choice of a's and b's leading to a convergent series defines a function, $f(x, y, z, t)$, which is periodic in the sense that it takes on the same value each time x, y, or z changes by L. For a given function, $f(x, y, z, t)$ it can be shown that the $a_{l,m,n}(t)$ and the $b_{l,m,n}(t)$ are given by the following formulas:

$$\left.\begin{aligned} a_{l,m,n}(t) + a_{-l,-m,-n}(t) &= \frac{2}{L^3} \int_0^L \int_0^L \int_0^L dx\, dy\, dz \cos \frac{2\pi}{L} (lx + my + nz) f(x, y, z, t) \\ b_{l,m,n}(t) - b_{-l,-m,-n}(t) &= \frac{2}{L^3} \int_0^L \int_0^L \int_0^L dx\, dy\, dz \sin \frac{2\pi}{L} (lx + my + nz) f(x, y, z, t) \end{aligned}\right\} \quad (16)$$

These formulas illustrate the fact that only the sum of the a's and the difference of the b's are determined by the function f.

From the above, we conclude that f may be specified completely in terms of the quantities $a_{l,m,n} + a_{-l,-m,-n}$ and $b_{l,m,n} - b_{-l,-m,-n}$, but we prefer to retain the specification in terms of the $a_{l,m,n}$ and $b_{l,m,n}$ because of the simpler mathematical expressions to which they lead.

* There will be, of course, the usual regularity conditions that prevent $\mathbf{a}$ from being infinite or discontinuous.

† The function must be piecewise continuous.

Equations (16) are obtained with the aid of the following orthogonality relations:*

$$\left.\begin{aligned}\int_0^L\int_0^L\int_0^L dx\,dy\,dz\cos\frac{2\pi}{L}(lx+my+nz)\\ \sin\frac{2\pi}{L}(l'x+m'y+n'z)&=0\\ \int_0^L\int_0^L\int_0^L dx\,dy\,dz\cos\frac{2\pi}{L}(lx+my+nz)\\ \cos\frac{2\pi}{L}(l'x+m'y+n'z)&=0\end{aligned}\right\}\quad(17a)$$

unless
$$\begin{pmatrix}l=l'\\ m=m'\\ n=n'\end{pmatrix}\quad\text{or}\quad\begin{pmatrix}l=-l'\\ m=-m'\\ n=-n'\end{pmatrix}$$

in which case it is $L^3/2$, except when $l = m = n = 0$, in which case it is L^3.

$$\int_0^L\int_0^L\int_0^L \sin\frac{2\pi}{L}(lx+my+nz)\sin\frac{2\pi}{L}(l'x+m'y+n'z)=0\quad(17b)$$

unless
$$\begin{pmatrix}l=l'\\ m=m'\\ n=n'\end{pmatrix}\quad\text{or}\quad\begin{pmatrix}l=-l'\\ m=-m'\\ n=-n'\end{pmatrix}$$

in which case it is $L^3/2$. [It is suggested that the reader prove (17a and b) as an exercise, and use the results to obtain (16).]

Fourier analysis, in the preceding form, enables us to represent an arbitrary function as a sum of standing plane waves of all possible wavelengths and amplitudes. The entire treatment is essentially the same as that used with waves in strings and organ pipes, except that it is three-dimensional.

Let us now expand the vector potential in a Fourier series. Because $\mathbf{a}$ is a vector, involving three components, each $a_{l,m,n}$ and $b_{l,m,n}$ also has three components and, hence, must be represented as a vector:

$$\mathbf{a}=\sum_{l,m,n}\left[\mathbf{a}_{l,m,n}(t)\cos\frac{2\pi}{L}(lx+my+nz)+\mathbf{b}_{l,m,n}(t)\sin\frac{2\pi}{L}(lx+my+nz)\right]$$

We assume that $\mathbf{a}_{0,0,0}$ is zero in the above series.†

* For the origin of the term "orthogonality" see Chap. 16, Sec. 10; also Chap. 10, Sec. 24.

† This follows from the fact that the part of $\mathbf{a}$ which is constant in space corresponds to no magnetic field, and to a spatially uniform electric field $\left(\mathcal{E} = -\frac{1}{c}\frac{\partial a}{\partial t}\right)$. Such a field requires a charge distribution somewhere to produce it, i.e., at the boundaries, and since we assume that no such distribution is present, we set $a_{0,0,0} = 0$.

We now introduce the propagation vector $\boldsymbol{k}$, defined as follows:

$$k_x = \frac{2\pi l}{L} \qquad k_y = \frac{2\pi m}{L} \qquad k_z = \frac{2\pi n}{L}.$$

$$k^2 = \left(\frac{2\pi}{L}\right)^2 (l^2 + m^2 + n^2) \tag{18}$$

By orienting our co-ordinate axes in such a way that the z axis is directed along the $\boldsymbol{k}$ vector, we obtain $l = m = 0$, and $k = 2\pi/L$. From the definition of k, it follows that $k/2\pi$ is the number of waves in the distance L; hence the wavelength is $\lambda = 2\pi/k$, or

$$k = 2\pi/\lambda \tag{19}$$

In this co-ordinate system a typical wave takes the form $\cos 2\pi nz/L$. Thus, the vector $\boldsymbol{k}$ is in the direction in which the phase of the wave changes. Going back to arbitrary co-ordinate axes, we conclude that $\boldsymbol{k}$ is a vector in the direction of propagation of the wave. Its magnitude is $2\pi/\lambda$, and it is allowed to take on only the values permitted by integral l, m, and n in eq. (18).

With this simplification of notation, we obtain

$$\boldsymbol{a} = \sum_k [\boldsymbol{a}_k(t) \cos \boldsymbol{k} \cdot \boldsymbol{r} + \boldsymbol{b}_k(t) \sin \boldsymbol{k} \cdot \boldsymbol{r}] \tag{20}$$

where the summation extends over all permissible values of $\boldsymbol{k}$.

6. Polarization of Waves. Let us now apply the condition div $\boldsymbol{a} = 0$ to (20). We have

$$\text{div } \boldsymbol{a} = \sum_k (-\boldsymbol{k} \cdot \boldsymbol{a}_k \sin \boldsymbol{k} \cdot \boldsymbol{r} + \boldsymbol{k} \cdot \boldsymbol{b}_k \cos \boldsymbol{k} \cdot \boldsymbol{r}) = 0$$

It is a well-known theorem that if a Fourier series is identically zero, then all of the coefficients, $\boldsymbol{a}_k$ and $\boldsymbol{b}_k$, must vanish.

Problem 1: Prove the above theorem, using the orthogonality relations (17).

From the above it follows that $\boldsymbol{k} \cdot \boldsymbol{a}_k(t) = \boldsymbol{k} \cdot \boldsymbol{b}_k(t) = 0$. Thus, $\boldsymbol{a}_k(t)$ and $\boldsymbol{b}_k(t)$ are perpendicular to $\boldsymbol{k}$, as are also the electric and magnetic fields belonging to the kth wave. Since the vibrations are normal to the direction of propagation, the waves are transverse. The direction of the electric field is also called the *direction of polarization.*

To describe the orientation of $\boldsymbol{a}_k$ let us return to the set of co-ordinate axes in which the z axis is in the direction of $\boldsymbol{k}$. The vector $\boldsymbol{a}_k$ can have only x and y components, and if we specify the values of these, we shall have specified both the magnitude and the direction of $\boldsymbol{a}_k$.

We designate the direction of the vector $\boldsymbol{a}_k$ by the subscript μ, writing $\boldsymbol{a}_{k,\mu}$, where μ is allowed to take on the values 1 and 2. For $\mu = 1$, $\boldsymbol{a}_{k,\mu}$ is in the x direction; but for $\mu = 2$, it is in the y direction. All possible

$\boldsymbol{a}_k$ vectors can then be represented as a sum of some $\boldsymbol{a}_{k,1}$ vector, and some other $\boldsymbol{a}_{k,2}$ vector. Hence, the most general vector potential, subject to the condition that div $\boldsymbol{a} = 0$, is given by

$$\boldsymbol{a} = \sum_{k,\mu} [\boldsymbol{a}_{k,\mu}(t) \cos \boldsymbol{k} \cdot \boldsymbol{r} + \boldsymbol{b}_{k,\mu}(t) \sin \boldsymbol{k} \cdot \boldsymbol{r}] \tag{21}$$

Here the summation extends over all permissible $\boldsymbol{k}$ vectors and over the two possible values of μ.

It can be verified from (14) and (21) that the $\boldsymbol{a}_{k,\mu}$ satisfies the following differential equation:

$$\frac{d^2\boldsymbol{a}_{k,\mu}}{dt^2} + k^2c^2\boldsymbol{a}_{k,\mu} = 0 \tag{22}$$

which shows that the $\boldsymbol{a}_{k,\mu}$ terms oscillate with simple harmonic motion and with angular frequency, $\omega = kc$.

7. Evaluation of the Electromagnetic Energy. The first step in evaluating the electromagnetic energy is to express $\mathcal{E}$ and $\mathcal{H}$ in terms of the Fourier series for $\boldsymbol{a}$. These expressions are:

$$\mathcal{E} = -\frac{1}{c}\sum_{k,\mu} (\dot{\boldsymbol{a}}_{k,\mu} \cos \boldsymbol{k} \cdot \boldsymbol{r} + \dot{\boldsymbol{b}}_{k,\mu} \sin \boldsymbol{k} \cdot \boldsymbol{r})$$

$$\mathcal{H} = \sum_{k,\mu} (-\boldsymbol{k} \times \boldsymbol{a}_{k,\mu} \sin \boldsymbol{k} \cdot \boldsymbol{r} + \boldsymbol{k} \times \boldsymbol{b}_{k,\mu} \cos \boldsymbol{k} \cdot \boldsymbol{r})$$

Problem 2: Derive the above expressions for $\mathcal{E}$ and $\mathcal{H}$.

Let us now evaluate the following over the cube of side L:

$$\frac{1}{8\pi}\int \mathcal{E}^2\, d\tau = \frac{1}{8\pi c^2}\sum_{k,\mu}\sum_{k',\mu'}\int_0^L\int_0^L\int_0^L dx\, dy\, dz$$
$$\begin{pmatrix} \dot{\boldsymbol{a}}_{k,\mu} \cdot \dot{\boldsymbol{a}}_{k',\mu'} \cos \boldsymbol{k} \cdot \boldsymbol{r} \cos \boldsymbol{k}' \cdot \boldsymbol{r} + \dot{\boldsymbol{b}}_{k,\mu} \cdot \dot{\boldsymbol{b}}_{k',\mu'} \sin \boldsymbol{k} \cdot \boldsymbol{r} \sin \boldsymbol{k}' \cdot \boldsymbol{r} \\ + \dot{\boldsymbol{a}}_{k,\mu} \cdot \dot{\boldsymbol{b}}_{k',\mu'} \cos \boldsymbol{k} \cdot \boldsymbol{r} \sin \boldsymbol{k}' \cdot \boldsymbol{r} + \dot{\boldsymbol{b}}_{k,\mu} \cdot \dot{\boldsymbol{a}}_{k',\mu'} \sin \boldsymbol{k} \cdot \boldsymbol{r} \cos \boldsymbol{k}' \cdot \boldsymbol{r} \end{pmatrix}$$

With the aid of eqs. (17) we see that all integrals vanish except when $\boldsymbol{k} = \boldsymbol{k}'$, and that all terms involving $\dot{\boldsymbol{a}}_{k,\mu} \cdot \dot{\boldsymbol{b}}_{k,\mu'}$ are zero. Furthermore, $\dot{\boldsymbol{a}}_{k,\mu} \cdot \dot{\boldsymbol{a}}_{k,\mu'} = 0$ unless $\mu = \mu'$. When $\mu \neq \mu'$, the two vectors are, by definition, perpendicular to each other. Thus, the above expression reduces to

$$\int \frac{\mathcal{E}^2\, d\tau}{8\pi} = \frac{L^3}{8\pi c^2}\sum_{k,\mu}\left[\frac{1}{2}(\dot{\boldsymbol{a}}_{k,\mu})^2 + \frac{1}{2}(\dot{\boldsymbol{b}}_{k,\mu})^2\right]$$

With a similar method, which involves somewhat more algebra, we obtain

$$\int \frac{\mathcal{H}^2\, d\tau}{8\pi} = \frac{L^3}{8\pi}\sum_{k,\mu} k^2\left[\frac{1}{2}(\boldsymbol{a}_{k,\mu})^2 + \frac{1}{2}(\boldsymbol{b}_{k,\mu})^2\right]$$

Problem 3: Derive the above expression for $\int \mathcal{H}^2\, d\tau$.

Thus, the electromagnetic energy in the cavity is (with $L^3 = V$)

$$E = \frac{V}{8\pi c^2} \sum_{k,\mu} \left\{ \frac{1}{2} [(\dot{a}_{k,\mu})^2 + c^2k^2(a_{k,\mu})^2] + \frac{1}{2} [(\dot{b}_{k,\mu})^2 + c^2k^2(b_{k,\mu})^2] \right\} \quad (23)$$

8. Meaning of Preceding Result for Electromagnetic Energy. The following are the most important properties of eq. (23):

(1) The energy is a sum of separate terms, one for each $a_{k,\mu}$, and one for each $b_{k,\mu}$. This means that different wavelengths and polarizations do not interact with each other, because the interaction of any two systems always requires that the energy of one should depend on the state of the other. Here we see that the energy in each wave of propagation vector $\boldsymbol{k}$ and polarization direction μ is proportional only to the square of $\dot{a}_{k,\mu}$ and $a_{k,\mu}$, and not to any of the other a's or b's. A similar result holds for each of the b's.

(2) The energy associated with each $a_{k,\mu}$ (or $b_{k,\mu}$) has the same mathematical form as that of a material harmonic oscillator. A harmonic oscillator of mass m, angular frequency ω, has energy

$$E = \frac{m}{2} (\dot{x}^2 + \omega^2 x^2)$$

By analogy, we can write

$$m = \frac{V}{8\pi c^2}, \qquad \omega = kc$$

The frequency is then $f = \omega/2\pi = kc/2\pi = c/\lambda$. We know, of course, that an electromagnetic wave of wavelength λ has just the above frequency.* This shows that our harmonic oscillator analogy gives the right description of the way in which the a's oscillate.

The analogy with a material oscillator can be carried further. For example, with material oscillators, we introduce a momentum $p = m\dot{x}$. Here the momentum is

$$p_{k,\mu} = \frac{V}{8\pi c^2} \dot{a}_{k,\mu}$$

We can then introduce a Hamiltonian function

$$H = \frac{p^2}{2m} + \frac{m\omega^2 x^2}{2}$$

For the $a_{k,\mu}$ we get

$$H = \frac{8\pi c^2}{L^3} \frac{(p_{k,\mu})^2}{2} + \frac{L^3}{8\pi} k^2 \frac{(a_{k,\mu})^2}{2} \quad (24)$$

Similar terms may be introduced for the $b_{k,\mu}$.

The correct equations of motion are obtained from the Hamiltonian equations

$$\dot{a}_{k,\mu} = \frac{\partial H}{\partial p_{k,\mu}} \qquad \text{and} \qquad \dot{p}_{k,\mu} = -\frac{\partial H}{\partial a_{k,\mu}}$$

* See also eq. (22).

which yield eq. (22), obtained originally by direct substitution into Maxwell's equations.

$$\ddot{a}_{k,\mu} + c^2k^2a_{k,\mu} = 0 \tag{25}$$

and similarly for the b's.

The $a_{k,\mu}$ and $b_{k,\mu}$ are, as we have seen, analogous to the co-ordinates of separate noninteracting harmonic oscillators. In a sense, the $a_{k,\mu}$ and $b_{k,\mu}$ may also be regarded as the co-ordinates of the radiation field. This is because, once they are given, the field is specified everywhere through eq. (20). There are an infinite number of these co-ordinates, because there are an infinite number of possible values of $\mathbf{k}$. But the infinity is discrete, or countable, as distinguished from the continuous infinity of points on a line. The main advantage of the Fourier series is that it enables us to describe the fields over a continuous region of space by means of a discrete infinity of co-ordinates.

How many independent co-ordinates are there for each permissible value of $\mathbf{k}$? First, there are two polarization directions; then we have, for each $\mathbf{k}$ and $\boldsymbol{\mu}$, an $a_{k,\mu}$ and a $b_{k,\mu}$. Thus, it would seem, at first sight, that we need four indepenaent co-ordinates for each value of $\mathbf{k}$. But, from eq. (16), we see that it is necessary to specify only the combinations $a_{k,\mu} + a_{-k,\mu}$ and $b_{k,\mu} - b_{-k,\mu}$, so that the number of variables necessary is reduced by a factor of two. This means that for each $\mathbf{k}$ there are two independent co-ordinates.

9. Number of Oscillators. We must now find the number of oscillators with frequencies between ν and $(\nu + d\nu)$. Since $\nu = kc/2\pi$, the problem is equivalent to that of finding the number between k and $k + dk$.

Now, for any reasonable value of k, the number of waves fitting into a box is usually very large. For example, at moderate temperatures, most of the radiation is in the infrared, with wavelengths $\sim 10^{-4}$ cm. Hence, when k changes in such a way that one more wavelength fits into the box, only a very small fractional shift of k results. It is possible, therefore, to choose the interval dk so small that no important physical quantity (such as the mean energy) changes appreciably within it, yet so large that very many radiation oscillators are included. This means that the number of oscillators can be treated as virtually continuous, so that we can represent it in terms of a density function.

We must now find the number of oscillators in the volume $dk_x\, dk_y\, dk_z$. If we imagine a space in which the co-ordinates are l, m, and n, there will be one oscillator every time l, m, and n take on separate integral values. Hence, there is one oscillator per unit cube of l, m, n space, so that the density in this space is unity. To go to k space, we use eq. (18), obtaining

$$\delta N_1 = dl\, dm\, dn = \frac{V}{(2\pi)^3}\, dk_x\, dk_y\, dk_z \tag{26}$$

It is now convenient to adopt polar co-ordinates in $\boldsymbol{k}$ space. We define $k^2 = k_x^2 + k_y^2 + k_z^2$. Then the element of volume becomes $k^2\,dk\,d\Omega$; where $d\Omega$ is the element of solid angle. Since we are not interested in the direction of $\boldsymbol{k}$, we integrate over $d\Omega$, obtaining $4\pi k^2\,dk$ for the element of volume, and

$$\delta N_1 = \frac{4\pi V}{(2\pi)^3} k^2\,dk \tag{27}$$

Writing $\nu = kc/2\pi$, we obtain

$$\delta N_1 = 4\pi V \frac{\nu^2\,d\nu}{c^3} \tag{28}$$

This gives the number of permissible values of $\boldsymbol{k}$ in the range between ν and $\nu + d\nu$. As shown in the section discussing the significance of the $\boldsymbol{a}$'s and $\boldsymbol{b}$'s, there are two independent coordinates for each $\boldsymbol{k}$, corresponding to the two directions of polarization. Thus, for the total number of oscillators between ν and $\nu + d\nu$, we find

$$\delta N = 2\delta N_1 = \frac{8\pi V}{c^3} \nu^2\,d\nu \tag{29}$$

10. Equipartition of Energy. To calculate the mean energy possessed by each oscillator when it is in thermodynamic equilibrium with the walls, we shall apply classical statistical mechanics to these oscillators. Although this theory was derived for material oscillators alone, the derivation involved only the formal properties of the equations of motion. Any other systems acting formally like material oscillators must, therefore, have the same equilibrium distribution of energy. It is shown in classical statistical mechanics* that in any assembly of independent, noninteracting systems (such as our assembly of radiation oscillators), the probability that a co-ordinate lies between q and $q + dq$, and that the corresponding momentum lies between p and $p + dp$, is equal to

$$A\,e^{-E/\kappa T}\,dp\,dq$$

E denotes total energy, kinetic and potential; and A denotes a normalizing factor, defined by the requirement that the total probability integrate out to unity or

$$\int_{-\infty}^{\infty}\int_{-\infty}^{\infty} A\,e^{-E/\kappa T}\,dp\,dq = 1$$

For a perfect gas, $E = p^2/2m$, and we obtain the familiar Maxwell-Boltzmann distribution of velocities

$$A\,e^{-p^2/2m\kappa T}\,dp\,dq$$

For the harmonic oscillator, we have

$$E = \frac{p^2}{2m} + m\omega^2\frac{q^2}{2}$$

* R. C. Tolman, *The Principles of Statistical Mechanics.*

It is convenient to transform these equations to new variables defined by

$$p = \sqrt{2m}\,P; \qquad q = \frac{Q}{\omega}\sqrt{\frac{2}{m}}$$

This yields $E = P^2 + Q^2$.

$$\frac{1}{A} = \int_{-\infty}^{\infty}\int_{-\infty}^{\infty} e^{-E/\kappa T}\frac{2}{\omega}\,dP\,dQ$$

The probability that the system lies between P and $P + dP$, and Q and $Q + dQ$, is

$$dW(P, Q) = \frac{e^{-(P^2+Q^2)/\kappa T}\,dP\,dQ}{\int_{-\infty}^{\infty}\int_{-\infty}^{\infty} e^{-(P^2+Q^2)/\kappa T}\,dP\,dQ}$$

Let us transform to polar co-ordinates R, ϕ, in phase space, where

$$P^2 + Q^2 = R^2 = E, \qquad \text{or} \qquad R\,dR = \frac{dE}{2}$$

The element of area is now

$$R\,dR\,d\phi = \frac{1}{2}d(R^2)\,d\phi = \frac{dE}{2}\,d\phi$$

Since we are not interested in the angle ϕ, we may integrate over it. We then obtain, for the normalized probability that the energy lies between E and $E + dE$:

$$W(E)\,dE = \frac{e^{-E/\kappa T}\,dE}{\int_0^{\infty} e^{-E/\kappa T}\,dE}$$

The mean value of the energy $\bar{E}$ is obtained by integration of $EW(E)$ over all energies. This means that we weight each energy according to its probability. We get

$$\bar{E} = \frac{\int_0^{\infty} E\,e^{-E/\kappa T}\,dE}{\int_0^{\infty} e^{-E/\kappa T}\,dE} = \kappa T\,\frac{\int_0^{\infty} e^{-\epsilon}\epsilon\,d\epsilon}{\int_0^{\infty} e^{-\epsilon}\,d\epsilon} = \kappa T \qquad (30)$$

where $\epsilon = E/\kappa T$. Thus, we prove that the average energy of each oscillator is κT. This is an example of the theorem of equipartition of energy.*

Collecting the information obtained from (29) and (30), we get the Rayleigh-Jeans law:

$$U(\nu)\,d\nu = \bar{E}\,\delta N = \frac{8\pi V}{c^3}\,\kappa T\nu^2\,d\nu \qquad (31)$$

Because this law disagrees with experiments, we conclude that the con-

* See Richtmeyer and Kennard, p. 161.

cepts of classical physics are in some way inadequate to describe the interaction of matter and radiation.

Planck's Hypothesis

11. The Quantization of the Radiation Oscillators. In searching for a modification of the above treatment that would reduce the contribution of the high frequencies to the energy, Planck was led to make an assumption equivalent to the following: The energy of an oscillator of natural frequency ν is restricted to integral multiples of a basic unit $h\nu$. This basic unit is not the same for all oscillators, since it is proportional to the frequency. The energy of an oscillator is, then, $E = nh\nu$, where n is any integer from 0 to ∞. With this assumption, Planck obtained an exact fit, within experimental error, to the observed distribution of radiation.

According to classical mechanics, there should be no restrictions whatever on the energy an oscillator may possess. Our experience with oscillators, such as radio waves, clock springs, and pendulums, seem to verify this prediction. How, then, can Planck's hypothesis be consistent with all these well-known results? The answer is that h is a very small quantity, equal to about 6.6×10^{-27} erg-sec. Hence, even for microwaves having a frequency as high as 10^{10} cps, the basic unit of energy is only 6.6×10^{-17} erg which is not detectable except by use of the most sensitive apparatus now available. With clock springs and pendulums of period of the order of 1 sec, the basic unit of energy is obviously so small that in such relatively gross observations as can now be made, the allowed values of energy seem to be continuous. With light waves, however, $\nu \sim 10^{15}$, and $h\nu \cong 10^{-12}$ erg, a value that can be detected with sensitive instruments. Hence, as we go to higher frequencies, where the basic unit becomes larger, quantization of the energy levels is easier to observe.

To obtain Planck's distribution of energy, we need to know what the probability is that the oscillator has an energy corresponding to its nth allowed value. Now, when n is very large, so that the discrete character of the energy becomes unimportant (as with radio waves, for example) we must obtain a result that is consistent with classical mechanics, which is known to be correct in this region. The simplest way of obtaining agreement is to choose a probability that is the same function of the energy as in the classical theory,* namely, $e^{-E/\kappa T}$. For a given energy, $E_n = nh\nu$, the probability is then

$$W(n) \sim e^{-E_n/\kappa T} = e^{-nh\nu/\kappa T}$$

* This choice involves an assumption that is justified in part by its success in accounting for the energy distribution in a blackbody. A systematic development of the theory of quantum statistics (see Tolman, *The Principles of Statistical Mechanics*) shows, however, that no other probability distribution would lead to thermodynamic equilibrium.

To normalize this, we write*

$$W(n) = \frac{e^{-nh\nu/\kappa T}}{\sum_{n=0}^{\infty} e^{-nh\nu/\kappa T}} = e^{-nh\nu/\kappa T}(1 - e^{-h\nu/\kappa T})$$

The mean energy is

$$\bar{E} = \sum_{n=0}^{\infty} E_n W(n) = (1 - e^{-h\nu/\kappa T}) \sum_{n=0}^{\infty} e^{-nh\nu/\kappa T} nh\nu$$

$$= h\nu(1 - e^{-h\nu/\kappa T}) \sum_{n=0}^{\infty} ne^{-nh\nu/\kappa T}$$

To evaluate this sum, we can write

$$\sum_{n=0}^{\infty} ne^{-n\alpha} = -\frac{d}{d\alpha} \sum_{0}^{\infty} e^{-n\alpha} = -\frac{d}{d\alpha} \frac{1}{1 - e^{-\alpha}} = \frac{e^{-\alpha}}{(1 - e^{-\alpha})^2}$$

We can apply this result, obtaining

$$\bar{E} = \frac{h\nu e^{-h\nu/\kappa T}}{1 - e^{-h\nu/\kappa T}} \tag{32}$$

Multiplying by δN, we find the Planck distribution

$$U(\nu) = \frac{8\pi V}{c^3} h\nu^3 \frac{e^{-h\nu/\kappa T}}{1 - e^{-h\nu/\kappa T}} \tag{32-a}$$

12. Discussion of Results. For small $h\nu/\kappa T$ the exponentials can be expanded, and retention of only the first terms yields $\bar{E} = \kappa T$, the classical result, in agreement with the fact that the Rayleigh-Jeans law is correct for small $h\nu/\kappa T$. As $h\nu/\kappa T$ becomes large, then $\bar{E} \to h\nu e^{-h\nu/\kappa T}$. This leads to the Wien law. In between, there is excellent agreement with experiment at all temperatures. Hence, despite the strangeness of Planck's hypothesis, there is evidently something to it.

The decrease in mean energy of the high-frequency oscillators arises because of the great amount of energy required to bring them to the first excited state, which is a state of rare occurrence. As ν is lowered or T raised, it becomes more likely that the oscillator will gain a quantum of energy. After the oscillator is excited to a high quantum number n, its behavior will be essentially classical, because the basic unit of energy is then much less than the mean available energy κT.

13. Material vs. Radiation Oscillators. Planck's original idea was not to quantize the radiation oscillators as we have done previously. Instead, he assumed that the radiation was in equilibrium with material oscillators in the walls of the container, and that these material oscillators

* We use the expansion $\frac{1}{1 - x} = \sum_{n=0}^{\infty} x^n$.

could give up or absorb radiant energy only in quanta with $E = nh\nu$. With this assumption, he obtained exactly the same distribution of radiant energy. The quantization of the radiation oscillators was a later idea that has, as we shall see, many far-reaching consequences. Moreover, the step of quantizing the radiation oscillators is almost imperative to explain the fact that the blackbody spectrum is independent of the materials of which the walls are composed.

The only alternative possibility is that *all* matter, and not only harmonic oscillators, can accept or emit radiation only in quanta of size, $E = h\nu$. But this means that all radiation ever emitted has energy restricted to $E = h\nu$; even if some were present with other energies it could not, by hypothesis, interact with matter and, hence, would be undetectable. This hypothesis is, then, equivalent to the statement that all radiation oscillators have their energies restricted to $E = nh\nu$.

14. Quantization of Material Oscillators. We can consider the specific heats of solids, to determine whether Planck's quantum hypothesis applies to material oscillators. In a crystal, for example, each atom is in equilibrium when it lies in its proper lattice position and, if disturbed, it can oscillate about the equilibrium position with a motion that is approximately simple harmonic for small oscillations.

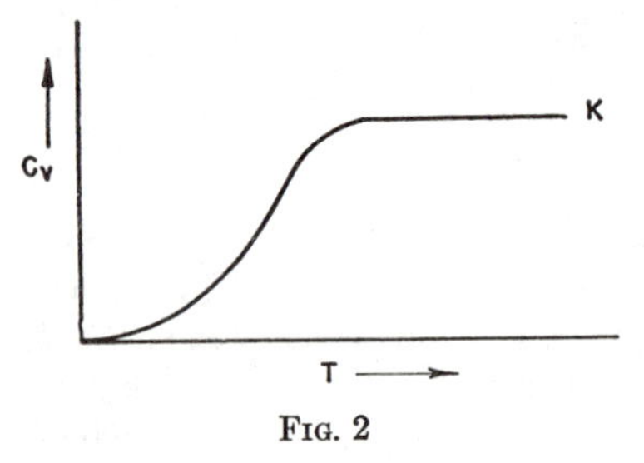

Fig. 2

To a first approximation, the oscillators can be regarded as independent. The frequency of the oscillation can be computed in terms of the mass of the atom and the elastic constants of the crystal (see Richtmeyer and Kennard). According to the classical equipartition theorem, each oscillator possesses energy κT and, therefore, makes a contribution to the specific heat of κ per atom. Experimentally, it is found that the specific heat approaches zero at absolute zero, and rises asymptotically to κ per atom at high temperatures, as shown in Fig. 2. Thus, the classical theory is certainly wrong at low temperatures.

Einstein proposed that this curve could be explained by assuming that the molecular oscillators are quantized with $E = nh\nu$. In contrast to the radiation oscillators, which can have all possible frequencies, the material oscillators have only *one* frequency, which is the characteristic frequency of the substance. Applying Planck's result for a given frequency, eq. 32, we obtain

$$\bar{E} = \frac{h\nu e^{-h\nu/\kappa T}}{1 - e^{-h\nu/\kappa T}} \quad \text{and} \quad \frac{\partial \bar{E}}{\partial T} = \frac{(h\nu)^2}{\kappa T^2} \frac{e^{-h\nu/\kappa T}}{(1 - e^{-h\nu/\kappa T})^2}$$

This formula clearly predicts a specific heat per molecule of κ at high temperatures, and approaching zero at very low temperatures as

$$\frac{h^2\nu^2}{\kappa T^2} e^{\frac{-h\nu}{\kappa T}}$$

It is in general agreement with experiment, except at very low temperatures ($\sim$10°K).

The reason for the discrepancy at very low temperatures was explained by Debye.* The oscillations of each atom are actually not independent of the oscillations of the others, but are coupled to them because of the forces between molecules. The description in terms of independent oscillations is, therefore, not completely accurate.

To describe the coupled oscillations of the molecules, we may consider, for example, a one-dimensional string of particles. Suppose that each particle interacts only with its two nearest neighbors. It can then be shown† that waves are propagated through this system resembling those propagated through a chain, except that here the waves are both longitudinal and transverse, whereas the waves in a chain are transverse only. When the wavelength is large compared with the distance between particles, the propagation differs very little from that in a continuous string; but as the length of the waves approaches the mean distance between particles, the law of propagation changes. For wavelengths shorter than the mean distance between particles, propagation becomes impossible.

Problem 4: In the one-dimensional string of particles specified above, let the equilibrium distance between particles be a. Suppose the force on the nth particle to be

$$F_n = -m\omega_0^2[(x_n - x_{n-1}) + (x_n - x_{n+1})] = m\ddot{x}_n$$

Here x_n is the deviation of the nth particle from its equilibrium position.

Find solutions of the form $x_n = A_n e^{i\omega t}$, and show that we can choose $A_n = e^{in\alpha}$, where α is a suitable constant whose relationship to ω is obtained by solving the equations.

Show that for low frequencies the oscillations resemble sound waves, and that $\omega \cong 2\pi v/\lambda$, where $v = \omega_0 a$ is the speed of sound in the system. Show also that there is a maximum possible frequency.

In three dimensions, a similar treatment can be given and, in this way, we can describe the propagation of sound waves through a crystal. As was done with the electromagnetic field, we can adopt the amplitudes of the possible sound waves as co-ordinates to describe the state of the system. Since these co-ordinates oscillate harmonically with the time, the energies of the associated oscillators must be quantized.‡ In computing the energy, however, we must take into account the fact that only a finite number of wavelengths is permissible, and also that the relation

* See Richtmeyer and Kennard, p. 450.

† F. Seitz, *The Modern Theory of Solids.* New York: McGraw-Hill Book Company, Inc., 1946, pp. 121 and 125.

‡ See Secs. 10 and 13.

between frequency and wavelength becomes more complex as we approach wavelengths comparable with interatomic spacing. When all these factors have been taken into account, the quantum hypothesis leads to excellent general agreement with experimental specific heats at all temperatures. Thus, in addition to quanta of electromagnetic energy, we now have evidence for the existence of quanta of sound energy.

15. Summary. We may conclude that all systems which oscillate harmonically are quantized with $E = nh\nu$, whether these systems be material oscillators, sound waves, or electromagnetic waves. Since we assume that all systems can interact with each other, the quantization of any one type of harmonic oscillator requires a similar quantization of all other types. If experiments had not verified the existence of this unity, the quantum theory would have had to be abandoned, or at least funda mentally modified.

CHAPTER 2

Further Developments of the Early Quantum Theory

The New Concepts of the Quantum Theory

THUS FAR, quantum restrictions on allowed energies have arisen only in connection with harmonic oscillators. We shall see, however, that the results of many experiments, together with the systematic and logical development of the quantum hypothesis, lead to the conclusion that all matter is subject to quantum restrictions. This conclusion thus enables us to explain correctly a wide variety of experimental data for which the results of the classical theory are either wrong or ambiguous. As examples, we shall deal with the photoelectric effect, the Compton effect, the energy levels of material systems, and the laws governing the emission and absorption of radiation. In all these examples, we shall also study in detail how the quantum laws approach the classical limit.

1. Photoelectric Effect. We begin with a discussion of the photoelectric effect. The study of blackbody radiation would enable one to deduce indirectly that electromagnetic waves can change their energy only in units of $h\nu$; it would certainly now seem desirable to verify directly whether this statement is true by a study of the emission and absorption of radiation. The earliest experimental investigations of this problem were concerned with the photoelectric effect. These experiments showed that electrons are emitted from a metal surface* that is irradiated with light or ultraviolet rays; also, that their kinetic energy is independent of the intensity of the radiation, but depends only on the frequency in the following manner:

$$\tfrac{1}{2}mv^2 = h\nu - W \tag{1}$$

Here ν is the frequency of the incident radiation, and W is the work function of the metal or, in other words, the energy needed to remove the electron from the interior of the metal.

Einstein was the first person to relate this result to Planck's hypothesis (1905). Perusal of the data showed that h was a universal constant, and equal to the h appearing in Planck's theory. This agreement is a strong confirmation of the hypothesis that the radiation field can change energy only in units of $h\nu$. If the constant had not been obtained, the theory would have been in serious difficulties.

* Some electrons are liberated from the layers below the surface; these lose energy as a result of having to penetrate the metal.

The next important task is to try to determine why the electron absorbs energy only in quanta, independently of the intensity of the radiation. In this connection, it is worth noting that with radiation of very low intensity we simply obtain a correspondingly low rate of emission of photoelectrons.

The simplest interpretation of this phenomenon is that light consists of particles* which, because they are localized objects, can transfer all their energy to the photoelectrons during a collision. This idea is strengthened by experiments in which very low-intensity beams are directed at a photographic plate;† we obtain dark spots at random positions, with an average density proportional to the intensity of the light. In the limit of a very intense beam, the distribution of spots gets so dense that it is practically continuous.

When the beam is so intense that it seems to be continuous, it must in some way become the equivalent of what is described as a *light wave* in classical physics. Such a "classical" wave has a certain rate at which energy is incident on any surface per unit area per unit time. Let us call the rate S. Then, when many quanta are present (intense beam or low frequencies), this rate must also be equal to the mean number, N, of incident quanta per unit area per second times their energy $h\nu$. Thus, we have

$$S = Nh\nu$$

If only a few quanta are present, N must be the probability that a quantum strikes a unit area per second.

Although the assumption that light is made up of localized particles enables us to explain the photoelectric effect in a very simple way, it cannot be made consistent with the extremely wide range of experiments leading to the conclusion that light is a form of *wave motion*. As an example of the type of experiment that calls for a wave interpretation, consider the measurement of the intensity pattern of the light that strikes a screen, after being diffracted through an arrangement of one or more slits. It very often happens that when two nearby slits are open, the intensity will be very small at certain points on the screen where either slit, separately, would produce a high intensity. This result is both qualitatively and quantitatively explained by the assumption that light is made up of waves, which can interfere either constructively or destructively so that, under some circumstances, the waves coming from each of the two slits may cancel each other.

It would be impossible, however, to explain interference if one assumed that light was made up of localized particles. Such particles would have to go through either one slit or the other, and the opening of a second slit

* A particle is an object that can always be localized within a certain minimum region, which we call its *size*.

† Ruark and Urey. (See list of references on p. 2.)

could hardly prevent a particle from reaching a certain point to which the particle would be free to go if this second slit were closed. On the other hand, the assumption of wavelike properties for light not only explains this particular experiment, but also a whole host of other experiments involving radiation—from radio waves to x rays. It is, therefore, certainly desirable to try to understand the appearance of quanta in terms of the wave theory of light, if possible.

To do this, we now consider the classical account of what happens in the photoelectric effect. When radiation strikes an electron vibrating within an atom, it transfers energy to the electron. If the electric field oscillates at a frequency that is resonant with the frequency of the electron in the atom, the electron will absorb energy from the light wave until it is liberated. One could try to explain the photoelectric effect by assuming that the properties of the atom are such that the electron would keep on gaining energy until it had picked up an amount equal to $h\nu$, after which it would be ejected. If atoms had these properties then, with very weak light, the photoelectric effect should not be observed for a long time, since it would take a long time to store the necessary quantum of energy. Experiments were conducted, however, with metallic-dust particles and very weak light. These dust particles were so small that it would have taken many hours to store $h\nu$ of energy; yet, some photoelectrons were found to appear instantaneously.

To explain the above result, we could suppose that the metal contained electrons with all sorts of energies and, when the metal was struck by a light wave, a few electrons of appropriate energy could be liberated immediately. If we consider a case in which $h\nu \gg W$, however, it seems unlikely that electrons with so large a surplus of energy would remain indefinitely inside the metal, until their release was triggered by a light wave of exactly the right frequency. Moreover, it has been found that no matter how we try to release an electron from a metal (for example, by bombardment of the metal by protons or by other electrons), we must always supply the same minimum energy, equal to the work function W. Similarly, it has been found that electrons cannot be liberated from atoms of a gas, unless a certain minimum energy equal to the ionization potential I is supplied (see the discussion of the Franck-Hertz experiments in Sec. 15). Yet, some electrons are liberated instantaneously by very weak light from gas atoms with a kinetic energy equal to $h\nu - I$. In view of all this evidence we must, therefore, rule out the possibility of explaining the photoelectric effect by assuming that some electrons initially possess nearly all the energy with which they escape.

If electrons in metals had such a range of energies, then it would be difficult to make the quantum hypothesis self-consistent, because only part of a quantum would have to be absorbed to liberate a typical electron. According to Planck's hypothesis, however, the radiation oscillators can supply a minimum of a full quantum in each absorption process. What would then happen to the rest of the quantum if only part of it were absorbed by the electron?

These particular efforts fail to explain the photoelectric effect in terms of a process of gradual accumulation of energy, and every similar attempt that has ever been carried out has also failed. This means that the wave theory is unable to account for the sudden appearance of finite amounts of energy on a single electron. We are, therefore, in a quandary. One set of experiments suggests that light is a *particle* that can be localized, and the other suggests, with equal emphasis, that it is a *wave.* Which approach leads to the correct picture? The answer is, neither.

Before we can obtain a correct theory of the wave-particle duality of the properties of light, we shall see that it is necessary to make radical changes in some of our most fundamental concepts dealing with the properties of matter and energy. These new concepts will be developed through the remainder of this book, but primarily in Chaps. 6, 8, and 22. For the present, however, we merely state that light must be regarded as existing in the form of *fundamental units,* or *quanta,* which can, in some circumstances, act like particles and, in other circumstances, like waves. We find a strong analogy here to the fable of the seven blind men who ran into an elephant. One man felt the trunk and said that "an elephant is a rope"; another felt the leg and said that "an elephant is obviously a tree," and so on. The question that we have to answer is: Can we find a single concept that will unify our different experiences with light, just as our concept of the elephant unifies the experiences of the seven blind men?

2. Differences between Classical and Quantum Laws of Physics. Our first step in the program of developing the new concepts needed in quantum theory will be to bring out two crucial differences between the kind of physical law obtained in classical theory and the kind suggested by experience with quantum phenomena. The first difference is that whereas classical theory always deals with *continuously varying quantities,* quantum theory must also deal with *discontinuous or indivisible processes.* The second difference is that whereas classical theory completely determines the relationship between *variables at an earlier time* and those at a *later time* (i.e., it is completely causal), quantum laws determine only *probabilities of future events* in terms of given conditions in the *past.*

3. The Indivisibility of Quantum Processes. Let us now consider some of the experimental evidence that indicates the need for introducing the concept of discontinuous or indivisible processes into the quantum theory. The first important piece of evidence comes from the photoelectric effect. We have already seen, for example, that while all efforts to explain the photoelectric effect as a process of gradual transfer of energy from radiation field to matter have failed, the assumption that the transfer of energy is a discontinuous process that takes place in jumps of size $\Delta E = h\nu$ is in agreement with all the experiments dealing with this phenomenon. Moreover, the same assumption is also required by Planck's hypothesis that the energy of the radiation oscillators is restricted to discrete values. That is, if the transfer of energy took place gradually, it would be necessary to consider states in which radiation oscillators had part of a quantum and, according to Planck's hypothesis, no such state is possible.

As we shall see later, there are many other experiments which demand the interpretation that the transfer of energy is a discontinuous process. For the present, we shall offer an experiment by Lawrence and Beams,* who tried to break up a light quantum into two parts by means of a very fast shutter, utilizing a Kerr cell that could be activated in 10^{-9} sec. If the light wave were continuous, as described by classical theory then, with the intensities of light used, it would have taken much longer than 10^{-9} sec for a full quantum of energy to come through. Thus, we should expect that the shutter would break up the quanta into smaller quanta. They found, however, that none of the quanta was ever broken up.

If we combine Planck's hypothesis with the fact that no one has ever been able to perform an experiment in which a part of a quantum has been detected, we are led to the conclusion that a quantum is an *indivisible unit of energy*. We may also see, from the failure of all attempts to follow the energy gradually, that the transfer of a quantum from one system to another is an *indivisible process*. The indivisibility of the quantum of energy, and the indivisibility of the process of transfer go together; they are necessary for each other's logical self-consistency. We should conclude, therefore, that in the transfer of a quantum, the system cannot be regarded as passing through a succession of intermediate states, in which the energy is exchanged in a continuous fashion. Instead, the quantum process must be regarded as discontinuous and as an indivisible unit. The transfer of a quantum is one of the basic events in the universe and cannot be described in terms of other processes. It may be called an *elementary process*, just as a proton or an electron is called an *elementary particle*, because it does not seem to be made up of other particles.

4. Probability and Incomplete Determinism in Quantum Laws. The indivisibility of quantum processes is totally at variance with classical physics, which describes all processes in a continuous fashion, each change being caused by the state of the system just before the change took place. Since classical laws presuppose the existence of continuous processes to which they apply, it is clear that discontinuous quantum jumps cannot be predicted by our classical laws. Our problem is, then, to find the new laws governing quantum transfers.

* Ruark and Urey, p. 83.

We come now to the second important difference between classical and quantum laws. It is an experimental fact, exemplified in the photoelectric effect and in a wide range of other experiments not yet studied here, that no law has been discovered which predicts exactly where and when an individual quantum will be transferred. Instead, only the probability of such a process may be predicted. For example, if only one quantum is directed at a metal surface, it is impossible to predict whether it will be absorbed and, if it is absorbed, exactly where and when. But if a beam contains many quanta, it is possible, from the intensity of the light used, to predict the mean number absorbed in any given region. Thus, in this case, quantum laws appear to control only the *probability* of an event and cannot predict its *occurrence* with certainty. We shall see that this behavior is not restricted to the photoelectric effect but is common to all quantum processes.

Thus, we can see that quantum laws are very different from their classical counterparts, which always imply that the behavior of the system is completely determined by exact causal laws. For example, all material particles obey Newton's equations of motion, $m\ddot{x} = \mathbf{F}$. Once the initial position and velocity of each particle are given, the future motion is determined exactly by the differential equations of motion. Thus, the trajectory of an electron is determined by three quantities:

(1) The position at any instant of time.
(2) The velocity at that time.
(3) The value of the force $\mathbf{F}$ at all times.

For an electrical particle, the force $\mathbf{F}$ is determined by the electric and magnetic fields. But these can be calculated exactly with the aid of Maxwell's equations and the initial values of electric and magnetic fields everywhere. Hence, according to classical physics, the motion of a charged particle (also of any other kind of particle) can be determined precisely for all time, once certain initial conditions are known. The same can be said about changes of the electromagnetic field. Classical theory may therefore be called *completely deterministic.*

Applying these general ideas, one concludes from classical theory that, in a light beam of a given intensity, electrons gain energy at a continuous rate, which is calculable from the light intensity and from the initial conditions of the electrons. On the other hand, experiments show that the process of energy transfer is discontinuous and apparently not governed exactly by deterministic laws, at least not by the deterministic laws of classical mechanics. Instead, so far as we can find out from experiment, only the probability of the process is determined.

At this point, it is worthwhile to go more deeply into the connection between the appearance of probability and the indivisibility of a quantum process. First, there is the previously mentioned fact that many classical laws (including Newton's equations of motion) which are essential for the operation of classical determinism must, by their very nature, refer to gradual and continuous processes. Hence, if only because this kind of law has no meaning in discontinuous processes, it cannot apply directly to quantum transfers. Some classical laws, however, do not require us to follow particles through a continuous path in space time, for example, conservation of energy, momentum, or angular momentum. Even in an impulsive collision in which we cannot follow the motion continuously, these laws apply for the collision as a whole. Such laws do have meaning even in discontinuous processes. It is an experimental fact that these laws can all be taken over directly into the quantum theory. For example, it has been shown experimentally that energy is always conserved in the photoelectric effect. Many other experiments also yield this result. Hence, not all classical deterministic laws must be abandoned, but only these requiring a description in terms of continuous processes.

5. Unlikelihood of Completely Deterministic Laws on a Deeper Level. One might wonder whether the appearance of probability in quantum processes is not a result of our ignorance of the correct variables to use in describing the system. In classical physics, probabilities often appear for just this reason. For example, in thermodynamics we measure the pressure, temperature, and volume of a given system. In very small regions of space, especially near the critical point, we find that these quantities no longer obey an equation of state exactly, but instead exhibit large random fluctuations about a mean value that is predicted by the equation of state. Hence, the deterministic laws of thermodynamics break down and are replaced by laws of probability. This is because the thermodynamic variables are no longer appropriate for the problem and must be replaced by the position and velocity of each molecule, which are, from the viewpoint of thermodynamics, hidden variables. The thermodynamic quantities are, then, merely averages of hidden variables that cannot be observed by thermodynamic methods alone. To find the underlying causal laws, we must accept a description in terms of the individual molecules.

The idea immediately suggests itself that probability in quantum processes arises in a similar way. Perhaps there are hidden variables that really control the exact time and place of a transfer of a quantum, and we simply haven't found them yet. Although this possibility cannot be absolutely ruled out, we can show that this is unlikely. The first point, of course, is that no experiment has yet shown the slightest trace of such hidden variables. The second point is that there are strong theoretical arguments which make it unlikely that such hidden variables exist. These will be discussed later (Chap. 22, Sec. 19). For the present, we shall merely assert, as a general principle, that only the probability of a quantum jump can be determined by the physical state of the system.

6. Correspondence Principle. Thus far, we have seen the need for introducing two nonclassical ideas. First, energy levels of harmonic oscillators are restricted to the values of $E = nh\nu$, with the result that energy transfers to and from such oscillators take place in quanta with $\Delta E = h\nu$. These quanta are indivisible, so that the energy involved is either all in the oscillator or all out of it. Second, only the probability of the transfer of a quantum is determined by the physical state of the system. How is it possible for these ideas to be consistent with the fact that, in the realm of ordinary experience, motion appears to be continuous and can be described by deterministic laws, such as Newton's equations of motion?

The apparent continuity of motion on a macroscopic scale is, of course, a result of the smallness of the quantum. When we have processes involving many quanta, the discontinuity is so small that it is not ordinarily visible. We must remember that most processes in classical physics involve rather low frequencies, not greater than 10^{10} cps. The energy jump $\Delta E = h\nu$ is about 10^{-16} erg at this frequency, a very small value. Suppose we want to see what a radio wave of this frequency does to an electron. Before the electron absorbs an energy corresponding to that which it would gain in falling through a potential drop of only 1 volt (1 ev $= 1.6 \times 10^{-12}$ erg), it must absorb 10^4 quanta. On the other hand, with light of frequency about 10^{15} cps, one quantum requires about 5 ev. Thus, at higher frequencies, quantization is much more important.

As for the appearance of apparently exact causal laws on a macroscopic scale, when only the probability of each elementary quantum transfer is determined, we merely note that, where many quanta are involved, the probability becomes *almost* a certainty (but not quite). This is very similar to the exact prediction, by insurance statistics, of the *mean* lifetime of a person within a large group, even though an *exact* prediction of the lifetime of a single individual in the group is not possible.

Let us, for example, consider again the interaction of an electron with a radio wave. Although there is only one electron it gains, as we have seen, many quanta from the radiation oscillators in a very short time. No one can predict exactly when and where any individual quantum will be transferred but, on the average, even in a microsecond, so many quanta are transferred that the mean energy given to the electron can be predicted within very narrow limits. In this way, the classical deterministic laws are still valid for all practical purposes, although only the probability of an elementary quantum process is determined.

A similar analysis may be made for other classical processes. For example, a planet in a gravitational field can be regarded as absorbing gravitational quanta that are emitted by the central star. These gravitational quanta carry gravitational momentum and energy, just as electromagnetic quanta carry electromagnetic momentum and energy. We visualize the sun as continually throwing out and reabsorbing gravitational quanta, producing a steady state of the mean number of quanta present in space. If we imagine that a planet can absorb quanta only when they are returning to the sun, we see that an inward force is produced by an enormous number of tiny impulses. The planet also emits quanta that are absorbed by the sun. Thus, the two bodies attract each other, and energy is conserved. Because there are so many of these impulses, even in a very short time, the average force will be practically constant.

The preceding description is merely a rough approximation of what occurs in this case; it should not be taken literally, but it is substantially accurate. To obtain a more accurate and detailed description, it is necessary to study, first, the general theory of the quantization of force fields, which is beyond the scope of this book.*

The ideas contained in the preceding examples are stated more generally in the form of the correspondence principle, which was first given by Bohr. This principle states that the laws of quantum physics must be so chosen that in the classical limit, where many quanta are involved, the quantum laws lead to the classical equations as an average. The problem of satisfying the correspondence principle is by no means trivial. In fact, the requirement of satisfying the correspondence principle, combined with indivisibility, the wave-particle duality, and incomplete determinism, will be seen to define the quantum theory in an almost unique manner.

7. Particle Properties of Light. Let us now consider the particle-like aspects of light in more detail. We have seen that a radiation oscillator can gain or lose energy only by the transfer of a whole quantum at once, i.e., with $\Delta E = h\nu$. If the oscillator is excited to the nth quantum state, it has energy $E = nh\nu$, which it can lose in n steps. Then, so far as the energy relations are concerned, there appear to be n equivalent particles present, each with energy $h\nu$. These equivalent particles are often called *photons*.

It is natural to ask, "What about the momentum of the equivalent particles?" Now, from electrodynamics it can be shown that the radiation field possesses momentum as well as energy. With the aid of Maxwell's equations, we can prove that this momentum is given by

$$p = \frac{1}{4\pi c} \int (\mathcal{E} \times \mathcal{H})\, d\tau \tag{2}$$

For a light wave in empty space, we know that $\mathcal{E}$ is normal to $\mathcal{H}$, and $|\mathcal{E}| = |\mathcal{H}|$. Hence, $\mathcal{E} \times \mathcal{H}$ is a vector normal to both $\mathcal{E}$ and $\mathcal{H}$, and therefore in the direction of propagation, $\boldsymbol{k}$. Its magnitude is

* G. Wentzel, *Quantum Theory of Fields*. New York: Interscience Publishers, Inc., 1948.

$$|\mathcal{E} \times \mathcal{H}| = \mathcal{E}^2 = \frac{\mathcal{E}^2 + \mathcal{H}^2}{2}$$

and since the energy of the wave is

$$E = \frac{1}{4\pi} \int \frac{\mathcal{E}^2 + \mathcal{H}^2}{2} \, d\tau$$

the momentum becomes

$$\boldsymbol{p} = \frac{E}{c} \hat{k} \tag{3}$$

where $\hat{k}$ is a unit vector in the direction of propagation.*

Definite evidence for the momentum of light is found in many places, for example, in the radiation pressure which is caused by the absorption of the momentum carried by light.

Problem 1: A radio antenna radiates 500 kw in a certain direction. What is the reaction on the antenna in dynes?

Let us now consider how the momentum of the radiation is affected by the quantization of energy. Since the energy comes in units of $h\nu$, the momentum ought to come in units of $h\nu/c$, or

$$\boldsymbol{p} = \frac{h\nu}{c} \hat{k} = \frac{h}{\lambda} \hat{k} = \hbar \boldsymbol{k} \tag{4}$$

where $\hbar = h/2\pi$. (Equation (4) is a special case of the de Broglie relation.†) Now, the energy and momentum of a particle of mass m are related by

$$\frac{E^2}{c^2} = m^2c^2 + p^2$$

Thus, the energy-momentum relation for a light quantum is the same as that for a particle of zero rest mass, traveling with the velocity of light.

We may, therefore, conclude that when a radiation oscillator is excited to its nth quantum state, it has energy $E = nh\nu$, and momentum $\boldsymbol{p} = n\hbar\boldsymbol{k}$, where n can change by one unit at a time. Thus, its energy and momentum behave like that of a collection of n particles, each with energy $h\nu$ and momentum to $\hbar\boldsymbol{k}$. We can, therefore, specify the state of excitation of the radiation field by specifying the number of equivalent particles corresponding to each $\boldsymbol{k}$.

We have seen that the electromagnetic field can interact with matter only by means of indivisible processes in which a full quantum is either emitted or absorbed. If we wish to describe these processes in terms of

* This relation can also be obtained from the theory of relativity. See R. C. Tolman, *Relativity, Thermodynamics, and Cosmology*. Oxford: Clarendon Press, 1934, Chap. 3.

† Chapter 3, eq. (19).

the language of equivalent particles, we must then say that interactions between matter and light take place only by means of the emission and absorption of photons.

8. Compton Effect. The Scattering of Electromagnetic Radiation. It is well known that electromagnetic radiation can be scattered by charged particles that are free to respond to the incident electromagnetic wave.* In the quantum theory, this process must be described as the absorption of a quantum from the incident light, and the emission of another quantum in a new direction. Insofar as energy-momentum relations are concerned, however, this process can equally well be described as the scattering of a single particle, which is not destroyed, but which merely suffers a change of energy and momentum.

Compton investigated the particle properties of light experimentally by scattering x rays from electrons. In this experiment, a beam of x rays of frequency ν was sent through matter. According to Sec. 7, the beam should act like a collection of particles, each with energy $h\nu$ and momentum $\hbar \mathbf{k}/c$. Occasionally a quantum is scattered by an electron,† deviating at an angle ϕ from the direction of the incident beam, and with a frequency ν'; the electron appears at an angle θ, as indicated in Fig. 1.

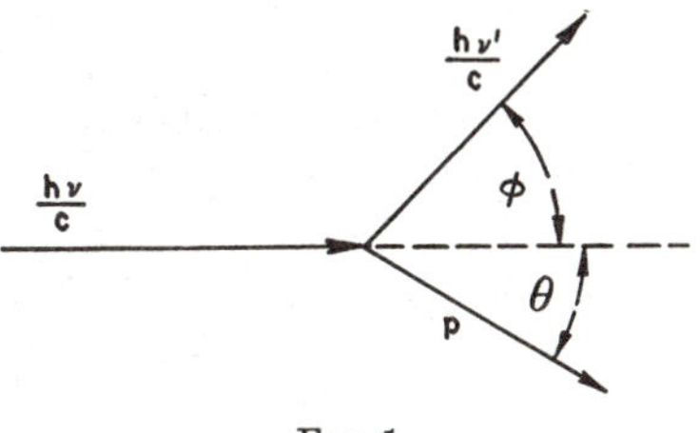

FIG. 1

All these quantities can be observed experimentally. If energy and momentum are conserved in individual scattering processes (as must happen if x-ray quanta are to act like particles), we can show that

$$\lambda' - \lambda = \frac{2h}{mc} \sin^2 \frac{\phi}{2} \tag{5}$$

m denoting the mass of the electron.

Problem 2: Prove the above result of eq. (5).

The quotient h/mc is called the *Compton wavelength* of the electron. (Its value is 2.42×10^{-10} cm.) Note that the scattered quantum always has a longer wavelength than the incident quantum.

Compton's experiments verified eq. (5) and thus demonstrated that the energy and momentum of light are quantized according to $E = h\nu$, $\mathbf{P} = \hbar \mathbf{k}$, and also that energy and momentum are conserved in individual scattering processes. The conservation laws can be more completely verified by studying the recoil electrons, and subsequent experiments

* Richtmeyer and Kennard, p. 476.

† The electronic energies resulting from motion in atomic orbits are much less than the energy of the x rays and, hence, can be neglected.

showed that the electron gained the same momentum and energy that the quantum lost.

9. Analysis of the Compton Effect. By means of a detailed analysis of this experiment, we can demonstrate the intricate and subtle relations between the classical and quantum theories, which are brought about through the correspondence principle.

To do this, let us first consider the classical interpretation of the scattering of light waves from an electron. According to classical theory, the incident light wave provides an oscillatory electric field, $\mathcal{E} = \mathcal{E}_0 \sin \omega t$, which sets the electron in oscillatory motion, and causes it to radiate symmetrically about the plane normal to the direction of the incident radiation. As a result, the total momentum radiated is zero. The momentum carried by that part of the light that was removed from the incident beam must, therefore, go into the electron. This produces a radiation pressure on the electron, which causes it to accelerate. To obtain the magnitude of the net radiation force on the electron, we can use a formula derived by Thomson,* for the rate at which energy is scattered out of an incident beam of intensity I (expressed in ergs per square centimeter per second). This result is $\frac{dW}{dt} = \frac{8\pi e^4}{3m^2c^4} I$. Since the momentum absorbed from the beam is W/c, the electron gains momentum at a rate given by

$$\frac{dP}{dt} = \frac{8}{3} \frac{\pi e^4}{m^2c^5} I \tag{6}$$

We can obtain a more direct picture of the mechanism producing radiation pressure by considering a wave incident in the z direction, with electric field in the x direction, and magnetic field in the y direction. Thus far, we have neglected magnetic forces, because the total force is $e\mathcal{E} + v/c \times \mathcal{H}$, and since $|\mathcal{E}| = |\mathcal{H}|$ in free space, the term involving v/c produces only a small effect on the motion, unless the electron moves with a speed close to that of light. Whenever $v/c \ll 1$, as is usually the case, we can then solve for the motion of the electron to a first approximation by neglecting the term, $v/c \times \mathcal{H}$, and then obtain a more accurate approximation by taking it into account as a perturbation. We find that, taken over a period, the average electric force on the electron vanishes, but that there is a component of magnetic force in the z direction, whose time average does not vanish. Calculation shows that it is equal to the value given in eq. (6).

Problem 3: It is found that an accelerated electron obeys the following equation of motion:

$$m\ddot{x} = \left(e\mathcal{E} + \frac{ev}{c} \times \mathcal{H}\right) - \frac{2}{3}\frac{e^2}{c^3}\dddot{x}$$

The latter term represents the force, arising from the reaction of the radiated field back on the electron.† Assuming that $\mathcal{E}_x = \mathcal{E}_0 \sin \omega t$, and $H_y = \mathcal{E}_0 \sin \omega t$, solve for the steady state of oscillation of the electron, neglecting magnetic forces. Show that with this motion the average electric force vanishes, but that the average mag-

* See J. J. Thomson, *Conduction of Electricity through Gases*. New York: Macmillan, 2nd ed., p. 321; also, Richtmeyer and Kennard, p. 477.

† H. A. Lorentz, *Theory of Electrons*. Leipzig: B. G. Teubner, 1909.

netic force has a nonvanishing component in the z direction, equal in value to that given in eq. (6). Use the fact that

$$\frac{e^2}{mc^3} \ll \tau = \frac{2\pi}{\omega}$$

As an electron gains speed, there is a resultant Doppler shift toward the lower frequencies. This shift occurs in two parts. First, the electron is receding from the light beam, so that it experiences a field of frequency lower than that of the incident light. Then, in the process of radiation, another Doppler shift comes in, which cancels the first one in the forward direction but doubles it in the backward direction. We can show, in fact, that the angular dependence of the Doppler shift is precisely that given by the eq. (5) for the Compton effect [see eq. (7a)].

Problem 4: Suppose that an electron initially at rest is exposed to a light beam of intensity I, wavelength λ_0, for a time T. Show that the Doppler shift is (in a nonrelativistic theory, where $v/c \ll 1$)

$$\lambda - \lambda_0 \cong 2\lambda_0 \frac{v}{c} \sin^2 \frac{\phi}{2} \tag{7a}$$

where v is the electron velocity. Show also that the Doppler shift is equal to

$$\lambda - \lambda_0 \cong \frac{2W\lambda_0}{mc^2} \sin^2 \frac{\phi}{2} \tag{7b}$$

where W is the total energy removed from the incident light beam.

According to classical theory, this Doppler shift will gradually increase with time, as the particle gains energy. Furthermore, it should be possible for the particle to gain any amount of energy, so that all possible values of the Doppler shift should be observable at a given angle as the intensity of radiation or the time of exposure are changed. Experimentally, it is found that only one value of the wavelength shift is observed at a given angle, independently of the intensity of the radiation and time of exposure. These facts indicate that the process of transfer of energy and momentum is not continuous as predicted by classical theory but is indivisible as suggested by quantum theory.

Let us now consider how the classical limit is described in terms of quantum processes. Suppose, for example, that an electron is struck by a radio wave, with many quanta in it. The electron keeps on scattering these quanta, but the maximum possible increase in wavelength in a single process is 10^{-11} cm, too small a value to be detected in a radio wave of length of the order of centimeters or more. On the other hand, the particle continues to gain energy, as it is struck by more and more quanta and, eventually, the velocity becomes so high that the Doppler shift becomes appreciable. In this way, we obtain a Doppler shift that appears to increase in a continuous fashion, as demanded by classical theory. (With x ray beams, however, the fractional shift in wavelength per quantum process is appreciable, so that the effects of indivisibility are important.)

At any particular stage, the frequency shift can be calculated quantum-mechanically from the conservation of energy and momentum, plus the Einstein-de Broglie relations* ($E = h\nu$, $p = h\nu/c$). To do this, we need only obtain the Compton shift for a photon of incident frequency ν_0, scattered from an electron moving initially with momentum p_0. To simplify the problem, let us suppose that the initial particle momentum is in the direction of the incident photon. The change of wavelength is then equal to

$$\lambda_0 - \lambda = \frac{2(h + \lambda_0 p_0)}{\sqrt{m^2c^2 + p_0^2} - p_0} \sin^2 \frac{\phi}{2} \tag{7c}$$

Problem 5: Verify eq. (7c).

To obtain the classical limit, we must assume that the change in momentum δp, occurring during the indivisible scattering process, is small compared with p_0. For the nonrelativistic case ($v/c \ll 1$) this momentum transfer is of the order of $h\nu_0/c = h/\lambda_0$, so that in the classical limit

$$\frac{h}{\lambda_0} \ll p_0 \qquad \text{or} \qquad \lambda_0 p_0 \gg h$$

Thus, eq. (7c) becomes

$$\lambda_0 - \lambda \cong 2\lambda_0 \frac{v_0}{c} \sin^2 \frac{\phi}{2}$$

in agreement with the classical value given in (7a).

It is significant that the values of the Doppler shift, calculated from the quantum theoretical assumption that the scattering process is *indivisible*, agree in the classical limit with the values obtained from classical theory on the basis of a totally different description involving, among other things, the assumption that the process is *continuous*. We should readily see that the origin of this agreement lies in the character of the Einstein-de Broglie relations that connect energy and momentum changes associated with indivisible quantum process with changes of frequency and wavelength associated with continuous classical process. Therefore, even at this early state of the development, the interpretation of the Compton and photoelectric effects in terms of indivisible quantum processes not only leads to agreement with the specific experiments on which this interpretation is based, but also leads to a theory having built into it the correct approach to the classical limit. This result is our first example, demonstrating that the application of the correspondence principle is far from trivial (see Sec. 6). As we proceed to develop the subject further, the close agreement between classical and quantum theories in regard to the correspondence limit will become more evident.

It is interesting to note that we obtain the correct quantum-mechanical frequency shift by setting $W = h\nu$ in the classically derived eq. (7b). On the basis of this

* The origin of the term "de Broglie relation" is explained in Chap. 3, Sec. 8

result, we might suggest from a naive point of view, that the electron seems to be scattering radiation as described classically, except that it is for some reason restricted to dealing with energy only in bundles of size $h\nu$. (This suggestion is very similar to an unsuccessful explanation of the photoelectric effect described in Sec. 1.) If the Compton effect occurred in this way, however, the frequency shift would still vary over a continuous range of values, from zero to a maximum, while the electron was being accelerated. Thus, we would not obtain agreement with experiment, which shows that at a given angle there is but one definite frequency shift, in agreement with the hypothesis that the scattering process is indivisible.

The fact that we obtain the correct Compton shift by setting $W = h\nu$ in eq. (7b) is related to the correspondence principle, but not in a rigorous way. This relation arises in the circumstance that if one takes the effects of indivisibility into account then, in the classical limit, the main further effect of quantization is to confine the changes of energy to integral multiples of $h\nu$. Of course, the Compton effect itself is normally observed when we are far from the classical limit, but it turns out somewhat by accident that, for this case, the results obtained by extrapolating the procedure suggested above into the quantum domain are correct even though the procedure is rigorously justifiable only in the classical limit. In most cases, as we shall see in later sections of this chapter, such crude extrapolations lead only to approximate formulas that are likely to be wrong by a factor of the order of 2 or 3, but which give a generally correct order-of-magnitude estimate of quantum-mechanical effects. In this connection, we point out also that the only rigorous expression of the correspondence principle for the Doppler shift is given in eq. (7c).

The correspondence principle applies not only for the frequency shifts, but also for the mean energy radiated. Thus, we know (see Secs. 3 and 4) that quantum laws yield only the probability of indivisible energy transfers from the radiation field to the electron, whereas classical laws yield deterministic expressions for the rate at which energy is transferred continuously. In the classical limit, where many quanta are involved, the average rate of energy transfer calculated from quantum probabilities, must agree with the definite rate of energy transfer, calculated from Newton's law of motion. The probability of scattering a quantum per unit time S must, therefore, be so chosen that the mean momentum absorbed per unit time is equal to the classical rate, so that

$$S\frac{h\nu}{c} = \frac{1}{c}\frac{dW}{dt} \qquad \text{or} \qquad S = \frac{8\pi e^4}{3h\nu m^2c^4} I$$

With this choice we obtain the correct classical causal laws applying to this case, where so many quanta are scattered that the deviation of the actual from the probable result becomes negligible.

Problem 6: Given a beam of x rays of frequency, $\nu = 10^{21}$ cps, with an intensity of 1 watt/cm^2. What is the probability that a quantum will be scattered from a single electron in one second? Compare the results with those obtained with an electromagnetic wave of frequency $\nu = 1$ cps of the same intensity. In which case do the causal laws apply?

The Quantization of Material Systems

It has already been shown, in connection with specific heats of solids, that material harmonic oscillators have their energies quantized in the same way as do the radiation oscillators. Furthermore, from the photoelectric and Compton effects, we have seen that the classical laws of conservation of energy apply in *each* transfer of a quantum between radiation and matter. If we then consider that the equilibrium distribution of blackbody radiation is independent of the material of which the walls are made, we are forced to the conclusion that *all* matter can absorb energy from radiation only in quanta with $\Delta E = h\nu$. The photoelectric effect and the Compton effect further bear out this idea in a more direct fashion. The simplest explanation for these facts is to assume that the energy levels of *all* matter are restricted to discrete values. When Bohr first proposed this idea, it seemed implausible, but we now have a great evidence in its support.

10. Evidence for Quantization of All Material Systems. In addition to the above arguments that make the idea plausible, there are some strong experimental grounds for believing in discrete energy levels for all material systems.

First, there is the problem of the stability of atoms. According to classical theory, accelerated electrons radiate energy at a rate equal to $\frac{2}{3}\frac{e^2}{c^3}|\ddot{x}|^2$. But electrons in atomic orbits are always accelerated and should, therefore, lose energy continuously until they fall into the nucleus. Actually, it is known that they stop radiating long before this happens. This fact strongly suggests that there is a minimum energy possible in the atom, corresponding to the lowest discrete quantized state of energy, and that radiation stops when this state is reached.

According to classical theory, an electron in a given orbit should radiate light having either the frequency of rotation in the orbit, or else some harmonic of this frequency. If, for example, the electron moves in a circular orbit with uniform speed, then only the fundamental should be radiated. In a highly elliptic orbit, however, the particle speeds up a great deal as it approaches the nucleus; and this produces a sharp pulse of radiation, which is repeated periodically. This sharp pulse produces corresponding harmonics in the radiated frequency.

Now, the frequency of rotation depends on the size and shape of the orbit, which are, according to classical mechanics, continuously variable. Hence, there should be a continuous distribution of frequencies in the spectrum, emitted by an excited atom. Actually, each type of atom emits a discrete group of frequencies,* that is characteristic of that atom. If there were a set of discrete energy levels then, according to the relation $\Delta E = h\nu$, one could readily explain the emission of discrete frequencies.

Furthermore, according to classical physics, if a given frequency ν is emitted, then various harmonics of this frequency may also appear,

* Since each frequency leads to a corresponding line in the spectrum of the atom, this result means that the spectrum is discrete, whereas classical physics predicts a continuous spectrum.

depending, as we have seen, on the nature of the orbital motion. Actually, it is found that observed groups of frequencies do tend to be produced together, but these do not stand to each other in the ratio of harmonics. Instead, experiments show that if two such lines with frequencies ν_1 and ν_2 occur, we are also likely to find the related frequencies $\nu_1 + \nu_2$ or $\nu_1 - \nu_2$. This rule of combination is known as the *Rydberg-Ritz principle.* It fits nicely into the idea that there are corresponding energy levels, as shown in Fig. 2.

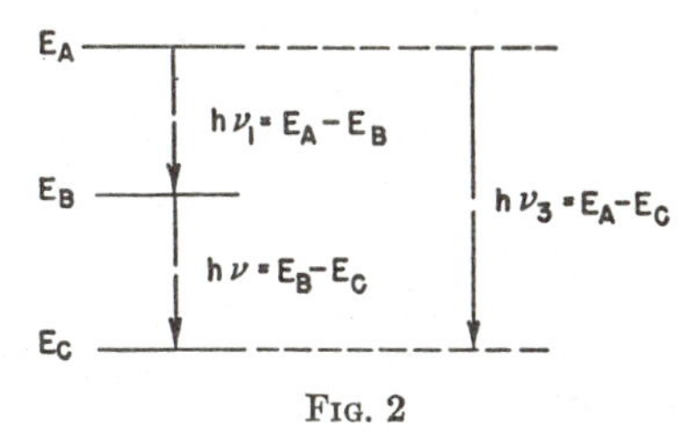

Fig. 2

We shall see later that the Rydberg-Ritz principle is the quantum-mechanical analogue of the appearance of harmonics in classical theory; in fact, in the classical limit of high quantum numbers, this principle leads to the prediction that harmonics are radiated.

There is a great deal of evidence for the idea that all kinds of motions are restricted to quantized energy levels. The spacing between the levels, however, need not in general be uniform as it is for the harmonic oscillator. In fact, evidence from spectral lines shows that it is not uniform.

11. Determination of Energy Levels. Our next problem is to discover how to calculate the spacing between energy levels. To study this question let us restrict ourselves for the present to a system with only one degree of freedom that undergoes anharmonic periodic motion. Such a system might, for example, be an anharmonic oscillator (in which the restoring force is not proportional to the displacement) or an electron in a hydrogen atom. The characteristic of anharmonic motions is that the period is a function of the amplitude and, therefore, of the energy. For example, the period of a pendulum increases with large amplitudes, but the period of rotation of an electron in an atom also increases as the orbit becomes larger. Hence, in general, $\nu = \nu(E)$, and only for a harmonic oscillator is ν independent of E.

Let us now consider this problem: What determines the energy levels? First, we know that the spacing between levels is $\Delta E = h\nu$, where ν is the actual frequency of the emitted light. But in the classical limit, ν is a definite function of E, which can be calculated from the equations of motion. If the correspondence principle is to be affirmed, these two frequencies must agree, at least in the classical limit of large quantum numbers. We obtain*

$$\Delta E = h\nu(E)$$

* Note that the present argument is very similar to the discussion of the Compton effect given in Sec. 9. In particular, the calculation of frequency shifts (and therefore changes of energy) from the classically derived eq. (7b), by setting the change of

The energy levels are, in principle, already determined by this formula. Starting from some arbitrary zero, the energy in the nth state is

$$E_n = \sum_0^n \Delta E_n = \sum_{n'}^n h\nu(E_0) + K \tag{8}$$

where n' is an integer large enough so that we remain in the classical region, and K is an arbitrary constant.

There is some ambiguity in eq. (8), because ΔE_n is associated with two energy levels, and we do not know exactly which of these to use in computing $\nu(E_n)$. Some average is perhaps the best value. In the classical limit, this ambiguity is negligible because $\Delta E_n \ll E_n$ and, in general, it will be important only when the change in frequency $\Delta\nu$, resulting from the change of energy $\Delta E = h\nu$, becomes comparable with ν, or when

$$h\frac{\partial \nu}{\partial E} \cong 1$$

For the harmonic oscillator $\partial\nu/\partial E = 0$, so that this method of quantization is correct for all energies. Even if $h(\partial\nu/\partial E) \cong 1$, the method should give at least an estimate of the energy.

A somewhat more elegant method exists for evaluating the energy levels, which is also instructive in showing the relation between classical and quantum mechanics. We first define a function

$$J_n = \sum_{n'}^n \frac{\Delta E_n}{\nu(E_n)} + J_{n'} \tag{9}$$

where n' is, again, a suitably large number.

By definition, J_n changes by h, when n changes by unity. In the classical limit, however, J_n undergoes only a small fractional change with a unit change of n and, as we have seen, $\nu(E_n)$ also undergoes only a small fractional change. The finite difference ΔE_n may, therefore, be regarded as a differential, and the sum may be approximated as an integral. We obtain

$$J(E) = \int_{E_0}^E \frac{dE}{\nu(E)} + J(E_0) \tag{10}$$

Differentiation gives

$$\frac{dJ}{dE} = \frac{1}{\nu(E)} = T(E)$$

where $T(E)$ is the period. Classically, $J(E)$ is a well-defined function capable of taking on continuous values. According to quantum theory, however, it can take on only discrete values, differing by h. This is the

---

energy (in this case W) equal to $h\nu$, presents a close analogy to the calculation of the energy difference ΔE from the classically calculated frequency $\nu(E)$.

general quantization condition. For the harmonic oscillator $\nu(E)$ is independent of E. Hence $J = E/\nu$, and $\Delta J = h = \Delta E/\nu$. This is the usual quantum condition for the harmonic oscillator. Although this method of obtaining energy levels is rigorous for the classical limit only, it yields approximately correct results, even when E_0 is allowed to go down into the quantum region.

12. The Action Variable. The quantity J was widely used in classical mechanics, even before the development of quantum theory. It is called the *action variable.* The formula usually given for J is

$$J = \oint p\, dq \tag{11}$$

where p is the momentum conjugate to the co-ordinate q. The integral is taken over the path actually covered by the particle during a single period of oscillation. Such an integral is known as a *phase integral.*

It is readily shown for the special case in which $p = \sqrt{2m[E - V(q)]}$, where $V(q)$ is the potential, that this definition is equivalent to the preceding one. To do this we note that q oscillates between limits that are functions of the energy. Since the particle goes from one limit of oscillation to the other and back, the phase integral is just twice the value of the integral taken between limits of oscillation, or

$$J = 2\int_{a(E)}^{b(E)} dq\, \sqrt{2m[E - V(q)]}$$

Let us now obtain $\partial J/\partial E$. By a well-known theorem of the calculus

$$\frac{\partial J}{\partial E} = 2\{\sqrt{2m[E - V(q)]}\}_{q=b}\frac{\partial b}{\partial E} - 2\{\sqrt{2m[E - V(q)]}\}_{q=a}\frac{\partial a}{\partial E} + 2\int_a^b \sqrt{\frac{m}{2[E - V(q)]}}\, dq$$

Now the limits of oscillation are determined by the fact that the kinetic energy is zero, i.e., $E = V$. Thus, the quantities in the brackets vanish and, since $\sqrt{\frac{2(E - V)}{m}} = \frac{dq}{dt}$, we obtain

$$\frac{\partial J}{\partial E} = 2\int_a^b \frac{dq}{dq/dt} = 2\int_a^b dt = T = \text{the period} \tag{12}$$

This shows that the J defined in eq. (11) is identical with the J first introduced in eq. (10), except for a constant of integration that is irrelevant for our purposes.

The quantization of the action, J, is usually referred to as "the Bohr-Sommerfeld quantum condition."

13. Quantization of Angular Momentum. Let us now apply our rule to the simple case of a particle moving in a plane with a polar angle ϕ and with a definite angular momentum p_ϕ. If the potential is spherically symmetrical, this angular momentum is known to be a constant of the motion. The co-ordinate conjugate to p_ϕ is ϕ itself. Now, during a period, ϕ goes from 0 to 2π, so that

$$J = \int_0^{2\pi} p_\phi \, d\phi = p_\phi \int_0^{2\pi} d\phi = 2\pi p_\phi \tag{13}$$

Then the quantization of J means that

$$\Delta J = h = 2\pi \, \Delta p_\phi \qquad \text{or} \qquad \Delta p_\phi = \frac{h}{2\pi} = \hbar$$

Angular momentum can therefore change only in units of $\hbar = h/2\pi$.

14. The Hydrogen Atom. Let us now investigate the effects of quantization on an electron moving in the field produced by a point charge, such as the nucleus of a hydrogen atom. This problem was first treated by Niels Bohr, who arrived at the quantization of angular momentum with the aid of the correspondence principle. We shall give here the general line of argument by which this can be done directly without reference to the quantization of action, J. Classically, the energy is

$$E = \frac{mv^2}{2} - \frac{Ze^2}{r} \tag{14}$$

For a circular orbit, the attractive force balances the centrifugal force. Thus, we obtain

$$\left.\frac{mv^2}{r} = \frac{Ze^2}{r^2} \qquad \text{or} \qquad mv^2 r = Ze^2 \qquad \text{and} \qquad \frac{Ze^2}{r} = mv^2 \right\} \tag{15}$$

With $mvr = p_\phi$, this becomes

$$vp_\phi = Ze^2$$

Furthermore, we obtain for the energy

$$E = -\frac{mv^2}{2} = -\frac{m}{2}\frac{(Ze^2)^2}{p_\phi^2} \tag{16}$$

Let us now consider a transition from an orbit where $p_\phi = p_{\phi_1}$ to one in which $p_\phi = p_{\phi_2}$. According to Einstein's relation, $\Delta E = h\nu$, we must have

$$\Delta E = h\nu = -\frac{m}{2}(Ze^2)^2\left(\frac{1}{p_{\phi_1}^2} - \frac{1}{p_{\phi_2}^2}\right) \tag{17}$$

Now, in a high quantum state, we must obtain agreement between the quantum mechanically calculated frequency and the classical frequency, which is that of rotation in the orbit, i.e., $\nu = v/2\pi r$. Furthermore, the fractional change of angular momentum is so small that we can replace the difference in eq. (17) by a differential, obtaining

$$\frac{\hbar v}{r} \cong \frac{m(Ze^2)^2 \, \Delta p_\phi}{(p_\phi)^3}$$

where Δp_ϕ is the allowed change in p_ϕ. Since $p_\phi = mvr$, this becomes

$$\frac{\hbar v}{r} = \frac{(Ze^2)^2}{m^2v^3r^3}\Delta p_\phi$$

From eq. (15), we readily obtain $\dfrac{v}{r} = \dfrac{(Ze^2)^2}{m^2v^3r^3}$

and $$\Delta p_\phi = \hbar \qquad \text{or} \qquad p_\phi = l\hbar + K \tag{18}$$

where l is an integer, and K is a constant. This result is in agreement with that obtained from the Bohr-Sommerfeld condition (which were actually later derived historically). Both results must, of course, agree since they are obtained from the correspondence principle.

Bohr then tentatively suggested that eq. (18) holds even for very small quantum numbers. Thus far, only changes in p_ϕ have been defined. We have, therefore, written $p_\phi = l\hbar + K$, where K is a constant. The allowed values of p_ϕ must, however, be the same for positive and negative values. This can be seen from the fact that in a right-handed co-ordinate system, the sign of the angular momentum is opposite to that calculated for a left-handed co-ordinate system. Thus, if we decide to alter the co-ordinate system from right- to left-handed, we reverse the sign of the angular momentum. But the final results of the theory must not depend on which type of co-ordinate system is adopted. This means that if a given value of the angular momentum is allowed, its negative must also be allowed. If we choose $K = 0$, then this criterion is satisfied, for the allowed values are $p_\phi = l\hbar$. With $K = \frac{1}{2}$, the allowed values are $p_\phi = (l + \frac{1}{2})\hbar$, or $(\ldots -\frac{5}{2}, -\frac{3}{2}, -\frac{1}{2}, \frac{1}{2}, \frac{3}{2}, \frac{5}{2}, \ldots)\hbar$, so that our requirements are also satisfied. With any other value of K, however, this condition cannot be met. For example, with $K = \frac{1}{4}$, we obtain for typical allowed values $(\ldots -1\frac{3}{4}, -\frac{3}{4}, \frac{1}{4}, 1\frac{1}{4}, 2\frac{1}{4}, \ldots)\hbar$, so that if we reverse the sign of the angular momentum, we do not obtain an allowed value.

To obtain agreement with observed spectra, we must take $p_\phi = l\hbar$. [Half-integral quantum numbers will, however, appear further ahead in other applications in connection with WKB approximation (Chap. 12) and spin, (Chap. 17).] Thus, from eq. (16) we obtain

$$E = -\frac{m}{2}\frac{(Ze^2)^2}{\hbar^2 l^2} = -\frac{Rch}{l^2} \tag{19}$$

where R is the Rydberg constant. The frequency of radiation emitted in a transition is

$$\nu = \frac{\Delta E}{h} = Rc\left(\frac{1}{l_2^2} - \frac{1}{l_1^2}\right) \tag{20}$$

This result not only explained the known spectra of hydrogen (and singly ionized helium), but also predicted new series of frequencies of emitted radiation, which were not known at the time. Thus, the quantum theory once again led to precise quantitative agreement with a wide range of experimental data, which classical physics could not even begin to explain.

The radius of an electronic orbit is

$$a = -\frac{Ze^2}{2E} = \frac{\hbar^2}{mZe^2}l^2 \tag{21}$$

The lowest circular orbit has $l = 1$. ($l = 0$ leads to absurd conclusions. We shall see that the Bohr theory does not treat the question of the labeling of the lower quantum states completely correctly, but that these difficulties are all resolved with the aid of Schrödinger's equation.) The radius of this orbit is known as a *Bohr radius* and is usually labelled "a_0." a_0 is equal to 0.528×10^{-8} cm. Thus, we have $a = l^2 a_0$, with $a_0 = \dfrac{\hbar^2}{mZe^2}$. The succeeding orbits increase in radius as the square of the quantum number.

In order to obtain a complete treatment of the hydrogen atom, we must also deal with the elliptical orbits. To do this, we note that in an elliptical orbit, the radius oscillates periodically with the frequency of rotation. Thus we can apply the Bohr-Sommerfeld conditions, quantizing the "radial action variable,"

$$J_r = \oint p_r\, dr \tag{22}$$

where p_r is the radial component of the momentum.

To evaluate p_r, we write

$$E = \frac{mv^2}{2} - \frac{Ze^2}{r} = \frac{p_r^2}{2m} + \frac{p_\phi^2}{2mr^2} - \frac{Ze^2}{r} \tag{23}$$

Because the force is spherically symmetrical, p_ϕ is a constant of the motion. Hence, the term $p_\phi^2/2mr^2$ acts like an added repulsive term in the potential, which tends to keep particles away from the origin. In fact, the "centrifugal force" can be obtained by differentiating this term with respect to r:

$$\frac{\partial}{\partial r}\left(\frac{p_\phi^2}{2mr^2}\right) = \frac{p_\phi^2}{mr^3} = mr\dot{\phi}^2 = \frac{mv_\phi^2}{r}$$

where v_ϕ is the component of the velocity in the ϕ direction.

We now solve for p_r, and obtain

$$J_r = 2\int_{r_{\text{min}}}^{r_{\text{max}}} \sqrt{2mE + \frac{Ze^2}{r} - \frac{p_\phi^2}{r^2}}\, dr \tag{24}$$

The range of integration goes from the minimum value of r to the maximum. Since these values occur where $p_r = m\dot{r} = 0$, this range lies between the points where the integrand is zero. The integral can be evaluated and the result is

$$J_r = -2\pi p_\phi + 2\pi Ze^2 \sqrt{\frac{m}{-2E}} \tag{25}$$

Problem 7: Prove the preceding result.

Solving for E, we obtain

$$E = -\frac{m}{2} \frac{(Ze^2)^2}{\left(\frac{J_r}{2\pi} + p_\phi\right)^2} = -\frac{m}{2} \frac{(2\pi)^2(Ze^2)^2}{(J_r + J_\phi)^2} \tag{26}$$

where $J_\phi = 2\pi p_\phi$.

Now we know that $J_\phi = lh$, and we must also have $J_r = sh$,* where s is an integer known as the "radial quantum number." The total energy is

$$E = -\frac{m}{2} \frac{(Ze^2)^2}{\hbar^2} \frac{1}{(l + s)^2} \tag{27}$$

We usually write

$$l + s = n = \text{principal quantum number} \tag{28}$$

and

$$E = -\frac{m}{2} \frac{(Ze^2)^2}{\hbar^2} \frac{1}{n^2} = -\frac{Rch}{n^2} \tag{29}$$

We observe that the allowed energy levels are precisely the same as those calculated with circular orbits. The energy depends explicitly only on n, and not on l or s separately. Thus, for each value of n, we can obtain a series of orbits of the same energy by giving l all possible values between 1 and n, whereas s takes on the corresponding values between $n - 1$ and 0.

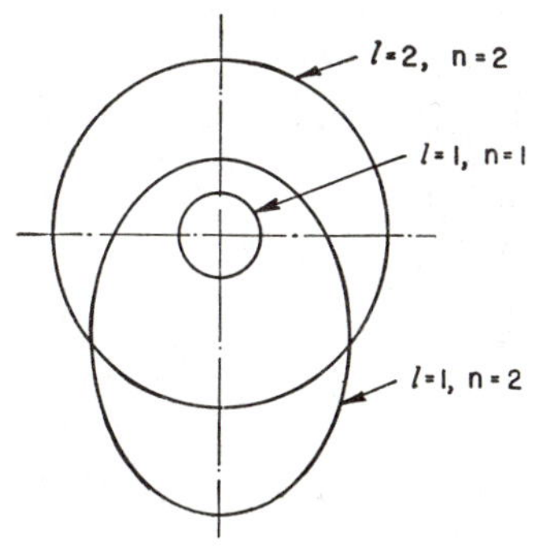

FIG. 3

To see what these orbits look like, we use the fact that the energy in an elliptical orbit is a function only of the length of the semi-major axis, and not of the eccentricity.† Thus, all orbits with the same energy are ellipses with the same semi-major axis. For the choice $l = n$, we have already seen that we obtain circular orbits. With $l < n$, we obtain ellipses, which become more eccentric for smaller values of l.

Let us enumerate the first few orbits. For $n = 1$, there is only one possibility, $l = 1$. This is a circular orbit. For $n = 2$, however, we can have $l = 2$ or $l = 1$. The former is circular, the latter elliptical. The

* Actually, we should take $J = s\hbar + k$, where k is a constant, but we choose $k = 0$ here. This derivation then applies rigorously only for high quantum numbers, but happens to lead to correct results for low quantum numbers also.

† Ruark and Urey, Chap. 5.

shape of the orbits for the first two energy levels is shown in Fig. 3. We see that for large n, the orbits rapidly become larger and more numerous. The quantum states are usually represented on an energy-level diagram, as shown in Fig. 4. The value of the energy relative to zero is given by

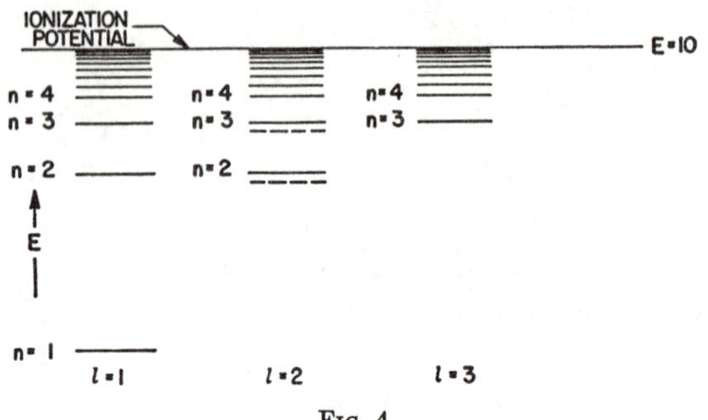

Fig. 4

the position of the line. This value is shown for hydrogen, for the first three values of l, by the solid lines. We note that as $E = 0$ is approached, the lines approach infinite density. For $E > 0$, the electron is free, so that we have, actually, an ion and electron. The energy needed to liberate an electron starting from the lowest energy level is called the *ionization potential.* For hydrogen, it is the negative of the result obtained by writing $n = 1$ in eq. (29), i.e., $E = Rch$.

The only laws of force such that orbits of the same n and different l always have the same energy are the laws of Coulomb attraction and the three-dimensional simple harmonic oscillator. In both cases, this result is closely connected with the fact that the frequency with which the radius returns to its original value is the same as that required for the angle to go through a change of 2π. To prove this for the Coulomb force, we observe that the frequency of radial oscillation is $\nu_r = \partial E/\partial J_r$, and that of going through an angle of 2π is $\nu_\phi = \partial E/\partial J_\phi$. From eq. (26), we verify that

$$\frac{\partial E}{\partial J_r} = \frac{\partial E}{\partial J_\phi}$$

so that $\nu_r = \nu_\phi$. For the allowed changes of energy, we then write

$$\Delta E \cong \frac{\partial E}{\partial J_r} \Delta J_r + \frac{\partial E}{\partial J_\phi} \Delta J_\phi = h(\nu_r \, \Delta s + \nu_\phi \, \Delta l)$$

or

$$\Delta E = h\nu(\Delta s + \Delta l) \tag{30}$$

Starting from the ground state at $l = 1$, $n = 1$ (or $s = 0$), we see that we come to the same value of the energy whether we increase s or l by 1. Similarly, the size of the next step is the same regardless of whether we increase s or l. In all cases, the total energy change then depends only on $n = s + l$. If, however, ν_r had been different from ν_ϕ, we would have obtained different results by changing s by 1 from those obtained by changing l by 1 and, in general, the energy would then depend on both s and l separately.

When $\nu_r = \nu_\phi$, we also obtain the result that the orbit closes, i.e., both r and ϕ return to their initial values at the same time. It is fairly clear that the elliptical orbits of hydrogen have this property. When, however, the radius does not return to its initial value at the same time that the angle goes through a change of 2π, the orbit does not close. If the difference in frequencies is not too great, the resulting motion may be described as a precession of the orbit. A precessing elliptical orbit is shown in Fig. 5.

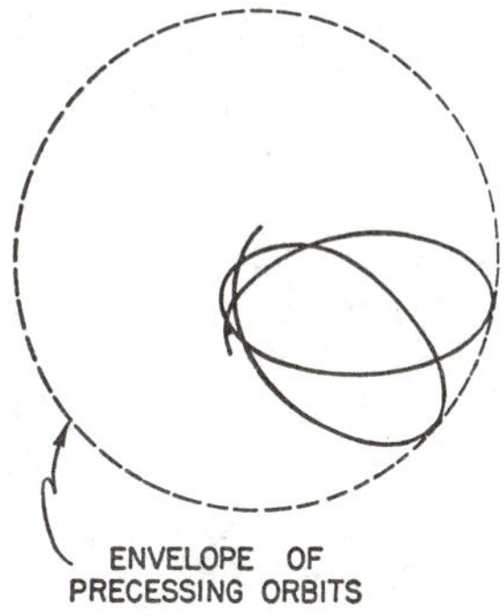

FIG. 5

In complex atoms, there will usually be deviations from the Coulomb law of force as a result of shielding.* In this case, ν_r and ν_ϕ will differ, the orbits will precess at a rate determined by the size of the deviation, and levels of the same n and different l will have different energies. This change of energies is illustrated for a few levels by the dashed lines in the energy-level diagram of Fig. 4. The effects of electron spin and relativistic corrections also produce similar, but usually much smaller, variations of the energy with l, even in hydrogen. This is known as the *fine structure*.‡ When all these effects are taken into account, we can obtain good agreement with the energy spectrum of hydrogen, and fair agreement with that of the alkali metals, such as sodium.

Problem 8: Suppose that we are given the potential $V = -\frac{Ze^2}{r} + \frac{K^2}{r^2}$, which is a modified Coulomb potential. The modifications are not in the right direction to describe shielding correctly, but are a good approximation to the modifications introduced by taking into account the relativistic corrections to the energy (see Ruark and Urey). Find the energy levels as a function of the radial quantum number and the angular momentum, and show that, for a given $n = l + s$, the energy obtained is not independent of l.

HINT: Evaluate the integral in eq. (24) by replacing p_ϕ^2 with $p_\phi^2 + 2mK^2$.

Although the early Bohr-Sommerfeld theory was successful in explaining the energy levels of a wide variety of systems, including rotating and vibrating molecules, as well as the applications discussed in connection with the theory of atomic spectra, it was, nevertheless, an incomplete and somewhat ambiguous theory. For example, there was no clear way to formulate the theory rigorously for complex atoms, or for nonperiodic motions, such as might be involved in the scattering of electrons on atoms. In some cases, the results were wrong for low quantum numbers, and various semiempirical efforts were made to improve agreement with experiments by using fractional quantum numbers. Where the theory was in error, however, it was seldom very much so; hence it was clearly on the right track.

Later, we shall see that all these ambiguities can be removed with the aid of wave mechanics, which is capable of yielding, in principle at least, a complete and quantitative theory of all phenomena in which relativistic effects are not important. Wave mechanics does not, how-

* We can often approximate the effects of other atomic electrons on a given electron by an average potential, in which the net effect of the other electrons is to tend to shield or "screen" the particle in question from the nucleus. This approximation, however, is not perfectly accurate, as it neglects the possibility of transfer of energy between the atomic electrons.†

† Ibid, p. 201.

‡ See Ruark and Urey, p. 135.

ever, contradict the general line of the Bohr-Sommerfeld theory but, instead, it shows just what the limitations of this earlier theory are and how they can be overcome.

15. Franck-Hertz Experiments. An important verification of Bohr's theory was provided by the Franck-Hertz experiment, and by a series of similar experiments that followed.* Until these experiments were finished, the only cases where energy transfers had been clearly proved to be quantized involved emission or absorption of radiation, or else transfers to material harmonic oscillators, such as those constituted by the atoms of a solid. The Franck-Hertz experiments were designed to determine if the transfers of energy were still quantized when the energy came from kinetic energy of free material particles.

The experiments consisted in passing a beam of electrons of controlled energy through a low-pressure gas, such as hydrogen. (Under normal conditions, the hydrogen is in its state of lowest possible energy.) It was found that the atoms were unable to absorb energy from the electron beam, unless the beam particles had an energy greater than or equal to that needed to raise the atomic electron from the lowest state to the first excited state,† as calculated from the relation $\Delta E = h\nu$. The fact that atoms were absorbing energy could be proved either from the sudden decrease of the beam current when the critical potential was exceeded, or from the simultaneous appearance of quanta that were radiated by the atoms which had absorbed the energy. Each time a critical energy corresponding to a known spectral line was exceeded, there was a new decrease in beam current, accompanied by the sudden appearance of the corresponding quantum in the emission spectrum of the gas. When the energy necessary to raise the electron from the ground state to $n = \infty$ was supplied, ions began to appear, but not before. (This result also shows clearly that we cannot hope to explain the photoelectric effect by assuming that there are electrons on the verge of being liberated.)

In subsequent experiments, it was possible to measure the energies of the electrons after passage through the gas, and it was found that the electrons which had struck atoms always lost exactly the same amount of energy that appears in a quantum. These experiments indicated that energy is conserved in individual quantum processes.

We must conclude that, in exchanges of energies between material particles, we are restricted to the same quantized values which appear in exchanges of energy between matter and radiation.

Correspondence Theory of Radiation

To study the radiation problem, we must consider what happens when an electron moves from one discrete orbit to another. This prob-

* See Ruark and Urey, p. 78.

† The first excited state is the first energy level above the lowest, or "ground," state.

lem is essentially the same as that of how a radiation oscillator changes from one energy state to another. We can easily see that the indivisibility of either one of these processes implies the indivisibility of the other. Consider, for example, a process in which an atom absorbs a quantum of radiation, going from the ground state to an excited state. We have seen that, in each individual quantum process, energy is conserved. Since the radiation oscillator gives up its energy in a single indivisible step, the atomic electron must likewise accept an equal amount of energy in the same indivisible step. Thus, the electron cannot be thought of as traversing intermediate states between the orbits. The apparently continuous motion of classical physics merely reflects the fact that, in the classical limit, the orbits are so close together that the indivisible and discontinuous nature of these transitions is not normally apparent. For example, by differentiating eq. (21), we find that the separation of orbits in hydrogen is $\Delta a = 2la$. The fractional change of radius between successive orbits is then $\Delta a/a = 2/l$, and this becomes very small as l becomes large.

The lack of complete determinism in the process of transfer of a quantum from radiation oscillators to the atom also implies a corresponding lack of complete determinism in the process by which the electron goes from one orbit to the next. Thus, if an atom is irradiated by a beam of light, we can predict only the probability that the atom goes into an excited state. Similarly, if the atom is in an excited state, we can predict only the probability that it will radiate a quantum and go to a state of lower energy. The classical deterministic laws apply only when the system is in such a high quantum state that many quanta are emitted before the radius undergoes an appreciable fractional change. Thus, in a classically describable radiation process, the electron actually emits a large number of quanta in a comparatively short time, and goes through many adjacent quantum states which are so closely spaced that the process appears to be continuous. So many quanta are emitted that the deviation between the actual number appearing and the probable number predicted by the theory becomes small. Thus, we obtain an almost deterministic result.

Problem 9: A simple harmonic oscillator of frequency 1000 cps has an energy of 1 erg. What is the mean statistical fluctuation in the number of quanta emitted when the oscillator changes its energy by 1 per cent?

16. Absorption of Radiation. Let us now consider the problem of absorption of radiation. We first return to the classical interpretation of absorption as a gradual process resulting from the electrical forces acting on a particle. In an electromagnetic wave, a charged particle experiences a force

$$F = e\left(\mathcal{E} + \frac{v}{c} \times \mathcal{H}\right) \tag{31}$$

Since $|\mathcal{E}| = |\mathcal{H}|$ in free space, and since $v/c \ll 1$ in most atoms, the second term is usually much smaller than the first. We also write (for a plane wave)

$$\mathcal{E} = \mathcal{E}_0 e^{i(\boldsymbol{k}\cdot\boldsymbol{x}-\omega t)} = \mathcal{E}_0 e^{i(\boldsymbol{k}\cdot\boldsymbol{x}_0-\omega t)} e^{i\boldsymbol{k}\cdot(\boldsymbol{x}-\boldsymbol{x}_0)}$$

where $\boldsymbol{x}_0$ is the position of the center of the atom.

In most cases, $1 \gg \boldsymbol{k}\cdot(\boldsymbol{x}-\boldsymbol{x}_0) = \dfrac{2\pi}{\lambda}|\boldsymbol{x}-\boldsymbol{x}_0|$, because λ, the wavelength, is of the order of 10^{-5} cm, but $|\boldsymbol{x}-\boldsymbol{x}_0|$ is of the order of atomic dimensions,

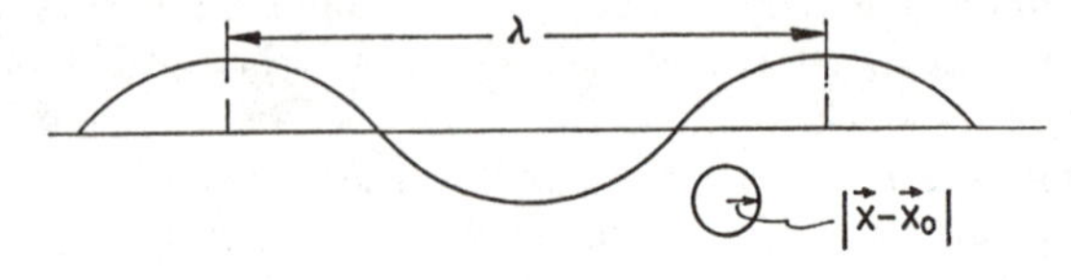

FIG. 6

which are of the order of 10^{-8} cm. This is illustrated in Fig. 6. Thus, we can usually write

$$\mathcal{E} = \mathcal{E}_0 e^{i(\boldsymbol{k}\cdot\boldsymbol{x}_0-\omega t)} = \mathcal{E}(\boldsymbol{x}_0)e^{-i\omega t} \tag{32}$$

where $\mathcal{E}(\boldsymbol{x}_0)$ is the value of the electric field at the center of the atom. This approximation is equivalent to replacing the atom by a point dipole of moment $\boldsymbol{M} = -e(\boldsymbol{x}-\boldsymbol{x}_0)$, located at the center of the atom.*

Let us study, for illustration, the rate at which a harmonic oscillator of natural frequency, ω_0, initially at rest in an equilibrium position, gains energy from an electromagnetic wave of angular frequency ω. This problem is studied because the results obtained with a harmonic oscillator are essentially the same as those obtained with any other system, and the mathematical expressions are simplified.

Suppose that we have an incident wave polarized in the x direction. Then, according to eq. (31), the equation of motion is

$$m(\ddot{x} + \omega_0^2 x) = e\mathcal{E}_0 \cos(\omega t + \phi_0) \tag{33}$$

where ϕ_0 represents the phase of the electric field at the point x_0, and the time $t = 0$. The boundary conditions are that $x = \dot{x} = 0$ at $t = 0$, and the solution corresponding to these conditions is

$$\begin{aligned} x &= \frac{e\mathcal{E}_0}{m(\omega_0^2-\omega^2)}\left[\cos(\omega t+\phi_0) - \cos(\omega_0 t+\phi_0) + \frac{(\omega-\omega_0)}{\omega_0}\sin\phi_0 \sin\omega_0 t\right] \\ &= \frac{e\mathcal{E}_0}{m(\omega_0+\omega)}\left\{2\sin(\omega_0-\omega)\frac{t}{2}\sin\left[(\omega_0+\omega)\frac{t}{2}+\phi_0\right] + \frac{\sin\phi_0}{\omega_0}\sin\omega_0 t\right\} \end{aligned} \tag{34a}$$

$$\begin{aligned} \dot{x} = \frac{e\mathcal{E}_0}{m(\omega_0+\omega)}\left\{\frac{2\omega}{\omega_0-\omega}\cos\left[(\omega_0+\omega)\frac{t}{2}+\phi_0\right]\sin(\omega_0-\omega)\frac{t}{2}\right. \\ \left. + \sin(\omega_0 t+\phi_0) - \sin\phi_0\cos\omega t\right\} \end{aligned} \tag{34b}$$

* See Chap. 18, Sec. 25.

Let us now describe the general character of the motion. We observe that the amplitude of oscillation increases and decreases with the "beat frequency" $\nu = (\omega_0 - \omega)/4\pi$. When ω_0 is far from ω, the beats are very rapid, and the maximum amplitude remains small for all time but, when ω_0 is close to ω, the time between beats becomes very long, and the maximum amplitude becomes large. The reason is that when the impressed frequency is far from the natural frequency of the oscillator, the external force rapidly gets out of phase with the oscillations that it produces, so that in a short time the forcing term begins to oppose the existing motion, and thus reduce the amplitude of oscillation. As ω approaches the "resonant frequency" ω_0, the impulses from the external field remain in phase with the oscillations for an increasingly longer time, so that an increasingly larger amplitude is built up, and the period of the beats becomes longer. To determine what happens when $\omega = \omega_0$, we find the limit of eq. (34) as ω approaches ω_0. This is

$$x = \frac{1}{2}\frac{e\mathcal{E}_0 t}{m\omega_0}\sin(\omega_0 t + \phi_0) + \frac{1}{2}\frac{e\mathcal{E}_0}{m\omega_0}\sin\phi_0 \sin\omega_0 t \tag{35}$$

It is readily verified by direct differentiation that eq. (35) is a solution of eq. (33), with ω set equal to ω_0, and that it satisfies correct boundary conditions.* This is the case of exact resonance, for which the amplitude of oscillation increases indefinitely with the time, because the forcing term never gets out of phase with the oscillations.

The energy of the oscillator is

$$W = \frac{m}{2}(\dot{x}^2 + \omega_0^2 x^2) \tag{36}$$

When ω is close to ω_0, we can approximate the value of x, after an appreciable time has elapsed, by neglecting the second term on the right side of eq. (34a) in comparison with the first. Similarly, we can approximate $\dot{x}$ by neglecting the second and third terms on the right side of eq. (34b) in comparison with the first. The result is

$$W \cong \frac{e^2\mathcal{E}_0^2}{8m}\frac{\sin^2(\omega_0 - \omega)t/2}{(\omega_0 - \omega)^2} \tag{37}$$

We conclude that the energy absorbed is proportional to $\mathcal{E}_0^2$, which is in turn proportional to the intensity of the radiation $I(x_0)$ at the center of the atom. For times so short that $(\omega_0 - \omega)t/2 \ll 1$, we can expand $\sin(\omega_0 - \omega)t/2$, and we find that the energy is proportional to t^2. For longer times, W goes through a maximum and returns to zero (because the forcing term gets out of phase with the oscillations that it produces).

The first of these results is reasonable, in the sense that it agrees with

* Note that the most general solution can be obtained by adding to eq. (35) the term $A\cos\omega_0 t + B\sin\omega_0 t$, where A and B are arbitrary constants.

general experience. The latter two results are not, because we expect to find that the energy gained by an oscillator should increase linearly with the time. We shall now show that a linear increase results from the fact that, in most applications, the frequency of the radiation is not perfectly defined but varies over a range. There is, in fact, an intensity function $dE = I(\nu)\,d\nu$, giving the energy to be found in the frequency range between ν and $\nu + d\nu$. In a blackbody, for example, $I(\nu)$ is given by the Planck distribution.*

To obtain the total energy transfer, we must integrate eq. (37) over all frequencies, noting that $\mathcal{E}_0^2$ is proportional to $I(\omega)$. Thus, we obtain

$$W \sim \int_0^\infty I(\omega) \frac{\sin^2 (\omega_0 \cdot \omega) \frac{t}{2}\, dw}{(\omega_0 - \omega)^2} \tag{38}$$

Let us now consider the function

$$F(\omega) = \frac{\sin^2 (\omega_0 - \omega)\frac{t}{2}}{(\omega_0 - \omega)^2}$$

The maximum value occurs at $\omega = \omega_0$, and it is equal to $t^2/4$. The function goes to zero, where $\omega_0 - \omega = 2\pi/t$. Thereafter, it behaves as shown in Fig. 7, decreasing rapidly with increasing $\omega_0 - \omega$. When t is large, $F(\omega)$ has a very sharp and narrow peak (of width $|\omega - \omega_0| \cong 2\pi/t$ and height $t^2/4$) located at $\omega = \omega_0$. Hence, the main contribution to the integral in eq. (38) comes from a very narrow frequency interval near $\omega = \omega_0$. Over this region $I(\omega)$, which is usually a continuous function, varies so little that it can be regarded as a constant and may, therefore, be taken out of the integral and evaluated at $\omega = \omega_0$. The integral then becomes

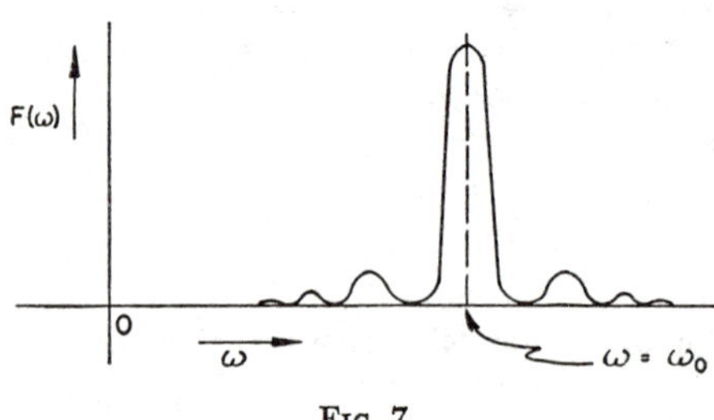

FIG. 7

$$W \sim I(\omega_0) \int_0^\infty \frac{\sin^2 (\omega - \omega_0) \frac{t}{2}\, d\omega}{(\omega_0 - \omega)^2} \tag{39}$$

The preceding integral can be approximated by noting that, for negative values of ω, the function $F(\omega)$ becomes negligible, when t is large. This is because the function is so sharply peaked at $\omega = \omega_0$ that, as can be seen from Fig. 7, it is negligible at $\omega = 0$, and so small for negative ω that

* See Chap. 1, eq. (32a).

its integral from $-\infty$ to 0 can be neglected. Thus we may, with small error, extend the range of integration to $-\infty$, obtaining

$$W \sim I(\omega_0) \int_{-\infty}^{\infty} \frac{\sin^2 (\omega - \omega_0) \frac{t}{2} \, d\omega}{(\omega_0 - \omega)^2} \tag{40}$$

This integral can be evaluated with the substitution

$$(\omega - \omega_0)t/2 = y \qquad \text{and} \qquad d\omega = (2/t)\, dy$$

This leads to

$$W \sim I(\omega_0) \frac{t}{2} \int_{-\infty}^{\infty} \frac{\sin^2 y \, dy}{y^2} = \frac{\pi}{2} I(\omega_0)t \tag{41}$$

We have now obtained the reasonable result that the energy absorbed is proportional to the product of the intensity and the time of exposure. Another important result of eq. (40) is, that during a time t, the major part of the transfer of energy comes from a frequency range

$$|\omega - \omega_0| \cong \pi/t$$

Thus, as t becomes larger, we find that absorption is confined to a narrower band of frequencies surrounding the resonance frequency. Although each frequency in this range contributes to the energy gain proportional to t^2, the net energy gain is proportional only to t, because the range of frequencies that can contribute decreases as $1/t$. If, however, we directed a wave of perfectly defined frequency at an absorber (for example, a radio wave into a resonant cavity), the energy would instead fluctuate with the beat frequency, as shown in eq. (37).

Let us now see how this problem is treated in the quantum theory. We know that the energy transfer is indivisible and goes in units of $h\nu$. According to the correspondence principle, however, we must choose the probability of transition such that, in the limit where there are many quanta, the classical rate of absorption of energy is obtained. To do this, we must choose for the probability per unit time of absorbing a quantum

$$S = \frac{1}{h\nu} \frac{dW}{dt} \tag{42}$$

where dW/dt is the classically calculated rate of absorption of energy from eq. (41). Since $dW/dt \sim I$, we obtain

$$S \sim I/h\nu \tag{43}$$

Let us now consider the question of whether these formulas, derived in the correspondence limit, will hold when only a few quanta are present. Experiments with the photoelectric effect show that the rate of absorption of quanta is proportional to the classically calculated intensity at the point of absorption. But the constant of proportionality does not, in general, agree exactly with that implied by eq. (42), when small quantum numbers are involved, although the results are seldom very much in error. To obtain the exact results, we must go to wave mechanics (see Chap. 18).

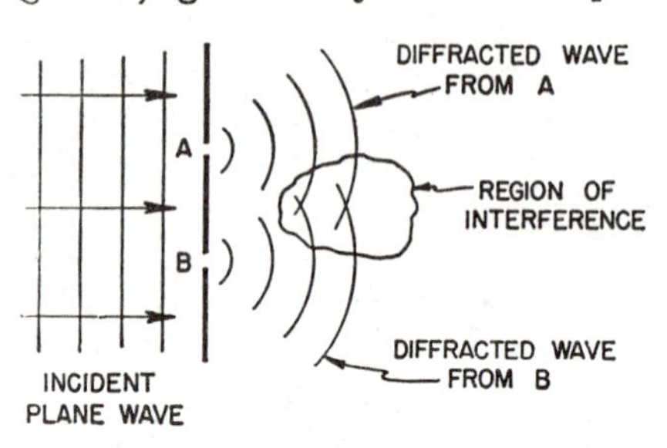

FIG. 8

Let us now investigate the relation of these results to the wave-particle duality. Consider, for example, a wave incident on A slit (Fig. 8), which produces an electric field to the right of the slit, denoted by $\mathbf{\mathcal{E}}_A(x, y, z, t)$. Let us now suppose that a second slit B is opened, and that the light passing through it produces the field $\mathbf{\mathcal{E}}_B(x, y, z, t)$. The total field is $\mathbf{\mathcal{E}} = \mathbf{\mathcal{E}}_A + \mathbf{\mathcal{E}}_B$, and the intensity of the light is proportional to

$$|\mathbf{\mathcal{E}}|^2 = |\mathbf{\mathcal{E}}_A + \mathbf{\mathcal{E}}_B|^2 = |\mathcal{E}_A|^2 + |\mathcal{E}_B|^2 + 2\mathbf{\mathcal{E}}_A \cdot \mathbf{\mathcal{E}}_B \tag{44}$$

The first two terms on the right represent the sum of the intensities of the separate beams, and the third represents interference effects. According to eq. (43), we see that, where destructive interference in the classically calculated wave pattern takes place, the theory yields zero probability of absorption of a quantum. We can conclude that the extrapolation of the theory obtained in the correspondence limit, down to the case of a few quanta, leads to a correct prediction of the wave-particle duality, as it is observed. Or conversely, we can base this result directly on observation and show that eq. (43) then leads, when many quanta are present, to the correct classical intensity pattern. Thus, we see again how closely the quantum laws are tied in with their classical limits.

17. Emission of Radiation. To calculate the probability of emission of quanta, we first study the classical theory of this process. The classical rate of radiation of energy by a moving particle is given by the well-known formula*

$$\frac{dW}{dt} = \frac{2}{3}\frac{e^2}{c^3}|\ddot{\mathbf{r}}|^2 = \frac{2}{3}\frac{e^2}{c^3}(\ddot{x}^2 + \ddot{y}^2 + \ddot{z}^2) \tag{45}$$

In evaluating the above quantities, it is convenient to evaluate by Fourier analysis the motion as a function of the time. In a periodic orbit of fundamental angular frequency ω_0, the variation of each co-ordinate can be represented by a Fourier series. For example,

* See Ruark and Urey, p. 762, eq. (31): Richtmeyer and Kennard, Chap. 2.

$$x = Rl \sum_n X_n e^{in\omega_0 t} = Rl \sum_n |X_n| e^{i(n\omega_0 t + \phi_n)} = \sum_n |X_n| \cos (n\omega_0 t + \phi_n) \tag{46}$$

where $X_n = |X_n| e^{i\phi_n}$. In a circular orbit, for instance, we have

$$x = \cos \omega_0 t \qquad y = \sin \omega_0 t$$

so that only the fundamental frequency is present. In an elliptical orbit, as seen in Sec. 14, higher harmonics are also present.

Generally, the motion may have more than one period. In fact, in the most general case, there are as many periods as there are independent degrees of freedom. In such a system, which is called "multiply periodic," we find that when the x co-ordinate, for example, returns to its original value, the y and z co-ordinates do not. Thus, the orbits do not close as they do when only one period is present. As seen in Sec. 14, an electron in a Coulomb force undergoes singly periodic motion. The treatment of radiation will be given here only for singly periodic systems, but the method of extension to multiply periodic systems* is fairly straightforward.

We shall calculate here only the rate resulting from the x component of the acceleration, noting that the effects of y and z motion can be added in a similar way.† We obtain

$$\begin{aligned}
\frac{dW_x}{dt} &= \frac{2}{3}\frac{e^2}{c^3}\omega_0^4 \left[\sum_n n^2 |X_n| \cos (n\omega_0 t + \phi_n)\right]^2 \\
&= \frac{2}{3}\frac{e^2}{c^3}\omega_0^4 \left\{\sum_n |X_n|^2 n^4 \cos^2 (n\omega_0 t + \phi_n)\right. \\
&\qquad \left. + 2 \sum_{n \neq m} [|X_n||X_m| n^2 m^2 \cos (n\omega_0 t + \phi_n) \cos (m\omega_0 t + \phi_m)]\right\} \\
&= \frac{2}{3}\frac{e^2\omega_0^4}{c^3} \left\{\sum_n \frac{[1 - \cos 2(n\omega_0 t + \phi_n)]}{2} |X_n|^2 n^4\right. \\
&\qquad + \sum_{n \neq m} |X_n||X_m| n^2 m^2 [\cos (n+m)\omega_0 t + \phi_n + \phi_m) \\
&\qquad\qquad \left. + \cos (n-m)\omega_0 t + \phi_n - \phi_m)]\right\}
\end{aligned} \tag{47}$$

Averaging over a period yields

$$\overline{\overline{\frac{dW_x}{dt}}} = \frac{e^2}{3c^3}\omega_0^4 \sum_n n^4 |X_n|^2 \tag{48}$$

We observe that the energy is a sum of separate terms, one for each

* For a discussion of multiply-periodic systems, see Born, *Mechanics of the Atom*. London: George Bell & Sons, Ltd., 1927.

† See eq. (45).

harmonic. Since the radius of the electronic orbit is usually small compared with the wavelength, however, we know that the radiation is the same as would occur from a point dipole for which the moment varied with time in the same way as x does. Furthermore, it can be shown that such a dipole radiates each harmonic independently of all the others. Hence, we can conclude that each of the terms in the series in eq. (48) represents the rate of radiation of energy $\overline{\overline{dW_n}}/dt$ at the frequency of the corresponding harmonic $\omega_n = n\omega_0$. We, therefore obtain

$$\nu_n = 2\pi n\omega_0 \tag{49}$$

$$\frac{\overline{\overline{dW_n}}}{dt} = \frac{e^2\omega_0^4}{3c^3}|X_n|^2 n^4 \tag{50}$$

Let us now go to the quantum theory. In the correspondence limit, the rate of emission of quanta of the nth harmonic must be such that the classical rate of radiation of energy is obtained. This means that

$$R_n = \frac{e^2\omega_0^3}{6\pi hc^3}|X_n|^2 n^3$$

where $$\Delta E_n = h\nu_n = 2\pi nh\omega_0 \tag{51}$$

This result is rigorously correct only in the classical limit, where the fractional change of electronic energy per quantum emitted is small. Yet it is seldom seriously wrong, even when the quantum numbers are small and may, therefore, be used as an approximation in this region. The exact probabilities can, as we shall see in Chap. 18, be found with the aid of the wave equation.

In order to obtain the nth harmonic, we must have a jump between quantum states for which the change of quantum number is n. To prove this, we write (in the classical limit) $\Delta E \cong \nu_0\,\Delta J$, where ν_0 is the fundamental frequency. But $\Delta E = h\nu = nh\nu_0$, so that $\Delta J = nh$. As long as we consider large quantum numbers, the spacing of the levels does not change much for a small change in n. Hence the frequencies emitted will be very nearly integral multiples of the fundamental. A better approximation is, however,

$$h\nu = \Delta E = \left(\frac{\partial E}{\partial J}\right)_{J=J_0}\Delta J + \frac{1}{2}\left(\frac{\partial^2 E}{\partial J^2}\right)_{J=J_0}(\Delta J)^2 + \cdots \tag{52}$$

Writing $\Delta J = nh$, we obtain

$$\nu = n\nu_0 + \frac{h}{2}\left(\frac{\partial^2 E}{\partial J^2}\right)_{J=J_0} n^2 + \cdots \tag{53}$$

In the classical limit, the second term becomes very small compared with the first. This can be seen by evaluating the ratio of the two terms, which is

$$\frac{\Delta J(\partial^2 E/\partial J^2)_{J=J_0}}{2(\partial E/\partial J)_{J=J_0}}$$

Now $\partial^2 E/\partial J^2 = \partial \nu_0/\partial J$. Thus, the ratio becomes $[\Delta J(\partial \nu_0/\partial J)]/2\nu_0$. But $\Delta J(\partial \nu_0/\partial J)$ is the change of frequency resulting from a change of action variable ΔJ, and this is very small in the classical limit. Thus, the spacing of frequencies corresponds to integral ratios of frequencies. When the quantum number becomes small, however, the spacing of energy levels ceases to be nearly uniform, and the frequencies emitted cease to be related to each other by integral ratios, although they are still related by the Rydberg-Ritz intercombination principle. In fact, we see that in the classical limit, the Rydberg-Ritz principle leads precisely to the classical harmonics.

From eq. (51), we conclude that large changes of quantum number are probable only when the classically calculated motion contains high harmonics. In hydrogen, this is the case for highly eccentric orbits. In transitions between circular orbits, however, only the fundamental is present, so that the quantum number can change only by unity.

Let us now consider what can happen to an electron that is known to be in the mth quantum state at time $t = 0$. If there are other states of lower energy, then the electron can emit a quantum and thus make a radiative transition to one of these lower states. Suppose that the probability per unit time that the electron go to the nth state is R_{mn}. (R_{mn} may be calculated for high quantum numbers from eq. (51). Note that in a transition involving the sth harmonic, the change of quantum number is $m - n = s$. For low quantum numbers, however, this result is only approximate so that if we want an accurate treatment, we must obtain R_{mn} from wave mechanics.) The total probability per unit time that the electron leaves the mth state is then $R_m = \sum_n R_{mn}$, where the summation is carried out only over states n, for which the energy is below that of m.

Now let $P_m(t)$ be the probability that after the time t the electron is still in the mth state. During the time interval between t and $t + dt$, the probability that an electron leaves the mth state is equal to the product of the probability $P_m(t)$, that it is in this state, times the probability $R_m\,dt$, that if it is in this state, it will leave during the time dt. Thus, we obtain

$$\frac{dP_m}{dt} = -R_m P_m$$

or, integrating and setting $P_m(0) = 1$, we get

$$P_m = e^{-R_m t} \tag{54}$$

Hence, the probability that the particle is in the mth state decays exponentially with time. Within a time, $\tau = 1/R_m$, P_m sinks to $1/e$ of its initial value and, soon after that, becomes negligible. Thus, the mean lifetime of an atom in an excited state is of the order of $\tau = 1/R_m$.

When an electron is in the lowest energy state (for example, the state with $l = n = 1$ in the hydrogen atom), further radiative transitions are impossible, since there is no other state to which the electron can go, unless energy is made available by incident quanta, or by other sources, such as a beam of fast particles. Hence, if an atom is left to itself, it tends to sink to the lowest quantum state, after which nothing more can happen to it. In this way, the stability of atoms is explained by quantum theory, whereas classically the prediction is that electrons keep on radiating until they fall into a nucleus.

Problems

10. Suppose that an electron is originally in the sth quantum state, and that it can make transitions to a whole range of states, among which is the mth state. The total probability per unit time that electron in the sth state makes a transition to any lower state is R_s, and the probability that it goes to the mth state is R_{sm}. The total probability that an electron in the mth state makes a transition to any lower state is R_m. Calculate the probability $P_m(t)$ that the electron is in the mth state.

11. Determine the energy levels of a particle of mass m, in a box of length L. Assume one dimension only.

12. Determine the energy levels of a particle bouncing in a gravitational field of acceleration g, off a level and perfectly elastic floor.

13. What is the mean lifetime of an electron in the first excited state of a hydrogen atom? Use the rough correspondence method indicated in previous discussion. Compare this with accurately calculated lifetimes from wave mechanics.

14. Find the energy levels of a rigid rotator (two-dimensional), of moment of inertia I.

Summary. As a result of the indivisibility and lack of complete determinism of elementary quantum processes, we have been led to far-reaching changes in our general concepts concerning the fundamental nature of matter and energy. Despite the fact that quantum and classical theories are very different, there is established, through the correspondence principle, a very close relation between them, from which the general form of the quantum theory can be determined by the requirement that it approach the correct classical limit.

CHAPTER 3

Wave Packets and De Broglie Waves

1. Introduction. After the development of the theory of the Bohr atom and of the Bohr-Sommerfeld quantum conditions, de Broglie was struck by the fact that the Einstein relation $E = h\nu$, coupled with the discrete character of energy levels, seemed to imply that each energy level was associated with a corresponding frequency. The appearance of a discrete set of allowed frequencies was, however, already a familiar phenomenon in classical physics in connection with the motion of waves in enclosures. For example, in the derivation of the Rayleigh-Jeans law, we found that the allowed wave vectors were $\boldsymbol{k} = \frac{2\pi}{L}(l_x, l_y, l_z)$, resulting in a spectrum of frequencies $\omega = ck = \frac{2\pi c}{L}(l_x^2 + l_y^2 + l_z^2)^{1/2}$. Other shapes of enclosures give more complicated, but similar, sets of allowed frequencies. De Broglie speculated that material particles were somehow associated with a hitherto undetected oscillatory phenomenon. In this way, a great unification between matter and light would be obtained. Both would be different forms of some new kind of system that could act sometimes like a wave and sometimes like a particle.

If matter is really comprised of waves, then how can we explain the particle-like properties that it has shown universally, until the appearance of quantum effects? Let us recall that several centuries ago it was thought that light, too, consisted of particles because, in common experience, its rays seemed to travel in straight lines. Since then it has been found that light diffracts around edges and shows interference phenomena, but that diffraction and interference are important only for distances comparable with a wavelength. De Broglie argued that if matter waves exist, then perhaps their wavelength is so short that we have only seen their motion as rays thus far, but that more sensitive experiments might indicate diffraction and interference effects.

In this chapter, we shall develop de Broglie's theory of matter waves, obtaining the so-called "de Broglie relationships," which define the wavelength in terms of the momentum, and the frequency in terms of the energy. We shall show that, in the classical limit, such waves move in a way that makes them resemble classical particles but they are, nevertheless, capable of accounting correctly for the existence of definite, allowed energy states at the quantum level in atoms. We shall then discuss the direct experimental evidence given by Davisson and Germer for the existence of electron waves. We shall see, however, that these waves must be interpreted in terms of the probability that a particle can be found at a given point. Finally, we shall derive a partial differential equation (Schrödinger's equation) that governs the propagation of de Broglie waves.

2. Motion of Pulses of Light. Before we develop the de Broglie theory of matter waves, we shall find it helpful to discuss the motion of light rays and to show in some detail the connection between the path of the ray and the underlying waves which make up the ray. This is of interest, not only for its own sake, but also because it provides a picture of the processes involved, on the basis of the relatively familiar light waves and also because it illustrates the necessary mathematical methods.

Let us begin with a pulse of light that might be defined, for example, by a shutter opened for a limited time τ. In general, such a pulse is three-dimensional, having a length $c\tau$ in the direction of motion of the pulse, and a diameter that depends on the narrowest aperture through which the light has gone and also on the divergence of the beam of rays as it passes through this aperture. If the dimensions of the pulse are large compared with a wavelength, but small compared with the dimensions of the apparatus, then the light beam acts like a particle (or a group of particles) localized in the pulse and moving with the speed of light.

We shall first consider a case in which a parallel beam of light strikes with normal incidence a shutter that remains open for a time so short that $c\tau$ is much less than the diameter of the pulse. This is essentially a one-dimensional case, since no important effects occur that involve the directions normal to the motion of the pulse. The pulse then simply travels with velocity c in its original direction of motion, which we take to be the x direction.

Now, an ordinary plane wave of definite wavelength λ is spread over all space and, therefore, cannot be used to describe the motion of a pulse, which is localized in a comparatively narrow region. To obtain a wave that is restricted to a definite region of space, we must construct what is known as a *wave packet*. A wave packet comprises a group of waves of slightly different wavelengths, with phases and amplitudes so chosen that they interfere constructively over only a small region of space, outside of which they produce an amplitude that reduces to zero rapidly as a result of destructive interference. The amplitude $\mathcal{E}$ of a one-dimensional wave packet (representing, for example, the Z component of the electric field) will, in general, resemble the curve shown in Fig. 1. We can construct a wave packet by taking a plane wave and integrating it over a small range of wavelengths. Let us take, for example,

$$E_Z(x) = \int_{k_0-\Delta k}^{k_0+\Delta k} dk e^{ik(x-x_0)} = 2\,\frac{\sin \Delta k(x - x_0)}{(x - x_0)}\, e^{ik_0(x-x_0)} \qquad (1)$$

When plotted as a function of $(x - x_0)$, the real part of $E_z(x)$ looks like the curve shown in Fig. 2 ($\Delta k \ll k_0$). We see that the amplitude of oscillation reaches a maximum at $x = x_0$, and goes down to zero where $x - x_0 = \pi/\Delta k$, after which it is a rapidly decreasing oscillatory function. We have thus obtained a wave function that is concentrated in a packet. In a function like the preceding, the real and imaginary parts each oscillate rapidly as a function of $(x - x_0)$. The intensity of the wave is proportional to the square of the maximum amplitude of oscillation. If, as is usually the case, the wavelength ($\lambda = 2\pi/k_0$) is much less than the width of the

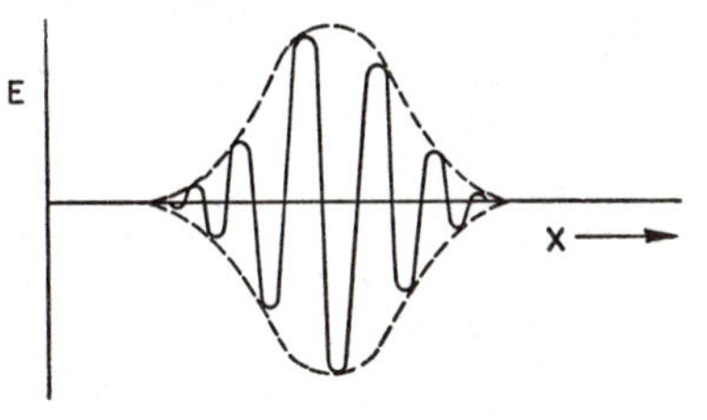

Fig. 1

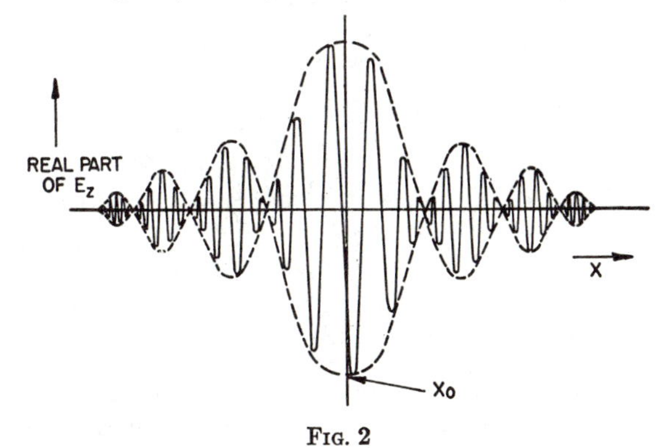

Fig. 2

packet Δk, this maximum is approximated very closely by the square of the absolute value of the complex function $E_z(x)$. Thus, we get

$$I \sim |E_z|^2 = \frac{4 \sin^2 \Delta k(x - x_0)}{(x - x_0)^2} \tag{2}$$

To obtain a more general type of packet, we multiply $e^{ik(x-x_0)}$ by a weighting function, $f(k - k_0)$, which is large near $k - k_0 = 0$ and dies out rapidly beyond some short distance Δk. For the present, we consider only functions which do not oscillate rapidly inside the region Δk, so that a graph of $f(k - k_0)$ resembles the curve shown in Fig. 3. Note that the use of a weighting function is, in a qualitative way, equivalent to integrating over a small range of wavelengths. (The effects of choosing rapidly oscillating functions for f

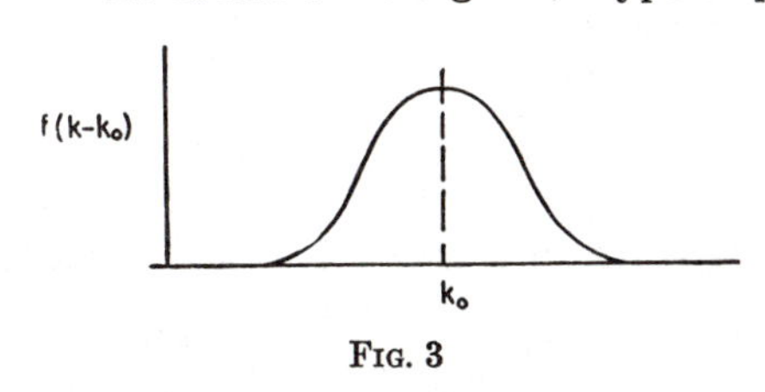

Fig. 3

will be discussed later.) We now assert that the following wave function is concentrated in a packet:

$$\psi = \int_{-\infty}^{\infty} f(k - k_0) e^{ik(x-x_0)}\, dk \tag{3}$$

To prove this, we note that at $x = x_0$, the argument of the exponential is zero for all k, hence all contributions to the integral coming from different values of k add up in phase, and the result is large. As $(x - x_0)$ becomes large, $e^{ik(x-x_0)}$ becomes a rapidly oscillating function of k, and its integral tends to cancel out. Thus, ψ is a function that is large only near $x = x_0$. At places far from $x = x_0$, the contributions of different k interfere destructively. Hence, any function defined in this way has the form of a wave packet.

As an example, let us choose $f(k) = \exp\left[-\dfrac{(k - k_0)^2}{2(\Delta k)^2}\right]$. This is chosen because it leads to simple mathematical results. We then get

$$\begin{aligned}\psi(x) &= \int_{-\infty}^{\infty} \exp\left[-\frac{(k-k_0)^2}{2(\Delta k)^2} + ik(x - x_0)\right] dk \\ &= \exp\left[ik_0(x - x_0) - \frac{(x - x_0)^2}{2}(\Delta k)^2\right] \int_{-\infty}^{\infty} \exp\left[-\frac{(k-k_0)^2}{2(\Delta k)^2}\right. \\ &\qquad\qquad \left. + i(k - k_0)(x - x_0) + \frac{(x - x_0)^2(\Delta k)^2}{2}\right] \\ &= \sqrt{2\pi}\,\Delta k \exp\left[ik_0(x - x_0) - \tfrac{1}{2}(x - x_0)^2(\Delta k)^2\right]\end{aligned} \tag{4}$$

We note that a Gauss function in k space leads to a Gauss function in x space. The Gauss function is the only one having this peculiar symmetry in x and k space. We note also that the resulting packet has a maximum at $x = x_0$ and clearly becomes negligible for large values of $(x - x_0)$.

3. The Width of a Wave Packet. At this point, we are ready to calculate what determines the width of a given wave packet. This problem is of special interest because the results are applied later in connection with the derivation of the uncertainty principle.

Let us begin with our two examples given above. In the first of these, the intensity is

$$I \sim \frac{\sin^2 (x - x_0)\,\Delta k}{(x - x_0)^2}$$

This quantity begins to become fairly small when $(x - x_0)$ takes on values appreciably larger than $1/\Delta k$ or when $(x - x_0) > \dfrac{1}{\Delta k}$. In a similar way, the intensity in our second example $\{I \sim \exp[-(x - x_0)^2(\Delta k)^2]\}$ begins to become small when $(x - x_0)^2 > \dfrac{1}{(\Delta k)^2}$. Since, in both cases, Δk is a measure of the range of wave numbers k present in the packet, we obtain the result that the product of the width of the packet in k space, and its width in x space is of the order of unity, or, mathematically speaking

$$\Delta x \, \Delta k \cong 1 \tag{5}$$

This means that a packet with a narrow range in k space must be very broad in x space and vice versa.

It is easy to show that this result holds for any packet of the form (3), where $f(k - k_0)$ is a smooth function that does not oscillate too rapidly. To do this, consider the function given in eq. (3). We see that contributions of difference k tend to interfere constructively as long as

$$k(x - x_0) < 1,$$

but that for larger values of k they tend to oscillate and get out of phase. Since f is large in only a limited region, Δk, we conclude that destructive interference begins to be important when $|x - x_0| > \frac{1}{\Delta k}$. Thus, we obtain once again the result that $\Delta x \, \Delta k \cong 1$.

We can summarize the preceding results in simpler terms by noting that in order to make up a packet that is small beyond a region Δx, it is necessary to add up a range of waves, involving functions such as $\cos k(x - x_0)$ and $\cos (k + \Delta k)(x - x_0)$, which can be in phase at $x = x_0$, but out of phase at $x = x_0 + \Delta x$, so that we must have $(x - x_0) \, \Delta k > 1$. Note, however, that this result required that we add up waves of the same sign. If $f(k - k_0)$ had been a rapidly oscillating function, it would not necessarily follow.

Similar results can be obtained relating the time Δt that is required for a pulse to pass a given point, to the range of angular frequencies $\Delta\omega$, needed to form such a pulse. Thus, the electric field at a given point can be expressed as

$$E = \int f(\omega - \omega_0) \exp [-i\omega(t - t_0)] \, d\omega \tag{6}$$

Such a function will be a pulse that is large only near $t = t_0$, and which has a width Δt obeying the relation

$$\Delta\omega \, \Delta t \cong 1 \tag{7}$$

The fact that a pulse requires a range of frequencies is responsible, for example, for the "band width" of a radio transmitter. To carry audio-frequency pulses on a radio wave, it is necessary to allow the frequency of the radio wave to shift by an amount of the order of magnitude of the audio frequencies that we wish to carry. If the receiver is tuned to accept a band width $\Delta\omega$, the shortest pulse that can be received has a duration $\Delta t \cong 1/\Delta\omega$.

4. Group Velocity. We now treat the problem of how a wave packet moves through space. To do this, we make use of the fact that, for light in free space, a wave of propagation vector k oscillates with frequency $\omega = ck$. Thus, we write

$$E(x, t) = \int_{-\infty}^{\infty} f(k - k_0) \exp [ik(x - x_0) - \omega t]\, dk$$
$$= \int_{-\infty}^{\infty} f(k - k_0) \exp [ik(x - x_0 - ct)] \quad (8)$$

Note that E is a function only of $(x - x_0 - ct)$; this means that the pulse travels at the velocity c without changing its shape; this is, of course, a well-known result.

The motion of the wave packet is caused by the change of phase of all of the different wavelengths, resulting from the multiplication by the term e^{-ickt}. Thus, when $t \neq 0$, the waves cease to add up in phase at $x = x_0$, but, instead, they are all in phase at the point $x = x_0 + ct$. The change of position of the wave packet is, therefore, caused by the change of conditions for constructive and destructive interference.

Suppose now that the packet enters a dispersive medium, which has an index of refraction $n(\lambda)$. The angular frequency, $\omega = 2\pi c/\lambda n(\lambda)$, is, in general, a fairly complicated function of λ and, therefore, of k. To denote this, we write $\omega = \omega(k)$. The electric field is then

$$E(x, t) = \int_{-\infty}^{\infty} f(k - k_0) \exp [ik(x - x_0) - i\omega(k)t]\, dk \quad (9)$$

The wave packet will change with time but, in general, the changes are not so simple as when $\omega = ck$, for now E cannot be written simply as $E(x - x_0 - ct)$. This means that not only the position of the center of the wave packet, but also the shape, will change with time. We shall discuss changes of shape later; let us now consider only how the packet as a whole moves. In order to find the position of the maximum of the packet, we note that, as in free space, there will be at each time one point where waves of different k do not tend to interfere destructively. This will happen wherever the phase of the exponential $\varphi = k(x - x_0) - \omega(k)t$ has an extremum. At this point, there will be a range of k where all waves have nearly the same phase; hence there will be constructive interference. To find this point, we set $\partial\varphi/\partial k = 0$. This gives

$$x - x_0 = t\frac{\partial \omega}{\partial k}$$

which means that the maximum of the wave packet moves through space with the velocity

$$V_g = \left(\frac{\partial \omega}{\partial k}\right)_{k=k_0} \quad (10)$$

V_g is called the *group velocity*, because it denotes the speed of motion of a group of waves collected together in the form of a packet. This is in contrast to the *phase velocity*, $V_p = \lambda\nu = \omega/k$, which is precisely the speed with which a point of constant phase moves when ω and k are defined. In general, the phase velocity has little physical significance; for example, the speed of transmission of a signal through a dielectric is given by the group velocity,* as is also the speed of transport of energy.

For the special case, $\omega = ck$, which holds in free space, one obtains $V_g = c = V_p$. Only if ω is proportional to k is the group velocity equal to the phase velocity.

5. Spread of Wave Packets. We have already seen that we cannot, in general, expect a wave packet to be transmitted through a dielectric without change in shape. The problem of solving for the change of shape by taking into account the specific dependence of ω on k is usually too complicated to be solved exactly. If $f(k - k_0)$ has a narrow enough peak, however, a good approximation is obtained by expanding $\omega(k)$ as a series of powers of $k - k_0$. This is because the main contributions to the integral come in a region of the order of the width of the peak in $f(k - k_0)$. Thus, we obtain

$$\omega(k) = \omega(k_0) + \left(\frac{\partial \omega}{\partial k}\right)_{k=k_0} (k - k_0) + \frac{1}{2}\left(\frac{\partial^2 \omega}{\partial k^2}\right)_{k=k_0} (k - k_0)^2 + \cdots \tag{11}$$

Setting $\quad \omega(k_0) = \omega_0, \quad \left(\frac{\partial \omega}{\partial k}\right)_{k=k_0} = V_g, \quad \left(\frac{\partial^2 \omega}{\partial k^2}\right)_{k=k_0} = \alpha,$

we get in eq. (9)

$$E = \exp \{i[k_0(x - x_0) - \omega_0 t]\} \int_{-\infty}^{\infty} f(k - k_0) \exp\left[i(k - k_0)(x - x_0 - V_g t) - \frac{i\alpha}{2}(k - k_0)^2 t\right] dk$$

With $k - k_0 = \kappa$, this becomes

$$E = \exp \{i[k_0(x - x_0) - \omega_0 t]\} \int_{-\infty}^{\infty} f(\kappa) \exp\left[i\kappa(x - x_0 - V_g t) - \frac{i\alpha}{2}\kappa^2 t\right] d\kappa \tag{12}$$

If we had $\alpha = 0$, then E would be a function only of $x - x_0 - V_g t$, and the pulse would not change its shape. In order to show how the α term affects the pulse, let us consider the special case already given in eq. (4)

$$f(\kappa) = e^{-\kappa^2/2(\Delta k)^2}$$

We get

$$E = \exp \{i[k_0(x - x_0) - \omega_0 t]\} \int_{-\infty}^{\infty} \exp\left[i\kappa(x - x_0 - V_g t) - \frac{\kappa^2}{2}\left(i\alpha t + \frac{1}{(\Delta k)^2}\right)\right] d\kappa \tag{13}$$

* This statement is true only if we are not too near a resonance. At such points, the pulse is so badly distorted that there is no clearly defined maximum. Instead, it is necessary to define a quantity called *signal velocity* (Stratton, *Electromagnetic Theory*, p. 338), which is the velocity of the front of the pulse. It is found that the signal velocity is never greater than c.

We can evaluate this integral by completing the square in the exponential, as with the simpler integral in eq. (4):

$$E = \exp\left\{i[k_0(x - x_0) - \omega_0 t] - \frac{(x - x_0 - V_g t)^2(\Delta k)^2}{2[1 + i\alpha t(\Delta k)^2]}\right\} \int_{-\infty}^{\infty} \exp\left\{-\frac{1}{2}\left[\frac{1 + i\alpha t(\Delta k)^2}{(\Delta k)^2}\right]\left[k - i\frac{(x - x_0 - V_g t)\,\Delta k}{1 + i\alpha t(\Delta k)^2}\right]^2\right\} dk$$

The integral multiplying the exponential is equal to

$$\sqrt{\frac{2\pi(\Delta k)^2}{1 + i\alpha t(\Delta k)^2}}$$

We can transform the argument of the exponential to a simpler form by multiplying numerator and denominator by $1 - i\alpha t(\Delta\kappa)^2$. We get

$$E = \exp\{i[k_0(x - x_0) - \omega_0 t]\}\sqrt{\frac{2\pi(\Delta k)^2}{1 + i\alpha(\Delta k)^2 t}} \exp\left[-\frac{(\Delta k)^2}{2}\frac{(x - x_0 - V_g t)^2}{1 + t^2(\Delta k)^4\alpha^2}\right]\exp\left[\frac{i\alpha t}{2}\frac{(\Delta k)^4(x - x_0 - V_g t)^2}{1 + t^2\alpha^2(\Delta k)^4}\right] \quad (14)$$

Since any quantity of the form $e^{i\lambda}$ has absolute value unity, and since the intensity of a wave is proportional to $|E|^2$, we conclude that the intensity in the above wave is

$$I \sim \exp\left[\frac{-(\Delta k)^2(x - x_0 - V_g t)^2}{1 + \alpha^2 t^2(\Delta k)^4}\right]$$

which is a Gaussian distribution centering at $x = x_0 + V_g t$, in agreement with our calculation from the discussion of group velocity. The mean width of the distribution (where I falls to $1/e$ of its maximum value) is

$$\delta x = \frac{1}{\Delta k}\sqrt{1 + \alpha^2 t^2(\Delta k)^4} = \delta x_0\sqrt{1 + \frac{\alpha^2 t^2}{(\delta x_0)^4}} \quad (15)$$

For times so short that $\alpha^2 t^2(\Delta k)^4 \ll 1$, we have $\delta x = 1/\Delta k \sim \delta x_0$ where δx_0 is the spread when $t = 0$. More generally, we see that the packet begins to spread appreciably only when $t > \dfrac{1}{\alpha(\Delta k)^2}$.

Problem 1: Consider a dielectric for which

$$n = 1 + \frac{k^2}{(\omega - \omega_0)^2 + \beta^2}$$

Take $\omega_0 = 10^{16}$, $\beta = 10^{12}$, and $k = 10^{16}$.

(a) Calculate the phase velocity and group velocity at $\omega = 10^{15}$ cps
(b) If we choose $\delta x_0 = 10^{-2}$ cm, how long will it take for a packet to double its dimensions, and how far will it go in this time?
(c) Repeat the calculations for $\omega = 10^{16} - 10^{14}$ cps.

6. More General Criteria for Widths of Packets. We note that for the packets derived in the previous section, we obtain

$$\Delta x \, \Delta k = \sqrt{1 + \alpha^2 t^2 (\Delta k)^4}$$

Hence, after a long time, $\Delta x \, \Delta k$ becomes very large. This shows that it is not necessary that in all packets the relation given by eq. (5) shall hold. We can readily see that the reason for the change of behavior is the presence of the multiplier $\exp\left[-i\frac{\alpha t}{2}(k - k_0)^2\right]$ in the integrand. When t is large, this oscillates very rapidly as a function of k, particularly for large k. Such oscillations prevent us from deducing that when $(x - x_0)\,\Delta k \cong 1$, the waves necessarily begin to interfere destructively, as was done in Sec. 2 for smooth functions, $f(k - k_0)$. The reason is that the changes of phase caused by the term in eq. (12) involving $\exp\,(-i\alpha t\kappa^2/2)$ can, in certain regions, cancel the changes produced by the term $\exp\, ik(x - x_0 - V_g t)$, so that we must go to much larger values of $(x - x_0)$ to get oscillation of the integrand than if $f(k - k_0)$ were a smoothly varying function.

It will be shown in Chap. 10, Sec. 9, that any modification of $f(k - k_0)$ from a smooth form will always result in an increase in the value of the product $\Delta k \, \Delta x$. Thus, we can generalize the results of Sec. 2 and state that

$$\Delta x \, \Delta k \geq 1 \tag{16}$$

This means that for any wave packet, the minimum possible value of $\Delta x \, \Delta k$ is of the order of unity, but that it is possible to construct wave packets for which this quantity is arbitrarily large. This reflects the fact that it is not necessary that every wave having a range of frequencies Δk must be put together in such a way that it interferes destructively beyond the distance $\Delta x \sim 1/\Delta k$. As an example, we may take a radio signal carrying noise, which covers a frequency range $\Delta\omega$. This noise can happen to have the right phase relations among its component waves, so that a pulse of width $1/\Delta t$ is created. But it is much more likely that the noise consists of a random series of much weaker pulses that are spread out over a much longer time.

7. Generalization to Three Dimensions. The general results of the previous work can be extended to three dimensions. To do this, we make up a packet from the integral

$$E = \int f(k_x - k_{x_0}, k_y - k_{y_0}, k_z - k_{z_0}) \exp\,[i(k_x x + k_y y + k_z z)]\, dk_x\, dk_y\, dk_z \tag{17}$$

where f is large only in a narrow region near $k_x = k_{x_0}$, etc. It can be shown that even in free space, there is a tendency for a three-dimensional light-wave packet to spread, but that as the diameter becomes much greater than the length, so that it approaches a one-dimensional packet, this rate of spread approaches zero. To the extent that the rate of spread is slow enough to be neglected, light-wave packets act like particles in three-dimensional space. For example, in free space they move in straight lines, at constant velocity and reflect specularly off surfaces, much like free particles bouncing elastically. In a dielectric, their speed changes. If we have a dielectric of variable density, for example, glass of nonuniform density, or a layer of air of nonuniform density, the packets are actually bent in curved paths. Thus, they follow curved orbits and look very much like particles under the influence of a force. We shall see later (Chap. 12, on WKB approximations) that this analogy can be carried very far indeed.

To the extent that the packet spreads, the analogy with a particle fails because a particle will never spread. We might, however, make a comparison with a collection of particles of slightly differing velocities that would gradually separate with the passage of time. We shall return later to this analogy in Chap. 5, Sec. 4.

Electron Waves

8. Motion of Electron Wave Packets. Let us now assume tentatively with de Broglie that matter is really comprised of waves, and that what we see in a gross observation is just the packet. Since the wave properties are essentially quantum mechanical, we may regard the path of the packet as a classical limit of the particle trajectory. It is necessary then that the group velocity of the wave packets be equal to the classical particle velocity, or that

$$v_g = \frac{\partial\omega}{\partial k} = \frac{p}{m} \tag{18}$$

where P is the particle momentum.

But we have the relation

$$\omega = \frac{2\pi E}{h} = \frac{E}{\hbar}$$

from which we obtain

$$\frac{\partial\omega}{\partial k} = \frac{1}{\hbar}\frac{\partial E}{\partial k}$$

But classically, $E = p^2/2m$, so that

$$\frac{\partial\omega}{\partial k} = \frac{p}{m\hbar}\frac{\partial p}{\partial k}$$

Equating this to the classically observed particle velocity, we obtain

$$\frac{p}{m\hbar}\frac{\partial p}{\partial k} = \frac{p}{m}$$

$$p = \hbar k \qquad \text{or} \qquad p = h/\lambda \tag{19}$$

This is the de Broglie relation. (Note that we could have added a constant of integration, but since we are merely seeking a way of describing particle motions by means of wave packets, we are at liberty to take the simplest possible case, which is obtained by setting this constant equal to zero.) The group velocity is then

$$v_g = \frac{p}{m} = \frac{\hbar k}{m} \quad \text{and} \quad \omega = \frac{E}{\hbar} = \frac{p^2}{2m\hbar} = \frac{\hbar k^2}{2m} \tag{20}$$

We observe that in contrast to light waves in free space, ω is not proportional to k.

The preceding treatment is not actually as given originally by de Broglie, who used arguments based on relativistic considerations.* The agreement between the relativistic and the correspondence treatment is not accidental, but is caused by the fact that it must be possible, in the correspondence limit, to obtain a relativistic description of the motion of wave packets, which reduces to the nonrelativistic one when $v/c \ll 1$. The advantage of the derivation based directly on the correspondence principle, however, is that it shows that we can construct a wave theory of matter without reference to the theory of relativity. De Broglie's derivation has the advantage, however, that it shows the relations $E = h\nu$ and $P = h/\lambda$ are relativistically invariant.†

9. Effects of Forces. Thus far, we have considered only the problem of representing the motion of a free particle by means of wave packets. When there is an external force, then the momentum of the particle changes as it moves from one place to another. According to the de Broglie relation $\lambda = h/P$, this means that the wavelength becomes a function of position. The precise form of the function is

$$\lambda = \frac{h}{\sqrt{2m[E - V(x)]}}$$

where E is the total energy of the particle, and $V(x)$ is the potential.

In optics, a similar change of wavelength with position occurs in a medium of a continuously variable index of refraction. In such media, we know that light rays follow a curved path, as is evidenced by the distortion of an image by a piece of glass of nonuniform composition, or by the production of a mirage in a layer of air having a temperature gradient. Similarly, as will be shown in Chap. 12, electron-wave packets with the wavelengths equal to the function defined above move along the (generally curved) classical particle orbits with the same speed that a classical particle would have, provided that the wavelength does not undergo a large fractional change within its own length. If the latter condition is not satisfied, however, characteristic wave effects such as diffraction and interference come in. These will be discussed in later work. We con-

* See Ruark and Urey, p. 516.

† The term "classical" refers always to a nonquantum theory. It is possible to give separately, for classical and quantum theories, a relativistic and a nonrelativistic formulation.

clude therefore that in all phenomena not requiring a description in terms of distances as small as a de Broglie length, the use of de Broglie wave packets leads to exactly the same results as does classical mechanics.

10. Effects of Quantization. With these notions, de Broglie was able to obtain the quantization of orbits in atoms. To do this, he assumed that the allowed orbits in hydrogen, for example, corresponded to a wave propagated in such a way as to circle the nucleus. A stationary wave, representing the electron in a stationary state, can be obtained only if the wave fits onto itself continuously after going around the nucleus. This requires that there be an integral number of waves in the circuit, or that $2\pi r/\lambda = n$, from which we conclude that $2\pi r p_\phi = nh$, and $p_\phi = n\hbar$, in agreement with Bohr's original condition (Chap. 2, Sec. 14).

Problem 2: Find the permissible energy levels of electron waves in a box of length L. (The wave function must be zero at the walls.*) Compare the result with that obtained from the Bohr-Sommerfeld conditions (Chap. 2, Problem 11).

One can show quite generally that the de Broglie relations always lead to the Bohr-Sommerfeld conditions. To do this, consider an arbitrary periodic motion that is to be quantized. (For simplicity, only one variable is taken here.) There will, in general, be some limits of oscillation, which we denote by $q_a = a(E)$ and $q_b = b(E)$. The classical particle is confined within these limits. If the particle is to be described in terms of the waves of de Broglie, then a steady state can be reached, only if the wave that reflects off the boundaries is in the correct phase to match that of the incident wave and to produce a standing wave. The precise effect of this requirement will depend on the nature of the boundaries, but in general, the result will be to define a discrete set of allowed frequencies, hence also of energies. When there are many wavelengths inside the region (which will always happen in the classical limit), one has roughly an integral number of waves in going from b to a and back again within a fractional error of the order of $\pm 1/n$, where n is the total number of wavelengths. But the total number of waves† is just

$$2\int_a^b \frac{dq}{\lambda(q)} = 2\int \frac{p}{h}\,dq = \oint p\,\frac{dq}{h}$$

Setting this equal to an integer leads exactly to the Bohr-Sommerfeld conditions. The possibility of a fractional number of waves agrees with the fact that the precise value of the quantum number is somewhat ambiguous in the Bohr-Sommerfeld theory. The wave theory leads, however, as we shall see in later work, to a precise value for this number, and is therefore unambiguous.

* The wave function will be zero at the walls because the electron is unable to penetrate into the walls. Thus, the wave function must be zero inside the walls themselves and continuity requires that it shall also be zero at the edge of the walls.

† Thus far, the derivation applies only to Cartesian co-ordinates, but it can be generalized in a straightforward way to arbitrary co-ordinates.

11. The Davisson-Germer Experiment. Thus far we have seen that for a free particle the wave packets can describe all classical motions, while for bound particles the conditions of continuity on the wave function lead to the correct quantization conditions. Whether or not the wave theory is really valid, however, can be tested only by looking for its characteristic new effects, namely diffraction and interference. Little attention was in fact paid to de Broglie's suggestions until several years after he first made them, when Davisson and Germer, while studying the scattering of electrons from metals, discovered that the electrons were diffracted in a manner very similar to that of light from a grating. That is, they found that a beam of electrons incident on a crystal came off only at definite angles, θ. From the grating equations $\lambda/a = \sin\theta$ where a is the space between elements of the grating, they calculated λ and obtained agreement with the value given by the de Broglie relation, i.e., $\lambda = h/p$. This showed that electrons were being diffracted from the crystal as if they were waves, with the length predicted by de Broglie.

The results of this experiment are of fundamental significance, since they demonstrate the wavelike properties of matter, which cannot be understood in terms of the ideas that matter is made up of elementary particles. Later experiments showed that other forms of matter, such as molecules, also exhibited diffraction properties. We are, therefore, led to the conclusion that all matter has the property that we previously met in connection with the electromagnetic field, namely, in some experiments, it seems to behave as if it were made up of particles, while in other experiments it shows equally good evidence of acting like waves. We have thus obtained a remarkable unification of two different branches of physics, but at the expense of introducing the paradoxical wave-particle duality.

12. Prediction of Electron Diffraction by Bohr-Sommerfeld Theory. At this point, it is worthwhile to point out a subsequent argument of Duane,* which shows that it is possible to interpret the previously existing Bohr-Sommerfeld theory in such a way as to obtain a prediction of the observed electron diffraction. To do this, let us consider an electron that is scattered by some kind of periodic structure, such as a grating, with spacing a (Fig. 4). According to classical theory, the electron has a component of velocity v_x in the direction of periodicity of the grating, which is constant until it strikes the grating. Every time the electron moves a distance a, the force between electron and grating repeats itself and, since the velocity is constant, the force will be periodic. Hence, in energy transfers to the grating, we can plausibly argue that the same

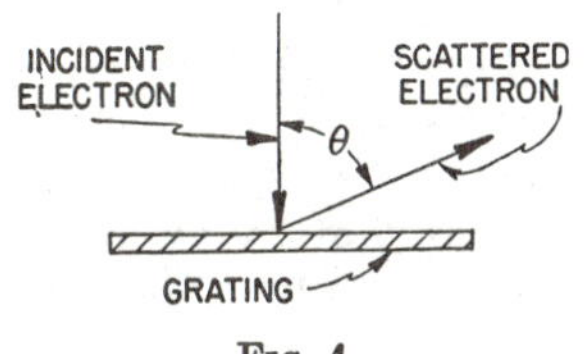

FIG. 4

* Heisenberg, W., *The Physical Principles of the Quantum Theory*, p. 77.

quantum conditions should apply as with any other periodic system, such as, for example, a harmonic oscillator. Thus, the action J can change only in units of h. To compute J, we note that the period is a/J_x. Thus $J = \oint p\,dq = pa$. The quantum condition is $a\,\Delta p = h$, or $\Delta p = h/a$. But, according to the wave theory, the momentum shift is

$$\Delta p = p \sin \theta = \frac{h}{\lambda} \sin \theta = \frac{h}{a}$$

Therefore, both the Bohr-Sommerfeld theory, as applied to the allowed orbits of the electron in the presence of the grating, and the wave theory, as applied to the diffraction of waves from the grating, lead to the same angle of deflection. In Duane's treatment, however, the appearance of definite angles comes from the quantization of momentum transfers, whereas in the wave theory it results from the interference of the electron wave. This result shows that the two methods are certainly closely related.

It should be noted, however, that the result of Duane's treatment is somewhat ambiguous, because in computing the period it is not clear whether the velocity before collision or after collision should be used. In the correspondence limit, where the angle of deflection is small, this ambiguity is not important. With the wave theory, however, there is no such ambiguity.

In view of the fact that we can explain the Davisson-Germer experiment with the aid of the Bohr-Sommerfeld theory, we might perhaps be tempted to refrain from taking so radical a step as to assert that electrons have some of the properties of waves. We must remember, however, that the Bohr-Sommerfeld theory can deal only with periodic motions, whereas the wave theory defines the effect of quantization even for aperiodic motions. Furthermore, from the logical point of view, the Bohr-Sommerfeld conditions are simply arbitrary restrictions on the possible motions of matter and are unable to describe what happens in the transition between allowed orbits. The wave theory, however, describes the quantization of allowed orbits naturally in terms of a spectrum of frequencies obtained from the boundary conditions that must be satisfied by the wave function. Moreover, the ambiguity at small quantum numbers, characteristic of the Bohr-Sommerfeld theory, is not present in the wave theory. We shall see, in fact, that the wave theory yields a complete and quantitatively correct treatment, which applies in principle to all phenomena and which agrees with experiment wherever a comparison has been made, over an enormous range of phenomena. It therefore seems preferable to take the wave theory as a basis and to derive the Bohr-Sommerfeld theory from it, as an approximation. We then come to the conclusion that, because of their wave properties, both electron and grating are restricted to certain quantized transfers of momentum between them, and the restrictions are such that the quanta of the one are of the same size as the quanta of the other, so that the whole theory fits together in a well integrated fashion.*

13. Interpretation of Wave Function in Terms of Probability. It is now necessary to try to obtain a more direct physical interpretation for the electron waves. These waves were first interpreted as representing the actual structure of the electron. In other words, it was suggested that the electron is, like light, a wave that spreads out, shows interference, etc. This interpretation, however, soon encountered serious difficulties when it was discovered that matter wave packets spread without limit and during moderate lengths of time are able to cover a very large region (i.e., billions of miles). To demonstrate this fact, we use eq. (9), setting

$$\omega = \frac{E}{\hbar} = \frac{\hbar k^2}{2m}$$

as obtained from eq. (20). This yields

$$\psi = \int f(k - k_0) \exp\left\{i\left[k(x - x_0) - \frac{\hbar k^2}{2m}t\right]\right\} dk \tag{21}$$

With $k - k_0 = \kappa$, we obtain

$$\psi = \exp\left\{i\left[k_0(x - x_0) - \frac{\hbar k_0^2}{2m}t\right]\right\} \int f(\kappa) \exp\left\{i\left[\kappa(x - x_0) - \frac{\hbar k_0 \kappa t}{m} - \frac{\hbar \kappa^2}{2m}t\right]\right\} d\kappa \tag{22}$$

This result is exactly the same as that obtained for light waves in eq. (12), provided that we set $\alpha = \hbar/m$, and $v_g = \hbar k_0/m$. Note, however, that whereas this expression is only an approximation for light waves, it is exactly correct for electrons.

We can conclude, then, that an electron wave packet will spread, and from eq. (15) we see that, in terms of the original width Δx_0, we get

$$\Delta x = \Delta x_0 \sqrt{1 + \frac{\hbar^2 t^2}{m^2(\Delta x_0)^4}} \rightarrow \frac{\hbar}{m\,\Delta x_0} t \qquad \text{as } t \rightarrow \infty \tag{23}$$

Problem 3:

(a) Suppose an electron wave packet is confined originally to 10^{-8} cm. How long will the packet take to spread to twice its original dimensions? How long until it exceeds the size of the solar system?

(b) Suppose that the wave packet representing the earth is confined to 1 m; how long will it take for the packet to double its original dimensions?

(c) How long would it take for a 1-gram object with a wave packet confined to 10^{-3} cm to double its dimensions?

The electron wave packet can spread very rapidly. Yet, whenever the position of an electron is observed, it is always found to be within a

* Further objections to replacing the wave theory by a generalization of the Bohr-Sommerfeld quantum conditions are given in Chap. 6, Sec. 11.

region of space that is as well-defined as we choose to make it. This fact would induce us to think of the electron as a particle; conversely, other experiments, such as those of Davisson and Germer, would have us consider it as a wave. Which is it? The problem is exactly the same as that encountered in connection with light waves and with the photoelectric effect (Chap. 2, Sec. 1).

The resolution of this problem must be deferred until Chaps. 6, 7, and 8 but, for the present, we shall say that, as in the case of light waves, the intensity of electron waves must be regarded as giving only the probability that a particle will be found at a given position. This idea constitutes a remarkable unification between the properties of light and those of matter, at the expense, however, of presenting us with a somewhat paradoxical dualism that requires the use of both wave and particle models to describe the same system.

We sum up our ideas on electron waves:

(1) The wavelength yields the electronic momentum according to $p = h/\lambda$.

(2) The wave intensity yields the probability that an electron will be found at a given position.

14. Comparison between Electron Waves and Electromagnetic Waves. Electron waves have two similarities with electromagnetic waves:

(1) The de Broglie relations $E = h\nu$ and $p = h/\lambda$ are satisfied by both of them.

(2) Each determines only the probability of a physical process.

There are, however, two important differences. The number of quanta in the electromagnetic field can change by emission and absorption, but the number of electrons cannot change. Hence, the integrated probability that an electron is somewhere in space must be unity and must remain unity for all time.* No such restriction applies to photons. We shall see in the next chapter that this restriction is closely connected with the form of Schrödinger's equation.

The other difference is that electromagnetic waves are expressed in terms of the vector potential $\boldsymbol{a}$. Electron waves, however, were first taken to be scalar functions. This is because no effects, such as polarization, were found that required the assumption of a directed field. Later, however, we shall see that the electron spin requires us to take two wave functions for ψ, which transform neither as vectors nor as scalars, but as an intermediate class of quantities called *spinors*. For the present,

* This requirement actually holds only to the extent that production of electron-positron pairs can be neglected. No pairs can be produced, however, unless adequate energy is available. This energy is $E = 2mc^2 \cong 1$ mev, where m is the mass of an electron. In this book, we shall not be interested in processes that involve so high an energy. For example, the energy of ionization of a hydrogen atom is only about 13.5 ev. For an account of pair production, see W. Heitler, *Quantum Theory of Radiation*. Oxford: Clarendon Press, 1936.

however, we neglect the effects of electron spin and regard the electron wave function as a scalar.

15. More Detailed Picture of Electron Waves. As shown in Sec. 8, the presence of a potential causes the index of refraction to vary continuously as a function of position and makes it possible for the wave to move in a curved path.* In an atom, the effective index of refraction varies in such a way that the parts of the wave at larger distances from the center move faster than those at smaller distances, so that the wave can, for this reason, circle the nucleus. In order that the wave have a definite frequency and, therefore, a definite energy, it is necessary that it fit onto itself continuously after going around the nucleus and that an integral number of wavelengths be present. If this condition were not satisfied, then the wave function could not take the simple form $\psi = e^{-iEt/\hbar}f(x, y, z)$, which is implied by the statement that its energy is definite, because the wave intensity $|f(x, y, z)|^2$ would be changed every time the wave made another circuit.

The exact form of the wave function can be obtained only by solving Schrödinger's equation, which we shall come to later. For the present, however, we shall use some of the results obtained from these solutions, in order to provide a general description of what these waves look like when they are inside an atom.* In a state of definite energy, the wave function is large only in a toroidal region, surrounding the radius predicted by the Bohr orbit for that energy level. A cross section of this region is shown in Fig. 5. Of course, the toroid is not sharply bounded, but the wave function reaches its maximum in this region and rapidly becomes negligible outside it. The next Bohr orbit would look very much the same, but would have a larger radius. In the classical limit, the width of the toroid is negligible in comparison with its diameter, so that we obtain something that looks exactly like a classical particle orbit. Waves with elliptical orbits are also possible.

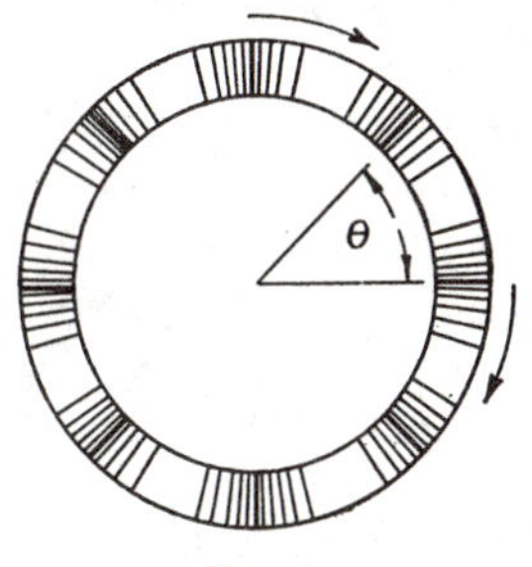

Fig. 5

The real and imaginary parts of ψ are propagated in wavelike fashion around the nucleus, with angular frequency $\omega = E/\hbar$ and with a wave vector $k = p/\hbar$. The probability of finding a particle in a given region is proportional to $|\psi|^2 = |f(x, y, z)|^2$.

Since the function f is more or less uniform in value over the toroid, it is likely that a particle will be found in a region that is fairly close to where the Bohr orbit theory says it should be. But we cannot predict at exactly what value of angle θ it will be found. This behavior is analogous to that obtained with free particles of a definite energy. For these,

* These results are obtained in detail in Chap. 15.

the wave function $\psi = e^{i(kx-\omega t)}$ yields a uniform probability that the particle can be found anywhere in space. In order to localize such a particle, it is necessary to make a packet to cover a range of energies. Similarly, in order to obtain a wave function in which an electron in an atom is certain to be in a fairly narrow range of angles θ, it is also necessary to make a packet containing a wide range of energies. If a particle has a definite energy it cannot, therefore, be localized within a definite range of angles; but if it is so localized, its wave function cannot have a definite frequency, but must contain a range of frequencies and, therefore, of energies. We shall return to this point later in connection with the uncertainty principle.

16. Transitions between Orbits. We have already seen that the motion of wave packets is governed by the equation of propagation (9), from which the group velocity was derived. From this equation it was also possible to calculate the spread of the wave packet. In fact, once we know the initial value of $f(k - k_0)$, it predicts exactly what happens to the wave amplitude as a function of time. The propagation of the wave function can therefore be called continuous and deterministic.

This result can be extended from the case of the free particle which we have treated thus far to the case of an electron in an atom. This will be done later in connection with Schrödinger's equation, but we shall quote some of the results of this treatment here. As long as the atom does not gain or lose energy from its surroundings, the wave continues to propagate around the nucleus and takes a form resembling that shown in Fig. 5, where the mean radius of the orbit is determined by the energy level in which the electron happens to be. If, however, the electron can gain energy from some other system, such as, for example, the electromagnetic field, then we shall see that the wave gradually flows from its original toroid into another one corresponding to a higher energy level.* While this process is taking place there is some probability that the particle can be found in either toroid. In fact, for neighboring energy levels, the toroids overlap to some extent so that the wave never goes from one region to another without crossing the intervening space, a necessary condition for continuity of flow.

The above description seems, at first sight, to furnish a continuous and deterministic account of how the electron gets from one quantum state to the next. This is in remarkable contrast to the Bohr-Sommerfeld theory, in which the transition between quantum states was discontinuous and indivisible, not passing through intermediate states, and for which only the probability of transition was determined by physical laws. Have we really eliminated the fundamental indivisibility and lack of determinism of quantum processes, discussed in connection with the

* See Chap. 4, eq. (4) and Chap. 9, eq. (49).

photoelectric effect and the Compton effect? Although the idea is very tempting, the answer is, no.

Let us consider, for example, an atom exposed to an electromagnetic wave, which, because it can supply energy, causes the electron wave to flow from an inner into an outer toroid. Experimentally, however, we know that there are cases when, after a very short time, a *full* quantum has been transferred to the atom. Since energy is conserved in each quantum process, the electron must go into an excited state in a correspondingly short time. Meanwhile, because the wave moves continuously, only a small part of it can have reached the outer ring.

To interpret this discrepancy, we use the fact that the time of irradiation of the atom by light determines only the *probability* of transfer of a quantum. We note that both the probability of this process and the wave intensity in the outer toroid increase at a rate that is proportional to the time. It seems inevitable that we shall apply our probability interpretation of the wave intensity here, and say that the steadily growing wave intensity in the outer ring corresponds to a steadily growing probability that an indivisible quantum of energy has been transferred and that the atom can therefore be found in an excited state.*

The need for retaining the indivisibility of the process of energy transfer can also be seen from the fact that the allowed frequencies of oscillation of the wave function, and therefore the allowed energies, are discrete. This means that the atom has no way of holding part of a quantum of energy so that the process of transfer must still be indivisible, even though the wave amplitude and the probability of finding the particle at a given point in space change in a continuous way. Thus, we conclude that the connection of the wave function with a real and observable event, such as a jump to a higher energy state, is only statistical, and the indivisibility and lack of complete determinism characteristic of the quantum theory are still present. Yet, as we shall see, the wave theory represents a tremendous advance, because it makes possible the quantitative calculation of energy levels and of the probabilities of transition.

Wave Equation

We now proceed to the derivation of the wave equation. The wave equation is, in general, a partial differential equation satisfied by the wave function, ψ, the nature of which will become clear in each case as the equation is developed and used. In this section, we consider only the special case of a free particle, but the results are generalized to an arbitrary system in Part II.

17. Fourier Analysis. Fourier Integrals. The first step in finding the wave equation is to treat the propagation of waves of arbitrary shape. In order to do this, it is convenient to use Fourier analysis. In the proof

* This interpretation is developed quantitatively in Chap. 18.

of the Rayleigh-Jeans law we used Fourier analysis to represent an arbitrary electromagnetic wave, which was confined within a large box. Here it is desirable to use waves which are not confined at all, because we want to represent free particles. To do this, we go from a Fourier series to what is called a Fourier integral.

The Fourier integral may be obtained in many ways, but the simplest is to take a Fourier series, expanded within a large box of side L, and allow the box to approach an infinite size. It is also a matter of convenience here to use the complex functions, e^{ikx}, rather than the real functions, $\cos kx$ and $\sin kx$. Thus, we can write the following Fourier series in one dimension, where the φ's are the coefficients.

$$\psi(x) = \frac{(2\pi)^{1/2}}{L} \sum_{n=-\infty}^{\infty} \exp\left(\frac{2\pi inx}{L}\right) \varphi\left(\frac{2\pi n}{L}\right) \tag{24}$$

If L is made very large, and if φ is a continuous function of $k = 2\pi n/L$, then the change in each term of the series resulting from a unit change of n will become very small. Hence the sum may be replaced by an integral. Since $\Delta n = 1$ we can simply write Δn in the above summation, then replace it by dn in the integral. Furthermore, we can write $dn = \dfrac{L}{2\pi}\, dk$, and we get

$$\psi(x) \to \frac{1}{\sqrt{2\pi}} \int_{-\infty}^{\infty} e^{ikx} \varphi(k)\, dk \tag{25}$$

The above is called a Fourier integral. In a manner analogous to that used in Chap. 1, eq. (17), we can show that if $\psi(x)$ is known, $\varphi(k)$ can be calculated. The result is

$$\varphi(k) = \frac{1}{\sqrt{2\pi}} \int_{-\infty}^{\infty} e^{-ikx} \psi(x)\, dx \tag{26}$$

By using the above equation to express $\varphi(k)$ in (25), we obtain

$$\psi(x) = \frac{1}{2\pi} \int_{-\infty}^{\infty} \int_{-\infty}^{\infty} \exp\,[ik(x - x')]\psi(x')\, dx'\, dk \tag{27}$$

This identity is called the Fourier integral theorem.

The important point is that with a suitable $\varphi(k)$ we can represent an arbitrary* $\psi(x)$ in an unbounded region, with the aid of a Fourier integral.

18. Wave Propagation for Free Particle. To find the way in which any wave function propagates, we now imagine that it is Fourier analyzed, so that at $t = 0$, we have

$$\psi(x) = \frac{1}{\sqrt{2\pi}} \int_{-\infty}^{\infty} [\varphi(k)]_{t=0} e^{ikx}\, dk$$

* The function is not totally arbitrary, but should be piecewise continuous, and the square of its absolute value should have a finite integral over all space.

Now we have already seen that in free space a wave with propagation vector k must oscillate with angular frequency, $\omega = \hbar k^2/2m$. Hence the value of ψ for all times is given by multiplying each $\varphi(k)$ by

$$\exp - i\hbar k^2 t/2m$$

$$\psi(x, t) = \frac{1}{\sqrt{2\pi}} \int_{-\infty}^{\infty} [\varphi(k)]_{t=0} \exp\left[i\left(kx - \frac{\hbar k^2}{2m} t\right)\right] dk \tag{28}$$

This tells us what happens to an arbitrary wave function as time passes, for the case of a free particle.

19. Wave Equation for Free Particle. We are now ready to obtain the partial differential equation satisfied by ψ. To do this, first differentiate eq. (28) with respect to time. We get

$$i\hbar \frac{\partial \psi(x, t)}{\partial t} = \frac{1}{\sqrt{2\pi}} \int_{-\infty}^{\infty} \varphi(k) \frac{\hbar^2 k^2}{2m} \exp\left[i\left(kx - \frac{\hbar k^2}{2m} t\right)\right] dk$$

Let us now evaluate

$$-\frac{\hbar^2}{2m} \frac{\partial^2 \psi}{\partial x^2}$$

$$-\frac{\hbar^2}{2m} \frac{\partial^2 \psi}{\partial x^2} = \frac{1}{\sqrt{2\pi}} \int_{-\infty}^{\infty} \varphi(k) \frac{\hbar^2 k^2}{2m} \exp\left[i\left(kx - \frac{\hbar k^2}{2m} t\right)\right] dk$$

Combining the above equations we obtain the following partial differential equation

$$i\hbar \frac{\partial \psi}{\partial t} = -\frac{\hbar^2}{2m} \frac{\partial^2 \psi}{\partial x^2} \tag{29}$$

The above equation follows from the de Broglie relations and from the classical relation, $E = p^2/2m$, which holds only for a free particle. To deal with the general problem of a particle acted on by forces, we shall have to use instead the classical relation $E = \frac{p^2}{2m} + V(x)$, where $V(x)$ is the potential energy. This will be done in Part II. The equation so derived was first obtained by Schrödinger, and is called Schrödinger's equation, of which eq. (29) is a special case.

Practically the entire quantum theory is contained in the wave equation, *once we know how to interpret the wave function* ψ. For example, one of its consequences is the conservation of energy and momentum of a free particle. To see this, we note that the function, $\psi = \exp i(kx - \omega t)$ is a solution, provided that $\hbar\omega = \hbar^2 k^2/2m$. But since $E = \hbar\omega$ and $p = \hbar k$, we get $E = p^2/2m$, the well-known classical relation. Because neither ω nor k changes with time, it follows that E and p also remain constant. We shall see later that if there are many particles which exert forces on each other, the wave equation applying for that case will still have as its consequence the conservation of the total energy and momentum of the system. Hence this example shows that those classical causal laws that are taken over directly into quantum mechanics are contained in the wave equation.

The wave equation gives a continuous and causal prediction of what happens to the wave function. But the wave function gives only the probability of where the electron can be found. In the classical limit, the observation is so rough that the difference between probable behavior and actual behavior is never detected. Hence the wave equation also determines the classical limit of the electronic motion. This can be seen more directly as follows. The group velocity of a wave packet is $v_g = \partial\omega/\partial k$, which depends on the relation between ω and k. But, as we have seen, the latter can be obtained by looking for solutions of the wave equation of the form $\exp i(kx - \omega t)$.

More generally, the wave equation determines the probable results of any process which the electron can undergo, so that it plays the same fundamental role in quantum theory that the equations of motion do in classical theory. It is not surprising, therefore, that the first step in solving any specific quantum theoretical problem, such as, for example, the hydrogen atom or the harmonic oscillator, is to find the correct expression of the wave equation for the system in question. In Part II, we shall show how this is done for the general case.

CHAPTER 4

The Definition of Probabilities

1. Introduction. Thus far we have been using the term probability in a rather loose way. We are now faced with the problem of obtaining a more precise definition of the following two probabilities, a knowledge of which is essential if we are to be able to apply the theory to an arbitrary experimental situation:

(1) $P(x)\,dx$, the probability that a particle can be found between x and $x + dx$.

(2) $P(k)\,dk$, the probability that the particle momentum lies between $\hbar k$ and $\hbar(k + dk)$.

We shall see that it is possible to obtain suitable definitions of both of these probabilities and that, in the nonrelativistic domain at least, the definitions so obtained must be regarded as the only ones consistent with all of the reasonable requirements that can be set up. Additional complications arising in the relativistic range of velocities ($v/c \sim 1$) will also be discussed. Finally, it will be shown that while $P(k)$ can be defined for light quanta, $P(x)$ cannot. This means that light quanta do not have all of the spatial properties of particles, since it is only to a limited extent that they can be attributed to a definite point in space.

2. Choice of Probability Function $P(x)$. Let us begin with the definition of $P(x)$. An acceptable definition of this quantity must satisfy at least the following requirements:

(1) The probability function $P(x)$ is never negative.

(2) The probability is large where $|\psi|$ is large and small where $|\psi|$ is small. (This is needed to give the de Broglie packets the proper interpretation so that they can lead to the description of the actual particle motion in the classical limit. If, for example, $P(x)$ were large when $|\psi|$ was small, we would not be justified in saying that the particle is likely to be somewhere near the maximum of the packet.)

(3) The significance of $P(x)$ must not depend in a critical way on any quantity which is known on general physical grounds to be irrelevant. For example, in a nonrelativistic theory $P(x)$ must not depend on where the zero of energy is taken, since we know that no significant results depend on this choice. In a relativistic theory, the integrated probability must be left unchanged if it is measured in a co-ordinate system moving with some other velocity.

(4) Since the electron is neither emitted nor absorbed anywhere in the system, the integrated probability of finding the electron somewhere in the system must be unity and remain unity for all time. (This requirement is certainly valid in a nonrelativistic theory; in a relativistic theory, however, we shall see later that it can be relaxed somewhat because of the possibility of creating electron-positron pairs when very high energy quanta are present.)

We shall tentatively choose the following function for the probability:

$$P(x) = \psi^*\psi \tag{1}$$

This function obviously satisfies requirements (1) and (2). That it satisfies (3) can be seen by noting that the addition of an arbitrary constant E_0 to the energy changes the frequency of oscillation of the wave function by $\Delta\omega = E_0/\hbar$. Thus, we obtain for the new wave function

$$\psi' = \psi \exp\left(\frac{-iE_0t}{\hbar}\right) \qquad \text{and} \qquad (\psi')^* = \psi^* \exp\left(\frac{iE_0t}{\hbar}\right)$$

so that

$$P'(x) = (\psi')^*\psi' = P(x)$$

Thus, $P(x)$ is not changed by this shift in the zero of energy.

We shall show below that $P(x)$ also satisfies (4); hence, we conclude that this definition of probability is certainly acceptable. Whether it is the most general definition satisfying these requirements will be discussed later.

3. Proof of Conservation of Probability. We wish to show that one can define $P(x)$ in such a way that $\int_{-\infty}^{\infty} P(x)\,dx = 1$. In order that this be possible, the following condition must be satisfied:

$$\frac{\partial}{\partial t}\int_{-\infty}^{\infty} P(x)\,dx = \frac{\partial}{\partial t}\int_{-\infty}^{\infty} \psi^*(x)\psi(x)\,dx$$

$$= \int_{-\infty}^{\infty}\left[\frac{\partial\psi^*}{\partial t}\psi + \psi^*\frac{\partial\psi}{\partial t}\right]dx = 0 \tag{2}$$

Now $\partial\psi/\partial t$ is determined in terms of ψ by the wave equation (29), Chap. 3. $\partial\psi^*/\partial t$ is given by the complex conjugate equation

$$-i\hbar\frac{\partial\psi^*}{\partial t} = \frac{-\hbar^2}{2m}\frac{\partial^2\psi^*}{\partial x^2}$$

We therefore obtain

$$\frac{\partial}{\partial t}\int_{-\infty}^{\infty} P(x)\,dx = \frac{\hbar}{2mi}\int_{-\infty}^{\infty}\left(\psi\frac{\partial^2\psi^*}{\partial x^2} - \psi^*\frac{\partial^2\psi}{\partial x^2}\right)dx$$

Now

$$\psi\frac{\partial^2\psi^*}{\partial x^2} - \psi^*\frac{\partial^2\psi}{\partial x^2} = \frac{\partial}{\partial x}\left(\psi\frac{\partial\psi^*}{\partial x} - \psi^*\frac{\partial\psi}{\partial x}\right)$$

The above equations lead to

$$\frac{d}{dt}\int_{-\infty}^{\infty} P(x)\,dx = -\frac{\hbar}{2mi}\int_{-\infty}^{\infty}\frac{\partial}{\partial x}\left(\psi^*\frac{\partial\psi}{\partial x} - \psi\frac{\partial\psi^*}{\partial x}\right)dx$$

$$= -\frac{\hbar}{2mi}\left(\psi^*\frac{\partial\psi}{\partial x} - \psi\frac{\partial\psi^*}{\partial x}\right)_{-\infty}^{\infty}$$

If we choose a bounded wave packet, ψ^* and $\psi \to 0$ as $x \to \pm\infty$, and we therefore obtain the conservation of probability. (It is always known that the electron under discussion is somewhere in a bounded region, which may in practically all cases be taken, for example, as the size of the solar system.)

4. Probability Current. It is possible to obtain still more information from the above equation, for we have

$$\frac{\partial P(x)}{\partial t} = \frac{\partial}{\partial t}(\psi^*\psi) = -\frac{\hbar}{2mi}\frac{\partial}{\partial x}\left(\psi^*\frac{\partial\psi}{\partial x} - \psi\frac{\partial\psi^*}{\partial x}\right)$$

If we define

$$S = \frac{\hbar}{2mi}\left(\psi^*\frac{\partial\psi}{\partial x} - \psi\frac{\partial\psi^*}{\partial x}\right) \tag{3}$$

we obtain

$$\frac{\partial}{\partial t}P(x) + \frac{\partial}{\partial x}S(x) = 0 \tag{4a}$$

This is a special case of the three-dimensional equation

$$\frac{\partial P}{\partial t} + \operatorname{div} \mathbf{S} = 0$$

and is analogous to the equation of continuity of flow in hydrodynamics, $\frac{\partial\rho}{\partial t} + \operatorname{div}\mathbf{j} = 0$, where ρ is fluid density, and $\mathbf{j}$ its current density. The meaning of this equation is that the changes in the amount of material in an element of volume can be regarded as due to the unbalanced flow of some current $\mathbf{j}$ across the boundaries. Similarly, we can regard changes in the probability as results of the flow of probability current $\mathbf{S}$.

It is easily shown that our definitions can be generalized to three dimensions. The wave equation is then

$$i\hbar\frac{\partial\psi}{\partial t} = -\frac{\hbar^2}{2m}\nabla^2\psi \tag{4b}$$

and the probability current vector is

$$\mathbf{S} = \frac{\hbar}{2mi}(\psi^*\nabla\psi - \psi\nabla\psi^*) \tag{5}$$

This idea that probability flows through space more or less like a fluid is very useful physically. For example, in the section on how de Broglie waves move from one energy level to another, we anticipated this idea by saying that the wave flowed from one ring to the next.

Problem 1: Take $\psi = \exp\left[i\left(\boldsymbol{k}\cdot\boldsymbol{x} - \frac{\hbar k^2 t}{2m}\right)\right]$ and show that $S = \frac{\hbar k P(x)}{m} = VP(x)$.
For this wave function, the current is, therefore, just the velocity times probability density, in analogy with the fluid equation $\boldsymbol{j} = \rho \boldsymbol{V}$, where ρ is the fluid density.

5. Is the Above Formulation the Most General One? The formulation of wave mechanics adopted here is based on three points, first, the de Broglie relations demanded by the Davisson-Germer experiments; secondly, the correspondence principle that is satisfied by having the wave packets move with the classical particle velocity p/m; and finally, the requirement that we can define a sensible probability function, which is conserved identically for arbitrary wave functions as a result of the wave equation. Let us now investigate whether it is possible to find other formulations that lead to the same results. We shall see that in the nonrelativistic domain, at least, all satisfactory formulations must be essentially equivalent to the one given here, and we shall indicate some of the difficulties which arise in the relativistic domain. The three questions that we shall investigate are

(1) Must the wave function be complex?
(2) What is the most general possible wave equation?
(3) What is the most general possible definition of probability?

We shall see that these three questions are closely related.

If we remember that light waves can be described by the vector potential, which is real, it seems at first sight strange that we must use complex wave functions for an electron. Of course, complex functions are often used as an auxiliary means of dealing with real quantities. For example, we can write for the vector potential in a plane wave

$$\boldsymbol{a} = \text{real part of } (e^{i(kx-\omega t)})$$

In order that this provide a correct description of the system, it is necessary that the equations, which are solved with the aid of complex functions, shall never couple real and imaginary parts; i.e., the two parts must be independent of each other. The fact that this is true for the vector potential is readily demonstrated from the fact that $\boldsymbol{a}$ satisfies the equation

$$\frac{\partial^2 \boldsymbol{a}}{\partial t^2} = c^2 \nabla^2 \boldsymbol{a}$$

If we write

$$\boldsymbol{a} = U + iV$$

we get

$$\frac{\partial^2 U}{\partial t^2} = c^2 \nabla^2 U \qquad \text{and} \qquad \frac{\partial^2 V}{\partial t^2} = c^2 \nabla^2 V$$

Thus U and V remain independent of each other, and the use of a complex function is merely an auxiliary device here. This will happen, in general, whenever the imaginary number i does not appear explicitly in the wave equation.

Let us now write for the electron wave function $\psi = U + iV$. Insertion into Schrödinger's equation yields

$$\frac{\partial U}{\partial t} = -\frac{\hbar}{2m}\frac{\partial^2 V}{\partial x^2} \quad \text{and} \quad \frac{\partial V}{\partial t} = \frac{\hbar}{2m}\frac{\partial^2 U}{\partial x^2} \tag{6}$$

Here we see that U and V are coupled, so that neither of them alone is a solution of Schrödinger's equation. Hence, in this case, it is essential to carry both functions U and V. The use of a complex number is, however, merely a shorthand notation for representing two real functions; hence, we can say that either a pair of real functions or their equivalent in the form of a single complex function, are needed to solve Schrödinger's equation. This will happen, in general, whenever the wave equation explicitly contains the imaginary number i, as does Schrödinger's equation in the term $i\hbar(\partial\psi/\partial t)$.

The fact that both U and V contribute to physical results can be seen from the definition of the probability, $P = \psi^*\psi = U^2 + V^2$. It is instructive to derive the conservation of probability with the use of the real functions U and V. Thus, we can write

$$\frac{\partial P(x)}{\partial t} = 2\left(U\frac{\partial U}{\partial t} + V\frac{\partial V}{\partial t}\right)$$

From eq. (6) we eliminate $\frac{\partial U}{\partial t}$ and $\frac{\partial V}{\partial t}$, obtaining

$$\frac{\partial P}{\partial t} = -\frac{\hbar}{m}\left(U\frac{\partial^2 V}{\partial x^2} - V\frac{\partial^2 U}{\partial x^2}\right) = -\frac{\hbar}{m}\frac{\partial}{\partial x}\left(U\frac{\partial V}{\partial x} - V\frac{\partial U}{\partial x}\right)$$

With $S = \frac{\hbar}{m}\left(U\frac{\partial V}{\partial x} - V\frac{\partial U}{\partial x}\right)$, we obtain $\frac{\partial P(x)}{\partial t} + \frac{\partial S(x)}{\partial x} = 0$. This demonstrates the conservation of probability. We see that the relation between U and V is essential to the proof. In fact, we find that the current S is a mutual property of U and V, which vanishes when either is identically zero.

By elimination, we can obtain from eq. (6) the equations satisfied separately by U and V. These are

$$\frac{\partial^2 U}{\partial t^2} = -\frac{\hbar^2}{4m^2}\frac{\partial^4 U}{\partial x^4}; \qquad \frac{\partial^2 V}{\partial t^2} = -\frac{\hbar^2}{4m^2}\frac{\partial^4 V}{\partial x^4}$$

Thus, we obtain the result that U and V separately satisfy second-order equations in the time. Note, however, that U and V are still coupled by the first-order eq. (6).

Problem 2: Consider the complex function $\exp[i(kx - \omega t)]$. Show that it satisfies a first-order differential equation, but that the lowest order linear equations satisfied by real and imaginary parts are of second order.

Since U and V separately satisfy second-order wave equations, the idea immediately suggests itself that we might be able to avoid complex functions by replacing Schrödinger's equation by the equivalent second-order equation

$$\frac{\partial^2\psi}{\partial t^2} = -\frac{\hbar^2}{4m^2}\frac{\partial^4\psi}{\partial x^4} \tag{7}$$

This leads to the frequency condition for plane waves, $\omega^2 = \hbar^2k^4/4m^2$, which is essentially the same as that given by the de Broglie relations. We shall first show that if this is the correct wave equation, then we are not even permitted to choose a complex function for ψ. This is because real and imaginary parts are now uncoupled, so that U, $\partial U/\partial t$, V, and $\partial V/\partial t$ may all be given arbitrarily initial values. In the calculation of the motion of the wave packet, however, all known classical motions are already described by the proper initial choice of $\psi = U + iV$ alone. Hence, the additional freedom in the choice of $\partial\psi/\partial t$ leads to new possible motions of the wave packets, which are not in accord with the correct classical limit.

To see this in greater detail, let us note that our second-order equation implies that $\omega = \pm\hbar k^2/2m$, while the first-order equation implies that only the $+$ sign is to be taken. The appearance of the $+$ or the $-$ sign indicates a greater freedom in the possible choice of wave functions, corresponding to the second-order character of the differential equation. Let us now construct a wave packet as was done before. Since we can take either sign, a wave of propagation vector k oscillates as $\exp ikx\left[a \exp\left(-\frac{i\hbar k^2t}{2m}\right) + b \exp\left(\frac{i\hbar k^2t}{2m}\right)\right]$, where a and b are arbitrary constants. The most general wave function is then

$$\psi(x, t) = \int_{-\infty}^{\infty} \phi(k) \exp(ikx)\left[a(k) \exp\left(-\frac{i\hbar k^2t}{2m}\right) + b(k) \exp\left(\frac{i\hbar k^2t}{2m}\right)\right] dk$$

But we already know that the correct classical limit is obtained when we choose $b = 0$ and $a = 1$. Nonzero values of b are easily shown to lead to the wrong result for motion of the wave packets. (For example, it would be possible to obtain packets moving in two directions at once, after the direction of the momentum had been determined.)

Let us now consider whether it is possible to set up an acceptable theory involving the second-order equation (7), but using only a single real wave function U. Since both U and $\partial U/\partial t$ can be given arbitrary initial values at any point x, this equation involves just as many arbitrary conditions as does Schrödinger's first-order equation with a complex ψ, so that the objections of the previous paragraph do not apply here. We shall see that the difficulty with this method is that one cannot obtain from it a suitable probability function.

We first show that no conserved probability function can be set up which depends on U only and not on $\partial U/\partial t$. To do this, suppose that we assume that we can write for the probability

$$P = P(U)$$

Then
$$\frac{\partial}{\partial t}\int_{-\infty}^{\infty} P(U)\,dx = \int_{-\infty}^{\infty} \frac{\partial P}{\partial t}\,dx = \int_{-\infty}^{\infty} \frac{\partial P}{\partial U}\frac{\partial U}{\partial t}\,dx \tag{8}$$

If the above expression is to vanish for arbitrary U, it is necessary that $\partial U/\partial t$ be defined in terms of U. But this implies a first-order wave equation. With a second-order differential equation, however, $\partial U/\partial t$ can be given an arbitrary initial value; hence, the above expression cannot vanish* for all U.

We now give an example in which we obtain a conserved function which depends, however, on $\partial U/\partial t$ as well as on the space derivatives of U. This function is

$$P = \frac{1}{2}\left(\frac{\partial U}{\partial t}\right)^2 + \frac{\hbar^2}{8m^2}\left(\frac{\partial^2 U}{\partial x^2}\right)^2 \tag{9}$$

Problem 3: Show that

$$\frac{\partial P}{\partial t} + \frac{\partial S}{\partial x} = 0$$

where
$$S = \frac{\hbar^2}{4m^2}\left(\frac{\partial U}{\partial t}\frac{\partial^3 U}{\partial x^3} - \frac{\partial^2 U}{\partial x^2}\frac{\partial^2 U}{\partial x \partial t}\right)$$

and hence prove that
$$\frac{\partial}{\partial t}\int_{-\infty}^{\infty} P\,dx = 0$$

We see from the problem that P is conserved. We see also from its definition that it can take on only positive values. Hence, it seems, at first sight, to be a perfectly acceptable function. The difficulty with this function is that it makes the probability depend on $\omega = E/\hbar$ and, therefore, on where we choose the zero point of energy. To see this, let us evaluate P for the special case of a plane wave

$$U = \cos(kx - \omega t)$$

We get
$$P = \frac{\omega^2}{2}\sin^2(kx - \omega t) + \frac{\hbar^2 k^4}{8m^2}\cos^2(kx - \omega t)$$

With $\omega = \dfrac{\hbar k^2}{2m}$, this reduces to

$$P = \frac{\omega^2}{2} = \frac{E^2}{2\hbar^2}$$

In a nonrelativistic theory, it should be possible to choose the zero of energy arbitrarily, and still obtain an equivalent theory. We saw, for example, that this was possible with the definition

$$P = \psi^*\psi$$

* If $\partial U/\partial t$ can be given an arbitrary value it can, for example, be chosen equal to $\partial P/\partial u$ itself. Insertion of this value into eq. (8) shows that $\partial P/\partial t$ is then the integral of a positive quantity, which cannot vanish unless $\partial P/\partial u$ is identically zero

With the definition of P given in eq. (9), however, we could for example make $P = 0$ by choosing the zero of energy suitably. Hence, this definition of probability will not do.

Thus far we have merely given one example of how the choice of a second-order wave equation and a real wave function does not lead to an acceptable definition of probability. It can be shown that this conclusion is generally valid.

We shall now show that wave equations of order higher than the second are inadmissible. Consider, for example, the fourth-order equation

$$\frac{\partial^4\psi}{\partial t^4} = \frac{\hbar^4}{(2m)^4}\frac{\partial^8\psi}{\partial x^8}$$

For a plane wave, this reduces to

$$\omega^4 = \left(\frac{\hbar k^2}{2m}\right)^4$$

which has four roots

$$\omega = \pm\frac{\hbar k^2}{2m} \qquad \text{and} \qquad \omega = \pm i\frac{\hbar k^2}{2m}$$

The imaginary roots correspond to inadmissible solutions; i.e., they take the form $\exp\left(\pm\frac{\hbar k^2 t}{2m}\right)$. Such wave functions become infinite as $|t| \to \infty$. In a similar way it can be shown that no other order of wave equation can be used, which satisfies all the requirements of Sec. 2.

We are thus required, in the nonrelativistic theory, to use an equation that is of first order with regard to the time, with complex wave function. It can also be shown that $P = \psi^*\psi$ is the most general probability function, which under these conditions will lead to conservation for arbitrary ψ and which also satisfies all the other conditions given in Sec. 2.

Problem 4: Consider, for example, the possible definition

$$P = \psi^*\psi + \frac{\partial}{\partial x}(\psi^*\psi)$$

Show that although this quantity is conserved, it can become negative for some wave functions, so that it is inadmissible.

HINT: Try $\psi = \cos kx$.

Finally, it should be pointed out that the present wave equation is unique only insofar as its classical limit is concerned, and that small changes which do not affect the classical limit can always be made. Of course, this is because the correspondence principle was used in deriving the wave equation. Such changes do have to be made, for example, when we wish to describe the electron spin. One therefore regards the formulation obtained in this section as something that is generally on the right track, but which may later be subjected to corrections demanded by more accurate experiments.

6. Relativistic Theories. In the attempt to extend the quantum theory to the relativistic domain, serious difficulties have arisen. We shall indicate the general nature of some of these difficulties.

The first step is to choose a relation between ω and k which from the classical relation between energy and momentum will lead to the correct motion of the wave packets in the classical limit. The simplest choice leading to this result is

$$\hbar^2\omega^2 = m^2c^4 + \hbar^2k^2c^2 \tag{10}$$

This is equivalent to the classical relation

$$E^2 = m^2c^4 + p^2c^2$$

It is readily shown that the above relation leads to the wave equation (in three dimensions).

$$\frac{\partial^2\psi}{\partial t^2} = c^2\nabla^2\psi - \frac{m^2c^4}{\hbar^2}\psi \tag{11}$$

Problem 5: Prove the above equation, also that the relation (10) leads to the correct classical limit for motion of wave packets.

The next problem is to try to define a probability function which satisfies all the requirements of Sec. 2, including the requirement that the integrated probability is invariant to a Lorentz transformation. We begin by noting that this theory must approach the usual nonrelativistic theory in the limit where $v/c \ll 1$. Since the latter involves a complex wave function, the relativistic theory must also have one, despite its second-order character, which would otherwise permit us to restrict ourselves to a real wave function. The fact that there are two frequencies for each k ($\omega = \pm\sqrt{(m^2c^4/\hbar^2) + k^2c^2}$) implies then that there will be a new variable. (The significance of this new variable will be discussed later. We shall see that it is important only in the relativistic region of velocities ($v/c \cong 1$); for the nonrelativistic range, its effects are completely negligible.)

The next problem is to define a probability function. Since the equation is of second order, P must involve both ψ and $\partial\psi/\partial t$, as shown in Sec. 5.

Two examples of functions that are conserved by the equations of motion are

$$P(x) = \frac{i}{2}\left(\psi^*\frac{\partial\psi}{\partial t} - \psi\frac{\partial\psi^*}{\partial t}\right) \tag{12}$$

$$P(x) = \hbar^2\left|\frac{\partial\psi}{\partial t}\right|^2 + \hbar^2c^2|\nabla\psi|^2 + m^2c^4|\psi|^2 \tag{13}$$

Problem 6: Prove that the above two quantities are conserved.

The first of these examples is inadmissible as a probability because it is not always positive, as can be seen by choosing $\psi \sim e^{i\omega t}$, which yields $P \sim 1$, while $\psi \sim e^{-i\omega t}$ yields $P \sim -1$. Thus, there will always be the possibility of negative values of P. The second example is always positive, but it is inadmissible because it does not lead to a relativistically invariant definition of the integrated probability. To show this, let us choose the wave function $\psi = \exp i\left(\frac{Et - p \cdot x}{\hbar}\right)$. We get

$$P(x) = E^2 + p^2c^2 + m^2c^4 = 2E^2$$

The probability density thus transforms like the square of an energy, which is like the 4–4 component of a tensor. The integrated probability can, therefore, be shown to transform like an energy, so that it is not invariant.

Problem 7: Prove the above statements.

More generally, it can be shown that we cannot construct a positive definite probability function satisfying the second-order equation (11), which yields a Lorentz invariant integrated probability. Several ways out of this difficulty have been tried:

(1) Dirac has developed a first-order relativistic wave equation,* by introducing four complex wave functions. The extra wave functions correspond to additional variables, which can be related to the spin and charge of the electron. In this way, he is able to obtain conserved probabilities, as well as an accurate description of many relativistic properties of the electron, not treated correctly by any other theory.

(2) Pauli and Weisskopf decided to give up the assumption that the number of particles is conserved.† In this way, they avoid the need of defining a conserved probability function. In doing this, they were guided by the fact that when a photon in the relativistic range of energies ($\sim$1 mev or more) is absorbed, its energy can be converted into an electron-positron pair, which did not previously exist. For nonrelativistic energies, this process is impossible; hence probability is conserved. The theory of Pauli-Weisskopf reduces to that of Schrödinger in the nonrelativistic limit.‡

The problem of making a relativistic quantum theory is still faced by grave difficulties.§ There is strong evidence that the method of Dirac is probably at least a very good approximation for electrons, while that of Pauli and Weisskopf may perhaps apply to a new type of particle called the meson. It is not worthwhile at this point to go into greater detail,

* P. A. M. Dirac, 2nd ed., Chap. 12. (See list of references given on page 2.)

† W. Pauli and V. Weisskopf, *Helv. Phys. Acta.*, 7, 7, 709 (1934).

‡ The new degrees of freedom corresponding to the second-order equation are related in this theory to the possibility of occurrence of both positive and negative charges. In the nonrelativistic limit, however, we know that the charge never changes. Hence, we can there restrict ourselves to one sign of charge and, with this restriction, we obtain two separate first-order equations, one for each type of charge.

§ A. Pais, *Positron Theory*. Princeton, N.J.: Princeton University Press, 1949.

but the main conclusion to be drawn from this section is that the problem of formulating quantum theory depends considerably on the nature of the systems that we wish to describe. What must be done depends partly on getting the results to agree with experiment, and partly on setting up a theory that is logically self-consistent.

7. The Probability Function for Light Quanta. At this point, it is worthwhile to make a few remarks about the electromagnetic wave equation. Since the wave equation is of second order, we conclude that no suitable probability function can be defined. In free space, however, there does exist at least one positive definite conserved function, namely, the energy density,

$$W = \frac{\mathcal{E}^2 + \mathcal{H}^2}{8\pi}$$

It is a well-known result of classical electrodynamics that, in terms of the Poynting vector, $S = \frac{\boldsymbol{\mathcal{E}} \times \boldsymbol{\mathcal{H}}}{4\pi} c$, we obtain

$$\frac{\partial W}{\partial t} + \operatorname{div} \boldsymbol{S} = -\boldsymbol{\mathcal{E}} \cdot \boldsymbol{j}$$

where $\boldsymbol{j}$ is the current density. In the presence of matter, we know that electromagnetic energy is absorbed and emitted; hence, we expect no conservation of energy. But in free space, $\boldsymbol{j} = 0$, and from the above equa tion, we see that W is conserved.

The classical theory postulates a continuous distribution of energy throughout space. In the quantum theory, however, one must take into account the fact that energy is possessed by the electromagnetic field in the form of indivisible quanta with $E = h\nu$. From the correspondence principle, we know that the quantum laws of probability must be so chosen that, in the classical limit, one obtains the classical result for the energy density. To do this, we might try, in a manner similar to that used in the correspondence theory of radiation, to define the probability that a quantum can be found in a given volume element $d\tau$, from the relation

$$h\nu(\boldsymbol{x})P(\boldsymbol{x})\, d\tau = W(\boldsymbol{x})\, d\tau$$

or

$$P(x) = \frac{W(x)}{h\nu(x)} = \frac{(\mathcal{E}^2 + \mathcal{H}^2)}{8\pi hc} \lambda(x)$$

where $P(x)$ is the probability density for light quanta, which is analogous to the function $\psi^*(x)\psi(x)$ for electrons.†

Strictly speaking, however, such a definition is meaningless because

† It seems reasonable to require that the probability associated with a quantum at a given point and at a given time shall depend at most on the field quantities and on their space and time derivatives, evaluated at that point and at that time.

the wavelength at a given point cannot even be defined. We can give this concept a rough meaning in terms of a wave packet, covering a region of space much larger than a wavelength because, as we have seen in eq. (5), Chap. 3,

$$\Delta x\,\Delta k \cong 1 \qquad \text{or} \qquad \frac{\Delta x\,\Delta\lambda}{\lambda^2} \cong 1$$

so that

$$\frac{\Delta\lambda}{\lambda} \cong \frac{\lambda}{\Delta x}$$

Thus the range of wavelength needed to define a packet of size $\Delta x \gg \lambda$ becomes so small that the definition of probability density for a light quantum as given above has in it only a small element of ambiguity.

This result is very different from that obtained with an electron, where we can define $P(x)$ in a given region dx, independently of how the wave function behaves outside this region. For radiation, however, only $W(x)$ can be defined and, to know the probability that a given region contains a light quantum, we must also know the wavelength, and this cannot be obtained from the values of the fields inside the region dx. Thus, the electron has more of the attributes of a classical particle than does the light quantum although neither, of course, has all the attributes of a classical particle, since they both show interference effects. In the chapter on the uncertainty principle we shall verify that, in an actual process of measurement of the position of a light quantum, it is impossible to localize its position within a region Δx, which is smaller than the wavelength of the photon.

In the light of these results, how do we interpret an experiment in which a light quantum strikes an atom of diameter of the order of 10^{-8} cm, whereas the light wavelength is much larger, of the order of 10^{-5} cm? Can we not say that the quantum was found within a region much smaller than its wavelength? The answer is that the quantum can be localized in this way only at the moment that it disappears by absorption. The concept of a particle is, therefore, of no help in interpreting the result of any other experiment; with an electron, however, we can say that immediately after it has been found at a given spot another observation will disclose the same electron at the same point. In this way, the concept of a particle unifies many different experimental results, whereas the idea that a light quantum exists at the point where it is absorbed explains only this one result. As we shall see, whenever the light quantum is observed under conditions in which it is not absorbed, it cannot be localized to a region smaller than λ.

8. Probability of a Given Momentum. Thus far, we have obtained a consistent mathematical formulation of the fact that electrons show both wave and particle properties with the aid of the assumption that the intensity of the wave at a given point $|\psi(x)|^2$ yields only the probability that a particle can be found at that point. In general, however, we are interested in measuring other properties of the electron besides its position; the most important are its momentum and energy. In classical physics, for example, a knowledge of the initial position and momentum (and, therefore, the energy) of every particle in the universe, plus the forces between these particles is, in principle, both necessary and sufficient to determine completely the future motion of the system. Since, in quantum theory, the wave equation takes the place of the equations of motion, we must now investigate the extent to which the momentum is defined by the wave function.

In doing this, we base our work on the observed fact that a wave having a length λ is always associated with a momentum $p = h/\lambda$, whereas if the frequency of the wave is ν, the associated energy is $E = h\nu$. But as we have seen in Chap. 3, all real waves take the form of packets, which have in them a range of values of the frequency and the wavelength. Yet, for an actual experiment designed to measure the momentum, it is always possible, as will be evident in several examples, to arrange conditions in such a way that we obtain some definite value for the momentum, even though the wave packet contains a range of values. This is similar to what happens in a measurement of the position, in which, likewise, some definite value is always found, even after the wave function has spread out over a large region of space.

It should be pointed out here that, insofar as classical physics is concerned, a particle is something that has simultaneously a definite position and a definite momentum. In quantum theory, however, we shall see that an electron, for example, can show either a definite position or a definite momentum, but not both at the same time. In a sense, the particle nature of the electron is inferred from its ability to show definite values of either position or momentum. We shall see that a light quantum can also have a definite momentum. As shown in Sec. 7, however, it does not have a position that can be defined to within better than a wavelength. We must conclude, then, that the light quantum shows a much less close resemblance to a classical particle than does an electron, but that its definite momentum still makes it worth-while for us to regard it as one sometimes.

Just as was done in interpreting $P(\boldsymbol{x}) = \psi^*(\boldsymbol{x})\psi(\boldsymbol{x})$ as the probability density in position space, it seems reasonable to assume tentatively that $P(\boldsymbol{k}) = \phi^*(\boldsymbol{k})\phi(\boldsymbol{k})$ is proportional to the probability density in $\boldsymbol{k}$ space and, therefore, in momentum space. The correctness of this tentative identification will be proved further ahead. For the present, we note that our interpretation of the intensity of Fourier coefficients $\phi(\boldsymbol{k})$ means that in a wave packet the exact value of the momentum can neither be predicted nor controlled, but only the probability of a given value of $\boldsymbol{k}$ is determined by the wave function. In a series of experiments with the same initial experimental conditions, there will be a statistical distribution of momenta over a range determined by the spread of $\boldsymbol{k}$ in momentum space, just as there will be a statistical distribution of observed positions over a range determined by the spread of $\boldsymbol{x}$ in position space. This is an extension of the wave-particle duality to include momentum space as well as position space.

To illustrate the preceding statements and to prove that the probability density in momentum space is proportional to $P(\boldsymbol{k}) = |\phi(\boldsymbol{k})|^2$, let us consider an experiment in which an electromagnetic wave is diffracted by a grating. If the incident wave has a definite wave vector $\boldsymbol{k}$,

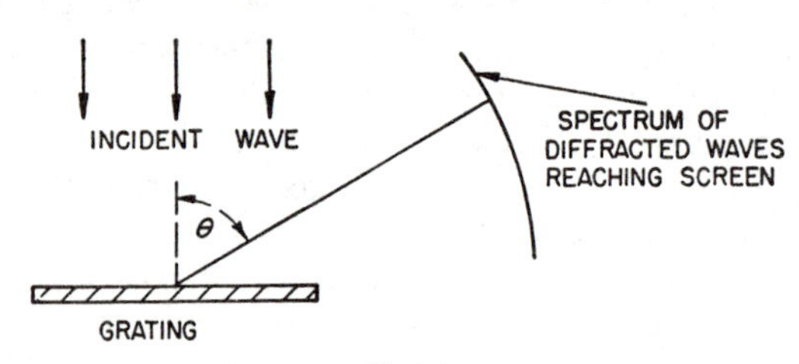

FIG. 1

then the diffracted wave will come off at a series of definite angles. For example, a wave which is incident normally on the grating comes off at the angles given by $\sin\theta = n\lambda/L$, where L is the space between rulings on the grating, and n is the order of the spectrum. If, however, the incident wave takes the form of a packet, then each Fourier component diffracts independently and comes off at the angle θ corresponding to its wave number as given in the preceding formula. Thus, the grating breaks up a packet into a spectrum as shown in Fig. 1. In a sense, the grating Fourier analyzes the packet in such a way that the amplitude of the wave $\mathcal{E}(\theta)$, appearing at a given angle θ, is proportional to the amplitude of the corresponding Fourier coefficient $\mathcal{E}_k$ in the incident wave. Similarly, the intensity $I(\theta)$ is proportional to $|\mathcal{E}(\theta)|^2$ and, therefore, to $|\mathcal{E}_{(k)}|^2$.

Let us now consider what happens when the incident packet contains only one quantum. Since the diffracted wave also has only one quantum in it, we conclude that the screen will be struck at only one point. The probability that a quantum strikes at a given angle θ is, as we have seen [eq. (43), Chap. 2] proportional to $I(\theta) = |\mathcal{E}(\theta)|^2$ and, therefore, to $|\mathcal{E}_k|^2$.

But now let us consider the fact that if a quantum strikes at an angle θ, its wave number is $k = 2\pi n/L \sin\theta$, so that its momentum must be $p = \hbar k = nh/L \sin\theta$. Since the wavelength does not change on diffraction, we conclude that the total momentum also remains constant (although its direction changes). As a result, a measurement of the angle θ also yields a measurement of the momentum that the particle had before it struck the grating. This shows that although there is a distribution of Fourier coefficients, $\mathcal{E}_k$, we still obtain in any one experiment only one value of the momentum at a time. Furthermore, the probability of a given momentum is proportional to $|\mathcal{E}_k|^2$.

Although there is a distribution of momenta, and therefore of energies, the *relation* between the energy and momentum for a light quantum, $E = pc$, remains exact and definite. This important fact follows from the de Broglie relations, plus the relation $\omega = kc$, which holds for an electromagnetic wave in free space.

A similar result can be obtained for electrons. This time we consider a Davisson-Germer experiment in which a beam of electrons is directed at a crystal. Because of the wave properties of electrons, they diffract in a manner similar to that of light waves. An electron of definite momentum, incident in a perpendicular direction, will arrive at a definite angle, given by $\sin\theta = n\lambda/L = nh/pL$. If the electron wave function is represented by a packet, there will be a spectrum of diffracted waves, and each Fourier component will diffract independently through its appropriate angle. Any one electron must, however, arrive at a definite angle at the detector, and the probability that it arrives at that angle is given by $|\psi(\theta)|^2$, where $\psi(\theta)$ is the wave function at that angle. But, as with the light quantum, we can show that $|\psi(\theta)|^2 \sim |\phi(k)|^2$ and that the absolute value of the momentum does not change on diffraction. Thus, a measurement of the angle of diffraction yields the momentum that the particle had before diffraction, so that a diffraction experiment can be used, if we choose, to measure momentum.* We conclude that the probability that the electronic momentum lies between $\hbar k$ and $\hbar(k + dk)$ must, therefore, be proportional to $|\phi(k)|^2$.

Although there is a distribution of momenta and energies, we point out that, as with the photons, the relation between these two quantities, in this case $E = p^2/2m$, is exactly true. This follows from the de Broglie relations, plus the frequency conditions, $\omega = \hbar k^2/2m$, which we can derive from Schrödinger's equation.

9. The Relation between $P(x)$ and $P(k)$. Let us now write the following expression for the probability that the momentum lies between $\hbar k$ and $\hbar(k + dk)$

$$P(k)\,dk = A|\phi(k)|^2\,dk \tag{14}$$

where A is a normalizing coefficient, defined so as to make

$$\int_{-\infty}^{\infty} P(k)\,dk = 1$$

Note the analogy to

$$P(x)\,dx = |\psi(x)|^2\,dx$$

Now, $P(k)$ and $P(x)$ are not independent of each other, but are related by

* Although this method of measuring momentum is somewhat unorthodox, it is as valid as any of the more familiar methods; for instance, the measurement of the potential drop needed to bring the particle to rest.

the fact that they are both determined from the same wave function. To demonstrate this relation, let us expand $P(k)$ in terms of $\psi(x)$ by means of a Fourier integral [eq. (26), Chap. 3]. We obtain

$$P(k) = A\phi^*(k)\phi(k) = \frac{A}{2\pi}\int_{-\infty}^{\infty}\int_{-\infty}^{\infty} \exp\,[i(x-x')k]\psi^*(x')\psi(x)\,dx\,dx' \quad (15)$$

Thus, $P(x)$ and $P(k)$ are *both* determined, once we know $\psi(x)$ at every point in space. Therefore it is not, in general, possible to give the two of them arbitrary sets of values independently of each other. We shall see that this has very important consequences, in connection with the uncertainty principle, treated in Chap. 5.

The preceding result shows that the wave function $\psi(x)$ determines *at least two* related probabilities. We shall see later that many more probabilities are determined by $\psi(x)$; in fact, the probabilities of all possible physical measurements. The wave function has often been called a "wave of probability," but a more accurate term is "a wave from which many related probabilities can be calculated." The peculiar complexity of the interrelation of probabilities can be seen by noting that, if we write the wave function, $\psi = R(x)e^{i\alpha(x)}$, where $R(x)$ and ϕ are real, then $P(x)$ is independent of $\alpha(x)$. Thus we might be tempted to say that only the absolute value of ψ is of physical significance. Although this is true if we are interested only in the position of the electron, we see from eq. (15) that the phase $\alpha(x)$ is important in determining the momentum distribution. We get

$$P(k) = \frac{A}{2\pi}\int_{-\infty}^{\infty}\int_{-\infty}^{\infty} \exp\,\{i[x - x' + \alpha(x) - \alpha(x')]\}R(x)R(x')\,dx\,dx'$$

Thus, every part of the wave function has significance for determining the probable results of some experiment.

10. Normalization Coefficient for $P(k)$. To obtain the normalizing coefficient A, we integrate eq. (14) over k, obtaining

$$\int_{-\infty}^{\infty} P(k)\,dk = \frac{A}{2\pi}\int_{-\infty}^{\infty}\int_{-\infty}^{\infty}\int_{-\infty}^{\infty} \exp\,[ik(x-x')]\psi^*(x')\psi(x)\,dx\,dx'\,dk$$

To evaluate this integral, we first note that it is defined as the limit, as $K \to \infty$, of the following integral

$$\frac{A}{2\pi}\int_{-\infty}^{\infty}\int_{-\infty}^{\infty} dx\,dx'\psi^*(x')\psi(x)\int_{-K}^{K} \exp\,[ik(x-x')]\,dk$$
$$= \frac{2A}{2\pi}\int_{-\infty}^{\infty}\int_{-\infty}^{\infty} \psi^*(x')\psi(x)\,\frac{\sin K(x-x')}{(x-x')}\,dx\,dx'$$

When K is large, the function $\sin K(x - x')/(x - x')$ has a large and narrow peak, of height equal to K and width $(x - x') \cong 1/K$. Outside this peak the function oscillates rapidly as a function of $(x - x')$ and soon becomes negligible. Thus, the main contribution to the integral comes from a very narrow region near $x - x' = 0$. If $\psi^*(x')$ is a continuous function, then it varies so little over this region that we can take it out of the integral over x' and evaluate it at $x' = x$. The result is

$$\frac{2A}{2\pi} \int_{-\infty}^{\infty} \psi^*(x)\psi(x)\, dx \int_{-\infty}^{\infty} \sin \frac{K(x - x')}{(x - x')}\, dx'$$

It is readily verified that the remaining integral over x' is equal to π. Thus, we get

$$\int_{-\infty}^{\infty} P(k)\, dk = \int_{-\infty}^{\infty} \psi^*(x)\psi(x)\, dx = \int_{-\infty}^{\infty} P(x)\, dx \qquad (16)$$

Hence, if $P(x)$ is normalized to unity, then $P(k)$ is automatically normalized by setting $A = 1$. Since $P(x)$ remains normalized for all time, we conclude that $P(k)$ remains so too. This feature of the theory is very satisfactory, demonstrating here, at least, self-consistency.

Summary on Probabilities

Let us now summarize our ideas on probability, contrasting the way that they apply to electrons and to light quanta.

For Electrons and Other Particles

1. There is a complex and scalar wave amplitude ψ, also called the *wave function*. It may be expressed either as $\psi(x)$ or, in Fourier analysis as a function of k, that is, $\phi(k)$.

2. From this wave function we can, in general, predict only the probability that a particle can be found with given position or momentum. In the classical limit, however, where we are not interested in an accuracy better than the size of a wave packet, this probability becomes, for all practical purposes, a certainty, so that we obtain the deterministic classical particle motion as an approximation.

For Light

1. There is a real wave amplitude that consists of a vector having only those components normal to the direction of propagation. It may be expressed either as $\mathcal{A}(x)$, or, in Fourier analysis, as $\boldsymbol{a_k}$.

2. The wave intensity determines only the probability that a quantum of energy will be absorbed when radiant energy is incident on matter; but in the classical limit, where many quanta are present, this probability becomes very nearly a certainty, so that we obtain the deterministic classical rate of absorption of energy as an approximation.

For Electrons and Other Particles

3. The probability that an electron can be found with positions between x and $x + dx$ is

$$P(x) = \psi^*(x)\psi(x)\,dx$$

4. The probability that an electron can be found with a momentum between $\hbar k$ and $\hbar(k + dk)$ is

$$P(k) = \phi^*(k)\phi(k)\,dk$$

(The problem of dealing with many electrons is deferred until some of the later chapters.)

5. The integrated probabilities $P(x)$ and $P(k)$ are conserved as a result of the wave equation.

6. There is a probability current

$$\mathbf{S} = \frac{\hbar}{2mi}(\psi^*\Delta\psi - \psi\Delta\psi^*)$$

which satisfies the relation

$$\frac{\partial P}{\partial t} + \text{div}\,\mathbf{S} = 0$$

Hence, we may think of the probability as a sort of fluid that flows from one point to another continuously and without loss or gain.

For Light

3. There is, strictly speaking, no function that represents the probability of finding a light quantum at a given point. If we choose a region large, compared with a wavelength, we obtain approximately

$$P(x) \cong \frac{\mathcal{E}^2(x) + \mathcal{H}^2(x)}{8\pi h\nu(x)}$$

but if this region is defined too well, $\nu(x)$ has no meaning. The term $\frac{\mathcal{E}^2 + \mathcal{H}^2}{8\pi}$ represents, in any case, the mean energy density. The probability per unit time that an atom at the point x will absorb a quantum is proportional to $W(x)$.

4. If there is only one quantum, the probability that its momentum lies between $\hbar k$ and $\hbar(k + dk)$ is proportional to $\frac{1}{8\pi}(\mathcal{E}_k^2 + \mathcal{H}_k^2)$. If there are many quanta, then $\frac{|\mathcal{E}_k|^2}{4\pi}$ is proportional to the mean number in the range from k to $k + dk$

5. $\int_{-\infty}^{\infty} P(k)\,dk$ is conserved, but only in free space, since light quanta can be absorbed or emitted by moving charges.

6. There is no corresponding quantity for light. There is, however, a current of energy $\mathbf{S} = \frac{c}{4\pi}(\boldsymbol{\mathcal{E}} \times \boldsymbol{\mathcal{H}})$, such that

$$\frac{\partial W}{\partial t} + \text{div}\,\mathbf{S} = 0$$

when there are no currents. This means that the mean energy also acts like a fluid, which flows continuously without loss or gain from one point to another.

CHAPTER 5

The Uncertainty Principle

1. Introduction. On the basis of the formulation of quantum theory obtained in the previous work, we now proceed to derive a very important expression yielding a quantitative estimate of the limitations on the possibility of giving a deterministic description of the world. This expression, which was first given by Heisenberg, is usually called *the uncertainty principle.*

We shall first give a statement of the uncertainty principle: If a measurement of position is made with accuracy Δx, and if a measurement of momentum is made *simultaneously* with accuracy Δp, then the product of the two errors can never be smaller than a number of order $\hbar$.* In other words

$$\Delta p\,\Delta x \geqq (\sim\hbar) \tag{1}$$

Since, in classical theory, a knowledge of the initial momentum and position of every particle is needed before the future orbits can be determined from the equations of motion, it is clear why this principle implies a quantum-mechanical limitation on the extent to which the deterministic description of classical theory can be applied.

In a similar way, it may be shown that if the energy of a system is measured to accuracy ΔE, then the time to which this measurement refers must have a minimum uncertainty given by

$$\Delta E\,\Delta t \geqq (\sim\hbar) \tag{2}$$

More generally, if Δq is the error in the measurement of any co-ordinate, and Δp is the error in its canonically conjugate momentum, we have

$$\Delta p\,\Delta q \geqq (\sim\hbar) \tag{3}$$

2. Proof of Uncertainty Principle for Electrons. To prove the uncertainty principle for electrons, we begin with eq. (16), Chap. 3, which gives the relation between the range of positions, Δx, and the range of wave numbers, Δk, appearing in a wave packet:

$$\Delta x\,\Delta k \geqq 1 \tag{4}$$

* We adopt the symbol $(\sim\hbar)$ to mean "a number of the order of $\hbar$." This is because the magnitude of an uncertainty is inherently a somewhat vague quantity for which there is considerable latitude in possible choice of definition. This latitude does not extend, however, further than a factor of 10 for any reasonable definition of the uncertainty in a measurement.

The above is a general property of waves and is not restricted to quantum theory. The uncertainty principle is obtained, however, when the following quantum-mechanical interpretations of the quantities appearing in the above equation are taken into account:

(1) The de Broglie equation $p = \hbar k$ creates a relationship between wave numbers and momentum, which is not present in classical waves. A classical electromagnetic wave with a given wave number k, for example, can have arbitrary amplitude and, therefore, arbitrary momentum.*

(2) Whenever either the momentum or the position of an electron is measured, the result is always some definite number.† Because of the de Broglie relation, a definite momentum implies a definite wave number k. On the other hand, a classical wave packet always covers a range of positions and a range of wave numbers.

(3) The wave function $\psi(x)$ determines only the probability of a given position, whereas the Fourier component $\phi(k)$ determines only the probability of a given momentum. This means that it is impossible to predict or control the exact location of the electron within the region Δx in which $|\psi(x)|$ is appreciable; and that it is impossible to predict or control the exact momentum of the electron within the region Δk, in which $|\phi(k)|$ is appreciable. Thus Δx is a measure of the minimum uncertainty, or lack of complete determinism of, the position that can be ascribed to the electron. Δk is, similarly, a measure of the minimum uncertainty, or lack of complete determinism of the momentum that can be ascribed to it.

From eq. (4), with the aid of the relation $\Delta p = \hbar\,\Delta k$, we now obtain the uncertainty principle, $\Delta p\,\Delta x \geqq \hbar$.

In a similar way, the energy-time uncertainty relation can be obtained, by starting with $\Delta\omega\,\Delta t \geqq 1$, where Δt is the range of time needed for a wave packet to pass a given point and $\Delta\omega$ is the range of angular frequencies in this packet. From the de Broglie relation, $\hbar\,\Delta\omega = \Delta E$ we obtain $\Delta E\,\Delta t \geqq \hbar$, where ΔE is the range of indeterminacy in the energy, and Δt is the range of indeterminacy during the time at which the electron passes a given point.

3. On the Interpretation of the Uncertainty Principle. An important question now arises in connection with the uncertainty principle: Can we think of the electron as something that has, simultaneously, well-defined values of position and momentum, which are uncertain to us because we cannot measure them with complete precision; or are we to think of the lack of complete determinism as originating in the very structure of matter itself? We shall see in Chap. 6, Sec. 11, and Chap. 22, Sec. 19 that the indeterminism is inherent in the very structure of matter and that the momentum and position cannot even exist with simultane-

* See Chap. 2, Sec. 7.

† See Chap. 3, Sec. 13; also Chap. 4, Sec. 8.

ously and perfectly defined values. The term "uncertainty principle" is, therefore, somewhat of a misnomer. A better term would be "the principle of limited determinism in the structure of matter." Because of its greater brevity, and because the term is already in common use however, we shall in this work, continue to refer to it as the uncertainty principle.

The idea that a particle has simultaneously well-defined values of position and momentum, which are uncertain to us, is equivalent to the assumption of hidden variables (see Chap. 2, Sec. 5) that actually determine what these quantities are at all times, but in a way that, in practice, we cannot predict or control with complete precision. We shall see in Chap. 22, Sec. 19, that the quantum theory is inconsistent with the assumption of such hidden variables.*

4. Relation of Spreading of Wave Packet to Uncertainty Principle. In eq. (23), Chap. 3, we have seen that a wave packet of initial width Δx_0 eventually spreads out to a width that approaches $\Delta x \cong \hbar t/m\,\Delta x_0$ as the time increases without limit. Thus, the narrower the wave packet to begin with, the more rapidly it spreads. We are now ready to see a simple physical reason for this spread in terms of the uncertainty principle. Because of the confinement of the packet within the region Δx_0, the Fourier analysis contains many waves of length of the order of Δx_0, hence momenta $p \cong \hbar/\Delta x_0$ and, therefore, velocities

$$\Delta v \cong \frac{p}{m} \cong \frac{\hbar}{m\,\Delta x_0}$$

Although the average velocity of the packet is equal to the group velocity, there is still a strong chance that the actual velocity will fluctuate about this average by the above amount, namely, by $\Delta v \cong \hbar/m\,\Delta x_0$. Because of this fluctuation (in direction as well as in magnitude), the distance covered by the particle is not completely determined, but can vary by as much as

$$\Delta x \cong t\,\Delta v \cong \frac{\hbar t}{m\,\Delta x_0}$$

This, however, is roughly what is predicted from the spread of the wave packet during this time. The spread of the wave packet may, therefore, be regarded as one of the manifestations of the lack of complete determination of the initial velocity necessarily associated with a narrow wave packet.

5. Relation of Stability of Atoms to Uncertainty Principle. We can see from the uncertainty principle that if an electron is localized, it must have, on the average, a high momentum, and hence a high kinetic energy. Thus, it takes energy to localize a particle. If there is nothing to hinder

* See also Chap. 5, Sec. 18.

the motion of the electron, the indefiniteness of momentum tends to destroy any initial localization as time passes. If, however, the electron is localized forcibly, for example, by putting it into a box, then the above-mentioned momentum will create a pressure on the box, very similar to the pressure created by the molecules of a gas. Thus, we may describe a permanently localized electron roughly as being under pressure. If this pressure is removed, the electron will begin to wander away, just as do the molecules of a gas, when the confining walls are removed.

This analogy is only partially accurate because it neglects the interference effects, arising from the wave properties of the electron, yet, it is often a very helpful way to picture some of the quantum properties of the electron. For example, we can use this picture to see why the electron in a hydrogen atom does not keep on radiating energy until it falls into the nucleus, as predicted by classical theory. The reason is that, according to the uncertainty principle, it takes a momentum $p \cong \hbar/\Delta x$, hence an energy $E = p^2/2m \cong \hbar^2/2m(\Delta x)^2$ to keep an electron localized within a region Δx. This momentum creates a pressure, which tends to oppose localization of the electron. In an atom, the pressure is opposed by the force attracting the electron back into the nucleus. The electron will come to equilibrium where the attractive force balances the effective pressure and, in this way, the mean radius for the lowest quantum state is determined. We can find the point of balance from the condition that the total energy, kinetic plus potential, must be a minimum.* The potential energy is of the order of $-e^2/\Delta x$ in a hydrogen atom. Thus we have

$$W \cong \frac{\hbar^2}{2m(\Delta x)^2} - \frac{e^2}{\Delta x}$$

$$\frac{\partial W}{\partial\, \Delta x} \cong -\frac{\hbar^2}{m(\Delta x)^3} + \frac{e^2}{(\Delta x)^2} \cong 0$$

$$\Delta x \cong \frac{\hbar^2}{me^2}$$

This result is just the radius of the first Bohr orbit. The argument is not exact, but only qualitative. Yet it shows how we can make a picture that gives, roughly, the right results. This picture is very useful in guessing approximately what happens in a complex atom, where calculations are either very difficult, or practically impossible. According to this picture, it is possible to force an electron into a region smaller than the first Bohr orbit, but this requires energy and will not happen if the electron is left to itself with an energy corresponding to the lowest Bohr orbit. If by some external means, however, an electron is initially localized in the nucleus, then the resulting kinetic energy would be so high that in a short time the electron would leave the atom altogether.

* We should actually give a three-dimensional treatment, but it is readily shown that the results would be the same as those given here.

This limitation on the localizability of the electron is inherent in the wave-particle nature of matter. Thus, in order to have an electron in a very small space, we must have very high Fourier components in its wave function and, therefore, the possibility of very high momenta. There is no way to force an electron to occupy a well-defined position *and still remain at rest.*

Problem 1:

(a) How much kinetic energy is required (on the average) to localize a 1-gram mass with an accuracy of 10^{-3} cm?

(b) How much to localize the earth to within 1 meter?

(c) How much to localize a proton within one atomic radius (take 10^{-8} cm)?

(d) How much to localize an electron within the same distance?

Can you draw any conclusions from this problem?

Problem 2:

Compute the mean pressure necessary to hold an electron within a nuclear radius (5×10^{-13} cm).

6. Theory of Measurements. So far, the limits on the possibility of simultaneous determination of position and momentum implied by the uncertainty principle are merely logical derivations from our assumptions about matter waves and their probability interpretation. Before we can be sure of the validity of these limits, it is necessary to construct a quantum theory of the processes of measurement, and to show that this theory leads to the same result. In other words, we must see in detail what happens in actual measuring processes that prevents us from making measurements that would permit us to determine momentum and position simultaneously with unlimited accuracy.

In any measurement, it is necessary to have some system that we regard as the measuring apparatus and from whose state we can draw inferences about the systems we are observing. In order that this be possible, it is necessary that the measuring apparatus interact with what is observed in a known and calculable fashion. For example, to use a camera to obtain a picture of the space-relations of objects, we must know how light is reflected by objects, how it gets to the lens, what the lens does to it, and how the record on the photographic plate is connected with the intensity of the light reaching it. With all these facts known, we can draw inferences from the photograph about the objects that were photographed. Of course, if we photograph a minute dust particle, the radiation pressure may change the motion of the dust particle. We try to use weak light to minimize such effects but, in any case, if we know the intensity of the light, we can always correct for them, as the radiation pressure is calculable.

7. Modification of Measurements by Quantum Effects. The above all applies to the classical theory which supplies, as far as it is valid, a deterministic theory of the interaction between observer and object, and thus makes possible the drawing of unique inferences about the object. This interaction must actually be treated, however, by the quantum theory, in which the determinism is limited, as we have seen, by the fact that the transfer of a *single* quantum is unpredictable and uncontrollable. Hence, if we wish to make observations that are accurate enough to reach the quantum level, an element of incomplete determinism enters into the interaction between the apparatus and what is observed. This behavior is totally different from that predicted by classical theory, which says that the disturbance resulting from the measuring apparatus can be made arbitrarily small, and can be corrected for by means of the deterministic classical laws involved, even if it is not made negligibly small.

Our general procedure, then, must be to study various devices used in making measurements of position and momentum. The results obtained with such devices are customarily interpreted with the aid of classical mechanics, but now we shall study the further limitations imposed by the quantum nature of the systems with which we are dealing. In this work, we shall restrict ourselves to showing, in a few specific cases, how quantum effects intervene to prevent measurements of unlimited precision. We shall only sketch the lines along which such a treatment can be generalized.*

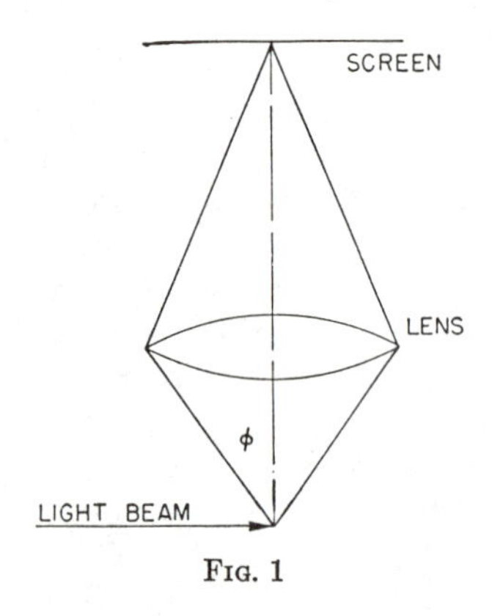

FIG. 1

8. Microscope. One way of measuring the position of an electron is with a microscope. To minimize the effect of radiation pressure on the electron, suppose the light is so weak that the electron scatters only one light quantum. (Of course, it must scatter at least one if we are to learn anything about it.)

After the quantum is scattered it passes through the lens and lands somewhere on a screen, which may, for example, be a photographic plate (Fig. 1). From the position of this spot, we try to deduce the position of the scattering electron. To do this, we must use the wave theory of light. Because of diffraction effects, we know that there is a region $\Delta x \cong \lambda/\sin\phi$, from which the light might have been scattered, if it is known to focus on a given spot (ϕ is the aperture of the microscope). The electron might have been anywhere in this region. To minimize this uncertainty, we may make λ small. But what does this do to the momentum of the electron? We know that the quantum has a momentum $p = -h/\lambda$. If the quantum is scattered through an angle ψ, it imparts to the electron

* For a fuller treatment of the theory of measurement see Chap. 22.

a momentum $\Delta p = p \sin \psi = \frac{h}{\lambda} \sin \psi$. There is no way to tell what the angle of scattering was; it might have been anything within the aperture ϕ of the lens. As a result, the momentum of the electron becomes uncertain by $\Delta p = \frac{h}{\lambda} \sin \phi = \frac{h}{\Delta x}$. This is in agreement with the uncertainty relation.

The reason for the uncertainty is not only that we had to use at least one quantum, which gives up some momentum, but that, to a certain extent, the size of this transfer is uncontrollable and unpredictable so that we cannot correct for it.

One way of trying to reduce the uncertainty in momentum transfer is to narrow the aperture of the lens so as to reduce the range of scattered angles that are accepted by the lens. This, however, decreases the resolving power of the lens because of increased diffraction effects, and there is a proportional decrease in the accuracy of the position measurement.

More generally we find that, no matter how we perform the experiment, a limit on the accuracy of the measurements corresponding to the uncertainty principle is always introduced at some point in the experiment. This limitation corresponds to the fact that the fundamental structure of matter is very different from that assumed by the classical theory.

9. Measurement of Momentum. Let us consider the method of measuring momentum by measuring the velocity with the aid of the Doppler shift of the light radiated by the particle. This is actually done in measuring the velocity of radiating atoms. (The velocity of a star, for example, is frequently obtained by measuring the red shift.) The connection between Doppler shift and velocity v is

$$\nu' \cong \nu\left(1 - \frac{v}{c}\right) \quad \text{or} \quad \frac{v}{c} = \frac{\nu - \nu'}{\nu} \tag{5}$$

where ν is the frequency radiated by the atom when at rest, ν' by the atom when in motion. Note that this is a nonrelativistic approximation.

Now the position at $t = 0$ can, in principle, be fixed arbitrarily with high accuracy. This could be done, for example, by having the particle come through a very fine slit at this time. Of course, we shall not then know the velocity of the particle, but that is what we are trying to measure. The uncertainty in the velocity depends on the accuracy with which we can measure ν', and it is well known that the uncertainty in ν' is $\Delta\nu' \cong 1/\tau$, where τ is the time duration of the wave train of radiated light. We therefore desire a long wave train. Now, the length of the wave train is determined by the time the atom takes to radiate its energy; in principle, this may be made arbitrarily long by choosing atoms that radiate slowly enough.

Because the radiation takes place in quanta, there is a minimum possible momentum transfer to the electron, $\Delta p = h\nu'/c$. To the extent that we can measure ν', we can calculate the transfer of momentum, so that no uncertainty would be introduced if we made a completely accurate measurement. This is an important point, because it shows that the existence of a minimum possible momentum transfer does not prevent us from making arbitrarily precise measurements of the momentum. Although we do change the momentum in this measurement, we know the magnitude of the change, and can, therefore, correct for it. An inherent lack of precision in a measurement can occur only when some crucial property of the system changes in an unpredictable and uncontrollable way during the course of the measurement. In this experiment, the unpredictable and uncontrollable quantity is the time of emission of the quantum. All we know is that the quantum was emitted at some time between 0 and τ. We are not allowed to measure this time more precisely, for its measurement would reduce the accuracy with which we can measure the frequency of the quantum.

But when the quantum is emitted, there will be an abrupt change of the electronic velocity given by $\Delta v = \Delta p/m = h\nu'/mc$. Hence, there will be an interval of time, anywhere between 0 and τ, during which it may have been traveling at a velocity different from the one that we measure. This will result in an uncertainty in the distance the particle covers, equal to

$$\Delta x \cong \tau\, \Delta v \cong \frac{h\nu'\tau}{mc}$$

Thus, we get

$$\Delta x\, \Delta p = m\, \Delta x\, \Delta v = \frac{h\nu'\tau\, \Delta v}{c}$$

But from eq. (5)

$$\Delta v \cong \frac{c\, \Delta\nu}{\nu} = \frac{c}{\nu\tau}$$

Now, for the nonrelativistic case which we are considering here, $\nu'/\nu \cong 1$. Thus, we obtain $\Delta x\, \Delta p \cong h$.

Here a rather interesting effect occurred as a result of the incompletely predictable and controllable time of emission of the quantum. Although the position before the measurement (at $t = 0$) was fairly well-defined, the deterministic relation between this position and the positions reached after the transfer of the quantum is destroyed while the measurement takes place. Thus, although the momentum is made more definite in the course of the measurement, the position is made less definite. A similar result would have been obtained in the experiment in which the position of an electron was measured by means of a microscope, if the momentum before the measurement had been well-defined. In this case, the deterministic relation between the momenta before and after the transfer of the quantum is likewise destroyed while the measurement takes place. Thus, the position is made more definite, but the momentum becomes less definite. (In this connection, see Chap. 8, Secs. 14 and 15.)

10. Energy Time Uncertainty. In Sec. 2 we have already demonstrated the energy-time uncertainty directly. We can show, however, that this relation also follows from $\Delta x\, \Delta p \geqq (\sim\hbar)$. To do this, let us suppose that we measure time by means of a particle moving at a known velocity that has been measured, for example, by the method discussed previously. To measure the time, we must simply know when the particle has covered a distance $x = vt$ relative to its original (accurately known) position. The uncertainty in the time is then given by $\Delta t \cong \Delta x/v$. But we have already seen that for Δx, the minimum uncertainty is $\Delta x \geqq (\sim\hbar/\Delta p)$. Thus, $\Delta t \cong \hbar/v\, \Delta p$. But $\Delta E \cong v\, \Delta p$; hence $\Delta E\, \Delta t \geqq (\sim\hbar)$.

11. Uncertainty Principle Applied to Light Quanta. Consider an electromagnetic wave packet made, for example, by opening a shutter for a length of time Δt. We obtain, in this way, a pulse of radiation that passes any given point in the time Δt. The electric field is large only during this time and is negligible at all other times.

Suppose that the pulse contains only one quantum. Now, if this pulse is allowed to strike a target containing many atoms, we know that some *one* of these atoms will absorb the quantum. The probability of absorption is proportional to $|\mathcal{E}|^2$, so that it is practically certain that the quantum will be absorbed during the time interval Δt when the electric field is large in the region containing the absorbing atoms. On the other hand, the exact time at which the transfer of a quantum from electromagnetic field to matter takes place can neither be predicted nor controlled; only the probability is known from the value of $|\mathcal{E}|^2$. The transfer may, therefore, take place at any time within the interval Δt. Thus, we can regard Δt as the uncertainty in the time of transfer of a quantum. We also know, however, that the pulse contains a range of angular frequencies $\Delta\omega \geqq 1/\Delta t$ and, therefore, a range of energies $\Delta E \geqq \hbar/\Delta t$. There is also no way to predict or control the value of the energy within the range ΔE. We therefore conclude that in any process in which a quantum is transferred from radiation field to matter (or vice versa), the product of the uncertainties in time of transfer and in the amount of energy transferred is $\Delta E\, \Delta t \geqq \hbar$.

In a similar way, we can show that the product of the uncertainty in the momentum transferred, Δp, and the uncertainty in the position, Δx, at which this transfer takes place satisfies the relation $\Delta p\, \Delta x \geqq \hbar$.

Note that in the preceding argument we have carefully avoided referring to light as made up of particles, or *photons*, as they are commonly called. In the process of emission or absorption of quanta, energy and momentum appear in one unit, just as if they had been supplied by a particle.* Yet, as we have seen on theoretical grounds† and as we shall see in more detail in the next section, it is impossible to ascribe a precise location to such a particle, except at the moment that it is annihilated. With an electron, on the other hand, we can always measure the position as accurately as we please without destroying the electron,‡ knowing, however, that the momentum becomes less definite in the process of measurement. For this reason we have thus far avoided talking about the location of the quantum, and have instead discussed only the uncertainty

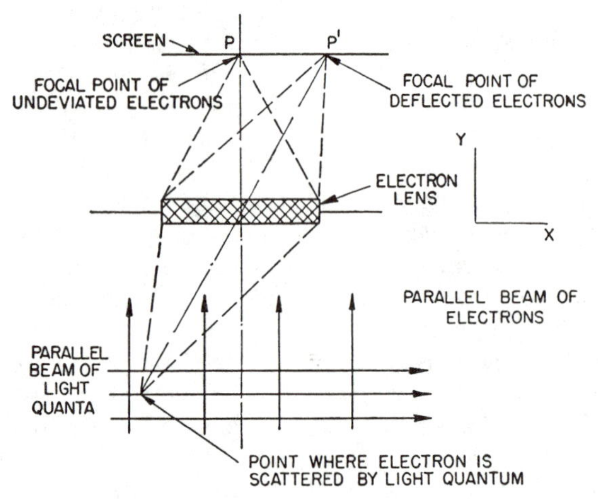

FIG. 2

in *time of transfer*, which is essentially what is really meant by the time of annihilation of the analogous particle. If we wish to refer to light quanta as "photons," one must use this word very cautiously, since it implies a more definite type of particle nature than is really possessed by light energy.

12. Observation of Light Quanta with Electron Microscope. To demonstrate directly the limitations on localizability of a light quantum (or of the equivalent photon, if we so choose to call it), let us try to observe its position with an electron microscope. We take advantage of the fact that light quanta and electrons scatter each other. A suggested arrangement is shown in Fig. 2.

We direct a beam of parallel-moving electrons normally at an electron lens, in such a way that the incident beam comes to a focus at the point P. A beam of quanta is allowed to cross the electron beam at right angles to the latter. Occasionally an electron is scattered and brought to a new

* See Chap. 2, Sec. 7.

† See Chap. 4, Sec. 7.

‡ This holds only in the nonrelativistic theory. See Chap. 4, Sec. 5.

focal point P'. From the position of this point, we try to draw some inferences about the x co-ordinate of the point at which the scattering took place and thus, if we imagine that the scattering is caused by an equivalent photon, we will have measured the position of this particle. (The x co-ordinate is taken in a direction normal to the electron beam and parallel to the beam of incident quanta.)

Let us first note that because of the wave nature of the electrons, the same general kind of limitation on accuracy of observation occurs as with the light microscope. Thus, the electron waves diffract around the lens edges in the same way that a light wave would. The main advantage of an electron microscope is that it is much easier to focus electrons of very short wavelength than to focus light of this kind. (Consider, for example, the focusing problems arising in dealing with an X ray microscope.) We therefore conclude that, as in the observation of the electron with photons, the best that we can do is to obtain $\Delta x \, \Delta p \geqq \hbar$.

With electrons we could, however, always make Δx as small as we pleased by using light of very short wavelength. We shall now see that, with light, there is a limitation on the minimum value of Δx, which is independent of the length of the electron waves. We shall work here only in the nonrelativistic limit and discuss the effects of relativity later. Now, in the nonrelativistic limit, the change of frequency of the quantum on scattering will be negligible [see Chap. 1, eq. (5)]. The maximum possible transfer of momentum from light quantum to electron will occur when the quantum is scattered through 180°, and this transfer will be $\Delta p_x \cong 2h\nu_q/c$, where ν_q is the frequency of the quantum. This means that the range of angles through which the electron can be scattered is of the order of

$$\Delta\theta \cong \frac{\Delta p_x}{p_{\text{el}}} = \frac{2h\nu_q}{cp_{\text{el}}}$$

Now, it is well known in optics that the smallest distance that can be resolved by a lens depends not on the angular aperture of the lens, but more directly on the angular aperture of the pencil of rays that is brought to a focus by the lens. Only if the pencil covers the whole lens is this distance determined by the aperture of the lens. More generally, the minimum resolvable distance is given either by $\Delta x \cong \lambda_{\text{el}}/\Delta\theta$, where $\Delta\theta$ is the angular width of the pencil of rays that enters the lens from any given point, or by $\lambda_{\text{el}}/\theta$, whichever is the larger. Thus, if $\Delta\theta$ is less than the aperture of the lens, the minimum resolvable distance is correspondingly increased. For this case, we obtain

$$\Delta x \cong \frac{\lambda_{\text{el}} c p_{\text{el}}}{2h\nu_q} \geqq \frac{\lambda_q}{2}$$

where we have used the expression given above for $\Delta\theta$ and the de Broglie relations.

From this result we conclude that, if we see a spot on the screen, the uncertainty in the point to which we can ascribe the origin of this spot is at least of the order of the wavelength that the light had before it was scattered. This means that unless we had some previous information about the length of the light wave, we cannot, from this experiment, draw any conclusions at all about the point at which the light was scattered. We also see that even if we do know this wavelength, we cannot ascribe a location to the light quantum that is more precise than this wavelength. On the other hand, for the electron, there was no need for any previous knowledge of its momentum and no limit on the possible accuracy with which it could be localized, as long as sufficiently energetic quanta were used in observing it. This distinction in behavior is in agreement with the results of Chap. 4, Sec. 7, where it was shown on theoretical grounds that the position of a light quantum cannot even be given a precise meaning, but that it can have a rough meaning provided that we do not try to define it better than to within a wavelength. The treatment given here has been incomplete in that we restricted ourselves to the nonrelativistic case, and to a case in which the beams of electrons and photons were initially perpendicular to each other. A more general treatment can be given, however, and it can be shown that the results are essentially the same.

We could have come to the same conclusion as in the previous paragraph by applying the uncertainty principle more directly. We know that the maximum uncertainty in momentum of the photon after it has scattered the electron is $2h\nu_q/c$. Now, if any process is to be used to make very accurate measurements of the position, it is necessary that it shall include some means of making large but unpredictable and uncontrollable transfers of momentum between observing apparatus and the system under observation. Because the magnitude of this transfer in the interaction between matter and light is limited to

$$\Delta p \cong \frac{2h\nu_q}{c}$$

the extent to which such means of observation can be used to localize a photon is also limited, according to the uncertainty principle, to

$$\Delta x \cong \frac{h}{\Delta p} = \frac{c}{2\nu_q} = \frac{\lambda_q}{2}$$

The method of analysis outlined in the last paragraph shows that in any measurement of position the accuracy is always limited by the largest possible momentum transfer between observing apparatus and the system under observation. The following problems will help show the importance of such limitations in a few specific cases.

Problem 3: Show that if electrons are observed by means of a proton microscope, the smallest distance within which the electron can be localized is either λ_{el} or $\frac{m_p}{m_e}\lambda_{pro}$, whichever is the smaller, where λ_{el} is the wavelength of the electron before the observation was made. (Use nonrelativistic theory.)

Problem 4: Show that if protons are observed by means of an electron microscope, the only limitation on the shortest distance that can be measured is the wavelength of the electrons. (Use nonrelativistic theory.)

Problem 5: Obtain the corresponding limitations applying to the observation of electrons by means of other electrons. (Use nonrelativistic theory.)

The above problems show that it is much easier to make accurate observations on heavy particles with light particles than vice versa. The maximum difficulties arise, therefore, when we try to observe a light quantum that has zero rest mass. The conclusions obtained from the problems, which were nonrelativistically formulated, cannot apply directly to the quantum, which goes at the speed of light, but the same general type of difficulty has been shown to arise in the effort to measure the position of the quantum.

In a completely relativistic theory of the electron and of other particles, difficulties arise that are similar to those met with in the case of the photon, but these are important only when the velocity is close to that of light. For small values of v/c, these theories approach the usual nonrelativistic theory, which we treat here.

13. Localization of Electromagnetic Energy and Momentum by Means of Slits and Shutters. We have already referred to a shutter as a means of forming a wave packet that is bounded in time. In a similar way, a slit provides a means of confining a wave within a definite region of space. In accordance with the uncertainty principle we may expect, then, that if a single quantum passes through a slit of width Δx, its momentum will be made uncertain by at least $\Delta p \cong \hbar/\Delta x$, but if it passes through a shutter in the time Δt, its energy will be made uncertain by at least $\Delta E \cong \hbar/\Delta t$.

We may ask, "What is the mechanism that produces these uncertainties?" First we note that, because of radiation pressure, even classical theory predicts the possibility of transfer of momentum from slit edges to the wave, and vice versa. Similarly, a moving shutter that opens and closes against the radiation pressure does work and can, therefore, exchange energy with the electromagnetic field. In the classical limit, this transfer is governed by the deterministic laws of radiation pressure, but on the quantum level the interaction must consist of indivisible transfers, which can neither be predicted nor controlled. Thus we obtain the possibility of uncertain transfers of momentum and energy.

Problem 6: Show that because of diffraction effects we can prove that when a single quantum passes through a slit of width Δx, it can receive an uncontrollable impulse $\Delta p \cong \hbar/\Delta x$ and thus demonstrate the uncertainty principle for this case. Show also that a shutter which is open for a time Δt can transfer an uncontrollable energy $\Delta E \cong \hbar/\Delta t$ to the quantum. Give a comprehensive discussion of the wave-particle duality in producing this uncertainty.

14. Application of Uncertainty Principle to Problem of Defining Orbits in Atoms. In the section on de Broglie waves, it was pointed out that when an atomic electron is in a state of definite energy, it can be found anywhere within a certain region which is near the orbit that a classical particle of the same energy will take. But the exact position of the electron in such an orbit cannot be predicted. In order to obtain a state in which the electron has a definite position, we have to make up a wave packet, containing waves of many possible energies. This means that an observation of the position would show the electron to be somewhere in a region corresponding to a spread over many possible energies, and that a measurement of the energy might disclose any one of a range of values with a probability depending on the intensity with which the wave corresponding to that energy appeared in the wave packet.

We can now show that this prediction corresponds to what would actually be observed if, for example, we attempted to use a microscope to measure where the electron was in its orbit. For simplicity, we shall restrict ourselves to the case of high quantum numbers, where very many orbits exist close to each other. Suppose that we wish to measure the time at which an electron passes a given position with accuracy Δt; from a series of such measurements we could then try to plot an orbit for the electron. To know the time at which the light was scattered to an accuracy Δt, we should have to use pulses of light of duration Δt or less. To make up such pulses, we need a range of frequencies $\Delta\omega \geqq 1/\Delta t$ and, therefore, a range of energies $\Delta E \geqq \hbar/\Delta t$. Now we certainly want to choose Δt considerably less than τ, the period of rotation of the electron in its orbit, or else we shall not be able to follow the course of its motion at all. But according to the Bohr-Sommerfeld theory, we have

$$\frac{1}{\tau} = \frac{\partial E}{\partial J} \cong \frac{\Delta E'}{\Delta J}$$

If we choose $\Delta J = h$, then $\Delta E'$ is the energy difference between adjacent orbits. We obtain, therefore, $\Delta E' \cong h/\tau$. Since $t \ll \tau$, we conclude that $\Delta E \gg \Delta E'$, so that the quantum has much more than enough energy to send the particle into the next orbit.

We conclude that it is impossible to follow a particle as it moves in a single Bohr orbit by watching it with a microscope, because the quanta used in observing it will not only send the electron into some other orbit, but into one that cannot be predicted or controlled. This result is in agreement with the fact that a wave packet with a definite position in the orbit must contain waves corresponding to many energies.

A rather interesting conclusion can be drawn from this hypothetical experiment. With the aid of the wave picture, we can follow the transition from one orbit to the next but cannot picture why the electron is always found in either one orbit or another and never in between. With the aid of the particle model, plus the Bohr-Sommerfeld quantum conditions, we can understand why the particle is always found in a definite orbit, but we cannot picture the process of transition between orbits. On the other hand, the uncertainty principle shows us that if we try to follow the particle by observing it in the process of transition, we impart such an uncertain energy to it that we do not know what orbit it is in. Hence, the transition between definite energies must not be followed continuously, or else it becomes a transition between unknown energies. Thus, although the particle model does not discuss the process of transition, it never gets into any inconsistencies, because within the framework of the particle model this process can never be observed anyway.

15. More General Application of Uncertainty Principle. All the examples given previously show that an uncontrollable and unpredictable transfer of a quantum in the interaction between observing apparatus and what is observed always intervenes to prevent us from inferring a unique connection between the state of the observing apparatus and the state of what is being observed. It might, at first sight, be thought that this difficulty could be avoided by considering the observing apparatus and what is being observed as part of a common system. For example, we might consider camera, photographic plate, light rays, and scenery as a combined system. The question of transfer of a quantum would not then arise because there is only one system to begin with. (The energy and momentum of this system are properties that are shared mutually by all of the parts; it is only when we try to isolate a given part that the problem of transferring quanta from one part to another will arise.)

The chief difficulty with the procedure outlined above is that it yields us no information. In order to obtain information from the system, we must interact with it somewhere, for example, by looking at the photographic plate, and in so doing, we will have to use light. Although the light used in observing the position of the plate will not, in general, alter the image on this plate to any significant extent, it will, nevertheless, transmit to the plate an unpredictable and uncontrollable momentum $\Delta p \cong \hbar/\Delta x$, in exactly the same way as occurred with the electron when its position was observed directly with a microscope.* Thus, when we use the plate in such a way as to provide information about the position of the electron, we inevitably make the momentum of the combined system (camera, plus plate, plus electron) indefinite. We conclude then that there is no indirect way to get around the uncertainty principle by

* See Sec. 8.

avoiding the step in which the transfer of an unpredictable and uncontrollable quantum takes place.

16. The Unity of the Quantum Theory. As shown in Sec. 2, the uncertainty principle was derived from three elements; the wave properties of matter, the indivisibility of energy and momentum transfers and the related particle properties of matter, and the lack of complete determinism. We then showed by analyzing various processes of measurement that the predicted limitation on determinism was actually verified. But equally important, one should notice, is the fact that, unless there had been an *unpredictable* transfer of an *indivisible* quantum of light having *both wave and particle properties*, we could have measured the position and momentum of an electron to an accuracy greater than that given by the uncertainty principle. Similar conclusions are obtained from an analysis of the functioning of the electron microscope, when it is used to make measurements on other particles. In fact, if there were anywhere in the universe a single system which did not combine the three elements of indivisibility, probability, and the wave-particle duality then this system could be used to make measurements on other systems which were more accurate than the limits of precision set by the uncertainty principle; and, as a result, one of the most fundamental predictions of the quantum theory could be contradicted. These three elements, therefore, work together to form a unit that would fall apart if any one of them were removed from any object in the universe. Thus, all parts of the quantum theory interlock in such a unified structure that it is very difficult to conceive of our giving up any one element, unless we give up the whole quantum theory.

17. Are there Hidden Variables Underlying the Quantum Theory? With this unity in mind, let us consider the possibility that quantum phenomena can be explained in terms of hidden variables that really determine where and when each quantum transfer takes place, so that the appearance of probability is merely an expression of our ignorance of the true variables in terms of which one can find causal laws (see Chap. 2, Sec. 5).*

Let us suppose, for the sake of argument, that such hidden variables exist. In order to observe them we must find some experimental result which depends on the state of the hidden variables; otherwise they can be of no real physical significance. Now, in all observations that have ever been made thus far, every conclusion of the quantum theory has been verified, including the one that we are not, in fact, able to predict or control the exact time and place of transfer of a quantum. Thus, even if there are hidden variables, we must conclude that no experiment made so far has ever depended on anything more than a random statistical average of these variables (analogous to the pressure and temperature in thermo-

* See also Chap. 5, Sec. 3.

dynamics) and that no experiment has yet, therefore, supplied any evidence for the existence of hidden variables. Moreover, we shall see in Chap. 22, Sec. 19 that the general conceptual framework of the quantum theory cannot be made consistent with the assumption of hidden variables that actually determine all physically significant events. In other words, no completely deterministic mechanism that could explain correctly the observed wave-particle duality of the properties of matter is even conceivable. Before we could justify the assumption of such a completely deterministic underlying theory, we would therefore have to prove first that the quantum theory is not in complete accord with experiments. But thus far quantum theory has been found to be in complete agreement with a very wide range of experiments, and in no case has it ever been found to contradict experiment. Of course, it is always possible that in some new range of experiment not yet studied, the predictions of the quantum theory may turn out to be wrong, and that here we will discover phenomena in which the hidden variables are not averaged out. If this ever happens, then we shall be forced to modify quantum theory in a fundamental way, but in such a way that in all phenomena with which we deal now, the new theory approaches the present quantum theory as a limit. At present, however, it seems extremely unlikely that we shall ever be able to obtain a totally deterministic description in terms of hidden variables. Although it is true that in the domain of relativistic quantum theory and in the study of the nature of the elementary particles the present theory is incomplete, every indication now points to a line of development in which the extent to which a causal description can be applied will be, if anything, even less than in the present quantum theory. Until we find some real evidence for a breakdown of the general type of quantum description now in use, it seems, therefore, almost certainly of no use to search for hidden variables. Instead, the laws of probability should be regarded as fundamentally rooted in the very structure of matter. In Chap. 8 we shall return to this question and try to show that such a point of view is, basically, just as reasonable as is the completely deterministic one, if not more so.

CHAPTER 6

Wave vs. Particle Properties of Matter

ONE OF THE MOST CHARACTERISTIC features of the quantum theory is the wave-particle duality,† i.e., the ability of matter or light quanta to demonstrate the wave-like property of interference, and yet to appear subsequently in the form of localizable particles, even after such interference has taken place. In this chapter, we shall consider the nature of these phenomena in greater detail, in order to show to what extent matter must be regarded as a wave and to what extent it must be regarded as a particle.

1. The Interference Pattern, and the Wave-particle Nature of Matter. The existence of interference patterns is the most important fact on which the assumption of wave properties for matter is based. Let us therefore begin by discussing the nature of the interference patterns that are met in connection with, for example, electron or photon diffraction. When a single electron (or photon) comes through a slit system, or a crystal, it leaves a single spot or track at the detector. If a second particle is *later* directed at the system, it too will produce a single track or spot at the detector. If many such particles, all having the same initial momentum, are *independently* sent through this system, then in time we obtain a statistical pattern of spots or tracks that shows maxima and minima of density very reminiscent of the interference patterns of optics. Yet, since the electrons clearly come through the slit system separately and independently, the interaction between electrons cannot cause the interference pattern.

If we regard an electron as nothing but a classical particle, then this phenomenon is indeed very difficult to understand. In quantum theory, however, we have seen that interference can be described quantitatively with the aid of a wave function, $\psi(x, t)$, which is associated with an individual electron in such a way that the probability that this particular electron can be found at a given spot is proportional to $\psi^*(x, t)\,\psi(x, t)$. If all of the electrons have the same initial momentum, $\mathbf{p}_0$, then associated with each of them must be the same incident wave function, $\exp(i\mathbf{p}\cdot\mathbf{x}/\hbar)$. This follows from the fact that an electron can have a given momentum $\mathbf{p}_0$ only when its wave vector‡ is $\mathbf{k} = \mathbf{p}_0/\hbar$. Since the wave functions

† See, for example, Chap. 2, Sec. 1; Chap. 3, Sec. 11; Chap. 5, Sec. 2.

‡ Strictly speaking, the plane wave is an approximation to a wave packet which

associated with each electron are all propagated in the same way, we conclude that even after diffraction each electron will continue to have associated with it a wave function that is the same as that of every other electron. This means, of course, that every electron has the same probability function.

After many such electrons have gone through the slit system, the resulting density of spots will therefore be proportional to $|\psi(x, t)|^2$. It may happen, however, that at certain points $\psi(x, t)$ vanishes as a result of interference between waves coming from several slits, whereas if only one of the slits had been open, there would have been a nonvanishing probability at these points. Alternatively at certain points the probability may be more than that which is the result of the sum of the contributions of the separate slits. We can thus obtain either destructive or constructive interference in the function determining the probability of arrival of an electron at a given point.

We conclude from the above that with a single electron one cannot really investigate the interference pattern, and that the wave properties of matter can be demonstrated clearly only when there are enough electrons to yield a statistical aggregate. We may ask why the individual electron is regarded as having any wave properties at all, if it is always found to arrive at the detector in a fairly definite location just as if it were a particle. The answer is that, to explain the appearance even of a statistical interference pattern, we must ascribe to matter certain wave properties, at least while it is in the process of going through the slit system. If we assumed that the electron *always* acted like a particle, then we would conclude that it could go through only one slit at a time. It is difficult to see, then, how the opening of another slit which, for example, may be millions of miles away, could make it unlikely that electrons would reach certain points to which they would otherwise have a high probability of going. Such long-range action of slits on particles is certainly contrary to all of our previous experience with particles. We might try to assume various modifications of the law of force between electrons and slit systems so as to try to explain this result but, as we shall see in Sec. 11, this effort would lead to all sorts of *ad hoc* hypotheses, which conflict with some of the most elementary requirements for a sensible theory. On the other hand, the wave interpretation of matter explains this result, as well as a whole host of other results, in a comparatively simple and yet quantitatively correct way. Thus, we conclude that even the individual electrons seem to be able to show certain wavelike properties.

The preceding discussion leads to the idea that an electron is neither a particle nor a wave, but is instead a third kind of object which has

---

is, however, usually so broad in practice that we can regard this width as infinite in interpreting most diffraction experiments.

some, but not all, of the properties of both particles and waves.* Under different circumstances, either the wave or the particle aspects of this object may manifest themselves more strongly. For this reason, the term *electron* will hereafter denote neither a wave nor a particle, but simply that object, whatever it is, which boils out of hot filaments, carries charge, demonstrates a certain ratio of charge to mass, shows certain deflections in electric and magnetic fields, shows certain diffraction properties in the Davisson-Germer experiment, certain energy levels in the hydrogen atom, etc. It will be the purpose of Chaps. 6, 7, and 8 to provide a better picture of this object.

2. Impossibility of Simultaneous Observation of Wave and Particle Properties of Matter. To find out whether an electron (or a photon) is more like a wave or more like a particle, we might try to see what happens to it while it is being diffracted. Let us consider, as an example, a hypothetical experiment, in which electrons (all having the same initial momentum and, therefore, as shown in Sec. 1, the same wave function) are sent one by one into a system consisting of two slits and a detecting screen to the right of those slits (see Fig. 1). Our objective in this experiment would be to discover to what extent an electron goes through one slit at a time, as if it were a particle, and to what extent it goes through both slits together, as if it were a wave. To do this, we could in each case observe the electron with the aid of a microscope, illuminating the region near the slits with plenty of light to insure the scattering of at least one quantum by each electron as it goes by. Then, to be able to find out whether an electron goes through one slit or through both, we shall have to use light of wavelength not greater than a, where a is the distance between slits. As shown in Chap. 5, Sec. 8, such a light quantum can deliver to the electron a momentum which is uncertain by $\Delta p \cong \hbar/a$. The angle of scattering will, in this way, be made uncertain by

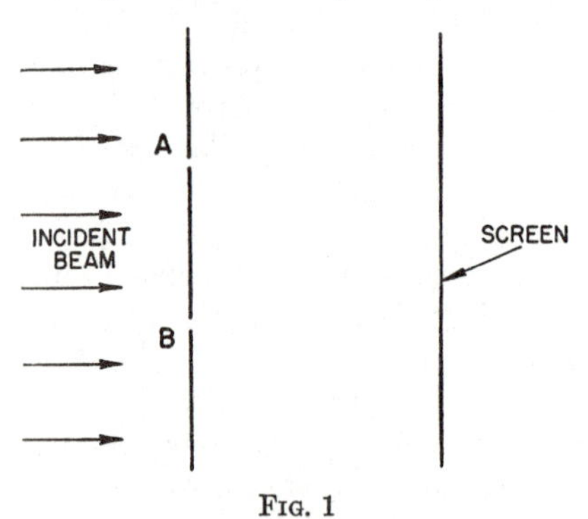

FIG. 1

$$\Delta\theta \cong \frac{\Delta p}{p} = \frac{h}{ap} = \frac{\lambda_{el}}{a}$$

where λ_{el} is the wavelength of the electron. This uncertainty, however, is as large as the angular difference between the minima of the interference pattern. The addition of the undetermined momentum, therefore, tends to destroy the interference pattern. In fact, we shall see in

* See Chap. 5, Sec. 2.

Sec. 4 that if the measurement is precise enough to define unambiguously the slit through which each electron passes, then no trace of the interference pattern will be left on the screen. On the other hand, if we had tried to avoid blotting out the interference pattern by using a quantum of longer wavelength, then the measurement would not have been precise enough to show unambiguously through which slit each electron had gone. We conclude that we cannot simultaneously observe through which slit each electron goes and also obtain an interference pattern. In other words, electrons seem to be able to go through one slit at a time as if they were particles, but only at the expense of losing their wavelike properties (i.e., the demonstration of interference). On the other hand, the wave-like property of interference can be demonstrated only under conditions in which the slit through which the electron passes is not defined.

To show that this conclusion does not depend on the particular method used to find through which slit the electron goes, let us consider, for example, the possibility of setting up a cloud chamber at the detecting screen. The cloud chamber indicates not only where the electron arrives at the screen but, because the electron leaves a visible track, it also tells in which direction the electron is going. By extrapolating the line of the track backward, we can then perhaps find out from which slit the electron came. The experiment is illustrated in Fig. 2.

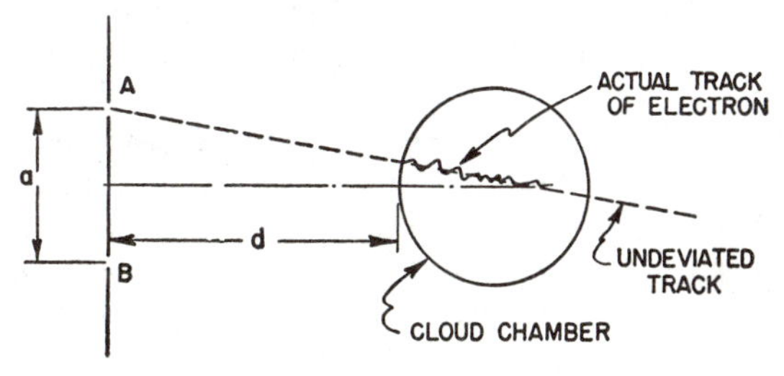

FIG. 2

We must remember, however, that the behavior of the electron in the cloud chamber is also limited by the uncertainty principle. It leaves a track by transferring quanta to neighboring atoms, which are therefore ionized and then serve as nuclei for the water droplets, which make the track visible. But in the transfer of a quantum from electron to atom, the electron suffers an uncontrollable change of momentum, so that it is deflected through an incompletely determined angle. According to the uncertainty principle, this change is $\Delta p \cong \hbar/\Delta x$, where Δx is the degree of uncertainty in the position measurement. The uncertainty in the angle of deflection is therefore $\Delta\phi \cong \dfrac{\Delta p}{p} \cong \dfrac{\hbar}{p\,\Delta x}$. This uncertainty in angle produces an uncertainty ΔX in the position* at which the electron crossed the slit system, equal to $\Delta X \cong d\,\Delta\phi \cong d\hbar/p\,\Delta x$, where d is the distance from the cloud chamber to the slit system. (This formula applies only when $a/d \ll 1$.)

Now, to determine whether or not an interference pattern existed, it is necessary

* This is a minimum uncertainty, which will be present if the direction is determined from two points of ionization; if more points have to be used, the uncertainty becomes larger.

to measure the position of the electron in the cloud chamber to an accuracy

$$\Delta x \cong d\,\Delta\theta \cong \frac{d\,\lambda_{el}}{a}$$

where $\Delta\theta$ is the angular separation between maxima and minima in the diffraction pattern. If we measure the position of the electron less accurately than this, then we have no way of knowing, for example, whether an electron arrived at a point where $|\psi|^2$ is a maximum or zero, and we are therefore unable to investigate the interference pattern. Writing $\lambda_{el} = h/p$, we obtain

$$\Delta x \cong \frac{dh}{ap} \qquad \text{and} \qquad \Delta X \cong \frac{d\hbar ap}{d\hbar p} \cong a$$

This result shows that, just as when we make our observations with a microscope, it is impossible to observe an interference pattern if we find out which slit the electron went through; and it is impossible to find out which slit the electron went through if we can obtain an interference pattern.

It is worth-while to state here that we have thus far considered only that part of the diffraction pattern arising from interference between waves coming from two different slits. There is also another part of the pattern arising from parts of the wave coming from different parts of the same slit. If, however, the slits are very narrow in comparison with their separation, variations in the pattern arising from this reason are negligible in comparison with variations arising from interference between slits. Thus we can, if we wish, neglect the effects of the finite width of each slit.

3. Effects of Process of Observation on the Wave Function. Let us now return to the experiment in which the position of the electron was observed with a microscope as it came through the slit system. Before the observation took place, the wave function certainly covered both slits, or else there could have been no interference. After the observation, however, the electron was found to be near either one slit or the other. The wave function corresponding to this new situation must then be a packet, which is near the slit where the electron is actually found.

What appears to have taken place is that when the position of the electron was observed, the wave function suffered a collapse from a broad front down to a narrow region. The exact region to which it collapses is not determined by the state of the wave function before collapse; only the probability of collapse to a given region is determined, and this is proportional to the value of $|\psi|^2$ in that region.

This type of collapse of the wave function does not occur in any classical wave theory. Why does it occur here? To answer this question, we must take into account the fact that, while an observation is taking place, there is an interaction between the particle and the observing apparatus. Thus far, Schrödinger's equation [Chap. 3, eq. (29)], which defines the wave function $\psi(x, t)$ has been derived only for a free particle.

The effect of any kind of force of interaction (for example, electrical, gravitational, electromagnetic, etc.) is to modify Schrödinger's equation. For example, while the observation is being made, the electromagnetic quantum used in connection with a microscope will change the wave equation for the electron. The precise way in which such modifications occur will be studied in detail in Chap. 22. For the present, however, we shall only describe some of the results obtained there.

We begin with the example previously used, in which an electron is directed at the two slit system, shown in Fig. 1. As in optics, the propagation of electron waves can be described by means of a Huyghens' principle.† This means that if we know the value of the wave function on a given wave front, then we can express its value elsewhere as the sum of contributions from different elements of that wave front, weighted with a phase factor, $\dfrac{\exp(2\pi ir/\lambda)}{r}$, where r is the distance from the point in question to the element of surface on the wave front.

In the two-slit experiment, all contributions to the wave function at the right of the slits comes either from slit A or from slit B. If we denote the wave function at slit A by $\psi_A^\circ(x_s)$ and that at slit B by $\psi_B^\circ(x_s)$, where x_s is the value of the co-ordinate at an arbitrary point in the plane of the slit, then according to Huyghens' principle the wave function at an arbitrary point x to the right of the slits is

$$\psi(x) \sim \int_A \frac{\exp[2\pi i(x - x_s)/\lambda]}{|x - x_s|}\psi_A^\circ(x_s)\,dx_s + \int_B \frac{\exp[2\pi i(x - x_s)]}{|x - x_s|}\psi_B^\circ(x_s)\,dx_s \quad (1)$$

where dx_s indicates integration over the plane of either slit A or slit B, as previously indicated. The preceding expression may be written more concisely as

$$\psi(x) = \psi_A(x) + \psi_B(x)$$

where $\psi_A(x)$ represents that part of the wave reaching the point x that has come from slit A, while $\psi_B(x)$ represents that part which has come from slit B.

If only slit A were open, then the probability that a particle reaches the point x would be equal to $P_A(x) = |\psi_A(x)|^2$, while if only slit B were open this probability would be $P_B(x) = |\psi_B(x)|^2$. When both slits are open, however, the probability is

$$P(x) = |\psi_A(x) + \psi_B(x)|^2 = P_A(x) + P_B(x) + \psi_A^*(x)\psi_B(x) + \psi^*(x)\psi_A(x) \quad (2)$$

Thus, in addition to the "separate-slit" terms, P_A and P_B, $P(x)$ contains the interference terms, $\psi_A^*\psi_B + \psi_B^*\psi_A$, which would not be present if the

† R. P. Feynman, *Rev. Mod. Phys.*, **20**, 377 (1948), Sec. 7.

experiment involved a probability distribution of classical particles, coming either through slit A or slit B. These interference terms constitute the characteristic effects coming from the wave properties of matter.

Let us now consider what happens to the wave function of the electron when the position is observed. As we shall see in Chap. 22, the interaction process involved in an observation always changes the wave function ψ in a way that cannot be predicted or controlled with complete accuracy. This change can be thought of in a rough manner as being caused by the unpredictable and uncontrollable quantum that is used in the process of measurement.* In general, this quantum can produce many different kinds of changes in the system under observation, and these changes will be reflected in corresponding changes in the wave function. It is always possible, however, to design the apparatus in such a way that the property that is being measured does not change in the course of the measurement. For example, if a microscope is used to measure the position of an electron, this position is not changed by the scattering of the quantum used in the measurement, and only the momentum is changed. (Of course, the position will change after the measurement is over, but this change is irrelevant for our discussion.) We shall see in Chap. 22 that, under these conditions, each part of the wave function corresponding to a definite position of the electron at the time of the measurement is changed during the course of the interaction between electron and observing apparatus in such a way that it is multiplied by an unpredictable and uncontrollable phase factor, $e^{i\alpha}$. For example, in the case considered, the wave function becomes

$$\psi = \psi_A(x)e^{i\alpha_A} + \psi_B(x)e^{i\alpha_B} \tag{3}$$

where α_A and α_B are different constants that can neither be predicted nor controlled.

In Chap. 22 it is shown exactly why these changes of phase are brought about. A very rough reason, however, can be given here. In any interaction between electron and observing apparatus, there is always some time Δt representing the duration of this interaction. During this time, the description of the electron as a separate system becomes inadequate, and the energy is determined not only by the state of the electron, but also by the quantum used in the process of interaction.

Now the wave function oscillates as $\exp(-iEt/\hbar)$. During the time of interaction, this energy is indefinite by some amount ΔE, and according to the uncertainty principle $\Delta E \cong h/\Delta t$. The uncertainty in the phase is, then, at least $\Delta E \, \Delta t/\hbar \cong 2\pi$. Thus, the phase of the wave function is made completely indefinite,† and there is no deterministic

* See Chap. 5, Sec. 8.

† Actually, the change of phase will be seen in Chap. 22 to be very much larger than 2π in all practical cases.

relation between the phase before interaction and the phase after interaction. Furthermore, we shall see also in Chap. 22 that there is no definite relation between α_A and α_B, so that the phase difference $\alpha_A - \alpha_B$ is also unpredictable and uncontrollable.

If the apparatus were such as to change the value of the quantity under observation, then the change of ψ occurring during the interaction with the apparatus would be more complicated, but we shall not discuss this possibility here because it can be shown that it does not alter in any essential way the conclusions that we shall now obtain.

To demonstrate the meaning of these changes of wave function, let us now compute the probability function

$$\psi^*(x)\psi(x) = P_2(x) = |\psi_A|^2 + |\psi_B|^2 + \psi_A^*\psi_B \exp[i(\alpha_B - \alpha_A)] + \psi_B^*\psi_A \exp[i(\alpha_A - \alpha_B)] \quad (4)$$

We see that the interaction with the observing apparatus has changed the interference terms, but not the "separate-slit" terms, $|\psi_A|^2$ and $|\psi_B|^2$. At those points where the wave functions, $\psi_A(x)$, and $\psi_B(x)$, do not overlap, and therefore do not interfere, the phase factors will produce no result whatever. Such points exist, for example, right at the slits themselves. We conclude, therefore, that the interaction with the observing apparatus did not change the probability of finding a particle at the slit system itself. This result is more or less to be expected, however, from the fact that we are considering only those methods of observing the position which do not change the position, so that the distribution of positions in the neighborhood of the slits is left unaltered in the course of the observation. The statistical distribution of particles elsewhere, however, can be changed considerably, because the "interference terms" in eq. (4) are altered by the factors $\exp[i(\alpha_B - \alpha_A)]$ and $\exp[i(\alpha_A - \alpha_B)]$. For example, at a point far enough to the right of the slits so that $\psi_A(x)$ and $\psi_B(x)$ overlap appreciably, the factor $\exp[i(\alpha_A - \alpha_B)]$ may be such as to change the character of the interference existing from destructive to constructive, thus increasing the probability of arrival of particles at this point.

In any particular experiment, α_A and α_B will be definite but unknown and uncontrollable constants. But, as pointed out in Sec. 1, $P_2(x)$ has meaning only insofar as it refers to a set of similar experiments carried out under equivalent initial conditions. The probability function that should be applied here is, therefore, the mean value of $\psi^*\psi$, averaged over many experiments. Because the phases $\alpha_A - \alpha_B$ fluctuate in a random and uncontrollable way from one experiment to the next, terms like $\exp[i(\alpha_A - \alpha_B)]$ will average out to zero, and the only terms remaining will be the contributions of the separate slits, $|\psi_A|^2$ and $|\psi_B|^2$. This means that after the electron has interacted with a device that enables us to tell which slit the electron went through, the waves coming through each slit cease to demonstrate observable interference effects, even though they continue to overlap in space.

4. Relationship of Destruction of Interference to Consistency of Wave-particle Duality. We shall now show that the statistical interpretation of the wave function, combined with the destruction of interference brought about by the interaction between the electron and the observing apparatus, are precisely what is needed to lead to a consistent formulation of the wave-particle duality. To do this let us suppose, for example, that the results of the measurement of the electronic position with the microscope are automatically recorded on a photographic plate. A spot will then be produced on the photographic plate with a position that depends on whether the electron has gone through slit A or B. If the apparatus functions properly, then an observer can, by looking at the plate, find out through which slit the electron went. Even before looking at the photograph, however, the observer knows that the apparatus will be able to show that the electron has gone through either one slit or the other, as if it were a particle, and not through both at once, as if it were a wave.

Let us now consider how these facts are to be described in terms of the electronic wave function. Before the apparatus has interacted with the electron, the wave function is given by $\psi = \psi_A(x) + \psi_B(x)$, but after interaction has taken place, it is given by $\psi_A(x)e^{i\alpha_A} + \psi_B(x)e^{i\alpha_B}$. Because of the unpredictable and uncontrollable changes of α_A and α_B, interference between $\psi_A(x)$ and $\psi_B(x)$ is destroyed and, as a result, the probability that a particle can be found at the point x becomes (see Sec. 3)

$$P = P_A(x) + P_B(x)$$

This function is, however, what would have been obtained from a distribution of classical particles coming through each slit separately. Thus, the electron acts, for all purposes, as if it had gone, like a particle, through a single distinct but unknown slit, with a probability

$$P_A = \int \psi_A^*(x)\psi_A(x)\,dx$$

that this slit was A, and

$$P_B = \int \psi_B^*(x)\psi_B(x)\,dx$$

that it was B. (The integration is to be carried out only over the region to the right of the slits.) Before the electron interacted with the measuring apparatus, however, it was capable of showing interference effects, which required the interpretation that it was able to go, like a wave, through both slits at the same time. We see, therefore, that when the effects of the measuring apparatus on the wave function are taken into account, we obtain what is effectively a transformation of the electron from a wavelike to a particle-like object. Such a transformation was also suggested in Sec. 2, in connection with the hypothetical experiment in which an effort was made to observe through which slit the electron went in the process of diffraction.†

We shall now show how the destruction of interference leads to a consistent account of what happens to the wave function when the observer looks at the photographic plate and finds out through which slit the electron actually went. As we have seen, the same results are predicted for all physical processes by the wave function

$$\psi = \psi_A(x)e^{i\alpha_A} + \psi_B(x)e^{i\alpha_B}$$

and by a wave function that is either *entirely* $\psi_A(x)e^{i\alpha_A}$ or *entirely* $\psi_B(x)e^{i\alpha_B}$, but with respective probabilities P_A and P_B that each of these is actually the correct wave function. When the observer finds out which slit the electron went through, he then replaces $\psi(x)$ either by $\psi_A(x)e^{i\alpha_A}$ or by $\psi_B(x)e^{i\alpha_B}$, depending on the results of the experiment. In this way, we describe the collapse of the wave function, discussed at the beginning of Sec. 3. Because of the destruction of interference, this collapse corresponds only to a choice of the two possible alternatives of the actual wave function and not to any real physical changes in the state of the electron itself.

Although the destruction of definite phase relations is deduced from the rest of the quantum theory (see Chap. 22), we wish to show now that this result is essential for the consistency of our probability interpretation of the wave function.

Suppose that in the hypothetical experiment described in Sec. 1, interference between $\psi_A(x)$ and $\psi_B(x)$ were not *completely* destroyed by the actions of the apparatus that was used to disclose the slit actually traversed by each electron. Then, by hypothesis, an interference pattern should be obtained on the screen to the right of the slits. But as soon as an observer consults the measuring apparatus (for example, the photographic plate on which the image of the electron in the microscope is recorded), he could find out through which slit each electron passed. Since each electron can in this way be shown unambiguously to have gone through a definite slit, its subsequent behavior must not depend on whether or not the other slit was open at that time. The probability of arrival of a given electron at any point on the screen should, therefore, be proportional to one of the "separate-slit" terms; i.e., either $|\psi_A(x)|^2$ or $|\psi_B(x)|^2$, depending on whether this electron traversed slit A or slit B. Since it is equally likely that a given electron shall traverse either slit, it follows that after many electrons have passed through the system, the pattern on the screen should be given by the sum of the separate-slit terms, and should not depend on the interference terms $\psi_A^*(x)\psi_B(x) + \psi_B^*(x)\psi_A(x)$. Thus, we have shown that if an observer consults the

† Note that the destruction of interference between ψ_A and ψ_B leads precisely to the blotting out of the interference pattern, as described in Sec. 2.

photographic plates, no interference pattern will be obtained on the screen. If interference between ψ_A and ψ_B were not *completely* destroyed by the actions of the observing apparatus we should, therefore, obtain a theory in which the statistical pattern of electrons striking the screen would depend on whether or not an observer chose to look at the record of the functioning of the measuring apparatus (in this case, the photographic plates). Such a theory could clearly make no sense. We conclude that complete destruction of interference between $\psi_A(x)$ and $\psi_B(x)$ is essential to the consistency of our interpretation of $|\psi_A(x)|^2$ and $|\psi_B(x)|^2$, respectively, as probabilities that the electron has passed either through slit A or slit B. This means that when the electron goes through the slit system under conditions in which it does not interact with a device that can be used to provide a measurement of its position, the wave function cannot consistently be regarded as undergoing a collapse down to $\psi_A(x)e^{i\alpha_A}$ or $\psi_B(x)e^{i\alpha_B}$, because interference between these functions still exists.

Abrupt changes in mathematical quantities analogous to the collapse of the wave function described above often occur in classical probability functions whenever new information is obtained. Thus, on the basis of insurance statistics, we can predict a certain life expectancy for a person of whom we know only that he is over 21 years old. Suppose, then, that we suddenly learn that he is actually 70 years old. At that moment, we immediately predict for him a much shorter life expectancy. The sudden change in life expectancy represents no change in the state of the person, but merely an improvement in our information about the person. Abrupt changes of this kind in life expectancy are permissible, because the life expectancy function merely tabulates statistical information and is, therefore, not in a one-to-one correspondence with the actual length of a given person's life. We may contrast such a statistical theory with a complete and deterministic theory, such as classical mechanics, which would in principle aim to predict a given person's life in terms of the motions of all his atoms and molecules. In this type of theory, the dynamical variables would be in a one-to-one correspondence with the system that is being described and, as a result, no changes in these variables could take place unless they reflected a real corresponding change in the system under description, and not simply an improvement of someone's information about this system.

Now, the variables that appear in the quantum theory bear some resemblance to classical statistical functions such as life expectancy but, as we shall see, they also differ from classical statistical functions in a very significant way. The similarity lies in the fact that the wave function predicts only the probabilities of actual events so that, like a classical statistical function, it is not in a one-to-one correspondence with the system that is being described. For this reason, the abrupt collapse of the wave function that occurs when an observer consults his apparatus (for instance, the photographic plate) represents no change in the object under observation, but merely a change of the statistical functions representing the observer's information about this system. On the other hand, the wave function differs from a classical probability function in the important respect that before interference has been destroyed by the actions of a suitable measuring apparatus, the wave function cannot consistently be interpreted in terms of a simple probability. This is because the phase relations between various parts of the wave function, as well as the amplitudes, have physical significance. Thus, in the hypothetical experiment in which an electron was sent through a two-slit system, the phase relations between the functions $\psi_A(x)$ and $\psi_B(x)$ determine the interference pattern that can be obtained on a screen to the right of the slits. As long as definite phase relations between $\psi_A(x)$ and $\psi_B(x)$ exist, the electron is capable of demonstrating the effects of interference and acting as if it passed wave-like through both slits simultaneously. A sudden collapse of the wave function would, therefore, at this time represent a real change in the physical state of the electron (from a wave-like to a particle-like behavior); as we have already seen, absurd results would follow if such abrupt changes in the wave function could be brought about simply by an improvement in an observer's information about the electron. It is only after definite phase relations between $\psi_A(x)$ and $\psi_B(x)$ have been destroyed by the actions of the observing apparatus that the collapse of the wave function ceases to imply a corresponding physical change in the state of the electron. This means that to the extent that definite phase relations exist between $\psi_A(x)$ and $\psi_B(x)$, the wave function is in a closer correspondence with the state of the electron than it would be if it were a simple classical probability function, specifying the likelihood that the electron goes through either of the slits. Nevertheless, the degree of correspondence between the wave function and the actual behavior of the electron is always less than that aimed for by the dynamical variables of classical mechanics. We shall see in Secs. 9 and 13 that this intermediate degree of correspondence of the wave function with the behavior of the electron provides the basis of a new physical picture of the quantum nature of matter.

5. Generalization of Previous Results. Let us now generalize the results of Sec. 4 to an arbitrary measurement of the position. To do

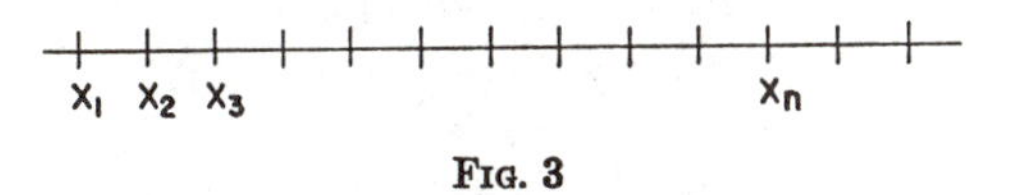

FIG. 3

this, we divide space up into blocks of width Δx, as shown in Fig. 3. We may think of this breakup as being produced by very fine wires, which absorb a negligible fraction of the incident wave. We then obtain, in essence, the problem of an infinite number of slits.

The wave function in the plane of the nth slit is denoted by $\psi_n^0(x_S)$. It is a function that is zero everywhere outside the nth slit, but inside this slit it is equal to $\psi(x_S)$, the actual value of the wave function in this plane. The wave function $\psi_n^0(x_S)$ therefore represents a state in which the electron is certain to go through the nth slit. According to Huyghens' principle, the complete wave function to the right of the slit system is

$$\psi(x) = \int \frac{\exp [2\pi i(x - x_S)/\lambda]}{|x - x_S|} [\psi_1^0(x_S) + \psi_2^0(x_S) + \psi_3^0 (x_S) + \ . \ . \ .] \, dx_S$$

More concisely, we write

$$\psi(x) = \psi_1(x) + \psi_2(x) + \psi_3(x) + \ . \ . \ .$$

where $\psi_n(x)$ represents that part of the wave function at the point x which has come from the nth slit.

The above is the wave function for a system which has not been disturbed by a measuring apparatus. If, however, one makes a measurement of position which is good enough to show which slit the electron goes through, then the process of interaction with the observing apparatus changes the wave function into the following:

$$\psi(x) = \psi_1(x)e^{i\alpha_1} + \psi_2(x)e^{i\alpha_2} + \psi_3(x)e^{i\alpha_3} + \ . \ . \ . \tag{4}$$

where each α is a different, but unpredictable and uncontrollable, constant phase factor.†

If we denote by $P_n(x) = \psi_n^*(x)\psi_n(x)$ the probability distribution that would be present if only the nth slit were open, we then obtain for the total probability that a particle reaches the point

$$P(x) = |\psi(x)|^2 = \sum_n P_n(x) + \sum_{n \neq m} (e^{i(\alpha_n - \alpha_m)}\psi_m^*\psi_n + e^{i(\alpha_m - \alpha_n)}\psi_n^*\psi_m) \tag{4a}$$

Since α_n and α_m are random phases, the interference terms (where $n \neq m$) will cancel out in a series of many experiments. Thus, the probability function is reduced to a set of noninterfering packets that may, as far as their relation to each other is concerned, be treated as classical probability functions.‡ This means that after the electron has interacted with the apparatus that measures its position, all subsequent processes undergone by the electron can have their probabilities calculated either with the wave function (7), or by the equivalent procedure of assuming that the wave function is *entirely* $\psi_1 e^{i\alpha_1}$, or $\psi_2 e^{i\alpha_2}$, or $\psi_3 e^{i\alpha_3}$, etc., with respective probabilities p_1, p_2, p_3, etc. Thus, the electron acts in every respect

† As in the two-slit problem, the changes may be more complicated if the process of observation of the position also changes the position.

‡ Note that the treatment is very similar to that of Sec. 4.

like a wave that has gone through *one* unknown slit of width Δx. To find out which slit the electron has actually traversed, we must consult the observing apparatus. Moreover, from the state of the system before the measurement took place, we can only predict the probability that a particular value of the position will be found.

Since the wave-like aspects of the electron are inferred from its ability to demonstrate the effects of interference over wide regions of space, we see that the destruction of definite phase relations accompanying a position measurement must also destroy all possibility of its demonstrating wave-like behavior over distances larger than the accuracy Δx of the measurement. Instead, the electron acts more like a particle that exists in a single distinct (but unknown) region of width Δx. Experiments that involved distances smaller than Δx would, however, be able to show the effects of interference and would, therefore, still require a wave interpretation. We may summarize these results with the statement that when an electron interacts with a device that can disclose its position, the particle-like aspects of the electron are emphasized at the expense of the wave-like aspects, although the electron is never completely identical with either a particle or a wave.

6. Measurement of Momentum. A rather similar result is obtained from any experiment that measures the momentum to accuracy

$$\Delta p = \hbar \, \Delta k$$

To describe this case, let us consider the Fourier component $\phi(k)$ of the wave function $\psi(x)$. The possible range of values that can occur as a

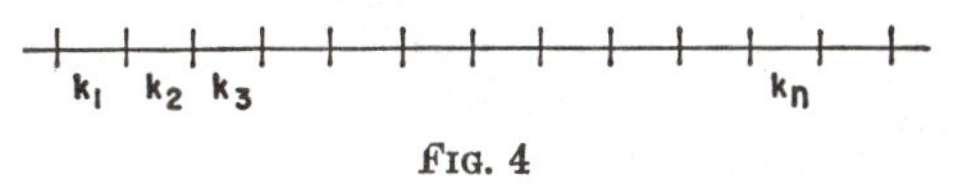

FIG. 4

result of a measurement of the momentum will be denoted by $k_1, k_2, \ldots, k_n \ldots$, and is indicated in Fig. 4. If the momentum of the system lies somewhere in the nth block, we can say that the system is represented by a corresponding wave packet in k space denoted by $\phi_n(k)$. Before an observation of the momentum, we can write*

$$\phi(k) = \phi_1(k) + \phi_2(k) +, \ldots, + \phi_n(k) +, \ldots \tag{5}$$

But after the electron has interacted with an apparatus that can measure its momentum, the wave function becomes

$$\phi(k) = \phi_1(k)e^{i\alpha_1} + \phi_2(k)e^{i\alpha_2} + \ldots + \phi_n(k)e^{i\alpha_n} + \ldots \tag{6}$$

* The above applies to a one-dimensional case. In a three-dimensional case, the functions $\phi_n(k)$ are zero everywhere, except in a block that lies in the neighborhood of k_n. These functions are not quite analogous to the $\psi_n(x)$ of Sec. 5, which are derived from a Huyghens' principle, taking into account the propagation of the wave after it passes through a given slit. In momentum space, however, there is no analogous propagation of waves through a Huyghens' principle.

where the α_n are unpredictable and uncontrollable phase factors. The probability is

$$p(k) = \sum_n |\phi_n(k)|^2 + \sum_{n \neq m} \phi_n^* \phi_m e^{i(\alpha_m - \alpha_n)} + \phi_m^* \phi_n e^{i(\alpha_n - \alpha_m)} \tag{7}$$

As with $p(x)$, the interference terms average out to zero, and we obtain a series of independent probabilities that the electronic momentum has a definite, but unknown, value. To find out what this value is, we must, as in the case of a position measurement, consult the observing apparatus. Moreover, from the state of the system before the measurement took place, we can predict only the probability of any given result.

When the electron obtains a comparatively definite momentum p, its wave function must obtain a correspondingly definite wave number, $k = p/\hbar$. Suppose that the momentum is left indefinite to the extent Δp, so that we have $\Delta k = \Delta p/\hbar$. This means that, even if the electron were initially localized in a very small region, its wave function would have to spread out to a width of $\Delta x \cong 1/\Delta k = \hbar/\Delta p$ after the electron interacted with any device that measures its momentum to this accuracy. The reason for the spread can easily be seen in terms of the uncontrollable phase shifts. Thus, before the interaction took place, the wave function was

$$\psi(x) = \frac{1}{(2\pi)^{1/2}} \sum_n \int \phi_n(k) e^{ik(x-x_0)}\, dk = \sum_n U_n(x)$$

where each $U_n(x)$ is a wave packet of width $\Delta x \cong 1/\Delta k$. The only way of obtaining a packet that is narrower than the individual packets $U_n(x)$ is to have destructive interference between the different $U_n(x)$, at points far from the center of the packet (see Chap. 3, Sec. 2). But after interaction with the apparatus has taken place, we obtain

$$\psi(x) = \left(\frac{1}{2\pi}\right)^{1/2} \sum_n e^{i\alpha_n} \int \phi_n(k) e^{ik(x-x_0)}\, dk = \sum_n e^{i\alpha_n} U_n(x)$$

Because of the appearance of the uncontrollable phase factors, destructive interference between the packets $U_n(x)$ can no longer occur, so that the resulting wave packet must be at least as wide as each $U_n(x)$. This means that the wave function as a whole has been transformed into a group of wave-like packets of fairly definite wavelength, which overlap in position space but do not interfere. The system acts, therefore, as if it had a fairly definite but unknown wavelength (the value of which can be found by consulting the apparatus and using the de Broglie relation $\lambda = h/p$). Thus, when an electron interacts with a device that measures its momentum, its wave-like aspects (definite wavelength) are emphasized at the expense of its particle-like aspects (definite position). An example of such a measurement is the interaction of an electron with a crystal, which would allow the electron wave function to spread out and also to obtain a definite wavelength, from which we could then compute the momentum (see Chap. 4, Sec. 8).

7. Relation of Phase Changes to Uncertainty Principle. It is of interest to note that the destruction of interference provides a simple description of the origin of the uncertainty principle. Thus, in the measurement of momentum, we have seen that the destruction of definite phase relations over parts of the wave function that are widely separated in k space prevents the formation of narrow packets in x space. Conversely, the destruction of interference over wide regions of x space, which accompanies an accurate position measurement, prevents the formation of narrow packets in k space. We see, therefore, that all the uncertainties which were ascribed in Chap. 5 to the transfer of uncontrollable quanta from the observing apparatus to the system under observation, may also be ascribed to uncontrollable changes in the phase of the wave function. But, since the uncontrollable phase changes and the transfers of uncontrollable quanta both originate in the interaction between observing apparatus and the system under observation, in accordance with the laws of quantum theory (see Chap. 22), these two methods of treating the problem must be equivalent ways of describing the same thing. (In fact, we shall see in Chap. 8, Sec. 13 that the treatment in terms of uncontrollable quantum transfers provides the so-called "causal" description; whereas the treatment in terms of uncontrollable phase changes in the wave function provides the complementary "space-time" description of the process of interaction between the two systems.)

8. Importance of Phase Relations. We have seen from the previous discussion that the phase relations between various parts of the wave function are as important as are the amplitudes in determining physically significant results. Thus, in the position representation, the phase relations between the $\psi(x)$ at different points in space controls the momentum distribution; but in the momentum representation the phase relations between the $\phi(k)$ control the position distribution. The phase relations are important even in the classical limit: For as we have seen in Chap. 3, Sec. 9, the motion of the center of the wave packet is determined by the changing phase relations among various $\phi(k)$. To see this in greater detail, note that the center of the packet occurs at the point where a wide range of $\phi(k)$ tend to interfere constructively, whereas some distance away they tend to cancel because of destructive interference. Since each $\phi(k)$ oscillates as exp $-i\hbar k^2t/2m$, the resulting changes of phase of the $\phi(k)$ with time change the positions of constructive and destructive interference and, therefore, govern the motion of the wave packet. The classical equations of motion are thus contained in the phase relations among the different $\phi(k)$.

9. Quantum Properties of Matter as Potentialities. On the basis of the results obtained thus far, we shall now show that the quantum theory leads us to a new concept of the inherent properties of an object to replace the classical concept. This new concept considers these properties as incompletely defined potentialities, the development of which depends on the systems with which the object interacts, as well as on the object itself. To demonstrate this concept we consider, first, an electron with a broad wave-like packet, of definite momentum and, therefore, of a definite wavelength. Such an electron is capable of demonstrating its wave-like properties when it interacts with a suitable measuring apparatus, such as a metal crystal. The same electron, however, is potentially capable of developing into something more like a particle when it interacts with a position-measuring device, at which time its wave-like aspects become correspondingly less important. But even while it is acting more like a particle, the electron is potentially capable of again developing its wave-like aspects at the expense of its particle-like aspects, if it is allowed to interact with a momentum-measuring device. Thus, the electron is capable of undergoing continual transformation from wave-like to particle-like aspect, and vice versa. At any particular stage of its development, it may further transform, while keeping its same general aspect; or it may emphasize the opposite aspect instead. The kind of apparatus with which the electron interacts determines which of these potential aspects prevails.

The quantum properties of the electron differ from those described in classical theory not only in that they are latent potentialities, but also in that these potentialities refer to developments, the precise outcome of which is not related completely deterministically to the state of the electron before it interacts with the apparatus. Consider, for example, a process in which an electron having, initially, a broad wave-like packet interacts with a device that can be used to measure its position. After interaction has taken place, the wave function is broken up into independent packets with no definite phase relations between them, each having a size of the order of magnitude of the error Δx in the measurement. But as we have seen, the electron exists in only *one* of these packets, and the wave function represents only the *probability* that a given packet is the correct one. This means that although the general direction of the development of the particle-like aspects of the electron is determined by the state of the system before interaction, the exact value of the position that will develop is not completely determined. Instead, there will be a range corresponding to the initial spread of the wave packet, over which the resulting position will fluctuate at random when the experiment is repeated many times and initial conditions are reproduced as accurately as the quantum nature of matter will allow (i.e., within the limits of precision set by the uncertainty principle).

The foregoing interpretation of the properties of the electron as incompletely defined potentialities finds its mathematical reflection in the fact that the wave function does not completely determine its own interpretation. Thus, before the electron has interacted with a measuring apparatus, the wave function defines two important kinds of probability; namely, the probability of a given position and the probability of a given momentum. But the wave function by itself does not tell us which of these two mutually incompatible probability functions is the appropriate one. This question can be answered only when we specify whether the electron interacts with a position-measuring device or with a momentum-measuring device. We conclude that, although the wave function certainly contains the most complete possible description of the electron that can be obtained by referring to variables belonging to the electron alone, this description is incapable of defining the general form (wave or particle) in which the electron will manifest itself. We are, therefore, again led to interpret momentum and position (and thus wave and particle aspects) as incompletely defined potentialities latent in the electron and brought out more fully only by interaction with a suitable measuring apparatus.

10. Inclusion of More General Interactions. Thus far, we have restricted ourselves to the consideration of interactions between an electron and a measuring apparatus. We shall see, however, in Chap. 22, Sec. 13, that transformations between wave and particle aspects of matter similar to those discussed in Sec. 9 can be brought about not only through interaction with a measuring apparatus, but also through interaction with any material system, whether it is part of a measuring apparatus or not.

This result is to be expected, because a measuring apparatus is nothing more than ordinary matter, so arranged that the results of interaction with the system of interest are subject to a comparatively simple and direct interpretation. If, for example, an electron is transformed into a wave-like object when it interacts with a metal crystal inside a piece of laboratory apparatus, it will also do the same if it interacts with a similar crystal at the bottom of the sea, or in interstellar space. Similarly, if an electron is transformed into a particle-like object when it interacts with a quantum of short wavelength, which happens to be associated with a microscope, it will react similarly if the quantum is generated in a spontaneous process without the intervention of any human being.

11. On the Reality of the Wave Properties of Matter. The ideas to which we have come in this chapter imply that the wave aspects of matter are just as real as the particle aspects. But we are so used to thinking in classical terms that we have an almost irresistible tendency to revert to making the implicit assumption that the electron is really a particle having a definite momentum and position that cannot be measured simultaneously. We tend to deemphasize the physical reality of the wave aspects, which show up in the importance of the phase relations in determining interference.

Because this classical concept is persistent, we shall now present some additional evidence to show that it leads to serious inconsistencies. Perhaps the most consistent formulation of such an idea is as follows: The electron is to be thought of as occupying a definite position and having a definite momentum that cannot be measured simultaneously with accuracies greater than those permitted by the uncertainty principle. The energy levels of atoms are to be explained by the Bohr-Sommerfeld theory, or perhaps by some refinement of this theory that may conceivably give better agreement with experiment. Electron diffraction is to be explained by arguments like those of Duane (Chap. 3, Sec. 12,) in which the appearance of definite angles is to be regarded as the result of the quantization of momentum transfers between electron and grating. Although Duane worked out his arguments only for a periodic structure, such as a grating, there are various ways in which we might conceivably extend this method to aperiodic structures, such as the two-slit system and the electron lens.

We shall not discuss any of these concepts in detail, but merely wish to point out that plausible theories can be worked out in which the allowed momentum transfers depend on the size and shape of the system, the number of holes, and so on, and in this way obtain effects that seem very much like those obtained from the wave theory of electron diffraction.

In any electron or photon diffraction experiment (as pointed out in Sec. 1), electrons or photons may be sent in one at a time, separated by such long intervals that they cannot possibly affect each other. To explain the resulting statistical interference patterns known to be produced after many particles have arrived at the detector, in terms of the model of a particle of uncertain position and momentum, we can assume that the probable range of angles of deflection of the particle is determined by certain restrictions (such as those suggested by Duane) on the quantized momentum transfers between particle and slit system. Such restrictions can depend at most, however, on the size and shape of the apertures and on the *actual* position and velocity with which the particle enters the system.

On the basis of this assumption, let us consider an experiment in which we observe the position of electrons with the aid of a proton microscope (see Chap. 5, Sec. 12 for details). We assume that the electrons are initially at rest, with a very well-defined momentum, and that there is a parallel beam of protons of well-defined momentum p incident on the microscope, as shown in Fig. 5. The position of the electron is made visible by the fact that it can scatter a proton, which then arrives at a different part of the image.

Now, if the protons are simply particles, the range of uncontrollable quantum deflections that they might obtain from the lens edge must be determined only by the size and shape of the lens and by the position and velocity of the proton as it enters the lens. But our general experience with diffraction of particles (such as photons and electrons) shows that the detailed nature of this phenomenon does not depend critically on the position and velocity of the particle. In other words, more or less the same *range* of momentum transfers can take place regardless of the direction of, or point of, origin of the particles. This is proved, for example, by the fact that the observed resolving power of an electron or photon lens is not strongly dependent on the direction in which the particles come through, or on the position of the object that is being viewed. Thus, we can say that the range of uncontrollable deflections should be determined mainly by the size and shape of the lens. Since this theory can be made to lead to more or less the same results as the wave theory, we conclude that the resolving power of the lens ought to be of the order of $\lambda/\sin\phi_0$, where ϕ_0 is the aperture of the lens, and λ is the de Broglie wavelength for the proton. If we choose $\phi_0 = \pi/2$, we obtain for the uncertainty in the position of the electron

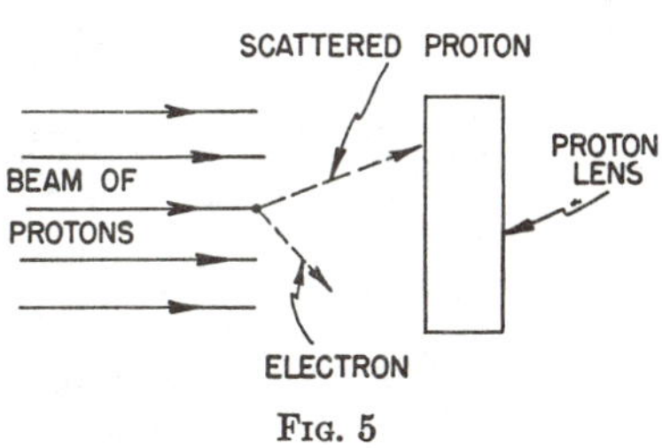

Fig. 5

$$\Delta x \cong \lambda \cong \frac{h}{p}$$

But because of conservation of momentum, we know that the electron cannot obtain from the proton a momentum larger than a value of the order of $\frac{m}{M}p$, where m is the electron mass, and M is the proton mass. Since the initial momentum of the electron was known to high accuracy, we obtain

$$\Delta p \cong \frac{m}{M}p \qquad \text{and} \qquad \Delta x\,\Delta p \cong \frac{m}{M}h$$

which is much less than the minimum permitted by the uncertainty principle. But we have previously seen in Chap. 5, Sec. 16 that a contradiction of the uncertainty principle at any point would make the entire wave-particle duality untenable. In this case, for example, we could define the momentum of the electron more accurately than the wave vector $\boldsymbol{k}$ is defined in its wave packet and thus contradict the de Broglie relation.

By using the treatment given in Chap. 5, Sec. 12 for a similar problem, this difficulty can be avoided, but only by making the assumption that, between the time it was scattered and the time it arrived at the detecting plate, the proton acted in every respect like a wave that originated at the point where it was scattered. Because of the small range of momenta that can be transmitted to it $\left(\Delta p \cong \frac{m}{M} p\right)$, the proton wave must have within it a correspondingly small range of wave vectors, so that it acts as a narrow pencil of rays does in optics.* Thus, the resolving power is not determined by the size of the lens, but by the unavoidable diffraction resulting from the small angular width of the pencil of rays

$$\Delta\theta \cong \frac{\Delta p}{p} \cong \frac{m}{M}$$

The resolving power of the lens is, therefore, only

$$\Delta x \cong \frac{\lambda}{\Delta\theta} \cong \frac{h}{p}\frac{M}{m}$$

Thus, we obtain $\Delta x\, \Delta p \cong h$, in agreement with the uncertainty principle.

To retain the model of a particle of uncertain position and momentum, we should have had to assume that the permissible range of momentum transfers from particle to lens was determined (but in a reciprocal way) by the range of momentum transfers from electron to proton. Thus, as it interacted with the lens, the proton would have to have some kind of "memory" that it had last interacted with an electron. If it had last interacted with a proton, its behavior would have been different.

It is clear that, to retain the concept of a particle in this experiment, we must adopt complicated, artificial, and implausible assumptions. That such assumptions can be made self-consistent is doubtful, more so that they can be made consistent with all known data concerning properties of matter. On the other hand, the same wave theory that explains so many other facts correctly can also deal with this problem in a simple and natural way without leading into any inconsistencies. We conclude, therefore, that the wave aspects of matter are as real as are the particle aspects and that, to obtain a complete and consistent theory, we must consider both aspects, each under its proper conditions. Thus, the conclusion reached in Sec. 1, that the individual electron must be regarded as having some wave-like properties, is given a more complete justification. (A qualitative picture of the connection between wave and particle is given in Sec. 12.)

* This behavior is an example of transformation between wave and particle aspects of matter. Thus, in its progress from the point of scattering to the point of detection, the proton acts like a wave; but in its interaction with the screen, it is transformed into a particle-like object.

12. Wave-mechanical Interpretation of a Track in a Cloud Chamber. It is of interest to apply our picture of transformations between wave and particle aspects of matter to show how a typical experimental situation is described, such as the detection of the track of a particle by means of a cloud chamber. (This problem has already been discussed to some extent in Sec. 2.) As the particle passes gas atoms, it excites (or ionizes) them, leaving a track of excited atoms and ions in the path of its trajectory. When the gas is expanded, the ions serve as condensing points for water droplets that make the track of the particle visible.

How can this process be understood on the basis of the wave theory? We use the fact that, if an atom is excited, or ionized, it is because the charged particle passed nearby and transferred a quantum of energy to the atom. Since ionization is very unlikely when the charged particle passes at a distance of more than a few atomic diameters from the atom, we conclude that an observation of the droplet resulting from the ion can serve, in principle, to localize the path taken by the particle within an accuracy of the order of a few atomic diameters. At the pressures used, a particle is practically certain to encounter an atom within a very short distance, say 10^{-5} cm. Thus, when the electron wave packet enters the chamber, it is quickly broken up into independent packets with no definite phase relations between them, each of the order of a few atomic diameters in size. As shown in Secs. 3 and 5, the electron exists in only *one* of these packets, and the wave function represents only the *probability* that any given packet is the correct one. Each of these packets can then serve as a possible starting point for a new trajectory, but each of these starting points must be considered as a separate and distinct possibility, which, if realized, excludes all others.

If the original momentum of the particle was very high, the uncertainty in momentum introduced as a result of the interaction with the atom results in only a small deflection, so that the noninterfering packets all travel with almost the same speed and direction as that of the incident particle. As each packet moves, it starts to spread, and the wave-like aspects of the electron begin to develop at the expense of the particle-like aspects. Before the packet can spread very far, however, it arrives near another atom, and once again it is broken up into noninterfering packets, each of which represents a distinct and separate possible location of a particle-like object, excluding all others. The process of continual interaction with gas atoms, therefore, prevents the appreciable development of the wave-like aspects of the incident "particle."

The actual trajectory, which is followed by watching the tracks of the ions, resembles Fig. 6. There will be many minute deflections occurring each time a packet arrives near an atom. These deflections are interpreted as the scattering of the particle by the atom. Since we cannot predict exactly where the packet strikes an atom or exactly how much momentum is transferred, the exact shape of the path cannot be predicted. But as long as the particle speed is high enough, large deflections are unlikely, and the path remains close to a straight line. Otherwise, it may deviate appreciably in a random way.

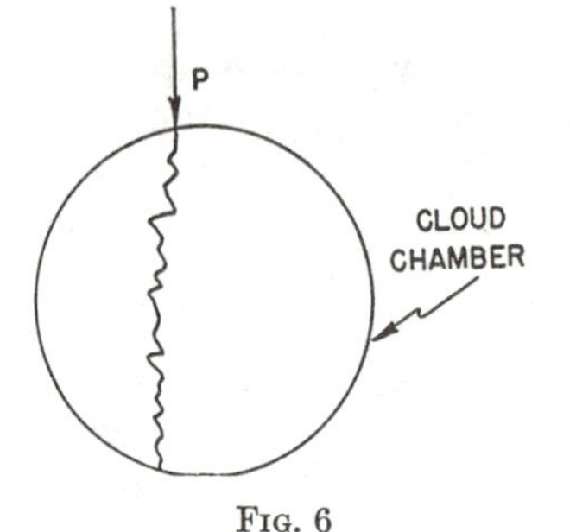

Fig. 6

If we compare this description with the classical description, using the idea that the chamber is struck by a particle, we obtain essentially the same result. Even classically, a series of deflections is expected, the size and distribution of which are determined exactly, in principle, by the location of each atom relative to the impinging particle. Since, in practice, we do not control this quantity, we obtain a series of random deflections.

Thus, we can understand the observed tracks of particles in a cloud chamber in terms of the wave function and its probability interpretation. All evidence for the particle nature of matter comes from experiments of this general type, in which the orbit is traced by a series of position measurements. But we have seen that the quantum theory predicts a particle-like behavior of the electron when it is treated in this way, because the continual interaction with position-measuring devices (in this case, the atoms that later serve as nuclei for water droplets) prevents the development of the wave-like aspects. To the extent that the measurements of position do not appreciably change the momentum, the system acts as if it had a continuous and fairly well-defined trajectory. On the other hand, as we approach the quantum level of accuracy, the uncontrollable deflections that occur in the interactions with gas atoms will prevent us from inferring a continuous and causally determined particle trajectory. If, for example, we were interested in a precise description of the motion of each electron while it scattered from a gas atom, the wave aspects of the electron would become significant.*

We conclude that the quantum concept of transformation between wave and particle aspects of matter is capable of explaining all forms of behavior manifested by matter.

13. Qualitative Picture of the Quantum Properties of Matter. We shall now summarize the material in this chapter by giving a preliminary qualitative picture of the quantum nature of matter.

The most important new concept to which we are led is that any given piece of matter (for instance, an electron) is not completely identical with either a particle or a wave but that, instead, it is something potentially capable of developing either one of these aspects of its behavior at the expense of the other. Which of the electron's opposing potentiali-

* See, for example, Chap. 21.

ties will actually be realized in a given case depends as much on the nature of the systems with which the electron interacts as on the electron itself. Because the electron continually interacts with many different kinds of systems, each of which develops different potentialities, the electron will undergo continual transformations between its different possible forms of behavior (i.e., wave or particle).*

The precise outcome of these transformations is not, however, completely deterministically related to the state of the system before the interaction takes place, but only statistically. In the classical limit, of course, the effects of these transformations can be ignored, because the wave-like properties of the electron produce a negligible effect. Similarly, for electromagnetic fields, the particle-like properties can likewise be ignored in the classical limit. In this way, we explain the fact that, classically, each type of system appears to take on a fixed "intrinsic" character (i.e., either always a particle or always a wave).

We must be careful not to imply that the electron is a complex object made up of many parts that simply rearrange themselves in response to the various forces that exist in the environment and thus transform from a wave-like to a particle-like object. Such a picture would be essentially equivalent to the assumption of hidden variables (in this case, the positions of the various "parts"), which really determine what the electron as a whole will do. But as we have already seen,† such an assumption of hidden variables cannot be made consistent with the present formulation of quantum theory. The transformations of the electron from wave-like to particle-like object, and vice versa, which are implied by the quantum theory refer to fundamental but not further analyzable changes in what would, in a classical theory, be called the "intrinsic" nature of the electron. In fact, quantum theory requires us to give up the idea that the electron, or any other object has, by itself, any intrinsic properties at all. Instead, each object should be regarded as something containing only incompletely defined potentialities that are developed when the object interacts with an appropriate system.

The conclusions of the previous paragraph contradict an assumption that has long been implicit in physics as well as in most other branches of science; namely, that the universe can correctly be regarded as made up of distinct and separate parts that work together according to exact causal laws to form the whole. In the quantum theory, we have seen that none of the properties of these "parts" can be defined, except in interaction with other parts and that, moreover, different kinds of interactions bring about the development of different kinds of "intrinsic"

* In this connection, let us recall the result of Sec. 10; namely, that transformations between wave and particle aspects of matter are not restricted to interactions with measuring apparatus but take place in interactions with all matter.

† Chap. 5, Secs. 3 and 17.

properties of the so-called "parts." It seems necessary, therefore, to give up the idea that the world can correctly be analyzed into distinct parts, and to replace it with the assumption that the entire universe is basically a single, indivisible unit. Only in the classical limit can the description in terms of component parts be correctly applied without reservations. Wherever quantum phenomena play a significant role, we shall find that the apparent parts can change in a fundamental way with the passage of time, because of the underlying indivisible connections between them. Thus, we are led to picture the world as an indivisible, but flexible and ever changing, unit.

We shall develop the further implications of the preceding qualitative description of the quantum nature of matter in Chaps. 8 and 22. It is suggested that the reader return to this chapter after reading Chaps. 8 and 22, to obtain a better understanding of the concepts involved.

CHAPTER 7

Summary of Quantum Concepts Introduced

STARTING WITH PLANCK'S HYPOTHESIS that the radiation oscillators are restricted to discrete energy states with $E = nh\nu$, we have come a long way. Although this hypothesis is totally at variance with all of classical physics, it gives a quantitative prediction of the electromagnetic spectrum emitted by a blackbody, which is in perfect agreement with experiments at all temperatures and frequencies that have ever been measured. Simply because radiation oscillators are in thermodynamic equilibrium with material oscillators in the walls, we are led now to expect that the material oscillators have their energies quantized in the same way. Einstein and Debye obtained a quantitative fit with the specific heats of a wide range of solids when they applied this idea to the vibrations of the atoms composing the solid.

The next step was to apply the idea of quantization of the energy of the radiation oscillators to exchanges of energy between electromagnetic fields and charged particles, such as electrons. This idea requires that these exchanges occur in quanta with $E = h\nu$. This is exactly what is observed in the photoelectric effect. On the other hand classical theory, which predicts a gradual transfer of energy, is clearly shown to be wrong. The fact that the energy of the oscillators is restricted to multiples of a definite unit suggests that the process of transferring energy to an electron is indivisible, since if it were not, there would be an intermediate state in which the oscillator had part of a quantum. Many experiments, in connection with the photoelectric effect, Compton effect, and other problems, have verified the indivisibility of all quantum processes.

Because the energy of a radiation oscillator can be transferred to an electron only in quanta, the electromagnetic wave takes on many particle-like properties. In particular, the sudden appearance of all the energy at one point suggests that light is made up of particles. Yet, even when only a single photon is present, light demonstrates certain interference properties which suggest equally strongly that it is a wave. Here we meet, for the first time, the wave-particle duality, characteristic of all material systems in quantum theory. Under some circumstances light shows interference properties and acts very much like a wave, while in other circumstances it acts like a particle. The two aspects are related by the fact that the wave intensity determines the probability that all the energy appears at one point, as if the system were a particle. Thus.

we see how probability, the wave-particle duality, and the indivisibility of quantum transfers are all related.

We then asked how it came about that, on a macroscopic scale, the continuous deterministic laws of classical physics seemed to be true, whereas on a fundamental level the basic processes were discontinuous and had only their probabilities determined. The answer lies in the correspondence principle, which is based on the idea that, first, the discontinuities are too small to be seen on a classical level and, second, that so many quantum processes take place in any classical process that the deviation of the actual result from the statistical average is negligible. In order that quantum laws lead to the correct classical laws, it is necessary to limit severely both the possible spacings of the quantum states and the probabilities of quantum processes. For example, with the aid of the correspondence principle, we derived the Bohr-Sommerfeld conditions for the quantization of action. These conditions then led to a large number of predictions that are in accord with experiment. In a similar way, the probability of radiation could be predicted roughly from the requirement that it must yield the correct classical rate of radiation in the correspondence limit.

The theory still suffered from three defects, however. First, it applied only to periodic motions; second, it gave no account of what happens in the transition from one energy level to another; and finally, it was unable to deal with complex atoms. All these defects were eventually removed by the wave theory of de Broglie and Schrödinger. The quantization of action follows naturally from boundary conditions on the waves in periodic systems; yet aperiodic systems are described by wave packets that move with an average velocity equal to that of the classical particles in their trajectories. The transitions between orbits are described as the gradual flow of the wave from one orbit to the next. Later, we shall see that with Schrödinger's equations we can treat all systems, however complex, and get quantitatively correct results in a tremendous number of applications, covering a range of fields from spectroscopy, chemistry and the theory of solids, to electrical conductivity, x rays, and the theory of atomic structure. In all these fields classical physics fails, and quantum physics gives the right results.

Although the wave theory is successful, it raises some paradoxes of its own. The wave theory successfully explains interference effects, such as those observed in the Davisson-Germer experiment. Yet, even after an electron, or an electromagnetic wave, has clearly suffered diffraction, it is always possible to find the electron or photon at some definite position. In a similar way, although the wave flows gradually from one orbit to the next, the process of energy transfer is still indivisible, because experiment shows that the atom either absorbs the full energy of a quantum, or absorbs none. If we recall that the quantum laws are known to be laws of probability only, the most natural interpretation of these results is that the wave yields the probability of finding a particle in a given region. In a similar way, we can show that the intensity of a given Fourier component $|\varphi(k)|^2$ yields only the probability that the momentum has the value $p = \hbar k$. This means that matter can exhibit, under different conditions, either wavelike or particle-like behavior, i.e., it shows a wave-particle duality in its properties.

The linkage between wave intensity and probability, on the one hand, combined with the linkage between wavelength and momentum, on the other hand, leads to the uncertainty principle, which is one of the most important results of the wave-particle duality. This uncertainty principle provides a more precise limitation on the applicability of the concepts of classical determinism than we were able to get without the aid of the wave picture. Finally, we saw that in any actual measurement process, there was always one stage where the uncontrollable and unpredictable transfer of an indivisible quantum intervened to prevent us from drawing inferences about the system under observation that were more accurate than the limitations given by the uncertainty principle. Because the uncertainty principle, which is predicted from the wave-particle duality, requires for its verification the elements of indivisibility of quantum transfers, and incomplete predictability of when and where the transfer takes place, we conclude that all three elements must be included to obtain a consistent quantum theory. Thus, the quantum theory possesses a very complete internal unity, such that each part works together with the others in an interlocking way, and such that the whole theory would fail unless each part were present.

Finally, we saw that even the wave function undergoes indivisible and uncontrollable changes when the object under observation interacts with a measuring apparatus. This behavior of the wave function leads to qualitative description of the properties of matter in terms of incompletely defined and mutually incompatible potentialities, which can be realized more fully only in interaction with a suitable system in the environment. For example, whether an electron shows more wave-like or more particle-like properties depends on whether it interacts with something that tends to bring out its wave-like or particle-like aspects. We are thus led to regard matter as something more fluid and dependent on the environment than classical physics would lead us to suppose. The further consequences of this new concept of matter will be discussed in Chaps. 8 and 22.

CHAPTER 8

An Attempt to Build a Physical Picture of the Quantum Nature of Matter

1. The Need for New Concepts.* We have traced, step by step, the chain of reasoning leading from classical to quantum physics. In so doing, we have obtained a theory that is in excellent quantitative agreement with a wide range of experiments, the results of which contradict even the qualitative predictions of classical theory. The new theory to which we are led represents, however, not only a far-reaching change in the content of scientific knowledge, but also an even more radical change in the fundamental concepts, in terms of which such knowledge is to be expressed. The three principal changes in these concepts are:

(1) Replacement of the notion of continuous trajectory by that of indivisible transitions.

(2) Replacement of the concept of complete determinism by that of causality as a statistical trend.

(3) Replacement of the assumption that the world can be analyzed correctly into distinct parts, each having a fixed "intrinsic" nature (for instance, wave or particle), by the idea that the world is an indivisible whole in which parts appear as abstractions or approximations, valid only in the classical limit.

Because most of our experience has been gained, thus far, in connection with phenomena that are described to an adequate degree of approximation by classical concepts, the new quantum concepts are unfamiliar. In this chapter, we shall try to make these new concepts more familiar and show that they are, basically at least, as reasonable as are those of classical theory. Our procedure will be, to provide a critical discussion of the classical concepts of continuity and complete determinism, in order to show that there is no *a priori* logical reason for their adoption. We shall also show that the quantum concepts of indivisible transitions and incomplete determinism are not only just as self-consistent from a logical standpoint, but also much more analogous to certain naive concepts that arise in many phases of common experience. We shall then be led to Bohr's principle of complementarity, which was the first qualitative state-

* Many of the ideas appearing in this chapter are an elaboration of material appearing in a series of lectures by Niels Bohr. (See N. Bohr, *Atomic Theory and Description of Nature.*)

ment of the new concepts needed for understanding the quantum properties of matter. After that, we shall discuss critically the classical concepts of analysis of a system into its component parts and the synthesis of these parts according to exact causal laws, in order to show that such procedures fail in the quantum domain. Thus, we shall be led to picture the world as an indivisible whole. Finally, we shall discuss certain analogies to quantum concepts that should help the reader understand in a more imaginative way some of the implications of the quantum theory.

2. Discussion of the Concept of Continuity. Let us begin with the problem of continuity of motion of particles in classical physics. The basic variables, describing a classical elementary particle, are its position and velocity (or momentum), both of which are assumed at each instant to have definite values varying continuously with the passage of time. We shall start by considering only our simplest ideas about these things, and later go on to the more sophisticated theories of continuity and the use of derivatives to describe the velocity of a particle.

3. Simple and Pictorial Ideas about Continuity of Motion. Our simplest ideas about the position of an object seem to imply that an object with a definite position is not moving. That is, if we try to picture the position of an object with *perfect* accuracy, we seem to imagine an object that is at one fixed position and at no other. We can try to represent motion as a succession of objects at slightly different positions, as is done in a motion picture, but a succession of fixed positions does not include all the properties that are usually associated with motion. In particular, it does not seem to include the idea that a real moving object is *continuously* covering space as time passes. To picture the actual process of motion taking place in a continuous manner, we must imagine an object that is covering some space during some interval of time. We may reduce the element of time to a very small value and thus reduce the indefiniteness of position to a correspondingly small value. But we cannot reduce the indefiniteness to zero and still obtain a picture of a moving object, for in picturing an object at an absolutely definite point in space we cannot seem to help picturing it as fixed. In other words, we cannot think of the position of an object and of its velocity simultaneously.

One might argue that we can define a continuous trajectory of a moving object and that at each instant of time it has some definite position. Whether this is true or not will be discussed later, but for the present we point out that this procedure merely specifies some of the results of motion after it has taken place. A picture of the object in the process of motion is not given. To obtain such a picture, we must allow our view of the position to blur slightly. For example, a blurred photograph of a speeding car suggests to us that the car is moving, because it implies the continuous covering of space during a period of time. On the other hand a sharp picture of a moving car, taken with a very fast camera, does not suggest motion. This residue of indefiniteness in our picture of a moving object suggests that we really think of such an object as being in a state of transition from one position to the next. When it is in this state of transition, our picture does not tell us quite where it is but shows us, instead, how its average position is changing with time. In this very concept of motion, we seem to have to include the idea that a continuously moving object has a somewhat indefinite range of positions.

4. Similarity of Simple Ideas of Motion and Quantum Concepts. The simple picture of motion has many points of similarity to the one suggested by quantum theory, although the two are, of course, not exactly the same. According to quantum theory, momentum, and hence velocity, can be given an exact meaning only when there is provision for a wave-like structure in space. When this provision is made, the wave packet* is in a state of transition through space, in which its average position moves from one point to the next at a fairly definite velocity. But the motion of the wave packet is analogous to our simple picture of a particle in motion because, in both, the particle is thought of as covering a range of positions at any instant, while the average position changes uniformly with time. Quantum theory, therefore, gives a picture of the process of motion that is considerably closer to our simplest concepts than does classical theory. We cannot visualize simultaneously a particle having a definite momentum and position. Quantum theory has shown that it is unnecessary to try, because such particles do not exist.

5. Similarity of Simple Ideas about Fixed Position and Quantum Concepts. In what way does our naive idea of motion seem to differ from the quantum picture? According to our simplest ideas, it is possible to have an object at rest, in a fixed position, whereas the uncertainty principle tells us that an object in a well-defined position has a highly indefinite momentum. Careful study shows, however, that this view is actually in close accord with our simple picture. What we can say is that, if we think of an object in a given position, we simply cannot think of its velocity at the same time.

Thus, if we forego the possibility of thinking of the motion of an object in a continuous fashion, we can begin by picturing it in a definite position, and then we can imagine that at a short, but finite, time later it is somewhere else. Where else, we cannot infer from our original picture; any position is equally consistent with it. This means that any velocity is also equally consistent with our picture of a particle in a definite position. This idea is remarkably close to the quantum theoretical description. The more precisely we define the wave packet, the more rapidly it spreads, and the less able are we to give an approximately continuous description of the motion. We conclude that our naive pictures and quantum theory are alike in that they both have the follow-

* See Chap. 3, Sec. 2 for a definition of a wave packet.

ing property: It is possible to give a continuous picture of the motion only if the position is blurred or made indefinite, and it is possible to give a picture of a particle in a definite position only if we forego the possibility of picturing it in continuous motion.

Although our naive pictures and the quantum theoretical description resemble each other closely, it must not be inferred that quantum theory is *identical* with the naive picture. We merely point out the close functional analogy in the way in which the two pictures deal with the problem of motion.

6. More Sophisticated Ideas, Including Concept of Continuous Trajectory. The objection might arise that our simple ideas are too naive to be taken seriously. Let us instead be satisfied with a description in terms of a continuous trajectory, for which the co-ordinate at each instant of time can be defined as accurately as we wish. We cannot picture the process of motion directly; we do not have to because the concept of the derivative can be used. To do this, we consider a small interval of time Δt and specify the distance Δx moved during this time. We then obtain the average velocity in this interval: $V_{av} = \Delta x/\Delta t$. If Δt is allowed to approach zero, and if the function describing the trajectory is smooth enough, V_{av} will approach a definite limit, which we define as the velocity at a definite point. Thus, although we are unable to imagine this velocity directly by some sort of mental picture, we can instead use the mathematical definition.

The question of whether this limit always exists is one that mathematicians have studied extensively. The limit certainly exists for common functions like $\sin \omega t$, but mathematicians can easily define functions that are everywhere discontinuous and that have no derivatives at any point. Consider, for example, a function that has a value of zero whenever the independent variable is a rational number, but unity when it is an irrational number. This function is completely discontinuous. In physics texts, however, no attention is given to such functions, because it is tacitly assumed that all functions describing the motion of actual material particles are continuous and differentiable. This is done, essentially, because it seems to be the most natural thing to do. But why does continuous motion seem so natural to us? Actually, many of the ancient Greeks were unable to grasp the idea of continuous motion, as those who have studied Zeno's paradoxes will know. One of the most famous of these paradoxes concerns an arrow in flight. Since at each instant of time the arrow is occupying a definite position, it cannot at the same time be moving. Zeno, therefore, concluded that motion is in some manner illusory. Many of the early Greek philosophers were unable to convince themselves that continuous motion is really such a natural thing.

Between ancient times and now our ideas of continuous motion developed through experience with planetary orbits, gun trajectories, etc., and by the associated theory of the differential calculus dealing with them. After studying such things for a while, succeeding generations gradually began to take the basic ideas for granted. But the only way to know whether the motion of particles can actually be described by functions with derivatives is to test the assumption by experiment. In other words, the classical idea that a particle has a trajectory for which the derivative can be defined at each point is based only on empirical evidence. Since the days of Newton, the great success of classical theory provided strong empirical evidence and it seemed inevitable and in the nature of things that a continuous trajectory was the only conceivable kind that real matter could follow. Yet, on a purely logical basis, there is no reason to choose the concept of a continuous trajectory in preference to that of a discontinuous trajectory. It is quite possible that $\Delta x/\Delta t$ shall approach a limit for a while as Δt is made smaller, and then cease to approach a limit as Δt is made smaller still. We may consider, for example, the experiment of measuring Δx for a real object, using a smaller and smaller Δt. For a while, this procedure gives us increasingly accurate information about the velocity. But eventually we reach intervals of time so small that the Brownian movement becomes important, and $\Delta x/\Delta t$ ceases to approach a definite limit. We could argue that this difficulty can be avoided by treating the motion of the individual molecules. But the experiments leading to the quantum theory have shown that if Δt is made too small, this effort will also fail. We conclude that, in a very accurate description, the concept of a continuous trajectory does not apply to the motion of real particles.*

7. Cause and Effect. Having seen from the previous discussion that the discontinuous aspects of quantum processes are not basically unreasonable, let us now consider the lack of complete determinism. The problems of determinism and causality have occupied a central position in all philosophical discussions since the time when man first tried to obtain a more complete general understanding of the world than is afforded by deductions from immediate experience. We shall, therefore, begin with a brief account of the kinds of ideas on causality that man has held and shall show on what basis the modern ideas on this subject have been developed.

8. Early Ideas on Cause and Effect. Some of the most primitive ideas on cause and effect probably arose when man noticed that, by exerting various forces on his surroundings and by doing work, he could produce desired effects or avoid undesirable effects. The most primitive notion of causality is, therefore, closely connected with the mechanical concepts of force and work. It is certainly true that human beings

* A further discussion of the relationship between the continuous and discontinuous aspects of the motion of matter is given in Chap. 22, Sec. 14. See also the discussion of the principle of complementarity in Chap. 8, Sec. 15.

can produce effects on other material systems only by exerting forces and doing work.

Later, the idea arose that the human body is made of the same kind of matter as inanimate objects. Perhaps, it was reasoned, such objects can also push on each other in the same way that people can. In this way, one could be led to the concept of inanimate causes. In this connection, we should note that the ultimate effect is not always in proportion to the cause. There are unstable systems; a boulder poised on the side of a hill can be made to produce tremendous effects as a result of comparatively small, causal forces.

Along with the idea of material forces as causes, there probably arose the idea of magic (which is essentially the production of effects without the intervention of material forces and the doing of work). This idea is both self-consistent and attractive, but experience has shown that such magical causes do not operate. Hence, this type of law of cause and effect has been discarded.

It is easy to guess where the concept of magic probably originated. Man discovered that he could affect others not only through material forces, but also by words and signs that did not seem to require such forces. He naturally extended this idea to inanimate objects and assumed that suitable magic words and signs would produce effects similar to those produced on people. This idea may be summarized generally in terms of the use of words, signs, symbols, and ideas as direct causes of events. Since then, we have found that sound and light exert material forces, and we now have the unified point of view that only through material forces can effects be produced on other objects whether they are alive or inanimate. In this connection, it must be remembered that men act like very unstable systems, so that the comparatively small forces involved in sound waves and light can produce big effects. By now we have such devices as photoelectric cells and microphones that, with the aid of vacuum-tube relays, can also respond to the small forces in sound and light and, eventually, produce large effects.

Another type of causal law put forward by men in early times was the teleological notion of "final causes," which conceives of events as being governed not so much by previous conditions or events as by a final goal toward which the whole universe is striving. This idea of cause was quite probably an extrapolation to all material systems of the feeling of purpose, which certainly plays an important role in governing the actions of human beings. This extrapolation, however, like that of signs and symbols as causes, has never obtained any reliable experimental verification, so that we do not now ascribe purposes to inanimate objects, except perhaps for the sake of a metaphor.

The conclusion to be drawn from this discussion is that our ideas on cause and effect, along with all the rest of our ideas, probably originated in the extrapolation of man's most immediate experiences to wider classes of phenomena. Of the many types of causal laws suggested in this way, only the concept of material forces as causes has thus far survived the test of general experience. We must remember, however, that the precise form that this concept has taken has been determined by long experience with classically describable systems and that, if reliable experiments in the quantum domain indicate the need for further changes in these laws, there is no fundamental reason why such changes should not be made.

9. Completely Deterministic vs. Causal Laws as Tendencies. At this point, we wish to call attention to the fact that, even in very early times, two alternative general types of causal laws appeared. One of these involved the notion of complete determinism; the other involved the notion of causes as determining general tendencies but not determining the behavior of a system completely. One of the earliest examples of complete determinism is the idea that the whole course of events is determined by fate—in a way that is beyond the power of man to change. The origin of such ideas cannot be definitely fixed, but it is not unlikely that they were rooted partially in the extent to which men felt themselves at the mercy of the forces of nature, that seemed far beyond the power of human control. Thus, we have the poetic simile of human life as a ship tossed about by the wind and waves.

Although some of the ancient philosophers did develop such ideas systematically, it is doubtful that the notion of complete determinism ever permeated very far into practical life. Instead, the idea most likely to have been used in connection with common experience is that a particular force or cause produces a tendency toward an effect, but that it does not guarantee the effect. At that time, work was done mostly by hand or with the aid of animals. The control of forces is not exact by these methods. To obtain the desired effects, one pushed in the right general direction, and pushed backward if one had gone too far. The force was used to produce a general tendency toward motion in a certain direction, without too precise an interest in what the results of these forces were. Before the advent of machinery, virtually all activities were of this general nature, involving at most the use of judgment and art rather than the precise control of motion.

It is very likely that the modern form of the idea of complete determinism was suggested, in part at least, by its resemblance to complex and precisely constructed machines, such as clocks. With the advent of astronomy and ballistics, and mechanics generally, where systems were obtained in which the workings of the causal laws could be traced in detail, the idea of exact causality, or complete determinism, began to grow rapidly and, with Newton's laws of motion, the idea obtained an exact and quantitative expression. Later it became a matter of common experience to deal with rapidly moving machinery, where a precise determination of the motions of all the parts was essential. As a result, practically everyone in the field was willing to admit that, on the atomic level, all processes taking place in the world might be understood in terms of such mechanical analogies and thus could be thought of as completely deterministic. At this stage of history (16th through 19th centuries) the view that the world resembled one huge machine replaced the earlier view that many of the inanimate objects resembled men and animals.*

10. Classical Theory Prescriptive and not Causal. It is a curiously ironical development of history that, at the moment causal laws obtained an exact expression in the form of Newton's equations of motion, the idea of forces as causes of events became unnecessary and almost meaningless. The latter idea lost so much of its significance because *both the past and the future* of the entire system are determined completely by the equations of motion of all the particles, coupled with their positions and velocities at any one instant of time. Thus, we can no more say that the future is caused by the past than we can say that the past is caused by the future. Instead, we say that the motion of all particles in space time is prescribed by a set of rules, i.e., the differential equations of motion, which involve only these space-time motions alone. The space-time order of events is, therefore, determined for all time, but this determination is not conceived of as being the result of the operation of anything like the primitive animistic notion of "causes."

Of course, the notion of forces as causes can be retained; in fact, this procedure appears to be the most convenient one to use in practice. From a purely logical point of view, however, the concept of force is redundant, because it is always possible, in principle, to express all classical physics in terms of the positions, velocities, and accelerations of all the particles in the universe. Thus, the law of gravitation can be expressed as follows: Two bodies suffer a mutual acceleration, in the direction of the line joining them, which is inversely proportional to the square of the distance between them. The acceleration of each body is also directly proportional to the mass of the other. In a similar way, all other laws can be expressed without the aid of the force concept. The force exerted by a spring balance, for example, can, in principle, be expressed in terms of the space co-ordinates of all the molecules in the balance.

The principle of enonomy of concepts would then suggest that, except

* There is a striking contrast between the mechanical-causal description of the motion of planets, given by Newton, and that given by many philosophers of ancient and medieval times. The latter said that planets moved in circles. This attitude was based on the assumption that a circle is the only perfect geometrical figure and such celestial bodies as planets must surely move only in perfect orbits. Here, they used the concept of final causes in their striving for perfection. In the Newtonian description, however, one uses the analogy of a huge machine, inexorably required to carry out certain motions.

as a convenient term lumping together the effects of many accelerations, the concept of force as the cause of acceleration ought to be discarded and replaced by the idea that particles simply follow certain trajectories determined by the equations of motion. Thus, classical theory leads to a point of view that is prescriptive and not causal.*

11. New Properties of Quantum Concepts: Approximate and Statistical Causality. With the advent of quantum theory, the idea of complete determinism was shown to be wrong and was replaced by the idea that causes determine only a statistical trend, so that a given cause must be thought of as producing only a tendency toward an effect. Once again, quantum theory was a step in the direction of the less sophisticated ideas that arise in ordinary experience, where one seldom encounters an exact relation between cause and effect, but instead usually thinks of a cause as producing a qualitative tendency in a given direction.

The complete determinism of the classical theory arose from the fact that, once the initial positions and velocities of each particle in the universe were given, their subsequent behavior was determined for all time by Newton's equations of motion. But in quantum theory, Newton's law of motion cannot be applied in this way to an individual electron, because the momentum and position cannot even exist under conditions in which they are both simultaneously defined with perfect accuracy. Suppose, for example, that we wish to aim an electron at a given spot. To do this, it is necessary for us first to find out where the electron is now, and then to give it that momentum which *causes* it to move to the spot desired. Since the uncertainty principle indicates that this cannot be done, we conclude that the concept of exact determinism does not apply in the quantum theoretical description of the electron.

We find that, although there are no exact deterministic laws in quantum theory, there are still statistical laws. In a series of many observations we can, for example, measure the position reached by a particle after the passage of a given time Δt, if the initial conditions are reproduced as completely as the quantum nature of matter permits. This position fluctuates from one measurement to the next, but remains in the neighborhood of a mean value determined by the momentum† according to the rule

$$\overline{\Delta x} = \frac{p}{m}\,\Delta t$$

If an electron is aimed at a certain point by suitably controlling its momentum, we obtain a fairly reproducible pattern of hits near this point. *In order to change the position of the center of this pattern, it is necessary to change the momentum of the system.* But even if the momentum is pre-

* We shall see in Sec. 14 that quantum theory leads to a quite different description of the motion of matter.

† This equation holds only to the extent that the momentum is defined.

cisely defined, we cannot predict or control the exact point at which the electron will actually strike. Thus, in quantum theory, as in common experience with the nonmechanical aspects of life, only a statistical trend in the course of events is determined and not the precise outcome in each case.

12. Energy and Momentum in Classical and Quantum Theories. We now proceed to a more detailed description of how the concept of causality as a statistical trend is to be applied in quantum theory. As a preliminary step, we shall undertake to define more carefully what is meant by energy and momentum, first in classical theory and then in quantum theory. More exact definitions are necessary because these terms play a key role in the precise expression of the causal aspects of the behavior of matter.

In classical mechanics, the energy of a body is defined in terms of its ability to do work on other bodies. (*Work* is defined as the product of the force exerted by the one body on the other and the distance through which this force moves.) Since the preceding definition determines only changes of energy, the zero of energy can be chosen arbitrarily at a convenient point.

If an object is able to do work because of a change in its state of motion, it is said to possess *kinetic energy* ($T = mV^2/2$). If it can do work because of a change in position, it is said to possess *potential energy* $V(x)$. Actually, however, all energy is, in a sense, a latent or potential property of matter, since it represents a potential ability to do work, which is realized only when matter changes its state in interaction with other matter. Because of the existence of radiant energy, we must, however, generalize the above definition to include the fact that so-called "empty space" can also have a potential ability to do work by virtue of its ability to support an electromagnetic field. Finally, in the theory of relativity, we must include in addition to all other types of energy, the so-called "rest energy," which is the potential ability of matter to do work in a process in which it is annihilated. Thus, most generally, we define changes of energy of any system (whether ordinary matter, electromagnetic field, or anything else) as a potential ability to do work on another system by a process in which both systems interact and undergo corresponding changes of state.

We may ask why energy plays a more important role in mechanics than is played by other functions, such as mv^3 or arc sinh (mv). The reason is that the total energy of any isolated system is conserved, whereas, in general, no such conservation laws can be found for most other functions.* This fact suggests that energy corresponds to a real physical attribute of matter. Nevertheless, it would be wrong to think of it as a substance that is added to matter, like sugar to water, because

* Momentum and angular momentum, besides energy, are conserved.

energy, as such, is never found in isolated form. Instead, it is better to retain the concept that energy is a potential ability of some system (such as matter or the electromagnetic field) to do work. This potential ability can be transferred to other systems in the process of interaction, but the total quantity never changes.

The momentum of a single particle is defined as $p = mv$. Because the total momentum of an isolated system does not change, we are likewise led to regard momentum as a real physical attribute of matter (and of the electromagnetic field). In fact, we can make a complete parallel between momentum and energy: As the change of energy in a given change of state can be defined in terms of the potential ability of a body to do work, the change of momentum can be defined in terms of its potential ability, in interaction with another body, to produce an impulse. (The impulse is defined as $I = Ft$, where t is the time during which the force, F, acts.) We can then choose an arbitrary zero of momentum for each body, which is most conveniently associated with the state in which each body is at rest.

In classical theory, the procedure of regarding energy and momentum as fundamental properties of matter is not absolutely necessary from a logical point of view, but merely a convenient and suggestive way of thinking of the subject, based on the fact that these quantities are conserved. For, after all, energy and momentum can be expressed as functions of the positions and velocities so that, as shown in Sec. 10, they are redundant concepts, since all the laws of motion can be expressed directly in terms of the space-time motions alone.

In quantum theory, however, the energies and momenta cannot be expressed in this way. Thus, classically, the momentum is defined as

$$p = \lim_{\Delta t \to 0} m \frac{\Delta x}{\Delta t}$$

But we have already seen that, in the quantum domain, this limit does not really exist when Δt is made too small. Yet, we cannot avoid regarding momentum as a real quantity, not only because it is important in controlling the statistical behavior of space-time motions (see Sec. 11), but also because the momentum can be defined in quantum theory through the de Broglie relation $p = h/\lambda$, even though it is no longer possible to describe the motion in terms of a well-defined orbit in space time. The only course that seems to be left open is to regard momentum as an independent physical property of matter that, in the classical limit, represents potential ability to produce an impulse but, more generally, is related uniquely to the de Broglie wavelength and statistically to the space-time motion of matter. Thus, when we say that an electron was observed to have a given momentum, this statement stands on the same footing as the statement that it had a given position.

Neither statement is subject to further analysis. We must, therefore, think of momentum and energy as properties residing within matter, properties that cannot be pictured directly but which are simply given the names *momentum* and *energy*. We know that they are there, because they produce effects which cannot be understood in terms of the classical assumption that the space-time motions are governed by rules involving the motions alone.

Pursuant to our procedure of regarding momentum (and energy) as fundamental rather than derived concepts, we now show that these quantities can be measured even without the use of a detailed space-time description of the motions of all the particles in the system. Thus, as shown in Chap. 5, we can measure momentum with the aid of a diffraction grating, or else measure the potential drop needed to bring a particle to rest. Neither of these methods requires a detailed space-time description. With the aid of such measurements, we can prove that energy and momentum are conserved even in the quantum domain. Hence, in every respect, the concepts of energy and momentum stand on a footing that is independent of the need for a precise space-time description of the motion of matter.

13. Momentum and Energy a Description of Causal Aspects of Matter. We now proceed to show that, both in classical and quantum theory, a precise description of the causal aspects of the motion of matter involves a specification of the energies and momenta of all the relevant parts of a system. Let us begin by defining more carefully what we mean by the "causal aspects" of the motion of matter. In a complete description of the behavior of any system, two distinct but related elements always enter. First, there is the simple space-time order of the events that describe this behavior. In other words, we must tell *what* happens. But we are not usually satisfied in science by such a description, since we also wish to know *why* these things happen. In other words, we seek a causal description of the relationship between events, as well as a space-time description of the events themselves.

Now, the very effort to provide such a causal description involves the tacit assumption that the relationships between events actually originate in some kind of causal factors that exist within matter and, in some way, are able to bring about the events in question. In classical physics, these causal factors are the forces acting on each particle in the system (although, as we have seen in Sec. 10, the concept of forces as causes is redundant in classical theory). The causal relationships are then contained in Newton's laws of motion; namely, that each particle tends to move uniformly in a straight line, except insofar as it is disturbed by forces producing proportionate accelerations. Forces may, therefore, be regarded as the *causes* of changes of velocity. These forces may either be internal, i.e. between parts of the same system, or else externally imposed. In any case, if the forces are specified for all time, and if the initial positions and velocities are given, then the future course of the motion is determined for all time. We can take advantage of this fact to predict this motion on the basis of our knowledge of initial conditions, and we can also use it to alter and control the future course of the motion by imposing suitable external forces on the appropriate parts of the system.

Now, in quantum theory, the concept of force is cumbersome to use. It is much easier to work in terms of momentum and energy, because the de Broglie relations yield a simple connection between wavelength and momentum. No such simple relations exist between the wave properties of matter and force. We can define force in a rough way as the *average rate of change of momentum.** This agrees with the classical definition in the correspondence limit and provides an extension that has meaning in the quantum domain. Except for the purpose of demonstrating the relation between classically definable forces and quantum theory, however, this definition is not particularly useful. Instead, it is much more convenient to follow the procedure outlined in Secs. 11 and 12, i.e., to regard momentum as a fundamental and not further analyzable property of matter, which, to the extent that it is defined, determines statistically only the mean distance covered by a particle in a given time according to the equation

$$\frac{d}{dt}(\bar{x}) = \frac{p}{m}$$

We are, therefore, regarding momentum as the direct cause of motion of matter, instead of following the procedure (equivalent in classical theory) of regarding force as the cause of changes of motion. When a system is left to itself, it undergoes, as in classical theory, some characteristic type of motion, except that, in quantum theory, the course of this motion is determined only statistically by the momenta of all of the relevant parts. If we wish to alter or control the statistical trend of this motion, we can only do so by changing the momentum (and energy) of the appropriate parts. Thus, we conclude that the relevant energies and momenta are the causal factors contained in matter; these control the relation between events at different times deterministically in classical theory and statistically in quantum theory.†

14. Relation between Space Time and Causal Aspects of Matter. We are now in a position to show that the quantum theory leads to a new concept of the relation between space time and causal aspects of matter. This concept presents both aspects as united and yet not so closely that the need for two distinct aspects can be eliminated.

* See Chap. 9, Sec. 26.

† Momentum will therefore frequently be called the "causal factor" or the "causal aspect," that is contained within matter.

To obtain this concept, we begin with the results of Secs. 10 and 12. These sections indicated that, whereas classical theory can be expressed in terms of a set of prescriptive rules relating space-time motions at different times, quantum theory cannot be so expressed. Energy and momentum (and, therefore, the causal factors) cannot be eliminated in terms of velocities and positions of the component particles. The quantum theoretical concept of causality, therefore, differs from its classical counterpart in that it must necessarily describe the relationships between space-time events as being "caused" by factors existing within matter (i.e., momenta), which are on the same fundamental and not further analyzable footing as that of space and time themselves. It is true that these causal factors control only a statistical trend in the course of space-time events, but it is just this property of incomplete determinism that prevents the causal factors from becoming redundant, and that thus gives a real content to the concept of causality in quantum theory.

The retention of both space-time and causal aspects of matter under conditions where neither can be precisely defined leads to a totally new concept of these properties. For, instead of regarding space-time and causal aspects as existing in simultaneously well-defined forms, we now regard them as opposing potentialities, either of which can be realized in a more precisely defined form in interaction with an appropriate system, but only at the expense of a corresponding loss in the degree of definition of the other.* Thus, if an electron interacts with a position measuring device, it will have a comparatively well-defined position, with a corresponding decrease in the degree of definition of its momentum. On the other hand, if an electron interacts with a momentum measuring device, it will have a comparatively well-defined momentum, with a corresponding loss in the degree of definition of its position. Moreover, as shown in Chap. 6, Sec. 10, such changes in the definition of various properties take place not only in interaction with a measuring apparatus but, more generally, in interaction with all matter. Thus, in terms of our new concept, matter should be regarded as having potentialities for developing either comparatively well-defined causal relationships between comparatively poorly defined events or comparatively poorly defined causal relationships between comparatively well-defined events, but not both together.† Which of these potentialities is more fully realized in a given case depends partially on the systems with which the object in question interacts so that, with the passage of time, it can manifest

* In Chap. 6, Secs. 9 and 13, we have already introduced the concept that, at the quantum level, the properties of matter are opposing potentialities, which can become more precisely defined only at each other's expense and in interaction with an appropriate environment.

† The events may, for example, be described in terms of the presence of particles in certain regions of space-time, while the causal relationships are described in terms of the momenta of all the relevant parts of the system. See Sec. 13.

either of its potentialities more strongly as it interacts with different systems. We are thus led to conceive of matter as something uniting these two aspects, space time and causal, which would be incompatible if precisely defined, but which exist together in incompletely defined forms and oppose each other in the sense that their degrees of definition are reciprocally related.

We might assume, at first sight, that because the wave function can be expressed either entirely as a function of the position or entirely as a function of the momentum that only one of these aspects is really needed for a complete description of the behavior of matter. But let us recall that the wave function is completely defined in a given representation only when both amplitude and phase have been specified. Now, as we have seen in Chap. 6, Secs. 4 to 10, the amplitudes in a given representation control the probabilities of a given value of the variable associated with that representation, but the phase relations are more closely associated with the probability distribution of the conjugate variable.

The physical meaning of the phase relations can, therefore, be apprehended only in terms of a measurement of the conjugate variable. For example, in a position representation, the phase relations of the wave function cannot be understood in terms of the space-time locations of a particle but require for their physical interpretation the introduction of the concept of momentum. We therefore conclude that the wave function contains a description of both space-time and causal aspects of matter implicitly within it (and each description is, of course, carried to an intermediate degree of accuracy, such that the uncertainty principle is not violated). Consequently, to obtain a qualitative account of the nature of matter as implied by quantum theory, we must likewise retain both descriptions, each with an intermediate and flexible degree of accuracy.

15. The Principle of Complementarity. In the previous section, we have seen that fundamental properties of matter, such as momentum and position, are compatible only when each is defined to an intermediate degree of accuracy, so that the uncertainty principle is not violated. Now, all theories preceding the quantum theory have tacitly assumed that the behavior of matter can be described completely in terms of suitable dynamical variables which are all, in principle, capable of being defined at the same time with arbitrarily high precision. The idea that the basic properties of matter do not, in general, exist in a precisely defined form therefore constitutes a far-reaching change in the kinds of concepts used for the expression of physical theories. This change is, in fact, so far-reaching that Bohr was led to enunciate it in terms of a general principle, which he called "the principle of complementarity." The full meaning of this principle can be appreciated only after we have seen how it works out in detail in a large number of cases. We shall, however, try here to indicate its significance in terms of a few simple examples and then state the principle in a more general form.

Let us begin with momentum and position. In classical physics, the momentum of a particle lies either within the range between p and $p + dp$, or else lies in some other range outside that between p and $p + dp$. But in quantum theory, when the wave packet is broader than the range dp, it is no longer correct to say that the particle definitely lies in any given range dp; nor is it correct to say that it definitely lies outside that range. Instead, we say that, under these conditions, momentum simply is not a well-defined property (although it could potentially be better defined at the expense of the degree of definition of the position, if the electron interacts, for example, with a momentum measuring device).

The blurring of definition of the momentum does not, however, exhaust the physical significance of the spread of the wave function over momentum space, for as we have seen in Chap. 6, Sec. 6, the phase relations in momentum space determine the position distribution. Similarly, the phase relations in position space determine the momentum distribution. Thus, we conclude that the incomplete definiteness of momentum and position is essential because, within the range of indefiniteness of each, exist the factors responsible for the definition of the other. Momentum and position might, therefore, perhaps be called "interwoven variables," although even this description is inadequate, since it does not include the idea that the very existence of either requires a certain degree of indefiniteness of the other. A more accurate description is obtained by calling them "interwoven potentialities," representing opposing properties that can be comparatively well defined under different conditions.

It might be argued that if we described matter solely in terms of a wave function, the need for indefinite or "potential" properties could perhaps be eliminated since, after all, the wave function can, in principle, become arbitrarily well defined. It must be remembered, however, that the wave function is not in one-to-one correspondence with the actual behavior of matter, but only in statistical correspondence.* Thus, the wave function would be meaningless without the prescription that it is to be interpreted in terms of the probability that the system will develop a definite position or a definite momentum, depending on the nature of the measuring apparatus with which it interacts. But this probability has a one-to-one correspondence with only the mean values of the variables that will be obtained in a series of experiments performed with equivalent initial conditions.† Insofar as each individual electron is concerned, it remains true that there is a limit to the precision with which it is appropriate for us to attribute simultaneously to it a definite position and a definite momentum. Thus, an individual electron must be regarded as being in a state where these variables are actually not well defined but exist only as opposing potentialities. These potentialities complement each other, since each is necessary in a complete description of the physical processes through which the electron manifests itself; hence, the name "principle of complementarity."

* Chap. 6, Sec. 4.
† Chap. 6, Sec. 1.

We now give a more general statement of the principle of complementarity: *At the quantum level, the most general physical properties of any system must be expressed in terms of complementary pairs of variables, each of which can be better defined only at the expense of a corresponding loss in the degree of definition of the other.* This principle is clearly in sharp contrast to the classical concept of a system that can be described by specifying all the relevant variables to an arbitrarily high precision. For, in the quantum theory, complementary pairs of variables are to some extent opposing potentialities, either of which can be made to develop a more precise value but only under conditions wherein the other develops a less precise value. This means, of course, that complementary variables are not actually incompatible, provided that they are not too precisely defined; it is only the *complete* precision of definition of each which is incompatible with that of the other.

The most common examples of complementary pairs of potentialities are the canonically conjugate variables of classical mechanics, such as momentum and position, energy and time. Since one of these is always related to the causal aspects of matter and the other to its space-time aspects, it follows that causal and space-time aspects are complementary. The principle of complementarity is, however, not restricted to dynamical variables, for it also applies to more general concepts. For example, we have seen in Chap. 6, Secs. 9 and 13, that wave and particle aspects of matter are opposing but complementary modes of realization of the potentialities contained in a given piece of matter, either of which may be emphasized more in interaction with an appropriate environment.

Another example of a complementary pair of concepts is continuity and discontinuity. Let us recall, for example, that in a transition between discrete energy levels in an atom, the electron jumps from one level to another, without covering intermediate values of the energy. On the other hand, the wave function moves continuously from the region of space corresponding to the initial orbit to that corresponding to the final orbit. We have not yet developed the mathematical apparatus needed to treat this problem in detail, but in Chap. 22, Sec. 14, it will be shown that the continuous and discontinuous aspects of the transition are complementary in the sense that both are needed for a complete description of the process, despite the fact that the complete precision of definition of either is incompatible with that of the other.

Further examples of the principle of complementarity will appear throughout the course of this book. We shall, however, anticipate here a few of the results of later chapters in qualitative terms. We shall see* that a given system is capable, in principle, of demonstrating an infinite variety of properties that cannot all exist in simultaneously well-defined forms. Thus, if we begin with a pair of properties (or categories) such

* Chap. 16, Sec. 25.

as momentum and position, we find that not only does neither of these exist in a precisely defined form, but that there is also an infinite number of new properties (or categories) that can become definite only when both momentum and position are somewhat indefinite. These properties will actually become definite only when the object in question interacts with an appropriate system, such as a suitable measuring apparatus that brings about the realization of this particular property in a definite form.

We see, then, that a given system is potentially capable of an endless variety of transformations in which the old categories figuratively dissolve, to be replaced by new categories that cut across the old ones. Thus, we are led to an exceptionally fluid and dynamic concept of the nature of matter, a concept in which a given object can always escape any well-defined system of categories that may be appropriate under a given set of conditions and that, according to classical lines of reasoning, would permanently limit its behavior in a definite way. A striking example of such transformation appears in connection with leakage of a "particle" through a potential barrier, where the so-called "particle" is able to traverse a classically impenetrable region of space, because the barrier brings out its wave-like potentialities.*

We conclude that the principle of complementarity represents a thoroughgoing change in the type of concept that is appropriate for the description of matter at the quantum level, as compared with the types of concepts appropriate at the classical level. (In this connection, see Chap. 23.)

16. The Indivisible Unity of the World. We now come to the third important modification in our fundamental concepts brought about by the quantum theory; namely, that the world cannot be analyzed correctly into distinct parts; instead, it must be regarded as an indivisible unit in which separate parts appear as valid approximations only in the classical limit. This conclusion is based on the same ideas that lead to the principle of complementarity; namely, that the properties of matter are incompletely defined and opposing potentialities that can be fully realized only in interactions with other systems (see Chap. 6, Sec. 13). Thus, at the quantum level of accuracy, an object does not have any "intrinsic" properties (for instance, wave or particle) belonging to itself alone; instead, it shares all its properties mutually and indivisibly with the systems with which it interacts. Moreover, because a given object, such as an electron, interacts at different times with different systems that bring out different potentialities, it undergoes (as we have seen in Sec. 14) continual transformation between the various forms (for instance. wave or particle form) in which it can manifest itself.

Although such fluidity and dependence of form on the environment have not been found, before the advent of quantum theory, at the level

* See Chap. 11, Sec. 4.

of elementary particles in physics, they are not uncommon in classical experience, especially in fields, such as biology, which deal with complex systems. Thus, under suitable environmental conditions, a bacterium can develop into a spore stage, which is completely different in structure, and vice versa. Yet we recognize bacterium and spore as different forms of the same living system. There is certainly similarity here to the quantum behavior of the electron, for we can also recognize wave and particle aspects of the electron as different "forms" of the same material entity.* In both cases, suitable environmental conditions can bring out one aspect or the other of two possible modes of behavior.

Yet, there is an important difference between the change from bacterium to spore and the change of the electron from a more wave-like to a more particle-like object. The change from bacterium to spore can probably be regarded as a rearrangement of the various parts of the bacterium and its environment (i.e., the atoms and molecules), brought about by the forces between these parts; whereas, as pointed out in Chap. 6, Sec. 13, the change of the electron cannot be described in this way. Instead, it is a fundamental change in what would classically be called the "intrinsic" nature of the electron, a change that is not further analyzable in terms of hypothetical component parts of the electron and its environment. This is the meaning of the statement that at the quantum level of accuracy, the universe is an indivisible whole, which cannot correctly be regarded as made up of distinct parts.

To clarify the implications of this point of view concerning the quantum nature of matter we shall, in the next few sections, present a fairly detailed analysis of the classical description of complex systems as made up of component parts. Then we shall show how this description breaks down in the quantum domain. We shall thus be led in another way to the conclusion, obtained more directly in Chap. 6, that the world must be regarded as an indivisible unit.

17. Distinction between Object and Environment on Classical Level. Whenever we meet with an object whose nature depends critically on the environment, whether in classical, quantum, or any other theory, we recognize that the description as a separate system is inadequate, and that we should study instead the combined system, consisting of object plus environment, as a unit. On a classical level, however, it is always assumed that, even when an object is strongly linked with its environment, a distinction between the two can be made at any instant of time on the basis of their separation in space. Thus, with the aid of a microscope for example, it can be seen that something is happening in a definite region of space, which, at any instant, calls for the interpretation that this particular region of space is occupied by a fairly definable object, which can be called the bacterium. (Although the physical line of sepa-

* See Sec. 15, and compare with the principle of complementarity.

ration between the bacterium and environment may not be perfectly sharp, it is still very narrow compared with the size of the bacterium.)

How are we to describe what happens in such a system with the passage of time? Clearly, there are strong interactions between bacterium and environment: first, because of the forces between them and, second, because of the exchange of matter between them. In fact, in a few hours most of the matter that was originally in the bacterium may have been expelled and replaced by matter from the surrounding medium. In the meantime, the bacterium may also have changed into a spore. How are we then justified in thinking of this as a continuation of the same living system that we saw originally? The justification lies partially in the continuity of the process of change undergone by the bacterium and partially in the fact that, at all times, the properties of bacterium and environment are determined by causal laws.

18. The Role of Continuity. The role played by continuity in making possible the identification of a changing object is fairly clear. If, for example, large discontinuous and erratic changes occurred in the bacterium, we could not then trace its identity with the passage of time. The continuity also insures that the bacterium will "stay put" long enough to allow it to be seen and recognized. That is, even if it is changing, the effect of changes can always be made arbitrarily small by choosing a sufficiently small interval of time in which to observe it.

19. The Role of Causal Laws. The role of causal laws in making possible the identification of an object, *whether it is changing or not,* is perhaps less obvious, but it is certainly no less important. The significance of causal laws in this problem can be demonstrated by a description of the procedures by which the bacterium is identified as such. For example, the bacterium may be seen by looking into a microscope. But unless the bacterium obeyed causal laws, at least to the extent of refracting, absorbing, and reflecting light in a systematic and reliable way, the microscope would be of no help in identifying it as a separate object. In another important test, the object must react to external disturbances in a known and reliable way. Thus, if a bacterium is prodded by means of a minute needle, it reacts more or less like a piece of jelly and not like a piece of glass. If certain dyes are inserted into the medium, then each type of bacterium shows its own characteristic staining reactions.

Many more such examples could be used, but we can sum up a wide range of general experience by saying that an object is identified by the way it reacts to forces of various kinds. These forces may be electromagnetic, mechanical, gravitational in origin. They may also arise from the forces of chemical interaction of molecules, or in still other ways not mentioned here. (Note that this criterion also includes seeing the object with the aid of light.) Since the statement that an object reacts in a definite way to forces implies that it obeys causal laws, we conclude that no object can even be identified as such unless it obeys causal laws.

The same type of criteria are used in recognizing elementary particles such as protons and electrons. The first convincing evidence for the existence of such particles was the appearance of apparently continuous tracks in a cloud chamber, tracks that were curved by electric and magnetic forces in exactly the way that the path of a charged particle would curve. It is from the reaction to electric and magnetic forces, and from the ionization of other atoms by the electric forces produced by a charged particle, that an electron or proton is identified.

20. Analysis and Synthesis. If a system moves continuously and obeys causal laws, we can continue, with the passage of time, to identify it as a separate object, even though it may interact strongly with its environment and suffer major changes as a result of this interaction. If such changes occur, they can then be understood in terms of the causal laws. Thus, the changes of structure of the bacterium when it goes over into the spore stage are thought to be caused by the electrical, magnetic, and chemical forces between the molecules constituting the bacterium and environment. And it is these forces which cause various parts of the system to move in such a way that, in time, the bacterium is transformed into a spore.

The preceding ideas can now be generalized. In practically all fields of science, as well as in much of everyday life, we tacitly make use of a program of analysis of the world into parts, and synthesis of these parts with the aid of causal laws. If this program is to have meaning, it is necessary that the parts have properties that enable them to be identified, in principle at least, and described as working together according to causal laws to form the whole.

The process of identification, as carried out in practice, always involves the tacit assumptions of continuity and causality. Thus we assume that, at any instant of time, each part occupies a definite region of space and has a definite shape and structure, all of which change continuously with the passage of time. Equally important, however, is the assumption that we can attribute definite and characteristic effects to each part. This means that in its interaction with the various types of forces used to probe its properties, the system is assumed to obey causal laws. Since, in principle, an object can be observed or probed by means of any kind of force, we conclude that if a system can be analyzed into identifiable parts, it must obey causal laws in all interactions. Otherwise, we would be led to doubt the identification of the parts, since all means of observation would not necessarily lead to the same results. The same general requirements of continuity and causality, which are needed to make a system analyzable into distinguishable parts, however, are also what are needed to make possible a description of how all the parts work together to form the whole. Thus, the programs of analysis and synthesis go together.

21. Applicability of Analysis and Synthesis to Classical Theory. It is immediately evident that, insofar as classical physics is valid, the requirements for an analysis of the world into parts and a synthesis of these parts into a whole, can all be satisfied. This follows from the fact that all parts of the world (for instance, atoms, molecules, electrons) are assumed to move continuously and to satisfy causal laws.

22. Classically Describable vs. Essentially Quantum-mechanical Systems. Let us now see what happens when we try to extend these ideas into the quantum domain. In doing this, it is convenient to make a distinction between classically describable processes and those essentially quantum-mechanical in nature. In the last analysis all processes are, of course, quantum-mechanical in nature, but there are many processes involving relatively large objects and, therefore, a great many quanta, where a precise description down to a quantum level of accuracy is not essential *because the interesting features of the system do not depend critically on the transfer of a few quanta more or less.* Such processes can most conveniently be described in terms of classical theory alone. Note that the distinction between classically describable and essentially quantum-mechanical systems is not on the basis of the accuracy with which we can make an observation but is, rather, on the basis of whether the objects of interest depend critically on the quantum properties of matter.

As an example, let us consider the bacterium again. By quantum standards, the bacterium is a fairly large object and it may, therefore, be expected that most of its actions can be understood with a classical description alone. Thus, the program of analysis of a cell into parts, with the ultimate objective of understanding how these parts work together to form the whole cell, can probably be justified directly on this basis of the applicability of classical theory to all the significant parts. It is not inconceivable, and perhaps not unlikely, that there may exist "chain reactions" in a cell, which can multiply the effects of certain crucial quantum processes to a classically observable level. If this should be true, then the program of analysis and synthesis would have to be reconsidered in the light of quantum theory, at least insofar as these crucial properties are concerned.

23. An Attempt to Analyze a Quantum System into Parts. When we come down to the quantum level of accuracy, serious difficulties appear in the effort to carry out a program of analysis and synthesis. These difficulties arise in the circumstance that the application of the causal laws requires a precise definition of the momentum of each part of the system; this is impossible when the system is localized in any way at all. On the classical level, this residue of lack of complete determinism is negligible, but as we try to deal with smaller and smaller objects, it becomes more and more difficult to probe their properties by means of their reaction to external forces. For example, they no longer reflect light continuously and in a definite way but, instead, begin to reflect it discontinuously (in the form of quanta) and somewhat erratically. Thus, when looked at in a microscope, such objects would appear to fluctuate in size, shape, and other properties, discontinuously and without much regularity of behavior. Their reaction to probing by mechanical or electrical forces would become equally erratic because of the rapid and uncontrollable exchanges of quanta between object and probe. Thus, it would be difficult to decide, for instance, whether the object was "hard" or "soft." The lack of continuity of motion, coupled with the rapidly and uncontrollably changing nature of all of the parts, would make it difficult for us to continue to identify each part with the passage of time, since between observations a part might change in a very fundamental way. For example, it might turn from something resembling a wave to something resembling a particle, but it would be impossible to follow the transition between the two in detail, as can be done in the transition from bacterium to spore. If there were many similar interacting parts (for example, elementary particles), it would soon become impossible to make certain that we were following the same part that we had started with.

24. The Indivisible Unity of Quantum Systems. From the above, it can be seen that as we try to improve the level of accuracy of description, the classical program of analysis into parts eventually becomes infeasible. The program of synthesis according to causal laws also becomes infeasible, since there are no exact causal laws. We are led, instead, to a new point of view, based on the idea that the quanta connecting object and environment constitute irreducible links that belong, at all times, as much to one part as to the other. Since the behavior of each part depends as much on these quanta as on its "own" properties, it is clear that no part of the system can be thought of as separate.

If, in a classical experiment, we discovered the presence of irreducible "links" between objects, we should then postulate a third object, the link, and thus re-establish the old type of description, this time in terms of *three* parts to the system. In quantum theory, however, these quanta do not constitute separate objects, but are only a way of talking about indivisible transitions of the objects already in existence. The fact that quanta are unpredictable and uncontrollable would, in any case, prevent their introduction as a third object from being of any use, since we could not in any definite way ascribe observed effects to them.

25. An Example: the Hydrogen Atom. Consider, for example, a hydrogen atom in the ground state interacting with an electromagnetic field carrying some energy. The atom can absorb a quantum but, during the process of transition, it is not in a definite energy state. Instead, it covers an indefinite range of energy states. The energy of the electromagnetic field is equally indefinite. During the process of transition, both systems are coupled because they are exchanging an indivisible quantum of energy belonging as much to the electron as to the electromagnetic field. It is, therefore, impossible to ascribe the future behavior of the system in a unique way, as can be done classically, to the state of each "part" (i.e., electron and electromagnetic fields), because the state of each part is indefinite and yet inextricably linked with that of the other part.

26. The Need for a Nonmechanical Description. The fact that quantum systems cannot be regarded as made up of separate parts working together according to causal laws means that we are now led to a fundamental change in our general methods of description of nature. Only in the classical limit, where the effects of individual quanta are negligible and where their combined effects can be approximated by a causal description, is it possible to separate the world into distinct parts. Even in the classical limit, we recognize that the separation between object and environment is an abstraction. But because each part interacts with the others according to causal laws, we can still give a correct description in this way. In a system whose behavior depends critically on the transfer of a few quanta, however, the separation of the world into parts is a non-permissible abstraction because the very nature of the parts (for instance, wave or particle) depends on factors that cannot be ascribed uniquely to either part, and are not even subject to complete control or prediction.

Thus, by investigating the applicability of the usual classical criteria for analyzing a system into distinct parts, we have been led to the same conclusion as that obtained directly in Chap. 6, Sec. 13: The entire universe must, on a very accurate level, be regarded as a single indivisible unit in which separate parts appear as idealizations permissible only on a classical level of accuracy of description. This means that the view of the world as being analogous to a huge machine, the predominant view from the sixteenth to nineteenth centuries, is now shown to be only approximately correct. The underlying structure of matter, however, is not mechanical.*

Summary of New Concepts in Quantum Theory

We have seen that the classical concepts of continuity, causality, and the analysis of the world into distinct parts are all necessary for each other's consistency; foregoing any one of them leads to the necessity for giving up all. Thus, as shown in Sec. 20, the analysis of a system into distinct parts has meaning only in a context where these parts move continuously and obey precisely defined causal laws. Similarly, it is easily seen that the concept of precisely defined causal laws has meaning only in a context where the world can be analyzed into distinct elements

* This means that the term "quantum mechanics" is very much of a misnomer. It should, perhaps, be called "quantum nonmechanics."

moving continuously. For without such elements, there will be no precisely definable variables to which the causal laws can be applied.

The entire system of classical concepts must, therefore, be replaced by a totally new system of quantum-theoretical concepts, each of which has meaning only in a context when all others are true. The system of quantum concepts involves the assumptions of incomplete continuity, incomplete determinism, and the indivisible unity of the entire universe. These may be summarized by saying the properties of matter are to be expressed in terms of opposing but complementary pairs of potentialities, either of which can be realized in a more definite form in an appropriate environment but only at the expense of a corresponding loss in the degree of definition of the other.*

The expression of the new quantum concepts is beset with severe difficulties, because much of our customary language and thinking is predicated on the tacit assumption that classical concepts are substantially correct. Such an assumption leads us to interpret quantum-theoretical results in a general classical context. Thus, when we say that there is an electron in a certain region of space, we tend to imply that there is, in this region, a separate object having intrinsic properties that are independent of the systems with which this object interacts. Yet, we know that an electron acts more like a wave or more like a particle, depending on what system it interacts with, as well as on the electron itself.

We anticipate that new ways of using language may ultimately be developed to avoid the previously mentioned tacit errors. For the present, however, we can only keep in mind that common scientific words such as "electron," "atom," "wave," and "particle" are already associated with classical concepts which cannot be applied without reservation in quantum theory. Thus, the word "electron' as used in quantum theory refers to something whose properties are much less fixed and independent of the environment than those contemplated in the classical concept of an electron. To avoid wrong interpretations of the quantum theory, arising from difficulties in the language, the reader should grasp the theory in terms of a whole new system of concepts. It has been the purpose of this chapter to indicate the scope of the changes in concept that are needed. This material should, however, also be read in close conjunction with Chaps. 6, 7, 22, and 23.

Analogies to Quantum Processes

There are wide ranges of experience in which occur phenomena possessing striking resemblances to quantum phenomena. These analogies will now be discussed, since they clarify the results of the quantum theory

* See Sec. 15 on the principle of complementarity.

Some interesting speculations on the underlying reasons for the existence of such analogies will also be introduced.

27. The Uncertainty Principle and Certain Aspects of Our Thought Processes. If a person tries to observe what he is thinking about at the very moment that he is reflecting on a particular subject, it is generally agreed that he introduces unpredictable and uncontrollable changes in the way his thoughts proceed thereafter. Why this happens is not definitely known at present, but some plausible explanations will be suggested later. If we compare (1) the instantaneous state of a thought with the position of a particle and (2) the general direction of change of that thought with the particle's momentum, we have a strong analogy.

We must remember, however, that a person can always describe approximately what he is thinking about without introducing significant disturbances in his train of thought. But as he tries to make the description precise, he discovers that either the subject of his thoughts or their trend or sometimes both become very different from what they were before he tried to observe them. Thus, the actions involved in making any single aspect of the thought process definite appear to introduce unpredictable and uncontrollable changes in other equally significant aspects.

A further development of this analogy is that the significance of thought processes appears to have indivisibility of a sort. Thus, if a person attempts to apply to his thinking more and more precisely defined elements, he eventually reaches a stage where further analysis cannot even be given a meaning. Part of the significance of each element of a thought process appears, therefore, to originate in its indivisible and incompletely controllable connections with other elements.* Similarly, some of the characteristic properties of a quantum system (for instance, wave or particle nature) depend on indivisible and incompletely controllable quantum connections with surrounding objects.† Thus, thought processes and quantum systems are analogous in that they cannot be analyzed too much in terms of distinct elements, because the "intrinsic" nature of each element is not a property existing separately from and independently of other elements but is, instead, a property that arises partially from its relation with other elements. In both cases, an analysis into distinct elements is correct only if it is so approximate that no significant alteration of the various indivisible connected parts would result from it.

There is also a similarity between the thought process and the classical limit of the quantum theory. The logical process corresponds to

* Similarly, part of the connotation of a word depends on the words it is associated with, and in a way that is not, in practice, completely predictable or controllable (especially in speech). In fact the analysis of language, *as actually used*, into distinct elements with precisely defined relations between them is probably impossible.

† See Secs. 24, 25, 26.

the most general type of thought process as the classical limit corresponds to the most general quantum process. In the logical process, we deal with classifications. These classifications are conceived as being completely separate but related by the rules of logic, which may be regarded as the analogue of the causal laws of classical physics. In any thought process, the component ideas are not separate but flow steadily and indivisibly. An attempt to analyze them into separate parts destroys or changes their meanings. Yet there are certain types of concepts, among which are those involving the classification of objects, in which we can, without producing any essential changes, neglect the indivisible and incompletely controllable connection with other ideas. Instead, the connection can be regarded as causal and following the rules of logic.

Logically definable concepts play the same fundamental role in abstract and precise thinking as do separable objects and phenomena in our customary description of the world. Without the development of logical thinking, we would have no clear way to express the results of our thinking, and no way to check its validity. Thus, just as life as we know it would be impossible if quantum theory did not have its present classical limit, thought as we know it would be impossible unless we could express its results in logical terms. Yet, the basic thinking process probably cannot be described as logical. For instance, many people have noted that a new idea often comes suddenly, after a long and unsuccessful search and without any apparent direct cause. We suggest that if the intermediate indivisible nonlogical steps occurring in an actual thought process are ignored, and if we restrict ourselves to a logical terminology, then the production of new ideas presents a strong analogy to a quantum jump. In a similar way, the actual concept of a quantum jump seems necessary in our procedure of describing a quantum system that is actually an indivisible whole in terms of words and concepts implying that it can be analyzed into distinct parts.*

28. Possible Reason for Analogies between Thought and Quantum Processes. We may now ask whether the close analogy between quantum processes and our inner experiences and thought processes is more than a coincidence. Here we are on speculative ground; at present very little is known about the relation between our thought processes and emotions and the details of the brain's structure and operation. Bohr suggests that thought involves such small amounts of energy that quantum-theoretical limitations play an essential role in determining its character.† There is no question that observations show the presence of an enormous amount of mechanism in the brain, and that much of this mechanism must probably be regarded as operating on a classically

* See, for example, Chap. 22, Sec. 14.

† N. Bohr, *Atomic Theory and the Description of Nature.*

describable level. In fact, the nerve connections found thus far suggest combinations of telephone exchanges and calculating machines of a complexity that has probably never been dreamed of before. In addition to such a classically describable mechanism that seems to act like a general system of communications, Bohr's suggestion involves the idea that certain key points controlling this mechanism (which are, in turn, affected by the actions of this mechanism) are so sensitive and delicately balanced that they must be described in an essentially quantum-mechanical way. (We might, for example, imagine that such key points exist at certain types of nerve junctions.) It cannot be stated too strongly that we are now on exceedingly speculative grounds.

Bohr's hypothesis is not, however, in disagreement with anything that is now known. And the remarkable point-by-point analogy between the thought processes and quantum processes would suggest that a hypothesis relating these two may well turn out to be fruitful. If such a hypothesis could ever be verified, it would explain in a natural way a great many features of our thinking.

Even if this hypothesis should be wrong, and even if we could describe the brain's functions in terms of classical theory alone, the analogy between thought and quantum processes would still have important consequences: we would have what amounts to a classical system that provides a good analogy to quantum theory. At the least, this would be very instructive. It might, for example, give us a means for describing effects like those of the quantum theory in terms of hidden variables. (It would not, however, prove that such hidden variables exist.)

In the absence of any experimental data on this question, the analogy between thought and quantum processes can still be helpful in giving us a better "feeling" for quantum theory. For instance, suppose that we ask for a detailed description of how an electron is moving in a hydrogen atom when it is in a definite energy level. We can say that this is analogous to asking for a detailed description of what we are thinking about while we are reflecting on some definite subject. As soon as we begin to give this detailed description, we are no longer thinking about the subject in question, but are instead thinking about giving a detailed description. In a similar way, when the electron is moving with a definable trajectory, it simply can no longer be an electron that has a definite energy.

If it should be true that the thought processes depend critically on quantum-mechanical elements in the brain, then we could say that thought processes provide the same kind of direct experience of the effects of quantum theory that muscular forces provide for classical theory. Thus, for example, the pre-Galilean concepts of force, obtained from immediate experience with muscular forces, were correct, in general.

But these concepts were wrong, in detail, because they suggested that the velocity, rather than the acceleration, was proportional to the force. (This idea is substantially correct, when there is a great deal of friction, as is usually the case in common experience.) We suggest that, similarly, the behavior of our thought processes may perhaps reflect in an indirect way some of the quantum-mechanical aspects of the matter of which we are composed.

PART II

MATHEMATICAL FORMULATION OF THE QUANTUM THEORY

CHAPTER 9

Wave Functions, Operators, and Schrödinger's Equation

ON THE BASIS of the physical theory developed in Part I, we are now ready to derive a mathematical formalism by which the quantum theory can be given a precise expression. More specifically, we shall first obtain formulas for the average value of any physical quantity. Then we shall develop the operator formalism, which is very convenient for the expression of these averages. Next we shall obtain Schrödinger's equation with the aid of the correspondence principle. Finally, we shall introduce the use of eigenvalues of operators, and their eigenfunctions. At this point, we shall be ready to go on to Part III, where the quantum theory is applied to various elementary problems.

1. Wave Formalism and Probability. We have seen in Part I that quantum theory, unlike classical theory, can, in general, predict only the probable and not the exact results of a measurement. These probabilities are determined by a wave function, $\psi(x)$. The probability that a particle be found with position between x and $x + dx$, is

$$P(x)\,dx = \psi^*(x)\psi(x)\,dx$$

The probability that a particle be found with momentum between $p = \hbar k$ and $p + dp = \hbar(k + dk)$ is

$$P(k)\,dk = \varphi^*(k)\varphi(k)\,dk$$

Since the only two properties of an elementary particle that we need treat at present are its position and momentum,† it is clear that insofar as any features of the behavior of the particle are predictable, all information about them is contained in the wave function. This is a very important point. In applications to systems that are more complex than a

† Spin and other properties exist, but these make only small corrections, which may, for our purposes, be neglected here (see Chap. 17).

single particle, this idea is generalized to the statement that there is a wave function which is a function of all the significant co-ordinates needed to describe the system, and from which all possible physical information about the system may be obtained. For example, in a two-body problem the wave function is $\psi(x_1, x_2)$, where x_1 and x_2 are the co-ordinates of the first and second particle, respectively. $\psi^*(x_1, x_2)\psi(x_1, x_2)\, dx_1\, dx_2$ is then equal to the probability that Particle 1 will be found between x_1 and $x_1 + dx_1$ at the same time Particle 2 is found between x_2 and $x_2 + dx_2$. Hence, for two particles the waves move in a six-dimensional space, and for N particles, in a $3N$-dimensional space.

In Chap. 17 we shall show that electrons have a spin, which requires the introduction of a spin co-ordinate s. The wave function for a single electron will then become $\psi(x, s)$. Hence, we see that, in general, as more variables are found to be needed to describe the system, we simply make the wave function a function of these new variables.

2. Hypothesis of Linear Superposition. A basic idea in any wave theory is that if ψ_1 and ψ_2 are possible wave functions, then any linear combination $a\psi_1 + b\psi_2$, where a and b are arbitrary constants, is also a possible wave function. This statement is known as the hypothesis of linear superposition. It is necessary to assume some such hypothesis to explain interference and the production of wave packets. For example, in optics interference patterns are often predicted with the aid of Huyghens' principle, which describes the wave intensity at any point as being determined by the linear superposition of waves starting from all possible points in a previous wave front. Whether this is the only hypothesis that can possibly explain interference is not known. It is, however, the simplest one that will do so, and it has been successful in explaining electromagnetic and acoustical interference phenomena. We tentatively extend this postulate to electron waves also. In fact, we have already done so without discussing the fact that it is a postulate in making up wave packets and in describing electron diffraction experiments in a manner analogous to the way that diffraction of light is treated. The great success of this interpretation then justifies its further application to a more general set of problems.

This kind of lack of uniqueness will appear quite frequently as we set up the mathematical formulation of quantum theory. We are adopting the point of view that by means of analogies and arguments that make our choices plausible, we can be led to fruitful ideas; in the last analysis, however, these must be checked by comparison with observation. We believe that this procedure is more suitable for developing quantum theory for a beginner than is the method in which one starts with a set of abstract postulates from which one makes a complete set of mathematical deductions that are compared with experiment. We believe also that the postulational approach has the further disadvantage of being too rigid, making it difficult to tell how the theory might be changed in case small disagreements with experiment should be found. The approach we have adopted may be called "heuristic," i.e., partially based on deduction and partially on intelligent guesses that may later have to be modified on the basis of more accurate experiments.

3. Concept of the State of a System in Quantum Theory. As pointed out in Sec. 1, the wave function has, in general, only a probability interpretation. It is, therefore, not in a one-to-one correspondence with the actual behavior of matter (see Chap. 6, Sec. 4). Yet, we are also assuming that the wave function contains all possible information relating to the system under description. How can we reconcile these two statements?

We do so in terms of the assumption that the properties of matter do not, in general, exist separately in a given object in a precisely defined form. They are, instead, incompletely defined potentialities realized in more definite form only in interaction with other systems, such as a measuring apparatus (see Chap. 6, Secs. 9 and 13; Chap. 8, Sec. 14). The wave function describes all these potentialities, and assigns a certain probability to each. This probability does not refer to the chance that a given property, such as a certain value of the momentum, actually exists at this time in the system, but rather to the chance that in interaction with a suitable measuring apparatus such a value will be developed at the expense of a corresponding loss in definiteness of some other variable, in this case the position. Therefore, when two similar systems have the same wave function, we cannot then deduce that they will necessarily behave the same in all processes in which they take part. We can merely state that they have the same range of potentialities within them and that, in each system, a given potentiality has the same probability of being developed, if both systems are treated in the same way (for instance, if the same variable is measured).

Now, in classical physics, two similar systems may be put into a state in which each has the same value of every significant variable (such as momentum and position). If this is done, then the subsequent behavior of both systems will be identical. Such systems could correctly be said to be in the same state. In quantum theory we have seen, however, that the incompletely definite nature of all significant variables prevents one from making any two systems so similar that they behave precisely the same in all subsequent processes. The best that can be done is to adjust conditions so that each system has the same probability for development of any of its various potentialities. To do this, we must obtain two systems having either the same wave function, or else wave functions differing at most by a constant phase factor $e^{i\alpha}$. (On the other hand, a phase factor depending on the position would imply different momentum distributions, as shown in Chap. 6, Sec. 7.) Under these conditions, the two systems are as similar as the incompletely definite character of their fundamental properties permits them to be. For this reason, they will be said to be "in the same quantum state."

How can two systems be made to have the same wave function and therefore go into the same quantum state? An example of such a process is described in Chap. 6, Sec. 1, where we saw that if electrons are directed into a slit system, one at a time, all with the same initial momentum, then all these electrons will have the same wave function (except for a constant phase factor) and will, therefore, be in the same quantum state. More generally, whenever all significant variables for two systems are defined as accurately as is consistent with the uncertainty principle, these systems will be in the same quantum state. Thus, to put two systems in the same quantum state. we try to reproduce initial conditions as accurately as the quantum nature of matter permits.

4. Statistical Significance of the Concept of Quantum State. Thus far, we have applied the concept of a quantum state only to individual systems. Also, we have seen that, even when two systems are treated as similarly as is consistent with their quantum natures, they may behave differently because within each system exists a range of incompletely defined potentialities. Thus, for individual cases, it would be difficult to prove that two systems were in the same quantum state, unless it happened that this quantum state could be described in terms of a definite value associated with some variable.

If, for example, the wave function is precisely $e^{ipx/\hbar}$, then we know that the momentum is certainly p. We could, therefore, be sure that two systems had the same wave function if we knew that each had definite and equal momenta. More generally, however, we have a wave packet, which implies a range of both potential momenta and positions which can be developed when either of these variables is measured. In this case, the fact that a system is in a given quantum state is in general manifested only in a statistical way.

For example, we may direct electrons with a given small range of momenta into a slit system as described in Chap. 6, Sec. 2. There will then be a whole range of potential positions at which an individual electron can arrive at a detecting screen to the right of the slits. But in any given case the wave function and, therefore, the quantum state do not determine precisely where the particle will actually arrive. Only after a large number of similar experiments where equivalent initial conditions are carried out will we obtain a statistical pattern of electronic positions, which is characteristic of the wave function and, therefore, of the quantum state.

Similarly, if the momenta of the electrons are measured after they pass through the slit system, we will obtain a statistical pattern of results, determined by the Fourier components of the wave function and, therefore, also by the quantum state. More generally, it is only the statistical pattern of results obtained under equivalent initial conditions that is determined by the quantum state. In some cases, this pattern may be so narrow that, to a first approximation, we may speak in terms of a well-defined result, particularly if the phenomena of interest do not depend critically on the precise value of the measured variable. Such a situation will always arise in the classical limit (i.e., when the spread of the wave packet can be neglected) so that we can speak approximately of well-defined values for *all* significant variables and thus obtain a specification of a classical state.

In general, statistical measurements at the quantum level must be carried out in a series of similar systems, each of which is subject to the same initial treatment. This is because, in the process of measurement, uncontrollable changes take place,† which result in a new quantum state that is not deterministically related to the quantum state existing before interaction with the measuring apparatus. Therefore, if we wish our data to refer to a given quantum state, we must discard each system after the measurement is over and start with a new system, prepared in an equivalent way.

5. Mathematical Expression for Averages. Let us now set up a mathematical formalism that expresses in a more precise way the general ideas outlined above. We begin with the problem of obtaining expressions for various important physical quantities.

Average Value of a Function of Position

The average value of x must be by definition

$$\bar{x} = \int_{-\infty}^{\infty} P(x)x\,dx \tag{1}$$

Since, as we have seen,

$$P(x) = \psi^*(x)\psi(x)$$

we may write‡

$$\bar{x} = \int_{-\infty}^{\infty} \psi^*(x)x\psi(x)\,dx \tag{2}$$

The x has been inserted between ψ^* and ψ for reasons of notational symmetry, the nature of which will become obvious later.

In a similar way, the average values of any function of x may be written

$$\bar{f}(x) = \int_{-\infty}^{\infty} \psi^*(x)f(x)\psi(x)\,dx \tag{3}$$

† See discussion of uncertainty principle in Chap. 5.

‡ Note that ψ is always assumed to be normalized because the total probability that the particle is somewhere in space must be unity. That is, $\int_{-\infty}^{\infty} \psi^*\psi\,dx = 1$. If ψ is not already normalized, then it can be normalized by multiplying it by a suitable constant A, such that $|A|^2\int\psi^*\psi\,dx = 1$. In Part I, it was shown that the total probability is conserved; hence if the wave function is initially normalized, it remains normalized for all time.

The generalization of this formalism to three dimensions is straightforward. Thus, we write

$$\bar{f}(x, y, z) = \int_{-\infty}^{\infty}\int_{-\infty}^{\infty}\int_{-\infty}^{\infty} \psi^* f(x, y, z)\psi \, d\tau \tag{3a}$$

where $d\tau$ represents the element of volume.

Hereafter, we shall restrict ourselves to a one-dimensional treatment in order to decrease the amount of notation needed since, in all cases, the generalization to three dimensions will be equally simple.

Average Value of a Function of the Momentum

The average value of the momentum is

$$\bar{p} = \int_{-\infty}^{\infty} pP(p)\, dp = \int_{-\infty}^{\infty} \Phi^*(p)p\Phi(p)\, dp \tag{4}$$

where $\Phi(p)$ is the normalized Fourier component of $\psi(x)$, with $p = \hbar k$.

In Chap. 4, Sec. 10, it was shown that if $\psi(x)$ is normalized, then $\varphi(k)$ is automatically normalized, so that

$$\int_{-\infty}^{\infty} \varphi^*(k)\varphi(k)\, dk = 1$$

It is often convenient, however, to introduce the functions $\Phi(p)$, which are normalized such that

$$1 = \int_{-\infty}^{\infty} \Phi^*(p)\Phi(p)\, dp$$

This condition will be satisfied if we write

$$\varphi(k) = (\hbar)^{1/2}\Phi(p)$$

Problem 1: Prove the above statement.

For any function of the momentum, the average is then given by

$$\overline{f(p)} = \int_{-\infty}^{\infty} f(p)P(p)\, dp = \int_{-\infty}^{\infty} \Phi^*(p)f(p)\Phi(p)\, dp \tag{4a}$$

Criterion for Acceptable Wave Functions

A basic requirement that any ψ must satisfy is that it be quadratically integrable, i.e., that

$$\int_{-\infty}^{\infty} |\psi|^2\, dx = \text{a finite number}$$

If this requirement is not satisfied, then we cannot even normalize the probability, so that it is impossible to give the wave function a meaning in terms of physically observable averages. A necessary (but not sufficient) requirement for ψ is, therefore, that $\psi \to 0$ as $x \to \pm\infty$, and $\Phi(p) \to 0$ as $p \to \pm\infty$.

We can, however, obtain more stringent physical requirements from the fact that the averages of all physically observable quantities must exist. Now x and p are clearly physically observable quantities, so that their averages must exist. The kinetic energy $T = p^2/2m$ is also an observable quantity. We can, therefore, set the condition on the wave function that $\int_{-\infty}^{\infty} \Phi^* p^2 \Phi \, dp$ must exist. A necessary (but not sufficient) condition for $\Phi(p)$ is, then, that $p\Phi(p) \rightarrow 0$ as $p \rightarrow \pm\infty$. If it is known that there are potential energies present of the form $V(x)$, then it is also necessary that $\bar{V}(x)$ shall exist. We repeat that, whenever we know a given function is physically important, we require of all acceptable wave functions that the average value of this quantity shall exist.

We shall see in the subsequent work that practically all wave functions likely to occur will have the property that $\overline{x^n}$ exists,† where n is an arbitrary positive number.

Further restrictions on the behavior of acceptable wave functions will be obtained in the next section of this chapter.

6. Operator Notation to Obtain Momentum Averages from Integrals in Position Space. It would be very useful to be able to compute averages of functions of the momentum directly from $\psi(x)$, without having to Fourier analyze the wave function. To find out how to do this, we express $\Phi(p)$ in eq. (4) as a Fourier integral

$$\Phi(p) = (\hbar)^{-\frac{1}{2}}\varphi(k) = \left(\frac{1}{2\pi\hbar}\right)^{\frac{1}{2}} \int_{-\infty}^{\infty} e^{-ikx}\psi(x)\, dx$$

We obtain (using $p = \hbar k$)

$$\bar{p} = \frac{\hbar}{2\pi} \int_{-\infty}^{\infty}\int_{-\infty}^{\infty}\int_{-\infty}^{\infty} e^{ikx'}\psi^*(x') k e^{-ikx}\psi(x)\, dx'\, dx\, dk \tag{5}$$

Let us now write

$$ke^{-ikx} = i\frac{\partial}{\partial x} e^{-ikx}$$

The integral then becomes

$$\bar{p} = \frac{\hbar}{2\pi} \int_{-\infty}^{\infty}\int_{-\infty}^{\infty}\int_{-\infty}^{\infty} e^{ikx'}\,\psi^*(x')\, dx' \left[i\frac{\partial}{\partial x} e^{-ikx}\right] \psi(x)\, dx\, dk \tag{5a}$$

Integration by parts over x, plus the fact that $\psi(\pm\infty) = 0$, yields

$$\bar{p} = \int_{-\infty}^{\infty} dx \int_{-\infty}^{\infty} \psi^*(x') \frac{\hbar}{i}\frac{\partial\psi(x)}{\partial x}\, dx' \int_{-\infty}^{\infty} \frac{dk}{2\pi} e^{ik(x'-x)} \tag{5b}$$

† This follows from the fact that for bound states $\psi(x) \rightarrow e^{-C|x|}$ as $x \rightarrow \pm\infty$, whereas, for free particles, we can always regard the system as equivalent to one that is contained in a very large box, so that the wave function vanishes outside the box, and all integrals involving x^n will converge (see Chap. 10, Sec 20).

With the aid of the Fourier integral theorem, we obtain

$$\bar{p} = \int_{-\infty}^{\infty} \psi^*(x) \frac{\hbar}{i} \frac{\partial \psi(x)}{\partial x} \, dx \tag{6}$$

We now have $\bar{p}$ expressed in terms of $\psi(x)$ and $\psi^*(x)$. Formally, the result looks somewhat similar to the result for $\bar{x}$, except that the number x appearing in the integral has been replaced by the differential operator, $\frac{\hbar}{i}\frac{\partial}{\partial x}$. Hence, whenever we wish to find $\bar{p}$, we can do so with the wave function expressed as a function of position, provided that we replace the number p by the operator $\frac{\hbar}{i}\frac{\partial}{\partial x}$ as in eq. (6). This replacement of numbers by operators is merely a formal device. It is, however, exceedingly useful, because it creates a formalism that greatly resembles that of taking averages in classical physics, except that operators replace certain types of numbers. It now becomes apparent why x was sandwiched between ψ^* and ψ in eq. (2).

With the aid of this formalism, we can see in more detail why $\psi(x)$ is more than a wave of probability. Not only is the average value of x determined by $\int_{-\infty}^{\infty} \psi^* x \psi \, dx$, but the average value of p is

$$\frac{\hbar}{i} \int_{-\infty}^{\infty} \psi^* \frac{\partial \psi}{\partial x} \, dx$$

Hence, the way in which the wave amplitude changes with position (i.e., its slope) also has physical significance. Even when $\psi^*\psi$ is constant, for example, if $\psi = e^{ikx}$, $\partial\psi/\partial x$ is by no means equal to zero. Thus, we see that the wave function includes more than a determination of the probability of a given position, for its slope determines the mean value of the momentum.

7. Functions of the Momentum. If we have any function of the momentum that can be expressed as a power series $f(p) = \Sigma C_n p^n$), then it is easy to show by reasoning similar to that used in dealing with $\bar{p}$ that

$$\overline{f(p)} = \int_{-\infty}^{\infty} \psi^*(x) \left[\sum C_n \left(\frac{\hbar}{i} \frac{\partial}{\partial x} \right)^n \right] \psi(x) \, dx \tag{6a}$$

This result requires for its validity that $\overline{p^n}$ exist for arbitrary n, and that the above series converge. These conditions are satisfied for most wave functions and operators with which we shall deal, but where they are not satisfied, it is not possible to express $\overline{f(p)}$ directly in terms of $\psi(x)$. Instead, we must use eq. (4a) for $\overline{f(p)}$.

The rule in evaluating any function of p is, therefore, to operate as many times with $\frac{\hbar}{i}\frac{\partial}{\partial x}$ as there are powers of p in the term which is being evaluated.

Problem 2: Prove the validity of eq. (6a) for $\overline{p^2}$, and extend the results by induction to $\overline{p^n}$.

In proving eq. (6a), it is necessary to assume that $\dfrac{\partial^n\psi}{\partial x^n} \to 0$ as $x \to \infty$ for arbitrary values of n. For all wave functions which have ever arisen thus far in connection with any real problems, this requirement is satisfied. If, however, wave functions ever appear in which this requirement is not satisfied, then the rule of obtaining $\overline{p^n}$ from $\int \psi^* \left(\dfrac{\hbar}{i}\dfrac{\partial}{\partial x}\right)^n \psi$ will no longer be applicable. Because the convergence of the integral $\int \psi^* \dfrac{\partial^n\psi}{\partial x^n}\,dx$ simplifies the theory a great deal, it seems reasonable here to assume that $\dfrac{\partial^n\psi}{\partial x^n} \to 0$ as $x \to \pm\infty$ as a postulate that will be retained unless strong experimental reasons for discarding it should arise.

8. Operators in Momentum Space. The Momentum Representation. When we work with $\psi(x)$, we have what is called a *position representation*. It is often more convenient to work with $\varphi(k)$ which, after all, defines the wave function just as effectively as does $\psi(x)$. If $\varphi(k)$ is given, then the wave function has what is called a momentum representation.

In momentum space, the momentum is represented as a simple number:

$$p = \hbar k$$

just as in position space, the co-ordinate x is represented as a number. Thus, $\bar{p} = \int_{-\infty}^{\infty} \Phi^*(p)p\Phi(p)\,dp$ (eq. 4). On the other hand, it is easy to show by Fourier analysis that the mean value of x is equal to the following integral:

$$\bar{x} = -\frac{\hbar}{i}\int_{-\infty}^{\infty} \Phi^*(p)\,\frac{\partial\Phi(p)}{\partial p}\,dp \tag{7}$$

(Note the negative sign.)

Problem 3: Prove the above statement.

Thus, in analogy with the evaluation of $\bar{p}$ in x space, we have

$$\bar{x} = \int_{-\infty}^{\infty} \varphi^*(k) i\frac{\partial}{\partial k}\varphi(k)\,dk \tag{8}$$

If $f(x) = \Sigma A_n x^n$, it can then be shown in a similar way that

$$\bar{f}(x) = \int_{-\infty}^{\infty} \varphi^*(k)\left[\sum A_n\left(i\frac{\partial}{\partial k}\right)^n\right]\varphi(k)\,dk \tag{8a}$$

Hence, whether x or p is represented as a differential operator depends on which space we are using, position or momentum. Which of these representations we use is entirely a matter of convenience.

Problem 4: Prove eq. (8a), and state the conditions under which it is valid.

9. Linearity of Operators. The operators introduced thus far have a property called linearity. An operator O is linear if the following is true:

(1) O operating on *any* wave function yields a new wave function, in general, not the same one; i.e., $O\psi_1 = \psi_2$.

(2) $O(\psi_1 + \psi_2) = O\psi_1 + O\psi_2$.

(3) $CO\psi = OC\psi$, where C is an arbitrary constant.

The reader will readily verify that all operators introduced thus far have this property of linearity.

10. The Co-ordinate x as an Operator. In the position representation, x is represented as a number. It may, however, also be regarded as an operator having the particularly simple property that it multiplies the wave function by a number. It is obvious that x is a linear operator.

In momentum space, $p = \hbar k$ has exactly the same properties as does x in co-ordinate space.

11. Multiplication of Operators. Commutators. We may now consider the multiplication of two operators together. We have already dealt with the use of powers of p, and powers of x. What about products like xp or $x^n p^m$?

The operator xp operating on $\psi(x)$ has the following meaning. We first take $\frac{\hbar}{i}\frac{\partial \psi}{\partial x}$, then multiply it by x. The operator $px\psi$ means that we first multiply ψ by x, then differentiate. It is clear that the two are not the same. In fact, we have

$$(xp - px)\psi = \frac{\hbar}{i}\left(x\frac{\partial \psi}{\partial x} - \frac{\partial}{\partial x}(x\psi)\right) = i\hbar\psi \tag{9}$$

$(xp - px)$ is called the *commutator* of the two operators x and p. The commutator of co-ordinates and momenta satisfies the simple relation $(xp - px) = i\hbar$. Note that because of their failure to commute, operators are not the same as numbers. They have all properties of numbers except this one. That is, they can be added, subtracted, multiplied by a constant, and multiply each other. But when they multiply each other, they do not, in general, satisfy the rule $ba = ab$, satisfied by numbers.

Problem 5: Find the commutators $(x^n p^m - p^m x^n)$ and $(e^{ikx}p - pe^{ikx})$.

12. General Functions Expressed as Operators. This far, we have obtained a method for evaluating the average value of any function of x and of any function of p. But suppose that we wish to obtain the average of some function, such as xp, which contains both x and p simultaneously.

One might guess that this could be done in the co-ordinate representation by extending our rule and replacing p by the operator $\frac{\hbar}{i}\frac{\partial}{\partial x}$, just as when we have only $f(p)$ to deal with. Thus, we write tentatively

$$\overline{xp} \stackrel{?}{=} \int_{-\infty}^{\infty} \psi^* x \frac{\hbar}{i}\frac{\partial \psi}{\partial x}\,dx \tag{10}$$

In a similar way, we might, in the momentum representation, replace x by $i\hbar\frac{\partial}{\partial p}$, writing

$$\overline{xp} \stackrel{?}{=} \int_{-\infty}^{\infty} \Phi^*(p)\left(i\frac{\partial}{\partial p}p\right)\Phi(p)\,dp \tag{10a}$$

A minimum requirement that must be satisfied by such a tentative rule is that it gives the correct averages in the classical limit; in other words, it must satisfy the correspondence principle. To show that it does satisfy the correspondence principle, we consider a wave function ψ, which takes the form of a wave packet. Insofar as all classical results are concerned, no important physical quantity can change appreciably within the packet. This is because, in the classical limit, the packet looks essentially like a particle, so that, if the system is to be described classically, the specific wave-like properties of the packet must not matter. Hence, we may neglect all changes of x within the packet, and replace x by $\bar{x}$, which may be regarded as essentially constant. This means that $\bar{p}$ may now be computed by the usual rule given in eq. (4). Hence we see that our tentative rule does give at least the correct classical limit. A similar argument may be made using the momentum representation, and the same conclusion is obtained.

The above tentative rule is readily generalized to any function of x and p that can be expressed as a series of powers of x and p. What we do (in the position representation) is to replace the number p, whenever it occurs, by the operator $\frac{\hbar}{i}\frac{\partial}{\partial x}$. Thus we write

$$f(x, p) = \sum_{m,n} A_{nm}x^n p^m \longrightarrow \sum_{n,m} A_{nm}x^n\left(\frac{\hbar}{i}\frac{\partial}{\partial x}\right)^m$$

and

$$\bar{f}(x, p) \stackrel{?}{=} \int_{-\infty}^{\infty} \psi^*(x)\sum_{n,m} A_{nm}x^n\left(\frac{\hbar}{i}\frac{\partial}{\partial x}\right)^m \psi(x)\,dx \tag{11}$$

The definition of operators not expansible in power series will be discussed later.

13. Reality of Average Values and the Order of Factors. Although the above rule gives the correct classical limit for averages of $f(x, p)$, it is somewhat ambiguous because the order in which the operators x and p appear is vital, whereas in the corresponding classical expression, this order is immaterial. We shall now show that this ambiguity is removed in part by the requirement that the mean value of any real function of x and p must be real for an arbitrary ψ.

It is easy to show, for example, that $\overline{xp}$ as defined above is not real. To do this, we write

$$\overline{xp} = \int_{-\infty}^{\infty} \psi^* x \frac{\hbar}{i} \frac{\partial \psi}{\partial x}\, dx \tag{11a}$$

Integration by parts yields (noting that the integrated part vanishes)

$$\overline{xp} = -\frac{\hbar}{i} \int_{-\infty}^{\infty} \psi \frac{\partial}{\partial x} (x\psi^*)\, dx = -\frac{\hbar}{i} \int_{-\infty}^{\infty} \left(\psi^*\psi + \psi x \frac{\partial \psi^*}{\partial x} \right) dx \tag{11b}$$

We note that the second term on the right-hand side of the above expression is equal to the complex conjugate of $\overline{xp}$. Hence $\overline{xp}$ is equal to its complex conjugate plus an additional term; this means that $\overline{xp}$ cannot be real.

14. Hermitean Operators. To avoid such complex averages for quantities which are basically real, we shall require, as has already been stated, that the mean value be defined such that it is real for arbitrary ψ. If $O(p, x)$ is the operator in question, we require that $\bar{O}$ be equal to its complex conjugate. Now we have

$$\bar{O} = \int_{-\infty}^{\infty} \psi^*(x) O \psi(x)\, dx \tag{12}$$

The complex conjugate of $\bar{O}$ is found by taking the complex conjugate of all parts of the integral. Hence the reality requirement is equivalent to the following:

$$\int_{-\infty}^{\infty} \psi^*(x) O \psi(x)\, dx = \int_{-\infty}^{\infty} \psi(x) O^* \psi^*(x)\, dx \tag{13}$$

O^* refers to the complex conjugate of the operator O. For example in the operator $p = \frac{\hbar}{i} \frac{\partial}{\partial x}$, we get $p^* = -\frac{\hbar}{i} \frac{\partial}{\partial x}$. Operators satisfying eq. (13) are said to be *Hermitean.*

It is readily shown that p is a Hermitean operator. To do this, we write

$$\bar{p} = \int_{-\infty}^{\infty} \psi^*(x) \frac{\hbar}{i} \frac{\partial \psi}{\partial x}\, dx \tag{14}$$

Integration by parts, with the vanishing of the integrated part, yields

$$\bar{p} = \int_{-\infty}^{\infty} \psi \left(-\frac{\hbar}{i} \frac{\partial \psi^*}{\partial x} \right) dx \tag{14a}$$

We see that $\bar{p}$ is equal to its complex conjugate, and hence, that p is a Hermitean operator.

Problem 6: Show that p^n is a Hermitean operator, hence that $f(p) = \Sigma A_n p^n$ is also Hermitean, provided that all the A_n are real. Show that if any of the A_n are complex, $f(p)$ is not Hermitean.

Problem 7: Show that $f(x)$ is a Hermitean operator, if $f(x) = \Sigma A_n x^n$, and all the A_n are real. Show that if any of the A_n are complex, $f(x)$ is not Hermitean.

Problem 8: Show that if $\frac{\partial^n \psi}{\partial x^n}$ does not approach zero as $x \to \pm \infty$, the operator $\left(\frac{\hbar}{i} \frac{\partial}{\partial x}\right)^{n+1}$ is not necessarily Hermitean.

From the problems, we see that the requirement of reality of averages is automatically satisfied for any real function of x or p. On the other hand, for functions of x and p together, it is not necessarily satisfied. To satisfy the reality condition, we shall now show that, in general, one must take the mean of two possible orders in which x and p may appear. Consider, for example

$$\left(\overline{\frac{xp + px}{2}}\right) = \frac{\hbar}{2i} \int_{-\infty}^{\infty} \psi^* \left(x \frac{\partial}{\partial x} + \frac{\partial}{\partial x} x\right) \psi \, dx \tag{15}$$

Integration by parts of $\int \psi^* x \frac{\partial \psi}{\partial x} dx$ yields (noting that the integrated part vanishes) $-\int \psi \frac{\partial}{\partial x} (x\psi^*) \, dx$; while integration of $\int \psi^* \frac{\partial}{\partial x} (x\psi) \, dx$ yields $-\int \psi x \frac{\partial \psi^*}{\partial x} dx$. Thus, we get

$$\left(\overline{\frac{xp + px}{2}}\right) = -\frac{\hbar}{2i} \int_{-\infty}^{\infty} \psi \left(x \frac{\partial}{\partial x} + \frac{\partial}{\partial x} x\right) \psi^* \, dx \tag{15a}$$

Hence, we have proved that $\left(\overline{\frac{xp + px}{2}}\right)$ is equal to its complex conjugate, and that the operator is therefore Hermitean.

Problem 9: Prove that the operator $\sum A_{nm} \left(\frac{p^n x^m + x^m p^n}{2}\right)$ is Hermitean, if all the A_{nm} are real.

15. Modified Rule for Average of $f(x, p)$. We can now give a more definite rule for getting the average value of any function of x and p. We not only replace p by $\frac{\hbar}{i} \frac{\partial}{\partial x}$, wherever it occurs, but we remove the ambiguity in order of factors by taking the mean between the two possible orders of x and p. In doing this, we always order the function in such a way that all factors involving p occur together, as do all factors involving x. Then, we replace $p^n x^m$ by $\frac{1}{2}(p^n x^m + x^m p^n)$. In this way, the operator is made Hermitean, and all averages computed with it are certain to be real. This process is called *Hermitization of the operator.*

The procedure adopted above in which all factors of p and all factors of x are grouped together before Hermitization is still somewhat arbitrary. Thus, to find the quantum-mechanical analogue of the classical product $(px)^2$, we could take either $\frac{p^2x^2 + x^2p^2}{2}$ or $\left(\frac{xp + px}{2}\right)^2$.

Problem 10: Show that the above two assumptions do not yield identical results, but that they differ by quantities of order $\hbar^2$.

There is still, therefore, some ambiguity in how we should define quantum-mechanical operators in the evaluation of averages. The results of the various definitions, however, differ by quantities of the order of $\hbar^2$, and therefore become important only at the quantum-mechanical level of accuracy. Since we are here trying to construct a consistent theory limited only by the requirement that it yield the correct classical behavior at high quantum numbers, it is clear that our present procedure is not definitive enough to remove this type of ambiguity. Instead, as mentioned before, we should regard this line of approach as somewhat heuristic, in the sense that it leads to a theory with the correct general form, but in which some of the details may later have to be filled in by direct reference to experiment. Further modifications, however, can only produce corrections of the order of some power of $\hbar$.

At present, there is no experimental basis for deciding which of the various alternative methods of ordering factors is right, simply because no systems have been found for which the predicted results depend on which method of ordering is adopted, as long as all observable quantities are calculated from averages of Hermitean operators. In the absence of any experimental data, we have chosen the order suggested in Problem 9, because it leads to the simplest mathematical expressions. Until some experiment is found for which the predicted results depend on the method of Hermitization, there will be no way to decide which is the correct method.

16. Hermitean Conjugate Operators. We have seen from the previous discussion that, in general, a non-Hermitean operator will yield a complex average value unless the operator is first Hermitized, that is, the orders in which x and p appear must be interchanged and half the sum of both orders taken. Nevertheless, it is often convenient to work in a purely mathematical way with non-Hermitean operators. Such a non-Hermitean operator may be regarded as a kind of operator analogue of a complex number. With any complex number, $C = a + ib$, it is always possible to define a complex conjugate number $C^* = a - ib$. Can we define a conjugate operator in an analogous way? It seems natural to require of the conjugate of any operator O, that its average value be the complex conjugate of the average value of O itself. More precisely stated, if we denote the conjugate of the operator O by $O^\dagger$, we require that

$$\int \psi^* O^\dagger \psi \, dx = \int (\psi O^* \psi^*) \, dx \tag{16}$$

From this definition, we see that if O is a Hermitean operator, then $O^\dagger = O$. In other words, a Hermitean operator is self-conjugate, according to our definition. This is in analogy with a real number, which is its own complex-conjugate.

It should be noted that $O^\dagger$ is not, in general, equal to O^*, the latter being obtained by simply replacing every i appearing in O by $-i$. For example, consider the operator,

$$p = \frac{\hbar}{i}\frac{\partial}{\partial x}$$

which is Hermitean. Thus, we have $p^\dagger = p$. Yet

$$p^* = -\frac{\hbar}{i}\frac{\partial}{\partial x} = -p$$

Hence, $p^\dagger \neq p^*$. To distinguish $O^\dagger$ from the complex conjugate of O, we refer to the former as the Hermitean conjugate of O. It may also be called the complex adjoint of O.

We may ask why one chooses to define the conjugate of an operator in this particular way. The answer is that the average value of an operator is the only thing having physical significance. Hence, the nearest quantum analogue of a complex function is an operator having complex averages. The appearance of complex numbers in the operator itself, however, is not particularly significant. For example, in the position representation the operator p is given by $p = \frac{\hbar}{i}\frac{\partial}{\partial x}$; yet its average is always real. Thus, the operator that approaches the complex conjugate function in the classical limit is not necessarily the complex conjugate operator but is, in general, the Hermitean conjugate.

One important question is whether it is always possible to find an operator that satisfies our definition of the Hermitean conjugate, if we start with an arbitrary operator O. The answer is that this can always be done. We shall not prove this here, but merely state that a proof can be given.

Problem 11: Show by integration by parts that $(xp)^\dagger = px$.

From an arbitrary operator O, we can always construct a Hermitean operator by taking the mean between the operator and its Hermitean conjugate. Thus, we write

$$\frac{O + O^\dagger}{2} = H \tag{17}$$

where H is a Hermitean operator. That this is true is fairly clear, from the fact that $(O^\dagger)^\dagger = O$. This provides a close analogy to finding the real part of a complex number c, from the formula $a = \frac{c + c^*}{2}$.

Is there an analogy to the imaginary part b of the complex number, which is equal to $b = \dfrac{c - c^*}{2i}$? To study this question, let us consider the operator A, which we define as

$$A = \frac{O - O^\dagger}{2} \tag{18}$$

and

$$A^\dagger = \frac{O^\dagger - (O^\dagger)^\dagger}{2} = \frac{O^\dagger - O}{2} \tag{19}$$

The operator A, therefore, possesses the property that $A^\dagger = -A$, or, in other words, that it is equal to the negative of its Hermitean conjugate. Such an operator is called *anti-Hermitean.*

From any anti-Hermitean operator A we can always construct a Hermitean operator by multiplying by i. To prove this, we first note that i is itself an anti-Hermitean operator, as can be seen by taking its average value.

$$\bar{i} = \int_{-\infty}^{\infty} \psi^* i \psi \, dx = -\int_{-\infty}^{\infty} \psi^* i^* \psi \, dx \tag{20}$$

(Note that $i^* = -i$.)

Problem 12: From the above results, prove that $i(0 - 0^\dagger)$ is a Hermitean operator.

Let us denote the Hermitean operator $\dfrac{O - O^\dagger}{2i}$ by the symbol B. Then we can write

$$O = \left(\frac{O + O^\dagger}{2}\right) + i\left(\frac{O - O^\dagger}{2i}\right) = H + iB$$

In this way, we have decomposed an arbitrary operator into the sum of two parts, one of which has a real average value and one of which has an imaginary average value. This is the complete analogue of the numerical expression $c = a + ib$. Note, however, that A and B do not necessarily commute and that, as a result, this decomposition is not entirely equivalent to what is done with numbers. For example, with numbers, we have

$$(a + ib)(a - ib) = a^2 + b^2$$

With operators, however, we have

$$(H + iB)(H - iB) = H^2 + B^2 + i(BH - HB)$$

17. Generalized Definition of a Hermitean Operator. Consider the Hermitean operator H, which satisfies the equation

$$\int_{-\infty}^{\infty} \psi^* H \psi \, dx = \int_{-\infty}^{\infty} \psi (H^* \psi^*) \, dx \tag{21}$$

for an arbitrary ψ. Let us write $\psi = \psi_1 + \psi_2$ where ψ_1 and ψ_2 are arbitrary functions. We get

$$\int_{-\infty}^{\infty} (\psi_1^* H\psi_1 + \psi_2^* H\psi_2)\,dx + \int_{-\infty}^{\infty} (\psi_1^* H\psi_2 + \psi_2^* H\psi_1)\,dx$$
$$= \int_{-\infty}^{\infty} (\psi_1 H^*\psi_1^* + \psi_2 H^*\psi_2^*)\,dx + \int_{-\infty}^{\infty} (\psi_1 H^*\psi_2^* + \psi_2 H^*\psi_1^*)\,dx \quad (22)$$

With the aid of (21), which enables us to cancel out the first integrals, we get

$$\int_{-\infty}^{\infty} (\psi_1^* H\psi_2 - \psi_2 H^*\psi_1^*)\,dx = \int_{-\infty}^{\infty} (\psi_1 H^*\psi_2^* - \psi_2^* H\psi_1)\,dx \quad (22a)$$

This relation must remain true for arbitrary ψ_1 and ψ_2; hence it must be true if we multiply ψ_1 by a constant factor, e^{ia}, and ψ_2 by e^{ib}. We then get

$$e^{i(b-a)} \int_{-\infty}^{\infty} (\psi_1^* H\psi_2 - \psi_2 H^*\psi_1^*)\,dx = e^{i(a-b)} \int_{-\infty}^{\infty} (\psi_1 H^*\psi_2^* - \psi_2^* H\psi_1)\,dx \quad (22b)$$

This relation can remain true for arbitrary a and b only if the integrals above are zero. Thus, we get

$$\int_{-\infty}^{\infty} \psi_1^* H\psi_2\,dx = \int_{-\infty}^{\infty} \psi_2 H^*\psi_1^*\,dx \quad (23)$$

This is an important result. It says that in any integral like

$$\int_{-\infty}^{\infty} \psi_1^* H\psi_2\,dx$$

we can obtain the same result if H is Hermitean by allowing H^* to operate on ψ_1^*, even when ψ_1 and ψ_2 are different. Our original definition, eq. (13), allowed us to do this only when ψ_1 and ψ_2 were the same.

18. Generalized Definition of Hermitean Conjugates. If an operator O is not Hermitean, we can generalize the definition of the Hermitean conjugate operator in a similar manner. To do this, we write

$$O = A + iB$$

where A, B are Hermitean operators. Then we have

$$O^\dagger = A^\dagger - iB^\dagger = A - iB$$

noting that $i^\dagger = -i$ and that i commutes with B.

Let us now consider the integral

$$\int_{-\infty}^{\infty} \psi_1^* O^\dagger \psi_2\,dx = \int_{-\infty}^{\infty} \psi_1^*(A - iB)\psi_2\,dx \quad (24)$$

Since A and B are Hermitean, we get

$$\int_{-\infty}^{\infty} \psi_1^*(A - iB)\psi_2\,dx = \int_{-\infty}^{\infty} \psi_2(A^* - iB^*)\psi_1^*\,dx = \int_{-\infty}^{\infty} \psi_2 O^*\psi_1^*\,dx_1 \quad (24a)$$

and we conclude that

$$\int_{-\infty}^{\infty} \psi_1^* O^\dagger \psi_2\,dx = \int_{-\infty}^{\infty} \psi_2 O^*\psi_1^*\,dx \quad (25)$$

This means that whenever $O^\dagger$ operates on the function to the right, we can evaluate the same integral by allowing O^* to operate on the function to the left.

19. Application to Finding Hermitean Conjugate of Product of Two Operators. Given two operators A and B and their Hermitean conjugates $A^\dagger$ and $B^\dagger$, what is the Hermitean conjugate of their product AB? To obtain this quanity, we write by definition (eq. 23) that

$$\int \psi_1^*(AB)^\dagger \psi_2 \, dx = \int \psi_2(A^*B^*)\psi_1^* \, dx \tag{26}$$

We now note that $B\psi_1$ is a new wave function, which we may call φ. Thus we get

$$\int \psi_1^*(AB)^\dagger \psi_2 \, dx = \int \psi_2 A^* \varphi^* \, dx \tag{26a}$$

Application of the definition of Hermitean conjugate yields

$$\int \psi_2 A^* \varphi^* \, dx = \int \varphi^* A^\dagger \psi_2 \, dx = \int (B^*\psi_1^*)(A^\dagger \psi_2) \, dx \tag{26b}$$

Writing $A^\dagger \psi_2 = f$, we have

$$\int (B^*\psi_1^*)(A^\dagger\psi_2) \, dx = \int f B^* \psi_1^* \, dx = \int \psi_1^* B^\dagger f \, dx = \int \psi_1^* B^\dagger A^\dagger \psi_2 \, dx \tag{26c}$$

Thus we get

$$\int \psi_1^*(AB)^\dagger \psi_2 \, dx = \int \psi_1^* B^\dagger A^\dagger \psi_2 \, dx \tag{27}$$

and we conclude that

$$(AB)^\dagger = B^\dagger A^\dagger \tag{27a}$$

If A and B are Hermitean, then

$$(AB)^\dagger = BA \tag{27b}$$

Note that even though A and B are separately Hermitean, their product is not necessarily Hermitean.

Problem 13: What relation must exist between B and A, in order that AB be Hermitean, if A and B are separately Hermitean?

20. Application to Commutators. If A and B are Hermitean, we see that the Hermitean conjugate of their commutator is

$$(AB - BA)^\dagger = (BA - AB) = -(AB - BA)$$

Thus, the commutator of two Hermitean operators is anti-Hermitean. To make the commutator Hermitean, we multiply by i. Thus, we write $i(BA - AB) =$ a Hermitean operator.

Problem 14: Show directly that $i(p^2x - xp^2)$ is Hermitean.

21. A Theorem on Hermitean Operators. We shall now prove the following theorem which will be very useful later: If the average value of the Hermitean operator H is zero for arbitrary ψ, then $H\psi$ must be identically zero for all ψ. This means that we may write $H \equiv 0$.

To prove this, we start out with the definition of $\bar{H}$

$$\bar{H} = \int \psi^* H \psi \, dx = 0 \tag{28}$$

We now write $\psi = \psi_1 + \psi_2$ where ψ_1 and ψ_2 are arbitrary.

$$\bar{H} = \int \psi_1^* H \psi_1 \, dx + \int \psi_2^* H \psi_2 \, dx + \int \psi_1^* H \psi_2 \, dx + \int \psi_2^* H \psi_1 \, dx = 0 \tag{29}$$

We note that, by definition, the first two terms are zero. Hence, we have

$$\int \psi_1^* H \psi_2 \, dx + \int \psi_2^* H \psi_1 \, dx = 0 \tag{29a}$$

Since this relation is true for arbitrary ψ_1, we may replace ψ_1 by $e^{ia}\psi_1$ where a is a constant. We obtain

$$e^{-ia} \int \psi_1^* H \psi_2 \, dx = -e^{ia} \int \psi_2^* H \psi_1 \, dx \tag{29b}$$

This can be true for arbitrary a only if each of the integrals vanish. Thus, we have

$$\int \psi_1^* H \psi_2 \, dx = 0 \qquad \text{for arbitrary } \psi_1 \text{ and } \psi_2 \tag{29c}$$

We can then choose $\psi_1 = H\psi_2$. Thus, we get

$$\int (H^* \psi_2^*)(H\psi_2) \, dx = 0 \tag{29d}$$

But the integrand of the above expression is the absolute value of the function $H\psi_2$; hence it is by definition either zero or positive everywhere. The integral can therefore be zero only if $H\psi_2 = 0$ for all ψ_2 or in other words, if $H \equiv 0$.

Summary on Operator Formalism

We have obtained a method of expressing average values of various quantities in terms of the wave function $\psi(x)$ or, alternatively, in terms of its Fourier component $\Phi(p)$. In doing this, we have found that it was convenient to introduce certain linear operators, which have some of the formal properties of numbers, but which do not commute. These operators have no direct physical significance, but have meaning only as mathematical auxiliaries, used in computing average values of physically observable quantities. Yet, they are extremely convenient to use, and greatly simplify the task of calculating these averages. These operators therefore find much use in the quantum theory.

Derivation of Schrödinger's Equation

22. General Form of Schrödinger's Equation. In Part I we have already seen that, for a free particle, the wave function satisfies the equation†

$$i\hbar \frac{\partial \psi}{\partial t} = -\frac{\hbar^2}{2m} \frac{\partial^2 \psi}{\partial x^2} \tag{30}$$

† The entire discussion will be given in one dimension only, but it is very easily generalized to three dimensions.

Further arguments were given which showed that this equation must always be of first order in the time, in order that we may obtain motions of wave packets that approach the classical limit correctly, and also in order that there may exist a conserved probability density function with generally sensible properties. This means that, in general, even when forces are present, we can write

$$i\hbar \frac{\partial \psi}{\partial t} = H(\psi) \tag{31}$$

where H is some function of ψ, not involving time derivatives of ψ. This is the general wave equation.

In Sec. (1) we showed that a fundamental postulate of quantum theory is the hypothesis of linear superposition of waves. This means that if ψ_1 and ψ_2 are possible wave functions, $a\psi_1 + b\psi_2$ is also a possible wave function. But since all permissible wave functions must be solutions of the wave equation, we conclude that the sum of two solutions is also a solution. The wave equation must therefore be a *linear* equation, and H must be a linear operator of the type which has already been discussed.

23. Conservation of Probability and Hermiticity of H. An additional requirement on H is that it must be Hermitean, in order to conserve probability; i.e., $\partial P/\partial t = 0$, where P is the integrated probability, i.e., $P = \int \psi^* \psi \, dx$. This means that we require that

$$\frac{\partial P}{\partial t} = \int \left(\frac{\partial \psi^*}{\partial t} \psi + \psi^* \frac{\partial \psi}{\partial t} \right) dx = 0$$

From the wave equation (31), we can express $\partial \psi / \partial t$ in terms of ψ, noting also that

$$-i\hbar \frac{\partial \psi^*}{\partial t} = H^* \psi^*$$

We obtain

$$\frac{\partial P}{\partial t} = \frac{i}{\hbar} \int (\psi H^* \psi^* - \psi^* H \psi) \, dx \tag{32}$$

From our definition of the Hermitean conjugate operator, eq. (25), this reduces to

$$\frac{\partial P}{\partial t} = \frac{i}{\hbar} \int \psi^* (H^\dagger - H) \psi \, dx \tag{33}$$

If the above is to be zero for arbitrary ψ then, according to the definition given in eq. (13), H must be a Hermitean operator. Conversely, if H is Hermitean, probability is always conserved.

24. Determination of *H* from Correspondence Principle. Further limitations on H will now be obtained from the correspondence principle. This can be done in a manner that is basically the same as that used in obtaining the wave equation for a free particle. In the case of the free particle we obtained the wave equation from the de Broglie relations, $E = h\nu$ and $p = h/\lambda$. But the latter were obtained from the requirement that wave packets move with the classical particle velocity, plus the requirement that $E = p^2/2m$. Let us now require that the average velocity of a wave packet be equal to the classical particle velocity, even when forces are present. This is essentially a requirement that our theory satisfy the correspondence principle or, in other words, that we get the classical result when we do not consider the finer details of the wave properties of matter, but only ask how the wave packet moves on the average.

25. General Formula for Time Derivative of Average Value of a Variable. In order to carry out our program of further limiting the form of H with the aid of the correspondence principle, we shall need a formula for the time rate of change of the average value of any operator O. Thus, we wish to evaluate

$$\frac{d}{dt}\bar{O} = \int \left(\frac{\partial \psi^*}{\partial t} O\psi + \psi^* O \frac{\partial \psi}{\partial t} \right) dx + \int \psi^* \frac{\partial O}{\partial t} \psi \, dx \tag{34}$$

Note that $\partial O/\partial t$ refers only to explicit time dependence of O. Thus, for the operators x and p, $\partial O/\partial t = 0$, but for $O = x + pt$ we get $\partial O/\partial t = p$. Note also that $\partial O/\partial t$ is Hermitean whenever O is Hermitean.

Once again, we express $\partial \psi/\partial t$ in terms of ψ, and $\partial \psi^*/\partial t$ in terms of ψ^*, obtaining

$$\frac{d}{dt}\bar{O} = \frac{i}{\hbar} \int [(H^*\psi^*)(O\psi) - (\psi^* O H \psi)] \, dx + \int \psi^* \frac{\partial O}{\partial t} \psi \, dx \tag{35}$$

From our definition of the Hermitean conjugate operator (16), we write

$$\frac{d}{dt}\bar{O} = \frac{i}{\hbar} \int \psi^*(H^* O - OH)\psi \, dx + \int \psi^* \frac{\partial O}{\partial t} \psi \, dx \tag{36}$$

Since H must be Hermitean, this reduces to

$$\frac{d}{dt}\bar{O} = \frac{i}{\hbar} \int \psi^*(HO - OH)\psi \, dx + \int \psi^* \frac{\partial O}{\partial t} \psi \, dx \tag{37}$$

This means that once we know the commutator of any operator O with H, we can always obtain the time rate of change of O.

Note that we have evaluated the net rate at which the average value of O changes. This change may be the result, in part, of changes of ψ and, in part, of changes of the operator O itself, arising from the explicit time dependence. It is, therefore, important to realize that $(d/dt)\bar{O}$ is very different from $\partial O/\partial t$.

26. Application to Evaluation of Average Motion of Wave Packet. Newton's laws of motion may be written (classically)

$$\frac{dp}{dt} = -\frac{\partial V}{\partial x} \quad \text{and} \quad \frac{dx}{dt} = \frac{p}{m} \tag{38}$$

In quantum theory, we cannot even define derivatives of x and p in the classical sense because there is no such thing as a continuous particle trajectory (see Chap. 8, Sec. 6). The nearest thing to derivatives is to be found by considering the time rates of change of the average values of x and p. In the classical limit, where the width of the wave packet can be neglected, these must become equal to the classically calculated values. This condition can most easily be satisfied by requiring that in the quantum theory, Newton's laws of motion be true when expressed in terms of the averages, $\bar{x}$ and $\bar{p}$. We therefore write

$$\left.\begin{aligned} \frac{d}{dt}\bar{p} &= -\int \psi^* \frac{\partial V}{\partial x} \psi\, dx \\ \frac{d}{dt}\bar{x} &= \int \psi^* \frac{p}{m} \psi\, dx = \int \psi^* \frac{\hbar}{im}\frac{\partial \psi}{\partial x}\, dx \end{aligned}\right\} \tag{39}$$

The first equation means that the rate of change of the average momentum is equal to the average force; the second, that the rate of change of the average position is equal to the average of p/m.

Let us evaluate $d\bar{p}/dt$ from eq. 36. We note that $\partial p/\partial t = 0$. It will be convenient to write $H = p^2/2m + g$, where g is an operator which vanishes when there are no forces.† We obtain

$$\frac{d\bar{p}}{dt} = \frac{i}{\hbar}\int \psi^* \left[\left(\frac{p^2}{2m}p - p\frac{p^2}{2m}\right) + (gp - pg)\right]\psi\, dx \tag{40}$$

Since p commutes with p^2, we are left with $\left(\text{writing } p = \frac{\hbar}{i}\frac{\partial}{\partial x}\right)$

$$\int \psi^* \left(g\frac{\partial \psi}{\partial x} - \frac{\partial}{\partial x} g\psi\right) dx = -\int \psi^* \frac{\partial V}{\partial x}\psi\, dx$$

or

$$\int \psi^* \left(\frac{\partial g}{\partial x} - \frac{\partial V}{\partial x}\right)\psi\, dx = 0 \tag{41}$$

Now, the above must be true for an arbitrary ψ. The simplest way to satisfy the relation is to choose $g = V$. More generally, we may write $g = V + f$, where

$$\int \psi^* \frac{\partial f}{\partial x}\psi\, dx = 0$$

† See Chap. 3, eq. (29) for Schrödinger's equation for a free particle

for arbitrary ψ. But this can be satisfied for arbitrary ψ only if $\partial f/\partial x = 0$, and hence, if $f = f(p)$.† Thus, we write in general

$$H = \frac{p^2}{2m} + V(x) + f(p) \tag{42}$$

We next show that $f(p) = 0$ by requiring the satisfaction of the equation

$$\frac{d}{dt}\bar{x} = \int \psi^* \frac{p}{m} \psi \, dx \tag{43}$$

To do this, we write (noting that $xV - Vx = 0$)

$$\frac{d}{dt}\bar{x} = \frac{i}{\hbar}\int \psi^* (Hx - xH)\psi \, dx$$

$$= \frac{i}{\hbar}\int \psi^* \left[-\frac{\hbar^2}{2m}\frac{\partial^2}{\partial x^2}(x\psi) + x\frac{\hbar^2}{2m}\frac{\partial^2 \psi}{\partial x^2}\right] dx + \frac{i}{\hbar}\int \psi^*[f(p)x - xf(p)]\psi \, dx$$

But $\frac{\partial^2}{\partial x^2}(x\psi) = 2\frac{\partial \psi}{\partial x} + x\frac{\partial^2 \psi}{\partial x^2}$. Hence we get

$$\frac{d}{dt}\bar{x} = \frac{1}{m}\int \psi^* \frac{\hbar}{i}\frac{\partial \psi}{\partial x}\, dx + \frac{i}{\hbar}\int \psi^* [f(p)x - xf(p)]\psi \, dx \tag{44}$$

To satisfy eq. (44) we require that

$$\int \psi^*[f(p)x - xf(p)]\psi \, dx = 0$$

for arbitrary ψ. This is possible only if $f(p)x - xf(p) = 0$. The reader will readily convince himself that no function of p, other than a constant, can satisfy this requirement. For example,

$$(p^n x - xp^n)\psi = \left(\frac{\hbar}{i}\right)^n \left[\frac{\partial^n}{\partial x^n}(x\psi) - x\frac{\partial^n \psi}{\partial x^n}\right] \neq 0$$

Hence $f(p)$ adds at most a constant to H, and we may absorb this into $V(x)$ if we wish. We have thus proved that Newton's equations of motion are satisfied on the average if ψ is a solution of the following wave equation:

$$i\hbar\frac{\partial \psi}{\partial t} = \left[-\frac{\hbar^2}{2m}\frac{\partial^2}{\partial x^2} + V(x)\right]\psi \tag{45}$$

Ehrenfest was the first to show that this wave equation leads to the satisfaction of Newton's equations of motion on the average; this result is therefore called Ehrenfest's theorem.

If we write the above result in operator notation, we get

$$i\hbar\frac{\partial \psi}{\partial t} = \left(\frac{p^2}{2m} + V\right)\psi = H\psi \tag{46}$$

† See theorem in Sec. 21.

The operator H is therefore just the classical Hamiltonian function, in which p is replaced by the operator $\frac{\hbar}{i}\frac{\partial}{\partial x}$.

27. General Rule for Obtaining H. It can be shown that this result may be generalized as follows: The wave equation satisfied by ψ is $i\hbar(\partial\psi/\partial t) = H\psi$ where H can be obtained from the classical Hamiltonian function by the replacement of each momentum p canonically conjugate to the co-ordinate* q by the operator $\frac{\hbar}{i}\frac{\partial}{\partial q}$. If p and q occur together as common factors of a term, the order must be symmetrized to make H a Hermitean operator.

28. Is the Above Equation the Most General One Possible? Does this prescription yield the most general wave equation consistent with the correspondence principle? The answer is that it does not. The wave equation was derived by making quantum-mechanical averages of the changes of x and p with time bear the same relation to each other that they do classically. But in any classical experiment, the energy is never measured to within a precision which is better than many units of $h\nu$. (The observation of spectral lines is a purely quantum-mechanical piece of data, because classical theory implies that spectra are continuous.) If terms were added to H, which led to contributions of the order of $h\nu$ to the mean energy, their results could not be observed in a purely classical experiment. The Hamiltonian operator is therefore not uniquely defined by the correspondence principle alone. For example, small corrections are introduced by spin and other relativistic effects which have been neglected, but which may be dealt with by a more careful treatment. (In this connection, see Sec. 2.)

The above derivation is, therefore, not a unique derivation of Schrödinger's equation, but one which leads to the main terms in the equation. Its purpose is to enable one to grasp the origin of the equation from the physical background, rather than requiring us to pick the equation out of the air, and then to deduce the physical background from the equation. The lack of uniqueness of the equation is also a useful thing to know because it shows us where modifications may be made, if necessary, to produce agreement with nonclassical experimental results.

29. Significance of Wave Equation. From the wave equation, we obtain the way in which the wave function changes with time. We have seen that, in this way, the classical motion is determined. Not only the classical averages, but all other averages change in a way that is determined by the wave equation. Schrödinger's equation is therefore analogous to Newton's equation of motion in classical physics, but unlike Newton's equation, it determines only the probability of real events.

* Strictly speaking, this rule is true only in rectangular coordinates. The operators can then be expressed in nonrectangular co-ordinates by means of an appropriate transformation. See, for example, Chaps. 14 and 15.

30. General Definition of Probability Current. We can now show that whenever the Hamiltonian takes the form

$$H = \frac{p^2}{2m} + V(x)$$

the probability current is the same as that given in Chap. 4, eq. (4). To see this we write (noting that V is always real)

$$\frac{\partial P(x)}{\partial t} = \frac{\partial \psi^*}{\partial t}\psi + \psi^* \frac{\partial \psi}{\partial t} = \frac{\hbar}{2mi}\left(\frac{\partial^2 \psi^*}{\partial x^2}\psi - \psi^* \frac{\partial^2 \psi}{\partial x^2}\right) \tag{47}$$
$$= -\frac{\hbar}{2mi}\frac{\partial}{\partial x}\left(\psi^* \frac{\partial \psi}{\partial x} - \psi \frac{\partial \psi^*}{\partial x}\right)$$

where we have used Schrödinger's equation to eliminate $\partial\psi/\partial t$ and $\partial\psi^*/\partial t$. Writing $S = \frac{\hbar}{2mi}\left(\psi^* \frac{\partial \psi}{\partial x} - \psi \frac{\partial \psi^*}{\partial x}\right)$, we get

$$\frac{\partial P}{\partial t} + \frac{\partial S}{\partial x} = 0 \tag{48}$$

in agreement with Chap. 4, eq. (4). It is easy to generalize the results to three dimensions. We obtain

$$H = -\frac{\hbar^2}{2m}\nabla^2\psi + V\psi \qquad \text{and} \qquad S = \frac{\hbar}{2mi}(\psi^* \nabla\psi - \psi \nabla\psi^*) \tag{49}$$

31. Interpretation of $\bar{H}$ as Average Energy. Does H have a physical meaning other than that of determining the way in which ψ changes with time? To see whether it does, let us consider its average value.

$$\bar{H} = \int \psi^*\left[-\frac{\hbar^2}{2m}\frac{\partial^2}{\partial x^2} + V(x)\right]\psi\, dx = \frac{\overline{p^2}}{2m} + \overline{V(x)} \tag{50}$$

Because H is Hermitean, its average value is certainly real. Furthermore, we see that $\bar{H}$ is equal to $\overline{\frac{p^2}{2m} + V(x)}$. In the classical limit, this is just the total energy of the system. Since quantum-mechanical averages are obtained by the rule of replacing p by the operator, $\frac{\hbar}{i}\frac{\partial}{\partial x}$, we conclude that $\bar{H}$ must therefore, in general, represent the average value of the energy.

32. Conservation of Energy. In classical physics, whenever the Hamiltonian function is not explicitly a function of the time, we can prove that H is a constant of the motion, so that energy is conserved. To prove this, we write

$$\frac{d}{dt}[H(p, q, t)] = \frac{\partial H}{\partial p}\dot{p} + \frac{\partial H}{\partial q}\dot{q} + \frac{\partial H}{\partial t} \tag{51}$$

According to the canonical equations,

$$\dot{p} = -\frac{\partial H}{\partial q} \qquad \text{and} \qquad \dot{q} = \frac{\partial H}{\partial p}$$

Thus, we get

$$\frac{dH}{dt} = -\frac{\partial H}{\partial p}\frac{\partial H}{\partial q} + \frac{\partial H}{\partial q}\frac{\partial H}{\partial p} + \frac{\partial H}{\partial t} = \frac{\partial H}{\partial t} \tag{52}$$

If $\partial H/\partial t = 0$ (as is usually the case), then $dH/dt = 0$, so that H equals a constant. In quantum theory, we obtain the mean rate of change of $\bar{H}$ with equation (37)

$$\frac{d\bar{H}}{dt} = \frac{i}{\hbar}\int \psi^* (HH - HH)\psi \, dx + \int \psi^* \frac{\partial H}{\partial t} \psi \, dx = \overline{\frac{\partial H}{\partial t}} \tag{53}$$

Thus, once again, if H is not explicitly a function of t, then $\bar{H}$ is a constant of the motion.

Thus we have shown that, just as the classical canonical equations of motion guarantee the conservation of energy in all cases in which H is not a function of time, Schrödinger's equation guarantees, in a similar way, the conservation of the mean energy in quantum theory.

CHAPTER 10

Fluctuations, Correlations, and Eigenfunctions

1. Statistical Fluctuations and Correlations. We have already seen that, in any measuring process, the observed value of a variable will, in general, fluctuate from one measurement to the next. It is useful to have a measure of this fluctuation. In classical physics, such fluctuations are often measured in terms of the mean square of the deviations of the actual value from the mean. Thus, the mean fluctuation in x is

$$\bar{F} = \overline{(x - \bar{x})^2} = \overline{[x^2 - 2x\bar{x} + (\bar{x})^2]} = \overline{x^2} - 2\bar{x}\bar{x} + (\bar{x})^2 = \overline{x^2} - (\bar{x})^2 \quad (1)$$

It is clear that if there is no fluctuation, i.e., if $x = \bar{x}$ in all measurements, then $\bar{F} = 0$. Since $(x - \bar{x})^2$ is necessarily a positive quantity, it is also clear that if any measurements at all occur in which x differs from $\bar{x}$, then $\bar{F}$ will not be zero. The larger the difference between x and $\bar{x}$, the larger will be its contribution to $\bar{F}$.

Of course, it must be remembered that a knowledge of $\bar{F}$ and $\bar{x}$ by no means defines the probability function $P(x)$. It merely defines the general way in which the value of x is distributed about the mean. In fact, it may be said that $\overline{(x - \bar{x})^2}$ yields a measure of the uncertainty in x, since it tells us roughly about how much its value will fluctuate from one measurement to the next. We may therefore write $\overline{(x - \bar{x})^2} = (\Delta x)^2$, where Δx is the uncertainty in x.

2. Extension to Quantum Theory. These ideas are easily extended to the quantum theory. If we know the wave function $\psi(x)$, then we know

$$P(x) = \psi^*(x)\psi(x)$$

and we therefore know the mean value of any function of x. In particular, the mean value of F is given by

$$\bar{F} = \int_{-\infty}^{\infty} \psi^*(x)(x - \bar{x})^2\psi(x)\, dx = (\Delta x)^2 \quad (2)$$

Very often, however, it is not convenient to discuss the wave function in all its precise detail, because this requires a solution of the wave equation. Sometimes, all we wish to know are a few crude properties of the distribution, such as $\bar{x}$ and $\bar{F}$, which tell us roughly the main general properties of the distribution. In particular, we shall see later that it is possible to draw certain conclusions about the magnitude of $(x - \bar{x})^2$, even when we do not know the precise form of ψ. The introduction of averages like $\bar{F}$ is therefore often very helpful.

We may, in a similar way, introduce the mean fluctuation in p, which is

$$\overline{(p - \bar{p})^2} = \int_{-\infty}^{\infty} \psi^* \left(\frac{\hbar}{i}\frac{\partial}{\partial x} - \bar{p}\right)^2 \psi \, dx = (\Delta p)^2 = \begin{pmatrix}\text{uncertainty} \\ \text{in } p\end{pmatrix}^2 \quad (3)$$

If we know the value of ψ, we can calculate the value of $(\Delta x)^2$ and $(\Delta p)^2$ from the preceding formulas. Later we shall investigate the general value of products like $\Delta x \, \Delta p$ and show that the uncertainty principle is always satisfied for any wave function. For the present, however, we can consider some special wave functions, as in the following problem, and show that it is satisfied in these cases.

Problem 1: Show that the uncertainty principle ($\Delta p \, \Delta x \geqq \hbar/2$) is satisfied for the following three wave packets:†

$$\psi = \alpha_1 e^{-\alpha x^2/2}$$

$$\psi = \alpha_2 e^{-\alpha|x|}$$

$$\psi = \frac{\alpha_3}{(\alpha^2 + x^2)^2}$$

In each case, α is chosen to normalize the total integrated probability.

3. Correlations between p and x. In any statistical distribution of two classical variables, such as p and x, an important question is whether or not the two variables are correlated. For example, among people there is no unique relation between height and weight, yet the two are statistically correlated in the sense that a taller person tends to be heavier than a shorter person. In a similar way, one may ask whether the distribution in p has any correlation with the distribution in x. In other words, does a large p tend to occur simultaneously with a large x or, vice versa, does it tend to occur with a small x? If either of these statistical relations exists, then it can be said that p and x are correlated. On the other hand, if no such relation exists, then the two are and may be said to be statistically independent.

Suppose that the height h and the weight w of people were statistically independent. This would mean that the distribution in height would be independent of the weight. Thus we could write that the probability of a given height between h and $h + dh$ is $R(h)\,dh$. In a similar way, the probability of any weight between w and $w + dw$ is independent of h, so that this probability is $S(w)\,dw$. Now the probability of two independent results is, by definition, the product of their separate probabilities. Thus, the probability that the height lies between h and $h + dh$ and the weight lies between w and $w + dw$ is given by the product,

$$P(h, w)\,dh\,dw = R(h)S(w)\,dh\,dw$$

† The exact size of the uncertainty depends on the way Δp and Δx are defined. For the definition we are now using, $\hbar/2$ is the correct value.

We can also see that if the distribution cannot be written as a product, then the two variables are not statistically independent. Consider, for example, $P(h, w) = 1/(h^2 + w^2)$. It is clear that the distribution function in h cannot be regarded as independent of w.

4. A Quantitative Measure of the Correlations in Classical Theory. A good quantitative measure of the extent of correlation of two classical quantities is the following mean value:

$$C_{1,1} = \overline{(x - \bar{x})(p - \bar{p})} = xp - \bar{x}\bar{p} \tag{4}$$

If the distribution in x is statistically independent of that in p, then $C_{1,1}$ must be zero, for in this case,

$$\overline{xp} = \int R(x)S(p)xp \, dx \, dp = \bar{x}\bar{p}$$

Hence, we get $C_{1,1} = 0$.

It is possible, however, for $C_{1,1}$ to be zero even when correlations are present. For example, large $|x|$ may be correlated with large $|p|$ but in such a way that for each value of x, it is equally likely that p is either positive or negative. Hence both $\overline{xp}$ and $\bar{x}\bar{p}$ vanish, even though some correlations are still present.

In order to obtain a measure of correlations of this more subtle type, we may consider the function

$$C_{2,2} = \overline{x^2p^2} - (\bar{x})^2(\bar{p})^2 \tag{5}$$

It is clear that $C_{2,2}$ vanishes if x and p are statistically independent, but does not vanish in the above case, where $C_{1,1}$ vanished.

In general, however, still more subtle types of correlation may exist, in which $C_{1,1}$ and $C_{2,2}$ both vanish. In order to test for all possible types of correlation, we may study the functions

$$C_{nm} = \overline{x^np^m} - (\bar{x})^n(\bar{p})^m \tag{6}$$

5. Specification of a Classical Statistical System through Mean Values of x^np^m. The above discussion shows the significance of the mean values of all kinds of terms of the form x^np^m. Terms like $\overline{x^n}$ and $\overline{p^m}$ can be interpreted in a way similar to that done with $\bar{x}$, $\overline{x^2}$, $\bar{p}$, and $\overline{p^2}$, but in terms of measurements of more complex and subtle properties of the fluctuation. Hence, if we know all about the fluctuations and correlations, we can calculate all the $\overline{x^np^m}$. From the products, x^np^m, we can construct an arbitrary function $f(x, p)$. Its average is

$$\bar{f}(x, p) = \overline{\Sigma A_{nm}x^np^m} = \Sigma A_{nm}\overline{x^np^m}$$

This means that the fluctuations and correlations determine the average of any physically observable quantity, so that they describe all features of the distribution that we have any need to know about. In statistics,

$x^n p^m$ is called the n, m moment of the distribution, in analogy with moments of momentum in mechanics.

If there were no fluctuations, we should have $\bar{f}(x, p) = f(\bar{x}, \bar{p})$. It is only because of fluctuations that the two differ. Each kind of function possesses sensitivity to certain kinds of fluctuations and correlations, depending on how large the coefficient of each $x^n p^m$ is in its power series expansion.

6. Quantum Definition of Correlations. In quantum theory, the correlation functions are obtained simply by replacing p by the operator, $\frac{\hbar}{i}\frac{\partial}{\partial x}$, and taking the mean between the two orders in which x and p occur. Thus

$$C_{nm} = \frac{1}{2}\int \psi^*\left[x^n\left(\frac{\hbar}{i}\frac{\partial}{\partial x}\right)^m + \left(\frac{\hbar}{i}\frac{\partial}{\partial x}\right)^m x^n\right]\psi\,dx - \left(\int \psi^* x^n \psi\,dx\right)\left[\int \psi^*\left(\frac{\hbar}{i}\frac{\partial}{\partial x}\right)^m \psi\,dx\right] \tag{7}$$

Problem 2: Find $C_{1,1}$ and $C_{2,2}$ for the function $\psi = \alpha e^{-\alpha x^2/2}$, where α is a normalizing factor.

Problem 3: Prove that $C_{1,1} = 0$ for any real wave function.

7. Application to Spreading Wave Packet for a Free Particle. Let us evaluate $C_{1,1}$ for the wave packet defined in Chap. 3, eqs. (14) and (22), which spreads out as time passes. (We assume as a special case that $\bar{p} = \bar{x} = 0$.) Since the wave function is initially real, $C_{1,1}$ vanishes at $t = 0$; hence there are no simple correlations between momentum and position at this time. Let us now see what happens when $t > 0$. To do this, we write the wave function as follows:

$$\psi = \alpha \exp\left[-(A - iB)\frac{x^2}{2}\right] \tag{8}$$

α is a normalizing factor, defined such that

$$\int_{-\infty}^{\infty} \psi^*\psi\,dx = 1 = \alpha^*\alpha \int_{-\infty}^{\infty} e^{-Ax^2}\,dx$$

$$A = \frac{(\Delta k)^2}{1 + \frac{\hbar^2 t^2}{m^2}(\Delta k)^4} \quad \text{and} \quad B = (\Delta k)^4 \frac{\hbar t}{m}\frac{1}{1 + \frac{\hbar^2 t^2}{m^2}(\Delta k)^4} \tag{9}$$

Now, $\bar{x} = 0$ since

$$(\alpha^*\alpha)\int_{-\infty}^{\infty} e^{-Ax^2/2} x e^{-Ax^2/2}\,dx = 0$$

Thus, we obtain

$$C_{1,1} = \frac{\hbar}{2i}\alpha^*\alpha \int_{-\infty}^{\infty} \exp\left[-(A + iB)\frac{x^2}{2}\right]\left(x\frac{\partial}{\partial x} + \frac{\partial}{\partial x}x\right) \exp\left[-(A - iB)\frac{x^2}{2}\right] dx \tag{10}$$

Let us integrate the term

$$\int_{-\infty}^{\infty} \exp\left[-(A + iB)\frac{x^2}{2}\right] \frac{\partial}{\partial x} x \exp\left[-(A - iB)\frac{x^2}{2}\right] dx$$

by parts, noting that the integrated part vanishes. Using the fact that

$$\frac{\partial}{\partial x} \exp\left[-(A + iB)\frac{x^2}{2}\right] = -x(A + iB) \exp\left[-(A + iB)\frac{x^2}{2}\right]$$

we then obtain

$$C_{1,1} = \frac{\hbar}{2i} \alpha^*\alpha \int_{-\infty}^{\infty} \exp\left[-(A + iB)\frac{x^2}{2}\right] x^2(A + iB - A + iB) \exp\left[-(A - iB)\frac{x^2}{2}\right] dx \quad (11)$$

or

$$C_{1,1} = \hbar\alpha^*\alpha B \int_{-\infty}^{\infty} e^{-Ax^2}x^2\, dx = \frac{\hbar B}{2A} = \frac{\hbar^2(\Delta k)^2}{2m} t = \frac{(\Delta p)^2}{2m} t \quad (12)$$

where $(\Delta p)^2$ is the initial uncertainty in momentum.

We see that although there are no correlations at $t = 0$, correlations begin to appear with the passage of time. The physical reason for these is simply that the faster particles move farther, so that a high momentum tends to become correlated with the covering of a large distance.

Another way of seeing the source of correlations is to note that the momentum operator, operating on the wave function $\exp[-(A - iB)x^2/2]$ is

$$\frac{\hbar}{i}\frac{\partial}{\partial x} \exp\left[-(A - iB)\frac{x^2}{2}\right] = \hbar x(B + iA) \exp\left[-(A - iB)\frac{x^2}{2}\right]$$

Because of the factor $\exp(iBx^2/2)$, which represents roughly a momentum of the order of Bx, it follows that the electron will tend to have a large momentum when x is large.

The results just obtained show clearly how the wave function is far more than a wave of probability. The probability $P(x)$ is just

$$P(x) = \psi^*(x)\psi(x) = \alpha^*\alpha\, e^{-Ax^2/2}$$

Yet, even though the B terms do not appear in the expression for $P(x)$, they have an effect on the correlation between momentum and position. Thus, the phase of the wave function [in this case, $\exp(iAx^2/2)$] contains an enormous amount of information of all types, much of it consisting of rather subtle interrelations between values of various quantities.†

8. Semi-classical Picture of Particle with Uncertain Position and Momentum. In Chap. 5, Sec. 4, it was pointed out that to a first approximation the effects of the quantum properties of matter can be pictured in

† In this connection, see Chap. 6, Secs. 6 and 8.

terms of a classical particle of uncertain momentum and position, provided that we assume that

$$\Delta p\, \Delta q \gtrsim (\sim\hbar)$$

In Chap. 6, Sec. 11, however, we saw that this picture must be used with caution, because it does not provide a completely correct account of the wave properties of matter. Yet, as long as its limitations are understood, it is often very helpful. For example, the result of the previous section was interpreted tacitly in terms of the idea that the electron acts, to some extent, like a classical particle with a probability distribution of momenta and positions. The spread of the wave packet is then related to the fact that particles of different velocity move different distances in the same time. This process introduces correlations between p and x, since those particles which move the fastest also cover the greatest distance.

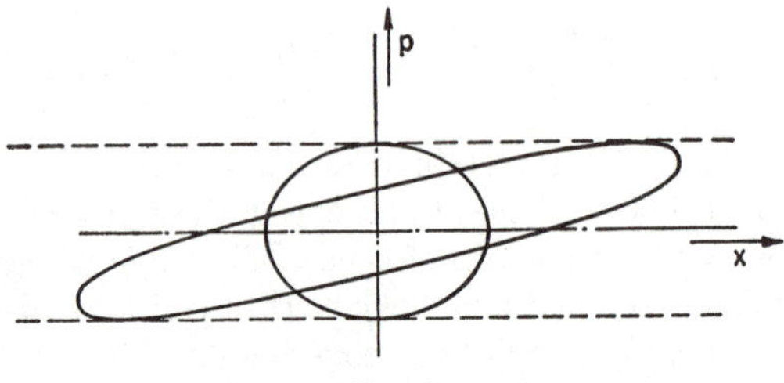

FIG. 1

It is of interest to represent the spread of the probability distribution in terms of a diagram in phase space. The original probability distribution was proportional to $\exp[-x^2/(\Delta x)^2]$ in position space, and by Fourier analysis (see Chap. 3, Sec. 2) it was shown to be proportional to $\exp[-p^2/(\Delta p)^2]$ in momentum space. A classical particle with this probability distribution would be most likely to be found in an elliptical region of phase space, centered at the origin with semiaxes Δx and Δp, as shown in Fig. 1. The area of this ellipse is roughly $\pi\, \Delta p\, \Delta x_0 \sim h/2$. With the passage of time, particles of positive momentum move to the right, whereas those of negative momentum move to the left. Thus, the ellipse is skewed, as shown in Fig. 1. The center of the ellipse remains unaltered. Δp also does not change, because each particle moves with constant velocity, but Δx is increased.

Problem 4: Prove that the area of the ellipse remains unchanged. (This is a special case of Liouville's theorem.)

Problem 5: Prove that the correlation function, $C_{1,1}$ obtained by assuming Gaussian classical distributions of particle momenta and positions, is the same as that obtained from the quantum theory. Show, however, that $C_{2,2}$ does not remain the same for both cases, but differs by quantities of the order of $\hbar$. (This is a special case of the general result that the wave properties of matter cannot be understood completely in terms of the classical concept of a particle of uncertain position and momentum.)

Problem 6: Analyze in detail how one can take advantage of correlations in this case in such a way as to permit the product $\Delta x \, \Delta p$ to become known within a minimum uncertainty of the order of $\hbar$, despite the spread of the wave packet.

9. A Generalization of the Uncertainty Principle. We have already seen that the values of x and p cannot be measured simultaneously and that the minimum uncertainties in these quantities satisfy the relation $\overline{(\Delta x)^2}\,\overline{(\Delta p)^2} \cong \hbar^2$. We shall now show that the minimum uncertainties for any two Hermitean operators, A and B, satisfy the rule

$$\overline{(\Delta A)^2}\,\overline{(\Delta B)^2} \cong \left[\overline{\frac{i}{2}(AB - BA)}\right]^2 \tag{13}$$

(Since $i(AB - BA)$ is a Hermitean operator, the quantity on the right is always positive.) We see that if we let

$$A = p = \frac{\hbar}{i}\frac{\partial}{\partial x} \qquad \text{and} \qquad B = x$$

we get the usual uncertainty relation, because

$$i(pq - qp) = \hbar$$

The uncertainty $\overline{(\Delta A)^2}$ is equal to $\int \psi^*(A - \bar{A})^2\psi \, dx$. For simplicity, let us write $A - \bar{A} = \alpha$ and $B - \bar{B} = \beta$. Then, what we wish to evaluate is

$$I = \left(\int \psi^*\alpha^2\psi \, dx\right)\left(\int \psi^*\beta^2\psi \, dx\right) \tag{14}$$

Because α is Hermitean, we can write

$$\int \psi^*\alpha^2\psi \, dx = \int \psi^*\alpha(\alpha\psi) \, dx = \int (\alpha^*\psi^*)(\alpha\psi) \, dx = \int |\alpha\psi|^2 \, dx \tag{15}$$

The same may be done with β. This gives us

$$I = \left(\int |\alpha\psi|^2 \, dx\right)\left(\int |\beta\psi|^2 \, dx\right) \tag{16}$$

At this point, it is convenient to represent the integrals as limits of sums. Thus,

$$I = \left[\sum_i |\alpha\psi_i|^2 \, \Delta x_i\right]\left[\sum_j |\beta\psi_j|^2 \, \Delta x_j\right] = \sum_{ij} |\alpha\psi_i|^2|\beta\psi_j|^2 \, \Delta x_i \, \Delta x_j \tag{17}$$

We now use a theorem called *Schwartz's inequality*. This theorem states that

$$\sum_{i,j} |A_i|^2|B_j|^2 \geqq \left|\sum_i A_i^*B_i\right|^2 \tag{18}$$

To prove this, we write

$$\left|\sum_i A_i^*B_i\right|^2 = \sum_{i,j} A_i^*B_iA_jB_j^*$$

and we then consider the following quantity:

$$Q = \sum_{i,j} |A_i|^2|B_j|^2 - \left|\sum_i A_i^*B_i\right|^2 = \sum_{i,j} (|A_i|^2|B_j|^2 - A_i^*B_iA_jB_j^*) \tag{19}$$

We note that if $i = j$, the contribution to Q in the sum vanishes. But if i and j are each given fixed values, not equal to each other, then the corresponding terms are:

$$|A_i|^2|B_j|^2 + |A_j|^2|B_i|^2 - A_i^*B_j^*A_jB_i - A_j^*B_i^*A_iB_j = |A_iB_j - A_jB_i|^2 \quad (20)$$

This quantity, however, is always either positive or zero. Hence, Q is made up of terms which are never negative, and we conclude that $Q \geqq 0$. This proves Schwartz's inequality. It is clear that $Q = 0$ only if each term in the series is zero, or if $A_iB_j - A_jB_i = 0$. This means that $A_i/A_j = B_i/B_j$, or that $A_i = CB_i$ where C is a constant.

Application of Schwartz's inequality to eq. (17) now yields

$$I \geqq \left|\sum_i (\alpha^*\psi_i^*)(\beta\psi_i)\ \Delta x_i\right|^2 = \left|\int \alpha^*\psi^*\beta\psi\ dx\right|^2 \quad (21)$$

Because α is Hermitean, we get

$$I \geqq \left|\int \psi^*\alpha\beta\psi\ dx\right|^2$$
$$= \left|\int \psi^*\left(\frac{\alpha\beta + \beta\alpha}{2}\right)\psi\ dx + \int \psi^*\left(\frac{\alpha\beta - \beta\alpha}{2}\right)\psi\ dx\right|^2 \quad (22)$$

The operator $(\alpha\beta + \beta\alpha)$ is a Hermitean operator; hence its average is always real and may be denoted by the number P. We also know that $i(\alpha\beta - \beta\alpha)$ is also Hermitean; hence its average is some real number Q.

Thus, we can write

$$I \geqq |P - iQ|^2 = P^2 + Q^2$$

Let us now note that

$$\frac{\alpha\beta + \beta\alpha}{2} = \frac{1}{2}[(A - \bar{A})(B - \bar{B}) + (B - \bar{B})(A - \bar{A})] \quad (23)$$

which is just the correlation function, $C_{1,1}$, for the two variables A and B. The best we can do about the correlation function is to make it zero. Whether $C_{1,1}$ is zero or not, however, the following relation holds:

$$(\Delta A)^2(\Delta B)^2 = I \geqq \left|\int \psi^*\left(\frac{\alpha\beta - \beta\alpha}{2}\right)\psi\ dx\right|^2 = \left|\frac{\overline{AB - BA}}{2}\right|^2 \quad (24)$$

This is a very important result. It means that whenever A and B do not commute, they cannot be measured simultaneously with perfect accuracy. If A and B do commute, however, then it can be proved that it is possible to measure them simultaneously, but we shall not prove it here.

When is $(\Delta A)^2(\Delta B)^2$ a minimum? In order that the equal sign hold in eq. (24), two requirements must be satisfied:

(1) $\alpha\psi = C\beta\psi$ (This means that the = sign holds in Schwartz's inequality.)

(2) $\overline{\dfrac{\alpha\beta + \beta\alpha}{2}} = 0$ (This means that α and β are uncorrelated.)

Let us consider the special case

$$\alpha = (x - \bar{x}), \qquad \beta = (p - \bar{p})$$

Let us also restrict ourselves to $\bar{x} = \bar{p} = 0$; the generalization to arbitrary values of these quantities is fairly easy. We then get

$$p\psi = Cx\psi \qquad \text{or} \qquad \frac{\hbar}{i}\frac{\partial\psi}{\partial x} = Cx\psi \tag{25}$$

Integration yields $\psi = \exp(iCx^2/2)$. Now, it was shown previously that with wave functions of the type, $\exp[-(A + iB)x^2/2]$, $C_{1,1}$ vanishes only if B is zero. Hence to satisfy condition (2), we must make C imaginary. But we must also guarantee that $\int_{-\infty}^{\infty} \psi^*\psi\, dx$ exists, i.e., that the total probability be normalized to unity. This can be satisfied only if we write $C = ia$ where a is positive. We then get $\psi = \exp(-ax^2/2)$, the well-known Gaussian distribution. More generally, when $\bar{x}$ and $\bar{p}$ are not zero, we get

$$\psi = \exp(i\bar{p}x) \exp\left[-a\frac{(x - \bar{x})^2}{2}\right] \tag{26}$$

This latter is the most general function for which the equal sign holds in the uncertainty principle.

Problem 7: Prove the above result for the most general ψ.

Problem 8: Calculate $(\Delta x)^2$ and $(\Delta p)^2$ for the wave function (8), which starts out as a Gaussian function at $t = 0$. Show that $(\Delta p)^2$ remains constant, but that $(\Delta x)^2$ increases with time. Hence $(\Delta x)^2(\Delta p)^2 > \hbar^2/2$ after $t = 0$, even though the equal sign holds at $t = 0$. Yet the probability function $P = \alpha^*\alpha\, e^{-Ax^2}$ is Gaussian. Explain why the equal sign does not hold, even though P is Gaussian. (This is another example of the physical significance of phase relations of the wave function.)

10. The Unusual Properties of the Gaussian Wave Function. We have seen already that the Gaussian wave function has the unusual property that it takes the same form in momentum space as in position space. It can be shown that the most general wave function with this property is $\exp[-ax^2 + ikx + ibx^2]$. Furthermore, as shown in Sec. 9, when $b = 0$, it has the property that it makes $\Delta x\, \Delta p$ a minimum and is the only type of function with this property. The peculiar properties of the Gauss function make it useful in many problems. The Gauss function also arises directly in connection with the wave functions of a harmonic oscillator (see Chap. 13). Gaussian wave functions are therefore of considerable importance in the quantum theory.

11. The Many-particle Problem. With the aid of the discussion of statistical correlation given previously, we can now show how to set up the wave equation for a system containing more than one particle. We have already indicated in Sec. 1 that when many particles are present, the wave function is a function of the co-ordinates of all of them. To show how probability can be defined for such a system, let us first consider the special case of two independent particles. Let the wave function of the first be $\psi_A(x_1)$ and that of the second $\psi_B(x_2)$. The probability that the first particle lies between x_1 and $x_1 + dx_1$ is

$$P_A(x_1)\,dx_1 = |\psi_A(x_1)|^2\,dx_1$$

whereas the probability that the second particle lies between x_2 and $x_2 + dx_2$ is $P_B(x_2)\,dx_2 = |\psi_B(x_2)|^2\,dx_2$. Because the probabilities are independent, the probability that the first particle lies between x_1 and $x_1 + dx_1$, and the second particle lies between x_2 and $x_2 + dx_2$ is, as shown in Sec. 3, just the product of the separate probabilities, or

$$P(x_1, x_2)\,dx_1\,dx_2 = \psi_A^*(x_1)\psi_A(x_1)\psi_B^*(x_2)\psi_B(x_2)\,dx_1\,dx_2$$

This result suggests a natural generalization of our formalism to the case of two particles. Thus, we are led to define a wave function that depends on both particle co-ordinates

$$\psi(x_1, x_2) = \psi_A(x_1)\psi_B(x_2)$$

The probability function is

$$P(x_1, x_2)\,dx_1\,dx_2 = \psi^*\psi\,dx_1\,dx_2$$

When the two particles are independent, the wave functions themselves, as well as the probabilities, are therefore expressible as a product of functions of each variable separately.

Problem 9: Prove that if $\psi_A(x_1)$ and $\psi_B(x_2)$ are separately normalized, then their product is also normalized (when integrated over x_1 and x_2).

If there are forces between the two particles, however, the probability distributions will cease to be independent. Thus, if one particle is an electron and the other a proton, they will attract each other and tend to form a hydrogen atom. Here it is most likely that both particles will always be found much closer together than they would be on the average in a random distribution. To designate this possibility, we write the general probability function as $P(x_1, x_2)$. In such circumstances, the wave function must also cease to be expressible as a product, so that we write it as $\psi(x_1, x_2)$. The formula for probability is still, however,

$$P(x_1, x_2)\,dx_1\,dx_2 = \psi^*(x_1, x_2)\psi(x_1, x_2)\,dx_1\,dx_2$$

It is easily shown that with those definitions for P and ψ, the formalism for operators and averages goes through in exactly the same way as for the one-particle problem.

As an example, let us obtain the wave equation for two interacting particles. The wave function is $\psi(x_1, y_1, z_1; x_2, y_2, z_2)$ where x_1 is the x co-ordinate of the first particle, while x_2 is that of the second particle, etc. According to the generalization to two particles of the rule given in Chap. 9, Sec. 27, the Hamiltonian operator is

$$H = \frac{p_1^2}{2m_1} + \frac{p_2^2}{2m_2} + V(x_1, y_1, z_1; x_2, y_2, z_2)$$

where $V(x_1, y_1, z_1; x_2, y_2, z_2)$ is the total potential energy of both particles. It includes the interaction energy between the two particles, as well as all other sources of potential energy. For example, if the particles each have a charge e, we obtain

$$V = \frac{e^2}{r_{1,2}}$$

where $$r_{1,2} = [(x_1 - x_2)^2 + (y_1 - y_2)^2 + (z_1 - z_2)^2]^{1/2}$$

Schrödinger's equation becomes

$$i\hbar \frac{\partial \psi}{\partial t} = \left[-\frac{\hbar^2}{2}\left(\frac{1}{m_1}\nabla_1^2 + \frac{1}{m_2}\nabla_2^2\right) + V(x_1, y_1 z_1; x_2, y_2, z_2)\right]\psi$$

Since H is a linear operator, the equation is still linear. It is an equation for a wave in a six-dimensional space and, therefore, it is, in general, very difficult to solve. The generalization to an arbitrary number of particles is obvious. Although these equations are often very difficult to treat, various approximate methods exist and a number of solutions for various many-body problems has been obtained which are in satisfactory agreement with experiment. Thus, we have what is, in principle at least, a method for dealing with an arbitrary system. This means that, for all cases, the wave equation may be regarded as playing the same fundamental role in quantum theory as Newton's laws of motion do in classical theory.

12. Eigenvalues and Eigenfunctions of Operators. In general, we have seen that when a system is in a given quantum state, i.e., when its wave function is given, the observed value of any variable cannot be predicted accurately but fluctuates about some mean, from one observation to the next. Can a system ever be put into such a quantum state that some variable has a definite, predictable, and reproducible value which never fluctuates? The answer is that it can.

In order that a variable, O, take on the same value in all observations, it is necessary, first, that its mean fluctuation vanish, or that

$$\overline{O^2} - (\bar{O})^2 = \int \psi^*[O^2 - (\bar{O})^2]\psi\, dx = 0 \qquad (27)$$

In a similar manner, it is necessary that the mean of any function of O be equal to that function of the mean, or that

$$\bar{f}(O) = f(\bar{O}) \tag{28}$$

If this requirement were not satisfied, we could infer that there must be instances when O fluctuates from its mean value.

It is clear that one method of satisfying this requirement is to choose ψ such that

$$O\psi_C = C\psi_C \tag{29}$$

where C is a constant. If this relation is satisfied, C is called an *eigenvalue*, or *characteristic value* of the operator O, and ψ_C is called an *eigenfunction*, or *characteristic function*, belonging to the eigenvalue C.

If eq. (29) is satisfied, then we have

$$\bar{O} = \int \psi_C^* O \psi_C \, dx = C \int \psi_C^* \psi_C \, dx = C \qquad \text{(since } \psi \text{ is normalized)}$$

Similarly, we have

$$\overline{O^n} = \int \psi_C^* O^n \psi_C \, dx = C^n \int \psi_C^* \psi \, dx = C^n$$

Hence, for an arbitrary function expressible as a power series, we get

$$\overline{f(O)} = \sum_n A_n \overline{O^n} = \sum_n A_n C^n = f(C) \tag{30}$$

Thus, we see that if ψ_C is an eigenfunction of the operator O, then the mean value of an arbitrary function of O is equal to that function of the mean value. We can conclude that there are no fluctuations in the value of O. This does not mean, however, that there are no fluctuations in the values of other operators. On the contrary, when an observable, such as the momentum p, is given a definite value, we already know from the uncertainty principle that the conjugate variable x must become completely indefinite.

It may be shown that if ψ is not an eigenfunction of the operator O, then the value of O must show some fluctuation.

Problem 9: Prove the above statement.

13. Examples of Eigenfunctions and Eigenvalues in Position Space. (1) *Momentum Operator.* The eigenfunctions of the momentum operator are found by solving the equation

$$\frac{\hbar}{i} \frac{\partial \psi}{\partial x} = p\psi \tag{31}$$

where p is a constant. We obtain

$$\psi = e^{ipx/\hbar} \tag{31a}$$

the well-known plane wave which, as we already know, represents a state of definite momentum, p. (p is the eigenvalue, $e^{ipx/\hbar}$ the eigenfunction.)

Strictly speaking, the above eigenfunctions cannot, in general, be normalized to unity, because the integrated probability $\int_{-\infty}^{\infty} \psi^*\psi \, dx$ diverges. Let us recall, however (Chap. 3, Sec. 2), that in any real problem, the wave function must take the form of a packet, since the particle is known to exist somewhere within a definite region, such as in the space surrounded by the apparatus. To obtain a bounded and therefore normalizable packet, we can integrate over momenta with an appropriate weighting factor, as is done in Chap. 3, eq. (3). In practice, however, this packet can be made so large relative to any physically significant dimensions that we can usually ignore its bounded character in computing almost any quantity other than the normalization coefficient. Similarly, the spread of momenta Δp in the packet can be made so small that it is usually a good approximation to ignore this also. We shall, therefore, quite frequently refer to wave functions like $\exp(ipx/\hbar)$, with the understanding that they really refer to packets which are very broad in position space and correspondingly narrow in momentum space.

(2) *Energy Operator*. For a free particle, the energy is $E = p^2/2m$. The equation for an eigenfunction is then

$$E\psi = -\frac{\hbar^2}{2m}\frac{\partial^2\psi}{\partial x^2} \tag{32}$$

The solution is

$$\psi = A \exp\left(i\,\frac{\sqrt{2mex}}{\hbar}\right) + B \exp\left(-i\,\frac{\sqrt{2mex}}{\hbar}\right) \tag{32a}$$

where A and B are arbitrary constants.

In other words, to each eigenvalue E belong two linearly independent eigenfunctions, which may be added in an arbitrary linear combination. Writing $p = \sqrt{2mE}$, we see that the two eigenfunctions are $\exp\left(\frac{ipx}{\hbar}\right)$ and $\exp\left(-\frac{ipx}{\hbar}\right)$, corresponding to the two possible directions of momentum which a particle of a given energy may have. This result holds in one dimension. In three dimensions, there are an infinite number of directions corresponding to a given energy and, therefore, an infinite number of eigenfunctions belonging to one eigenvalue E.

14. Degenerate Operators. If more than one independent eigenfunction belongs to a given eigenvalue, the operator is said to be *degenerate* for that eigenvalue. For example, the eigenvalues of the operator E have a two-fold degeneracy in a one-dimensional problem, and an infinite-fold degeneracy in three dimensions.

If only one linearly independent eigenfunction belongs to each eigenvalue of an operator, the operator is said to be *nondegenerate.* p is a nondegenerate operator.

(3) *The Eigenfunction of the Operator* x. We have seen that the operators p and E have fairly simple eigenfunctions in position space What about the operator x? It is clear that if we take any smooth. function of x, then the operator x multiplies it by a variable number and, therefore, by definition, no continuous function of x can be an eigen-function of x. What we need is a function that is multiplied by the same number regardless of the value of x; that is, we want

$$x\psi = C\psi \tag{33}$$

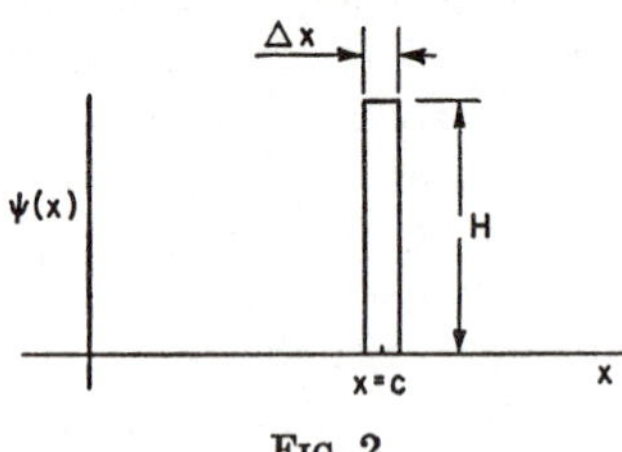

FIG. 2

where C is a constant. The only kind of a function that will satisfy this requirement is one which is zero everywhere except at $x = C$. Such a function is highly singular; it is better to think of it as the limit of a nonsingular function. We might, for example, take the function plotted in Fig. 2. The function is zero everywhere, except in the region Δx. In order to normalize it to unity, we write

$$\int_{-\infty}^{\infty} \psi^*\psi\, dx = \int_{C-\Delta x/2}^{C+\Delta x/2} H^2\, dx = H^2\,\Delta x = 1$$

or

$$H = \frac{1}{\sqrt{\Delta x}}$$

As $\Delta x \to 0$, $H \to \infty$. In the limit, we therefore obtain a function which is zero everywhere, except at $x = C$, and which is infinite at $x = C$, but which approaches infinity in such a way that the wave function remains normalized.

There are many different kinds of functions that approach eigenfunctions of the operator x. For example, consider the limit of a normalized Gaussian function

$$\psi = \lim_{\Delta x \to 0} \frac{\exp\left[-\dfrac{x^2}{2(\Delta x)^2}\right]}{\sqrt{2\pi}\,\Delta x} \tag{34}$$

As $\Delta x \to 0$, $\psi \to \infty$, but $\int_{-\infty}^{\infty} \psi^*\psi\, dx = 1$ independently of Δx. For small Δx, the above approaches a sharply peaked function of width Δx and height $1/\sqrt{2\pi}\,\Delta x$.

Problem 10: Show that $\dfrac{A}{(x - x_0)^2 + (\Delta x)^2}$ approaches an eigenfunction of x as Δx approaches zero. Evaluate the constant A so as to normalize the probability.

15. The Dirac Delta Function. The previous eigenfunctions of the operator x are normalized in such a way that the integrated probability is unity. It is often convenient mathematically to use eigenfunctions which are instead normalized so that

$$\int_{-\infty}^{\infty} \psi(x)\,dx = 1 \tag{35}$$

Thus, in the function defined in eq. (1), we require that

$$\int_{x_0-\Delta x/2}^{x_0+\Delta x/2} H\,dx = H\,\Delta x = 1 \qquad \text{or} \qquad H = \frac{1}{\Delta x} \tag{36}$$

With Gaussian-type functions,

$$\psi = A \exp\left[\frac{-x^2}{2(\Delta x)^2}\right]$$

we require that

$$A\int_{-\infty}^{\infty} \exp\left[\frac{-x^2}{2(\Delta x)^2}\right] dx = 1 \qquad \text{or} \qquad A = \frac{1}{\sqrt{2\pi}\,\Delta x} \tag{37}$$

More generally, we consider any sharply peaked function $S_{\Delta x}(x - x_0)$, which is appreciable only in a region of width Δx centering about $x = x_0$ and which has the property that

$$\int_{-\infty}^{\infty} S_{\Delta x}(x - x_0)\,dx = 1.$$

In the limit, as Δx approaches zero, such a function approaches what Dirac has called a *delta function*, denoted by $\delta(x - x_0)$. The only two important properties of the δ function are: (1) It is zero everywhere except at one point, and (2) it is infinite at this one point, but approaches infinity in such a way that its integral is unity.

Strictly speaking, the δ function cannot be given a meaning in the customary mathematical sense, because it must be infinite at the point $x = x_0$. Whenever we refer to the δ function, we always mean a function $S_{\Delta x}(x - x_0)$, which can be made as sharply peaked as necessary by making Δx small enough. Considerable writing and discussion is saved by using the δ function as if it were a proper mathematical function of the ordinary type, but keeping in mind its real definition as the limit of $S_{\Delta x}$ as Δx approaches zero.

The most important property of the δ function is obtained by considering the integral

$$I = \int_{-\infty}^{\infty} f(x) S_{\Delta x}(x - x_0)\,dx \tag{38}$$

where $f(x)$ is an arbitrary continuous function. If Δx is made small enough, then the variation of $f(x)$ in the region in which the integrand is appreciable can be made as small as we please. Thus, the function $f(x)$ can be replaced by the constant $f(x_0)$ and taken outside of the integral. We are left with

$$I \cong f(x_0) \int_{-\infty}^{\infty} S_{\Delta x}(x - x_0)\, dx = f(x_0) \tag{39}$$

As Δx approaches zero, the error involved in this procedure becomes arbitrarily small. Thus, we obtain

$$f(x_0) = \lim_{\Delta x \to 0} \int_{-\infty}^{\infty} f(x) S_{\Delta x}(x - x_0)\, dx = \int_{-\infty}^{\infty} f(x)\, \delta(x - x_0)\, dx \tag{40}$$

Problem 11: Show that $\lim\limits_{\Delta x \to 0} \dfrac{A}{(x - x_0)^2 + (\Delta x)^2}$ can be regarded as a δ function, provided that A is given a suitable value. Obtain the required value of A and explain why A differs from that obtained in Problem 10.

16. Eigenfunctions of the Momentum Operator in Momentum Space. In momentum space, the eigenfunction of the operator p, belonging to the eigenvalue p_0, is a δ function, namely,

$$\varphi(p) = \delta(p - p_0) \tag{41}$$

In position space, however, the eigenfunction of this operator was $\exp(ip_0x/\hbar)$. Thus, whether an eigenfunction of a given operator is a continuous function of a δ function may depend on the representation that we are using. The above eigenfunctions are, of course, not normalized to yield an integrated probability of unity. The same things can be said about normalization of eigenfunctions of p in momentum space as were said about eigenfunctions of x in position space.* Thus, we actually work with functions covering a small range of momenta Δp, whose range, however, can be made so small that it can in most applications be ignored altogether.

17. Momentum Representation of Eigenfunctions of *x*. Thus far, we have confined ourselves to the co-ordinate representation but we know that, in the momentum representation, the variable x can be represented by the operator $i\hbar \dfrac{\partial}{\partial p}$. The eigenfunctions of x are, therefore, given by

$$i\hbar \frac{\partial \varphi}{\partial p} = x_0 \varphi \tag{42}$$

where x_0 is now some constant value of the co-ordinate, independent of p. The solution is

$$\varphi_x(p) = \exp\left(-\frac{ipx_0}{\hbar}\right) \tag{42-a}$$

The eigenfunctions of x in momentum space are therefore plane waves, just like the eigenfunctions of p in co-ordinate space. The same remarks about normalization apply as in Sec. 13. We must integrate over a small range in x_0 to obtain a packet bounded in momentum space. Note also that, in position space, this small range of x_0 corresponded to the width of the peak in the function $S_{\Delta x}(x - x_0)$.

* See Sec. 14.

18. Connection Between Eigenfunctions of x in Position Space and in Momentum Space. Are the plane-wave eigenfunctions of x in momentum space consistent with the δ function eigenfunction of x in co-ordinate space? The answer is that they are. This can be seen in two ways. First, we may compute $\varphi(k)$ for a δ function in position space. We get

$$\varphi(k) = \frac{1}{\sqrt{2\pi}} \int_{-\infty}^{\infty} \exp\,(-ikx)\delta(x - x_0)\,dx = \frac{\exp\,(-ikx_0)}{\sqrt{2\pi}} = \frac{\exp\,(-ipx_0/\hbar)}{\sqrt{2\pi}} \tag{43}$$

This is in agreement with our previous result, obtained from the definition in p space of the operator x.

Another way to treat this problem is to take the wave function $\varphi(k)$ and find $\psi(x)$. Using the value of $\varphi(k)$ obtained from eq. (43) for the δ function, we get

$$\psi(x) = \frac{1}{\sqrt{2\pi}} \int_{-\infty}^{\infty} \varphi(k)e^{ikx}\,dk = \frac{1}{2\pi} \int_{-\infty}^{\infty} e^{ik(x-x_0)}\,dk \tag{44}$$

This integral does not exist in a rigorous treatment. Yet we can find its limit as the limits of integration approach infinity. Thus we write

$$\psi(x) = \lim_{K\to\infty} \int_{-K}^{K} \exp\,[ik(x - x_0)]\,\frac{dk}{2\pi} = \frac{\sin K(x - x_0)}{\pi(x - x_0)} \tag{44a}$$

A plot of $\psi(x)$ is given in Fig. 3. ψ reaches its peak value, K/π, when $x = x_0$ and begins to decrease rapidly when $|x - x_0| > 1/K$. There-

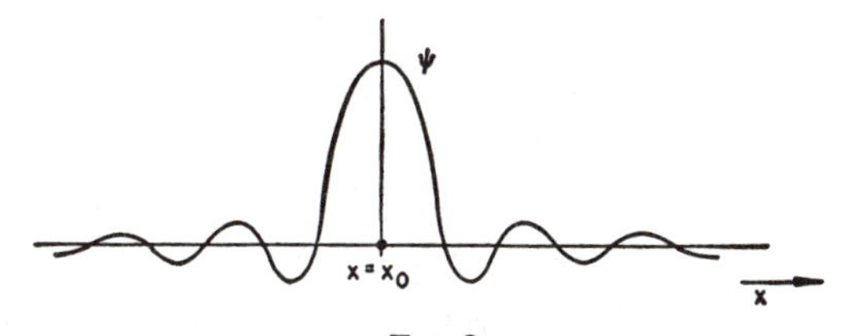

FIG. 3

after, it oscillates rapidly with a period equal to $2\pi/K$. The main contribution to $\psi(x)$ therefore comes from the narrow region near $|(x - x_0)| \leqq 1/K$. As $K \to \infty$, the region in which the function is large becomes narrower and narrower, and therefore takes on the character of a δ function. This means that we can write

$$f(x) = \lim_{K\to\infty} \int_{-\infty}^{\infty} \int_{-K}^{K} \exp\,[ik(x - x_0)]f(x_0)\,dx_0\,\frac{dk}{2\pi} = \int_{-\infty}^{\infty} \int_{-\infty}^{\infty} \exp\,[ik(x - x_0)]f(x_0)\,dx_0\,\frac{dk}{2\pi} \tag{45}$$

But the above is just the Fourier integral theorem.* Thus, we see that if we regard $\int_{-\infty}^{\infty} \exp [ik(x - x_0)]\, dk$ as a δ function, then the Fourier integral theorem becomes a special case of the use of δ functions.

19. Differentiation of the δ Function. At first sight, it would seem impossible to differentiate a function as discontinuous as the δ function. If we remember the definition in terms of the limit, however, we can give the derivative a meaning. Noting that $S_{\Delta x}(x - x_0)$ is a function with a sharp, high peak at $x = x_0$, we see that its derivative is a function with a sharp positive peak when x is a little less than x_0 and a sharp negative peak when x is a little greater than x_0 (Figs. 4 and 5).

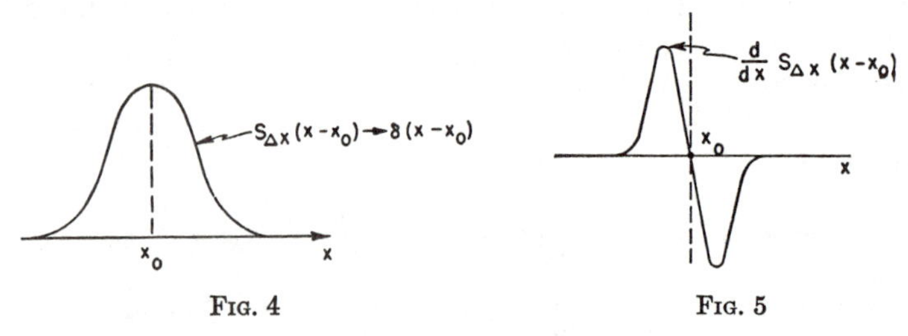

FIG. 4 FIG. 5

To see what it means to differentiate $\delta(x - x_0)$ in an actual application, we write

$$f(x) = \lim_{\Delta x \to 0} \int_{-\infty}^{\infty} S_{\Delta x}(x - x_0) f(x_0)\, dx_0 \tag{46}$$

$$\frac{df(x)}{dx} = \lim_{\Delta x \to 0} \int_{-\infty}^{\infty} \frac{dS_{\Delta x}(x - x_0)}{dx} f(x_0)\, dx_0 \tag{47}$$

We now note that

$$\frac{dS_{\Delta x}}{dx} = -\frac{d}{dx_0} S_{\Delta x}$$

$$\frac{df(x)}{dx} = -\lim_{\Delta x \to 0} \int_{-\infty}^{\infty} \frac{dS_{\Delta x}(x - x_0)}{dx_0} f(x_0)\, dx_0 \tag{47a}$$

We now integrate by parts with respect to x_0, noting that the integrated part vanishes. We get

$$\frac{df}{dx} = \lim_{\Delta x \to 0} \int_{-\infty}^{\infty} S_{\Delta x}(x - x_0) \frac{df(x_0)}{dx_0} = \int_{-\infty}^{\infty} \delta(x - x_0) \frac{df(x_0)}{dx_0} = \frac{df}{dx} \tag{47b}$$

We therefore conclude that the correct meaning to give to $\frac{d}{dx}\delta(x - x_0)$ whenever it occurs in an integral is the following:

$$\frac{df}{dx} = \int_{-\infty}^{\infty} \frac{d\delta(x - x_0)}{dx} f(x_0)\, dx_0 \tag{47c}$$

* See Chap. 3, Sec. 17.

$\dfrac{d\delta(x - x_0)}{dx}$ is a function which, like $\delta(x - x_0)$, is zero except at $x = x_0$, but because it takes this double-peaked form, shown in Fig. 5, it means roughly that we subtract the value of $f(x - \Delta x)$ from that of $f(x + \Delta x)$ and divide by $2\Delta x$; thus we obtain df/dx.

Problem 12: Prove by successive differentiation that

$$\frac{d^n f(x)}{dx^n} = \int_{-\infty}^{\infty} \frac{d^n \delta(x - x_0)}{dx^n} f(x_0)\, dx_0$$

and explain what this equation means in terms of the $S_{\Delta x}(x - x_0)$ function.

20. Discrete vs. Continuous Eigenvalues of Operators. Suppose that we have a particle confined in a box of length L. Then, (in one dimension), the wave function must vanish at $x = 0$ and $x = L$. The only eigenfunctions of the operator $p^2/2m$ for which this happens are

$$\psi = \sin \frac{n\pi x}{L} \tag{48}$$

where n is an integer. Thus, we have

$$\frac{p^2}{2m} \sin \frac{n\pi x}{L} = -\frac{\hbar^2}{2m} \frac{\partial^2}{\partial x^2} \sin \frac{n\pi x}{L} = \left(\frac{n\pi\hbar}{L}\right)^2 \frac{1}{2m} \sin \frac{n\pi x}{L} \tag{48a}$$

and the eigenvalues of $\dfrac{p^2}{2m}$ are limited to $\left(\dfrac{n\pi\hbar}{L}\right)^2 \dfrac{1}{2m}$. There is thus a discrete spectrum of eigenvalues, one for each integral value of n.

In general, whenever there is some boundary condition to satisfy, the eigenvalues will turn out to be discrete, as they did in the above example. On the other hand, in free space, with no boundaries, all positive values of $p^2/2m$ are permissible, and the spectrum is continuous. This is also a general rule; if there are no boundary conditions limiting the region where the wave function is large, then operators will usually have a continuous spectrum. In some cases, such as the hydrogen atom, we shall see that part of the spectrum is discrete and part is continuous. The discrete part of the spectrum corresponds to various quantum states of the hydrogen atom; the continuous part corresponds to states in which the atom is ionized.

Any continuous spectrum may always be regarded as the limit of a discrete spectrum in which the containing walls are allowed to recede to infinity. For example, in the case of a free particle in a box, if we let $L \to \infty$ then the spacing between energy levels approaches zero, and the spectrum approaches a continuous range of values.

21. The Expansion of an Arbitrary Function as a Series of Eigenfunctions. A mathematical theorem, which we shall present without proving, states that an arbitrary function can be expanded as a series involving all of the eigenfunctions of any Hermitean operator satisfying certain regularity conditions which will not be given here. More precisely stated, the theorem is as follows:

Let A be a sufficiently regular Hermitean operator, with eigenvalues a and eigenfunctions ψ_a. Then, by means of a suitable choice of coefficients, we can express an arbitrary (reasonably regular) function $f(x)$ in terms of the following series:

$$f(x) = \sum_a C_a \psi_a(x) \tag{49}$$

This is summed over all possible values of a. If A possesses a continuous set of eigenvalues, the sum should be replaced by an integral

$$f(x) = \int C(a) \psi_a(x)\, da \tag{50}$$

If A possesses both discrete and continuous eigenvalues, we must sum over all the discrete values, and integrate over all the continuous values.

Examples: (a) $\sin px/\hbar$ is an eigenfunction of the Hermitean operator $p^2/2m$. In a box (see Sec. 20), the only allowed values of p are $p = \hbar\pi n/L$, where L is the size of the box, and n is an integer. Our theorem says that we can expand an arbitrary wave function that is zero at the walls of the box as follows:

$$\psi(x) = \sum_{n=0}^{\infty} C_n \sin \frac{n\pi x}{L} \tag{50a}$$

But this is just a Fourier series in which the value of ψ is restricted to be zero at the boundaries at $x = 0$ and $x = L$. We already know that such a Fourier series can represent an arbitrary function of this kind. Fourier analysis is, therefore, a special case of the expansion theorem.

(b) $\exp ipx/\hbar$ is an eigenfunction of the Hermitean operator p. In free space, all values of p are permissible, and the allowed values are therefore continuous. The expansion theorem says that

$$\psi(x) = \int_{-\infty}^{\infty} C(p') \exp\left(\frac{ip'x}{\hbar}\right) dp' \tag{50b}$$

But this is just the Fourier integral, which is also a special case of the expansion theorem.*

(c) The eigenfunctions of the operator x are $\delta(x - x_0)$. These form a continuous set of eigenfunctions, one for each value of x_0. The expansion theorem states that we can write

$$\psi(x) = \int \psi(x_0) \delta(x - x_0)\, dx_0 \tag{50c}$$

We have already seen from the definition of the δ function that the above equation is true. The above equation may, therefore, be regarded as a special case of the expansion theorem.

* One can generalize these results and say that a function satisfying certain boundary conditions can be expanded in terms of eigenfunctions which satisfy the same boundary conditions, provided that these conditions are linear, as they always are in quantum theory.

Later, we shall come to eigenfunctions of more complicated operators and the expansion theorem will allow us to form new types of series, analogous to the Fourier series and integrals, but involving new functions instead, like Bessel's functions, Legendre polynomials, Hermite polynomials, and others.

22. The Expansion Postulate. Because the specification of the conditions under which such an expansion is possible is rather complicated and not thoroughly worked out, it is perhaps better to replace the expansion theorem by the following postulate:

Every Hermitean operator representing some observable quantity in the quantum theory will be tentatively assumed to have the property that an arbitrary acceptable wave function can be expanded as a series of its eigenfunctions.

It is a fact that all operators of this kind that are now known have this property. This property is, as we shall see, so closely bound up with the interpretation of quantum theory that if it were ever found not to be satisfied, fundamental changes in the theory would probably be needed. Thus, it seems reasonable to postulate its validity here.†

23. A Theorem: The Eigenvalues of a Hermitean Operator Are Real. The proof of this theorem is simple. Since the average value is real for an arbitrary function, it must be real for an eigenfunction. Thus, we say that if O is Hermitean

$$\int \psi_a^* O \psi_a \, dx = a \int \psi_a^* \psi_a \, dx = a \tag{51}$$

must be real, if ψ_a is an eigenfunction.

24. The Orthogonality of Eigenfunctions of a Hermitean Operator. Suppose that ψ_a and ψ_b are each different eigenfunctions of some Hermitean operator O, belonging respectively to the eigenvalues a and b. Then, we can show that if a and b are different eigenvalues, it follows that

$$\int \psi_a^* \psi_b \, dx = 0 \tag{52}$$

Any functions that satisfy the above relation are said to be orthogonal. We wish, therefore, to prove the orthogonality of eigenfunctions belonging to different eigenvalues.

To do this, let us consider the following integral

$$I = \int \psi_a^* O \psi_b \, dx = \int \psi_a^* b \psi_b \, dx \tag{53}$$

Because O is Hermitean, we also have

$$I = \int \psi_b O^* \psi_a^* \, dx \tag{53a}$$

Because the eigenvalues of O are real, we have

$$O^* \psi_a^* = a^* \psi_a^* = a \psi_a$$

We then obtain

$$I = a \int \psi_a^* \psi_b \, dx \tag{53b}$$

† For further justification of this postulate see Sec. 29.

Thus, equating the two values of I, we get

$$(a - b)\int \psi_a^* \psi_b \, dx = 0 \tag{53c}$$

If $a \neq b$, then $\int \psi_a^* \psi_b \, dx = 0$ so that ψ_a and ψ_b are orthogonal. If $a = b$, we cannot draw any definite conclusions.

Examples: (a) In a Fourier series, we have seen that each term sin $(n\pi x/L)$ is an eigenfunction of the Hermitean operator, $p^2/2m$. Our theorem states that $O = \int_O^L \sin(n\pi x/L) \sin(m\pi x/L)\, dx$ unless $m = n$. But we already know this from a study of Fourier series. The orthogonality of terms in a Fourier series is therefore a special case of our general theorem.†

(b) Two possible eigenfunctions of $p^2/2m$ (for the case of a free particle) are exp $(ipx/\hbar)$ and sin $(px/\hbar)$. These belong to the same eigenvalue of $p^2/2m$. A brief calculation shows that they are not orthogonal. This is an example of how orthogonality can fail when both eigenvalues are the same. It is not necessary, however, that two eigenfunctions with the same value be nonorthogonal. For example, exp $(ipx/\hbar)$ and exp $(-ipx/\hbar)$ belong to the same value of $E = p^2/2m$, and yet they are orthogonal, as a simple calculation shows. We therefore see that equality of eigenvalues merely prevents us from drawing conclusions about the orthogonality of eigenfunctions.

Problem 13: Consider a box of side L, with periodic boundary conditions. The allowable eigenfunctions of p are then $\exp\left(-\frac{i2\pi nx}{L}\right)$, with $p = \frac{2\pi n\hbar}{L}$. Show that eigenfunctions belonging to different p are orthogonal.

25. Calculation of Expansion Coefficients. Let us assume an expansion in terms of the normalized eigenfunctions of the operator A. For simplicity, suppose that the allowed eigenvalues are discrete, although the same methods may be applied to a continuous spectrum. Thus, we have

$$\psi = \sum_a C_a \psi_a \tag{54}$$

Now multiply by ψ_b^* and integrate over all space. From the fact that $\int_{-\infty}^{\infty} \psi_b^* \psi_a \, dx = 0$ if $a \neq b$ and unity if $a = b$, we obtain

$$C_b = \int \psi_b^* \psi \, dx \tag{55}$$

If ψ is given, the preceding equation enables us to calculate the expansion coefficient C_b. The calculation of Fourier coefficients is a special case of this result.

26. Expansion of Dirac δ Function in Terms of Eigenfunctions of an Arbitrary Hermitean Operator. We can make an immediate application of the previous result to obtain a more general expression for the Dirac δ function than we already have. By the expansion theorem, we write

$$\delta(x - x_0) = \sum_a f(a)\psi_a(x) \tag{56}$$

† See Chap. 1, Sec. 5.

where ψ_a is a normalized eigenfunction of the Hermitean operator A, corresponding to the eigenvalue a.

To solve for $f(a)$, we multiply by $\psi_a^*(x)$ and integrate over x. From the orthogonality relations, and from $\int \psi_a^* \psi_a \, dx = 1$, we obtain

$$f(a) = \int \psi_a^*(x) \delta(x - x_0) \, dx = \psi_a^*(x_0) \tag{57}$$

and

$$\delta(x - x_0) = \sum_a \psi_a^*(x_0) \psi_a(x) \tag{58}$$

For a continuous spectrum of eigenvalues, we have

$$\delta(x - x_0) = \int \psi_a^*(x_0) \psi_a(x) \, da \tag{58a}$$

This is a very useful result, which we shall frequently find occasion to apply. We may give as an example the expansion in terms of the eigenfunctions of p, which, as we have seen, leads to Fourier analysis of the wave function. By applying (58) to wave functions defined such that they are periodic in a container of side L, for example, we obtain

$$\delta(x - x_0) = \frac{1}{L} \sum_{-\infty}^{\infty} \exp\left[\frac{2\pi i n(x - x_0)}{L}\right] \tag{59}$$

Problem 14: By considering

$$y_N(x - x_0) = \frac{1}{L} \sum_{-N}^{N} \exp\left[\frac{2\pi i n(x - x_0)}{L}\right]$$

show that

$$\lim_{N \to 0} \int_{-\infty}^{\infty} y_N(x - x_0) \psi(x_0) \, dx_0 = \psi(x)$$

Hence justify the use of the infinite sum as a δ function.

27. Representation of an Operator in Terms of Its Eigenfunctions. If ψ is expanded in terms of the eigenfunctions of the operator A, we can then obtain a very simple expression for the function $A\psi$. To do this, we note that the effect of A on each eigenfunction ψ_a is merely a multiplication by the corresponding value of a. Thus if $\psi = \sum_a C_a \psi_a$, we obtain

$$A\psi = \sum_a a C_a \psi_a \tag{60}$$

$$A^n \psi = \sum_a a^n C_a \psi_a \tag{60a}$$

For an arbitrary function of A, we then write

$$f(A)\psi = \sum_a f(a) C_a \psi_a \tag{60b}$$

Examples: Let ψ be expanded in terms of the eigenfunctions of the momentum operator p.

$$\psi(x) = \int_{-\infty}^{\infty} \varphi(p) \exp\left(\frac{ipx}{\hbar}\right) dp \tag{61}$$

$$p\psi(x) = \int_{-\infty}^{\infty} p\varphi(p) \exp\left(\frac{ipx}{\hbar}\right) dp \tag{61a}$$

$$f(p)\psi(x) = \int_{-\infty}^{\infty} f(p)\varphi(p) \exp\left(\frac{ipx}{\hbar}\right) dp \tag{61b}$$

Equation (60b) enables us to show in a very simple way what the operator does to an arbitrary wave function ψ, and thus to "represent" the operator by means of a set of numerical operations. This procedure is a generalization of what is done, for example, in the momentum representation, where the operator p is represented simply by the number p that multiplies the eigenfunction $\exp(ipx/\hbar)$. Similarly, in the position representation, the operator x is represented simply by a number, x, that multiplies the wave function $\psi(x)$. Every Hermitean operator, A, has a representation in terms of its eigenfunctions, ψ_a, such that its effect is simply to multiply ψ_a by a number a. Thus, we have generalized the idea of a representation from that of position and momentum spaces alone to the possibility of using a space involving the eigenfunctions of any Hermitean operator. We shall find this generalization very useful later,† in connection with the matrix formulation of the quantum theory.

One of the most important advantages of the procedure of representing an operator in the way outlined above is that it enables us to generalize the definition of a function of an operator to cases in which this function is not expressible as a power series. Thus, we can regard eq. (60b) as a definition of an arbitrary function of an operator and, in this way, avoid the specialization to functions that can be represented as power series. Our purpose in starting with the power series method, however, was to motivate the formulation of quantum theory in a more natural way than would have been possible if we had started out immediately with the most general case.

28. Mean Value of $f(A)$ in Terms of Expansion into Eigenfunction of A. The mean value of $f(A)$ is, by definition,

$$\bar{f}(A) = \int_{-\infty}^{\infty} \psi^*(x) f(A)\psi(x)\, dx \tag{62}$$

Let us expand ψ and ψ^* as a series of eigenfunctions of ψ_a, using the fact that

$$f(A)\psi_a = f(a)\psi_a$$

$$\bar{f}(A) = \int_{-\infty}^{\infty} \sum_a \sum_{a'} C_{a'}^* C_a \psi_{a'}^* f(a)\psi_a\, dx \tag{62a}$$

Using the fact that

$$\int_{-\infty}^{\infty} \psi_{a'}^* \psi_a\, dx = \begin{cases} 0, & \text{if } a \neq a' \\ 1, & \text{if } a = a' \end{cases}$$

† See Chap. 16.

we obtain

$$\bar{f}(A) = \sum_a C_a^* C_a f(a) \tag{62b}$$

29. Physical Interpretation of Expansion Coefficients in Terms of Probabilities. Equation (62b) provides the basis for a simple physical interpretation of the expression $C_a^* C_a$. To obtain this, we note that an alternative expression for $\bar{f}(A)$ can be obtained in terms of $P(a)$, where $P(a)$ is the probability that the variable A will be found in a measurement to have the numerical value a, corresponding to a particular eigenfunction ψ_a. This expression is

$$\bar{f}(A) = \sum_a P(a) f(a) \tag{63}$$

If the two expressions are to be equal *for arbitrary functions* $f(a)$, then it is necessary that

$$P(a) = C_a^* C_a \tag{64}$$

Thus, we have shown that $C_a^* C_a$ is the probability that in a measurement the system can be found in a state in which the variable A has the exact numerical value a.

Example: In the expansion $\psi = \int_{-\infty}^{\infty} \varphi(p) \exp(ipx/\hbar)\, dp$ the probability that the momentum is between p and $p + dp$ is $P(p) = \varphi^*(p)\varphi(p)\, dp/2\pi$. [Note that $(2\pi)^{-1/2} \varphi(p)$ is the expansion coefficient for ψ into a series of eigenfunctions of the operator p.]

This is an exceedingly important result, as it enables us to extend the definition of probability to a wider class of observable quantities than position and momentum. Thus, if A is set equal to the Hamiltonian operator, ψ_a represents a wave function corresponding to a definite energy state, and $C_a^* C_a$ is the probability that the system has this energy.†

The role of the expansion postulate (Sec. 22) in making possible our present interpretation of $|C_a|^2$ is clearly a key one. If it were not possible to expand an arbitrary ψ as a series of ψ_a, an integral part of our method of interpreting the wave function would then become untenable. The general requirements of consistency and unity of the theory would therefore suggest that in the absence of contradictions with experiment, we can safely regard the expansion postulate as a definition, or as a criterion which must be satisfied by an operator before we accept it as a suitable observable for use in the quantum theory. The fact that all observables now known satisfy this criterion is then experimental proof of the validity of this postulate.

Further Interpretation of Probabilities in Quantum Theory

30. Interference of Probabilities. Suppose that the system is in such a state that the operator A has an eigenvalue a, so that the wave function

† This result justifies the qualitative discussion of transitions between orbits given in Chap. 3, Sec. 16.

is $\psi_a(x)$. The probability that a particle can be found at the point x is then

$$P_a(x) = \psi_a^*(x)\psi_a(x) \tag{65}$$

This means that if a large number of equivalent systems are prepared in such a way that the observable A has the same definite value a, then if the position x is subsequently measured, $P_a(x)$ will yield the probability that the result turns out to be x. Similarly, if systems are prepared with the observable A equal to another definite numerical value b so that the wave function is equal to $\psi_b(x)$, then the probability of finding a definite value of x will be

$$P_b(x) = \psi_b^*(x)\psi_b(x) \tag{66}$$

Now, let us consider a new situation in which the system is prepared in such a way that the observable A may have either the value a or the value b with respective probabilities Q_a and Q_b (with $Q_a + Q_b = 1$). According to classical physics, the probabilities should add, as given below:

$$P(x) = Q_aP_a(x) + Q_bP_b(x) \tag{67}$$

In quantum theory, however, the new probability is not related to the old one in such a simple way. Instead of adding probabilities, we must, according to the hypothesis of linear superposition, add wave functions. The combined wave function is

$$\psi = C_a\psi_a(x) + C_b\psi_b(x) \tag{68}$$

where C_a and C_b are constants which must be determined. According to eq. (64),

$$Q_a = C_a^*C_a, \qquad \text{and} \qquad Q_b = C_b^*C_b$$

Hence, we may write

$$C_a = (Q_a)^{1/2}e^{i\phi_a} \qquad \text{and} \qquad C_b = (Q_b)^{1/2}e^{i\phi_b}$$

where ϕ_a and ϕ_b are phase factors which *cannot be determined from a knowledge of* Q_a *and* Q_b *alone*. What really determines these phase factors will be discussed later.

We then obtain

$$P(x) = \psi^*(x)\psi(x) = C_a^*C_a\psi_a^*(x)\psi_a(x) + C_b^*C_b\psi_b^*(x)\psi_b(x) \\ + C_a^*C_b\psi_a^*(x)\psi_b(x) + C_b^*C_a\psi_a(x)\psi_b^*(x) \tag{69}$$

The above may be rewritten as follows:

$$P(x) = Q_aP_a(x) + Q_bP_b(x) \\ + (Q_aQ_b)^{1/2}[e^{i(\phi_b-\phi_a)}\psi_a^*(x)\psi_b(x) + e^{-i(\phi_b-\phi_a)}\psi_a(x)\psi_b^*(x)] \tag{69a}$$

We see that besides the terms which we should expect classically, there are additional terms in $P(x)$ which result from the interference of $\psi_a(x)$

and $\psi_b(x)$. The phase difference between two different parts of the wave function controls the magnitude of the interference terms in a way that could never occur with classical probabilities. Throughout the rest of this book, we shall see in more detail just how these phase differences are determined by the precise physical situation. For the present, we merely point out that since a physically observable probability distribution depends on the phase difference $\phi_a - \phi_b$, then this phase difference is, in general, physically observable, even though the absolute value of the phase itself has no physical significance. The phase difference between different parts of the wave function is therefore an all important quantity, the nature of which we shall have to study later in more detail. In this connection, see Chap. 6, where we were led to similar conclusions.

31. Eigenfunctions of the Hamiltonian Operator. Since the Hamiltonian operator is Hermitean, we can, according to the expansion postulate, expand an arbitrary function as a series of its eigenfunctions. Thus, we may write

$$\psi(x) = \sum_E C_E \psi_E(x) \tag{70}$$

where $\psi_E(x)$ is an eigenfunction of H, belonging to the eigenvalue E. Because the Hamiltonian function is equal to the total energy of the system, it is clear that an eigenstate of the operator H is one in which the energy is perfectly defined, and equal to E.

32. Change of Eigenfunctions of H with Time. Each eigenfunction of H must satisfy the wave equation

$$i\hbar \frac{\partial \psi_E}{\partial t} = H\psi_E \tag{71}$$

But

$$H\psi_E = E\psi_E$$

Thus, we have

$$i\hbar \frac{\partial \psi_E}{\partial t} = E\psi_E \tag{71a}$$

for which the solution is

$$\psi_E = (\psi_E)_{t=0}\, e^{-iEt/\hbar} \tag{71b}$$

Hence, we see that the eigenfunctions of H oscillate harmonically with a frequency given by $2\pi\nu = E/\hbar$ or $\nu = E/h$, which is just the de Broglie relation, in agreement with what was used in the free-particle problem.†

33. Change of Probability with Time. Stationary States. For an eigenstate of the Hamiltonian, the probability, $P(x)$, is

$$P(x) = \psi_E^*(x)\psi_E(x) = (\psi_E^*)_{t=0}(\psi_E)_{t=0} \tag{72}$$

Note that $P(x)$ is not a function of the time. Similarly, it can be shown by Fourier analysis that $P(k)$ is also constant. A state in which the energy is well-defined is one in which all probabilities remain constant

† Equation (28), Chap. 3.

with time. It is, therefore, a *stationary state.* Thus, when an atom is in a state of a given energy, the probability of any experimental result is independent of the time at which the experiment is done. If, for example, we measure the position of an electron that is in a stationary state, the value we obtain will fluctuate from one experiment to the next. But the probability of a given value will be independent of the amount of time which has elapsed since the state was prepared. This is in contrast to what happens, for example, if we make up a wave packet, for in this case the wave packet moves through space and spreads out, so that the probability of a given position changes with time.

We see from the above that when an electron is in a state of definite energy, corresponding to some Bohr orbit, the probability of any experimental result remains constant. Actually, however, we know that if an atom is in an excited state, it radiates and goes to a state of lower energy, so that it is not really in a completely stationary state. The excited states are stationary only to the extent that radiation is neglected. Later, when we have formulated a more complete theory, we shall show how to take into account precisely the changes of ψ resulting from the possibility of radiation.* For the present, we merely note that an excited state lasts some mean time, τ, which depends on the rate of radiation. As shown in eq. (54), Chap. 2, the probability that the system has not radiated is $P = e^{-t/\tau}$, which becomes negligible soon after $t = \tau$.

34. Relation of Time-dependent Probabilities to Uncertainty Principle. If an atom in a definite Bohr orbit emits light within a time of the order τ, then the frequency of the light must be indefinite to the extent $\Delta\omega \cong 1/\tau$. (See Chap. 5, Secs. 2 and 11.) The energy, therefore, becomes indefinite by the amount $\Delta E = \hbar\,\Delta\omega \cong \hbar/\tau$. This is another example of the uncertainty principle. To see that this is so, let us note that because the quantum is practically certain to have been radiated within a time of the order of τ, one could know the time of emission to within this accuracy from the mere fact that emission has occurred. Thus, one has a rough measure of time, and the energy must become correspondingly uncertain. This means that merely because it can radiate, the energy of an excited state of an atom is made intrinsically uncertain to the extent $\Delta E \cong \hbar/\tau$.

If there are many atoms that are in excited states, then the energy radiated by each will be different and will fluctuate through the range $\Delta E \cong \hbar/\tau$. As a result, the spectral line will have a finite width $\Delta\nu \cong 1/\tau$. This is called the *natural line breadth;* there is no way to make the line any sharper than this, even after all effects, such as Doppler shift, collision broadening, etc., have been eliminated.

Problem 15: (a) What is the line breadth for the first excited state of the hydrogen atom? (See Problem 13, Chap. 2.) Give it in terms of energy and frequency.

* Chap. 18.

(b) What temperature would be needed to make the Doppler broadening smaller than this number for hydrogen? For mercury?

(c) When an excited atom collides with an unexcited atom, the energy of excitation is usually transferred to the other atom. In this way, the lifetime of the excited state of the first atom is shortened. If l is the mean free path, and v is the mean atomic velocity, then the time between collisions is l/v. For hydrogen at atmospheric pressure, the free path for this transfer is about 10^{-4} cm.

How low a pressure is needed to make this type of collision broadening smaller than the natural line breadth at room temperature? At 10°K? Assuming the same free path for mercury, what pressure is needed for mercury at room temperature?

35. The Importance of Eigenfunctions of *H*. Eigenstates of the energy are particularly important, not only because we must deal with many systems that do have a definite energy, but also because an arbitrary time-dependent system can have its wave function expanded as a series of eigenfunctions of H. Thus, we can put $\psi_E = (\psi_E)_{t=0}\, e^{-iEt/\hbar}$ into eq. (70), obtaining

$$\psi(x, t) = \sum_E C_E[\psi_E(x)]_{t=0} \exp\left(-\frac{iEt}{\hbar}\right) \tag{73}$$

Thus, if we know the expansion of $\psi(x)$ at $t = 0$, the preceding equation gives the general value of $\psi(x)$ at any time t. The eigenfunctions of the Hamiltonian operator are, therefore, particularly significant in solving the problem of the time rate of change of a wave function, once the initial value is known. For these reasons, the problem of obtaining the eigenfunctions of H is one of the basic problems in all of quantum theory, and once we have obtained all of them, we have obtained a general solution of the wave equation as a function of time, given by eq. (73).

36. Change of Probability with Time for a General Wave Function. From eq. (72), we write

$$P(x) = \psi^*(x)\psi(x) = \sum_E \sum_{E'} (\psi^*_{E'})_{t=0}(\psi_E)_{t=0} C^*_{E'} C_E \exp\left[-\frac{i(E - E')t}{\hbar}\right] \tag{74}$$

We see that, in general, $P(x)$ is a function of time. Only those terms with $E = E'$ do not contain the time. As an example, we consider the case where all C_E are zero except two, namely, C_{E_1} and C_{E_2}. We have

$$\psi = C_{E_1}(\psi_{E_1})_{t=0} \exp\left(-\frac{iE_1t}{\hbar}\right) + C_{E_2}(\psi_{E_2})_{t=0} \exp\left(-\frac{iE_2t}{\hbar}\right) \tag{75}$$

$$\begin{aligned} P(x) = {} & C^*_{E_1}C_{E_1}[\psi^*_{E_1}(x)\psi_{E_1}(x)]_{t=0} + C^*_{E_2}C_{E_2}[\psi^*_{E_2}(x)\psi_{E_2}(x)]_{t=0} \\ & + C^*_{E_1}C_{E_2}[\psi^*_{E_1}(x)\psi_{E_2}(x)]_{t=0} \exp\left[-\frac{i(E_2 - E_1)t}{\hbar}\right] \\ & + C^*_{E_2}C_{E_1}[\psi^*_{E_2}(x)\psi_{E_1}(x)]_{t=0} \exp\left[-\frac{i(E_1 - E_2)t}{\hbar}\right] \end{aligned} \tag{76}$$

We see that there is a time-constant part, and a part which oscillates with the frequency $\nu = (E_1 - E_2)/h$.

It is very significant that the way quantum theory describes changes of probability with time is through the terms involving the interference of the contributions of different stationary states. Motion is, therefore, described in an essentially nonclassical way. The change of any particular probability distribution is produced simply by the changing phase relations between different components of the wave function corresponding to different stationary states. Here we see a simple case of how the phase difference between two stationary states has physical significance; namely, it controls the change of probability with time. Because the process of motion is described in terms of the interference of wave functions belonging to different energies, we conclude that changing probabilities will exist only when there is a range of energies present or, in other words, when the energy is made somewhat indefinite. In this way, the uncertainty principle between energy and time is automatically contained in the theory.

A similar result was obtained in Chap. 3, Secs. 4 and 13, where it was shown that the motion of wave packets is caused by the change of position of constructive and destructive interference of waves of different k, brought about by the changing phase relations introduced by the time-dependent phase factor $\exp(-i\hbar k^2t/2m)$. [See eq. (21) Chap. 3.] More generally, the way in which the wave function changes with time is determined by the form of the Hamiltonian operator. Thus, the Hamiltonian operator may be said to contain the causal laws, insofar as they have meaning.

PART III

APPLICATIONS TO SIMPLE SYSTEMS. FURTHER EXTENSIONS OF QUANTUM THEORY FORMULATION

CHAPTER 11

Solutions of Wave Equation for Square Potentials

1. Introduction to Part III. We shall apply in Part III, the physical ideas developed in Part I and the mathematical ideas developed in Part II to the solution of various elementary problems, starting from the simplest cases and gradually working up to more complex systems. We shall begin with a one-dimensional problem in which space is divided into a finite number of regions, in each of which the potential is constant but different in value from the potential of the others. With this simple problem we will be able to illustrate many important specifically quantum-mechanical effects, such as penetration of a potential barrier, reflection of electron waves by a sharp change in potential, and the binding of particles into a narrow region by an attractive force.

The next problem will be to show how Schrödinger's equation leads to results approaching those of classical physics in the correspondence limit of high quantum numbers. This will be done with the aid of the WKB approximation (Wentzel–Kramers–Brillouin). In this problem, we shall see more clearly the precise connection between classical and quantum theory. Applications of this approximation will also be made to the problem of the lifetime of excited states of a nucleus.

Throughout this treatment an effort will be made to give a simple and pictorial method for thinking of the qualitative effect of various kinds of forces on the wave function. In this way, it is hoped that the student can learn to make a qualitative picture enabling him to estimate the general form of the wave function in more complex problems, without actually solving the equations exactly. In the simple cases of the harmonic oscillator and the hydrogen atom, we shall compare the approximate results with the exact solutions.

Finally, we shall introduce the matrix formulation of quantum theory, and apply it to the case of electron spin.

2. Eigenfunctions of the Energy. Analogy to Index of Refraction in Optics. In Chap. 10, Sec. 35, it was shown that solutions of Schrödinger's equation which are eigenfunctions of the Hamiltonian operator are particularly significant, not only because many systems met with in practice do have a definite energy, but also because the time variation of these eigenfunctions takes the particularly simple form

$$\psi = \psi_E(x) \exp\left(-\frac{iEt}{\hbar}\right) \tag{1}$$

When a system has a definite energy, all probabilities are constant, so that the state is stationary. Furthermore, an arbitrary solution of Schrödinger's equation can be formed from suitable linear combinations of the above solutions.

In Part III, we shall concern ourselves mainly with the problem of calculating the eigenfunctions of the Hamiltonian operator. In other words, we wish to solve the equation

$$H\psi = \frac{-\hbar^2}{2m}\nabla^2\psi + V(x)\psi = E\psi$$

and in so doing find out which values of E are permissible, in that the associated ψ_E satisfies all boundary conditions that we place on the wave function.

We may write our equation as follows:

$$\nabla^2\psi + \frac{2m}{\hbar^2}[E - V(x)]\psi = 0 \tag{2}$$

In optics, the wave equation for a wave of definite angular frequency ω may be written

$$\nabla^2 A + \frac{\omega^2}{c^2}n^2 A = 0 \tag{3}$$

where n is the index of refraction. Hence the wave equation for ψ now resembles that for light in a medium in which

$$n^2 = \frac{2m}{\hbar^2}[E - V(x)]\frac{c^2}{\omega^2} \tag{4}$$

or, in other words, a medium in which n is a function of the position. This is a very useful analogy, and one which we shall frequently find occasion to apply.

3. Square Potentials. In general, $V(x)$ may take any conceivable functional form. A form which leads to equations that are particularly easy to solve is to have V constant everywhere in a certain region (say, from $x = a$ to $x = b$), then to have it equal to another value in the next region (say, from $x = b$ to $x = c$), then to still another value in the next region, etc. Such a potential might look like the graph shown in Fig. 1. It is called a "square potential," because of the appearance of square corners in its graph. In nature there are no potentials which are actually square, for these imply an infinite force at the points of discontinuity in the potential. Yet, the square potential represents many actual systems roughly, and its mathematical simplicity enables us to use it to draw

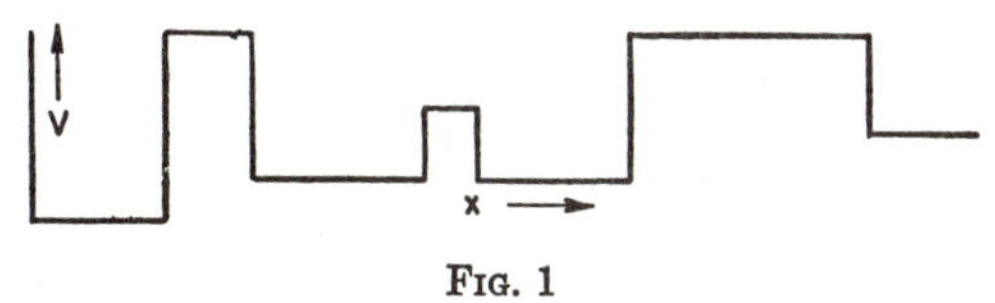

FIG. 1

conclusions that are at least qualitatively applicable to such systems. For example, the mutual potential energy of two molecules has the general form shown in Fig. 2. Many properties of molecular wave functions can be understood qualitatively by means of the square potential shown in Fig. 3, which includes two essential properties of the force; namely, attraction when the molecules are at a moderate distance, and repulsion when they are very close. It must be noted, however, that those properties of the molecule which depend on the precise shape of the curve in

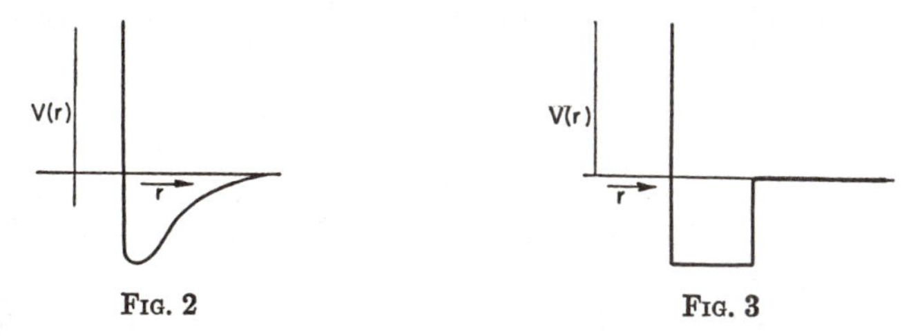

FIG. 2 FIG. 3

Fig. 2 (for example, coefficient of thermal expansion) cannot be treated at all by this simplified potential. On the other hand, this method will give a rough approximation to the energy levels.

Another set of forces which may be represented fairly well by the square potential is the force between nuclear particles, such as neutrons and protons. The force between a proton and a neutron, for example, is characterized by two properties:

(1) It is appreciable only over a very short distance, of the order of 2×10^{-13} cm. That this is indeed small can be seen by comparing it with atomic radii, which are of the order of 2×10^{-8} cm.

(2) In the range where the forces are appreciable, they are very large—much larger than the forces holding atoms together.

From scattering experiments, one can get a rough idea of the shape of the potential energy of interaction between a neutron and a proton.*

* See Chap. 21, Sec. 56. See also H. Bethe, *Elementary Nuclear Theory*. New York: John Wiley & Sons, Inc., 1947; Chap. 4.

It is more or less as shown in Fig. 4. To a first approximation, however, the potential may be represented by the square potential of Fig. 5. The range of the potential turns out to be 2.8×10^{-13} cm and the depth about 20 mev. This depth contrasts with molecular interaction energies of the order of 2 ev.

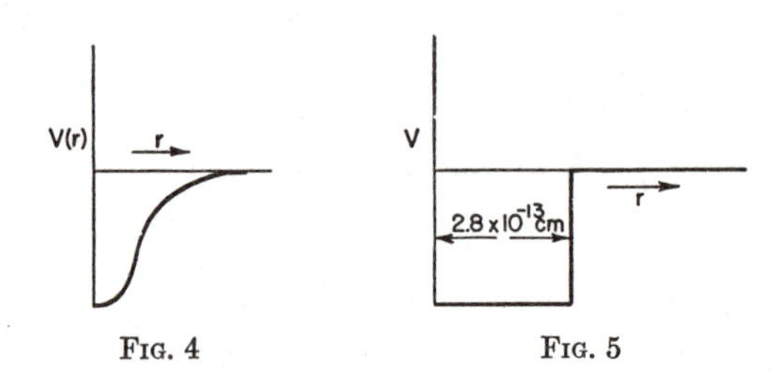

FIG. 4 FIG. 5

4. Solution of Problem of Square Potential. In any region where V is constant, the solution of the wave equation is

$$\psi_E = A \exp\left[i\sqrt{2m(E-V)}\frac{x}{\hbar}\right] + B \exp\left[-i\sqrt{2m(E-V)}\frac{x}{\hbar}\right] \quad (5)$$

where A and B are arbitrary constants. The time-dependent solution is

$$\psi = A \exp\left[\frac{i(px - Et)}{\hbar}\right] + B \exp\left[\frac{-i(px + Et)}{\hbar}\right] \quad (6)$$

where $p = \sqrt{2m(E-V)}$. It is clear that the first term represents a wave moving to the right, while the second term represents a wave moving to the left.

As we go from one region to the next, V changes, so that the length of the wave also changes. At the boundary between regions, certain boundary conditions must be satisfied. Because the differential equation is of second order in x, it is necessary that both ψ and its first derivative be continuous at the boundaries. This follows from the fact that ψ, E, and V are all assumed to be finite. ψ must be finite if its physical interpretation in terms of probability is to have meaning, whereas E and V must be finite, because infinite energies do not occur in nature. From the differential eq. (2), we then conclude that $d^2\psi/dx^2$ is everywhere finite (but not necessarily continuous). $d^2\psi/dx^2$ can be finite, however, only if $d\psi/dx$ is continuous. Thus, we obtain the first of our boundary conditions. In order that $d\psi/dx$ exist everywhere, however, as is implied by the mere use of a differential equation, it is also necessary that ψ be continuous. This gives us the second boundary condition.

Let us illustrate the application of these boundary conditions with the aid of a simple problem in which the potential undergoes only one discontinuous change, as shown in Fig. 6.

Case A: $(E > V)$

Suppose that electrons with some energy E are sent in from the left and that $E > V$. Classically we should expect that no electrons would be reflected at $x = 0$, since all of them have enough energy to enter the region $x > 0$. What is predicted by quantum theory for this problem? To answer this question, let us use the optical analogy. The electron acts, to some extent, like a wave coming in from the left, striking a sudden shift in potential at $x = 0$, where it experiences what is effectively a sudden shift in index of refraction. Just as with light striking a sheet of glass, we may expect part of the wave to be reflected and part to be transmitted.

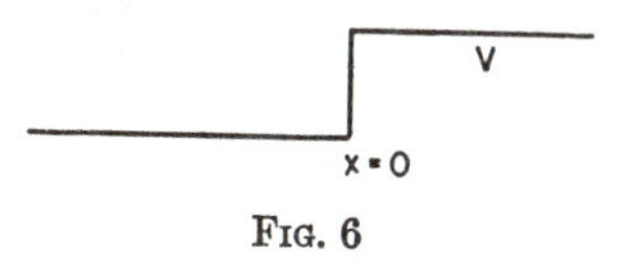

FIG. 6

In a complete quantum treatment of this problem we would actually have to start with an incident wave packet, representing the electron coming in initially from the left. This packet would come up to the barrier and part of it would be reflected and part transmitted. The reflected part of the wave packet would yield the probability that the electron was reflected, while the transmitted part would yield the probability that the electron was transmitted. We shall actually carry out this procedure in Sec. 17. Meanwhile, however, we shall adopt a procedure that is more abstract, but which leads to the same results in a mathematically simpler way. We shall assume that the packet is so broad that the incident wave can be approximated by the wave function $B \exp(ip_1x/\hbar)$ where $p_1 = \sqrt{2mE}$. The incident wave will then represent a situation in which the probability density remains constant with time, but in which there is a steady stream of electrons moving to the right. The mean probability current density will be $j = |B|^2 p_1/m$. (In order to maintain a constant probability despite this flow of current, it would be necessary to supply electrons from the left at a steady rate.)

There will also be a reflected wave, which we represent by

$$C \exp(-ip_1x/\hbar)$$

The complete wave function to the left of the barrier is

$$\psi_1 = B \exp\left(\frac{ip_1x}{\hbar}\right) + C \exp\left(\frac{-ip_1x}{\hbar}\right)$$

The transmitted wave amplitude is denoted by

$$\psi_2 = A \exp\left(\frac{-ip_2x}{\hbar}\right) \qquad \text{where} \qquad p_2 = \sqrt{2m(E - V)}$$

The constants A, B, and C must now be determined from the boundary conditions that the wave function and its first derivative are continuous at $x = 0$.

Noting that

$$\frac{d\psi_2}{dx} = \frac{ip_2}{\hbar} A \exp\left(\frac{ip_2x}{\hbar}\right)$$

and
$$\frac{d\psi_1}{dx} = \frac{ip_1}{\hbar}\left[B \exp\left(\frac{ip_1x}{\hbar}\right) - C \exp\left(\frac{-ip_1x}{\hbar}\right)\right]$$

we obtain, by setting $x = 0$,

$$A = B + C \tag{7}$$

$$p_2A = p_1(B - C) \tag{8}$$

Solution for A and C yields

$$A = \frac{2p_1B}{p_1 + p_2} \tag{9}$$

$$C = \frac{(p_2 - p_1)}{p_1 + p_2} B \tag{10}$$

We have thus obtained the amplitudes of the reflected and transmitted waves, which are respectively A and C, in terms of B, the amplitude of the incident wave. The fraction of electrons which are transmitted, T, is equal to the ratio of the transmitted current to the incident current. The transmissivity is therefore

$$T = \frac{|A|^2}{|B|^2}\frac{p_2}{p_1} = \frac{4p_1p_2}{(p_1 + p_2)^2} \tag{11}$$

The reflectivity, R, is then just the ratio of the intensities of reflected and incident waves

$$R = \frac{|C|^2}{|B|^2} = \frac{(p_1 - p_2)^2}{(p_1 + p_2)^2} \tag{12}$$

The sum of the reflectivity and transmissivity ought, by definition, to be unity. To verify that it is, we write

$$T + R = \frac{(p_1 - p_2)^2 + 4p_1p_2}{(p_1 + p_2)^2} = \frac{(p_1 + p_2)^2}{(p_1 + p_2)^2} = 1 \tag{13}$$

Problem 1: Compute the probability current for this problem (a) when $x < 0$, (b) when $x > 0$. Show that the two are the same, and thus prove that probability is conserved. Show that the current is $S = v_t\rho$, where v_t is the velocity of the transmitted particles, ρ is the probability density for this wave.

Problem 2: Show that the continuity of ψ and its derivative implies the conservation of probability current at $x = 0$.

We note that the reflectivity approaches zero as p_2 approaches p_1, but that it approaches unity as p_2 approaches zero. Since

$$p_2 = \sqrt{2m(E - V)}$$

the reflection coefficient becomes large only when V is comparable in size with E. Yet, some reflection exists no matter how small V is.

It must be emphasized again that this property of reflection from a sharp change in potential is a purely quantum-mechanical effect; it arises from the wave nature of matter and does not exist in classical theory. We shall see later, in studying the WKB approximation,* that if the change in potential is not sharp within a wavelength of the electron wave, there will be practically no reflection. The classical result will therefore be right only in a slowly changing potential. As soon as the potential begins to change appreciably within an electron wavelength, $\lambda = h/p$, the wave properties of matter begin to manifest themselves, and one of them is this property of reflection from a potential that is not great enough in numerical value to stop the particle and turn it around.

Case B: $(E < V)$

If electrons are sent into this system with $E < V$, then according to classical physics, they will all be turned around at $x = 0$ and none will ever penetrate to positive values of x. What does quantum theory say about this problem?

To study this question, we begin by investigating the nature of the solutions of the wave equation when $E < V$. In this region, the wave equation is

$$-\frac{\hbar^2}{2m}\frac{\partial^2\psi}{\partial x_2} + (V - E)\psi = 0$$

and the solution is

$$\psi = A \exp\left(\sqrt{2m(V - E)}\,\frac{x}{\hbar}\right) + B \exp\left(-\sqrt{2m(V - E)}\,\frac{x}{\hbar}\right) \tag{14}$$

Note that the solutions are *real* exponentials rather than complex exponentials. In order that the probability remain finite as $x \to \infty$, it is necessary that we choose only the negative exponential, i.e., that we choose $A = 0$.

When $x < 0$, we do as before and write the most general solution as

$$\psi = C \exp\left(i\sqrt{2mE}\,\frac{x}{\hbar}\right) + D \exp\left(-i\sqrt{2mE}\,\frac{x}{\hbar}\right) \tag{14a}$$

If the function is to be continuous at $x = 0$, we must have

$$C + D = B \tag{15}$$

If the derivative of ψ is to be continuous at $x = 0$, it is readily verified that

$$\frac{i}{\hbar}\sqrt{2mE}\,(C - D) = -\frac{B}{\hbar}\sqrt{2m(V - E)}$$

or

$$C - D = iB\sqrt{\frac{V - E}{E}}$$

* Chap. 12.

Hence, we obtain

$$C = \frac{B}{2}\left(1 + i\sqrt{\frac{V - E}{E}}\right) \tag{16}$$

$$D = \frac{B}{2}\left(1 - i\sqrt{\frac{V - E}{E}}\right) \tag{17}$$

The ratio of the intensity of the reflected wave to that of the incident wave is

$$R = \frac{|D|^2}{|C|^2} = 1 \tag{18}$$

It is also of interest to calculate the phase of the waves. To do this, we write $\sqrt{(V - E)/E} = \tan \varphi$. Then we have

$$C = \frac{B}{2}(1 + i \tan \varphi) = \frac{B}{2 \cos \varphi}(\cos \varphi + i \sin \varphi) = \frac{B}{2 \cos \varphi} e^{i\varphi} \tag{19}$$

$$D = \frac{B}{2}(1 - i \tan \varphi) = \frac{B}{2 \cos \varphi}(\cos \varphi - i \sin \varphi) = \frac{B}{2 \cos \varphi} e^{-i\varphi} \tag{20}$$

Writing

$$\frac{B}{2 \cos \varphi} = \frac{I}{2} = \text{a new constant}$$

we obtain for the wave function when $x < 0$,

$$\begin{aligned}\psi &= \frac{I}{2}\left[\exp\left(i\sqrt{2mE}\frac{x}{\hbar} + i\varphi\right) + \exp\left(-i\sqrt{2mE}\frac{x}{\hbar} - i\varphi\right)\right] \\ &= I \cos\left(\sqrt{2mE}\frac{x}{\hbar} + \varphi\right)\end{aligned} \tag{21}$$

For a typical case, the wave function looks more or less like Fig. 7.

We see from eq. (18) that the entire wave is reflected because the reflected intensity is equal to the incident intensity. Because the wave

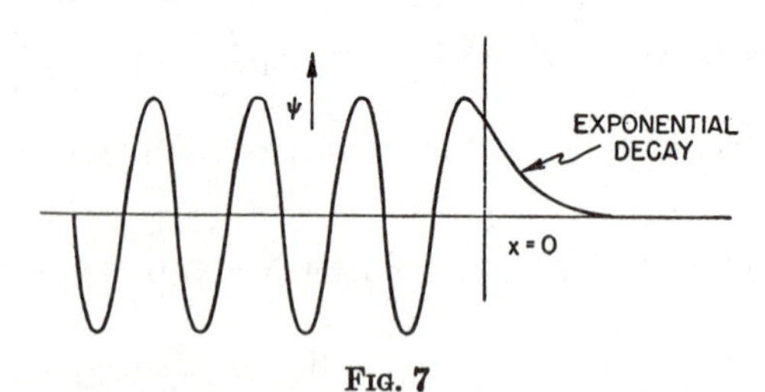

FIG. 7

equation implies the conservation of probability, we conclude that no electrons are transmitted.

Problem 3: Prove that the probability current is zero for Case B, that is, for $E < V$.

This result is in agreement with the classical prediction for this case. Yet there is a new feature, not present in the classical theory, coming from the penetration of the exponentially decaying part of the wave function into the region $x > 0$. This implies that an electron can be found in the region where $V > E$, whereas classically it could never enter this region, because it does not have enough energy.

To understand this phenomenon, we must remember that matter is not identical with the classical particle model, but that the electron also has wave properties, which can be just as important as the particle properties (see Chap. 6, Sec. 11). The region where $V > E$ corresponds to an imaginary index of refraction

$$n^2 = \left(\frac{2m}{\hbar^2}\right)(E - V)\frac{c^2}{\omega^2}$$

We already know of one case in optics where n is imaginary, namely that of total internal reflection of light. In this case there is exactly the same type of exponential penetration of the wave from the more dense into the less dense medium.* This new property of penetration into classically inaccessible regions must therefore be thought of in terms of the wave model; it is really one effect for which the particle model gives no description at all.

Suppose that we set up a device in the classically inaccessible region which actually measures the position of the electron; for example, we might use a microscope. Would this not involve a contradiction with the law of conservation of energy, since we would have in this region a particle with negative kinetic energy? Actually, we would find that to do an experiment which proved that the electron was definitely in this region, we would have to use such energetic light that the electron would be given a positive kinetic energy, and no contradiction would then arise from its being present in this region. To prove this, we note that any observation which guaranteed that the electron was in the region where $V > E$ would have to put the electron into such a state that the wave function was represented by a wave packet, practically all of which was in the region where $V > E$. But to form a wave packet, we need wave functions that oscillate; otherwise, they cannot interfere destructively far from the center of the packet. The solutions with $V > E$ are real exponentials in the region $x > 0$, so that they do not oscillate and no wave packet can be made from them. The only wave functions that oscillate are those for which $E > V$. We conclude that if the electron is ever in a state in which it is certain to be in the region $x > 0$, then it must have been given so much energy that it could have gotten into this region even classically.

* J. A. Stratton, *Electromagnetic Theory*. New York: McGraw-Hill Book Company, Inc., 1941, p. 497.

The penetration of the electron into regions where $V > E$ is paradoxical only if we try to hold onto the idea that matter consists of classical particles. Because of the wave properties of matter, however, an electron of definite energy is a different sort of thing from what it is classically. In fact, an electron can have a definite energy only when its wave function is an eigenfunction of the Hamiltonian operator, and therefore only when the electron is spread over a broad region of space. The electronic kinetic energy is just such a property that it must become positive whenever the electron is localized in a definite region. The statement that a particle penetrates into regions of negative kinetic energy is therefore meaningless, since the electron cannot have the localizability that leads us to attribute to it particle-like properties when it is in a region which would classically lead to negative kinetic energies. It would be just as wrong to talk of particles of negative kinetic energy as to talk of interference of particles in a Davisson-Germer experiment. Instead, we must say that both of these effects result from situations in which the wavelike aspects of matter are emphasized. In fact, from the point of view expressed in Chap. 6, Secs. 4 to 9, the process of measurement of position literally transforms the electron from a wave-like object into a particle-like object. In other words, interaction with a potential for which $V > E$ leads to a fuller realization of the electron's wave-like potentialities, while interaction with a position-measuring device leads to a fuller realization of its particle-like potentialities.

5. Penetration of a Barrier. Are there any cases in which the penetration of the particle into a classically inaccessible region produces physically important results? The answer is that if the region where $V > E$ is only of finite extent, then a particle may "leak" through a potential barrier which is so high that it could never get through classically. Suppose, for example, that the potential looked like that shown in Fig. 8.

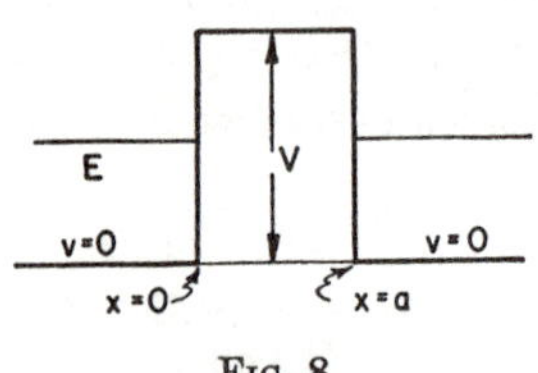

FIG. 8

In the region from $x = 0$ to $x = a$, $V > E$. According to classical theory, a stream of particles coming from the left would therefore be totally reflected. However, because of the wave nature of matter, we know that there is some probability that the particle penetrates out to the other side of the barrier and, as we shall show, it can actually escape into the region $x > a$, where $E > V$.

To treat this problem, we start from the right-hand side of the barrier, $x > a$. We know that there are no particles coming in from the right, but that there are particles streaming from the barrier toward the right. The wave function in this region is, therefore,

$$\psi = A \exp\left(\frac{ip_1 x}{\hbar}\right) \quad \text{where} \quad p_1 = \sqrt{2mE} \tag{22}$$

Within the barrier, the most general solution is

$$\psi = B \exp\left(\frac{p_2 x}{\hbar}\right) + C \exp\left(-\frac{p_2 x}{\hbar}\right) \quad \text{where} \quad p_2 = \sqrt{2m(V - E)} \tag{23}$$

We note that there is no reason now to throw out the exponentially increasing solution, because the region where $V > E$ has only a finite extent. To make ψ and $d\psi/dx$ continuous at $x = a$, we must have

$$B \exp\left(\frac{p_2 a}{\hbar}\right) + C \exp\left(-\frac{p_2 a}{\hbar}\right) = A \exp\left(\frac{ip_1 a}{\hbar}\right) \tag{24}$$

$$B \exp\left(\frac{p_2 a}{\hbar}\right) - C \exp\left(-\frac{p_2 a}{\hbar}\right) = \frac{Aip_1}{p_2} \exp\left(\frac{ip_1 a}{\hbar}\right) \tag{25}$$

Solving for B and C, we get

$$B = \frac{A}{2}\left(1 + \frac{ip_1}{p_2}\right) \exp\left[(ip_1 - p_2)\frac{a}{\hbar}\right] \tag{26}$$

$$C = \frac{A}{2}\left(1 - \frac{ip_1}{p_2}\right) \exp\left[(ip_1 + p_2)\frac{a}{\hbar}\right] \tag{27}$$

Let us consider the case where $p_2 a/\hbar \gg 1$; in other words, we suppose that the exponentials change a great deal from one side of the barrier to the other. We then notice that $|C| \gg |B|$. At the other side of the barrier (where $x = 0$), the main term in the wave function is the one involving $C \exp(-p_2 x/\hbar)$.

When $x < 0$, the wave function is

$$\psi = D \exp\left(\frac{ip_1 x}{\hbar}\right) + E \exp\left(-\frac{ip_1 x}{\hbar}\right)$$

In order that ψ and $\partial\psi/\partial x$ be continuous at $x = 0$, we must have

$$D + E = C + B \tag{28}$$

$$D - E = \frac{ip_2}{p_1}(C - B) \tag{29}$$

$$D = \frac{C}{2}\left(1 + \frac{ip_2}{p_1}\right) + \frac{B}{2}\left(1 - \frac{ip_2}{p_1}\right) \tag{30}$$

$$E = \frac{C}{2}\left(1 - \frac{ip_2}{p_1}\right) + \frac{B}{2}\left(1 + \frac{ip_2}{p_1}\right) \tag{31}$$

If the barrier is thick, we may, to a first approximation, neglect B. We then obtain

$$D = \frac{A}{4}\left(1 + \frac{ip_2}{p_1}\right)\left(1 - \frac{ip_1}{p_2}\right) \exp\left(\frac{ip_1 a}{\hbar}\right) \exp\left(\frac{p_2 a}{\hbar}\right) \tag{32}$$

$$E = \frac{A}{4}\left(1 - \frac{ip_2}{p_1}\right)\left(1 - \frac{ip_1}{p_2}\right) \exp\left(\frac{ip_1 a}{\hbar}\right) \exp\left(\frac{p_2 a}{\hbar}\right) \tag{33}$$

It is of interest to solve for the ratio of the intensity of the transmitted wave to that of the incident wave, i.e., the transmission coefficient T.

$$T = \frac{|A|^2}{|D|^2} = \frac{16 \exp(-2p_2a/\hbar)}{[1 + (p_2/p_1)^2][1 + (p_1/p_2)^2]} \tag{34}$$

The preceding result shows that there is a small probability that an object can penetrate a potential barrier which it could not even enter according to classical theory. This probability decreases rapidly as the barrier gets thicker and also as it gets higher.

As pointed out in Sec. 4, this property of barrier penetration is entirely due to the wave aspects of matter and is, in fact, very similar to the total internal reflection of light waves. If two slabs of glass are placed close to each other, but not touching, then light will be transmitted from one

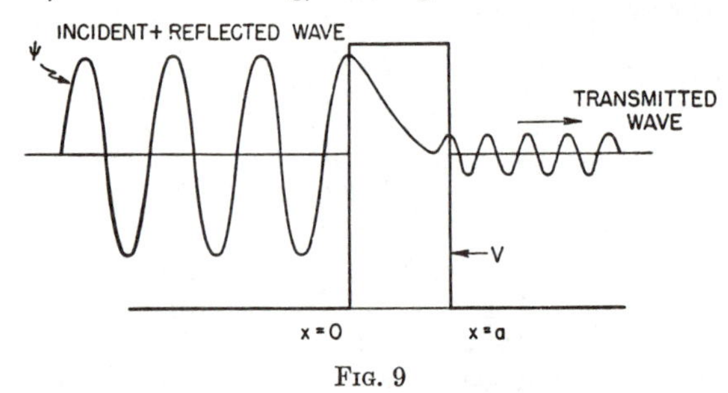

FIG. 9

slab to the second, even if the angle of incidence is greater than the critical angle. The intensity of the transmitted wave, however, decreases exponentially with the thickness of the layer of air. The reason for the transmission is exactly the same as with electron waves, namely, the exponential penetration of the wave into the region of imaginary index of refraction.

The wave function looks more or less as in Fig. 9. Most of the incident wave is reflected, but a small part is transmitted.

In order to compute the reflection coefficient, it is necessary to use the exact solution, which does not neglect B. The reflection coefficient is

$$R = \frac{|E|^2}{|D|^2} \tag{35}$$

Problem 4: Prove (with the exact solution) that $T + R = 1$.

Compute the probability current inside the barrier and show that it is equal to the current in the transmitted wave. Hence verify the conservation of probability for this case. (Note that the current for this case is contributed to by the effects of interference of the exponentially increasing and exponentially decreasing solutions. As a result, the neglect of the smaller solution in this region is not permitted if we wish to compute the current.)

6. Applications of Barrier Penetration. The principal example of barrier penetration is the α decay of nuclei. It is known that certain nuclei can emit α particles, but the mean time needed to emit such particles varies over an enormous range from one radioactive nucleus to another. The theory of α decay is based on the idea that the α particles are held inside the nucleus by tremendous attractive forces, very similar to those involved in the attraction of neutrons for protons. These forces, however, have a very short range, so that they are completely negligible unless the α particle is inside the nucleus. The α particles and the nucleus are both positively charged. This means that the electrical forces tend to make them repel each other. When the α particle is inside the nucleus, this electrical repulsion is much less than the nuclear attractive forces, but outside the nucleus it is the only force present. If, therefore, an α particle is brought toward a nucleus from a long distance, it will at first be repelled electrically and will have a potential energy

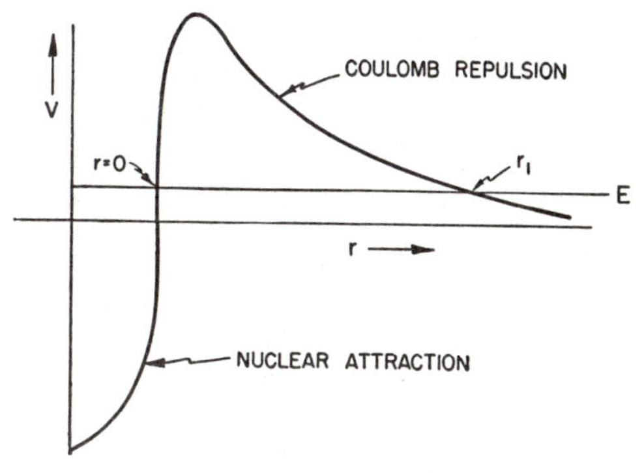

FIG. 10

$2Ze^2/r$, where Ze is the charge on the nucleus and $2e$ is the charge on the α particle. When it reaches the nucleus, this repulsion is rapidly overbalanced by the nuclear attraction. The potential curve as a function of the distance r of the α particle from the center of the nucleus looks more or less like the curve shown in Fig. 10. If the α particle has an energy E that is not great enough to carry it over the repulsive Coulomb barrier, then, according to classical physics, the α particle would be trapped inside the nucleus, once it got in. But because of its wave properties, the α particle actually has a small probability of leaking through the barrier.

To find the mean rate of emission, we assume that the α particle moves back and forth more or less freely inside the nucleus. There is independent evidence that it does so at a speed of about 10^9 cm/sec.* Since the heavy radioactive nuclei, such as uranium, have radii of about 10^{-12} cm, the α particle strikes the barrier about 10^{21} times per second. Each time it strikes the barrier, the probability that it penetrates is equal to

* H. Bethe, *Elementary Nuclear Physics*, p. 110.

the transmissivity T, of the barrier, given by eq. (34). Hence, the probability that it comes out in one second is given by

$$P = 10^{21}T \quad \text{per sec}$$

The mean lifetime of the nucleus is just the reciprocal of this, or

$$\tau = \frac{10^{-21}}{T} \quad \text{sec}$$

To compute T, we need to know the quantities $E - V$ and a, the thickness of the barrier. Actually, the barrier is far from rectangular, as can be seen from Fig. 4; hence, the present treatment is not very good for this case. A better treatment will be given later with the aid of the WKB approximation. Here we shall merely attempt to obtain an order of magnitude for T. For uranium, the mean value of $V - E$ is about 12 mev,* the mean width about 3×10^{-12} cm.

The factor $[1 + (p_2/p_1)^2][1 + (p_1/p_2)^2]$ is so close to unity that we can neglect it in comparison with the exponential. For the α particle, $m = 6.4 \times 10^{-24}$ gram. Noting that 1 ev $= 1.6 \times 10^{-12}$ ergs, we obtain

$$2\sqrt{2m(V - E)}\frac{a}{\hbar} = \frac{2\sqrt{12.8 \times 10^{-24} \times 12 \times 1.6 \times 10^{-12}}}{1.07 \times 10^{-27}} \times 3 \times 10^{-12} \cong 90$$

The result is that

$$P = 10^{21}\, e^{-90} = 10^{-18} \text{ per sec} = 10^{-11} \text{ per year}$$

It is clear that this number is sensitive to the exact value of $(V - E)$ and a, since these appear in the exponential. As a result this treatment gives merely a crude estimate. We can also see that the lifetimes for different elements may be expected to vary widely, since $V - E$ and a will vary, and since the exponential is sensitive to these quantities. In Chap. 12, however, with the aid of the WKB approximation, we shall give a treatment that is in closer agreement with experiment and that gives a better idea of how the lifetime for a decay varies for different elements.

7. The Square Well Potential. Let us now consider a square potential that is attractive, rather than repulsive, as shown in Fig. 11. Let this potential be $-V_0$ in the region from $x = a$ to $x = -a$, and zero elsewhere. Now, suppose that a stream of electrons is directed at it from the left. According to classical physics, no electrons would ever be turned back but, as we have already seen, the wave theory tells us that electrons will be reflected from the sharp edges at $x = a$ and $x = -a$.

* H. Bethe, *Elementary Nuclear Physics.*

As a result, there will be a reflected and a transmitted wave, as well as an incident wave.*

To solve this problem, we start in the region $x > a$, where there is only a transmitted wave. The wave function is, therefore,

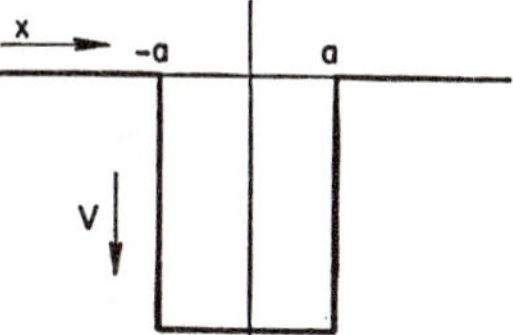

Fig. 11

$$\psi = A \exp\left(\frac{ip_1x}{\hbar}\right) \tag{36}$$

where p_1 is the momentum in the region $p_1 = \sqrt{2mE}$.

In the region of the well, from $x = -a$ to $x = +a$, the wave function is

$$\psi = B \exp\left(\frac{ip_2x}{\hbar}\right) + C \exp\left(-\frac{ip_2x}{\hbar}\right) \tag{37}$$

where $p_2 = \sqrt{2m(E + V_0)}$.

To solve for B and C in terms of A, we must make ψ and $d\psi/dx$ continuous at $x = a$:

$$A \exp\left(\frac{ip_1a}{\hbar}\right) = B \exp\left(\frac{ip_2a}{\hbar}\right) + C \exp\left(-\frac{ip_2a}{\hbar}\right) \tag{38}$$

$$\frac{p_1}{p_2} A \exp\left(\frac{ip_1a}{\hbar}\right) = B \exp\left(\frac{ip_2a}{\hbar}\right) - C \exp\left(-\frac{ip_2a}{\hbar}\right) \tag{39}$$

Solution of these equations yields

$$B = \frac{A}{2}\left(1 + \frac{p_1}{p_2}\right) \exp\left[\frac{i(p_1 - p_2)a}{\hbar}\right] \tag{40}$$

$$C = \frac{A}{2}\left(1 - \frac{p_1}{p_2}\right) \exp\left[\frac{i(p_1 + p_2)a}{\hbar}\right] \tag{41}$$

In the region $x < -a$, the wave function is

$$\psi = D \exp\left(\frac{ip_1x}{\hbar}\right) + E \exp\left(-\frac{ip_1x}{\hbar}\right) \tag{42}$$

To make ψ and $d\psi/dx$ continuous at $x = -a$, we have

$$D \exp\left(-\frac{ip_1a}{\hbar}\right) + E \exp\left(\frac{ip_1a}{\hbar}\right) = B \exp\left(-\frac{ip_2a}{\hbar}\right) + C \exp\left(\frac{ip_2a}{\hbar}\right) \tag{43}$$

* Note that this treatment will also apply to the potential barrier, provided that $E > V_0$, i.e., provided that the kinetic energy inside the barrier remains positive.

$$D \exp\left(-\frac{ip_1a}{\hbar}\right) - E \exp\left(\frac{ip_1a}{\hbar}\right) = \left(\frac{p_2}{p_1}\right)\left[B \exp\left(-\frac{ip_2a}{\hbar}\right) - C \exp\left(\frac{ip_2a}{\hbar}\right)\right] \tag{44}$$

$$D = \left(\frac{B}{2}\right)\left(1 + \frac{p_2}{p_1}\right) \exp\left[\frac{i(p_1 - p_2)a}{\hbar}\right] + \left(\frac{C}{2}\right)\left(1 - \frac{p_2}{p_1}\right) \exp\left[\frac{i(p_1 + p_2)a}{\hbar}\right] \tag{45}$$

$$E = \left(\frac{B}{2}\right)\left(1 - \frac{p_2}{p_1}\right) \exp\left[\frac{-i(p_1 + p_2)a}{\hbar}\right] + \left(\frac{C}{2}\right)\left(1 + \frac{p_2}{p_1}\right) \exp\left[\frac{i(p_2 - p_1)a}{\hbar}\right] \tag{46}$$

$$D = \left(\frac{A}{4}\right) \exp\left(\frac{2ip_1a}{\hbar}\right)\left[\left(1 + \frac{p_2}{p_1}\right)\left(1 + \frac{p_1}{p_2}\right) \exp\left(-\frac{2ip_2a}{\hbar}\right) + \left(1 - \frac{p_2}{p_1}\right)\left(1 - \frac{p_1}{p_2}\right) \exp\left(\frac{2ip_2a}{\hbar}\right)\right] \tag{47}$$

$$E = \left(\frac{A}{4}\right)\left[\left(1 - \frac{p_2}{p_1}\right)\left(1 + \frac{p_1}{p_2}\right) \exp\left(-\frac{2ip_2a}{\hbar}\right) + \left(1 + \frac{p_2}{p_1}\right)\left(1 - \frac{p_1}{p_2}\right) \exp\left(\frac{2ip_2a}{\hbar}\right)\right] \tag{48}$$

The transmissivity is

$$T = \frac{|A|^2}{|D|^2} = \left[\cos\left(\frac{2p_2a}{\hbar}\right) + \frac{i}{2}\left(\frac{p_1}{p_2} + \frac{p_2}{p_1}\right) \sin\left(\frac{2p_2a}{\hbar}\right)\right]^{-2} = \left[\cos^2\left(\frac{2p_2a}{\hbar}\right) + \frac{1}{4}\left(\frac{p_1}{p_2} + \frac{p_2}{p_1}\right)^2 \sin^2\left(\frac{2p_2a}{\hbar}\right)\right]^{-1} \tag{49}$$

Noting that

$$\cos^2 = 1 - \sin^2 \qquad \text{and} \qquad \left(\frac{p_1}{p_2} + \frac{p_2}{p_1}\right)^2 - 4 = \left(\frac{p_1}{p_2} - \frac{p_2}{p_1}\right)^2$$

we obtain

$$T = \frac{1}{1 + \frac{1}{4}(p_1/p_2 - p_2/p_1)^2 \sin^2(2p_2a/\hbar)} \tag{50}$$

This result is very interesting. First, we see that for $p_1 = p_2$, $T = 1$. This is very natural, because there is then no potential well at all. If $p_1 \neq p_2$, the transmissivity is, in general, less than unity, indicating that some reflection has taken place. This reflection from an attractive potential is a result of the wave nature of matter; it resembles the reflection of sound waves from the open end of an organ pipe. There is, however, one case in which $T = 1$ even though $p_1 \neq p_2$, namely, when $\sin^2(2p_2a/\hbar) = 0$, or $p_2 = N\pi\hbar/2a$, where N is an integer.

How can we understand this result? To see what it means, we note that this problem is very similar to that of the Fabry-Perot interferometer in optics.* In our problem, the wave is reflected at the sharp edges of the potential, which correspond to the edges of a piece of glass in optics, where, likewise, a sharp change of index of refraction takes place. This problem therefore resembles that of two sheets of glass, separated by a distance of $2a$. In the treatment of the Fabry-Perot interferometer, it is shown that if the wave, which reflects from the surface at $x = +a$,

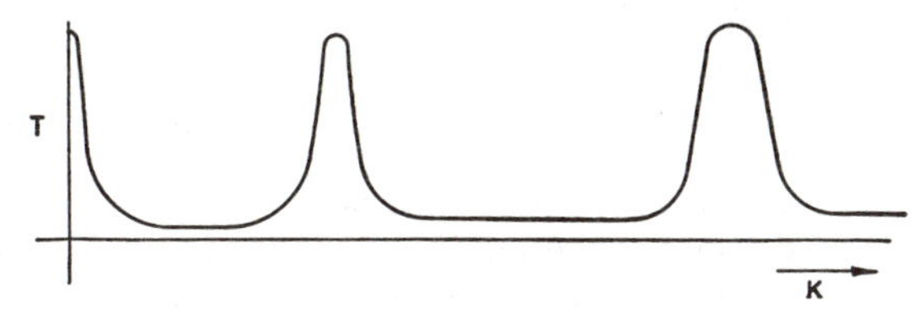

FIG. 12

arrives back at that surface after reflecting from $x = -a$ with a phase shift of $2\pi N$, then it will interfere constructively with the next wave coming in and, as a result, the transmitted wave is reinforced. Thus, for certain wavelengths, the transmission coefficient is unity. As a function of wave number, the transmission coefficient resembles the curve given in Fig. 12. The sharpness and breadth of the peaks depend on the reflection coefficient and, in our problem, this depends on the ratio p_1/p_2.

Problem 5: Compute the reflectivity, $R = |E|^2/|D|^2$ and show that $T + R = 1$.

8. Width of Peak in Transmission Resonances. To compute the width of the peak, we first assume that p_2/p_1 is large, so that the peak will be sharp. Then we shall ask how far from $p_2 = N\pi\hbar/2a$ we will have to be to make T drop to $\frac{1}{2}$. This will occur where

$$\frac{1}{4}\left(\frac{p_1}{p_2} - \frac{p_2}{p_1}\right)^2 \sin^2\left(\frac{2p_2a}{\hbar}\right) = 1 \tag{51}$$

or where

$$\sin\left(\frac{2p_2a}{\hbar}\right) = \pm\frac{2}{|p_1/p_2 - p_2/p_1|} \tag{52}$$

If the denominator on the right is large, then $2p_2a/\hbar$ will differ only slightly from $N\pi$ at this point, and we can write

$$\frac{2p_2a}{\hbar} \cong N\pi \pm \frac{2}{|p_1/p_2 - p_2/p_1|} \tag{53}$$

$$p_2 \cong \frac{N\pi\hbar}{2a} \pm \frac{\hbar/a}{|p_1/p_2 - p_2/p_1|} \tag{54}$$

$$\delta p_2 \cong \frac{\hbar}{a}\frac{1}{|p_1/p_2 - p_2/p_1|} \tag{55}$$

* F. A. Jenkins and H. E. White, *Fundamentals of Physical Optics*. New York McGraw-Hill Book Company, Inc., 1937, p. 93.

Writing $p_2 = \sqrt{2m(E + V_0)}$, we get, by differentiation (assuming $\delta p_2/p_2$ to be small),

$$\delta p_2 \cong \sqrt{\frac{m}{2(E + V_0)}}\,\delta E \tag{55a}$$

and

$$\delta E \cong \sqrt{\frac{2(E + V_0)}{m}}\,\delta p_2 \cong \sqrt{\frac{2(E + V_0)}{m}}\,\frac{\hbar}{a}\,\frac{1}{|p_1/p_2 - p_2/p_1|}$$
$$= \frac{v_2\hbar/a}{|p_1/p_2 - p_2/p_1|} \tag{56}$$

where v_2 is the velocity of the particle inside the well.

It is easy to explain the width of the transmission resonances in terms of the process of reflection of the wave back and forth between the edges of the potential well. If T is the transmission coefficient for a wave striking the sharp potential edge at $x = a$, then the wave will reflect back and forth approximately $1/T$ times before most of it has been transmitted. According to eq. (11), the transmission coefficient is

$$T = \frac{4p_2p_1}{(p_1 + p_2)^2}$$

where p_1 is the momentum of the transmitted particle and p_2 that of the incident particle.

The phase shift suffered by a wave as it crosses the well and returns is $4p_2a/\hbar$, which is equal to $2N\pi$ for the case of exact resonance. The total phase shift after $1/T$ reflections is $\varphi = 4p_2a/T\hbar$, which is equal to $\varphi_R = 2N\pi/T$ for exact resonance. Constructive interference will begin to fail when $\varphi - \varphi_R \cong 1$, or when

$$\varphi - \varphi_R = \frac{4\Delta p_2 a}{T\hbar} \cong 1$$

and

$$\Delta p_2 \cong \frac{\hbar T}{4a} \cong \frac{p_1p_2}{(p_1 + p_2)^2}\frac{\hbar}{a}$$

In order that a sharp resonance occur, it is necessary that T be small, so that many reflections can take place. This will happen only if $p_1 \ll p_2$. As a result, $(p_1 + p_2)^2 \cong p_2^2$, and we obtain $\Delta p_2 \cong \frac{p_1}{p_2}\frac{\hbar}{a}$. This is the same as obtained in eq. (55), using the same approximation, i.e., $p_1 \ll p_2$. Note that the whole argument is only approximate and qualitative.

9. The Ramsauer Effect. An interesting example of these transmission "resonances" occurs in the scattering of electrons from atoms of noble gases, such as neon and argon. The potential energy of an electron inside such an atom looks somewhat as shown in Fig. 13. To a first approximation, it may be represented as a square well of radius about

2×10^{-8} cm and uniform depth V_0. Now, it turns out that for very slow electrons, having a kinetic energy of the order of 0.1 electron volt, the effective depth and radius of the well are such that there is a transmission resonance for electrons. Thus, the atom seems to be practically transparent to electrons of this speed. Also, the probability of scattering, for example, is much less than is obtained with atoms for which this resonance does not exist, or for the same atoms at higher electronic energies, for which the resonance also does not exist. This effect was first observed experimentally by Ramsauer and was later explained in terms of quantum theory. We shall study this effect in greater detail in the theory of scattering, where the complications resulting from the three-dimensional nature of the problem are taken into account.*

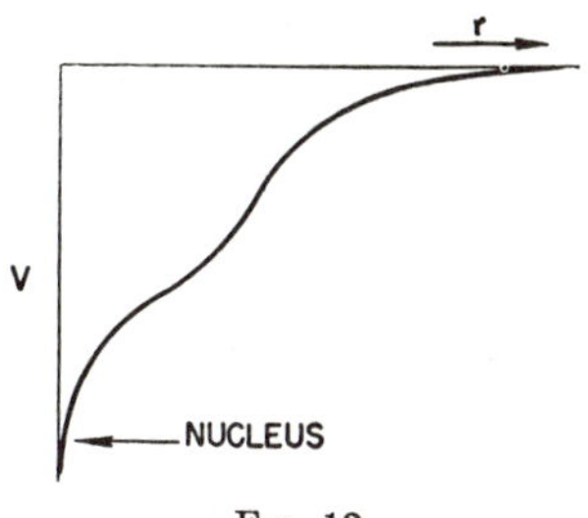

FIG. 13

Problem 6: Assuming a square well of range 2×10^{-8} cm, how deep would the well have to be to provide a transmission resonance for electrons of 0.1-ev kinetic energy?

10. Bound States. According to classical theory, a particle for which $E < 0$ would be bound inside the potential well, because it would not have the energy to escape. Are there any bound states in the quantum treatment of this problem? We shall see that there may be such bound states but that, in general, the possible energies of bound states are not continuous as in classical theory, but discrete. This is also in contrast to quantum results for positive energies, where we have seen that there was no restriction on the values of E for which solutions to the wave equation existed, so that a continuous spectrum is obtained in this case.

We begin the solution for the bound-state eigenfunctions and eigenvalues by noting that in the region where $x > a$, the solution is a linear combination of real exponentials; hence to make ψ finite as $x \to \infty$, we must choose the exponential that decreases with increasing x. Thus we write $\psi = A \exp(-p_1 x/\hbar)$, where $p_1 = \sqrt{2m|E|}$ and E is the energy of the bound state, which is negative.†

Within the square well, the wave function is

$$\psi = B \exp\left(\frac{ip_2 x}{\hbar}\right) + C \exp\left(-\frac{ip_2 x}{\hbar}\right) \tag{57}$$

where $p_2 = \sqrt{2m(V_0 - |E|)}$. The continuity conditions lead to (for $x = a$)

* See Chap. 21, Sec. 51. See also N. F. Mott and H. S. W. Massey, *The Theory of Atomic Collisions*. Oxford: Clarendon Press, 1933, p. 133.

† V_0 is by definition a positive number, as defined in connection with Fig. 12.

$$B\exp\left(\frac{ip_2a}{\hbar}\right) + C\exp\left(-\frac{ip_2a}{\hbar}\right) = A\exp\left(-\frac{p_1a}{\hbar}\right)$$

$$B\exp\left(\frac{ip_2a}{\hbar}\right) - C\exp\left(-\frac{ip_2a}{\hbar}\right) = \frac{iAp_1}{p_2}\exp\left(-\frac{p_1a}{\hbar}\right)$$

Solving for B and C, we get

$$\left.\begin{aligned} B &= \frac{A}{2}\left(1+\frac{ip_1}{p_2}\right)\exp\left[-\frac{a}{\hbar}(p_1+ip_2)\right] \\ C &= \frac{A}{2}\left(1-\frac{ip_1}{p_2}\right)\exp\left[-\frac{a}{\hbar}(p_1-ip_2)\right] \end{aligned}\right\} \tag{57a}$$

It is now necessary that at $x = -a$, the solution fit smoothly onto an exponential which decreases as $x \to -\infty$. This will not happen, in general, unless the binding energy, $|E|$, has a suitable value. To find out when such a solution is possible, we write (for $x < -a$)

$$\psi = D\exp\left(\frac{p_1x}{\hbar}\right) \tag{58}$$

The continuity conditions are

$$D\exp\left(\frac{p_1a}{\hbar}\right) = B\exp\left(-\frac{ip_2a}{\hbar}\right) + C\exp\left(\frac{ip_2a}{\hbar}\right) \tag{59}$$

$$\frac{p_1}{p_2}D\exp\left(\frac{p_1a}{\hbar}\right) = i\left[B\exp\left(-\frac{ip_2a}{\hbar}\right) - C\exp\left(\frac{ip_2a}{\hbar}\right)\right] \tag{60}$$

By dividing the second of these by the first, we obtain

$$\begin{aligned} -i\frac{p_1}{p_2} &= \frac{B\exp(-ip_2a/\hbar) - C\exp(ip_2a/\hbar)}{B\exp(-ip_2a/\hbar) + C\exp(ip_2a/\hbar)} \\ &= \frac{[1+(ip_1/p_2)]\exp(-2ip_2a/\hbar) - [1-(ip_1/p_2)]\exp(2ip_2a/\hbar)}{[1+(ip_1/p_2)]\exp(-2ip_2a/\hbar) + [1-(ip_1/p_2)]\exp(2ip_2a/\hbar)} \end{aligned} \tag{60a}$$

To simplify this expression, we write

$$\frac{p_1}{p_2} = \tan\varphi, \qquad 1+\frac{ip_1}{p_2} = \frac{1}{\cos\varphi}(\cos\varphi + i\sin\varphi) = \frac{e^{i\varphi}}{\cos\varphi}$$

We get

$$\begin{aligned} \tan\varphi &= i\,\frac{\exp[i(\varphi - 2p_2a/\hbar)] - \exp[-i(\varphi - 2p_2a/\hbar)]}{\exp[i(\varphi - 2p_2a/\hbar)] + \exp[-i(\varphi - 2p_2a/\hbar)]} \\ &= \tan\left(\frac{2p_2a}{\hbar} - \varphi\right) \end{aligned} \tag{61}$$

The above equation implies that $\varphi = 2p_2a/\hbar - \varphi + N\pi$, where N is any integer, positive or negative. Solution for φ yields

$$\varphi = p_2 \frac{a}{\hbar} + \frac{N\pi}{2}$$

$$\tan \varphi = \frac{p_1}{p_2} = \tan\left(\frac{p_2 a}{\hbar} + \frac{N\pi}{2}\right) = \begin{cases} \tan \dfrac{p_2 a}{\hbar} & N \text{ even} \\ -\cot \dfrac{p_2 a}{\hbar} & N \text{ odd} \end{cases} \tag{62}$$

Expressing p_1 and p_2 in terms of E and V_0, we then obtain

$$\sqrt{\frac{|E|}{V_0 - |E|}} = \begin{cases} \tan\left[\sqrt{2m(V_0 - |E|)}\,\dfrac{a}{\hbar}\right] & N \text{ even} \\ -\cot\left[\sqrt{2m(V_0 - |E|)}\,\dfrac{a}{\hbar}\right] & N \text{ odd} \end{cases} \tag{63}$$

The above is a transcendental equation defining $|E|$. Wherever it has a solution, we have a possible energy level. The equation must, in general, be solved numerically or graphically. We can, however, obtain an approximate idea of the location of the energy levels. To do this, we rewrite the equations with the substitution

$$\sqrt{2m(V_0 - |E|)}\,\frac{a}{\hbar} = \xi; \qquad 2m|E| = 2mV_0 - \left(\frac{\hbar}{a}\right)^2 \xi^2 \tag{64}$$

These yield

$$\frac{\hbar}{a} \frac{\xi}{\sqrt{2mV_0 - (\hbar/a)^2\xi^2}} = \begin{cases} \cot \xi & N \text{ even} \\ -\tan \xi & N \text{ odd} \end{cases} \tag{65}$$

After we have solved for ξ, then we can obtain $|E|$ from eq. (64)

Case A: N odd.

It is necessary to find the intersection of the curve $y_1 = \tan \xi$ with the curve

$$y_2 = -\frac{\hbar}{a}\xi\left[2mV_0 - \left(\frac{\hbar}{a}\right)^2\right]^{-1/2}$$

(see Fig. 14). We note that the curve for y_2 extends only as far as

$$\xi = \pm\sqrt{2mV_0}\,\frac{a}{\hbar}$$

since, by definition, $|E|$ must be positive, and larger values of ξ would lead to negative values of $|E|$ in eq. (64). The curve for y_2 goes through the origin, with a slope depending on V_0 and a. and finally becomes infinite at

$$\xi = \pm\sqrt{2mV_0}\,\frac{a}{\hbar}$$

The intersection of y_1 and y_2 at $\xi = 0$ is an extraneous root and does not lead to a true solution of Schrödinger's equation.

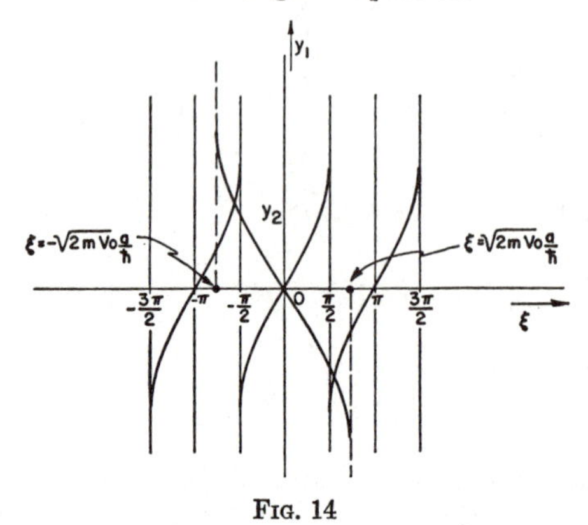

Fig. 14

Problem 7: Prove by substitution of eq. (57) into Schrödinger's equation that the root $\xi = 0$ does not lead to a solution.

If $\sqrt{2mV_0}\frac{a}{\hbar} < \frac{\pi}{2}$, there will be no additional intersections between y_1 and y_2, and therefore no bound-state solutions. It is readily verifiable that the condition for N bound-state solutions is

$$\sqrt{2mV_0}\frac{a}{\hbar} > \left(N + \frac{1}{2}\right)\pi$$

or

$$V_0 > \frac{1}{2m}\left(\frac{\hbar}{a}\right)^2\left(N + \frac{1}{2}\right)^2\pi^2$$

Note that we always obtain positive and negative roots in pairs. Since the value of $|E|$ depends only on ξ^2 [see eq. (64)], each pair leads, however, to only one value of $|E|$.

Case B: N even.

A similar treatment can be given for N even. We plot $y_1 = \cot \xi$, and find its intersection with

$$y_2 = \frac{\hbar}{a}\frac{\xi}{\sqrt{2mV_0 - (\hbar/a)^2\xi^2}}$$

(see Fig. 15).

The first solution occurs when $\xi < \pi/2$, the next one when $\xi > \pi$, and so on. At least one solution of this type (N even) can therefore exist, no matter how small V_0 is. For two solutions to exist, it is necessary that $\xi > \pi$, or that $V_0 > \frac{1}{2m}\left(\frac{\hbar}{a}\right)^2\pi^2$. As V_0 is increased, more and more solutions eventually become possible.

Problem 8: Suppose V_0 is 20 mev, $a = 2.8 \times 10^{-13}$ cm. Find (numerically or graphically) for a proton ($m = 1.6 \times 10^{-24}$ gram) Find them for an α particle of mass $m = 6.4 \times 10^{-24}$ gram.

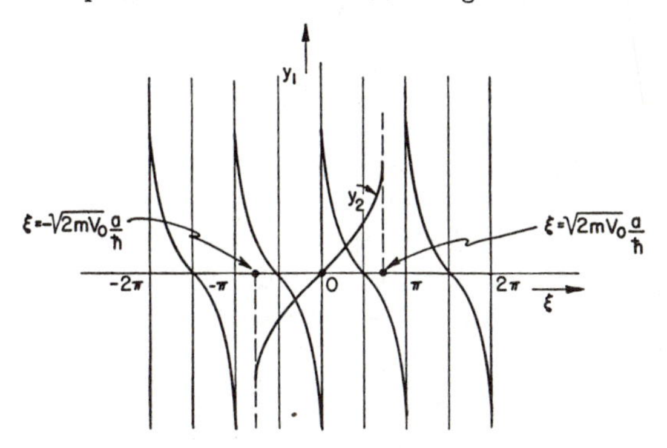

FIG. 15

11. Limit of an Infinitely Deep Well. If a well is infinitely deep, the solution in the classically inaccessible region, $\exp\left[-\sqrt{2m(V_0 - |E|)}\,\frac{x}{\hbar}\right]$, dies out with infinite speed, so that the wave function must be zero at each edge of the well. The solution must then be $\psi = \sin\left(N\frac{\pi}{2}\frac{x}{a}\right)$, where N is any integer.* Since the solution can also be written

$$\psi = \sin\sqrt{2m(V_0 - |E|)}\,\frac{x}{\hbar}$$

we have

$$\sqrt{V_0 - |E|} = \frac{\hbar}{a}\left(\frac{N\pi}{2}\right)\frac{1}{\sqrt{2m}} \qquad \text{or} \qquad V_0 - |E| = \frac{1}{2m}\left(\frac{\hbar}{a}\right)^2\left(\frac{N\pi}{2}\right)^2$$

We can readily verify that eqs. (63) lead to the same solution since, as $V_0 \to \infty$, we have $\frac{|E|}{V_0 - |E|} \to 0$; hence

$$\left.\begin{aligned} N \text{ odd:} \qquad & \tan\sqrt{2m(V_0 - |E|)}\,\frac{a}{\hbar} \to 0 \\ N \text{ even:} \qquad & \cot\sqrt{2m(V_0 - |E|)}\,\frac{a}{\hbar} \to 0 \end{aligned}\right\}$$

This leads to

$$\sqrt{V_0 - |E|} = \frac{1}{\sqrt{2m}}\left(\frac{\hbar}{a}\right)\frac{N\pi}{2}$$

which is in agreement with the result obtained directly.

* We have, for convenience, shifted the origin to one side of the well. We retain this notation only in this section.

12. Graphical Interpretation of Solutions. There is a simple graphical point of view that enables us to understand readily the general nature of all these different kinds of solutions. Let us consider the wave equation

$$\frac{d^2\psi}{dx^2} + \frac{2m}{\hbar^2}(E - V)\psi = 0 \tag{1}$$

This equation defines the second derivative of the wave function ψ, in terms of ψ and $E - V$. When $E > V$ (positive kinetic energy), the second derivative is opposite in sign to ψ itself. ψ is, therefore, concave toward the axis, so that the wave function will tend to oscillate. (This result is in agreement with the exact solution

$$\psi = A \cos \sqrt{2m(E - V)}\,\frac{x}{\hbar} + B \sin \sqrt{2m(E - V)}\,\frac{x}{\hbar}$$

The bigger $E - V$ is, the more rapidly does ψ curve, and the more rapidly it oscillates.)

When $V > E$, however, $d^2\psi/dx^2$ has the same sign as ψ, so that ψ is convex toward the axis. This means that if ψ is already increasing, it will increase even more rapidly, because the slope must be always increasing. (This is in agreement with the exact solution,

$$\psi = A \exp\left[-\sqrt{2m(V - E)}\,\frac{x}{\hbar}\right] + B \exp\left[\sqrt{2m(V - E)}\,\frac{x}{\hbar}\right]$$

The bigger $V - E$ is, the more rapidly will the exponential change.)

Let us now consider the bound states of the square well. When $x < -a$ (see Fig. 16), we start with an exponential solution that is increasing with increasing x and curving upward. At $x = -a$, the kinetic energy becomes positive and, since ψ is positive, the curvature becomes negative. The wave function then begins to curve back toward $\psi = 0$, at a rate depending on $V_0 - |E|$. If $V_0 - |E|$ is large enough, the slope will be negative by the time we reach $x = a$. When $x > a$, the function begins to curve back upward again, because $V_0 - |E|$ is negative. For a general choice of $|E|$, it will eventually increase without bounds and, therefore, become an inadmissible solution. Only if $|E|$ is such that the slope at $x = a$ exactly matches the required slope of a decreasing exponential

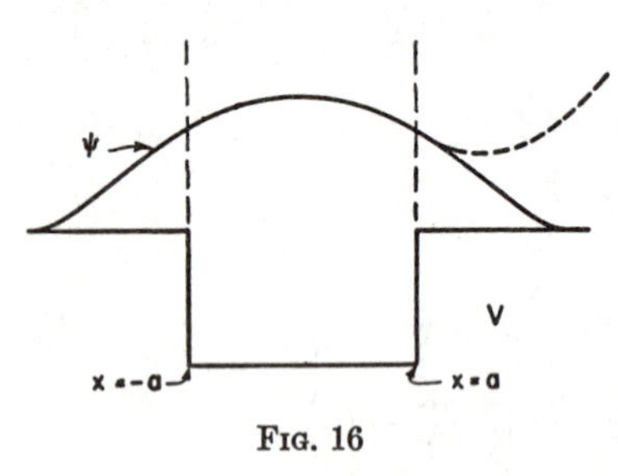

FIG. 16

$$\left[\psi = \exp\left(-\frac{x}{\hbar}\sqrt{2m|E|}\right)\right]$$

will the solution remain bounded as $x \to \infty$. Thus, only certain values of $|E|$ will lead to bound states. These will be the eigenvalues.

If V_0 is very large, then ψ can fit onto the decaying exponential at $x = a$ after one or more oscillations. These will be additional bound states. Such possibilities are illustrated in Fig. 17. The larger V_0 is, the greater, in general, will be the number of such possibilities.

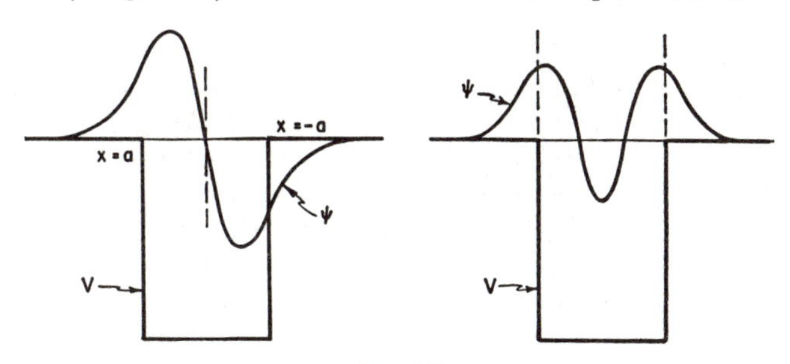

FIG. 17

Each solution may be described in terms of the number of zeros (or nodes) that the wave function has. For example, the first solution mentioned has no nodes, the second solution has one, the third two, etc. Generally, the number of nodes in the solution is equal to the number N, appearing in eq. (64).

Solution for Wave Function

For each value of N for which there is a solution to eq. (63), we can now solve for the wave function. To do this, we note that once $|E|$ is known, $p_1 = \sqrt{2m|E|}$ and $p_2 = \sqrt{2m(V_0 - |E|)}$ are also known. This means that (57) and (60a), defining the wave function inside the well, can now be solved, so that the entire wave function can be expressed in terms of the single constant D, defined in eq. (58). The constant D can be evaluated by normalizing the wave function.

Problem 9: Show by obtaining the wave function that the number of nodes is equal to N.

13. Application of Expansion Theorem. In Chap. 10, Sec. 22, it was pointed out that an arbitrary function can be expanded as a series of eigenfunctions of any Hermitean operator. Let us now apply this theorem to the Hamiltonian operator for the square well potential. The eigenfunctions must include the continuous spectrum of eigenvalues appearing when $E > 0$ and also all bound states with $E < 0$.

At first sight, it may not be clear why the bound states are needed. The reason is that within the well the eigenfunctions for $E > 0$ are so distorted by the potential that they are unable to express certain types of functions at all. The functions which cannot be expressed as an integral of continuum eigenfunctions are, in fact, just the bound-state wave functions. To see this in greater detail, let us note that the bound-state eigenfunctions are orthogonal to the continuum functions (see Chap. 10, Sec. 24). It is, therefore, impossible to expand the bound-state functions in terms of the continuum functions for, according to Chap. 10, eq. (55) the expansion coefficient is just

$$C_E = \int \psi_E^*(x)\psi_B(x)\,dx$$

which is zero when ψ_B is a bound-state wave function, and ψ_E belongs to the continuum. Thus, to express all possible functions, we must sum over bound states, as well as integrate over the continuous spectrum.

14. Application to Deuteron. So far, we have considered only a one-dimensional problem, whereas all actual problems are three-dimensional. But we shall see in Chap. 15 that in terms of the radius r, the wave equation for ψ is similar to the one-dimensional wave equation that we have given here. In fact, for the special case that ψ is a function of r only, and not of the spherical polar angles ϑ and φ, the equations will be shown to be identical with the one-dimensional case.† There is, however, one important new restriction, namely, that the wave function must always be zero at the origin. This arises, as we shall see, from the requirement that certain functions remain finite as $r \to 0$. For the present, let us merely accept this requirement.

To find out which bound-state wave functions satisfy the requirement that $\psi = 0$ at $x = 0$, we refer to eq. 57, which gives the value of ψ within the potential well. At $x = 0$, we have

$$\psi = B + C = 0$$

The additional requirement is, therefore, that $B = -C$. From eq. (57) this is seen to be the equivalent of

$$\left(1 + \frac{ip_1}{p_2}\right)\exp\left(\frac{-iap_2}{\hbar}\right) = -\left(1 - \frac{ip_1}{p_2}\right)\exp\left(\frac{iap_2}{\hbar}\right)$$

Writing $p_1/p_2 = \tan\varphi$, we obtain

$$\exp\left[-i\left(\frac{ap_2}{\hbar} - \varphi\right)\right] = -\exp\left[i\left(\frac{ap_2}{\hbar} - \varphi\right)\right]$$

or

$$\cos\left(\frac{ap_2}{\hbar} - \varphi\right) = 0$$

Hence

$$\varphi = \frac{ap_2}{\hbar} + \frac{N\pi}{2}$$

where N is an odd integer. Comparing this with eq. (62), we see that if N is restricted to odd values in eq. (62), then the two equations are

† See Chap. 15, Sec. 3.

equivalent. We therefore conclude that all bound solutions of the three-dimensional problem must have N odd. As shown in Sec. 10, no such bound solution is possible unless $V_0 \geq \frac{1}{2m}\left(\frac{\hbar}{a}\right)^2\left(\frac{\pi}{2}\right)^2$ and, in general, a bound solution with a given value of N is possible only when $V_0 \geq \left(\frac{\hbar}{a}\right)^2\left(\frac{1}{2m}\right)\left(\frac{N\pi}{2}\right)^2$. The number of bound states depends, therefore, on the depth of the potential well, its radius, and the mass of the particle.

The deuteron consists of a neutron and a proton bound together by a force that can be represented by a square well potential (see Sec. 3). Experimentally it has been found that the binding energy* is 2.237 mev. Using the radius given in Sec. 3, $r = 2.8 \times 10^{-13}$ cm, we can calculate the depth V_0 necessary to yield this binding energy. It is known that no levels exist below this level.† That is, there is only one bound state. In eq. (63), we therefore set $N = 1$. This gives

$$\sqrt{\frac{|E|}{V_0 - |E|}} = -\cot\sqrt{2m(V_0 - |E|)}\,\frac{a}{\hbar}$$

$$\tan\sqrt{2m(V_0 - |E|)}\,\frac{a}{\hbar} = -\sqrt{\frac{V_0 - |E|}{|E|}} \tag{66}$$

Let us write

$$\sqrt{2m(V_0 - |E|)}\,\frac{a}{\hbar} = \xi$$

We obtain

$$\tan\xi = -\frac{\hbar}{a}\frac{\xi}{\sqrt{2m|E|}} \tag{67}$$

Since $|E|$, m, and a are known, we can solve for ξ graphically and use this result to solve for V_0. The result is $V_0 = 21.2$ mev. Note that we must use the reduced mass $m = M/2$, where M is the proton mass, which is also practically equal to the neutron mass. This is because the wave equation really refers to the relative co-ordinates of the neutron and proton. We shall discuss this point in greater detail in connection with the hydrogen atom. (See Chap. 15, Sec. 5.)

Problem 10: Obtain V_0 in the manner suggested in the preceding section.

Note that eq. (66) really determines the product $\sqrt{V_0 - |E|}\,a$, and, therefore, also $(V_0 - |E|)a^2$. Since $|E|/V_0$ is small, the knowledge of the deuteron binding energy enables us to determine the approximate product of V_0a^2.

* The binding energy of a bound state is that energy needed to raise the energy to $E = 0$, at which point the particles are no longer bound together. It is clear that the binding energy is equal to $|E|$ in eq. (64).

† H. Bethe, *Elementary Nuclear Theory*, Chap. 7.

15. Interpretation of Energy Levels in Terms of Uncertainty Principle. The fact that no bound states are possible unless

$$V_0 > \frac{1}{2m}\left(\frac{\hbar}{a}\right)^2\left(\frac{\pi}{2}\right)^2$$

is easily understood in terms of the uncertainty principle. To have a bound state, a particle must be localized roughly within the radius of the well. To have a wave function large only in a region of the size of the well, there must also be a range of momenta $\sim \frac{\hbar}{a}$ and, therefore, energies $\sim \frac{1}{2m}\left(\frac{\hbar}{a}\right)^2$.* Before a particle can be trapped within the well, the potential energy given up when the particle enters the well must be greater than the kinetic energy that the particle obtains merely because it is localized within the radius a. Thus, no bound states at all are possible unless $V_0 > \frac{1}{2m}\left(\frac{\hbar}{a}\right)^2$. If V_0 is barely great enough to provide the kinetic energy necessary to localize the particle within the well, then the binding energy $|E|$ will be very small. If V_0 is increased, the binding energy becomes greater, and eventually V_0 becomes so great that it can supply the kinetic energy necessary to make the wave function oscillate once within the well. At this point, a new bound state becomes possible. If V_0 is made greater still, eventually a third oscillation becomes possible, then a fourth, etc. Thus, the number of bound states depends on how much deeper the well is than the minimum amount needed to contain the particle within the well.

16. Use of Observed Energy Levels to Provide Information about the Potential. In atomic theory, the usual procedure in quantizing is to start with the classical Hamiltonian function and to form the Hamiltonian operator by replacing the number p, wherever it occurs, by the operator, $\frac{\hbar}{i}\frac{\partial}{\partial x}$. But in many cases, we do not know the classical Hamiltonian function, because our only experience with the system has been on a purely quantum-mechanical level. This is especially true in nuclear physics, since nuclear forces have a very short range. In order for nuclear forces to act in a classical fashion, it would be necessary to have particles for which the de Broglie wavelength $\lambda = h/p$ was much less than the range of the forces, which is about 2.8×10^{-13} cm. We should, therefore, need momenta much greater than

$$p \cong \frac{h}{2.8 \times 10^{-13}} = \frac{6.6 \times 10^{-27}}{2.8 \times 10^{-13}} = 2.4 \times 10^{-14}$$

* See Chap. 5, Sec. 5, where a similar discussion is given in connection with the lowest bound state of a hydrogen atom.

The energy for protons would have to be greater than

$$E = \frac{p^2}{2m} \cong 100 \text{ mev}$$

Most experiments in nuclear physics involve much smaller energies ($\sim$1 to 20 mev). Furthermore, at energies of 100 mev and higher, there is evidence that the idea whereby the system can be described by a wave equation involving some definite potential function is breaking down. In other words, at very high energies it is likely that quantum theory may have to be seriously modified. As a result, in the nuclear domain, the entire formulation of the theory in terms of a Hamiltonian operator is a tentative procedure, which can be justified only to the extent that it is successful. It must be emphasized, however, that in the domain of atomic physics, which involves distances not shorter than 10^{-12} cm, the concepts of ordinary quantum theory are known by experiment to be on a very solid foundation. Even here, however, it is often necessary to correct the Hamiltonian operator by small terms such as those involving the spin,* which are not contained in the classical Hamiltonian function.

The net result of this situation is that in certain problems, especially in nuclear physics, it is necessary to guess the potential function and try to verify our guesses by seeing whether they predict results agreeing with experiment. One of the most important types of results is the energy levels of the system in question: for example, in the case of the deuteron, we saw that because there was only one energy level with a depth of 2.237 mev, we could solve for the product of the depth of the potential V_0 and the square of its range, a^2. If there had been more energy levels, we should have come to different conclusions about the nature of the potential. We must keep in mind, therefore, the possibility of using the observations of energy levels as a tool to investigate potential functions which we cannot measure directly in a classical fashion. We shall return to this type of consideration many times in the future and we shall also stress the role of the study of scattering as a means of probing into the nature of atomic and nuclear systems.†

17. Wave Packets Made up from Eigenfunctions in the Continuum. Thus far, in discussing the continuum eigenfunctions ($E > 0$) for the square well potential, we used plane-wave solutions, which spread over all space. Such solutions actually represent an abstraction never realized in practice, because all real waves are bounded in one way or another. A wave packet more closely represents what happens in a real experiment.

For example, we start out with an incident packet, far from the well. This packet moves toward the well, spreading slowly as it moves. When

* See Chap. 17.
† See Chap. 21, Sec. 11.

it strikes the well, part of it is reflected, and part enters the well.* The part inside the well reflects back and forth, and part of it goes out to form a transmitted wave, while part returns to contribute to the reflected wave. The reflected wave from inside the well interferes with the wave reflected directly from the well. When the phase relations are right, the reflected wave is canceled by interference of these two parts so that only a transmitted wave is present, and there is a transmission resonance, as described after eq. (50). In general, however, both reflected and transmitted waves are present. If we start with an incident wave packet, the reflected and transmitted waves will also take the form of packets. In this manner, the progress of the wave through the potential can be described as a function of the time. After a long time, no part of the wave will remain inside the potential well.

It is instructive to carry out in detail the solution of the wave equation corresponding to the boundary condition of an incident wave packet as $t \to -\infty$. Let us start with the solution in the region $x < -a$ (see Fig. 12).

We have for the time-dependent solution

$$\psi(p_1) = \left[D(p_1) \exp\left(\frac{ip_1x}{\hbar}\right) + F(p_1) \exp\left(-\frac{ip_1x}{\hbar}\right)\right] \exp\left[-\frac{iE(p_1)t}{\hbar}\right] \tag{68}$$

where D and F† are, in general, functions of p_1, and $E(p_1) = p_1^2/2m$. To form a packet, it is necessary to integrate ψ over p_1 with a weighting factor $f(p_1 - p_0)$, peaked near some value, which we denote as p_0. We obtain

$$\psi(x, t) = \int dp_1 f(p_1 - p_0) \exp\left[-\frac{iE(p_1)t}{\hbar}\right] \left[D(p_1) \exp\left(\frac{ip_1x}{\hbar}\right) + F(p_1) \exp\left(-\frac{ip_1x}{\hbar}\right)\right] \tag{69}$$

In general, D and F are fairly smooth functions of p_1, defined by eqs. (47) and (48). For convenience, we can choose A such that $D(p_1) = 1$. This choice yields

$$\frac{A}{4} = \exp\left(-\frac{2ip_2a}{\hbar}\right)\left[\left(1 + \frac{p_2}{p_1}\right)\left(1 + \frac{p_1}{p_2}\right) \exp\left(-\frac{2ip_2a}{\hbar}\right) + \left(1 - \frac{p_2}{p_1}\right)\left(1 - \frac{p_1}{p_2}\right) \exp\left(\frac{2ip_2a}{\hbar}\right)\right]^{-1} \tag{70}$$

* In any given instance, the electron is either reflected or transmitted, but the intensities of the respective waves yield the probabilities that each of these processes takes place.

† We are using F here instead of E, which was used in the same place in eq. (42).

and, after rearranging,

$$F = -i(p_1^2 - p_2^2) \exp\left(-\frac{2ip_2a}{\hbar}\right) \sin\left(\frac{2p_2a}{\hbar}\right)\left[2p_1p_2\cos\left(\frac{2p_2a}{\hbar}\right)\right.$$
$$\left. + i(p_1^2 + p_2^2)\sin\left(\frac{2p_2a}{\hbar}\right)\right] \Big/ \left[4p_1^2p_2^2 + (p_1^2 - p_2^2)\sin^2\left(\frac{2p_2a}{\hbar}\right)\right] \quad (71)$$

It is often convenient to write

$$F(p_1) = R(p_1)e^{-i\varphi_1} \quad (72)$$

where $R(p_1) = |f(p_1)|$.

(Note that p_2 is expressed in terms of p_1.) We obtain

$$\varphi_1 = \frac{2p_2a}{\hbar} + \tan^{-1}\left(\frac{2p_1p_2}{p_1^2 + p_2^2}\cot\frac{2p_2a}{\hbar}\right) \quad (73)$$

Insertion of these values into eq. (69) for ψ now yields

$$\psi(x, t) = \int dp_1 f(p_1 - p_0)\exp\left(-\frac{iE(p_1)t}{\hbar}\right)$$
$$\left\{\exp\left(\frac{ip_1x}{\hbar}\right) + R(p_1)\exp\left[-i\left(\frac{p_1x}{\hbar} + \varphi_1(p_1)\right)\right]\right\} \quad (74)$$

To find the place where ψ is a maximum, we look for the place where the phase of the wave has an extremum when differentiated with respect to p_1. This insures that many waves of different p_1 will add up in phase, thus producing a peak (see Chap. 3, Sec. 2).

For an incident wave, the phase has an extremum when

$$\left.\begin{aligned} &\frac{\partial}{\partial p_1}\left(p_1\frac{x}{\hbar} - E(p_1)\frac{t}{\hbar}\right)_{p_1=p_0} = 0 \\ &\text{or when} \\ &x = \left(\frac{\partial E}{\partial p_1}\right)_{p_1=p_0} t = \frac{p_0}{m}t \end{aligned}\right\} \quad (75)$$

As $t \to -\infty$ we see that this point recedes indefinitely to the left. But as $t \to +\infty$, this point would have to be at $x \to +\infty$. Since the incident wave function has meaning only for x negative, it is clear that after $t = 0$, the incident wave disappears altogether, as it should.

Let us now look at the reflected wave. The condition for an extremum of its phase is

$$\left.\begin{aligned} &\frac{\partial}{\partial p_1}\left(p_1\frac{x}{\hbar} + \varphi_1(p_1) + E(p_1)\frac{t}{\hbar}\right)_{p_1=p_0} = 0 \\ x &= -\left(\frac{\partial E}{\partial p_1}\right)_{p_1=p_0} t - \hbar\left(\frac{\partial \varphi_1}{\partial p_1}\right)_{p_1=p_0} = -\frac{p_0}{m}t - \hbar\left(\frac{\partial \varphi_1}{\partial p_1}\right)_{p_1=p_0} \\ &= -v_0t - \hbar\left(\frac{\partial \varphi_1}{\partial p_1}\right)_{p_1=p_0} \end{aligned}\right\} \quad (76)$$

As $t \to +\infty$, we see that $x \to -\infty$. Hence, a reflected wave packet appears after the incident wave packet has struck the well. The significance of the term involving $\partial\phi_1/\partial p_1$ will be discussed in Sec. 19.

18. Wave Packet for Transmitted Wave. The transmitted wave amplitude is

$$\psi(x, t) = \int dp_1 f(p_1 - p_0) \exp\left[-\frac{iE(p_1)t}{\hbar}\right] A(p_1) \exp\left(\frac{ip_1 x}{\hbar}\right)$$

As in eq. (72) we write

$$A = |A|\, e^{i\varphi_2} \tag{77}$$

But, according to eq. (70),

$$A = \frac{\exp(-2ip_2a/\hbar)[\cos(2p_2a/\hbar) + i/2(p_1/p_2 + p_2/p_1)\sin(2p_2a/\hbar)]}{\cos^2(2p_2a/\hbar) + \frac{1}{4}(p_1/p_2 + p_2/p_1)^2 \sin^2(2p_2a/\hbar)} \tag{78}$$

The phase φ_2 can then be written

$$\varphi_2 = -\frac{2p_2a}{\hbar} + \tan^{-1}\left[\frac{1}{2}\left(\frac{p_1}{p_2} + \frac{p_2}{p_1}\right)\tan 2p_2\frac{a}{\hbar}\right] \tag{79}$$

The wave function becomes

$$\psi(x, t) = \int dp_1 f(p_1 - p_0)|A| \exp\left\{i\left[p_1\frac{x}{\hbar} + \varphi_2 - E(p_1)\frac{t}{\hbar}\right]\right\} \tag{80}$$

The maximum of the wave pocket occurs where the derivative of the argument of the exponential is zero, or where

$$x = \left(\frac{\partial E}{\partial p_1}\right)_{p_1=p_0} t - \hbar\left(\frac{\partial\varphi_2}{\partial p_1}\right)_{p_1=p_0} = \frac{p_0}{m}t - \hbar\left(\frac{\partial\varphi_2}{\partial p_1}\right)_{p_1=p_0} \tag{81}$$

As $t \to +\infty$ the maximum appears in the region where $x > a$. Thus, after sufficient time, the transmitted wave appears and travels with the group velocity $v_0 = p_0/m$.

19. Time Delay of Wave as It Crosses Potential Well. If there were no potential well causing the wave to be reflected, we should expect the transmitted wave to move with its center at $x = p_0t/m$. The additional term in eq. (81) represents a time delay, as can be seen by noting that it causes a given value of x to be reached later than if this term were not present.

Let us now evaluate the time delay of the transmitted wave:

$$\Delta t = \frac{\hbar}{v_0}\left(\frac{\partial\varphi_2}{\partial p_1}\right)_{p_1=p_0} \tag{82}$$

In differentiating φ_2, we must use the fact that

$$p_2^2 = 2m(E - V_0) = p_1^2 - 2mV_0$$

Hence

$$p_2 \frac{\partial p_2}{\partial p_1} = p_1 \qquad \text{or} \qquad \frac{\partial p_2}{\partial p_1} = \frac{p_1}{p_2}.$$

From eq. (79), we then obtain

$$\hbar \frac{\partial \varphi_2}{\partial p_1} = -2a \frac{p_1}{p_2} + \frac{\hbar}{2} \frac{\left\{\left(\frac{2}{p_2} - \frac{p_2}{p_1^2} - \frac{p_1^2}{p_2^3}\right) \tan 2p_2 \frac{a}{\hbar} + \frac{2a}{\hbar}\left[\left(\frac{p_1}{p_2}\right)^2 + 1\right] \sec^2\left(2p_2 \frac{a}{\hbar}\right)\right\}}{1 + \frac{1}{4}\left(\frac{p_1}{p_2} + \frac{p_2}{p_1}\right)^2 \tan^2\left(2p_2 \frac{a}{\hbar}\right)} \tag{83}$$

It is readily verified that when $p_1 = p_2$ (no barrier), $\Delta t = 0$. If $p_1 \neq p_2$, the result is rather complex, because there are two effects operating in opposite directions. First, the particle is speeded up as it enters the well; this should tend to make Δt negative. Second, the particle is reflected back and forth inside the well. This should tend to make Δt positive. Near a transmission resonance, the latter effect will win out, particularly if $p_1 \ll p_2$, because the reflection coefficient is then very high. Let us now calculate Δt. We note that at a transmission resonance

$$\tan \frac{2p_2 a}{\hbar} = 0 \qquad \text{and} \qquad \sec^2 \frac{2p_2 a}{\hbar} = 1$$

We get

$$v_0 \Delta t = -\frac{2ap_1}{p_2} + a\left[1 + \left(\frac{p_1}{p_2}\right)^2\right] \tag{84}$$

We note when p_1/p_2 is small, Δt is positive and approximately equal to a/v_0, where v_0 is the velocity of the particle *outside* the well. Since the particle actually goes faster inside the well in the ratio of p_2/p_1, it must suffer a number of reflections of the order of p_2/p_1. According to eq. (11), this is proportional to the inverse of the transmission coefficient; hence the delay is seen to be caused solely by the process of reflection.

Problem 11: Find the time delay for the reflected wave at a transmission resonance and interpret the result in terms of reflection of the wave inside the well.

20. Metastable (or Virtual) States of Trapping an Object within a Well. The previous discussion indicates that even when an object has enough energy to escape, it may, after entering the well, be reflected back and forth many times before it manages to get back out. This will happen if $p_1 \ll p_2$, i.e., if the depth of the well is much greater than the kinetic energy of the particle outside the well, and if the conditions are such that there is a transmission resonance ($2p_2a/\hbar = N\pi$). (From eq. 84, one can easily show that if we are not near such a resonance, the time delay is not very great, so that there is little likelihood of trapping the particle.) If the number of reflections of the wave is very great, the system appears to be in an almost stationary state, which, however, gradually decays, as the wave is slowly transmitted after many internal reflections. Such a state is called a virtual, or a metastable, level. Its energy is positive, in contrast to that of a true bound state, which is always negative. Its lifetime is given by Δt, as calculated in eq. (84).

Because such a metastable wave function constitutes, in effect, a wave packet which passes through the nucleus in a time Δt, its energy must, according to the uncertainty principle, fluctuate by

$$\Delta E \cong \frac{\hbar}{\Delta t}$$

Another way of obtaining the same result is to note that a metastable state can have physical significance only when the incident wave packet is so narrow that it passes a given point in a time less than the delay occurring inside the well. If this condition is not satisfied, then the time delay will be blotted out by the initial width of the packet itself, and no time delay will be observable. But to form a packet narrower than Δt, we need a range of energies greater than $\hbar/\Delta t$. Thus, a metastable state can exist only under conditions in which the energy is left undefined by this amount.

21. Metastable Singlet State of Deuteron. An important case of a metastable state, in which a particle is bound temporarily by reflection from a sharp edge, occurs in what is called the *singlet state* of the deuteron. Previously we stated that the neutron attracted the proton with a potential energy of 21.2 mev. Actually this is the potential only when the neutron spin is parallel to the proton spin. If the spins are antiparallel, the potential is less* and, in fact, equal to 11.85 mev. Remembering that in a three-dimensional problem we must have $\psi = 0$ at the origin, we can show that this reduction in potential is sufficient to prevent the wave function from curving downward to meet a decaying exponential at the edge of the well, so that if the spins are antiparallel, there are no bound states. In fact, for $E = 0$, it turns out that the wave starting out zero at the origin does not quite reach a phase of $\pi/2$ at the edge of the well. This result is illustrated in Fig. 18. But with a small positive energy ($\cong$ 40 kev), the phase becomes $\pi/2$ at the edge of the well. According to the discussion following eq. (50), this is the condition for a transmission resonance and, therefore, for a virtual level. As a result, there should be a metastable singlet level at a very low positive energy. The lifetime is

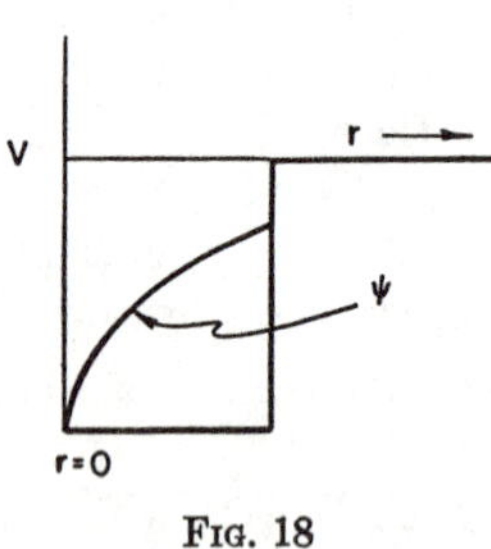

FIG. 18

$$\Delta t = \frac{a}{v_0} \tag{85}$$

* H. Bethe, *Elementary Nuclear Physics*, p. 43.

where v_0 is the velocity outside the well. The number of reflections is of the order of $\frac{v_{(\text{inside})}}{v_0} = \sqrt{\frac{E + V_0}{E}}$. With $V_0 = 20$ mev and $E \cong 40$ kev, this number is of the order of 20.

We shall see in Chap. 12, Sec. 18, that much longer-lived metastable states are possible, as a result of reflection of a particle from a potential barrier. In fact, metastable states are extremely common in nuclear physics and are one of the most important nuclear phenomena now being studied.

CHAPTER 12

The Classical Limit of Quantum Theory. The WKB Approximation

1. Introduction. In the previous chapter, we investigated systems where the potential changes sharply as a function of position. Let us now consider the opposite extreme, where the potential energy changes very slowly as a function of position. A discontinuity in the potential corresponds to a discontinuity in the refractive index and, as we have seen, electron waves are reflected by such discontinuities, just as light waves are. Since the slowly changing potential is analogous to a slowly changing index of refraction, we may learn what behavior to expect for electrons in this case from the corresponding problem with light.

In a medium where the index of refraction changes in a continuous way, light is not reflected, although its path may be curved as a result of refraction.* What is the critical rate of change of index of refraction with position, which determines whether or not a light wave will be reflected? We shall see that a great deal of reflection occurs only when the wavelength $\lambda = c/n\nu$ changes by a large fraction of itself within the distance of a wavelength. Now, the change of wavelength $\delta\lambda$ occurring in a distance δx is

$$\delta\lambda = \frac{\partial\lambda}{\partial x}\,\delta x \tag{1}$$

Setting $\delta x = \lambda$, we find that the condition for no appreciable reflections of the wave is that

$$|\delta\lambda| = \left|\frac{\partial\lambda}{\partial x}\lambda\right| \ll \lambda \qquad \text{or} \qquad \left|\frac{\partial\lambda}{\partial x}\right| \ll 1 \tag{2}$$

Exactly the same treatment applies to electron waves. Since λ is given by the de Broglie wavelength $\lambda = h/p$, the condition for no reflection is that

$$\left|\frac{\partial\lambda}{\partial x}\right| = \left|\frac{h}{p^2}\frac{\partial p}{\partial x}\right| \ll 1 \tag{3}$$

Substituting $p^2 = 2m(E - V)$, we obtain

$$\frac{hm\left|\dfrac{\partial V}{\partial x}\right|}{[2m(E-V)]^{3/2}} \ll 1 \qquad \text{or} \qquad \frac{\lambda\left|\dfrac{\partial V}{\partial x}\right|}{2(E-V)} \ll 1 \tag{4}$$

* See Chap. 3, Sec. 9, for a further qualitative discussion of this problem.

Hence, the requirement for the absence of a reflected wave is that the potential energy be a slowly changing function of position, and that $E - V$ shall not be too small.

In a three-dimensional problem the wave would still be deflected, even though it is not reflected. A wave packet would, therefore, follow a curved trajectory, rather than a straight line, but this curved trajectory must be exactly the same path as is predicted classically for the particle in a force field since, as we have shown in Chap. 9, Schrödinger's equation for the wave function leads to Newton's equations of motion in the classical limit. Thus, whenever the change of potential energy within a de Broglie wavelength is small compared with the kinetic energy, the specifically quantum-mechanical features arising from the wave properties of matter will not make themselves felt, and the classical description will be adequate.

We can see from eq. (3) that the applicability of classical concepts requires that $\hbar$ be small in comparison with $\left(\frac{1}{p^2}\frac{\partial p}{\partial x}\right)^{-1}\cdot$ The wide range of validity of classical physics is therefore really a reflection of the fact that, by ordinary standards, $\hbar$ is a rather small number. We can imagine a world in which $\hbar$ is much larger; such a world would show quantum-mechanical effects on a macroscopic scale.

2. The WKB Approximation. Whenever eq. (3) is satisfied, so that the classical limit is being approached, we can use what is called the WKB approximation (Wentzel–Kramers–Brillouin). This approximation takes advantage of the fact that the wavelength is changing slowly, by assuming that the wave function is not changed much from the form it would take if V were constant, namely,

$$\psi = \exp\left(\frac{ipx}{\hbar}\right) \qquad \text{where} \qquad p = \sqrt{2m(E - V)} \tag{5}$$

This suggests that it will be convenient to write the wave function in the form

$$\psi = \exp\left(\frac{iS}{\hbar}\right) \tag{6}$$

where S is a function of x. In general, S may be complex. If V is very nearly constant, we may expect that S is roughly equal to px, but to obtain a better value, we must solve for it from Schrödinger's equation. To do this for a general potential V, we approximate S as a series of powers of $\hbar$.

$$S = S_0(x) + \hbar S_1(x) + \frac{\hbar^2}{2} S_2(x) + \cdots \tag{7}$$

The first few terms of this series will be a good approximation only when $\frac{\hbar S_1}{S_0}, \frac{\hbar}{2}\frac{S_2}{S_1},$ etc. are all small. Since we already know that for a constant potential $S_0 = px$ and S_1, S_2, etc. are all zero, we may expect that this requirement can be satisfied when V changes very slowly as a function of x.

In one sense, it may be said that this approximation requires the smallness of h. If, for example, we imagine a series of worlds where h becomes progressively smaller, this expansion will become progressively better. Since the classical description gets better as h is made smaller, it is clear that this expansion is good only in the classical limit.

To solve for S, we insert eq. (6) into Schrödinger's equation, and obtain

$$0 = -\frac{\hbar^2}{2m}\frac{\partial^2\psi}{\partial x^2} + (V - E)\psi$$
$$= \left\{\frac{1}{2m}\left[\left(\frac{\partial S}{\partial x}\right)^2 - i\hbar\frac{\partial^2 S}{\partial x^2}\right] + (V - E)\right\}\exp\left(\frac{iS}{\hbar}\right)$$

or

$$\frac{1}{2m}\left(\frac{\partial S}{\partial x}\right)^2 + (V - E) - \frac{i\hbar}{2m}\frac{\partial^2 S}{\partial x^2} = 0 \tag{8}$$

We now insert the expansion (7) for S into the above equation, and collect all terms according to the power of $\hbar$ that they multiply. The result is (up to second order in $\hbar$)

$$0 = \frac{1}{2m}\left(\frac{\partial S_0}{\partial x}\right)^2 + (V - E) + \frac{\hbar}{m}\left(\frac{\partial S_0}{\partial x}\frac{\partial S_1}{\partial x} - \frac{i}{2}\frac{\partial^2 S_0}{\partial x^2}\right)$$
$$+ \frac{\hbar^2}{2m}\left[\frac{\partial S_0}{\partial x}\frac{\partial S_2}{\partial x} + \left(\frac{\partial S_1}{\partial x}\right)^2 - \frac{i\partial^2 S}{\partial x^2}\right] \tag{9}$$

Since this equation must be satisfied independently of the value of $\hbar$, it is necessary that the coefficient of each power of $\hbar$ be separately equal to zero. This requirement leads to the following series of equations

$$\frac{1}{2m}\left(\frac{\partial S_0}{\partial x}\right)^2 + V - E = 0 \tag{10}$$

$$\frac{\partial S_0}{\partial x}\frac{\partial S_1}{\partial x} - \frac{i}{2}\frac{\partial^2 S_0}{\partial x^2} = 0 \tag{11}$$

$$\frac{\partial S_0}{\partial x}\frac{\partial S_2}{\partial x} + \left(\frac{\partial S_1}{\partial x}\right)^2 - i\frac{\partial^2 S_1}{\partial x^2} = 0 \tag{12}$$

and so on.

These equations can be solved successively. That is, the first equation defines S_0 in terms of $V - E$, the second defines S_1 in terms of S_0, the third defines S_2 in terms of S_1 and S_0, etc. Solving, we obtain

$$\frac{\partial S_0}{\partial x} = \pm\sqrt{2m(E - V)} \tag{13}$$

$$S_0 = \pm\int_{x_0}^{x}\sqrt{2m(E - V)}\,dx$$

We assume here that $E > V$; the case $E < V$ will be treated in Sec. 7.

$$\frac{\partial S_1}{\partial x} = \frac{i}{2}\frac{1}{(\partial S_0/\partial x)}\frac{\partial^2 S_0}{\partial x^2} = \frac{i}{2}\frac{\partial}{\partial x}\ln\frac{\partial S_0}{\partial x} \tag{14}$$

$$S_1 = \frac{i}{2}\ln\frac{\partial S_0}{\partial x}$$

$$\exp(iS_1) = \frac{1}{\sqrt{\partial S_0/\partial x}} = \frac{1}{\sqrt[4]{2m(E-V)}}$$

Similarly,

$$S_2 = \frac{1}{2}\frac{m(\partial V/\partial x)}{[2m(E-V)]^{3/2}} - \frac{1}{4}\int\frac{m^2(\partial V/\partial x)^2\,dx}{[2m(E-V)]^{5/2}} \tag{15}$$

Because S_1 is a logarithm of $\partial S_0/\partial x$, it is not, in general, small compared with S_0. Both S_0 and S_1 must, therefore, be retained. On the other hand, from eq. (15), we can see that S_2 will be small whenever $\partial V/\partial x$ is small and $E - V$ is not too close to zero. We can also show that the smallness of the higher approximations (S_3, S_4, etc.), requires the smallness of all derivatives of V. Thus, the WKB approximation will be good whenever V is a sufficiently smooth and slowly varying function.

To obtain a more precise criterion for the applicability of the WKB approximation, we require that the absolute value of total phase shift coming from the second approximation, namely $|\hbar S_2/2|$, be small compared with unity. Inspection of the integral appearing in eq. (15) shows that it is of the same order of magnitude as the unintegrated term to the left of it. Our criterion therefore becomes

$$\frac{\hbar m\,(\partial V/\partial x)}{[2m(E-V)]^{3/2}} \ll 1 \tag{16}$$

But this is exactly the result obtained in eq. (4) by asking that the fractional change in wavelength be small over a distance of a wavelength.

Similar criteria involving the higher derivatives of V can be obtained, but we shall not do so here.

The solution, according to the WKB approximation, is then (absorbing the factor of $\sqrt{m}$ in the constants A and B)

$$\psi = \frac{A}{\sqrt[4]{E-V(x)}}\exp\left[i\int_{x_0}^{x}\sqrt{2m(E-V)}\,\frac{dx}{\hbar}\right] + \frac{B}{\sqrt[4]{E-V(x)}}\exp\left[-i\int_{x_0}^{x}\sqrt{2m(E-V)}\,\frac{dx}{\hbar}\right] \tag{17}$$

where A and B are arbitrary constants. The positive exponential corresponds to a wave moving in the positive direction, the negative exponential, to a wave moving in the negative direction. For the special case where V is a constant, these reduce respectively to the plane waves, $\exp(ipx/\hbar)$ and $\exp(-ipx/\hbar)$.

3. WKB Approximation—an Asymptotic Expansion. It can be shown that the series (7) does not converge but is, instead, an asymptotic expansion for S. This means that if we take a finite number of terms, it is always possible to find a value of $\hbar$ so small that the difference between this finite sum and the true value of S is less than any number that we care to choose. Yet, if we take more terms in the series, the expansion may begin to diverge away from the true value of S. Thus it is, in general, best for such an expansion to take only one or two terms, and then to apply it only to those cases where the remaining terms are small.

4. Physical Interpretation of Solutions in Terms of Classical Distribution of Particles. Let us choose the special case where $B = 0$. The probability that a particle lies between x and $x + dx$ is then

$$P(x) = \psi^*\psi = \frac{|A|^2}{\sqrt{E - V}} = \frac{|A|^2 \sqrt{2/m}}{v(x)} \tag{18}$$

where v is classical particle velocity. The current density is

$$j = \frac{\hbar}{2mi}\left[\psi^* \frac{\partial\psi}{\partial x} - \psi \frac{\partial\psi^*}{\partial x}\right] = v(x)P(x) \tag{19}$$

This wave function then corresponds to a distribution of particles with a probability density proportional to the reciprocal of the classical velocity, and with a mean velocity equal to the classical velocity. But this is exactly what is to be expected in a classical statistical ensemble, because the time spent by a particle in any region is inversely proportional to the velocity in that region. Hence, $\psi^*\psi$ is in this approximation the same as the classical probability distribution function. The phase S also has physical significance, in that its rate of change with position $\partial S/\partial x$ is equal to the mean momentum. Since the absolute value of S cannot be determined, we conclude that the classical distribution to which our WKB approximate wave function corresponds is one in which the phase $S(x) = \int_{x_0}^{x} \sqrt{2m(E - V)}$ is totally unknown, whereas the energy E is known exactly. The first effect of quantum theory (in the WKB approximation only) is, therefore, to leave the motion of the particles unchanged but to replace a description involving individual particles by a description involving a statistical distribution of particles, distributed uniformly over the phase S. For the special case of a free particle, $S = p(x - x_0)$. For this case, a uniform distribution over S means a uniform distribution over x. The distribution over x remains, even when V is not constant, but as we have seen, the probability is no longer uniform, but varies as $1/v(x)$.

It should be noted that having $P(x) \sim 1/v$ is characteristic only of the WKB approximation and that this is not, in general, the way that P varies.

Problem 1: For the square well potential, with $E > 0$, show that $\psi^*\psi$ is not proportional to $1/v$. (See Chap. 11, Secs. 10 and 12.)

5. Wave Packets. The Time-dependent Solution. The time-dependent approximate solution of the wave equation for a given energy is

$$\psi = \frac{A}{\sqrt{p}} \exp\left\{\frac{i}{\hbar}[S_0(x, E) - Et]\right\} \tag{20}$$

Let us write $S_0(x, E) - Et = S(x, t, E)$. The function S, which is the phase of the wave function, is also equal to a function appearing in classical mechanics, namely, Hamilton's principal function.† To verify this, we note that S satisfies the following differential equations:

$$\text{(a)}\ \frac{\partial S}{\partial t} = -E \qquad \text{(b)}\ \frac{\partial S}{\partial x} = p$$

$$\text{(c)}\ \frac{\partial S}{\partial x_0} = -p_0 \qquad \text{(d)}\ -\frac{\partial S}{\partial t} = \frac{1}{2m}\left(\frac{\partial S}{\partial x}\right)^2 \tag{21}$$

These are just the equations that define Hamilton's principal function. Thus, in the classical limit, the phase of the wave function approaches a function that had already been studied by Hamilton, during the nineteenth century, in an effort to obtain an analogy between mechanics and geometric optics. In fact, Hamilton showed that the equations for the trajectories of particles in classical mechanics were the same as those defining light rays in geometric optics, provided that these rays were obtained from waves by means of a Huyghen's construction, in which the phase of the wave was taken to be the function S. It is not surprising, therefore, that the phase S also appears as an intermediate function in the derivations of the connection between the wave theory of quantum mechanics and the particle theory of classical mechanics.

In order to show up this connection more closely, let us form a wave packet by integrating over a small range of energies:

$$\psi(x, t) = \int \exp\left[\frac{i}{\hbar} S(x, t, E)\right] f(E - E_0) \frac{dE}{\sqrt{p}} \tag{22}$$

The center of the packet will occur where waves of different energy tend to remain in phase, or where $\partial S/\partial E = 0$. But

$$\frac{\partial S}{\partial E} = \frac{\partial S_0}{\partial E} - t \tag{23}$$

Thus, we obtain

$$t = \frac{\partial S_0}{\partial E} = \frac{\partial}{\partial E}\int_{x_0}^{x} \sqrt{2m(E - V)}\, dx = \int_{x_0}^{x} \sqrt{\frac{m}{2(E - V)}}\, dx = \int_{x_0}^{x} \frac{dx}{v(x)} \tag{24}$$

† Born, *Mechanics of the Atom.*

The center of the wave packet, therefore, passes through the point x at the time $t = \int_{x_0}^{x} \frac{dx}{v}$. But this is just the time needed for the particle to cover the distance from x_0 to x, according to classical physics. The centers of wave packets therefore move with the classical velocity. This result was to be expected from the fact that Schrödinger's equation was so chosen as to lead to Newton's laws of motion in the classical limit. It should be noted, however, that in order to make the time of passing a given point defined to within an accuracy Δt, it is necessary to choose a range of energies $\Delta E \cong \hbar/\Delta t$.

Problem 2: Prove the above statement.

6. Time-dependent Three-dimensional WKB Approximation. As in the one-dimensional case, we write

$$\psi = e^{if/\hbar} \tag{25}$$

The three-dimensional time-dependent Schrödinger's equation is

$$i\hbar \frac{\partial \psi}{\partial t} = -\frac{\hbar^2}{2m} \nabla^2 \psi + V\psi \tag{26}$$

The equation for f becomes

$$-\frac{\partial f}{\partial t} = \frac{1}{2m} (\nabla f)^2 + \frac{\hbar}{2mi} \nabla^2 f + V \tag{27}$$

Writing

$$f = f_0 + \frac{\hbar}{i} f_1 + \ldots \tag{28}$$

we obtain

$$-\frac{\partial f_0}{\partial t} = \frac{1}{2m} (\nabla f_0)^2 + V \tag{29}$$

$$-\frac{\partial f_1}{\partial t} = \frac{1}{m} \nabla f_0 \cdot \nabla f_1 + \nabla^2 \frac{f_0}{2m} \tag{30}$$

The first of these equations is known in classical mechanics as the Hamilton-Jacobi equation.† It is the equation which defines Hamilton's principal function, which we have previously called S. Writing $f_0 = S_0 - Et$, we get

$$\frac{1}{2m} (\nabla S_0)^2 + V = E \tag{31}$$

This is the three-dimensional generalization of the equation for the function S_0, which we obtained in eq. (21).

The meaning of eq. (30) is also easily shown. We first write

$$P(x) = \psi^* \psi = e^{2f_1} \tag{32}$$

† *Ibid.*

Then we note that the probability current is

$$S(x) = \frac{\hbar}{2mi}(\psi^*\nabla\psi - \psi\nabla\psi^*) = \frac{e^{2f_1}}{m}\nabla f_0 = \frac{P(x)}{m}\nabla f_0 \tag{33}$$

Since $S(x)$ is also $vP(x)$, where v is the mean velocity, we have

$$v = \frac{\nabla f_0}{m} \tag{34}$$

We therefore write

$$\text{(a)}\ \frac{1}{2P}\frac{\partial P}{\partial t} = \frac{\partial f_1}{\partial t} \qquad \text{(b)}\ \frac{1}{2P}\nabla P = \nabla f_1$$

$$\text{(c)}\ \frac{\nabla^2 f_0}{m} = \nabla \cdot \frac{\nabla f_0}{m} = \nabla \cdot v \tag{35}$$

With these substitutions, eq. (30) becomes

$$-\frac{1}{2P}\frac{\partial P}{\partial t} = \frac{v \cdot \nabla P}{2P} + \frac{\nabla \cdot v}{2} \tag{36a}$$

or

$$\frac{\partial P}{\partial t} + v \cdot \nabla P + P\nabla \cdot v = \frac{\partial P}{\partial t} + \nabla \cdot (vP) = 0 \tag{36b}$$

The latter equation states that the change of probability in a given region is caused by the unbalanced probability current vP. It therefore shows that, in the WKB approximation, the probability may be regarded as flowing in a purely classical way, since a classical distribution of probability $P(x)$ would also have a probability current $S(x) = vP(x)$.

7. Penetration of a Barrier. As in the case of a square potential, the WKB approximation possesses real exponential solutions when $V > E$. The solution is

$$\psi = \frac{1}{\sqrt[4]{2m(E-V)}}\left\{A \exp\left[\int_{x_0}^{x} \sqrt{2m(V-E)}\,\frac{dx}{\hbar}\right] + B \exp\left[-\int_{x_0}^{x} \sqrt{2m(V-E)}\,\frac{dx}{\hbar}\right]\right\} \tag{37}$$

where A and B are arbitrary constants. Thus, the property of barrier penetration is not restricted to square potentials, but clearly exists for all kinds of potentials.

8. Connection Formulas. To treat the problem of barrier penetration where the WKB approximation is valid, we must find how to connect solutions in the region where $V > E$ with those where $E > V$. Consider, for example, the potential barrier shown in Fig. 1. Suppose that the energy E of the particle is such that $E = V$ at the point $x = a$. Classically, the particle would slow down to zero velocity at this point, and

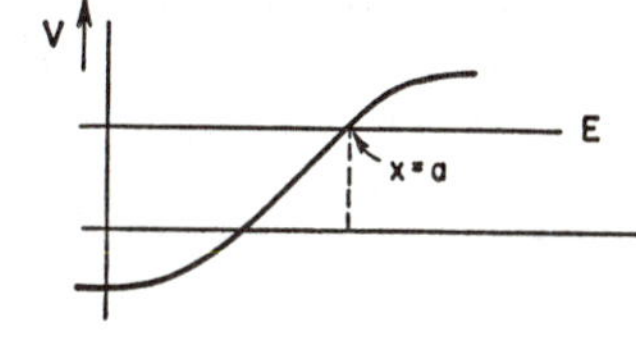

FIG. 1

then turn back. Quantum-mechanically, however, we know that the wave penetrates some distance further into the barrier.

Unfortunately, we cannot use the WKB approximation in the region near $x = a$, because when $E = V$, the conditions for its applicability break down (see eq. 4). Thus, if we start with a given solution, say

$$\psi \sim \frac{1}{\sqrt{p_1}} \exp\left(-\int_a^x p_1 \frac{dx}{\hbar}\right) \qquad (\text{with } p_1 = \sqrt{2m(V-E)})$$

which is a good approximation to the true solutions at some distance to the right of $x = a$, all that we know is that at a sufficient distance to the left of $x = a$, the approximate solution will be

$$\psi \sim \frac{A}{\sqrt{p_2}} \exp\left(i\int_a^x p_2 \frac{dx}{\hbar}\right) + \frac{B}{\sqrt{p_2}} \exp\left(-i\int_a^x p_2 \frac{dx}{\hbar}\right)$$

where A and B are unknown constants and $p_2 = \sqrt{2m(E-V)}$. The values of A and B cannot be found with the aid of the WKB approximation alone because they are determined by the nature of the solution in the region where this approximation does not apply. To obtain the values of A and B we need to have an expression for the solution in the region near $x = a$. In general, this is too complex a problem to be solved, except by numerical techniques. If the WKB approximation is applicable at some distance from $x = a$, however, we need a better solution only over a small region near $x = a$, extending out to where the WKB approximation is valid. If this region is small enough, then the potential function can be represented approximately by a straight line within the region, with a slope equal to that of the potential curve at the classical turning point $x = a$. Since $E = V$ at $x = a$, we can write

$$V - E = C(x - a)$$

where C is a constant, equal to $(\partial V/\partial x)_{x=a}$. Thus, in this region, Schrödinger's equation reduces approximately to

$$\frac{-\hbar^2}{2m}\psi'' + C(x-a)\psi = 0 \tag{38}$$

This equation is still fairly difficult to solve, but it can be solved with the aid of Bessel's functions of order $\frac{1}{3}$. After it is solved, one must carry the solution far enough from $x = a$, so that the WKB approximation becomes applicable; and then fit the solution in each region to the suitable WKB approximation. In this way, the constants A and B can be determined. We shall not go through the details of this procedure here, but merely quote the results.*

* See R. E. Langer, *Phys. Rev.*, **51**, 669 (1937). For a treatment by a different method, see E. C. Kemble, *The Fundamental Principles of Quantum Mechanics.* New York: McGraw-Hill Book Company, Inc., 1937, p. 95.

9. Connection Formulas.

Case A: Barrier to the Right.

Suppose that $V > E$ to the right of $x = a$, and let $p_2 = \sqrt{2m(E - V)}$, $p_1 = \sqrt{2m(V - E)}$. Then, far to the right of $x = a$, let us consider the approximate solution, which is a decaying exponential, namely:

$$\psi \cong \frac{1}{\sqrt{p_1}} \exp\left(-\int_a^x p_1 \frac{dx}{\hbar}\right) \tag{39a}$$

Far to the left of $x = a$, the connection formula states that this solution approaches

$$\psi_2 \cong \frac{2}{\sqrt{p_2}} \cos\left(\int_x^a p_2 \frac{dx}{\hbar} - \frac{\pi}{4}\right) \tag{39b}$$

Similarly, it may be shown that the following connections hold for solutions which approach an increasing exponential to the right of $x = a$:

$$\frac{1}{\sqrt{p_2}} \sin\left(\int_x^a p_2 \frac{dx}{\hbar} - \frac{\pi}{4}\right) \rightleftarrows -\frac{1}{\sqrt{p_1}} \exp\left(\int_a^x p_1 \frac{dx}{\hbar}\right) \tag{40}$$

Case B: Barrier to the Left.

It is convenient to write down the formulas for the case where the classically forbidden region is to the left of $x = a$.

For the solution which decays exponentially to the left, we find the following connection formula:

$$\frac{1}{\sqrt{p_1}} \exp\left(-\int_x^a p_1 \frac{dx}{\hbar}\right) \rightleftarrows \frac{2}{\sqrt{p_2}} \cos\left(\int_a^x p_2 \frac{dx}{\hbar} - \frac{\pi}{4}\right) \tag{41}$$

If the wave function increases exponentially to the left, we obtain

$$\frac{1}{\sqrt{p_2}} \sin\left(\int_a^x p_2 \frac{dx}{\hbar} - \frac{\pi}{4}\right) \rightleftarrows -\frac{1}{\sqrt{p_1}} \exp\left(\int_x^a p_1 \frac{dx}{\hbar}\right) \tag{42}$$

It should be noted that the solution in the region $x < a$ is not simply the continuation of the solution in the region $x > a$. For example, the continuation of $\frac{1}{\sqrt{p_2}} \exp\left(-\int_a^x p_2 \frac{dx}{\hbar}\right)$ is just $\frac{1}{\sqrt{p_2}} \exp\left(-i \int_a^x p_2 \frac{dx}{\hbar}\right)$, whereas the actual solution in the region $x < a$ involves

$$\cos\left(\int_a^x p \frac{dx}{\hbar} - \frac{\pi}{4}\right)$$

which has waves running in both directions.

It should also be noted that the connection formulas enable us only to obtain the relation between the solutions in a region at some distance to the right of the turning point, $x = a$, with those in a region some distance to the left. In order to obtain the form of the wave function in the intermediate region, we should have to refer to the exact solution, which involves Bessel's functions of order $\frac{1}{3}$. In practice, however, it turns out that a knowledge of the exact form of the solution in the intermediate region is, for most purposes, unnecessary, and that the connection formulas are all that we need.

The exact mathematical conditions under which the connection formulas can be proved rigorously to apply are fairly complex.* We shall restrict ourselves here to the statement that in all practical applications which are ever made, the requirements are the following:

(1) There exist regions on either side of the turning point containing many wavelengths, in which the WKB approximation applies.

(2) In the region near the turning point (at $x = a$), over which the WKB approximation does not apply, the kinetic energy can be represented approximately by a straight line, $E - V = C(x - a)$. In other words, the potential should not undergo a large fractional change in slope within this region. The region in which the WKB approximation does not apply should be regarded as covering at least the distance to the first node of the wave function, and preferably should include the first few oscillations. Inside the barrier, the WKB approximation will begin to be reliable after $\int_a^x \sqrt{2m(V - E)}\, \frac{dx}{\hbar}$ becomes appreciably greater than unity.

There are two kinds of problems in which the connection formulas break down. The first of these arises when the particle has an energy such that its classical turning point occurs near the top of a barrier, where the slope of the potential is small. As a result, the straight-line approximation to the potential breaks down, and the connection formulas must be altered in a way which we shall only mention here.† Such a potential is shown in Fig. 2. For energies nearly enough to carry the particle over the barrier, the connection formulas are not valid. But for lower energies, they are valid because the potential can be represented approximately by a straight line.

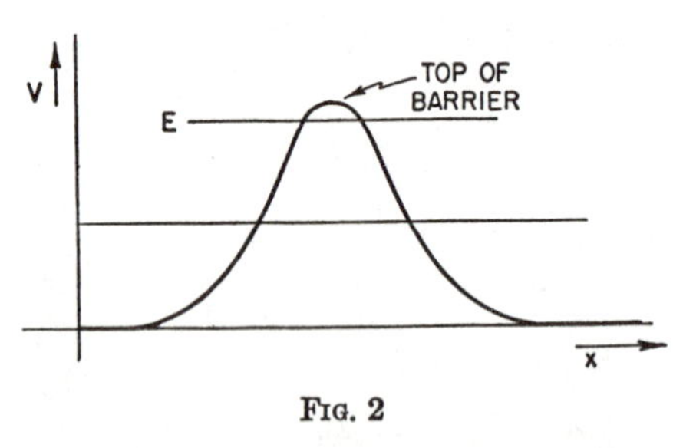

Fig. 2

The second kind of problem arises when the potential changes too rapidly in slope; for example, in the case of a square well potential. For such a potential, the slope is everywhere zero, except at the points of

* For a detailed discussion of these conditions, see Kemble, *Fundamental Principles of Quantum Mechanics*, pp. 103–112.

† *Ibid.*

discontinuity, where it is infinite. The connection formulas also break down for the type of potential shown in Fig. 3. To solve such a problem, one must use the exact solution of the wave equation in each region, and fit these solutions smoothly at the boundaries.

A few remarks on the direction of the arrow in the connection formulas are in order here. Strictly speaking, the arrow should point only in the direction in which the real exponential is increasing. The reason is that because of a slight failure of the WKB approximation, there is always a possibility that with any given solution, some of the other solution is introduced. If we are connecting in the direction of the increasing exponential, then the other solution will decrease exponentially in this direction and, thus, introduce a negligible correction to the wave function. On the other hand, if we are connecting in the direction of the decreasing exponential, the other solution will increase exponentially and may therefore become much larger than the decreasing exponential, even though its coefficient is very small. It can easily be shown, however, that in the calculation of energy levels (virtual or real), this effect produces only a very small error, so that, in practice, we can usually run the arrows both ways, even though the rigorous justification of such a procedure may be rather difficult. If, for a given energy, one wanted to know the wave function very accurately, however, then it would be permissible to run the arrow only in the direction in which the real exponential is increasing.

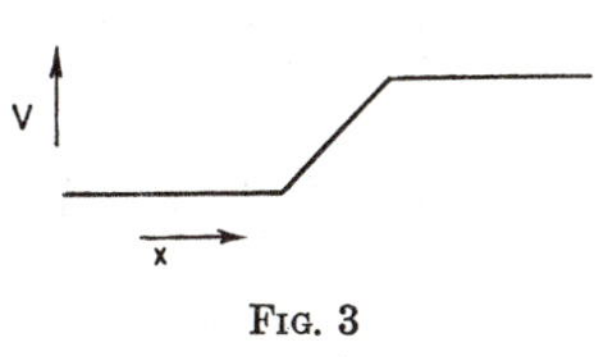

FIG. 3

10. Probability of Penetration of a Barrier. One of the most important problems to which the connection formulas apply is that of penetration of a potential barrier. In order that the WKB approximation apply within a barrier, it is necessary that the potential function not change too rapidly. In order that the connection formulas apply, it is necessary that the barrier be thick enough and high enough so that $\int_b^a \sqrt{2m(V-E)}\,\frac{dx}{\hbar}$ be considerably greater than unity. If these conditions apply, we can then easily calculate the probability of penetration of the barrier.

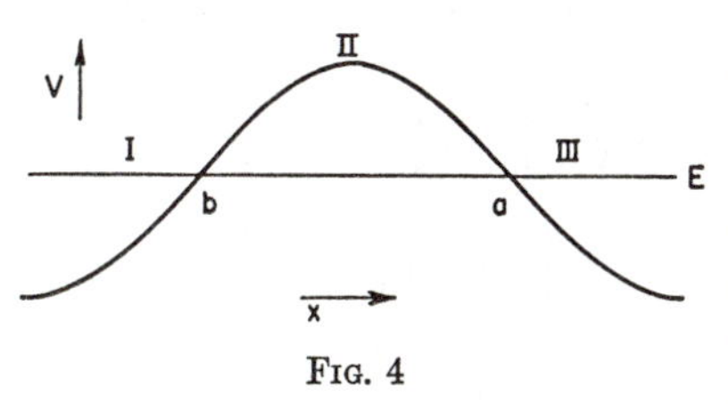

FIG. 4

The barrier is represented in Fig. 4, and the energy is such that the turning points are at $x = a$ and $x = b$. Suppose that particles are incident from the left. Some are reflected, and some transmitted. To the right, in region III, there is therefore only a transmitted wave. This we may represent by

$$\psi_{III} \sim \frac{A}{\sqrt{p}} \exp\left(i \int_a^x p \frac{dx}{\hbar} - \frac{\pi}{4}\right)$$

with

$$p = \sqrt{2m(E - V)} \tag{43}$$

The phase factor of $(-\pi/4)$ is included in the exponential for reasons of convenience in applying the connection formulas. Since A is complex, such a phase factor may be absorbed in it. To apply the connection formula, we first write

$$\psi_{III} \cong \frac{A}{\sqrt{p}}\left[\cos\left(\int_a^x p \frac{dx}{\hbar} - \frac{\pi}{4}\right) + i \sin\left(\int_a^x p \frac{dx}{\hbar} - \frac{\pi}{4}\right)\right] \tag{43a}$$

We now apply the connection formula, for the case in which the barrier is to the left, obtaining

$$\psi_{II} \cong \frac{A}{\sqrt{p_1}}\left[\frac{1}{2}\exp\left(-\int_x^a p_1 \frac{dx}{\hbar}\right) - i \exp\left(\int_x^a p_1 \frac{dx}{\hbar}\right)\right]$$

where

$$p_1 = \sqrt{2m(V - E)} \tag{44}$$

The next step is to apply the connection formulas to find the wave functions in region I. In this region, the barrier is to the right. We must therefore first put ψ_{II} in a form in which it is convenient to apply the formulas for this case. We obtain

$$\psi_{II} \cong \frac{A}{\sqrt{p_1}}\left[\frac{1}{2}\exp\left(-\int_b^a p_1 \frac{dx}{\hbar} + \int_b^x p_1 \frac{dx}{\hbar}\right) - i \exp\left(\int_b^a p_1 \frac{dx}{\hbar} - \int_b^x p_1 \frac{dx}{\hbar}\right)\right] \tag{44a}$$

Equations (39b) and (40) then yield

$$\psi_I \cong -\frac{A}{\sqrt{p}}\left[\frac{1}{2}\exp\left(-\int_b^a p_1 \frac{dx}{\hbar}\right) \sin\left(\int_x^b p \frac{dx}{\hbar} - \frac{\pi}{4}\right) + 2i \exp\left(\int_b^a p_1 \frac{dx}{\hbar}\right) \cos\left(\int_x^b p \frac{dx}{\hbar} - \frac{\pi}{4}\right)\right] \tag{45}$$

or

$$\psi_I \cong -\frac{iA}{\sqrt{p}}\left\{\exp\left[-i\left(\int_b^x p \frac{dx}{\hbar} + \frac{\pi}{4}\right)\right]\left[\exp\left(\int_b^a p_1 \frac{dx}{\hbar}\right) - \frac{1}{4}\exp\left(-\int_b^a p_1 \frac{dx}{\hbar}\right)\right] + \exp\left[i\left(\int_b^x p \frac{dx}{\hbar} + \frac{\pi}{4}\right)\right]\left[\exp\left(\int_b^a p_1 \frac{dx}{\hbar}\right) + \frac{1}{4}\exp\left(-\int_b^a p_1 \frac{dx}{\hbar}\right)\right]\right\} \tag{45a}$$

The transmission coefficient is just the ratio of the transmitted intensity times the velocity after transmission to the incident intensity times the velocity of the incident particles. Noting that the ratio of the velocities is equal to the ratio of the momenta, we obtain

$$T = \left[\exp\left(\int_b^a p_1 \frac{dx}{\hbar}\right) + \frac{1}{4}\exp\left(-\int_b^a p_1 \frac{dx}{\hbar}\right)\right]^{-2} \tag{46}$$

If the WKB approximation is to be applicable, then $\int_b^a p_1 \frac{dx}{\hbar} \gg 1$, so that the negative exponential may be neglected in comparison with the positive exponential. This gives us

$$T \cong \exp\left(-2\int_b^a p_1 \frac{dx}{\hbar}\right) \tag{46a}$$

We see that except for some factors which are usually fairly close to unity, this is essentially the same result as we obtained with a square barrier* for which $\int_b^a p_1 \frac{dx}{\hbar} = (b - a)\frac{p_1}{\hbar}$. When the height of the barrier is variable, however, we need merely integrate over it in the manner shown above.

Problem 3: Prove that the sum of the transmission and reflection coefficients is unity.

Problem 4: Evaluate the current in regions I, II, III, and show that it is the same in all three regions, thus demonstrating the conservation of probability. Why is there a current in region II, even though the wave function does not oscillate?

11. Applications of Barrier Penetration Probability.

(1) *Cold Emission of Electrons from Metals.* The electrons in a metal move in a more or less constant potential, but when they reach the edge, they are attracted back into the metal by a potential energy of the order of 5 to 10 electron volts. The force pulling the electron back into the metal is that arising from the "image" charge induced in the metal as the electron leaves the surface. Most of the force is experienced within a very short distance of the edge of the metal, perhaps one or two atomic diameters, or 3 to 5 $\times$ 10^{-8} cm. The potential function resembles the graph given in Fig. 5. The energy W necessary to liberate an electron is called the *work function.*

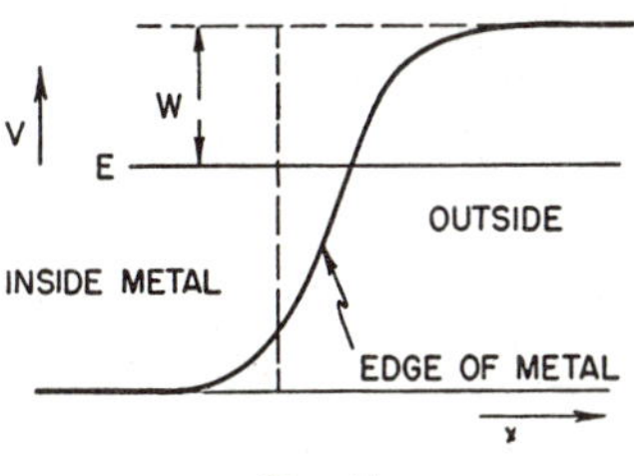

Fig. 5

Suppose now that the metal is placed in a strong electric field which is in such a direction as to tend to pull electrons out of the metal. The potential function will now be represented by a curve like that shown in Fig. 6, because to the original potential will have been added the electrical potential $-e\mathcal{E}x$, where $\mathcal{E}$ is the electric field and x the distance from the edge of the surface of the metal. Thus, there will always be a position $x = a$, where the electron will have a positive kinetic energy,

* Eq. (34), Chap. 11.

even though it is outside the metal, and there will be a finite probability that it leaks through the barrier and leaves the metal permanently. This process is called *cold emission* of electrons, to contrast it with thermal emission, which takes place when the electron acquires enough energy from random thermal motions to go over the barrier.

To calculate the transmission coefficient, we need to know how V changes in the region near the edge of the metal. Because the distance a is usually much greater than an interatomic distance, the contribution to the result of the region in which the potential function curves (near

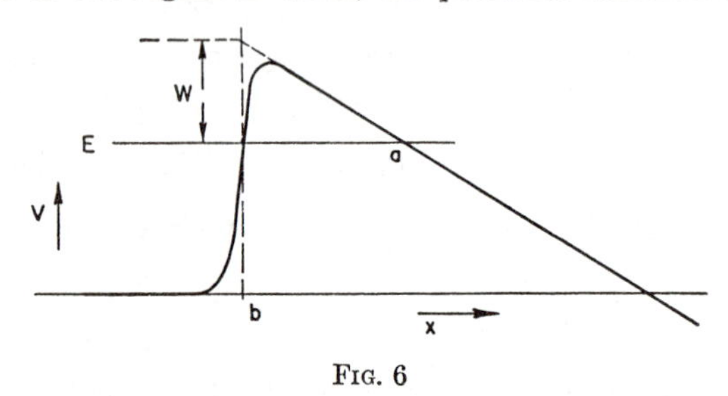

FIG. 6

$x = 0$) will be small in comparison to that of the rest of the region between $x = 0$ and $x = a$. This means that the precise way in which the potential curves is not very important and that we can approximate the potential everywhere in the region from $x = 0$ to $x = a$ by the straight line $V = -e\mathcal{E}x$. Although the WKB approximation may break down in the region near $x = 0$ because the potential curves over rather sharply, this, too, will introduce no important errors, because the contribution of this region to the factor determining the transmissivity is only a small fraction of the total effect. This means that we can use the WKB approximation throughout, and set $V - E = W - e\mathcal{E}x$. From eq. (46a) we then obtain (noting that $W - e\mathcal{E}x = 0$ at $x = a$)

$$T = \exp\left[-2\int_0^a \sqrt{2m(W - e\mathcal{E}x)}\,\frac{dx}{\hbar}\right] = \exp\left[-\frac{4}{3}\sqrt{2m}\,\frac{W^{3/2}}{\hbar e\mathcal{E}}\right] \quad (47)$$

From the transmission coefficient, we can compute the current by multiplying T by the number of electrons that strike the edge of the metal per second. We see that the current should increase rapidly with field strength and should also be greatest for materials of the lowest work function, W. This is what is observed experimentally. There is one discrepancy, however, in that the observed currents are much greater numerically than those calculated from eq. (47). This is because the metal surface is not flat, but has microscopic irregularities, which cause the electric field near the surface to be much greater than the field far from the surface. Since T is very sensitive to $\mathcal{E}$, an enormous increase in current can result even if $\mathcal{E}$ is only doubled or tripled.

(2) *Radioactive Decay.* We have already pointed out that a charged particle such as a proton or an α particle is bound in the nucleus by a strong attractive force that has a very short range.* When the particle leaves the nucleus, it is repelled by the Coulomb force.† As a result, the potential energy consists of a well with a repulsive barrier at the edge, as shown in Fig. 10, Chap. 11. A particle can therefore be trapped inside the nucleus for a long time, even though it has a positive energy, provided that this energy is less than the maximum height of the barrier. The mean lifetime for emission of the particle is then given in Chap. 11, Sec. 6, and is $\tau = 10^{-21}/T$ sec, where T is the transmissivity of the barrier. To calculate T, we shall use the WKB approximation.‡ If the nucleus has a charge Ze, then an α particle of charge $2e$ is repelled electrostatically with an energy $V = 2(Z-2)\dfrac{e^2}{r}\cdot$ To apply the WKB approximation, we should rigorously have to know just how the potential curves over near $r = r_0$. But because the nuclear forces vanish in a distance that is much less than $r_1 - r_0$ we may, as in the case of cold emission of electrons, consider only the electrostatic energy and set r_0 equal to the radius of the nucleus. One can use for r_0 the formula

$$r_0 = 2 \times 10^{-13} Z^{1/3} \text{ cm}$$

which has been obtained by independent methods.§

Let us now obtain T. We have

$$T = \exp\left\{-\frac{2}{\hbar}\int_{r_0}^{r_1}\sqrt{2m\left[2(Z-2)\frac{e^2}{r} - E\right]}\,dr\right\} \tag{48}$$

r_1 is the place where

$$E = 2(Z-2)\frac{e^2}{r_1} \qquad \text{or} \qquad r_1 = 2(Z-2)\frac{e^2}{E}$$

The integral under the exponential may be simplified with the substitutions

$$U = \sqrt{\frac{r_1}{r} - 1} \qquad r = \frac{r_1}{1+U^2} \qquad v = \sqrt{\frac{2E}{m}}$$

$$2\int_{r_0}^{r_1}\sqrt{2m\left[2(Z-2)\frac{e^2}{r} - E\right]}\,\frac{dr}{\hbar} = \frac{4r_1 vm}{\hbar}\int_0^{\sqrt{\frac{r_1}{r_0}-1}} \frac{U^2\,dU}{(1+U^2)^2}$$

$$= \frac{2vr_1 m}{\hbar}\left(\tan^{-1} U - \frac{U}{1+U^2}\right)_0^{\sqrt{\frac{r_1}{r_0}-1}} \tag{49}$$

* Chap. 11, Sec. 3.

† Chap. 11, Sec. 6.

‡ The actual problem is three-dimensional, but as in the case of the deuteron, (Chap. 11, Sec. 14), the form of the wave equation is the same as that of the one-dimensional case.

§ See F. Rasetti, *Elements of Nuclear Physics.* New York: Prentice-Hall, Inc., 1936, p. 220.

Let us set $\frac{r_0}{r_1} = \cos^2 W$ and $\sqrt{\frac{r_1}{r_0} - 1} = \tan W$. Then we get (eliminating r_1 in terms of E)

$$T = \exp\left[-4\frac{(Z-2)e^2}{\hbar v}(2W - \sin 2W)\right] \tag{50}$$

Problem 5: Calculate the lifetime for uranium α-particles. (Look up energy of α-particles.) Compare the result for polonium α-particles. Note how sensitive the lifetime is to the energy. Compare the results with the observed values and explain whatever discrepancies may exist.

12. Probability of Penetration into Nucleus from Outside. If high-speed charged particles, such as protons or α particles, are directed at a nucleus, there will, in general, be some chance that they enter. To do so, however, they must either be energetic enough to go over the barrier or else they must "leak" through in the manner already described. Now, the nucleus is a three-dimensional object, so that the previous formulas cannot be applied rigorously. Yet, one can see that two conditions determine the probability of entry into a nucleus of a charged particle which does not have enough energy to go over the barrier.

(1) The particle must make a fairly direct collision with the nucleus, i.e., it must not strike in such a direction that it will tend to glance off.

(2) It must penetrate the barrier.

The probability of making a collision with the nucleus can be estimated, if one knows the area of the nucleus, by the same methods used in kinetic theory to estimate mean free paths of gas molecules.* If N is the number of molecules per cubic centimeter, and A is the cross-sectional area of the nucleus, then in passing through l centimeters of matter, the probability of such a collision is $P = NAl$.

To find the probability of penetration, we multiply P by T. For protons, we must evaluate the potential energy from the formula $V = Ze^2/r$. Hence, we have $r_1 = Ze^2/E$, and

$$T = \exp\left[-\frac{2Ze^2}{\hbar v}(2W - \sin W)\right] \tag{51}$$

with

$$\cos^2 W = \frac{r_0}{r_1}$$

Problem 6: For $Z = 92$, what is the transmission coefficient for entry into the nucleus by a proton of energy 3 mev? For a proton of energy 8 mev?

It is found that the WKB approximation yields fairly good agreement with experiment for the way in which the transmissivity of a barrier varies with Z and W, but that it does not yield a very good estimate of the precise value of the numerical factor in front of the exponential. The failure of exact agreement is not surprising, first, because the exact

* See Chap. 21, Sec. 3; see also E. H. Kennard, *Kinetic Theory of Gases.* New York: McGraw-Hill Book Company, Inc., 1938, pp. 97–126.

shape of the potential near the edge of the nucleus is not known, second, because exactly what is happening in the nucleus is not known, and third, because the WKB approximation itself probably fails in a small region near the edge of the nucleus, where the potential curves over rather sharply. Yet, the main variations in the transmissivity result from the Coulomb barrier, which extends a long way from the edge of the nucleus, and which is fairly correctly described in our present treatment. A more detailed theory must, however, await a more detailed knowledge of what is happening inside the nucleus and at its edge.

13. Bound States of a Potential Well. Let us now consider the problem of a potential well in the WKB approximation. Such a well can be represented by the curve shown in Fig. 7.

For $E < 0$, the particle will, according to classical theory, oscillate back and forth between the limits at $x = a$ and $x = b$, where the kinetic energy vanishes. The period of oscillation depends on the shape of the

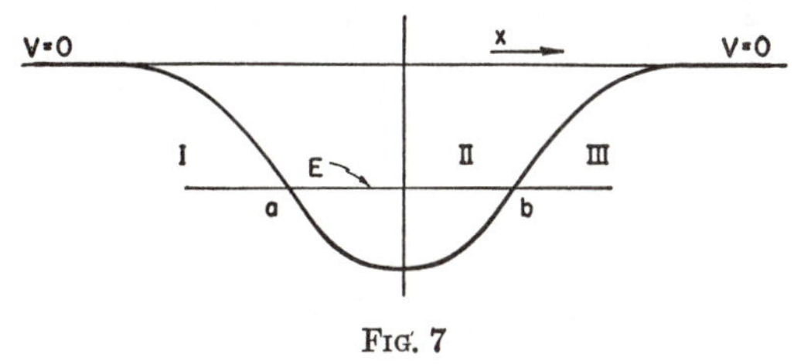

FIG. 7

potential and on the position of the limits of oscillation, the latter of which depends on the energy. Such an oscillation will be anharmonic, unless $V = kx^2/2$, in which case the system is a harmonic oscillator. We have seen in eq. 12, Chap. 2, that the period of oscillation is given by

$$\tau = \frac{\partial J}{\partial E}$$

where
$$J = \oint p\, dq = 2\int_a^b \sqrt{2m(E - V)}\, dx$$

Solution for Wave Function. We shall now obtain the wave functions for this potential, using the WKB approximation.

According to quantum theory, we know that the wave function penetrates exponentially into the region where $V > E$. Let us start in region I, where $x < a$. We must choose that solution which decays exponentially to the left. This is

$$\psi_{\text{I}} = \frac{A}{\sqrt{p_1}} \exp\left(-\int_x^a p_1 \frac{dx}{\hbar}\right)$$

where
$$p_1 = \sqrt{2m(V - E)}$$

The connection formula for the barrier to the left [eq. (41)] then yields for the wave function inside the well

$$\psi_{II} = \frac{2A}{\sqrt{p}} \cos\left(\int_a^x p\,\frac{dx}{\hbar} - \frac{\pi}{4}\right) = \frac{2A}{\sqrt{p}} \cos\left(\int_x^a p\,\frac{dx}{\hbar} + \frac{\pi}{4}\right) \tag{52}$$

In order to find what happens to this wave function in region III, we must rewrite it in a form suitable for the application of the formulas for the barrier at the right. We obtain

$$\psi_{II} = \frac{2A}{\sqrt{p}} \cos\left(\int_x^b p\,\frac{dx}{\hbar} - \int_a^b p\,\frac{dx}{\hbar} + \frac{\pi}{4}\right)$$
$$= \frac{2A}{\sqrt{p}} \cos\left(\int_x^b p\,\frac{dx}{\hbar} - \frac{\pi}{4} - \int_a^b p\,\frac{dx}{\hbar} + \frac{\pi}{2}\right) \tag{53}$$

In region III we want the solution to be a decaying exponential. Applying eq. (39b) for the barrier to the right, and noting that the connection formulas would be equally good if multiplied by a minus sign, we see that a decaying exponential is obtained only if the phase of the trigonometric function in ψ_{II} is such that the latter takes the form $\cos\left(\int_x^b p\,\frac{dx}{\hbar} - \frac{\pi}{4}\right)$. We can easily see that this condition will be satisfied if, and only if,

$$\int_a^b p\,\frac{dx}{\hbar} = \left(N + \frac{1}{2}\right)\pi \tag{54}$$

where N is any integer.

Writing
$$J = \oint p\,dq = 2\int_a^b p\,dx$$
we get
$$J = (N + \tfrac{1}{2})h \tag{55}$$

where N is any integer. The above is just the same as the Bohr-Sommerfeld quantum condition,* except for the $\frac{1}{2}$ added to N. Thus, the classical limit of the wave theory leads to the older quantum conditions. The correcting term of $\frac{1}{2}$ was already guessed before wave theory, because it was needed to fit the observed energy levels.

Form of Wave Function. The connection formulas indicate that the wave function seems to approach the turning points at $x = a$ and $x = b$ with phases of $N\pi + \frac{\pi}{4}$ and $-\frac{\pi}{4}$ respectively, where N is some integer. Actually, the WKB approximation breaks down in this region, but roughly the above will be near the right phase. The wave function with $N = 0$ then has a phase of $\frac{\pi}{4}$ at $x = a$ and $-\frac{\pi}{4}$ at $x = b$. As a result, the lowest state has only a quarter wavelength, inside the region of

* See Chap. 2, Sec. 12.

positive kinetic energy. (The wave function resembles the curve shown in Fig. 8.) This is the reason for the half-quantum number. The lowest state has no nodes.

The next state has a single node, the following state two, etc. The classification of bound states according to the number of nodes, therefore, holds in the WKB approximation, as well as for the square well.* In fact, it is a classification that holds for an arbitrary potential function. That is, *the number of the quantum state is equal to the number of nodes in the wave function.*

We can show that if the connection formulas do not apply, the quantum conditions may be slightly altered. For example, in a very deep potential well, which is square, the wave function vanishes at the turning points, and we must fit an integral number of half waves into the well, so that we get $J = (N + 1)h$. The appearance of $N + \frac{1}{2}$ in the WKB approximation is caused by the fact that the wave can penetrate exponentially into the classically forbidden region. In other words, it has a little more room than it would have if the barrier were infinitely high. As a result, only a quarter wavelength need be fitted into the classically accessible region, instead of the half wavelength that is necessary with the infinite barriers. On the other hand, in a very shallow well, the turning points may be in a region where the curvature of the potential over a wavelength is appreciable, so that there is also the connection-formulas breakdown, and the phase of the wave at the turning point is altered.† The quantum conditions for such a case are complex, and we shall not treat them here.

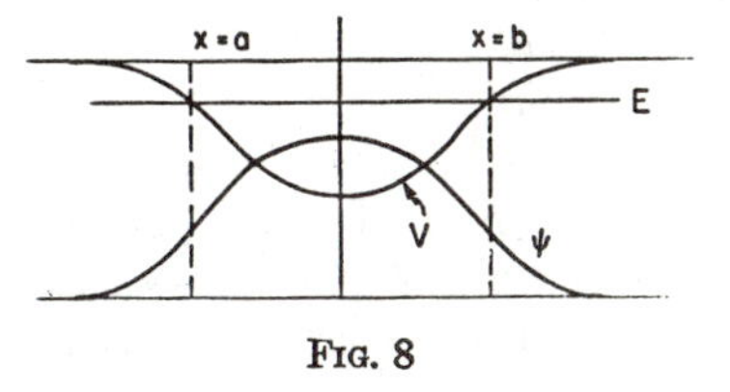

FIG. 8

Problem 7: Apply the above to calculate the energy levels of a harmonic oscillator and obtain the result $E = (n + \frac{1}{2})h\nu$. Note the fact that the lowest energy level is $E = \frac{1}{2}h\nu$ and not $E = 0$ as given by the older Bohr-Sommerfeld theory.‡ Explain this in terms of the uncertainty principle.

Plot the general form of the wave functions as given by the WKB approximation for the first four quantum states.

14. Virtual or Metastable States in the WKB Approximation. If the potential well is surrounded by a barrier, as is the case for charged particles in the nucleus, then besides the true bound states, there may exist virtual or metastable bound states, made possible by the fact that an electron wave of positive energy can reflect back and forth from the barrier many times before it penetrates. Such a potential might resemble that shown in Fig. 9. For the sake of simplicity, we shall take the

* Chap. 11, Sec. 12.

† See Sec. 8.

‡ See Chap. 2, Sec. 12.

potential to be symmetric about $x = 0$ although this is not essential. The turning points are then assumed to be at $x = a, b, -a$, and $-b$.

This problem will resemble very much that of metastable states of the square well without barriers, already treated,* but here the states will have a much longer life-time because the transmissivity of a barrier is, in general, much less than that of a sharp edge in the potential.

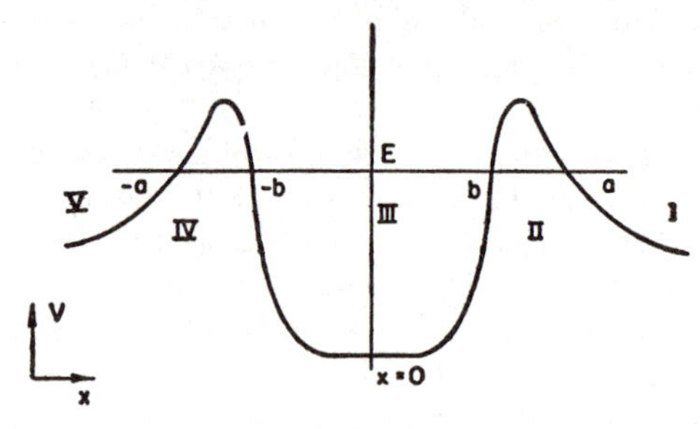

FIG. 9

As in the case of the square well, we shall assume that particles are incident from the left. Some will be reflected and some transmitted. To the right of $x = a$, however, only a transmitted wave will be present. In region I, we therefore write

$$\psi_{\mathrm{I}} = \frac{A}{\sqrt{p}} \exp\left(i \int_a^x p \frac{dx}{\hbar} - i\frac{\pi}{4}\right) \tag{56}$$

The problem of finding the wave function in regions II and III is exactly the same as the one which has already been treated, namely that of barrier penetration. From eq. (45) we then obtain

$$\psi_{\mathrm{III}} = -\frac{A}{\sqrt{p_w}}\left[\frac{1}{2} e^{-\int_b^a p_1 \frac{dx}{\hbar}} \sin\left(\int_x^b p_w \frac{dx}{\hbar} - \frac{\pi}{4}\right) + 2ie^{\int_b^a p_1 \frac{dx}{\hbar}} \cos\left(\int_x^b p_w \frac{dx}{\hbar} - \frac{\pi}{4}\right)\right] \tag{57}$$

where p_w is the absolute value of the momentum inside the well. This solution must now be carried across the well and into region IV, with the aid of the connection formulas. To use these formulas, we first rearrange the arguments of the trigonometric functions. We write

$$\int_x^b p_w \frac{dx}{\hbar} = -\int_b^x p_w \frac{dx}{\hbar} = -\left(\int_b^{-b} p_w \frac{dx}{\hbar} + \int_{-b}^x p_w \frac{dx}{\hbar}\right) \tag{57a}$$

Putting $\int_{-b}^b p_w\, dx = J/2$, where J is the action variable, and

$$\exp\left(\int_b^a p_1 \frac{dx}{\hbar}\right) = \Theta$$

* Chap. 11, Sec. 20.

we obtain

$$\psi_{\mathrm{III}} = \frac{A}{\sqrt{p_w}}\left\{-2i\Theta\cos\left[\int_{-b}^{x} p_w\frac{dx}{\hbar} - \frac{\pi}{4} + \left(\frac{\pi}{2} - \frac{J}{2\hbar}\right)\right] + \frac{1}{2\Theta}\sin\left[\int_{-b}^{x} p_w\frac{dx}{\hbar} - \frac{\pi}{4} + \left(\frac{\pi}{2} - \frac{J}{2\hbar}\right)\right]\right\} \tag{58}$$

We now expand the sines and cosines, and collect terms:

$$\psi_{\mathrm{III}} = \frac{A}{\sqrt{p_w}}\left\{\begin{aligned} &\cos\left(\int_{-b}^{x} p_w\frac{dx}{\hbar} - \frac{\pi}{4}\right)\left[-2i\Theta\cos\frac{1}{2}\left(\pi - \frac{J}{\hbar}\right) + \frac{1}{2\Theta}\sin\frac{1}{2}\left(\pi - \frac{J}{\hbar}\right)\right] \\ &+ \sin\left(\int_{-b}^{x} p_w\frac{dx}{\hbar} - \frac{\pi}{4}\right)\left[2i\Theta\sin\frac{1}{2}\left(\pi - \frac{J}{\hbar}\right) + \frac{1}{2\Theta}\cos\frac{1}{2}\left(\pi - \frac{J}{\hbar}\right)\right]\end{aligned}\right\} \tag{59}$$

The next step is to obtain the solution in the region IV. We use the connection formulas for the barrier to the left. Furthermore, we write

$-\int_{x}^{-b} = \int_{-a}^{x} - \int_{-a}^{-b}$, and, noting that $\Theta = \exp\left(\int_{-a}^{-b} p_1\frac{dx}{\hbar}\right)$, obtain

$$\psi_{\mathrm{IV}} = \frac{A}{\sqrt{p_1}}\left\{\begin{aligned} &\exp\left(\int_{-a}^{x} p_1\frac{dx}{\hbar}\right)\left[-i\cos\frac{1}{2}\left(\pi - \frac{J}{\hbar}\right) + \frac{1}{4\Theta^2}\sin\frac{1}{2}\left(\pi - \frac{J}{\hbar}\right)\right] \\ &- \exp\left(-\int_{-a}^{x} p_1\frac{dx}{\hbar}\right)\left[2i\Theta^2\sin\frac{1}{2}\left(\pi - \frac{J}{\hbar}\right) + \frac{1}{2}\cos\frac{1}{2}\left(\pi - \frac{J}{\hbar}\right)\right]\end{aligned}\right\} \tag{60}$$

The next step is to obtain the wave function in region V. To do this, we apply the formulas (39b) and (40) for the barrier to the right, obtaining

$$\psi_{\mathrm{V}} = \frac{-A}{\sqrt{p}}\left\{\begin{aligned} &\sin\left(\int_{x}^{-a} p\frac{dx}{\hbar} - \frac{\pi}{4}\right)\left[-i\cos\frac{1}{2}\left(\pi - \frac{J}{\hbar}\right) + \frac{1}{4\Theta^2}\sin\frac{1}{2}\left(\pi - \frac{J}{\hbar}\right)\right] \\ &+ \cos\left(\int_{x}^{-a} p\frac{dx}{\hbar} - \frac{\pi}{4}\right)\left[4i\Theta^2\sin\frac{1}{2}\left(\pi - \frac{J}{\hbar}\right) + \cos\frac{1}{2}\left(\pi - \frac{J}{\hbar}\right)\right]\end{aligned}\right\} \tag{61}$$

We now rewrite the above in terms of exponentials,

$$\psi_{\text{V}} = \frac{-A}{2\sqrt{p}} \left\{ \begin{aligned} &\exp\left[i\left(\int_{-a}^{x} p\frac{dx}{\hbar} + \frac{\pi}{4}\right)\right]\left[2\cos\frac{1}{2}\left(\pi - \frac{J}{\hbar}\right)\right. \\ &\qquad\qquad \left. + i\left(4\Theta^2 + \frac{1}{4\Theta^2}\right)\sin\frac{1}{2}\left(\pi - \frac{J}{\hbar}\right)\right] \\ &+ i\exp\left[-i\left(\int_{-a}^{x} p\frac{dx}{\hbar} + \frac{\pi}{4}\right)\right]\left(4\Theta^2 - \frac{1}{4\Theta^2}\right) \\ &\qquad\qquad \sin\frac{1}{2}\left(\pi - \frac{J}{\hbar}\right) \end{aligned} \right\} \tag{62}$$

We see that ψ_{V} includes an incident and a reflected wave. The transmissivity T is equal to the ratio of transmitted to incident intensities, or

$$T = 4\left[4\cos^2\frac{1}{2}\left(\pi - \frac{J}{\hbar}\right) + \left(4\Theta^2 + \frac{1}{4\Theta^2}\right)^2 \sin^2\frac{1}{2}\left(\pi - \frac{J}{\hbar}\right)\right]^{-1} \tag{63}$$

Writing $\cos^2 = 1 - \sin^2$, we obtain

$$T = \left[1 + \frac{1}{4}\left(4\Theta^2 - \frac{1}{4\Theta^2}\right)^2 \sin^2\frac{1}{2}\left(\pi - \frac{J}{\hbar}\right)\right]^{-1} \tag{64}$$

Problem 8: Prove that $T + R = 1$.

15. Discussion of Eq. (64) for Transmissivity. The WKB approximation should be applied only when the barrier is high and thick; in this case Θ is large, and T is usually very small. Comparing the results with those for a square well without barriers [eq. (34), Chap. 11], we see that since Θ^2, which is $\exp\left(2\int_b^a \frac{p_2\,dx}{\hbar}\right)$, is usually much bigger than p_2/p_1, the transmissivity is, in general, much smaller than for an attractive square well. Yet, just as with the square well, there are points where T is unity. Such transmission resonances occur where

$$\pi - \frac{J}{\hbar} = -2N\pi \qquad \text{or} \qquad J_N = \left(N + \frac{1}{2}\right)h$$

N being 0, 1, 2, ... It is interesting that the condition for a transmission resonance is exactly the same as that for a bound state.* We shall presently show that at transmission resonance, one has a metastable energy level, just as with the square well potential.

The reason for the transmission resonance is exactly the same as with the square well;† i.e., waves which have reflected back and forth are in phase with those just coming in. But the phase of the reflected wave is slightly different than for the square well, because a slowly varying potential reflects in a somewhat different way than does a sudden change.

* See eq. (55).

† Chap. 11, Secs. 8 and 20.

It is especially interesting that, although a single high and thick barrier has a very small transmissivity, two such barriers in a row can be completely transparent for certain wavelengths. This behavior can be understood only in terms of the wavelike aspects of matter. The high transmissivity arises because, for certain wavelengths, the reflected waves from inside interfere destructively with those from outside, so that only a transmitted wave remains.

16. Width of Resonance. Transmission Coefficient near Resonance. Near the resonant point, we may expand J, writing

$$J - J_N = \frac{\partial J}{\partial E}\,\delta E = \tau_0(E - E_N)$$

where τ_0 is the classical time needed to cross the well and return. Let us note that

$$\sin \frac{1}{2}\left(\pi - \frac{J}{\hbar}\right) \cong -\frac{1}{2}\tau_0 \frac{(E - E_N)}{\hbar}$$

near a resonance. We then obtain from eq. (64), when Θ is large,

$$T \cong \frac{1}{1 + \frac{\tau_0^2}{\hbar^2}(E - E_N)^2\Theta^4} \tag{65}$$

(Note that this formula is a good approximation only near a resonance.) The graph of T versus E will show a value of T which is generally small, but which becomes large near $E = E_N$, as illustrated in Fig. 10. The half width (where $T = \frac{1}{2}$) is obtained by setting

$$\Theta^4(E - E_N)^2\frac{\tau_0^2}{\hbar^2} = 1 \qquad \text{or} \qquad E - E_N = \frac{\hbar}{\tau_0\Theta^2} \tag{65a}$$

Near a transmission resonance, the rapid increase in probability of transmission is formally very similar to the increase of response of a damped

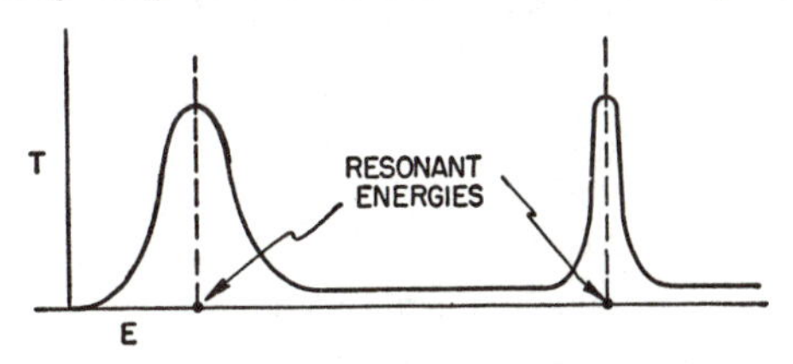

Fig. 10

harmonic oscillator to an impressed force that is near the resonant frequency. We shall show later that the reasons for the resonance phenomena in these two problems are analogous.

Note that if Θ is large, the resonance will be very sharp. The reason for the sharp resonance with large Θ is that there are many reflections.

Even at a point only slightly off resonance, where the waves suffer a small phase change each time that they reflect back and forth, the waves inside the barriers will eventually get out of phase with the incident wave, as a result of the cumulative changes of phase occurring after many reflections.

17. Intensity of Wave Inside Well. It is of interest to compute the ratio of the probability density inside the well to the probability density in the incident wave. Using eqs. (57) and (62),

$$\frac{|\psi_{\mathrm{III}}|^2}{|\psi_{\mathrm{inc}}|^2} = \frac{p}{p_w} \frac{\left[4\Theta^2 \cos^2\left(\int_x^b p\,\frac{dx}{\hbar} - \frac{\pi}{4}\right) + \frac{1}{4\Theta^2}\sin^2\left(\int_x^b p\,\frac{dx}{\hbar} - \frac{\pi}{4}\right)\right]}{\left[1 + \frac{1}{4}\left(4\Theta^2 - \frac{1}{4\Theta^2}\right)^2 \sin^2 \frac{1}{2}\left(\pi - \frac{J}{\hbar}\right)\right]} \tag{66}$$

where p_w is the momentum in the well and p is that in region V.

For a high quantum state, the trigonometric functions oscillate many times inside the well. Since each oscillation takes only a small part of the well, it is useful to consider the average density in a given region. If we do this, we can replace the $\cos^2$ and $\sin^2$ expressions each by $\frac{1}{2}$. Making the assumption that Θ is large, we then obtain

$$\frac{|\psi_{\mathrm{III}}|^2}{|\psi_{\mathrm{inc}}|^2} = \frac{p}{p_w} \frac{2\Theta^2}{1 + 4\Theta^4 \sin^2 \frac{1}{2}(\pi - J/\hbar)} \tag{67}$$

We note that far from a resonance, this ratio is usually small (for large Θ), so that the wave function has a shape resembling that shown in Fig. 11. At a resonance, however, the ratio $|\psi_{\mathrm{III}}|^2/|\psi_{\mathrm{inc}}|^2$ is large, and as a result, the wave function inside the well is also large. Since the transmitted

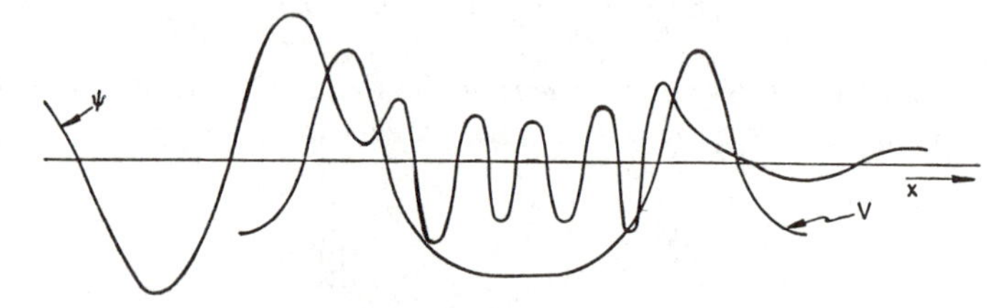

FIG. 11

wave has for this case the same intensity as the incident wave ($T = 1$), the wave function now has a shape resembling that shown in Fig. 12. Thus a very intense wave is trapped in the well, reflecting back and forth between the barriers in such a phase as to continually re-enforce itself, and leaking out very slowly.

The build-up of the wave inside the barriers near resonance involves a process very similar to that of building up a strong standing wave in an organ pipe, or in a resonant cavity undergoing electromagnetic oscillation.

In the latter examples, a small periodic impulse supplied externally can build up a large wave inside, provided that this impulse has a frequency near to that of the resonant system. The smaller the losses in the system resulting from friction, radiation, etc., the larger the wave amplitude, and the sharper the resonance. The quantum-mechanical problem is very similar, as the wave coming in from outside behaves rather like the "forcing term" in the harmonic oscillator. If this has the same frequency as that of the wave that is reflecting back and forth across

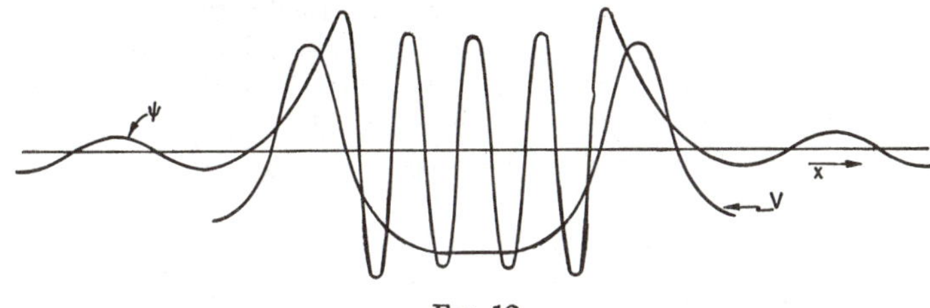

FIG. 12

the potential, a strong wave is built up inside. The smaller the losses caused by transmission through the barriers, the stronger the wave, and the sharper the resonance. Thus, we see that the analogy with mechanical and electrical resonance phenomena is very close.

The large transmissivity at resonance is produced by the fact that the wave is so big inside the well that even if only a small fraction leaks through, it produces a large result. The large amplitude inside also makes possible a large probability of entry into the region between the barriers. This is because the probability current across the barrier is proportional to $\psi^*\nabla\psi - \psi\nabla\psi^*$, so that if ψ gets large enough, the effect of the small barrier transmissivity is cancelled. The dependence of the transmissivity on the intensity of the wave *inside* the barriers is characteristically a wave phenomenon; it is hard to imagine, for example, how the transmission of a particle through a barrier could depend on what it was going to do after it got in. An analogy to the greater ease of penetration of the wave near resonance arises in a pendulum undergoing simple harmonic motion. If a given periodic force is in resonance with the pendulum, the rate of transfer of energy to the pendulum is proportional to the amplitude of vibration *already* in existence.

To a first approximation, the wave inside the barrier resembles a bound-state wave function, because it is large in such a restricted region. When we form a wave packet, moreover, we shall see that as a function of time the wave enters the well, stays in for a long time, and slowly leaks out through the barriers so that during the time that the wave packet is in the well it is very difficult to distinguish it from a bound-state wave function. In fact, the metastable states of the well with barriers resemble bound states much more closely than do those of the well without barriers, mainly because their lifetimes are much longer as a result of the very small transmissivities of the barriers.

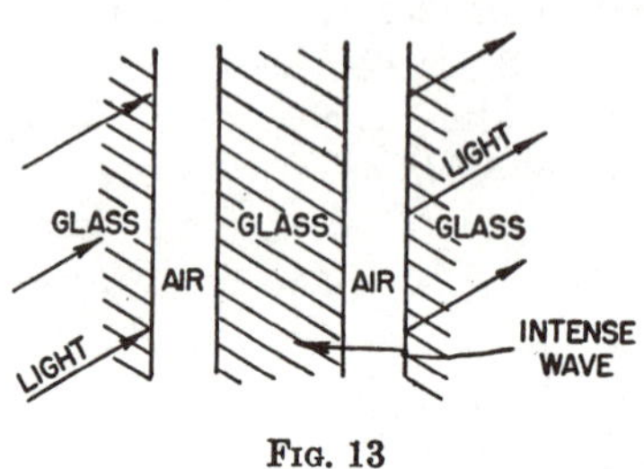

FIG. 13

A similar very intense wave can be built up by total internal reflection of light inside a thin sheet of glass, which is placed very close to two slabs of glass, but not quite in contact, as shown in Fig. 13. For certain wavelengths complete transmission will result, and the light inside the middle glass sheet will build up to a very great intensity. This could be detected by means of a small imperfection in the middle sheet, which would glow brilliantly for certain colors.

18. Formation of Wave Packets. Lifetimes of Virtual States. We shall now form a wave packet, just as we did with the square well problem. To do this, it is convenient to choose A so that the incident wave is just $\frac{1}{\sqrt{p}} \exp\left[i\left(\int_{-a}^{x} p\,\frac{dx}{\hbar} + \frac{\pi}{4}\right)\right]$. With the aid of eq. (62) we see that we must then choose

$$A = -2\left[2\cos\frac{1}{2}\left(\frac{\pi - J}{\hbar}\right) + i\left(4\Theta^2 + \frac{1}{4\Theta^2}\right)\sin\frac{1}{2}\left(\pi - \frac{J}{\hbar}\right)\right]^{-1} \tag{68}$$

Writing $A = -Re^{i\varphi}$, we obtain

$$\tan\varphi = -\frac{1}{2}\left(4\Theta^2 + \frac{1}{4\Theta^2}\right)\tan\frac{1}{2}\left(\pi - \frac{J}{\hbar}\right) \tag{69}$$

To form a wave packet, we integrate over a small range of energies. The incident wave will be

$$\psi_i = \int f(E - E_0)\,\frac{dE}{\sqrt{p}}\exp\left[i\left(\int_{-a}^{x} p\,\frac{dx}{\hbar} + \frac{\pi}{4} - \frac{Et}{\hbar}\right)\right] \tag{70}$$

The transmitted wave will be

$$\psi_t = \int f(E - E_0)R(E)\,\frac{dE}{\sqrt{p}}\exp\left[i\left(\int_{a}^{x} p\,\frac{dx}{\hbar} - \frac{\pi}{4} + \varphi - \frac{Et}{\hbar}\right)\right] \tag{71}$$

The center of the incident wave packet will be found where the phase has an extremum, or where

$$t = \frac{\partial}{\partial E}\int_{-a}^{x} p\,dx = \frac{\partial}{\partial E}\int_{-a}^{x}\sqrt{2m(E - V)}\,dx = \int_{-a}^{x}\frac{dx}{\sqrt{2(E - V)/m}}$$
$$= -\int_{x}^{-a}\frac{dx}{v} \tag{72}$$

Thus, the incident wave crosses the point x at a time t, which is the negative of that taken to go from x to the turning point at $-a$.

The center of the transmitted packet is then given by

$$t = \frac{\partial}{\partial E}\int_a^x p\,dx + \hbar\frac{\partial\varphi}{\partial E} = \int_a^x \frac{dx}{v} + \hbar\frac{\partial\varphi}{\partial E} \tag{73}$$

The time of crossing the point x is just that needed to move from the point a to the point x, plus $\hbar\frac{\partial\varphi}{\partial E}$. Hence $\hbar\frac{\partial\varphi}{\partial E}$ is roughly equal to the time delay of the wave as it reflects back and forth inside the well. By differentiating (69), we obtain

$$\begin{aligned}\hbar\sec^2\varphi\frac{\partial\varphi}{\partial E} = &+\frac{1}{4}\left(4\Theta^2+\frac{1}{4\Theta^2}\right)\left[\sec^2\frac{1}{2}\left(\pi-\frac{J}{\hbar}\right)\right]\frac{\partial J}{\partial E}\\ &-\frac{\hbar}{\Theta}\left(4\Theta^2-\frac{1}{4\Theta^2}\right)\left[\tan\frac{1}{2}\left(\pi-\frac{J}{\hbar}\right)\right]\frac{\partial\Theta}{\partial E}\end{aligned}$$

$$\hbar\frac{\partial\varphi}{\partial E} = \frac{+\left\{\begin{aligned}&\left[\sec^2\frac{1}{2}\left(\pi-\frac{J}{\hbar}\right)\right]\left(\Theta^2+\frac{1}{16\Theta^2}\right)\frac{\partial J}{\partial E}\\ &\quad-\frac{4\hbar}{\Theta}\left(\Theta^2-\frac{1}{16\Theta^2}\right)\left[\tan\frac{1}{2}\left(\pi-\frac{J}{\hbar}\right)\right]\frac{\partial\Theta}{\partial E}\end{aligned}\right\}}{1+4\left(\Theta^2+\frac{1}{16\Theta^2}\right)^2\tan^2\frac{1}{2}\left(\pi-\frac{J}{\hbar}\right)} \tag{74}$$

We see that if Θ is large, this time delay will be small, unless $\frac{1}{2}(\pi - J/\hbar) \cong N\pi$, or, in other words, unless we are near a virtual energy level. At a virtual energy level, we have

$$\Delta t = \hbar\frac{\partial\varphi}{\partial E} = \left(\Theta^2+\frac{1}{16\Theta^2}\right)\frac{\partial J}{\partial E} \cong \tau\Theta^2 \tag{74a}$$

for large Θ, where $\partial J/\partial E = \tau =$ classical period. Thus, when Θ is large, the appearance of the particle on the other side of the well is delayed for a time much longer than that needed to cross the well and return. We note that the explanation of this time delay is the same as for the delay obtained with the square well near resonance (see Chap. 11, Secs. 8 and 20). To prove this, we note that $1/\tau$ is just the number of times the particle strikes the barrier per second, while Θ^{-2} is the transmission coefficient for the barrier [eq. (46a)]. Thus, the mean time for getting through the barriers is of the order of $\tau\Theta^2$.

19. Wave Packet Inside Well (near Resonance). In order to demonstrate more clearly the fact that the particle stays inside the well for a long time, let us calculate the wave packet inside region III. To do this, we must integrate ψ_{III} obtained from eq. (57) over a small range of energies near a resonant point E_N. We write

$$\psi = \int_{-\infty}^{\infty} \psi_{\text{III}}(E) \exp\left(-\frac{iEt}{\hbar}\right) f(E - E_N)\, dE \tag{75}$$

To demonstrate the existence of a time delay Δt, we must choose a wave packet which is narrow enough so that it passes a given point in a time less than the delay that we are trying to discuss; otherwise, the time delay could not be distinguished from possible fluctuations in the time which were just caused by the width of the packet. According to the uncertainty principle, in order to define the time of passing a given point to an accuracy Δt, we need a range of energies $\Delta E \sim \hbar/\Delta t$. Hence $f(E - E_N)$ is assumed to be large only within this range.

We obtain ψ_{III} from eq. (57), neglecting terms involving $1/\Theta$, which are assumed to be small. We also note that at a resonance, $\sin \frac{1}{2}(\pi - J/\hbar)$ is zero, and $\cos \frac{1}{2}(\pi - J/\hbar)$ is unity. Near a resonance, we expand these quantities, keeping only first-order terms. Thus $\cos \frac{1}{2}(\pi - J/\hbar) \cong \pm 1$,

$$\sin \frac{1}{2}\left(\pi - \frac{J}{\hbar}\right) \cong \mp \frac{1}{2\hbar}\frac{\partial J}{\partial E}(E - E_N) = \mp \frac{\tau_0}{2\hbar}(E - E_N)$$

where $\tau_0 = \partial J/\partial E$. Obtaining A from eq. (68) we are led to

$$\psi_{\text{III}} \cong -\frac{2i}{\sqrt{p_w}}\Theta \frac{\cos\left(\int_x^b p\frac{dx}{\hbar} - \frac{\pi}{4}\right)}{1 - i\Theta^2\left(\frac{\tau_0}{\hbar}\right)(E - E_N)} \tag{76}$$

Because Θ is big, ψ_{III} becomes small for values of $E - E_N$, which are still so small that the expansion is valid. Thus, the main contribution to ψ will come from comparatively small values of $E - E_N$. We can therefore replace p_w by p_N, its value at resonance, since p_w will not change much in the region in which the denominator is small.

$$\psi_{\text{III}} \cong -\frac{2i\Theta}{\sqrt{p_N}}\cos\left(\int_x^b p_N\frac{dx}{\hbar} - \frac{\pi}{4}\right)\frac{1}{1 - \frac{i\Theta^2}{\hbar}\tau_0(E - E_N)} \tag{77}$$

and $$\psi \cong -\frac{2i\Theta}{\sqrt{p_N}}\cos\left(\int_x^b p_N\frac{dx}{\hbar} - \frac{\pi}{4}\right)\int \frac{f(E - E_N)\exp(-iEt/\hbar)\, dE}{1 - i(E - E_N)\frac{\Delta t}{\hbar}} \tag{78}$$

where we have written

$$\Delta t = \tau_0\Theta^2 \tag{78a}$$

Now, we have seen that $f(E - E_N)$ was chosen so that it was large in a region much bigger than $\hbar/\Delta t$; hence the reciprocal of the denominator becomes small at much smaller values of $E - E_N$ than does the numerator. To a first approximation, we may therefore regard $f(E - E_N)$ as a constant, which we shall, for the sake of convenience, take to be unity.

The remainder of the integral is then easily evaluated by standard methods, which yield*

$$\int_{-\infty}^{\infty} \frac{e^{-iEt/\hbar}\,dE}{1 - i(E - E_N)\dfrac{\Delta t}{\hbar}} = \begin{Bmatrix} \dfrac{2\hbar\pi}{\Delta t} e^{-iE_Nt/\hbar} e^{-t/\Delta t} & t > 0 \\ 0 & t < 0 \end{Bmatrix} \tag{79}$$

Thus, we obtain

$$\psi \cong \left[\frac{\hbar}{\Delta t}\frac{4\pi i\Theta}{\sqrt{p_N}} \cos\left(\int_x^b p_N \frac{dx}{\hbar} - \frac{\pi}{4}\right) e^{-iE_Nt/\hbar}\right] \begin{Bmatrix} e^{-|t|/\Delta t} & t > 0 \\ 0 & t < 0 \end{Bmatrix} \tag{80}$$

The discontinuous change in the wave function at $t = 0$ arises from our approximation of $f(E - E_N)$ by a constant. This is equivalent to assuming an infinitely sharp packet. Just as soon as the packet strikes the barrier at $t = 0$, the wave function inside therefore suddenly rises from zero to some definite value. If the width of the packet had been taken into account, and if the expansions in $(E - E_N)$ had not been used, the rise time for the wave function inside would have differed slightly from zero.

The interesting part of the result, however, is that the wave function after $t = 0$ is the same as that of a true stationary state, except for the exponential decay with time.

20. Uncertainty Principle. It is clear from the definition of Δt [eq. (78a)] and of ΔE, the width of the resonance [eq. (65a)], that $\Delta E\,\Delta t \sim \hbar$. This means that to make the virtual level, we must use a wave packet of width $\Delta E \sim \hbar/\Delta t$. As a result, when the state decays, particles with this range of energies will appear. Also, this range of energies would have to be used in forming the virtually bound state.

21. Application to Radioactive Systems. The application of these conclusions to radioactive systems is fairly direct.† We assume that the radioactive nucleus is formed at some time, which we denote as $t = 0$. The exact manner of formation is unimportant for the subsequent behavior, but it could have been formed, for example, by bombardment, as is implied by our use of an incident wave packet, which enters the region between the barriers. After the metastable state is formed, it decays exponentially, and a wave appears outside the region between the barriers. This describes exactly the probability of the decay process.

* See, for example, Whittaker and Watson, *Modern Analysis*, 3d ed., London: Cambridge University Press, 1920, p. 123, Problem 15.

† Actually, the process of decay of metastable nuclei has been highly idealized in this treatment, because many important factors have been neglected. The treatment given here is intended less as a complete theory of nuclear decay than as an illustration of the wave aspects of matter in the quantum theory. For a more complete treatment, see Bethe, *Elementary Nuclear Theory*.

To demonstrate the existence of the metastable state, one would have to form it, as we have seen, from an incident wave packet which was narrow enough to pass a given point in a time less than the lifetime Δt. In a typical case, for example, one might bombard a given nucleus with α-particles. The time at which an α-particle passes a given point can easily be controlled by electronic equipment to within 10^{-6} sec. The resultant (radioactive) metastable state might in some cases last for periods of the order of seconds. Thus, one could very easily demonstrate the existence of a metastable state. But if the time of entry into the nucleus had not been defined within an hour, then the experiment could not have been used to demonstrate the existence of such a state.

22. Application to Nuclear Reactions. In a typical nuclear reaction, a proton without enough energy to go over the Coulomb barrier may strike a nucleus, and occasionally enter by leaking through the barrier.* If the energy is such as to form a long-lived virtual state, it may stay inside the nucleus a long time, and then be re-emitted. Furthermore, we have seen that near a resonance, the probability of entry into the nucleus through the barrier is large. Such a nuclear reaction might be, for example, the following:

$$\mathrm{Li}^7 + \mathrm{p}^1 \rightarrow \mathrm{Be}^8 \xrightarrow[\text{metastable}]{} \mathrm{Li}^7 + \mathrm{p}^1$$

Actually, we have overidealized the situation in the following ways:

(1) The problem is three-dimensional. The three-dimensional treatment is more or less the same as the one-dimensional case, and when it is carried out, one finds that near a virtual level, there is also a high probability of entry into the nucleus by penetration of the barrier, as well as a long-lived metastable state. When the particle leaves the nucleus, however, it may be thrown out in any direction. Thus, the net result is to scatter the incident proton. One therefore finds that near a virtual level, there is a sharp increase in the probability of scattering, which resembles in functional form the sharp increase in transmissivity shown in Fig. 10.

(2) While the proton is in the nucleus, additional reactions may take place.† For example, if there are true bound states, the proton may radiate some energy, and a stable nucleus may be formed, so that the proton is not re-emitted. In the case of the $\mathrm{Li}^7 + \mathrm{p}^1$ reaction, a γ ray can be emitted and another state of Be^8 can be formed. In the case of Be^8, the system can also dispose of its energy by breaking up into two α particles. With other nuclei, still other possibilities exist. For example, the incoming proton may give up its energy to a neutron already in the nucleus, which latter particle is then emitted, leaving the proton bound inside the nucleus. Such a reaction, the net effect of

* See Sec. 12.

† See, for example, Bethe, *Elementary Nuclear Physics*, Chap. 17.

which is to replace a free proton by a free neutron, is called a p-n reaction.

In general then, there are many competing reactions, only one of which is re-emission of the original bombarding particle without loss of energy. If R_i is the rate at which the ith process goes on, then the total rate at which the metastable state is destroyed is $R = \sum_i R_i$. The net lifetime is $\Delta t = 1/R$. It can then be shown that, in accordance with the uncertainty principle, the width of the state is correspondingly increased to $\delta E \cong \hbar/\Delta t$. The resonance is therefore broadened, if the metastable nucleus can decay in any way other than by re-emission of the incident particle.*

* For a more complete discussion of resonance, see H. Bethe, *Reviews of Modern Physics*, **9,** 75 (1937), also E. P. Wigner, *Phys. Rev.*, **70,** 607 (1946) and H. Feshbach, D. C. Peaslee, and V. F. Weisskopf, *Phys. Rev.*, **71,** 145 (1947).

CHAPTER 13

The Harmonic Oscillator

1. Introduction. We shall now take up the problem of the harmonic oscillator. This problem is important in itself, especially because the radiation field acts like a collection of such oscillators.* Besides, many systems can be represented approximately as harmonic oscillators. For example, potential energy of two atoms as a function of their separation is usually a curve of the type shown in Fig. 1. There is usually some distance, $x = a$, at which the potential has a minimum. This is a point of stable equilibrium. Near this point, the potential can be expanded as a series of powers of $x - a$, and since $\partial V/\partial x = 0$ at this point, we have

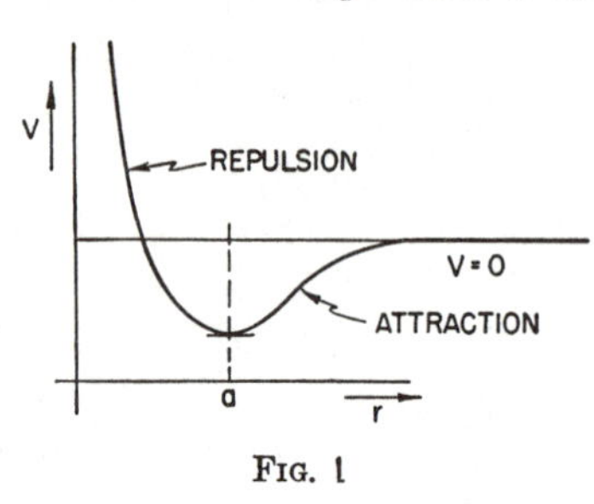

FIG. 1

$$V \cong \frac{k}{2}(x - a)^2 \tag{1}$$

This is just the harmonic oscillator potential.

In general, any system in stable equilibrium can be represented near the equilibrium position by means of a harmonic oscillator.

2. Wave Equation. For an oscillator with force constant k, the potential is $V = kx^2/2 = m\omega^2x^2/2$, where ω is the angular frequency of oscillation. The wave equation is then

$$-\frac{\hbar^2}{2m}\psi'' + \left(\frac{m\omega^2x^2}{2} - E\right)\psi = 0 \tag{2}$$

It is convenient to make the substitutions,

$$x = \sqrt{\frac{\hbar}{m\omega}}\, y, \qquad E = \frac{\omega\hbar}{2}\epsilon$$

The wave equation becomes

$$\frac{d^2\psi}{dy^2} + (\epsilon - y^2)\psi = 0 \tag{3}$$

* Chap. 1, Sec. 8.

3. General Form of Solutions. The potential well has, as shown in Fig. 2, a parabolic shape. For a large enough $|x|$ the potential energy is always greater than the total energy, so that the solution of the wave equation is a linear combination of real exponentials. To find the form of the exponential, we can use the WKB approximation.* The solution for large $|x|$ will involve

$$\exp\left[\pm\int_a^x \sqrt{2m(V-E)}\,\frac{dx}{\hbar}\right]$$

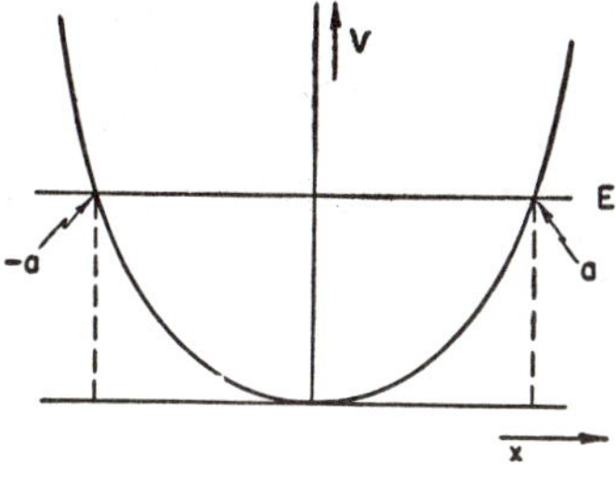

FIG. 2

Now for large $|x|$, $V \gg E$, so that

$$\sqrt{V-E} \cong \sqrt{V} = \sqrt{\frac{m}{2}}\,\omega x$$

The solutions are therefore of the order of

$$A\exp\left(\frac{m\omega}{\hbar}\frac{x^2}{2}\right) + B\exp\left(-\frac{m\omega}{\hbar}\frac{x^2}{2}\right) \tag{4}$$

We must choose that solution which dies out exponentially at a large $|x|$. Thus, if we start at a large negative x, with a solution varying as

$$\exp\left(-\frac{m\omega}{\hbar}\frac{x^2}{2}\right)$$

we want to end up at large positive x with the same type of solution. It is therefore necessary that the solution curve over in the region of positive kinetic energy in such a way as to fit a decaying exponential when

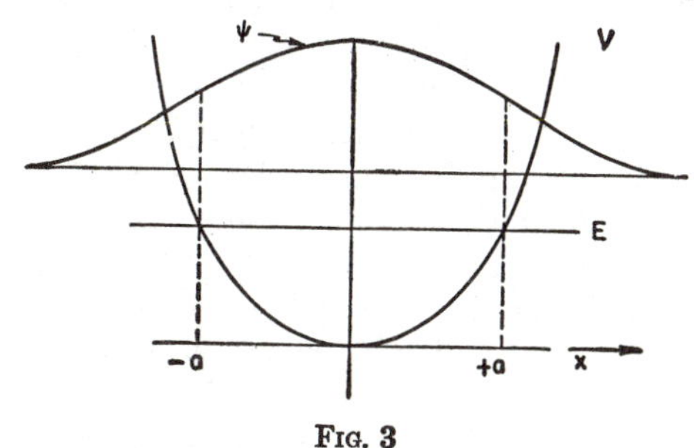

FIG. 3

$|x| > a$. The problem is very similar to that of the bound states of the square potential well.† The lowest state has no nodes, and this is shown in Fig. 3. The next state, which has one node, is represented in Fig. 4. It should be noted that not only does the wave function curve more rapidly when the energy is higher, but that there is also a larger

* Eq. (37), chap. 12.
† Chap. 11, Sec. 12.

region in which the kinetic energy is positive. For a very high quantum state, there are very many oscillations, and the WKB approximation* will be good. There is no limit to the number of possible bound states, because the potential becomes infinitely high as $x \to \infty$. It is therefore always possible to raise the energy, and, in this way, to cause the wave function to make another oscillation before it reaches the region of negative kinetic energy. If, however, the potential ceases to increase indefinitely at large values of x, as is, for example, true of interatomic forces (see Fig. 1), then the number of bound states will be finite.

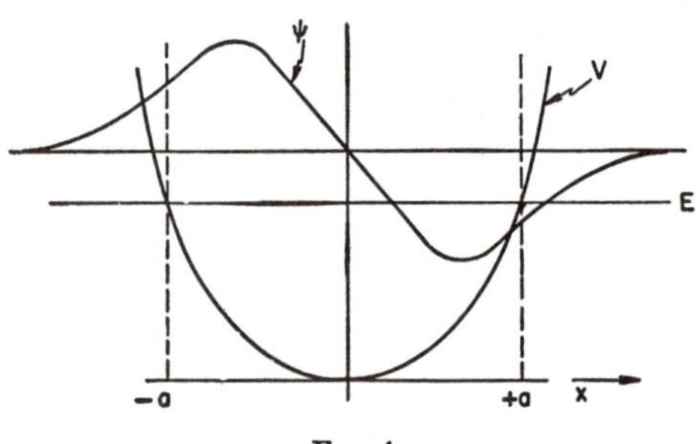

FIG. 4

4. Methods of Exact Solution. The general method of finding the eigenvalues and eigenfunctions of this type of equation is first, to represent the solution near the origin by a power series in which there are enough terms to carry the solution into the exponentially decaying region. If the series does not converge over the whole region, it may be necessary to use numerical integration, or else expansions about several different points in succession. In any case, one must choose the power series to be such that it fits smoothly to the decaying exponential. In general, as shown in the case of the square well, such a fit does not occur unless the energy is given one of a discrete set of possible values. These values are the eigenvalues, and the associated solutions are the eigenfunctions.

5. Schrödinger's Method of Factorization. Although the procedure outlined above is commonly used for the harmonic oscillator problem,† as well as for many other similar problems, we shall use here a simple method, developed by Schrödinger,‡ which, however, is not so general in its application. Nevertheless, as we shall see, it can be applied in several of the problems which we shall study in this course, among them the problem of quantization of angular momentum (see Chap. 14).

The basis of this method is to "factor" the Hamiltonian operator into two operators, each involving only first derivatives. In this problem, we do so by noting that

* Eq. (17), Chap. 12.
† See Pauling and Wilson, *Introduction to Quantum Mechanics*, pp. 67–72.
‡ E. Schrödinger, *Proc. Roy. Irish Acad.*, **A47**, 53 (1942); see also Dirac, *The Principles of Quantum Mechanics*, 3rd ed., Chap. 6.

$$\left(\frac{d^2}{dy^2} - y^2\right)\psi = \left[\left(\frac{d}{dy} - y\right)\left(\frac{d}{dy} + y\right) - 1\right]\psi \tag{5}$$

Equation (3) may therefore be written as follows:

$$\left(\frac{d}{dy} - y\right)\left(\frac{d}{dy} + y\right)\psi_\epsilon = -(\epsilon - 1)\psi_\epsilon \tag{6}$$

where ψ_ϵ is the eigenfunction belonging to the eigenvalue ϵ.

The next step is to operate on this equation from the left with the operator $\left(\frac{d}{dy} + y\right)$. In doing this, we note that

$$\left(\frac{d}{dy} + y\right)\left(\frac{d}{dy} - y\right) = \frac{d^2}{dy^2} - y^2 - 1$$

We therefore obtain

$$\left(\frac{d}{dy} + y\right)\left(\frac{d}{dy} - y\right)\left(\frac{d}{dy} + y\right)\psi_\epsilon = \left(\frac{d^2}{dy^2} - y^2 - 1\right)\left(\frac{d}{dy} + y\right)\psi_\epsilon$$

$$= -(\epsilon - 1)\left(\frac{d}{dy} + y\right)\psi_\epsilon \tag{7}$$

Writing $\left(\frac{d}{dy} + y\right)\psi_\epsilon = \varphi_\epsilon =$ a new wave function, we obtain

$$\left(\frac{d^2}{dy^2} - y^2\right)\varphi_\epsilon = -(\epsilon - 2)\varphi_\epsilon \tag{8}$$

We conclude that if ψ_ϵ is an eigenfunction of Schrödinger's equation corresponding to the eigenvalue ϵ, then $\varphi_\epsilon = \left(\frac{d}{dy} + y\right)\psi_\epsilon$ is an eigenfunction of the same equation corresponding to an eigenvalue of $\epsilon - 2$. Thus, given any one solution, we can always derive another. Furthermore, if ϵ is an eigenvalue, then $\epsilon - 2$ must also be an allowed value.

We can repeat this procedure indefinitely, and we are thus led to the conclusion that if ϵ is an eigenvalue, then $\epsilon - 2n$ is also an eigenvalue, where n is an integer. But if n is made big enough, the eigenvalue (and consequently the energy) will eventually become negative, since ϵ is proportional to the energy. But, we can easily see that the energy of a harmonic oscillator is always positive. We write for the mean value of E,

$$\bar{E} = \int \psi^* H \psi \, dx = -\int \frac{\hbar^2}{2m}\psi^* \frac{\partial^2 \psi}{dx^2}\, dx + \int \psi^* \frac{m\omega^2 x^2}{2}\, dx \tag{9}$$

Integration of the first integral by parts (noting that the integrated part vanishes since $\psi \to 0$ as $x \to \infty$) yields

$$\bar{E} = \frac{\hbar^2}{2m}\int \frac{\partial \psi^*}{\partial x}\frac{\partial \psi}{\partial x}\, dx + \frac{m\omega^2}{2}\int \psi^* x^2 \psi \, dx \tag{9a}$$

Both integrals are by definition positive; hence $\bar{E} > 0$. But, for an eigenfunction,

$$\bar{E} = \int \psi_E^* H \psi_E \, dx = E \int \psi^* \psi \, dx = E \tag{9b}$$

since ψ is assumed to be normalized. Thus, we conclude that all eigenvalues of E, and hence of ϵ, must be positive.

How can we avoid this contradiction? This can be done only if the lowest positive value of ϵ is such that $\left(\frac{d}{dy} + y\right)\psi_\epsilon = 0$. For in this case, no solutions of negative ϵ can be obtained. Multiplication of this equation by the operator $\left(\frac{d}{dy} - y\right)$ yields

$$\left(\frac{d}{dy} - y\right)\left(\frac{d}{dy} + y\right)\psi_\epsilon = \frac{d^2\psi_\epsilon}{dy^2} - y^2\psi_\epsilon + \psi_\epsilon = 0 \tag{10}$$

From (10) we see that the lowest value of ϵ must be $\epsilon = 1$. The only allowed values of ϵ must then be such that $\epsilon - 2n = 1$, where n is an integer. The eigenvalues therefore are

$$\epsilon = 2n + 1 \tag{11}$$

and the eigenvalues of E are

$$E = (2n + 1)\frac{\hbar\omega}{2} = \left(n + \frac{1}{2}\right)\hbar\omega \tag{12}$$

Thus, we have proved that the eigenvalues of the harmonic oscillator are exactly equal to those obtained from the WKB approximation,† even including the half integral quantization of energy.

Problem 1: Explain why the lowest state cannot have a zero energy.

6. Solution for Wave Functions. We can easily solve for the wave function of the lowest state. The equation defining it is given above as

$$\frac{d\psi_0}{dy} + y\psi_0 = 0 \qquad \text{or} \qquad \frac{d\psi_0}{\psi_0} + y\,dy = 0 \tag{13}$$

The solution is

$$\psi_0 = A\,e^{-y^2/2} \tag{13a}$$

where A is a constant. To normalize the wave function we must choose $A = 1/\sqrt{\pi}$. The lowest state is therefore simply a Gauss function.

The remaining wave functions can all be obtained from ψ_0. To do this, we first write Schrödinger's equation as follows:

$$\left(\frac{d}{dy} + y\right)\left(\frac{d}{dy} - y\right)\psi_\epsilon = -(\epsilon + 1)\psi_\epsilon \tag{14}$$

† Eq. (55), Chap. 12.

Multiplication from the left by $\left(\frac{d}{dy} - y\right)$ yields

$$\left(\frac{d}{dy} - y\right)\left(\frac{d}{dy} + y\right)\left(\frac{d}{dy} - y\right)\psi_\epsilon = \left(\frac{d^2}{dy^2} - y^2 + 1\right)\left(\frac{d}{dy} - y\right)\psi_\epsilon \quad (15)$$

$$= -(\epsilon + 1)\left(\frac{d}{dy} - y\right)\psi_\epsilon$$

or

$$\left(\frac{d^2}{dy^2} - y^2\right)\left(\frac{d}{dy} - y\right)\psi_\epsilon = -(\epsilon + 2)\left(\frac{d}{dy} - y\right)\psi_\epsilon \quad (16)$$

We see that the function $\varphi = \left(\frac{d}{dy} - y\right)\psi_\epsilon$ satisfies Schrödinger's equation, but corresponds to an eigenvalue of $\epsilon + 2$.* Hence, if we have any eigenfunction ψ_ϵ, we can always generate the next higher eigenfunction by operating on it with the operator $\left(\frac{d}{dy} - y\right)$. In this way we can obtain all eigenfunctions from ψ_0. The nth eigenfunction is

$$\psi_n = C_n(-1)^n\left(\frac{d}{dy} - y\right)^n \psi_0 = C_n(-1)^n\left(\frac{d}{dy} - y\right)^n e^{-y^2/2} \quad (17)$$

C_n is a normalizing constant, which we shall determine in Sec. 8.

By carrying out these operations, we obtain the first few eigenfunctions.

$$\psi_0 \sim e^{-y^2/2} \quad (18a)$$

$$\psi_1 \sim 2ye^{-y^2/2} \quad (18b)$$

$$\psi_2 \sim (2y^2 - 1)e^{-y^2/2} \quad (18c)$$

7. Hermite Polynomials. In general, we see that ψ_n is equal to $e^{-y^2/2}$ times a polynomial of the nth degree. Thus, we may write

$$\psi_n = C_n\, e^{-y^2/2}h_n(y) \quad (19)$$

where h_n is the polynomial in question. The h_n are called Hermite polynomials. We may also write

$$h_n(y) \sim e^{y^2/2}\psi_n(y) \quad (19a)$$

A somewhat more convenient expression for h_n can be obtained with the aid of the fact that for an arbitrary function φ the following relations hold:

$$\left(\frac{d}{dy} - y\right)\varphi = e^{y^2/2}\frac{d}{dy}\left(e^{-y^2/2}\varphi\right) \quad (20)$$

* Note that $\epsilon = 2n + 1$, where n is the quantum number.

We therefore obtain from eq. (17)

$$\psi_1 = -C_1 e^{y^2/2} \frac{d}{dy} e^{-y^2} \qquad \psi_2 = C_2 e^{y^2/2} \frac{d^2}{dy^2} e^{-y^2} \tag{21}$$

$$\psi_n = (-1)^n C_n e^{y^2/2} \frac{d}{dy} e^{-y^2/2} e^{+y^2/2} \frac{d^{n-1}}{dy^{n-1}} e^{-y^2} \tag{22}$$

$$= (-1)^n C_n e^{y^2/2} \frac{d^n}{dy^n} e^{-y^2} = C_n e^{-y^2/2} h_n(y)$$

8. Normalization Factor. To evaluate the normalizing factor, we set

$$\psi_n^* = C_n^* e^{-y^2/2} h_n$$

and

$$\psi_n = C_n (-1)^n e^{y^2/2} \frac{d^n}{dy^n} e^{-y^2}$$

The normalization condition becomes

$$\int_{-\infty}^{\infty} \psi_n^* \psi_n \, dy = (-1)^n |C_n|^2 \int_{-\infty}^{\infty} h_n(y) \frac{d^n}{dy^n} e^{-y^2} \, dy = 1 \tag{23}$$

Let us now integrate by parts n times, noting that the integrated parts always vanish. Each time we integrate by parts, we introduce the factor -1. This gives us $(-1)^n$, which cancels the $(-1)^n$ appearing in front of the integral. We obtain

$$|C_n|^2 \int_{-\infty}^{\infty} e^{-y^2} \frac{d^n}{dy^n} h_n(y) \, dy = 1 \tag{24}$$

Since $h_n(y)$ is a polynomial of the nth degree, the differentiation will wipe out all terms except that involving y^n. Writing

$$h_n(y) = \sum_{\gamma=0}^{n} A_\gamma y^\gamma$$

and noting that $\frac{d^n}{dy^n} y^n = n!$, we obtain

$$\frac{d^n}{dy^n} h_n(y) = n! \, A_n \tag{25}$$

To evaluate A_n, we note that the coefficient of y^n in $(-1)^n e^{y^2} \frac{d^n}{dy^n} e^{-y^2}$ is just 2^n. Thus, we get $A_n = 2^n$. Equation (24) then becomes

$$|C_n|^2 2^n n! \int_{-\infty}^{\infty} e^{-y^2} \, dy = 1 \qquad \text{or} \qquad C_n = \frac{1}{\sqrt{2^n n! \sqrt{\pi}}} \tag{26}$$

As a function of y, the normalized wave function is then

$$\psi_n(y) = \frac{(-1)^n e^{y^2/2}}{\sqrt{2^n n! \sqrt{\pi}}} \frac{d^n}{dy^n} e^{-y^2} = \frac{e^{-y^2/2} h_n(y)}{\sqrt{2^n n! \sqrt{\pi}}} \tag{27}$$

In order to normalize the wave function over $x = \sqrt{\frac{\hbar}{m\omega}}\, y$, we must multiply it by $\sqrt[4]{\omega m/\hbar}$. As a function of x, the normalized wave function is therefore

$$\psi_n(x) = \exp\left[-\left(\omega\frac{m}{\hbar}\right)\left(\frac{x^2}{2}\right)\right] h_n\left(\sqrt{\frac{\omega m}{\hbar}}\,x\right)\frac{\sqrt[4]{\omega m/\hbar}}{\sqrt{2^n n!\sqrt{\pi}}} \tag{28}$$

9. Generating Function. A very useful relation satisfied by Hermite polynomials can be obtained by multiplying $h_n(y)$ by $\frac{t^n}{n!}$ and summing over n. We obtain

$$\sum_0^\infty h_n(y)\frac{t^n}{n!} = e^{y^2}\sum_0^\infty \frac{d^n}{dy^n}e^{-y^2}\frac{(-t)^n}{n!} \tag{29}$$

If we expand the function $e^{y^2}e^{-(t-y)^2} = e^{-t^2+2ty}$ as a series of powers of t, however, we get precisely the above series. Hence, we may write

$$e^{-t^2+2ty} = \sum_0^\infty h_n(y)\frac{t^n}{n!} \tag{30}$$

e^{-t^2+2ty} is called *a generating function* for the Hermite polynomials, because from it one can generate all the Hermite polynomials by expansion in a series of powers of t.

10. Recurrence Relations. The generating function may be used to derive a number of useful relations between different Hermite polynomials. For example, if one differentiates eq. (30) with respect to y, one obtains

$$2te^{-t^2+2ty} = \sum_0^\infty \frac{dh_n}{d_y}\frac{t^n}{n!} = 2\sum_0^\infty h_n(y)\frac{t^{n+1}}{n!} \tag{31}$$

Since this must be true for all t, the coefficients of equal powers of t must be equal. The above equation may be written

$$\sum_1^\infty \frac{t^n}{n!}\left(\frac{dh_n}{dy} - 2nh_{n-1}\right) = 0 \tag{32}$$

We then obtain

$$\frac{dh_n}{dy} = 2nh_{n-1} \tag{33}$$

Another relation is obtained by differentiation with respect to t:

$$2(y-t)e^{-t^2+2yt} = \sum_0^\infty h_n(y)\frac{t^{n-1}}{(n-1)!} = 2(y-t)\sum_0^\infty h_n(y)\frac{t^n}{n!} \tag{34}$$

This equation can be rewritten as

$$\sum_{0}^{\infty} \frac{t^n}{n!} (2yh_n - 2nh_{n-1} - h_{n+1}) = 0 \tag{35}$$

We conclude that

$$2yh_n(y) = 2nh_{n-1}(y) + h_{n+1}(y) \tag{36}$$

11. A Few Auxiliary Mathematical Relations. We have already seen that, given an eigenfunction ψ_n, we can always construct an eigenfunction belonging either to the next higher or lower eigenvalue by multiplying respectively by the operators $\left(\frac{d}{dy} - y\right)$ or $\left(\frac{d}{dy} + y\right)$ It will be helpful later to express the effect of this operator in terms of the normalized eigenfunctions ψ_n. We first note that $-\left(\frac{d}{dy} - y\right)\psi_n = C\psi_{n+1}$, where C is a constant to be chosen so as to make ψ_{n+1} also a normalized function. Since ψ_n is assumed to be normalized, we write

$$\psi_n = \frac{(-1)^n e^{y^2/2}}{\sqrt{2^n n! \sqrt{\pi}}} \frac{d^n}{dy^n} e^{-y^2} \tag{37}$$

Noting that $-\left(\frac{d}{dy} - y\right)\psi_n = -e^{y^2/2}\frac{d}{dy}(e^{-y^2/2}\psi_n)$, we obtain

$$\begin{aligned}
-\left(\frac{d}{dy} - y\right)\psi_n &= \frac{(-1)^{n+1}}{\sqrt{2^n n! \sqrt{\pi}}} e^{y^2/2} \frac{d^{n+1}}{dy^{n+1}} e^{-y^2} \\
&= \frac{\sqrt{2(n+1)}(-1)^{n+1} e^{y^2/2}}{\sqrt{2^{n+1}(n+1)! \sqrt{\pi}}} \frac{d^{n+1}}{dy^{n+1}} e^{-y^2} \\
&= \sqrt{2(n+1)}\,\psi_{n+1}
\end{aligned} \tag{38}$$

To obtain $\left(\frac{d}{dy} + y\right)\psi_n$, we write*

$$\left(\frac{d}{dy} + y\right)\left(\frac{d}{dy} - y\right)\psi_{n-1} = \left(\frac{d^2}{dy^2} - y^2 - 1\right)\psi_{n-1} = -2n\psi_{n-1}$$

We also use the fact proved above that $\left(\frac{d}{dy} - y\right)\psi_{n-1} = -\sqrt{2n}\,\psi_n$. This gives us

$$\left(\frac{d}{dy} + y\right)\psi_n = \sqrt{2n}\,\psi_{n-1} \tag{39}$$

12. General Form of Solution. The nth eigenfunction consists of an nth order polynomial multiplying a factor of $e^{-y^2/2}$. The latter factor makes the wave function approach zero as y approaches infinity. The

* See eq. (14).

polynomial h_n has n roots; hence the wave function has n nodes. In this way, it resembles qualitatively the WKB wave functions,* since, with n nodes, it will undergo a corresponding number of oscillations. We see once again that the number of the quantum state is equal to the number of nodes in the solution. In general, the wave function oscillates within the classically accessible region ($E > V$) and dies out in a Gaussian fashion where $E < V$.

13. Orthogonality of Hermite Polynomials. According to Chap. (10), Sec. 24, the eigenfunctions of a Hermitean operator belonging to different eigenvalues are orthogonal. This means that

$$\int_{-\infty}^{\infty} \psi_n^* \psi_m \, dy = \int_{-\infty}^{\infty} \frac{e^{-y^2} h_m(y) h_n(y) \, dy}{\sqrt{2^{n+m} n! m! \pi}} = 0 \tag{40}$$

when $n \neq m$.

Let us suppose that $m > n$. Then we can prove the above statement directly by writing $e^{-y^2/2} h_m(y) = (-1)^m e^{y^2/2} \dfrac{d^m}{dy^m} e^{-y^2}$. We get

$$\sqrt{2^{n+m} n! m! \pi} \int_{-\infty}^{\infty} \psi_n^* \psi_m \, dy = (-1)^m \int_{-\infty}^{\infty} h_n(y) \frac{d^m}{dy^m} e^{-y^2} \, dy \tag{41}$$

Integration by parts m times, noting that the integrated part vanishes, yields

$$(-1)^{m+n} \int_{-\infty}^{\infty} e^{-y^2} \frac{d^m}{dy^m} h_n(y) \, dy \tag{42}$$

But differentiation of the polynomial m times yields zero, if $m > n$. Thus, we have demonstrated the orthogonality of eigenfunctions of the Hamiltonian for the harmonic oscillator.

14. Expansion Postulate. According to the expansion postulate,† it is possible to expand an arbitrary function as a series of eigenfunctions of the Hamiltonian operator for a harmonic oscillator. Thus, we write for an arbitrary function, φ,

$$\varphi(y) = \sum_{n=0}^{\infty} \frac{C_n e^{-y^2/2}}{\sqrt{2^n n! \sqrt{\pi}}} h_n(y) \tag{43}$$

To solve for C_n, we multiply by $e^{-y^2/2} h_m(y) / \sqrt{2^m m! \sqrt{\pi}}$ and integrate over y. Using the properties that the wave functions are orthogonal and normalized, we obtain

$$C_m = \frac{1}{\sqrt{2^m m! \sqrt{\pi}}} \int_{-\infty}^{\infty} e^{-y^2/2} h_m(y) \varphi(y) \, dy \tag{44}$$

* Chap. 12, Sec. 13; see also, Pauling and Wilson, *Introduction to Quantum Mechanics*, pp. 73–82.

† Chap. 10, Sec. 22.

Example: Expansion of the δ function.
Setting $\varphi(y) = \delta(y - y_0)$, we obtain

$$C_m = \frac{1}{\sqrt{2^m m! \sqrt{\pi}}} e^{-y_0^2/2} h_m(y_0) \tag{45}$$

and

$$\delta(y - y_0) = \sum_{n=0}^{\infty} e^{-(y^2+y_0^2)/2} \frac{h_n(y)h_n(y_0)}{2^n n! \sqrt{\pi}} \tag{46}$$

We note that to form a δ function we need to go to highly excited states of the oscillator. In other words, if a particle is highly localized, no matter where, the energy becomes very indefinite.

The formation of a δ function from Hermite polynomials may be illustrated by considering the first three polynomials alone.

$$\psi_0^*\psi_0 \sim e^{-y^2} \tag{47a}$$

This is symmetrical around the origin.

$$\psi_1^*\psi_1 \sim y^2 e^{-y^2} \tag{47b}$$

This is also symmetrical around the origin, but it is zero at the origin.

$$\psi_2^*\psi_2 \sim (2y^2 - 1)^2 e^{-y^2} \tag{47c}$$

This is also symmetric about the origin. Let us now consider linear combinations of ψ_0 and ψ_1.

$$\psi = a_0\psi_0 + a_1\psi_1 \sim e^{-y^2}(a_0 + a_1 y)$$

Thus, it is possible to form a function which is small for negative y and large for positive y. (For example, let $a_0 = a_1 = 1$.) Such a function is shown in Fig. 5.

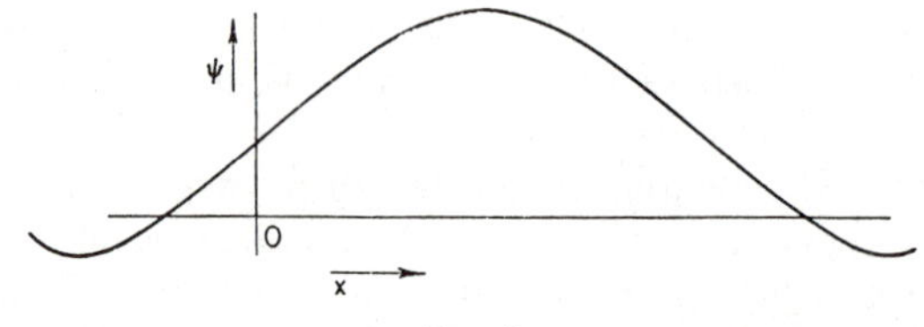

FIG. 5

As more and more polynomials are included, a packet is formed that is more and more localized. The essential point is that in order to make a localized packet, we need many energy states.

15. Wave Packets. The use of a δ function provides a wave packet that is infinitely well localized and therefore an abstraction. It is of interest to follow the motion of a wave packet that is initially of finite extent. Consider, for example, a particle that was localized initially with the packet

$$\psi(y) = e^{-(y-y_0)^2/2} = e^{-y^2/2} e^{yy_0 - y_0^2/2} \tag{48}$$

This is a packet centering about the position $y = y_0$.

In order to find how ψ changes with time, we must first expand ψ as a series of eigenfunctions of E, then multiply the nth eigenfunction by exp $(-iE_nt/\hbar)$ [see equation (73), Chap. 10]. We could use eq. (43) to expand ψ into a series of Hermite polynomials. Fortunately, however, the generating function gives us a ready-made expansion of this particular function. We obtain, setting $y_0 = 2\lambda$ and using eq. (30),

$$[\psi(y)]_{t=0} = \exp\left[-\left(\frac{y^2}{2} + \lambda^2\right)\right] \exp(2y\lambda - \lambda^2)$$

$$= \exp(-\lambda^2) \exp\left(-\frac{y^2}{2}\right) \sum_{n=0}^{\infty} h_n(y) \frac{\lambda^n}{n!} \quad (49)$$

The functions $e^{-y^2/2}h_n(y)$ are eigenfunctions of the Hamiltonian, belonging to an energy $E = (n + \frac{1}{2})\hbar\omega$. The wave function therefore becomes

$$\psi(y, t) = \exp\left(-\frac{i\omega t}{2}\right) \exp\left[-\left(\lambda^2 + \frac{y^2}{2}\right)\right] \sum_{n=0}^{\infty} h_n(y) \frac{(\lambda e^{-i\omega t})^n}{n!} \quad (50)$$

We now rewrite the above function, obtaining (with the aid of the definition of the generating function)

$$\psi(y, t) = \exp\left(-\frac{i\omega t}{2}\right) \exp\left[-\left(\lambda^2 + \frac{y^2}{2}\right)\right]$$
$$\exp(-\lambda^2 e^{-2i\omega t} + 2\lambda e^{-i\omega t} y) \quad (51a)$$

Writing $\lambda = y_0/2$, we obtain

$$\psi(y, t) = \exp\left(-\frac{i\omega t}{2}\right) \exp\left\{-\left[\frac{y^2}{2} + \frac{y_0^2}{4}(1 + e^{-2i\omega t}) - 2y\frac{y_0}{2}e^{-i\omega t}\right]\right\} \quad (51b)$$

$$= \exp\left(-\frac{i\omega t}{2}\right) \exp\left\{-\frac{1}{2}\left[y^2 - 2yy_0 \cos \omega t + \frac{y_0^2}{2}(1 + \cos 2\omega t)\right]\right\}$$
$$\exp\left[\frac{i}{2}\left(y_0^2 \frac{\sin 2\omega t}{2} - 2yy_0 \sin \omega t\right)\right] \quad (51c)$$

The probability density is

$$P = \psi^*\psi = \exp[-(y^2 - 2yy_0 \cos \omega t + y_0^2 \cos^2 \omega t)]$$
$$= \exp[-(y - y_0 \cos \omega t)^2] \quad (52)$$

We see that the center of the wave packet moves in the path of the classical motion (i.e., $y = y_0 \cos \omega t$). This result is more or less to be expected from the fact that Schrödinger's equation leads to the classical equations of the motion on the average. But there is an unusual feature of the motion of this packet, namely, it does not change its shape with time. Normally, we expect wave packets to spread out with time, but this particular packet does not. We cannot go completely into the reasons for this unusual behavior, but it can be said that this is so because of a peculiarity of the harmonic oscillator wave functions that is not duplicated in any other system. Some insight into the reason for this behavior can be gained by considering an arbitrary time-dependent harmonic oscillator wave function that has been expanded in terms of the eigenfunctions ψ_n

$$\psi = \sum_{n=0}^{\infty} C_n\psi_n(x)e^{-in\omega t}e^{-i\omega t/2} \tag{53}$$

Let us now form $P = \psi^*\psi$,

$$P = \sum_{m=0}^{\infty}\sum_{n=0}^{\infty} C_n^*C_m\psi_n^*(x)\psi_m(x)e^{i(n-m)\omega t} \tag{54}$$

We note that all parts of P are either constant in time or else oscillate harmonically with a frequency that is some multiple of the basic frequency, $\nu = \omega/2\pi$. After the passage of a basic period, the wave function therefore repeats itself. As a result, no wave packet ever spreads out indefinitely, because after a period it must return to its original shape.

It is clear that the periodicity of P (and also of ψ) arises from the fact that for a harmonic oscillator, the frequency of oscillation of each term in eq. (54) is a multiple of ν. For any system other than a harmonic oscillator, some parts of the wave function would oscillate with frequencies that were not multiples of the basic frequency, and the function would not be completely periodic.

Thus, we conclude that for a harmonic oscillator and only for a harmonic oscillator can we expect periodic wave packets. Yet, we may expect that, in general, even in a harmonic oscillator the shape of the wave packet will not remain absolutely constant with time. The particular wave packet that we have chosen is unusual, in that it has the same wave function as does the lowest state of the oscillator, except that its center has been displaced by an amount y_0. It may be shown that this property is the cause of the constant shape of P for this case, but we shall not show it here.

16. Mean Values of Kinetic and Potential Energies. To obtain the mean value of the potential energy, we use the identity

$$y = \frac{1}{2}\left(y + \frac{d}{dy}\right) + \frac{1}{2}\left(y - \frac{d}{dy}\right) \tag{55}$$

$$y^2 = \frac{1}{4}\left(y + \frac{d}{dy}\right)^2 + \frac{1}{4}\left(y - \frac{d}{dy}\right)^2 + \frac{1}{4}\left(y + \frac{d}{dy}\right)\left(y - \frac{d}{dy}\right) + \frac{1}{4}\left(y - \frac{d}{dy}\right)\left(y + \frac{d}{dy}\right) \tag{55a}$$

Note that we must be careful of order of factors in the latter part. We then obtain

$$y^2 = \frac{1}{4}\left(y + \frac{d}{dy}\right)^2 + \frac{1}{4}\left(y - \frac{d}{dy}\right)^2 + \frac{1}{2}\left(y^2 - \frac{d^2}{dy^2}\right) \tag{55b}$$

and $\int \psi_n^* y^2 \psi_n \, dx$

$$= \int \psi_n^* \left[\frac{1}{4}\left(y + \frac{d}{dy}\right)^2 + \frac{1}{4}\left(y - \frac{d}{dy}\right)^2 + \frac{1}{2}\left(y^2 - \frac{d^2}{dy^2}\right) \right] \psi_n \, dy \quad (56)$$

We now use the fact that

$$\left(y + \frac{d}{dy}\right)^2 \psi_n = 2\sqrt{n(n-1)}\, \psi_{n-2}$$

$$\left(y - \frac{d}{dy}\right)^2 \psi_n = 2\sqrt{(n+1)(n+2)}\, \psi_{n+2}$$

[See eqs. (38) and (39)]. From the fact that eigenfunctions corresponding to different n are orthogonal, we see that the first two terms vanish. The third term is just proportional to the energy, so that

$$\bar{V} = \frac{\hbar\omega}{2} \overline{x^2} = \frac{1}{2} \int_{-\infty}^{\infty} \psi_n^* \left[\frac{-\hbar^2}{2m} \frac{d^2}{dx^2} + \frac{m\omega^2 x^2}{2} \right] \psi_n \, dx = \frac{E_n}{2} \quad (57)$$

We conclude that the mean value of the potential energy is half the total energy, and since $E = T + V$, the mean kinetic energy must also be half the total energy. This is a result that also occurs with the classical harmonic oscillator, provided that we average over a period

Problem 2: Prove that $\bar{T} = \bar{V}$ for a classical harmonic oscillator.

Problem 3: Prove that the mean value of any odd power of x or p is zero, for any eigenstate of the oscillator. Prove that it is not necessarily zero for a linear combination of the first two eigenstates.

Problem 4: The wave packet in eq. (48) was so chosen that its mean momentum vanished at $t = 0$. Suppose we had chosen

$$[\psi(y)]_{t=0} = \exp\left(\frac{ip_0 x}{\hbar}\right) \exp\left[-\frac{1}{2}(y - y_0)^2\right]$$

where $x = \sqrt{\frac{\hbar}{m\omega}}\, y$. Evaluate the time-dependent wave function for this case.

Problem 5: By methods similar to those used for x^2, evaluate $\overline{x^4}$ and $\overline{p^4}$ for the case that $\psi = \psi_n(x)$, where n represents an arbitrary eigenvalue.

CHAPTER 14

Angular Momentum and Three-dimensional Wave Equation

IN THIS CHAPTER, we shall consider the problem of how to treat the three-dimensional wave equation. The method of separation of variables will be used, and in the process of solving the problem we shall investigate the properties of the angular-momentum operators and of the spherical harmonics, which are their eigenfunctions. Particular attention will be paid to the question of measurability of various components of the angular momentum and to the problem of describing orbits.

1. Separation of Variables. We shall begin with the description of the procedure used in separating variables. In three dimensions the wave equation takes the form

$$\nabla^2\psi + \frac{2m}{\hbar^2}(E - V)\psi = 0 \tag{1}$$

It very often happens that the potential is a function only of the distance, r, from some center of symmetry, which might, for example, be the center of an atom. In this case it is convenient to express Schrödinger's equation in terms of the spherical polar co-ordinates, r, ϑ, and φ. In any text in theoretical physics* it is shown that the Laplacian operator expressed in this way is

$$\nabla^2\psi = \frac{1}{r}\frac{\partial^2}{\partial r^2}(r\psi) + \frac{1}{r^2}\left[\frac{1}{\sin\vartheta}\frac{\partial}{\partial\vartheta}\sin\vartheta\frac{\partial}{\partial\vartheta} + \frac{1}{\sin^2\vartheta}\frac{\partial^2}{\partial\varphi^2}\right]\psi \tag{2}$$

Let us denote the operator in the brackets by Ω.

If V is a function only of r, we shall show that solutions can be obtained that are products of two functions, one of which involves only the radius, and the other of which involves ϑ and φ. To obtain such a solution, let us write tentatively

$$\psi = v(r)Y(\vartheta, \varphi) \tag{3}$$

Schrödinger's equation becomes

$$\left\{\frac{1}{r}\frac{\partial^2}{\partial r^2}[rv(r)] + \frac{2m}{\hbar^2}[E - V(r)]v(r)\right\}Y(\vartheta, \varphi) + \frac{v(r)}{r^2}\Omega Y(\vartheta, \varphi) = 0 \tag{4}$$

* See, for example, Slater and Frank, *Introduction to Theoretical Physics*. New York: McGraw-Hill Book Company, Inc., 1933.

Division by $\psi/r^2 = vY/r^2$ yields

$$r^2 \left\{ \frac{1}{vr} \frac{\partial^2}{\partial r^2} [rv(r)] + \frac{2m}{\hbar^2} [E - V(r)] \right\} = - \frac{\Omega Y(\vartheta, \varphi)}{Y(\vartheta, \varphi)}$$

In the above equation we ask that a function of r alone be equal to a function of ϑ and φ alone, *for all values of* r, ϑ, φ. This is possible only if each function is a constant. Let this constant be denoted by $-c$ We then obtain the following two equations:

$$\frac{1}{r} \frac{d^2}{dr^2} (rv) + \frac{2m}{\hbar^2} [E - V(r)]v = - \frac{cv}{r^2} \tag{5a}$$

$$\Omega Y(\vartheta, \varphi) = cY(\vartheta, \varphi) \tag{5b}$$

If we can obtain physically allowable solutions of the above two equa tions, then the product vY will be a solution of Schrödinger's equation. We shall see that, in general, well-behaved solutions will be possible only for certain values of c and of E, and furthermore we shall show that an arbitrary function can be expanded as a series of the products $v(r)Y(\vartheta, \varphi)$.

It should be noted, however, that the separation of the wave function into a product of a function of r and of a function of ϑ and φ depended on the fact that V was a function of r only. If V had involved r and ϑ together in an inextricable way, no such separation would have been possible.

Problem 1: Show that a separation into products involving spherical polar coordinates is possible if $V = f(r) + R(r)G(\vartheta, \varphi)$ provided that $R(r) = K/r^2$.

It is clear that in every problem in which V is a function of the radius only, the same angular functions, Y, will be required. We shall therefore defer considering the radial equation for a while, until we have solved for the allowed values of c, and the corresponding allowed eigenfunctions, $Y(\vartheta, \varphi)$. This latter may be done in many ways. For example, one can solve eq. (5b) in a manner similar to that used for the harmonic-oscillator equation, and one can thereby show that only certain values of c will lead to physically admissible functions, which have been named "spherical harmonics." We shall, however, use a somewhat more roundabout method, which has the advantages of giving a better physical picture, and of using simpler mathematical techniques. For the more direct method the reader is advised to consult texts on quantum theory,* electrodynamics, or mathematical analysis.

2. Angular Momentum. Let us begin by noting that the separation of Schrödinger's equation in spherical co-ordinates is very analogous to what is done classically when we write the Hamiltonian in spherical co-ordinates as follows:

* For a treatment of this method, see, for example, Pauling and Wilson, p. 118.

$$H = \frac{1}{2m}\left(p_r^2 + \frac{L^2}{r^2}\right) + V(r),$$

where p_r is the radial momentum $(m\dot{r})$ and L is the angular-momentum vector. By comparison with eq. (4), we note that L^2 enters into the classical Hamiltonian function in a way that is very analogous to that of Ω in the quantum-mechanical Hamiltonian operator. This suggests that Ω is the operator corresponding to L^2. In order to see whether or not this is true, we shall first make a brief study of the general properties of the angular-momentum operators.

The three components of angular momentum are the following:

$$\left.\begin{aligned} L_z &= xp_y - yp_x \qquad & L_x &= yp_z - zp_y \\ L_y &= zp_x - xp_z \qquad & L &= r \times p \end{aligned}\right\} \tag{6}$$

Note that L_y and L_z can be derived from L_x by cyclic permutation of the variables x, y, z and p_x, p_y, p_z. Quantum mechanically, we replace p_x by $\frac{\hbar}{i}\frac{\partial}{\partial x}$; similar replacements are made for p_y and p_z, obtaining

$$L_z = \frac{\hbar}{i}\left(x\frac{\partial}{\partial y} - y\frac{\partial}{\partial x}\right) \qquad L_x = \frac{\hbar}{i}\left(y\frac{\partial}{\partial z} - z\frac{\partial}{\partial y}\right) \qquad L_y = \frac{\hbar}{i}\left(z\frac{\partial}{\partial x} - x\frac{\partial}{\partial z}\right) \tag{7}$$

It is clear that the above operators are all Hermitean as they stand.

3. Commutation Rules for Angular Momentum. By direct computation, we obtain

$$\begin{aligned} (L_x, L_y) &= (L_xL_y - L_yL_x) \\ &= -\hbar^2\left[\left(y\frac{\partial}{\partial z} - z\frac{\partial}{\partial y}\right)\left(z\frac{\partial}{\partial x} - x\frac{\partial}{\partial z}\right) - \left(z\frac{\partial}{\partial x} - x\frac{\partial}{\partial z}\right)\left(y\frac{\partial}{\partial z} - z\frac{\partial}{\partial y}\right)\right] \\ &= -\hbar^2\left[y\frac{\partial}{\partial x} - x\frac{\partial}{\partial y}\right] = i\hbar L_z \end{aligned} \tag{8a}$$

By cyclical permutation, we obtain

$$(L_y, L_z) = i\hbar L_x \qquad (L_z, L_x) = i\hbar L_y \tag{8b}$$

The above commutation rules can be combined symbolically as follows:*

$$L \times L = i\hbar L \tag{9}$$

Note that two different components of the angular momentum do not commute. Hence it will not be possible, in general, to measure L_x, L_y, and L_z simultaneously, because, as shown in eq. (13), Chap. 10, the

* If L were a numerical vector, $L \times L$ would vanish. But the components of L are operators which do not commute, so that $L \times L$ need not vanish.

product of the uncertainties of two quantities is proportional to the mean value of their commutator. We shall return to this point later.

4. Total Angular Momentum. The absolute value of the angular momentum (or the total angular momentum) $|L|$ is defined by the relation

$$L^2 = L_x^2 + L_y^2 + L_z^2 \tag{10}$$

It is of interest to obtain the commutation relations of L^2 with the components, L_x, L_y, L_z. Let us take, for example,

$$\begin{aligned}(L^2, L_z) = (L^2L_z - L_zL^2) &= (L_x^2 + L_y^2)L_z - L_z(L_x^2 + L_y^2) \\ &= L_x(L_xL_z - L_zL_x) + (L_xL_z - L_zL_x)L_x \\ &\qquad + L_y(L_yL_z - L_zL_y) + (L_yL_z - L_zL_y)L_y\end{aligned}$$

Applying the commutation relations (8), we obtain

$$-i\hbar(L_xL_y + L_yL_x - L_yL_x - L_xL_y) = 0 = (L^2, L_z) \tag{11}$$

Thus we conclude that L^2 commutes with L_z. By symmetry, we conclude that it also commutes with L_x and L_y. In other words, it is possible to measure simultaneously L^2 and any *single* component of L. Because the components, L_x, L_y, and L_z do not commute with each other, however, not more than one of these at a time can be specified independently.

5. Angular Momentum in Polar Co-ordinates. At this point it is convenient to represent the Cartesian components of L in terms of spherical polar co-ordinates. To do this, we note that

$$\left.\begin{array}{lll} x = r\sin\vartheta\cos\varphi & y = r\sin\vartheta\sin\varphi & z = r\cos\vartheta \\ r^2 = x^2 + y^2 + z^2 & \cos\vartheta = \dfrac{z}{r} & \tan\varphi = \dfrac{y}{x} \end{array}\right\} \tag{12}$$

We wish to express $\partial/\partial x$, $\partial/\partial y$, $\partial/\partial z$ in terms of $\partial/\partial r$, $\partial/\partial\vartheta$, $\partial/\partial\varphi$. We shall need the following expressions, the verification of which will be left as an exercise.

$$\left.\begin{array}{lll} \dfrac{\partial r}{\partial x} = \sin\vartheta\cos\varphi & \dfrac{\partial r}{\partial y} = \sin\vartheta\sin\varphi & \dfrac{\partial r}{\partial z} = \cos\vartheta \\ \dfrac{\partial\vartheta}{\partial x} = \dfrac{1}{r}\cos\vartheta\cos\varphi & \dfrac{\partial\vartheta}{\partial y} = \dfrac{1}{r}\cos\vartheta\sin\varphi & \dfrac{\partial\vartheta}{\partial z} = -\dfrac{1}{r}\sin\vartheta \\ \dfrac{\partial\varphi}{\partial x} = -\dfrac{1}{r}\dfrac{\sin\varphi}{\sin\vartheta} & \dfrac{\partial\varphi}{\partial y} = \dfrac{1}{r}\dfrac{\cos\varphi}{\sin\vartheta} & \dfrac{\partial\varphi}{\partial z} = 0 \end{array}\right\} \tag{13}$$

With these relations, we obtain

$$\begin{aligned}\frac{\partial f}{\partial x} &= \frac{\partial f}{\partial r}\frac{\partial r}{\partial x} + \frac{\partial f}{\partial\vartheta}\frac{\partial\vartheta}{\partial x} + \frac{\partial f}{\partial\varphi}\frac{\partial\varphi}{\partial x} \\ &= \sin\vartheta\cos\varphi\frac{\partial f}{\partial r} + \frac{1}{r}\cos\vartheta\cos\varphi\frac{\partial f}{\partial\vartheta} - \frac{1}{r}\frac{\sin\varphi}{\sin\vartheta}\frac{\partial f}{\partial\varphi}\end{aligned} \tag{14}$$

Similar expressions follow for $\partial f/\partial y$ and $\partial f/\partial z$. Finally, we obtain

$$\left.\begin{aligned} L_z &= \frac{\hbar}{i}\frac{\partial}{\partial\varphi} \\ L_x &= -\frac{\hbar}{i}\left(\sin\varphi\frac{\partial}{\partial\vartheta} + \cot\vartheta\cos\varphi\frac{\partial}{\partial\varphi}\right) \\ L_y &= \frac{\hbar}{i}\left(\cos\varphi\frac{\partial}{\partial\vartheta} - \cot\vartheta\sin\varphi\frac{\partial}{\partial\varphi}\right) \end{aligned}\right\} \tag{15}$$

Problem 2: Verify the above relations for $\partial r/\partial x$, $\partial\vartheta/\partial x$, etc., and also obtain the above results for $\boldsymbol{L}$. Also, verify directly by differentiating that $(L_x, L_y) = i\hbar L_z$.

By using the above results for $\boldsymbol{L}$, we can compute L^2 in spherical polar co-ordinates, and obtain the result

$$L^2 = -\hbar^2\Omega \tag{16}$$

where Ω is the operator defined in eq. (2). The problem of obtaining the allowed values of c is therefore equivalent to that of obtaining the eigenvalues of the (Hermitean) operator L^2. Furthermore, we can rewrite Schrödinger's equation in the following form

$$\left[-\frac{\hbar^2}{2m}\left(\frac{\partial^2}{\partial r^2} + \frac{2}{r}\frac{\partial}{\partial r}\right) + \frac{L^2}{2mr^2} + V\right]\psi = H\psi = E\psi \tag{17}$$

6. Constants of the Motion. Now, if V is not a function of ϑ and φ, L^2 and $\boldsymbol{L}$ will commute with the Hamiltonian operator. To see this, note from eq. (17) that the Hamiltonian now contains ϑ and φ only in the form of the operator L^2. This operator commutes with $\boldsymbol{L}$ and, of course, with L^2, as does everything else in the Hamiltonian. One can therefore define the following three quantities simultaneously:

(1) The eigenvalue of H (i.e., the energy).

(2) The eigenvalue of L^2 [i.e., the total angular momentum (squared)].

(3) The eigenvalue of any component of $\boldsymbol{L}$ (e.g., the z component of the angular momentum).

Furthermore, since $\boldsymbol{L}$ and L^2 commute with H, their average values remain constant with time [see Chap. 9, eq. (37)]. $\boldsymbol{L}$ and L^2 are therefore quantum-mechanical constants of the motion when $V = V(r)$, just as they are classically. Once we fix the eigenvalues of L^2 and of some component of $\boldsymbol{L}$, these values will not change with time.

7. Eigenvalues of L_z. One can choose the z axis to be in any desired direction. Having done so, we now try to find the eigenfunctions of L_z. That is, we want to satisfy the equation

$$L_z\psi = \frac{\hbar}{i}\frac{\partial\psi}{\partial\varphi} = c\psi \tag{18}$$

The solution is

$$\psi = e^{ic\varphi/\hbar}f(r, \vartheta) \tag{19}$$

where $f(r, \vartheta)$ is an arbitrary function of r and ϑ.

Now ψ must be a single-valued function of x, y, z. It must, therefore, be periodic in φ, with period 2π. This is possible only if $c/\hbar = m$, where m is an integer. Thus, the eigenvalues of L_z are*

$$L_z = m\hbar \tag{20}$$

The eigenfunctions are

$$\psi = e^{im\varphi} f(\vartheta, r) \tag{21}$$

where $f(\vartheta, r)$ is an arbitrary function of ϑ and r.

8. Expansion Postulate. The expansion postulate (see Chap. 10, Sec. 22) states that an arbitrary wave function can be expanded as a series of eigenfunctions of the Hermitean operator L_z. We see that this is indeed true, since an arbitrary single-valued function of x, y, z can be expressed as a Fourier series, with the aid of the functions in eq. (21).

9. Simultaneous Eigenfunctions of L_z and L^2. Let us now take the above eigenfunctions of L_z and insert them into the equation defining an eigenfunction of L^2. We obtain [using eqs. (2) and (16)]

$$\begin{aligned} L^2 e^{im\varphi} f(r, \vartheta) &= -\hbar^2 \Omega e^{im\varphi} f(r, \vartheta) \\ &= -\hbar^2 \left[\frac{1}{\sin \vartheta} \frac{\partial}{\partial \vartheta} \sin \vartheta \frac{\partial}{\partial \vartheta} + \frac{1}{\sin^2 \vartheta} \frac{\partial^2}{\partial \varphi^2} \right] e^{im\varphi} f(r, \vartheta) \\ &= -\hbar^2 \left[\frac{1}{\sin \vartheta} \frac{\partial}{\partial \vartheta} \left(\sin \vartheta \frac{\partial}{\partial \vartheta} \right) - \frac{m^2}{\sin^2 \vartheta} \right] e^{im\varphi} f(r, \vartheta) \\ &= c e^{im\varphi} f(r, \vartheta) \end{aligned}$$

This leads to a differential equation defining f:

$$-\hbar^2 \left[\frac{1}{\sin \vartheta} \frac{\partial}{\partial \vartheta} \left(\sin \vartheta \frac{\partial}{\partial \vartheta} \right) - \frac{m^2}{\sin^2 \vartheta} \right] f = cf \tag{22}$$

The eigenfunctions of this equation will be denoted by $f_c^m(\vartheta)$. It is clear then that the products $\psi_c^m = f_c^m(\vartheta) e^{im\varphi}$ are simultaneous eigenfunctions of L^2 and L_z. According to the expansion postulate, an arbitrary function of ϑ can be expanded as a series of eigenfunctions f_c^m, so that the products ψ_c^m can then represent an arbitrary function of ϑ and φ.

10. Determination of Simultaneous Eigenvalues and Eigenfunctions of L_z and L^2. We must now determine the allowed values of L^2 and the associated expressions for the allowed eigenfunctions $f_c^m(\vartheta)$, when

* In Chap. 17, Sec. 3, we shall see that, in general, the eigenvalues of the operator corresponding to a particular component of the angular momentum may be either integral or half-integral multiples of $\hbar$. The restriction of orbital angular momenta to integral multiples of $\hbar$ arises from the requirement of a single-valued wave function. The angular momentum arising from electron spin is, however, $\hbar/2$. The requirement of a single-valued wave function does not extend to the spin variables, which are concerned with "internal" properties of the electron, rather than with its spatial location.

$$L_z = m\hbar\dagger$$

We begin by noting that if L^2 has a definite numerical value, this value is not changed, if we operate on ψ either with L_x, L_y, or L_z. To prove this, we use the fact that any component of L commutes with L^2. Thus, we begin with

$$L^2\psi = c\psi \qquad \text{(for an eigenfunction)} \tag{23}$$

Operating with any component of L, say L_z, we obtain

$$L^2L_z\psi = L_zL^2\psi = L_zc\psi = cL_z\psi$$

or

$$L^2(L_z\psi) = c(L_z\psi) \tag{24}$$

As a result, if ψ is an eigenfunction of L^2, $(L_z\psi)$ is also an eigenfunction, belonging to the *same eigenvalue of* L^2. The same is true of $(L_x\psi)$ and $(L_y\psi)$.

Let us now suppose that $L_z\psi_m = \hbar m\psi_m$, i.e., that ψ_m is simultaneously an eigenfunction of L^2 and L_z. On multiplying the above equation by the operator $(L_x + iL_y)$, we obtain

$$(L_x + iL_y)L_z\psi_m = \hbar m(L_x + iL_y)\psi_m \tag{25}$$

We now use the commutation rules (8) and are led to the following:

$$(L_x + iL_y)L_z - L_z(L_x + iL_y) = -\hbar(L_x + iL_y) \tag{26}$$

Equation (25) can then be rewritten

$$L_z(L_x + iL_y)\psi_m = \hbar(m + 1)(L_x + iL_y)\psi_m \tag{27}$$

From this, we conclude that if ψ_m is an eigenfunction, in which $L_z = m\hbar$, then the function $(L_x + iL_y)\psi_m$ is an eigenfunction of L_z belonging to $L_z = (m + 1)\hbar$, but to the same value of L^2. Thus, if we start with a given eigenfunction, we can always generate new eigenfunctions of L_z belonging to the same value of L^2.

In a similar way, one can show that

$$L_z(L_x - iL_y)\psi_m = \hbar(m - 1)(L_x - iL_y)\psi_m \tag{28}$$

The operator $L_x - iL_y$ therefore reduces the value of m by unity, but also leaves L^2 unchanged.

Repeated application of the operator $L_x + iL_y$ will enable us to generate eigenfunctions of a fixed L^2 belonging to indefinitely large eigenvalues of L_z unless there is some value of m for which $(L_x + iL_y)\psi_m$ vanishes. Similarly, repeated application of $L_x - iL_y$ will lead to indefinitely large negative values of m unless $(L_x - iL_y)\psi_m$ vanishes for some ψ_m. The situation is rather similar to the one met in the problem of the harmonic oscillator.

† $L_z = m\hbar$ is a standard abbreviated way of writing $L_z\psi_m = m\hbar\psi_m$. Because L_z has a definite value when ψ_m is an eigenfunction of L_z, the statement has also the meaning that the variable L_z is equal to $m\hbar$.

Is there any reason why indefinitely large $|m|$ cannot be associated with a fixed value of L^2? That there is such a reason can be seen from the definition

$$L^2 = L_x^2 + L_y^2 + L_z^2 = L_x^2 + L_y^2 + m^2\hbar^2 \tag{29}$$

Now, it is readily shown that the mean value of L_x^2 and of L_y^2 must always be positive.

Problem 3: Prove that the mean value of L_x^2 is always positive.
HINT: Rotate the axes so that the new z axis is parallel to the old x axis. Then

$$L_x^2 = -\hbar^2 \frac{\partial^2}{\partial\varphi^2}$$

In a state in which L^2 and L_z have definite values, the following must therefore hold:

$$\hbar^2 m^2 \leq L^2 \tag{30}$$

This means that $|\hbar m|$ must not be allowed to grow larger than $\sqrt{L^2}$. Therefore, in the state in which $|m|$ is as large as is consistent with a given L^2, we must have either

$$(L_x + iL_y)\psi_{m_1} = 0 \qquad \text{or} \qquad (L_x - iL_y)\psi_{m_2} = 0$$

where m_1 is the maximum positive value of $L_z/\hbar$ consistent with a given L^2 and m_2 is the maximum negative value.

In order to find the relation between L^2 and m_1 and m_2, we consider the expression

$$\begin{aligned} L^2\psi_{m_1} = (L_x^2 + L_y^2 + L_z^2)\psi_{m_1} &= [(L_x - iL_y)(L_x + iL_y) + L_z^2 + \hbar L_z]\psi_{m_1} \\ &= [(L_x - iL_y)(L_x + iL_y) + \hbar^2(m_1^2 + m_1)]\psi_{m_1} \end{aligned} \tag{31}$$

Since

$$(L_x + iL_y)\psi_{m_1} = 0 \tag{32}$$

it follows that

$$L^2\psi_{m_1} = \hbar^2 m_1(m_1 + 1)\psi_{m_1} \tag{33}$$

It is readily shown in a similar way that when $(L_x - iL_y)\psi_{m_2} = 0$, we obtain

$$L^2\psi_{m_2} = \hbar^2 m_2(m_2 - 1)\psi_{m_2} \tag{34}$$

If eqs. (33) and (34) are to be true simultaneously, then we must have* $m_2 = -m_1$. In other words, the maximum negative value of m is just the negative of the maximum positive value.

It is customary to designate m_1 by the integer l. With this notation, we obtain

$$L^2 = \hbar^2 l(l + 1) \tag{35}$$

For any given value of l, the allowed value of m is any positive or negative integer between $\pm l$, including zero.

* Another solution is $m_2 = m_1 + 1$, but this is inadmissible, because by hypothesis m_1 is the largest positive value of $L_z/\hbar$.

11. Vector Representation of Allowed Angular Momenta. Consider a case in which l is some fixed number. Then the total angular momentum may be represented by a vector of length

$$\frac{L}{\hbar} = \sqrt{l(l+1)} \tag{36}$$

The component m in the z direction may be any integer, up to and including $\pm l$. The number m is called the "azimuthal quantum number," while l is the "total angular momentum quantum number." The possible values of the component m can be represented schematically as the projections of $L/\hbar$ on the z axis. The vector diagram in Fig. 1 illustrates the possible states for $l = 2$ and $L/\hbar = \sqrt{6} \cong 2.5$.

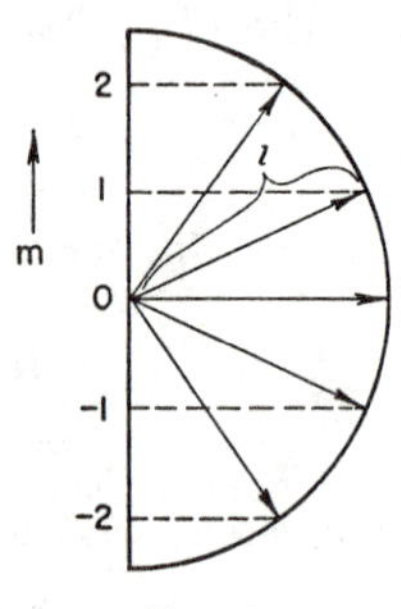

FIG. 1

Since we know that L_x and L_y are undefined within the limitation that $L_x^2 + L_y^2 = L^2 - m^2\hbar^2$, one should think of this vector as being distributed at random over all azimuthal angles consistent with the known value of the projection of L on the z axis. The vector L should therefore be thought of as covering a cone, with vector angle given by

$$\cos\vartheta = \frac{m}{\sqrt{l(l+1)}}$$

12. Effect of Fluctuation in Direction of L. It is important to note that even when $m = \pm l$, the angular momentum does not point exactly in the z direction, but that it has residual x and y components, which are not completely definable. This arises from the fact that L_x and L_y do not commute with L_z, so that they cannot be fixed at zero in a state in which L_z is definite. It is therefore unavoidable that there be some fluctuation of L_x and L_y, which contributes to $L^2 = L_x^2 + L_y^2 + L_z^2$. To describe this fluctuation quantitatively, we first note that whenever the wave function is an eigenfunction of L^2 and L_z, the mean values of L_x and L_y are zero. This corresponds to the fact that L covers a cone of directions, the axis of which is in the z direction. Thus, the fluctuations in L_x and L_y are respectively

$$\overline{(\Delta L_x)^2} = \overline{(L_x - \bar{L}_x)^2} = \overline{L_x^2} - (\overline{L_x})^2 = \overline{L_x^2}$$
$$\overline{(\Delta L_y)^2} = \overline{(L_y - \bar{L}_y)^2} = \overline{L_y^2} - (\overline{L_y})^2 = \overline{L_y^2}$$

We then obtain

$$\overline{L^2} = \overline{L_x^2} + \overline{L_y^2} + \overline{L_z^2} = \overline{(\Delta L_x)^2} + \overline{(\Delta L_y)^2} + \hbar^2 m^2 = \hbar^2 l(l+1)$$

or

$$\overline{(\Delta L_x)^2} + \overline{(\Delta L_y)^2} = \hbar^2(l^2 + l - m^2) \tag{37}$$

From the above, we see that the fluctuation in the components of $\boldsymbol{L}$ which are normal to z will be a minimum when $m = l$. We therefore obtain

$$[\overline{(\Delta L_x)^2} + \overline{(\Delta L_y)^2}]_{\min} = l\hbar^2$$

This means in a rough manner of speaking that the minimum angle between the direction of the $\boldsymbol{L}$ vector and the z axis is given by

$$\sin \vartheta_{\min} = \frac{\sqrt{l}}{\sqrt{l(l+1)}} = \frac{1}{\sqrt{l+1}} \tag{38}$$

It is not correct, however, to imagine that the angular momentum points in some definite direction which we do not happen to be able to measure with complete precision. Instead, whenever L^2 and L_z have definite values, one should imagine that the entire cone of directions corresponding to those values of L_x and L_y consistent with given L^2 and L_z are covered simultaneously because as we shall see in Secs. 17 and 18, important physical consequences may follow from the effects of interference of wave functions corresponding to different components of angular momentum.

In the limit of high quantum numbers, the angle $\vartheta_{\min}$ becomes very small, and the angular momentum vector can then point in a fairly well-defined direction.

13. Eigenfunctions of L^2 and L_z. We can find the eigenfunctions by a method similar to that used for the harmonic oscillator. We note that if we obtain a single eigenfunction of these two operators, then all other eigenfunctions corresponding to the same value of L^2 can be obtained by repeated application of the operators $(L_x + iL_y)$ and $(L_x - iL_y)$. The first eigenfunction, corresponding to $m = l$, can be obtained, however, from the condition (eq. 32) that*

$$(L_x + iL_y)\psi_l^l = 0 \tag{39}$$

From eqs. (15) we express L_x and L_y in terms of ϑ and φ, obtaining

$$L_x + iL_y = \hbar e^{i\varphi}\left(\frac{\partial}{\partial\vartheta} + i \cot \vartheta \frac{\partial}{\partial\varphi}\right) \tag{40}$$

Note that for $m = l$, $\psi_l^l = f_l^l(\vartheta)e^{il\varphi}$, so that $i\dfrac{\partial\psi_l^l}{\partial\varphi} = -l\psi_l^l$. Equation (40) then becomes

$$\frac{\partial\psi_l^l}{\partial\vartheta} = l \cot \vartheta \psi_l^l$$

Integration yields

$$\ln \psi_l^l = l \ln (\sin \vartheta) + k(\varphi) \qquad \text{or} \qquad \psi_l^l = g(\varphi)(\sin \vartheta)^l \tag{41}$$

* The notation, ψ_l^m means that $L_z/\hbar = m$, $L^2/\hbar^2 = l(l+1)$.

$g(\varphi)$ is an arbitrary function of φ, but it must be chosen to make ψ_l^l an eigenfunction of L_z with eigenvalue $\hbar l$. This means that $g(\varphi) = e^{il\varphi}$. Thus, we obtain

$$\psi_l^l = e^{il\varphi} (\sin \vartheta)^l \tag{42}$$

In order to obtain values of ψ_l^m corresponding to smaller values of m, we must operate with $(L_x - iL_y)$. From eq. (15), we obtain:

$$-(L_x - iL_y) = \hbar e^{-i\varphi} \left(\frac{\partial}{\partial \vartheta} - i \cot \vartheta \frac{\partial}{\partial \varphi} \right) \tag{43}$$

With the aid of the relation $-i \dfrac{\partial \psi_l^l}{\partial \varphi} = l\psi_l^l$

we obtain

$$-(L_x - iL_y)\psi_l^l = \hbar e^{-i\varphi} \left(\frac{\partial}{\partial \vartheta} + l \cot \vartheta \right) \psi_l^l \tag{44}$$

Using the relation

$$\left(\frac{\partial}{\partial \vartheta} + l \cot \vartheta \right) \psi_l^l = \frac{1}{(\sin \vartheta)^l} \left[\frac{\partial}{\partial \vartheta} (\sin \vartheta)^l \psi_l^l \right]$$

we get*

$$\psi_l^{l-1} = \frac{1}{(\sin \vartheta)^l} e^{i(l-1)\varphi} \frac{\partial}{\partial \vartheta} (\sin \vartheta)^{2l} \tag{45}$$

To obtain ψ_l^{l-2}, we note that

$$\begin{aligned} (L_x - iL_y)\psi_l^{l-1} &= e^{-i\varphi} \left[\frac{\partial}{\partial \vartheta} + (l-1) \cot \vartheta \right] \psi_l^{l-1} \\ &= \frac{e^{-i\varphi}}{(\sin \vartheta)^{l-1}} \frac{\partial}{\partial \vartheta} (\sin \vartheta)^{l-1} \psi_l^{l-1} \end{aligned}$$

Thus, we get

$$\psi_l^{l-2} = \frac{1}{(\sin \vartheta)^{l-1}} e^{i(l-2)\varphi} \frac{\partial}{\partial \vartheta} \left[\frac{\frac{\partial}{\partial \vartheta} (\sin \vartheta)^{2l}}{\sin \vartheta} \right]$$

Repeated application of $L_x - iL_y$ yields

$$\psi_l^{l-s} = \frac{e^{i(l-s)\varphi}}{(\sin \vartheta)^{l-s}} \left(\frac{1}{\sin \vartheta} \frac{\partial}{\partial \vartheta} \right)^s (\sin \vartheta)^{2l} \tag{46}$$

This result can be simplified with the substitutions,

$$\cos \vartheta = \zeta \qquad \frac{\partial}{\partial \vartheta} = - \sin \vartheta \frac{\partial}{\partial \zeta}$$

We obtain

$$\psi_l^{l-s} = \frac{e^{i(l-s)\varphi}}{(1 - \zeta^2)^{(l-s)/2}} \frac{\partial^s}{\partial \zeta^s} (1 - \zeta^2)^l \tag{47}$$

* We are neglecting factors of $\hbar$ and $-$ signs, since these will eventually be absorbed in the normalization coefficient.

These are the unnormalized eigenfunctions of L_z and L^2.

14. Legendre Polynomials. Let us begin to study these eigenfunctions for the special case $l = s$. In this case, the z component of the angular momentum is zero.

$$\psi_l^0 = \frac{\partial^l}{\partial \zeta^l} (1 - \zeta^2)^l \tag{48}$$

Because L_z is zero, these functions do not involve φ. Since $(1 - \zeta^2)^l$ is a polynomial of degree $2l$, and since each differentiation lowers the degree of a polynomial by one, it can be seen that ψ_l^0 is a polynomial of degree l. These polynomials are called Legendre polynomials (except for a constant factor), and will hereafter be denoted by $P_l(\zeta)$. Thus we get

$$\psi_l^0 \sim P_l(\zeta) \sim \frac{d^l}{d\zeta^l} (1 - \zeta^2)^l \tag{48a}$$

The following are a few of the properties of Legendre Polynomials:

(1) *Orthogonality.* Because Legendre polynomials are eigenfunctions of the Hermitean operators L_z and L^2, belonging to different eigenvalues of L^2, it may be expected that polynomials with different l are orthogonal.* Since the polynomials are not functions of the radius, it is necessary merely to integrate over the solid angle. Thus (writing $\zeta = \cos \vartheta$)

$$\int P_l(\cos \vartheta) P_m(\cos \vartheta)\, d\Omega = \int_0^{2\pi} d\varphi \int_0^{\pi} P_l(\cos \vartheta) P_m(\cos \vartheta) \sin \vartheta \, d\vartheta$$

Noting that

$$\sin \vartheta \, d\vartheta = -d\zeta$$

we obtain

$$\int_{-1}^{1} P_l(\zeta) P_m(\zeta)\, d\zeta = 0 \tag{49}$$

if $l \neq m$. This result can be proved directly from the definition of P_l [eq. (48a)].

Problem 4: Prove the orthogonality of different Legendre polynomials directly. HINT: Suppose $l > m$ and integrate by parts l times.

(2) *Differential Equation for Legendre Polynomials.* We have seen that the Legendre polynomial of degree l satisfies the equation

$$L^2 P_l = \hbar^2 l(l + 1) P_l$$

L^2 is given by eqs. (2) and (16). Since P_l is not a function of φ, the term involving $\partial/\partial\varphi$ vanishes. We obtain

$$-\frac{1}{\sin \vartheta} \frac{\partial}{\partial \vartheta} \left(\sin \vartheta \frac{\partial P_l}{\partial \vartheta} \right) = l(l + 1) P_l \tag{50}$$

* Chap. 10, Sec. 24.

Writing $\cos \vartheta = \zeta$, we get

$$\frac{d}{d\zeta}\left[(1-\zeta^2)\frac{dP_l}{d\zeta}\right] + l(l+1)P_l = 0 \tag{51}$$

This is Legendre's equation. The Legendre polynomials are usually obtained by solving this equation* and choosing only those solutions which are regular in the region $-1 \leq \zeta \leq 1$.

(3) *Normalization of Legendre Polynomials.* The Legendre polynomials can be normalized by methods similar to those used for the Hermite polynomials. We shall only quote the results here. $P_n(\zeta)$ is usually defined as follows:†

$$P_n(\zeta) = \frac{1}{2^n n!}\frac{d^n}{d\zeta^n}(\zeta^2-1)^n \quad \text{(Formula of Rodrigues)} \tag{52}$$

With this definition, the normalization can be obtained from the relations

$$\int_{-1}^{1} [P_n(\zeta)]^2\, d\zeta = \frac{2}{2n+1} \tag{52a}$$

(4) *A Generating Function for Legendre Polynomials.* Just as with Hermite polynomials, we can obtain a generating function for Legendre polynomials.‡ We shall not, however, carry out this work here, but merely quote the result.

$$V = \sum_0^\infty t^n P_n(\zeta) = \frac{1}{[1-2t\zeta+t^2]^{1/2}} \tag{53}$$

(5) *Recurrence Formulas.* From this function, one can get recurrence formulas. For example, let us differentiate with respect to t. We obtain, after a little rearranging, the following equation

$$(1-2\zeta t+t^2)\sum_{n=0}^{\infty} nt^{n-1}P_n(\zeta) = (\zeta - t)\sum_0^\infty t^n P_n(\zeta)$$

Equating coefficients of t^n yields

$$(n+1)P_{n+1}(\zeta) - (2n+1)\zeta P_n(\zeta) + nP_{n-1}(\zeta) = 0 \tag{54a}$$

Similarly, from the relation

$$t\frac{\partial V}{\partial t} = (\zeta - t)\frac{\partial V}{\partial \zeta}$$

we obtain

$$\zeta\frac{dP_n}{d\zeta} - \frac{dP_{n-1}}{d\zeta} = nP_n \tag{54b}$$

Still other formulas can be obtained in a similar way.

* See, for example, E. T. Copson, *Theory of Functions of a Complex Variable.* London: Clarendon Press, 1935, p. 273.

† *Ibid.*, p. 275.

‡ *Ibid.*, p. 277.

(6) *First Few Legendre Polynomials.* From eq. (53) we find

$$P_0(\zeta) = 1 \qquad P_2(\zeta) = \frac{3\zeta^2 - 1}{2} \tag{55}$$
$$P_1(\zeta) = \zeta$$

(7) *Expansion Postulate.* The Legendre polynomials are all eigenfunctions of L_z and L^2 belonging to the eigenvalue $L_z = 0$. From the expansion postulate* it then follows that an arbitrary function of

$$\zeta = \cos\vartheta$$

can be expanded as a series of Legendre polynomials. Thus,

$$f(\zeta) = \sum_n c_n P_n(\zeta) \tag{56}$$

To obtain c_m, when $f(\zeta)$ is given, multiply by $P_m(\zeta)$ and integrate over ζ from -1 to 1, using orthogonality and normalization of the $P_n(\zeta)$. We get

$$c_m = \frac{2m+1}{2} \int_{-1}^{1} P_m(\zeta) f(\zeta)\, d\zeta \tag{57}$$

(8) *Nodes.* Because $P_l(\zeta)$ is a polynomial of degree l, it must have l zeros. One can show† that these zeros all lie between $\zeta = -1$ and $\zeta = +1$. The wave function therefore oscillates l times in the range of angles between $\vartheta = 0$ and $\vartheta = \pi$.

15. Associated Legendre Functions. The Legendre polynomials are the eigenfunctions of L^2 with $L_z = m\hbar = 0$. The eigenfunctions for other values of m are given in eq. (47). Let us denote the eigenfunctions belonging to $L^2 = l(l+1)\hbar$ and $L_z = m\hbar$ by the symbol $P_l^m(\zeta)$; the complete wave function is then $P_l^m(\zeta)e^{im\varphi}$.

An alternative expression for $P_l^m(\zeta)$ can be obtained from the expansion given in eq. (47). We found that ψ_l^0 corresponded to $L_z = m\hbar = 0$. We now substitute $m = l - s$ in eq. (47),

$$\psi_l^m = \frac{e^{im\varphi}}{(1-\zeta^2)^{m/2}} \frac{\partial^{l-m}}{\partial\zeta^{l-m}} (1-\zeta^2)^l \tag{58}$$

If we consider negative values of m, then we can use eq. (48) and obtain

$$\psi_l^{-m} \sim e^{-im\varphi}(1-\zeta^2)^{m/2} \frac{d^m}{d\zeta^m} P_l(\zeta) \tag{59}$$

This means that

$$P_l^{-m}(\zeta) = (1-\zeta^2)^{m/2} \frac{d^m}{d\zeta^m} P_l(\zeta) \tag{60}$$

* Chap. 10, Sec. 22.

† By reference to (48a), we see that $P_l(\zeta)$ is the lth derivative of a function which has zeros of the lth order at $\zeta = +1$ and $\zeta = -1$. The first derivative must then have a zero between $+1$ and -1, the next derivative two, as the reader will readily verify, and we see that the lth derivative has l zeros.

Since we could equally well have started at $m = -l$ in Sec. 13, and worked up, it is clear that we can also obtain

$$\psi_l^m \sim e^{im\varphi} P_l^m(\zeta) \tag{61}$$

$P_l^m(\zeta)$ are known as the "associated Legendre functions."

Note that $$P_l^{-m}(\zeta) = P_l^m(\zeta)$$

The following are a few of the significant properties of the associated Legendre functions:

(1) *Normalization.* The normalizing coefficient for the $P_l^m(\zeta)$ can be obtained from a relation which we shall quote here without proof:†

$$\int_{-1}^{1} [P_l^m(\zeta)]^2 \, d\zeta = \frac{(l + |m|)!}{(l - |m|)!} \frac{2}{2l + 1} \tag{62}$$

(2) *Surface Harmonics.* According to the expansion postulate, the most general function of ϑ and φ can be expressed in terms of a series of the functions

$$P_l^m(\cos \vartheta) e^{im\varphi}$$

These functions are simultaneous eigenfunctions of the commuting operators L^2 and L_z, and since they are not degenerate (i.e., either l or m is different for each function), they are all orthogonal. The normalized functions are denoted by $Y_l^m(\vartheta, \varphi)$. We can therefore write for an arbitrary function

$$\psi(\vartheta, \varphi) = \sum_{l,m} c_l^m Y_l^m(\vartheta, \varphi) \tag{63}$$

To evaluate c_r^s, we multiply by Y_r^{s*} and integrate over the solid angle, using the orthogonality and normalization properties:

$$c_r^s = \int \psi(\vartheta, \varphi) Y_r^{s*}(\vartheta, \varphi) \, d\Omega \tag{64}$$

where $$d\Omega = \sin \vartheta \, d\vartheta \, d\varphi$$

(3) *Differential Equation Satisfied by Associated Legendre Functions.* Since $P_l^m(\zeta)e^{im\varphi}$ is an eigenfunction of $L^2/\hbar^2$ with eigenvalue $l(l + 1)$, and of $L_z/\hbar$ with eigenvalue m, it must satisfy the equation

$$L^2 P_l^m(\zeta) e^{im\varphi} = \hbar^2 l(l + 1) P_l^m(\zeta) e^{im\varphi}$$

The operator L^2 is obtained from eqs. (2) and (16), noting that

$$\frac{\partial^2}{\partial \varphi^2} e^{im\varphi} = -m^2 e^{im\varphi}$$

The resulting equation is

$$\frac{d}{d\zeta}\left[(1 - \zeta^2) \frac{dP_l^m(\zeta)}{d\zeta}\right] + \left[l(l + 1) - \frac{m^2}{1 - \zeta^2}\right] P_l^m(\zeta) = 0 \tag{65}$$

† Copson, p. 281.

The associated Legendre functions are often obtained by solving this equation and finding the eigenvalues.*

(4) *General Form of the* $P_l^m(\zeta)$. We have already seen that the $P_l(\zeta)$ are simply polynomials, which have l roots in the region from $\zeta = -1$ to $\zeta = +1$. From eq. (58) we see that the $P_l^m(\zeta)$ consist of a polynomial $\left[\frac{d^m}{d\zeta^m} P_l(\zeta)\right]$ of degree $l - m$, multiplying the factor $(1 - \zeta^2)^{m/2} = (\sin \vartheta)^m$. The reader will readily verify† that this polynomial has $l - m$ roots in the range from $\zeta = -1$ to 1; hence the function does not oscillate as often as when $m = 0$. The factor $(\sin \vartheta)^m$ tends to make the function bigger near $\vartheta = \pi/2$, that is, in the equatorial plane. The larger m is, the sharper is this maximum. In fact, when $m = l$ we see from eq. (58) that

$$P_l^m(\zeta) \sim (1 - \zeta^2)^{m/2} = (\sin \vartheta)^m \tag{66}$$

When m is large, this has a sharp maximum at $\vartheta = \pi/2$. The physical meaning of this maximum is that the particle tends to be found near the equatorial plane. As we approach the classical limit of large m, this maximum becomes sharper and sharper so that the state $l = m$ approaches a situation in which the particle seems to be circulating in an orbit that is almost exactly in the equatorial plane. There remain, however, small fluctuations in the position or the orbit, which reflect the fact that if the z component of the angular momentum is defined, the x and the y components cannot be controlled exactly.

As m is decreased for a given l, the factor $(\sin \vartheta)^m$ becomes less important, the maximum becomes less sharp, and finally, for $m = 0$, we have seen that the wave function covers all values of ϑ about equally. What is happening as m is reduced while l remains fixed is that the direction of the total angular momentum is being shifted away from the z axis. In the classical limit, we can say that the orbit is being tilted out of the equatorial plane (the orbit is always normal to L). But in the quantum description no particular directions of L_x and L_y are preferred; we should therefore think of the system as if it were distributed over all possible L_x and L_y consistent with the given values of l and m. This clearly means that the particle will be found over a range of latitude angles, ϑ, which increases as m is decreased. When we reach $m = 0$, the orbital plane is normal to the equatorial plane, and the particle covers the full range of ϑ.

As has been pointed out in Sec. 12, the direction of the orbital plane and of the angular-momentum vector (which in classical physics is perpendicular to this plane) must be regarded as incompletely defined, in the sense that the particle covers all directions simultaneously. This

* See, for example, Pauling and Wilson, p. 131.
† See Sec. 14, footnote belonging to subsection 8.

follows from our general interpretation of the wave function, in which interference properties depending on the wave functions at different angles are shown to determine results of physical significance.

(5) *Number of Nodes of the* $Y_l^m(\vartheta, \varphi)$. We have seen that $P_l^m(\zeta)$ has $l - m$ zeros. Each of these zeros defines a nodal cone, corresponding to constant latitude. If we consider the real part of the complete angular wave function, $P_l^m(\zeta) \cos m\varphi$, we see that $\cos m\varphi = 0$ defines m nodal planes. The total number of nodal surfaces is then equal to

$$l - m + m = l$$

This is a useful fact, to which we shall refer later.

16. Measurement of Angular Momentum. Stern-Gerlach Experiment. One way of measuring the angular momentum is by means of a Stern-Gerlach experiment.* Suppose that we wish to measure the angular momentum of the electrons in a given type of atom, let us say, magnesium. A beam of these atoms is prepared, for example, by

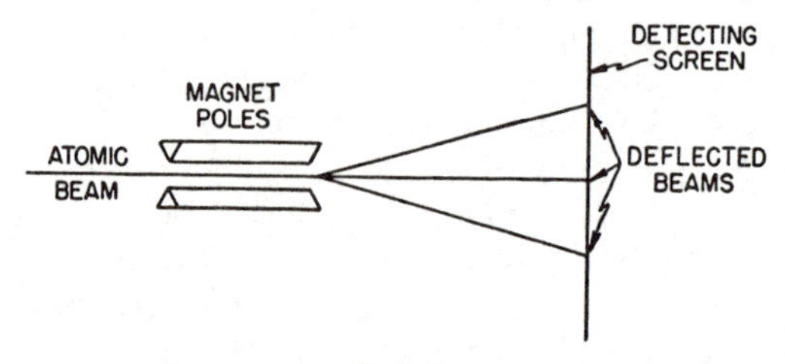

FIG. 2

evaporation from the solid, and passing the evaporated atoms through a set of collimating slits. This beam then enters a region in which there is an inhomogeneous magnetic field that is normal to the direction of motion of the atoms. The apparatus† is shown schematically in Fig. 2. One can show that if electrons are circulating in orbits in which their angular momentum is $\boldsymbol{L}$, then their magnetic moment is‡

$$\boldsymbol{\mu} = \frac{e\boldsymbol{L}}{2mc} \tag{67}$$

Now in an inhomogeneous magnetic field, the atoms experience a force given by

$$\boldsymbol{F} = \nabla(\boldsymbol{\mu} \cdot \boldsymbol{\mathcal{H}}) \qquad \text{(where } \boldsymbol{\mathcal{H}} \text{ is the magnetic field)} \tag{68}$$

Since, as is usually true, $\partial\mathcal{H}_z/\partial y$ is small, we obtain

$$F_z \cong \mu_z \frac{\partial\mathcal{H}_z}{\partial z} = \frac{e}{2mc}\frac{\partial\mathcal{H}_z}{\partial z} L_z \tag{69}$$

* Richtmeyer and Kennard, p. 407.

† The length of the magnet (in the x direction) is much greater than the distance between pole faces (z direction). Thus, one can neglect edge effects, and the problem becomes two-dimensional.

‡ See Chap. 15, eq. (52).

Thus, each atom experiences a force which is proportional to the z component of its electronic angular momentum (taken about the center of the atom). As a result, it picks up a corresponding momentum, and the beam is given a corresponding deflection. The beam is collected some distance from the magnet at a point that is far enough away so that atoms of different L_z have been separated. By measuring the deflection, one can calculate L_z.

Since there is no preferred direction for L_z before the atoms enter the magnetic field, it is clear that atoms of each value of L_z will occur with equal frequency, in a random fashion, as they boil out of the metal. Hence one will obtain a single spot on the detecting screen for each permissible value of L_z. Since the total number of permissible components of L_z is $2l + 1$, one can measure l simply by counting the number of spots on the screen. (This treatment actually applies only to those atoms for which the total electron spin is zero; the effect of spin will be to change the number and distribution of spots.* In fact, this experiment provides a way of showing that there is an electron spin because in many cases, the distribution of spots does turn out to be different from that predicted above.)

It should be noted that the Stern-Gerlach experiment yields a direct proof of the quantization of angular momentum, since, according to classical theory, there should be a continuous range of angular momenta and, therefore, a continuous range of positions at which atoms arrive on the screen.

17. Transformation to a Rotated System of Axes. In Sec. 7 it was pointed out that one could choose for the z axis an arbitrary direction. It would seem at first sight that the chosen axis would have some special significance, just because the angular momentum in its direction—and in no other direction—has been quantized. We wish to show, however, that despite the quantization of only one component of the angular momentum, no result of physical significance will depend on which axis happens to have been chosen. In order to prove this, we shall show that one can obtain the same wave function and, therefore, the same probabilities for all physical quantities, by working in a system of co-ordinates in which the axes have been rotated by an arbitrary amount relative to the original axes.

Let us begin with the case of zero angular momentum ($L^2 = 0$). For this case, we must also have $L_z = 0$, and the wave function is not even a function of ϑ or φ [see eq. (16)]. This means that if we rotate our co-ordinate axes, the wave function is left unchanged, so that in the new co-ordinate system, it still corresponds to $L^2 = 0$ and $L_z = 0$. It is clear therefore, that for this case no physical results will depend on what axes have been chosen.

* For a treatment of spin, see Chap. 17.

For higher angular momenta, it will still remain true that L^2 is left unchanged by a rotation. This follows from the fact that L^2 is a scalar. Thus, the value of l is left unchanged. The components of the angular momentum can, however, be expected to change. In order to illustrate the nature of these changes, let us consider the case, $l = 1$. In the original co-ordinate system, the three normalized wave functions are [see eqs. (47) and (62)]

$$\begin{aligned} m &= 1 \qquad & \psi_1 &= \sqrt{\frac{3}{8\pi}} \sin \vartheta \, e^{i\varphi} \\ m &= 0 \qquad & \psi_0 &= \sqrt{\frac{3}{4\pi}} \cos \vartheta \\ m &= -1 \qquad & \psi_{-1} &= \sqrt{\frac{3}{8\pi}} \sin \vartheta \, e^{-i\varphi} \end{aligned} \tag{70}$$

For the sake of illustration, consider a rotation through an angle β about the y axis. In doing this, it will be convenient to write the wave functions as

$$\psi_1 = \sqrt{\frac{3}{8\pi}} \frac{(x + iy)}{r} \qquad \psi_0 = \sqrt{\frac{3}{4\pi}} \frac{z}{r} \qquad \psi_{-1} = \sqrt{\frac{3}{8\pi}} \frac{(x - iy)}{r} \tag{71}$$

The old co-ordinates are related to the new by the relations

$$\begin{aligned} y &= y' \qquad z = z' \cos \beta - x' \sin \beta \\ x &= z' \sin \beta + x' \cos \beta \qquad r' = r \end{aligned}$$

We then obtain

$$\left.\begin{aligned} \psi_1 &= \sqrt{\frac{3}{8\pi}} \frac{1}{r} (x' \cos \beta + z' \sin \beta + iy') \\ &= \frac{1}{2} \psi_1'(1 + \cos \beta) + \frac{\psi_0' \sin \beta}{\sqrt{2}} + \psi_{-1}' \frac{(\cos \beta - 1)}{2} \\ \psi_0 &= \sqrt{\frac{3}{4\pi}} \frac{1}{r} (z' \cos \beta - x' \sin \beta) \\ &= -\psi_1' \frac{\sin \beta}{\sqrt{2}} + \psi_0' \cos \beta - \psi_{-1}' \frac{\sin \beta}{\sqrt{2}} \\ \psi_{-1} &= \sqrt{\frac{3}{8\pi}} \frac{1}{r} (x' + z' \sin \beta - iy') \\ &= \psi_1' \frac{(\cos \beta - 1)}{2} + \frac{\psi_0' \sin \beta}{\sqrt{2}} + \psi_{-1}' \frac{(\cos \beta + 1)}{2} \end{aligned}\right\} \tag{72}$$

This means that in the new co-ordinate system, each ψ_m becomes a linear combination of eigenfunctions of $L_{z'}$. If, for example, $L_z = 0$ in the old co-ordinate system, then in the new system, it will be possible for the $L_{z'}$ to be $+1$, 0, or -1. The respective probabilities that these values will be obtained are given by the coefficients of the corresponding wave functions. Thus, we obtain

$$P_{+1} = \frac{\sin^2 \beta}{2} \qquad P_0 = \cos^2 \beta \qquad P_{-1} = \frac{\sin^2 \beta}{2} \tag{73}$$

The sum of these probabilities is, as it ought to be, unity.

Problem 5: Prove that the eigenvalue of L^2 is not changed in the transformation (72).

It is of interest to take the special case where $\beta = 90°$. Here, we obtain $P_{+1} = \frac{1}{2}$, $P_{-1} = \frac{1}{2}$, so that a particle with $L_z = 0$ in the old system will have equal probabilities that if $L_{z'}$ in the new system is measured, one will find $L_{z'} = +1$ or $L_{z'} = -1$. But since z' (in the new system) is the same as x in the old system, we can also conclude that a particle with $L_z = 0$ will have equal probabilities that if L_x is measured, the result will be $+1$ or -1.

Let us note that the eigenfunction for $L_z = 0$ is made up by interference of two different eigenfunctions of L_x. We therefore conclude that a zero value of L_z is a mutual (or interference) property of the two states $L_x = 1$ and $L_x = -1$. Therefore, a particle with $L_z = 0$ cannot be thought of as having a definite value of L_x, but must instead be thought of as covering both states at once. This is just another way of expressing the fact that the observables L_x and L_z cannot be measured simultaneously, which was already obtained from the noncommutativity of the operators L_x and L_z.

We may now inquire into what happens to the wave function of a particle for which $L_z = 0$ when L_x is measured. In accordance with Chap. 6, Sec. 3, and Chap. 22, Sec. 9, the wave function will be broken into two parts, each corresponding to a definite value of L_x, and each multiplied by an uncontrollable phase factor $e^{i\alpha}$, which destroys coherent interference. Thus, before the measurement of L_x, the wave function is [see eq. (72)], setting $\beta = \pi/2$

$$\psi_0 = -\frac{1}{\sqrt{2}}(\psi_1' + \psi_{-1}') \tag{74}$$

After the measurement, it is

$$\psi_0 = -\frac{1}{\sqrt{2}}(\psi_1' e^{i\alpha_1} + \psi_{-1}' e^{i\alpha_2}) \tag{75}$$

Although the value of L_x has been made definite (i.e., either $+1$ or -1) by the functioning of the measuring apparatus, the value of L_z cannot now be definite because, for example, $L_z = 0$ only when the wave function is $\psi_0 = -\frac{1}{\sqrt{2}}(\psi_1' + \psi_{-1}')$. Thus, after the measurement we will have a mixture of two values of L_z. More generally, it is clear from the above that a wave function corresponding to a definite L_z has just such a structure that L_x cannot be made definite when L_z has an eigenvalue, and vice versa.

18. Double Stern-Gerlach Experiment. The measurement described in the previous paragraph can be realized experimentally by means of a double Stern-Gerlach experiment. Suppose that we send a beam of atoms through an inhomogeneous magnetic field, which is in the z direction, and thus separate the particles according to the values of L_z. We

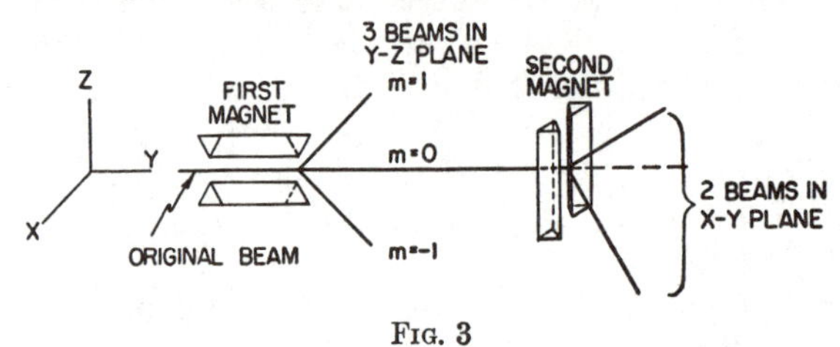

FIG. 3

could then choose a particular beam, for example, the one with $L_z = 0$, and send it through a second magnet, which is at right angles to the first. According to the results of the previous paragraph, this beam would then split into two, with $L_x = +1$, $L_x = -1$ (but for each of which L_z was indefinite). The experiment is illustrated in Fig. 3.

Similar results can be derived for $m = \pm 1$.

Problem 6: Discuss what happens to particles with $m = 1$, in the double Stern-Gerlach experiment, when $\beta = 90°$; also when $\beta = 45°$.

19. Physical Equivalence of All Co-ordinate Systems. In order to demonstrate the physical equivalence of all choices of direction of the z axis, it is necessary only to show that the same wave function can be expressed by means of an equivalent procedure in all co-ordinate systems. Consider, for example, an arbitrary wave function corresponding to $l = 1$.

$$\psi = C_1\psi_1 + C_0\psi_0 + C_{-1}\psi_{-1}$$

where the subscripts refer to the values of m.

Problem 7: Prove that to normalize ψ, one must have $|C_1|^2 + |C_0|^2 + |C_{-1}|^2 = 1$.

In the rotated system, we obtain

$$\begin{aligned}\psi = &\left[\frac{C_1(1+\cos\beta)}{2} - \frac{C_0\sin\beta}{\sqrt{2}} + \frac{C_{-1}(\cos\beta - 1)}{2}\right]\psi'_{+1} \\ &+ \left[\frac{C_1\sin\beta}{\sqrt{2}} + C_0\cos\beta + \frac{C_{-1}\sin\beta}{\sqrt{2}}\right]\psi'_0 \\ &+ \left[\frac{C_1(1-\cos\beta)}{2} + \frac{C_0\sin\beta}{\sqrt{2}} + \frac{C_{-1}(1+\sin\beta)}{2}\right]\psi'_{-1} \\ = &\ C'_1\psi'_{+1} + C'_0\psi'_0 + C'_{-1}\psi'_{-1} \end{aligned} \tag{76}$$

where the primed quantities refer to axes that have been rotated through an angle β about the old y axis.

Problem 8: Prove that $|C'_1|^2 + |C'_0|^2 + |C'_{-1}|^2 = |C_1|^2 + |C_0|^2 + |C_{-1}|^2 = 1$.

We conclude then that the same function ψ can be expanded in terms of eigenfunctions of L_z, with z taken in an arbitrary direction. Each co-ordinate system simply requires different expansion coefficients, but the same general procedure. Since the state of the system depends only on the wave function, we see that the quantization of L in a definite direction does not really give that direction any special properties and that the same results for all physical quantities could have been obtained with any other direction.*

As an example, we consider the Stern-Gerlach experiment, which we have thus far described by quantizing the angular momentum along the direction of the magnetic field. Although this is without doubt the most convenient direction to choose, it is easily shown that the same results would have been obtained in any other system.

In order to do this, we consider the operator $\boldsymbol{\mu} \cdot \boldsymbol{\mathcal{H}}$ appearing in eq. (68), choosing our z axis in a direction not necessarily the same as that of $\boldsymbol{\mathcal{H}}$. For simplicity, however, we choose $\mathcal{H}_y = 0$,† but the generalization to arbitrary $\mathcal{H}_y$ will be fairly straightforward. Thus we write

$$\boldsymbol{\mu} \cdot \boldsymbol{\mathcal{H}} \cong \mathcal{H}_x L_x + \mathcal{H}_z L_z = \frac{\hbar}{i}\left[\mathcal{H}_z \frac{\partial}{\partial \varphi} - \mathcal{H}_x \left(\cot \vartheta \cos \varphi \frac{\partial}{\partial \varphi} + \sin \varphi \frac{\partial}{\partial \vartheta}\right)\right]$$

Now a particle can experience a definite deflection only if its wave function is an eigenfunction of the operator $\boldsymbol{\mu} \cdot \boldsymbol{\mathcal{H}}$. It is readily shown, however, that the eigenfunctions of the operator $\boldsymbol{\mu} \cdot \boldsymbol{\mathcal{H}}$ are precisely the eigenfunctions of the operator $L_{z'}$ with the z' axis taken along the direction of the magnetic field.

Problem 9: Prove the above statement.

This means that although we used an arbitrary co-ordinate system, we still obtained the result that only those particles with a definite component of L in the direction of $\boldsymbol{\mathcal{H}}$ will obtain a definite deflection. An arbitrary wave function can always be expanded as a series of the three eigenfunctions in question (for $l = 1$), and the coefficients in this expansion yield the probability of a given deflection in this direction. Thus, the physical results are independent of the co-ordinate system in which the problem is set up.

This result can be understood physically in terms of the concept of the properties of matter as incompletely defined potentialities, which are

* This is a special case of a canonical transformation. See Chap. 16, Sec. 15.

† Starting with the co-ordinate system used in Sec. 16, we make a rotation about the y axis.

realized only in interaction with other systems. (See Chap. 6, Sec. 13, and Chap. 8, Sec. 15.) Thus, when an atom has a definite value of L_z, it has indefinite values of L_x and L_y, but it also has the latent ability to develop a definite, but incompletely predictable, value for either L_x or L_y, provided, for example, that it interacts with a suitably oriented Stern-Gerlach apparatus. In such a process, it would, of course, develop an indefinite value for L_z. This concept of noncommuting variables as mutually incompatible potential properties of matter provides a qualitative description of the invariance of the physical significance of the quantum theory of angular momentum to changes of the direction in which the axis is chosen. For the concept of invariance to rotation is contained in the statement that *any* axis may serve as the direction in which the potentialities for developing a definite component of the angular momentum can be realized, provided that the electron interacts with an appropriate system.

20. Generalization to Arbitrary Rotations and Arbitrary l. These results can easily be extended to arbitrary rotations (i.e., to those not necessarily taken about the y axis). Furthermore, similar results can be obtained for larger values of l. One can show, in fact, that on rotation, any given spherical harmonic, $Y_l^m(\vartheta, \varphi)$, goes into a linear combination of spherical harmonics of the same l, with coefficients that depend on the angle and direction of rotation.* Thus, one obtains

$$Y_l^m(\vartheta, \varphi) = \sum_{m'} C^l_{m,m'} Y_l^{m'}(\vartheta', \varphi') \tag{77}$$

21. Application to Construction of Orbits. We can apply some of these results to the problem of constructing wave functions that represent orbits in the case of large l. We have already seen (Sec. 15) that the case $l = m$ represents an orbit approximately in the equatorial plane, at least, insofar as the uncertainty principle permits us to define such an orbit. We now ask the question: what wave functions represent an orbit that is tilted away from the equatorial plane? We already know that the $Y_l^m(\vartheta, \varphi)$ cannot represent such an orbit, since they represent a situation in which the relative magnitudes of L_x and L_y are totally unknown. We can, however, readily construct a tilted orbit by taking an orbit that is originally in the equatorial plane, and by then rotating the co-ordinate axes through some angle γ. Then, according to eq. (75), the rotated wave function is

$$\psi_l^l(\vartheta, \varphi) = \sum_m C^l_{l,m} Y_l^m(\vartheta', \varphi')$$

Thus, to represent a tilted orbit, we need a linear combination of spherical harmonics. The exact linear combination depends on the $C^l_{l,m}$. We

* Kramers, p. 166. (See list of references on p. 2.)

shall not consider the values of the C's in detail here, however, but instead shall merely point out that when the C's are calculated for the case of large l, one obtains a wave packet of spherical harmonics that has a sharp maximum for that value of m which corresponds to the projection of $\mathbf{L}$ on the z axis. But to define the direction of the plane of the orbit, we have to make m somewhat uncertain, in the sense that we combine a range of values of m. In the classical limit, however, this range is negligible so that the system seems to have a definite value of each component of angular momentum and, therefore, a definite direction of its orbital plane.

Problem 10: By using the normalization and orthogonality properties of the $P_l(\cos \vartheta)$, obtain the expansion of the function $\delta(\vartheta - \vartheta_0)$ in Legendre polynomials. Obtain the corresponding expansion of $\delta(\vartheta - \vartheta_0)\delta(\varphi - \varphi_0)$ in spherical harmonics.

We can summarize by saying that an electron of definite angular momentum is something very different in quantum theory from what it is in classical theory. In quantum theory, it can have a definite angular momentum only when its wave function has the appropriate dependence on angle, i.e., $Y_l^m(\vartheta, \varphi)$. This is analogous to the fact that it can have a definite momentum only when its wave function has the appropriate dependence on the position, i.e., $e^{ipx/\hbar}$. Thus, it would be meaningless to consider an electron that had simultaneously a definite angle and a definite angular momentum. This is again an illustration of the wave aspects of matter, not understandable on the basis of the particle model.

When the angle is measured, then the actions of the apparatus change the system from a wavelike object with definite angular momentum and indefinite angle to a particle-like object with definite angle and indefinite angular momentum. On the other hand, if the angular momentum were subsequently measured, the system would be changed back into a wavelike object. (See Chap. 6, Secs. 4 to 10.)

CHAPTER 15

Solution of the Radial Equation; the Hydrogen Atom; the Effect of a Magnetic Field

IN THIS CHAPTER we continue the program of solving the three-dimensional wave equation and apply the results to a number of problems.

1. The Radial Equation. Since an arbitrary function of ϑ and φ can be expanded as a series of surface harmonics, an arbitrary function of r, ϑ, and φ can be expressed by allowing the coefficients of the surface harmonics to be functions of r. Thus

$$\psi(r, \vartheta, \varphi) = \sum_{l,m} f_{l,m}(r) Y_l^m(\vartheta, \varphi) \tag{1}$$

The above expansion in terms of $Y_l^m(\vartheta, \varphi)$ can be carried out for any fixed value of r. The resulting coefficients in the expansion will then depend on r. This dependence is noted by referring to the coefficients as $f_{l,m}(r)$.

According to eqs. (17), and (35), Chap. 14, Schrödinger's equation may now be written

$$\sum_{l,m}\left[-\frac{\hbar^2}{2m}\left(\frac{d^2}{d^2r^2}+\frac{2}{r}\frac{d}{dr}\right) - E + V(r) + \frac{\hbar^2}{2m}\frac{l(l+1)}{r^2}\right] f_{l,m}(r) Y_l^m(\vartheta, \varphi) = 0 \tag{2}$$

In order that this equation hold for an arbitrary value of ϑ and φ, it is necessary that

$$-\frac{\hbar^2}{2m}\left[\frac{d^2}{dr^2}+\frac{2}{r}\frac{d}{dr}-\frac{l(l+1)}{r^2}\right] f_{l,m} + [-E + V(r)] f_{l,m} = 0 \tag{2a}$$

[This can be proved by multiplying by $Y_r^s(\vartheta, \varphi)$ and integrating over ϑ and φ.]

Problem 1: Prove the above equation.

We note that the equation for $f_{l,m}(r)$ is independent of m and depends only on l. Hereafter, we shall write $f_{l,m}(r) = f_l(r)$.

The above equation may be simplified by writing $f_l(r) = g_l(r)/r$. We then obtain

$$\frac{d^2 g_l(r)}{dr^2} + \frac{2m}{\hbar^2}\left[E - V(r) - \frac{\hbar^2}{2mr^2} l(l+1)\right] g_l(r) = 0 \tag{3}$$

2. Normalization of g_l. The element of volume in spherical polar co-ordinates is $r^2\,dr\,d\Omega$. But the $Y_l^m(\vartheta, \varphi)$ are already normalized over $d\Omega$. The $f_l^{(r)}$ should, therefore, be normalized over the radius, as follows:

$$\int_0^\infty |f_l(r)|^2 r^2\,dr = 1 \tag{4a}$$

Writing $rf_l = g_l$, we obtain

$$\int_0^\infty |g_l(r)|^2\,dr = 1 \tag{4b}$$

3. Special Case: $l = 0$ (s waves). For the special case, $l = 0$, the above equation for $g(r)$ is exactly the same as the one-dimensional Schrödinger equation for ψ. But there is one important qualifying condition: Since ψ must be everywhere finite, and since $g = rf$, it is clear that g must approach zero at the origin at least as rapidly as r. This is a new boundary condition, not present in the one-dimensional problem. It has already been applied in the deuteron problem (see Chap. 11, Sec. 14).

States of zero angular momentum are called *s* states. This terminology dates from the early days of spectroscopy.* In this notation, states of various angular momentum are labeled as follows:

l	*Name*	
0	s	Sharp
1	p	Principal
2	d	Diffuse
3	f	Fundamental
4	g	

From here on, the letters increase in alphabetical order.

4. Centrifugal Potential. When $l \neq 0$, the equation for g_l is the same as a one-dimensional Schrödinger equation in which the potential function is $V(r) + \dfrac{\hbar^2}{2mr^2}l(l+1)$. The system thus acts as if there were a repulsive potential, $\hbar^2 l(l+1)/2mr^2$, in addition to the usual potential. As pointed out in Chap. 2, Sec. 14, this repulsive potential may be thought of as the term responsible for the centrifugal force that tends to keep particles of nonzero angular momentum away from the origin. Suppose, for example, that $V(r)$ is $-e^2/r$ (a Coulomb potential). The effective potential for $l \neq 0$ might then have a shape resembling that shown in Fig. 1. At large distances from the origin, the Coulomb potential is the main term, but for small r this is more than overbalanced by the repulsive centrifugal term. The equilibrium point occurs where the derivative of the effective potential is zero, or where

* See H. E. White, *Introduction to Atomic Spectra*. New York: McGraw-Hill Book Company, Inc., 1934, p. 13.

$$\frac{\partial V}{\partial r} - l(l+1)\frac{\hbar^2}{mr^3} = 0 \tag{5a}$$

For an attractive Coulomb force $\partial V/\partial r = e^2/r^2$, so that we obtain

$$r_{\text{equilibrium}} = \frac{l(l+1)\hbar^2}{me^2} \tag{5b}$$

We note that the equilibrium radius increases with the angular momentum, as we should expect. This radius is the one for which the attractive force just balances the centrifugal force. It is therefore the classical radius for a circular orbit.

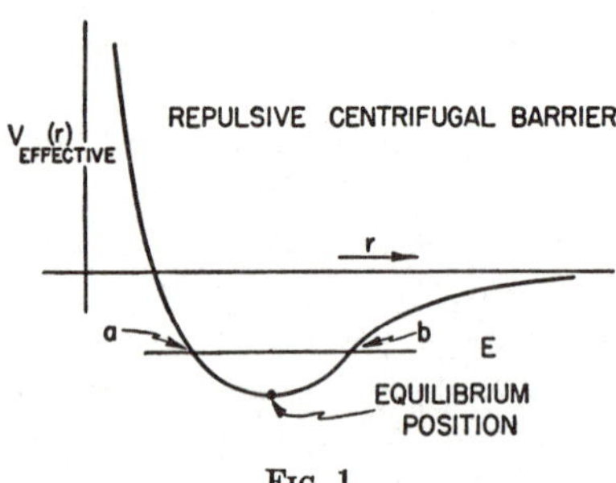

FIG. 1

In general, if a particle is bound ($E < 0$), it will oscillate (classically) between some limits $r = a$, and $r = b$, as shown in Fig. 1. For example, in an elliptic orbit of a hydrogen atom, the radius oscillates periodically between inner and outer limits. Only for a circular orbit is there no oscillation.

5. Separation into Relative Co-ordinates. Thus far, we have been assuming that the potential has been a function of the distance r from a fixed point. In many problems, such as, for example, the hydrogen atom, the potential is a function of the relative distance, $\mathbf{r}_1 - \mathbf{r}_2$, between the electron and the proton ($\mathbf{r}_1$ is the radius vector of the electron, and $\mathbf{r}_2$ that of the proton). Nevertheless, as is possible in classical theory, the equations can be separated into two sets, one involving only $\mathbf{r}_1 - \mathbf{r}_2$ and the other involving only the position of the center of mass. To do this, we write the Hamiltonian for the two particles as follows:*

$$-\frac{\hbar^2}{2m_1}\nabla_1^2 - \frac{\hbar^2}{2m_2}\nabla_2^2 + V(\mathbf{r}_1 - \mathbf{r}_2) = H \tag{6}$$

Let us now make the substitution

$$\boldsymbol{\xi} = \mathbf{r}_1 - \mathbf{r}_2 \qquad \boldsymbol{\eta} = \frac{(m_1\mathbf{r}_1 + m_2\mathbf{r}_2)}{(m_1 + m_2)}$$

$\boldsymbol{\xi}$ represents the separation between the two particles, $\boldsymbol{\eta}$ the position of the center of mass.

It will be left as an exercise for the reader to prove that

$$H = \frac{-\hbar^2}{2(m_1 + m_2)}\nabla_\eta^2 - \frac{\hbar^2(m_1 + m_2)}{2m_1m_2}\nabla_\xi^2 + V(\boldsymbol{\xi}) \tag{7}$$

Problem 2: Prove the above statement.

* See Chap. 10, Sec. 11.

Let us tentatively write the wave function as a product $\psi = F(\xi)G(\mathbf{n})$. We shall see later than an arbitrary wave function can be expressed as a sum of these products. The eigenvalues of H are then determined by the equation

$$\frac{-\hbar^2}{2(m_1+m_2)}F(\xi)\nabla_\eta^2 G(\mathbf{n}) - \frac{\hbar^2(m_1+m_2)}{2m_1m_2}G(\mathbf{n})\nabla_\xi^2 F(\xi) + V(\xi)F(\xi)G(\mathbf{n}) = EF(\xi)G(\mathbf{n}) \quad (8)$$

Division by $F(\xi)G(\mathbf{n})$ yields

$$\frac{-\hbar^2}{2(m_1+m_2)}\frac{\Delta_\eta^2 G(\mathbf{n})}{G(\mathbf{n})} + \left[-\frac{\hbar^2(m_1+m_2)}{2m_1m_2}\frac{\nabla_\xi^2 F(\xi)}{F(\xi)} + V(\xi)\right] = E \quad (9)$$

This equation can have a solution for arbitrary ξ and $\mathbf{n}$ only if the part involving $\mathbf{n}$ and the part involving ξ are each separately and identically constant. Thus, we obtain

$$\frac{-\hbar^2}{2(m_1+m_2)}\frac{\nabla_\eta^2 G(\mathbf{n})}{G(\mathbf{n})} = E_0 = \text{constant} \quad (10)$$

The above equation, however, is exactly the same as Schrödinger's equation for a free particle of mass $m_1 + m_2$. The wave function for the center of mass, therefore, behaves exactly as if the system were a single particle with kinetic energy E_0 and mass equal to the total mass of the system. This is just the quantum analogue of the classical result that the center of mass of a system of particles moves at a constant rate, independent of the forces between the particles. The quantum result can also be generalized to an arbitrary number of particles. The function $G(\mathbf{n})$ is given by

$$G(\mathbf{n}) = A\, e^{ip\cdot\eta/\hbar} + B\, e^{-ip\cdot\eta/\hbar} \quad (11)$$

where A and B are arbitrary constants and $|p| = \sqrt{2(m_1+m_2)E_0}$.

Let us denote the difference between the total energy E and the energy associated with the center of mass E_0 by the symbol E_r (relative energy). Then the equation for F becomes

$$\frac{\hbar^2}{2\mu}\nabla^2 F + (E_r - V)F = 0 \quad (12)$$

where

$$E_r = E - E_0 \qquad \text{and} \qquad \mu = \frac{m_1m_2}{m_1+m_2}$$

μ is known as the *reduced mass*. The above equation is clearly the same as that of a particle with energy E_r and the reduced mass μ in a potential arising from a fixed center.

Usually, in solving problems like that of the hydrogen atom, we are not interested in the energy of the motion of the atom as a whole, but merely the energy resulting from the relative motion of the electron and the proton. It is the relative energy, for example, that appears in the form of radiation when an electron jumps from a given stationary state to one of lower energy. In practice, we shall therefore usually solve only the equation for $F(\xi)$, and obtain the possible values of E_r, which will subsequently be denoted by E. This procedure has already been adopted in solving for the deuteron energy levels (Chap. 11, Sec. 14) where the two particles, neutron and proton, have practically the same mass, so that $\mu = m/2$.

The possibility of using a series of functions like $F(\xi)G(\mathbf{n})$ to express an arbitrary function of ξ and $\mathbf{n}$ arises from the fact that F and G are each eigenfunctions of a Hermitean operator. [G is the eigenfunction of $\frac{-\hbar^2}{2(m_1 + m_2)} \nabla\xi^2$ and F is the eigenfunction of the operator appearing in eq. (11).] According to the expansion theorem, an arbitrary function can therefore be expanded as a series of these products. Hereafter, unless otherwise specified, we shall restrict ourselves to solving for $F(\xi)$, remembering, however, that the complete wave function is a sum of products, $F(\xi)G(\mathbf{n})$.

6. Preliminary Discussion of General Form of Solution for Hydrogen Atoms. We now proceed to solve Schrödinger's equation for the hydrogen atom. We shall first, however, discuss the general form of the solutions in a qualitative way, in order to illustrate the technique of showing what a wave function looks like without solving the problem exactly. We restrict ourselves to the relative co-ordinates, $\mathbf{r}_1 - \mathbf{r}_2$, which will hereafter be denoted by $\mathbf{r}$. We are also going to seek only those solutions for which E is negative; these correspond to bound states. The solutions with E positive represent electrons that come in from an infinite distance, undergo scattering by the potential, and then go back out for an infinite distance. This type of solution will be studied in Chap. 21, in connection with scattering problems. In classical physics, positive E results in a hyperbolic orbit, negative in an elliptic orbit.

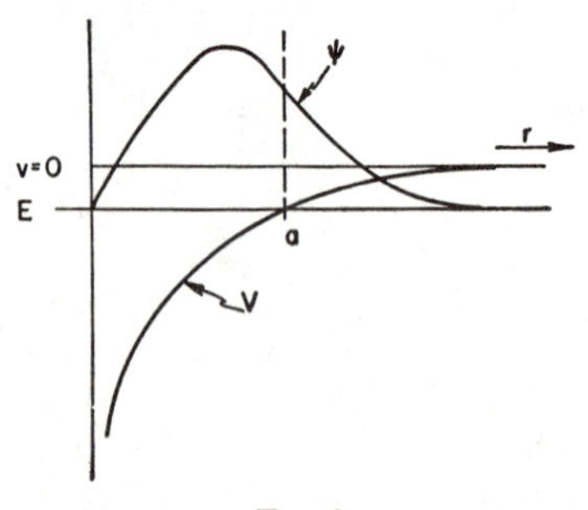

FIG. 2

7. General Form of Solution for *s* Waves. For *s* waves, the centrifugal potential vanishes, and the actual potential takes the shape shown in Fig. 2. If the energy is negative, then there is a point $r = a$, beyond which $E - V$ is negative; classically, the particle would never reach radii larger than a. The value of this radius is given by $|E| = e^2/a$ or $a = e^2/|E|$. Beyond this point, the solutions do not oscillate, but have a generally exponential behavior. As $r \to \infty$, the potential becomes negligible, and one can approximate g in eq. (3) by a solution of the equation

$$\frac{d^2g}{dr^2} - \frac{2\mu}{\hbar^2}|E|g = 0$$

$$g = A \exp\left(-\sqrt{\frac{2\mu|E|}{\hbar^2}}\,r\right) + B \exp\left(\sqrt{\frac{2\mu|E|}{\hbar^2}}\,r\right) \tag{13}$$

In order that the wave function remain finite as $r \to \infty$ the coefficient B of the increasing exponential must vanish. We shall see that this requirement determines the allowed values of $|E|$.

At the origin, g must start out with the value zero. The general form of the solution can be seen with the aid of the discussion in Chap. 11, Sec. 12. When $r < a$, $E - V$ is positive; hence if g is positive, the wave function has a negative curvature. If $E - V$ is sufficiently large, the solution may curve enough to make the slope of g negative at $r = a$. Beyond $r = a$ the curvature is positive. In general, g will ultimately approach an increasing exponential as $r \to \infty$, but for a certain value of $|E|$, it will fit a decaying exponential exactly. This value of $|E|$ will be an eigenvalue of the energy. The wave function so obtained is shown in Fig. 2; it has no nodes, except the unavoidable one at the origin. It must therefore be the lowest energy state, because a wave function which oscillates has a higher kinetic energy than one which does not.

The next state will be one in which the wave function fits a decaying exponential after it goes through a node. Because E is greater ($|E|$ is less, but E is negative), the wavelength inside the potential,

$$\lambda = h/p = h/\sqrt{2m(E - V)}$$

is less,* so that this wave function oscillates more rapidly than does the lower energy solution. There is, furthermore, more room to oscillate, because the turning point occurs at a larger radius when $|E|$ is smaller ($a = e^2/|E|$). The wave function for this state is shown schematically in Fig. 3.

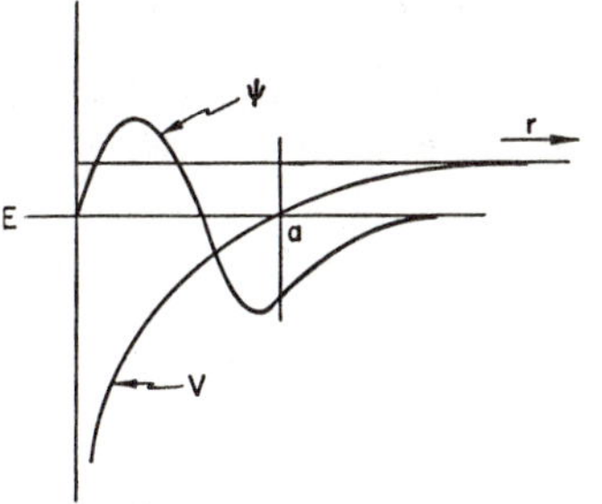

FIG. 3

Still higher energy states would involve wave functions with still more nodes. In the case of the square potential, the number of possible bound states depended on the depth of the potential and the radius. We shall see, however, that the Coulomb force has an indefinitely large number of bound states. This is because the Coulomb force dies out

* This argument is somewhat rough, because when the potential is a function of position, the wavelength at a given point has no precisely definable meaning. One can however give an approximate meaning to this wavelength whenever V does not change too rapidly, as was done in connection with the WKB (Chap. 12).

comparatively slowly as a function of the distance, so slowly in fact, that it is always possible to get more oscillations into the wave function by decreasing $|E|$ and, hence, increasing a.

Application of WKB Approximation to Determine Approximate Energy Levels. To see in greater detail how this works out, let us use the WKB approximation, which is certainly good in the case where the wave function oscillates many times and has some significance, even for the lower quantum states. We must be careful, however, to choose only those solutions which vanish at the origin. We therefore write for the WKB approximate solution

$$g \sim \frac{1}{\sqrt[4]{E-V}} \sin\left[\int_0^r \sqrt{2m(E-V)}\,\frac{dr}{\hbar}\right] \tag{14a}$$

In order to find out how g behaves as r approaches infinity, we apply the connection formulas for the barrier at the right [eq. (39), Chap. 12]. Note that the turning point is $r = a$. To do this, we rewrite the above equation for g, obtaining

$$g \sim \frac{1}{\sqrt[4]{E-V}} \sin\left[\int_0^a \sqrt{2m(E-V)}\,\frac{dr}{\hbar} - \int_r^a \sqrt{2m(E-V)}\,\frac{dr}{\hbar}\right]$$
$$= \frac{-1}{\sqrt[4]{E-V}} \cos\left[\int_r^a \sqrt{2m(E-V)}\,\frac{dr}{\hbar} - \int_0^a \sqrt{2m(E-V)}\,\frac{dr}{\hbar} + \frac{\pi}{2}\right]$$

The connection formulas, Chap. 12, eq. (39), show that g will fit a decaying exponential if, and only if,

$$\frac{\pi}{2} - \int_0^a \sqrt{2m(E-V)}\,\frac{dr}{\hbar} = -\frac{\pi}{4} - N\pi$$

where N is an integer, or

$$J = 2\int_0^a \sqrt{2m(E-V)}\,dr = \left(N + \frac{3}{4}\right)h \tag{14b}$$

Note the appearance of the $\frac{3}{4}$ in the quantum condition. This originates in the requirement that the wave function be zero at the origin, as distinguished from the one-dimensional case,* where no such requirement is made.

We must now evaluate the integral

$$J = 2\sqrt{2m}\int_0^a \sqrt{-|E| + \frac{e^2}{r}}\,dr$$

where $a = e^2/|E|$. Let us first make the substitution $r = ay = e^2y/|E|$. This gives

$$J = \frac{2e^2}{\sqrt{|E|}}\sqrt{2m}\int_0^1 \sqrt{\frac{1}{y} - 1}\,dy$$

* Chap. 12, eq. (55).

The integral is readily evaluated and yields $\pi/2$. The quantum condition (14) then becomes

$$\frac{\pi e^2}{2\hbar}\sqrt{\frac{2m}{|E|}} = \left(N + \frac{3}{4}\right)\pi \tag{15a}$$

Solving for E, we obtain

$$E = \frac{-me^4}{2\hbar^2[N + (\frac{3}{4})]^2} \tag{15b}$$

where N is any integer from zero on up. This formula disagrees with the exact formula (Sec. 12, Eq. (22)) in that $N + \frac{3}{4}$ should be replaced by $N + 1$. Yet, it is very nearly right, and, in the correspondence limit, the difference between it and the correct formula is too small to be detected. The reason that it fails for small N is that the WKB approximation is not strictly applicable.

Problem 3: Investigate the validity of the WKB approximation for the wave functions for different values of N in the case of the Coulomb potential.

Eigenvalues Approach to a Series Limit. As $N \to \infty$, we see that the energy levels get closer and closer to the continuum, which begins at $|E| = 0$. The levels become denser and denser. Thus, in a high quantum state, the levels are so close together that it is hard to tell the difference between the discrete quantized levels and the continuous range of energies predicted by classical theory. The energy level diagram is shown in Chap. 2, Fig. 4.

It is interesting to study in greater detail the reason for the appearance of an infinite number of levels near the continuum for the hydrogen atom, as contrasted with the finite number of levels for the square well. As has already been pointed out, the reason is that in the hydrogen atom the potential extends out to infinity. One might ask the question, what will happen with some other potential that dies out smoothly as $r \to \infty$, for example, as $1/r^n$ or e^{-br}? Will it show an infinite number of levels as $|E| \to 0$, like the hydrogen atom, or will it behave more like the square well and show a finite number of levels? We shall not work out the answer here, but shall merely quote the result that if $V \to 0$ as r^{-n} and if $n \leq 2$, then there are an infinite number of energy levels, but if $n > 2$, there are only a finite number. For e^{-br}, there are also only a finite number. The proof of these statements is left as an exercise for the reader.

Definition of Principal Quantum Number. We obtain a new energy level each time the function g has a node. The number of nodes therefore provides a convenient system of ordering the different states. The number of nodes in g (including the one at the origin) is called the principal quantum number of the state and is usually denoted by n. This definition holds only for s states; for higher angular momenta, we shall modify it in a way that will be discussed later.

It is clear that each node (except for the one at $r = 0$) defines a surface, on which the wave function vanishes. In this case, the surface is spherical. In more general problems, it is convenient to define the principal quantum number as the total number of nodal surfaces (which need not be spherical in the general case). We shall return to this question after discussing the solutions of higher angular momentum.

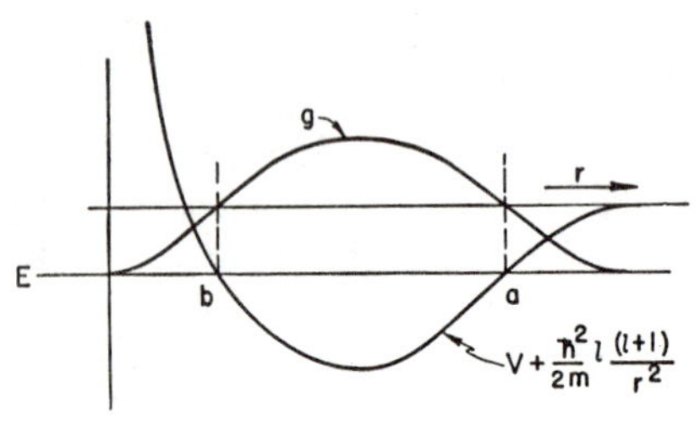

FIG. 4

8. General Form of Solution When $l > 0$. When $l \neq 0$, the addition of the repulsive centrifugal potential creates the effective potential resembling that shown in Fig. 4. Classically, the particle oscillates between the limits a and b, at which the radial component of the velocity vanishes. The general form of the solution can easily be seen. We know first, that $g = 0$ at $r = 0$. In fact, we can show that $g \cong r^{(l+1)}$ is a good approximation near the origin. To prove this, let us note that for small r the main term in the effective potential is $\hbar^2(l)(l+1)/2mr^2$. Thus the differential equation becomes approximately

$$\frac{d^2g}{dr^2} - \frac{l(l+1)}{r^2} g \cong 0 \tag{16a}$$

It is readily verified by direct substitution that the most general solution is

$$g = Ar^{l+1} + Br^{-l} \tag{16b}$$

where A and B are arbitrary constants. The solution involving r^{-l} is inadmissible because g must vanish at the origin. Thus, near the origin, an approximate solution is

$$g \cong Ar^{l+1} \tag{16c}$$

We can readily see, by plotting this function, that when $l > 0$, it curves upward as one goes to larger values of r. Another way of seeing the reason for this upward curvature is to note that the effective kinetic energy is negative; so that if g is taken to be positive, the slope of the wave function must increase with increasing r. This increase of slope continues from the origin until $r = b$, where the effective kinetic energy becomes positive, so that the wave function begins to curve back downward. The first bound state will occur at that energy for which g curves downward just sufficiently to meet a decaying exponential when $r > a$. The second bound state occurs when it meets the decaying exponential after passing through a node; the third, after passing through two nodes, etc.

Problem 4: Find the energy levels when $l > 0$ by the WKB approximation. Note that when $l \neq 0$, there is no need to impose the boundary condition that $g = 0$ at the origin, because the effective centrifugal barrier accomplishes this result automatically. The usual WKB treatment (Sec. 13, Chap. 12), defining the energy levels, may be used here.

The small value of the wave function near the origin when $l > 0$ is the result, of course, of the repulsive effects of the centrifugal potential. One can see that it is very unlikely that a particle of high angular momentum can be found near the origin. In fact, the particle is unlikely to be found until we go out to a radius large enough to make the effective radial kinetic energy positive. This will occur where

$$\frac{\hbar^2}{2m}\frac{l(l+1)}{r^2} \cong \frac{p^2}{2m}$$

where p is the total momentum. The particle is therefore unlikely to go to radii smaller than that given by $pr = \hbar\sqrt{l(l+1)}$. A rough way of thinking about this result is to say that a particle with momentum p is not likely to get closer to the origin than a distance at which a classical particle moving with this momentum in a circular orbit would have the appropriate angular momentum.

9. Definition of Quantum Numbers. In atomic problems it has become customary to designate bound states by the following three quantum numbers:

n = the principal quantum number.
l = orbital angular momentum quantum number.
m = azimuthal quantum number = z component of angular momentum, in units of $\hbar$.

The principal quantum number is defined as one plus the total number of nodal surfaces in the wave function. If one uses r times the wave function, which is also equal, in our case, to $g_l Y_l^m(\vartheta, \varphi)$, the principal quantum number is then equal to the total number of nodal surfaces in this function, provided that we regard the node at the origin as a surface. Since there are l nodal surfaces* in a spherical harmonic of degree l, the principal quantum number is then $l + N$, where N is the number of nodes in the radial function g (including the one at the origin).

The reason for this definition of n will become clear later, when we obtain the exact energy levels of the hydrogen atom.

If electron spin is taken into account, the quantum numbers must be further modified, in a way that will be treated in Chap. 17.

10. Physical Interpretation of Wave Functions of Different n, l, m.

Case 1: $l = 0$. These states have zero angular momentum. A classical orbit of zero angular momentum is one in which the particle moves in a radial direction, oscillating back and forth, plunging into the

* Chap. 14, Sec. 15

center of the atom and back out again periodically. Quantum-mechanically, one cannot speak of an exact orbit, but, instead, a wave packet must be made up. In order to make up a packet that moves on a definite radial line (i.e., definite angles, φ and ϑ), we must include many different angular momenta (see Chap. 14, Sec. 21). We can therefore no longer say that for a state of zero angular momentum the particle goes exactly through the center of the atom, just because the exact definition of its path requires many angular momenta. Yet, one can say that for an s state the particle comes closer to the nucleus on the average than for any other state. This is because of the absence of the centrifugal barrier. This fact is reflected in the form of the wave function near the origin $g \sim r^{l+1}$. The greater l is, the smaller is g near the origin.

Case 2: $l > 0$. These states have nonzero angular momentum, so that they correspond to circular or elliptical orbits in the classical limit. A circular classical orbit is one in which r is defined exactly and remains constant for all time. Of course, this is impossible in the quantum theory, because of the uncertainty principle. Yet, the most nearly circular orbit would be the one for which the radial part of the wave function was most localized and had the least number of nodes. If the wave function changes its sign many times in a given region, which it will do if there are many nodes, then there will be a correspondingly large momentum in the direction in which these changes of sign take place. This is because the mean value of p is just $\frac{\hbar}{i}\int \psi^* \nabla \psi \, dx$, so that if ψ changes sign rapidly, the resulting large contributions to $\nabla\psi$ will also produce large momenta.† If the radial wave function has several nodes, then it will correspond to an orbit that is farther from being circular than one which has only one node, simply because a circular orbit requires zero radial momentum.

Since the states with $l = n - 1$ have the minimum radial oscillation of the wave function, they correspond most nearly to circular orbits. To see this in greater detail, let us consider the complete wave function,

$$\psi = \frac{g_l(r)}{r} Y_l^m(\vartheta, \varphi)$$

We have already seen that the different values of m describe different orientations of the plane of the (approximate) orbit.‡ To get an orbital plane that is approximately normal to the z axis, we choose $l = m$ (see

† Actually, the radial momentum operator is not $\frac{\hbar}{i}\frac{\partial}{\partial r}$, but $\frac{\hbar}{i}\left(\frac{\partial}{\partial r} + \frac{1}{r}\right)$. (See Dirac, 3d. ed., p. 153.) The additional term does not, however, alter the argument appreciably.

‡ Chap. 14, Secs. 12, 15, and 21.

Chap. 14 Sec. 15). Then if we choose $n - 1 = l$ (i.e., one node in the radial wave function), the wave will be large only within a toroidal region centered in the xy plane and about the radius for which the effective potential, shown in Fig. 4, is a minimum. Thus, we justify the picture of the wave function suggested in Chap. 3, Sec. 15.

11. Formation of Wave Packets. For states in which $n - 1 > l$, there will be nodes in the radial wave function besides the one at the origin. These correspond, as we have seen, to additional radial momentum. In the classical limit, these wave functions represent elliptic orbits. It may seem at first sight that since ψ is a product of radial and angular wave functions, there can be no correlation between r and φ. (φ is necessary to describe an elliptic orbit.) To obtain this correlation, one must make a wave packet, using a range of values of l. We shall not carry this procedure out here, but shall only quote the result that in this way elliptical toroidal wave packets can be formed, when l and n are large. The reason that l and n have to be large is that in order to form a packet one needs functions that oscillate a great deal; otherwise one could not get destructive interference a long way from the center of the packet.

12. Exact Solution for Hydrogen Atom. We shall now solve the hydrogen atom problem exactly. In doing this, it is convenient to make the following substitution in eq. (3):

$$r = \frac{\hbar}{\sqrt{2\mu}}\, x \qquad \sqrt{2\mu}\,\frac{e^2}{\hbar} = k \qquad E = -W \tag{17a}$$

where W is the binding energy. Eq. (3) then becomes

$$\frac{d^2g}{dx^2} + \left[\frac{k}{x} - \frac{l(l+1)}{x^2} - W\right] g = 0 \tag{17b}$$

We have already seen that for large x the solution is approximately $g = e^{-\sqrt{W}x}$ [see eq. (13)]. This fact suggests that it will be convenient to write the solution as

$$g = Ue^{-\sqrt{W}x}$$

Insertion of this value of g into eq. (17) yields

$$\frac{d^2U}{dx^2} - 2\sqrt{W}\,\frac{dU}{dx} + \left[\frac{k}{x} - \frac{l(l+1)}{x^2}\right] U = 0 \tag{18}$$

Our boundary condition on U is now that as $x \to \infty$, $Ue^{-\sqrt{W}x}$ must approach zero rapidly enough so that the integrated probability is finite or, in other words, that $\int_0^\infty U^2 e^{-2\sqrt{W}x}\, dx$ converges.

There are many ways to solve this problem. For example, the method of factoring the differential equation used with the harmonic oscillator*

* See Chap. 13.

will also work here. We shall, however, take this opportunity to illustrate a more common, but less elegant, method, namely, expansion in a power series.

Let us therefore try to obtain a solution of the form

$$U = \sum_{N=0}^{\infty} C_N x^{N+s}$$

We use the form x^{N+s} because we already know from eq. (16c) that the solution will, in general, start out with some power of x; the value of s will be determined from the differential equation.

Substitution of the above expression into the differential eq. (18) yields

$$\sum_N C_N[(N+s)(N+s-1)x^{N+s-2} - 2(N+s)\sqrt{W}\,x^{N+s-1} - l(l+1)x^{N+s-2} + kx^{N+s-1}] = 0 \quad \text{(19a)}$$

We now collect all terms with equal powers of x, obtaining

$$\sum_N x^{N+s-2}\{C_N[(N+s)(N+s-1) - l(l+1)] - C_{N-1}[2\sqrt{W}(N+s-1) - k]\} = 0 \quad \text{(19b)}$$

In order that this equation be true for arbitrary x, the coefficient of each power of x must vanish. This leads to the following set of equations:

$$\frac{C_N}{C_{N-1}} = \frac{2(N+s-1)\sqrt{W} - k}{(N+s)(N+s-1) - l(l+1)} \quad \text{(19c)}$$

Since, by hypothesis, $C_{-1} = 0$ and $C_0 \neq 0$, it follows that

$$s(s-1) = l(l+1)$$

The above is known as the indicial equation; it determines the lowest power of x appearing in the expansion. Its solutions are

$$s = l + 1 \qquad s = -l$$

This result is in agreement with eq. (16b), obtained by solving the equation by an approximation good only at small x. Because g must vanish at the origin, $s = -l$ leads to an inadmissible wave function. We must therefore start with $s = l + 1$.

For any choice of C_0, C_1 is then determined by eq. (19c), by setting $N = 1$. C_2 can be obtained from C_1 by setting $N = 2$, and so on. Thus, we can obtain a complete solution, provided that the series converges.

It is not hard to prove that the series converges for all values of x. To do this, we note that the ratio of successive terms in the series is

$$\frac{C_N x}{C_{N-1}} = \frac{[2(N + s - 1)\sqrt{W} - k]x}{(N + s)(N + s - 1) - l(l + 1)} \tag{19d}$$

For large N, this ratio is asymptotic to $2\sqrt{W}\,x/N$; this is the same ratio as for the exponential series $e^{2\sqrt{W}x} = \sum_N \frac{(2\sqrt{W}\,x)^N}{N!}$. Now, it is shown in mathematics* that two series for which this limiting ratio is the same and not equal to unity converge or diverge together. Since the exponential series converges for all values of x, so does the series that we have derived.

The next question is to see how the solution behaves as $x \to \infty$. To do this, we use an extension of the above theorem, which states that as $x \to \infty$, two series for which the ratio in eq. (19d) is the same and not equal to unity have the same type of approach to infinity. In other words, as $x \to \infty$ our wave function will show the same behavior as $e^{2\sqrt{W}x}$. Thus, we write

$$g = Ue^{-\sqrt{W}x} \sim e^{+\sqrt{W}x} \qquad \text{as} \qquad x \to \infty$$

In general, the above series therefore leads to an inadmissible wave function. An exception arises, however, if the series terminates at a finite value of N, for then g will be just $e^{-\sqrt{W}x}P(x)$, where $P(x)$ is a polynomial in x. This function is clearly quadratically integrable. The condition for termination is that C_N shall vanish for some finite value of N, which must be at least unity if the solution itself is to be nonvanishing. According to eq. (19c) (setting $s = l + 1$), this will happen when

$$2(N + l)\sqrt{W} = k \qquad \text{or} \qquad W = \frac{k^2}{4(N + l)^2} \tag{19e}$$

From the definition of k, eq. (17a), we obtain

$$E = -W = \frac{-\mu e^4}{2\hbar^2(N + l)^2} \tag{20}$$

These are the energy levels of the hydrogen atom. Note that the reduced mass should be used here.

Since the series terminates after N terms, it is clear that $U(x)$ is equal to x^{l+1}, multiplying a polynomial of degree $N - 1$; hence it can have at most N real zeros. It can be shown that it must always have N real zeros, but we shall not do so here. We therefore conclude that the wave function has N nodes (including the one at the origin produced by the factor x^{l+1}). One can compare this wave function with the general form discussed in Secs. 7 and 8 and see that the two are the same. Thus,

* See, for example, E. T. Whittaker and G. N. Watson. *A Course of Modern Analysis*, London: Cambridge University Press, 1920, 3d. ed., p. 18.

we have

$$g = x^{l+1}p_N^l(x)e^{-\sqrt{W}x} \tag{21}$$

where $p_N^l(x)$ is a polynomial of degree N.* The wave function starts out as x^{l+1}, goes through $N - 1$ nodes, besides the one at the origin, and finally dies out exponentially.

From our definition of the principal quantum number (see Sec. 9), we have $n = N + l$. The energy levels of a hydrogen atom are therefore given by

$$E = -\frac{\mu e^4}{2\hbar^2 n^2} \tag{22}$$

These are the same as those derived by Bohr from his early quantum theory [Chap. 2, eq. (19)].

13. Degeneracy of Hydrogen Energy Levels. We note that the energy levels of hydrogen depend only on the principal quantum number n, and not on l or m. This statement means that, in order to know the energy, it is sufficient to know only the value of n, and that after n is specified, the energy does not depend on the values of l or m. Hence, there are, in general, many different quantum states that have the same energy, and the system is therefore degenerate.† The lowest state occurs with $n = 1$; there is then only a single node at the origin in the radial wave function. Since $n = N + l$, where N is the number of nodes in the radial wave function, we must have $l = 0$ for this state. This state is therefore nondegenerate, since there is only one wave function which has $n = 1$, namely the one for which $l = m = 0$. The next state has $n = 2$; here there are four possible states. We may have one node in the radial function and $l = 1$, in which case m can take on the values $-1, 0, 1$. We can also have two nodes in the radial function ($N = 2$) and $l = m = 0$. As we go higher, the degree of degeneracy increases.

The property that the energy is not a function of l is possessed only by the hydrogen atom and the three-dimensional isotropic harmonic oscillator.‡ For example, in atoms other than hydrogen, the energy is a function not only of n, but also of l. This is because the potential energy of a given electron in non-hydrogenic atoms is not $-Ze^2/r$, but is modified by the screening effects of other electrons. The greater the deviation from a Coulomb potential, the bigger will be the energy difference of levels of the same n and different l. Since the largest deviations occur for the heaviest atoms, there will be a general tendency toward increasing the separation of energy levels of the same n and different l as the atomic number is increased. Even in hydrogen, the

* See Sec. 14 for a precise definition of $p_N^l(x)$.

† Chap. 10, Sec. 14.

‡ For a discussion of classical and early quantum treatments of degeneracy, see Chap. 2, Sec. 14.

degeneracy of levels of the same n can be removed by impressing an external electric field, which causes each level to change its energy by an amount that depends on l. This is known as the first-order Stark effect, and we shall discuss it later in connection with perturbation theory.* The effects of spin and relativity also produce a small splitting of these levels, called the fine structure.

The degeneracy of levels of the same l and n, but different m, is common to all central fields, i.e., to all potentials that are functions of the radius only. This can be seen from the fact that the radial equation (2a), which determines the allowed energies, does not contain m, but

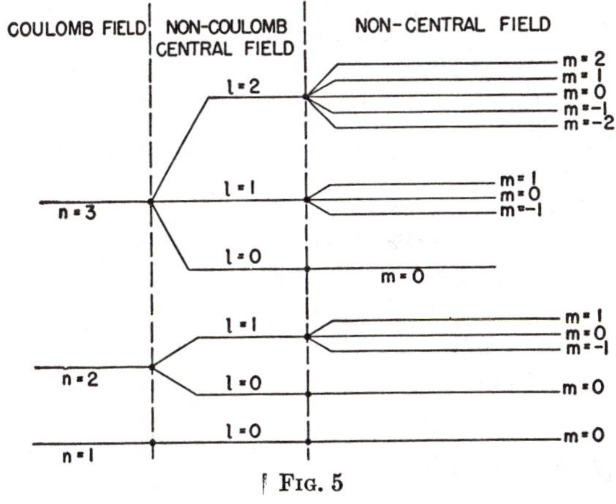

FIG. 5

contains only the total angular momentum l. This degeneracy is removed, however, when the field is noncentral. Such a noncentral field may be supplied, for example, by an external magnetic field, which then causes levels of different m to have different energy. This splitting of energy levels in an external magnetic field gives rise to the Zeeman effect, which we shall study in Sec. 27.

In Fig. 5 is shown a schematic diagram indicating the degeneracy of the various energy levels of hydrogen and the manner in which this degeneracy is removed.

It should be noted that both the signs and the magnitudes of the different shifts of energy levels can vary, depending on the type of force that is causing the splitting of the levels.

14. The Laguerre Polynomials and the Associated Laguerre Polynomials. The polynomials which are solutions to eq. (18) had already been studied independently in a mathematical way by Laguerre long before the Schrödinger wave equation was discovered. The Laguerre polynomials are special cases of a class of functions called confluent hypergeometric functions.

* Chap. 19, Sec. 11.

The Laguerre polynomials are defined as follows:*

$$L_r(\rho) = e^{\rho} \frac{d^r}{d\rho^r} (\rho^r e^{-\rho}) \tag{23}$$

The associated Laguerre functions are obtained by differentiating the Laguerre polynomials. Thus,

$$L_r^s(\rho) = \frac{d^s}{d\rho^s} L_r(\rho) \tag{24}$$

It is easily verified by direct substitution that these functions satisfy the equation

$$L_r^{s\prime\prime}(\rho) + \left(\frac{s+1}{\rho} - 1\right) L_r^{s\prime}(\rho) + \frac{(r-s)}{\rho} L_r^s(\rho) = 0 \tag{25}$$

If one writes $U = x^{l+1}v$, one obtains from eq. (18)

$$v'' + \left[\frac{2(l+1)}{x} - 2\sqrt{W}\right] v' + \frac{[k - 2\sqrt{W}(l+1)]}{x} v = 0 \tag{26a}$$

We eliminate k by using (19e). Thus $k = 2\sqrt{W}(N + l) = 2\sqrt{W}\,n$. With this relation, and with the substitution $z = 2\sqrt{W}\,x$, eq. (26a) becomes

$$\frac{d^2v}{dz^2} + \left[\frac{2(l+1)}{z} - 1\right]\frac{dv}{dz} + \frac{n-(l+1)}{z} v = 0 \tag{26b}$$

If we choose $r = n + l$, $s = 2l + 1$, $\rho = z = 2\sqrt{W}\,x$, eq. (26b) becomes identical with (25). We therefore obtain

$$v = L_{n+l}^{2l+1}(2\sqrt{W}\,x) \tag{27}$$

The complete solution of the wave equation is

$$\psi_{n,l}^m(x) \sim e^{-\sqrt{W}x} x^l L_{n+l}^{2l+1}(2\sqrt{W}\,x) Y_l^m(\vartheta, \varphi) \tag{28}$$

The wave function can be normalized with the aid of the relation†

$$\int_0^\infty e^{-\rho}\rho^{2l}\,[L_{n+l}^{2l+1}(\rho)]^2\rho^2\,d\rho = \frac{2n[(n+l)!]^3}{(n-l-1)!} \tag{29}$$

Let us now go back to r as the independent variable. We use the substitutions

$$x = \sqrt{2\mu}\,\frac{r}{\hbar} \qquad \sqrt{W} = \frac{k}{2n} = \sqrt{2\mu}\,\frac{e^2}{2n\hbar}$$

We also write $a_0 = \hbar^2/\mu e^2$ = radius of first Bohr orbit. The result is

$$\psi_{n,l}^m \sim e^{-r/na_0}\left(\frac{r}{a_0}\right)^l L_{n+l}^{2l+1}\left(\frac{2r}{na_0}\right) Y_l^m(\vartheta, \varphi) \tag{30}$$

* See Pauling and Wilson, pp. 130–132.
† Pauling and Wilson, p. 451.

For the special case in which $n = l + 1$, the above wave function becomes particularly easy to interpret, because this case corresponds most nearly to the classical circular orbit. From eq. (24), we see that we must differentiate a polynomial of $(2l + 1)$th degree $2l + 1$ times, thus obtaining a constant. The final result is

$$\psi_{l+1,l}^m \sim e^{-r/na_0} \left(\frac{r}{a_0}\right)^l Y_l^m(\vartheta, \varphi) \tag{31}$$

A graph of $g(r) = f(r)/r$ is shown schematically in Fig. 6 (compare with Fig. 4). The maximum value occurs where

$$r = n(l + 1)a_0 = n^2 a_0$$

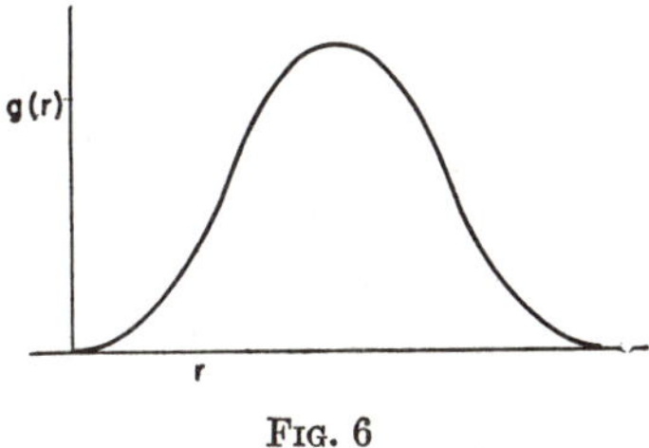

Fig. 6

But this is exactly where the nth Bohr orbit occurs in the early quantum theory. Thus, we see how the wave function tends to center around the old Bohr orbits. When $n > l + 1$, the wave function will have a polynomial in front of e^{-r/na_0}. It will therefore show a few oscillations, just as we were led to believe in the qualitative discussion of the shape of the wave function.

Problem 5: Show that the radial wave functions $g_n^l(r)$ are orthogonal, in the sense that

$$\int g_n^l(r) g_{n'}^l(r)\, dr = 0 \qquad \text{when} \qquad n \neq n'$$

Should they be orthogonal for different values of l? Explain your answer.

Problem 6: Express an arbitrary function as a series of hydrogen atom eigenfunctions and show how to calculate the coefficients in the expansion. Note that we must integrate over the continuum wave functions obtained when $E > 0$, as well as sum over the discrete bound state levels. For a discussion of continuum levels, see Chap. 21, Secs. 58 and 59.

15. Three-dimensional Harmonic Oscillator. Thus far, we have treated only the one-dimensional harmonic oscillator.* It is instructive to extend this treatment to three dimensions, not only because the three-dimensional oscillator is of some importance in itself, but also because the problem can be solved, as we shall see, in two different ways, each of which illustrates some important quantum-mechanical principles.

It is shown in mechanics that it is always possible to obtain a frame of co-ordinates in which the potential energy of a three-dimensional harmonic oscillator is

$$V = \frac{m}{2}(\omega_x^2 x^2 + \omega_y^2 y^2 + \omega_z^2 z^2) \tag{32}$$

ω_x, ω_y, ω_z are respectively the angular frequencies of the x, y, and z components of the oscillation. Note that, in general, all three may be differ-

* See Chap. 13.

ent. The co-ordinate axes with respect to which the potential takes this particularly simple form are called the principal axes. More generally, when the axes are other than the principal axes, the potential takes the form $\Sigma A_{ij}x_ix_j$, where $x_i = x, y, z$ for $i = 1, 2, 3$, respectively.

An example of a three-dimensional harmonic oscillator is supplied by an atom in a crystal. Such an atom has an equilibrium position in the lattice, about which it executes simple harmonic motion when it suffers a small disturbance. If the crystal is anisotropic, the angular frequencies of oscillations along the three principal axes of the crystal are all different. For an isotropic crystal, the three ω's are the same, and we obtain

$$V = \frac{m\omega^2}{2}(x^2 + y^2 + z^2) = \frac{m\omega^2 r^2}{2} \tag{33}$$

Thus, in general, V is not a radially symmetrical function, except when all three ω's are the same.

Schrödinger's equation becomes

$$\nabla^2\psi + \frac{2m}{\hbar^2}\left[E - \frac{m}{2}(\omega_x^2x^2 + \omega_y^2y^2 + \omega_z^2z^2)\right]\psi = 0 \tag{34}$$

This equation can be solved by separation of variables. Let us write

$$\psi = X(x)Y(y)Z(z) \tag{35}$$

Schrödinger's equation can then be written

$$\left[\frac{1}{X}\frac{d^2X}{dx^2} - \left(\frac{m\omega_x x}{\hbar}\right)^2\right] + \left[\frac{1}{Y}\frac{d^2Y}{dy^2} - \left(\frac{m\omega_y y}{\hbar}\right)^2\right] + \left[\frac{1}{Z}\frac{d^2Z}{dz^2} - \left(\frac{m\omega_z z}{\hbar}\right)^2\right] = -\frac{2mE}{\hbar^2} \tag{36}$$

To obtain a solution, we must have each of the above three brackets identically equal to constants, which we denote by

$$-\frac{2m}{\hbar^2}E_x \qquad -\frac{2m}{\hbar^2}E_y \qquad -\frac{2m}{\hbar^2}E_z$$

respectively. The equations then become

$$\left.\begin{aligned}
&\frac{d^2X}{dx^2} + \left[\frac{2m}{\hbar^2}E_x - \left(\frac{m\omega_x}{\hbar}x\right)^2\right]X = 0\\
&\frac{d^2Y}{dy^2} + \left[\frac{2m}{\hbar^2}E_y - \left(\frac{m\omega_y}{\hbar}y\right)^2\right]Y = 0\\
&\frac{d^2Z}{dz^2} + \left[\frac{2m}{\hbar^2}E_z - \left(\frac{m\omega_z}{\hbar}z\right)^2\right]Z = 0\\
&E = E_x + E_y + E_z
\end{aligned}\right\} \tag{37}$$

Each of the equations is the same as that of the one-dimensional harmonic oscillator. The energies are therefore

$$E_x = \hbar\omega_x(n_x + \tfrac{1}{2}) \qquad E_y = \hbar\omega_y(n_y + \tfrac{1}{2}) \qquad E_z = \hbar\omega_z(n_z + \tfrac{1}{2}) \tag{38}$$

16. Possibility of Degeneracy of Energy Levels. If ω_x, ω_y, ω_z are all different, then no two levels will coincide, unless there exists a relation among the ω's such that

$$\gamma_x\omega_x + \gamma_y\omega_y + \gamma_z\omega_z = 0$$

where γ_x, γ_y, γ_z are suitable integers (which may be either positive or negative). If such a relation exists, the ω's are said to be linearly dependent; otherwise, they are linearly independent.

It is clear that if the ω's are linearly dependent, then one can always find a new level which has the same energy as a given one by adding γ_x to n_x, γ_y to n_y, and γ_z to n_z, for in this case, the energy is

$$E = (E_x + E_y + E_z) = \hbar[(n_x + \gamma_x)\omega_x + (n_y + \gamma_y)\omega_y + (n_z + \gamma_z)\omega_z] + \frac{\hbar}{2}(\omega_x + \omega_y + \omega_z)$$

$$= \hbar(n_x\omega_x + n_y\omega_y + n_z\omega_z) + \frac{\hbar}{2}(\omega_x + \omega_y + \omega_z)$$

Thus, if the ω's are linearly dependent, the system will be degenerate.

17. Spherically Symmetric Case. The most degenerate possible case occurs where all the ω's are equal. Here we obtain

$$E = \hbar\omega(n_x + n_y + n_z + \tfrac{3}{2}) \tag{39}$$

If we define $n_x + n_y + n_z = N$, then the degree of degeneracy of a level is equal to the number of ways that N can be written as the sum of three nonnegative integers. For example, for $N = 0$, the level is nondegenerate; for $N = 1$ it is triply degenerate (either n_x, n_y, or n_z may be 1); for $N = 2$ it is six-fold degenerate, etc.

18. Form of Wave Functions for Spherically Symmetric Case. The wave functions are simply the products of the three eigenfunctions of the one-dimensional oscillators. Thus, we obtain for the unnormalized functions [see eq. (28), Chap. 13]

$$\psi_{n_x,n_y,n_z}(x, y, z) = \exp\left[-\frac{m\omega}{2\hbar}(x^2 + y^2 + z^2)\right] h_{n_x}\left(\sqrt{\frac{m\omega}{\hbar}}\,x\right) h_{n_y}\left(\sqrt{\frac{m\omega}{\hbar}}\,y\right) h_{n_z}\left(\sqrt{\frac{m\omega}{\hbar}}\,z\right) \tag{40}$$

The above is the eigenfunction corresponding to the quantum numbers n_x, n_y, and n_z.

19. An Important Property of Degenerate Eigenfunctions. If a given set of ψ's all belong to the same energy level, then they possess the following important property: any linear combination of this set of ψ's also belongs to the same energy level. For example, let $\psi_{i,n}$ represent the ith member of a set of eigenfunctions belonging to the energy level E_n. Then it follows that the wave function $U = \sum_i A_i\psi_{i,n}$ also belongs to the level E_n, where the A_i are arbitrary constants. The proof of this statement is fairly obvious. One simply writes

$$HU = \sum_i A_i H\psi_{i,n} = \sum_i A_i E_n \psi_{i,n} = E_n \sum_i A_i \psi_{i,n} = E_n U$$

20. Relation of Hermite Polynomials to Spherical Harmonics. At this point, one can note that if V is a function of the radius only, it should be possible to express the solution as the product of a radial function and a spherical harmonic, as was done, for example, with the hydrogen atom. This could be done by solving the radial equation, but we shall obtain this expression directly from the solution given in eq. (40). To do this, let us begin with the simplest cases first.

Case 1: $n_x = n_y = n_z = 0$. For this case, the wave function is just

$$\exp\left(-\frac{m\omega}{2\hbar}r^2\right)$$

Since ψ is not a function of ϑ and φ, we see that the lowest state is an s state; thus, it is already expressed as a product of a radial function and the zeroth spherical harmonic.

Case 2: *First Excited State.* This level is, as we have seen, triply degenerate. One can have either n_x, n_y, or $n_z = 1$, while the others are zero. The three unnormalized eigenfunctions are, respectively [see eq. (28), Chap. 13, for $h_n(x)$, and so on]

$$\left.\begin{aligned}
n_x = 1:\quad & \exp\left[-\left(\frac{m\omega}{2\hbar}r^2\right)\right]x = r\exp\left[-\left(\frac{m\omega}{2\hbar}r^2\right)\right]\sin\vartheta\cos\varphi \\
& \sim r\exp\left[-\left(\frac{m\omega}{2\hbar}r^2\right)\right][Y_1^1(\vartheta,\varphi) + Y_1^{-1}(\vartheta,\varphi)] \\
n_y = 1:\quad & \exp\left[-\left(\frac{m\omega}{2\hbar}r^2\right)\right]y = r\exp\left[-\left(\frac{m\omega}{2\hbar}r^2\right)\right]\sin\vartheta\sin\varphi \\
& \sim r\exp\left[-\left(\frac{m\omega}{2\hbar}r^2\right)\right][Y_1^1(\vartheta,\varphi) - Y_1^{-1}(\vartheta,\varphi)] \\
n_z = 1:\quad & \exp\left[-\left(\frac{m\omega}{2\hbar}r^2\right)\right]z = r\exp\left[-\left(\frac{m\omega}{2\hbar}r^2\right)\right]\cos\vartheta \\
& \sim r\exp\left[-\left(\frac{m\omega}{2\hbar}r^2\right)\right]Y_1^0(\vartheta,\varphi)
\end{aligned}\right\} \quad (41)$$

[See eq. (71), Chap. 14, for definition of $Y_1^m(\vartheta, \varphi)$]. By forming suitable linear combinations of the three degenerate eigenfunctions, we can obtain wave functions that are simply products of radial functions and spherical harmonics. Thus,

$$\left.\begin{aligned}
\exp\left[-\left(\frac{m\omega}{2\hbar}r^2\right)\right](x+iy) &\sim r\exp\left[-\left(\frac{m\omega}{2\hbar}r^2\right)\right]\sin\vartheta\exp(i\varphi)\\
&\sim r\exp\left[-\left(\frac{m\omega}{2\hbar}r^2\right)\right]Y_1^1(\vartheta,\varphi)\\
\exp\left[-\left(\frac{m\omega}{2\hbar}r^2\right)\right](x-iy) &\sim r\exp\left[-\left(\frac{m\omega}{2\hbar}r^2\right)\right]\sin\vartheta\exp(-i\varphi)\\
&\sim r\exp\left[-\left(\frac{m\omega}{2\hbar}r^2\right)\right]Y_1^{-1}(\vartheta,\varphi)\\
\exp\left[-\left(\frac{m\omega}{2\hbar}r^2\right)\right]z &\sim r\exp\left[-\left(\frac{m\omega}{2\hbar}r^2\right)\right]Y_1^0(\vartheta,\varphi)
\end{aligned}\right\} \tag{42}$$

For the higher excited states, similar methods may be applied. For example, in the second excited state, we can either include $h_2(x)$, $h_2(y)$, $h_2(z)$, or products like $h_1(x)h_1(y)$. All of these polynomials can be expressed in terms of r, ϑ, and φ. When this is done, we find that in the second excited state the angular factors contain $Y_2^m(\vartheta, \varphi)$ and Y_0. The six degenerate states can then be re-expressed in terms of the five states in which $l = 2$, and one state in which $l = 0$.

If the three ω's had not been equal, then V would not have been a function of r alone, and the expression of the wave function as simple products of radial functions and spherical harmonics would have been impossible. For example, the lowest state wave function would then be

$$\exp\left[-\frac{m}{2\hbar}(\omega_x^2x^2 + \omega_y^2y^2 + \omega_z^2z^2)\right]$$

There is no way to write this as a function of r times a spherical harmonic.

One can give a simple physical interpretation to the various wave functions. For example, the case $n_z = 1$, which corresponds to oscillation only in the z direction, has $Y_1^0(\vartheta, \varphi)$ for its angular factor; hence there is, as one would expect, no z component of angular momentum. The case $n_y = 1$, corresponding to oscillation in the y direction, has an equal probability* that L_z is $\pm\hbar$. This means that although the average value of L_z is zero, as we would expect, this zero value is achieved by having an equal probability that L_z is ± 1. The above result reflects the fact that even though the particle is moving in only the y direction, it may be on either side of the origin, and may, therefore, have either a positive or a negative component of the angular momentum.

21. The Hamiltonian for a Charged Particle in a Given Electromagnetic Field. We wish now to extend the previous theory to the treatment of a charged particle in an electromagnetic field that is specified exter-

* See Chap. 14, Sec. 17.

nally. In other words, we assume that the electromagnetic field is produced entirely by charges and currents other than the one that we are considering, and neglect fields produced by the charge that we are studying. Such a problem might arise, for example, if we had an atom in an external magnetic field or if the atom were illuminated by light produced by other atoms. In this section, our objective will be to show how this problem can be given a quantum-mechanical formulation.

The first step is to obtain the classical Hamiltonian function. In terms of the vector potential, $\boldsymbol{a}(x, y, z, t)$, and the scalar potential, $\phi(x, y, z, t)$, we shall see in Problem 7 that this Hamiltonian is

$$H = \frac{\left(\boldsymbol{p} - \frac{e}{c}\boldsymbol{a}\right)^2}{2m} + V(\boldsymbol{x}) + e\phi \tag{43}$$

where V is that part of the potential energy which is of nonelectromagnetic origin. Note that the only new step has been to replace $\boldsymbol{p}$ by

$$\boldsymbol{p} - \frac{e}{c}\boldsymbol{a}$$

The equations of motion are derived from the canonical equations

$$\dot{q}_i = \frac{\partial H}{\partial p_i} \qquad \dot{p}_i = -\frac{\partial H}{\partial q_i}$$

Using the above Hamiltonian, we obtain for the velocity

$$\dot{\boldsymbol{x}} = \frac{1}{m}\left[\boldsymbol{p} - \frac{e}{c}\boldsymbol{a}(\boldsymbol{x})\right] \qquad \text{or} \qquad \boldsymbol{p} = m\dot{\boldsymbol{x}} + \frac{e}{c}\boldsymbol{a}(\boldsymbol{x})$$

Note that when there is a vector potential, the *canonical* momentum, $\boldsymbol{p}$, is no longer equal to its customary value of $m\dot{\boldsymbol{x}}$.

Problem 7: Show that the above Hamiltonian leads to the correct classical equations of motion, which are

$$m\frac{d^2\boldsymbol{r}}{dt^2} = -\nabla V + e\boldsymbol{\mathcal{E}} + \frac{e}{c}\boldsymbol{v} \times \boldsymbol{\mathcal{H}}$$

where $\boldsymbol{\mathcal{E}}$ is the electric field and $\boldsymbol{\mathcal{H}}$ is the magnetic field.

HINT: Write

$$\frac{d\boldsymbol{a}}{dt} = \frac{\partial \boldsymbol{a}}{\partial t} + (\boldsymbol{v}\cdot\nabla)\boldsymbol{a}$$

and note that

$$(\boldsymbol{v}\cdot\nabla)\boldsymbol{a} = -\boldsymbol{v} \times (\nabla \times \boldsymbol{a}) + \nabla(\boldsymbol{v}\cdot\boldsymbol{a})$$

22. Quantum-mechanical Hamiltonian. To obtain the quantum-mechanical Hamiltonian operator, we follow the usual procedure of replacing $\boldsymbol{p}$ by the operator $(\hbar/i)\nabla$ wherever it occurs. The Hamiltonian is then

$$\frac{\left(\frac{\hbar}{i}\nabla - \frac{e}{c}a\right)^2}{2m} + e\phi + V = \frac{-\hbar^2}{2m}\nabla^2 + e\phi + V$$
$$- \frac{\hbar e}{2mci}(a \cdot \nabla + \nabla \cdot a) + \frac{e^2}{2mc^2}a^2 \quad (44)$$

23. Conservation of Probability. Probability Current. The above Hamiltonian is clearly Hermitean, so that probability is conserved. Nevertheless, it is useful to calculate the change of probability, in order to obtain an expression for the probability current. Using the expression $i\hbar(\partial\psi/\partial t) = H\psi$, we obtain

$$\frac{\partial P}{\partial t} = \frac{\partial}{\partial t}(\psi^*\psi) = \frac{\partial\psi^*}{\partial t}\psi + \psi^*\frac{\partial\psi}{\partial t}$$
$$= -\frac{\psi}{2mi\hbar}\left(-\frac{\hbar}{i}\nabla - \frac{e}{c}a\right)\left(-\frac{\hbar}{i}\nabla - \frac{e}{c}a\right)\psi^*$$
$$+ \frac{\psi^*}{2mi\hbar}\left(\frac{\hbar}{i}\nabla - \frac{e}{c}a\right)\cdot\left(\frac{\hbar}{i}\nabla - \frac{e}{c}a\right)\psi$$

With the aid of a little algebra, the above can be combined to yield

$$\frac{\partial P}{\partial t} + \nabla \cdot \left[\frac{\hbar}{2mi}\left(\psi^*\nabla\psi - \psi\nabla\psi^*\right) - \frac{e}{mc}a\psi^*\psi\right] = 0 \quad (45)$$

If we write

$$S = \frac{\hbar}{2mi}(\psi^*\nabla\psi - \psi\nabla\psi^*) - \frac{e}{mc}a\psi^*\psi \quad (46)$$

we obtain

$$\frac{\partial P}{\partial t} + \nabla \cdot S = 0$$

Thus, in order to obtain a conserved charge, we must modify the definition of current when a vector potential is present.

Problem 8: Prove eqs. (45) and (46).

24. Classical Limit. It should be recalled that we did not prove that Schrödinger's equation approaches Newton's laws of motion when a vector potential is present. This may be done, however, in a manner similar to that used in the absence of a vector potential, but this will be left as a problem.

Problem 9: Show that the average values of x and p satisfy the classical equations of motion when a vector potential only is present.

25. Gauge Invariance. Let us now see whether our theory is invariant to a gauge transformation (see Chap. 1, Sec. 3). In other words, we require that no physical result shall change when the potentials undergo a gauge transformation. If this requirement were not satisfied the equations of motion in the classical limit would in general be changed by a gauge transformation, in contradiction to the known fact that they are not changed under such a transformation. In this connection, it must be noted that even classically, the canonical momentum,

$$p = mv + \frac{e}{c}a$$

depends on the choice of gauge. The only physically significant quantities are those which are gauge invariant. In this case, the gauge invariant quantity is the velocity, $v = \frac{1}{m}\left(p - \frac{e}{c}a\right)$.

Problem 10: Prove that the velocity is invariant to a classical gauge transformation.

In quantum theory, there is no such thing as a velocity.* Instead, one has only an average velocity, which is the average of the operator $\frac{1}{m}\left(p - \frac{e}{c}a\right)$. As in the case of zero vector potential, this operator can be defined only when the position of the electron is not too well defined.

In order to demonstrate the property of gauge invariance of physically significant quantities, let us write the complete Schrödinger equation

$$i\hbar\frac{\partial\psi}{\partial t} = \frac{1}{2m}\left(\frac{\hbar}{i}\nabla - \frac{e}{c}a\right)^2\psi + (e\phi + V)\psi \tag{47}$$

Let us now make a gauge transformation.† Schrödinger's equation becomes

$$i\hbar\frac{\partial\psi}{\partial t} = \frac{1}{2m}\left(\frac{\hbar}{i}\nabla + \frac{e}{c}\Delta f - \frac{e}{c}a\right)^2\psi + (e\phi' + V)\psi + \frac{e}{c}\frac{\partial f}{\partial t}\psi$$

One can easily show that the above is equivalent to

$$i\hbar\frac{\partial}{\partial t}(e^{ief/\hbar}\psi) = \frac{1}{2m}\left(\frac{\hbar}{i}\nabla - \frac{e}{c}a'\right)^2(e^{ief/\hbar}\psi) + (e\phi' + V)(e^{ief/\hbar}\psi) \tag{48}$$

Thus, in terms of the new potentials, a' and φ', a new wave function, $\psi' = e^{ief/\hbar}\psi$, satisfies the same wave equation as was satisfied formerly by ψ itself. $e^{ief/\hbar}\psi$ is therefore a new solution. We note that the probability is the same with this new function as with the old. Furthermore, the probability current is the same expression in terms of ψ' and a' as it is in terms of ψ and a.

Problem 11: Prove the preceding statement.

* See Chap. 8, Sec. 6.

† The gauge transformation is $a \to a' - \nabla f$, $\phi \to \phi' + \frac{1}{c}\frac{\partial f}{\partial t}$. See Chap. 1, Sec. 3.

We can show that the expressions for all physically observable quantities are in a similar way left unchanged by a gauge transformation. We therefore conclude that in quantum theory, as well as in classical theory, a gauge transformation leads to no new physical consequences.

26. Special Case: Uniform Magnetic Field. One can readily verify that the following choice of vector potential leads to a uniform magnetic field, $\mathcal{H}$, directed in the z direction:

$$a_x = \frac{\mathcal{H}y}{2} \qquad a_y = \frac{\mathcal{H}z}{2} \qquad a_z = 0$$

Problem 12: Prove the above statement. Prove also that the potentials $a_x = \mathcal{H}_y$, $a_y = a_z = 0$, lead to the same magnetic field. Show that the two are related by a gauge transformation and find the gauge transformation.

Another example of a gauge transformation is the elimination of the scalar potential ϕ for radiation in free space. (This was done in Chap. 1 in connection with the radiation oscillators.*)

Hamiltonian With Constant Magnetic Field. With the above potentials, the Hamiltonian becomes

$$H = \frac{-\hbar^2}{2m}\nabla^2 + \frac{e\mathcal{H}}{2mc}\frac{\hbar}{i}\left(x\frac{\partial}{\partial y} - y\frac{\partial}{\partial x}\right) + \frac{e^2}{8mc^2}\mathcal{H}^2(x^2 + y^2) + V \quad (49)$$

27. The Zeeman Splitting of Levels of Different *m*. With the above Hamiltonian, one can treat a number of problems, for example, the Zeeman splitting of energy levels in a magnetic field. To do this. we express H in spherical polar co-ordinates,

$$H = \frac{-\hbar^2}{2\mu}\left(\frac{\partial^2}{\partial r^2} + \frac{2}{r}\frac{\partial}{\partial r} + \frac{L^2}{\hbar^2 r^2}\right) + \frac{e\mathcal{H}}{2\mu c}\frac{\hbar}{i}\frac{\partial}{\partial\varphi} + \frac{e^2\mathcal{H}^2}{8\mu^2c^2}\rho^2 + V(r) \quad (50)$$

where $\rho = x^2 + y^2$. Note that $V(r)$ is a spherically symmetric potential, which is assumed to be the type generally present in atoms. μ is the reduced mass. The above Hamiltonian leads to a wave equation that differs from the equation holding in the absence of a magnetic field in two respects.

(1) There is a field proportional term, $\frac{e\hbar}{2\mu c}\frac{\hbar}{i}\frac{\partial}{\partial\varphi}$, added to H.

(2) The effective potential is changed by the addition of $\frac{e^2\mathcal{H}^2}{2\mu c^2}\rho^2$.

The latter effect involves $\mathcal{H}^2$; thus for weak magnetic fields, it produces a second-order correction that may be neglected. To the first order, then, the only effect of the magnetic field is to change the Hamiltonian by $\nabla H = \frac{e\mathcal{H}}{2\mu c}\frac{\hbar}{i}\frac{\partial}{\partial\varphi}$. Note, however, that in this approximation the Hamiltonian still commutes with L^2 and with L_z, so that L^2, L_z, and H

* See Chap. 1, Sec. 3.

can be specified simultaneously. Let us assume that $L^2 = l(l+1)\hbar^2$ and $L_z = m\hbar$. We then obtain

$$H = \frac{-\hbar^2}{2\mu}\left[\frac{\partial^2}{\partial r^2} + \frac{2}{r}\frac{\partial}{\partial r} - \frac{l(l+1)}{r^2}\right] + \frac{e\mathcal{H}}{2\mu c}\hbar m + V(r) \tag{51}$$

The only effect of the magnetic field is then to add a constant to the energy, proportional to the azimuthal quantum number, m. This means that some of the degeneracy is removed because levels of different m now have different energies. This behavior is illustrated in Fig. 5. To a first approximation, the wave functions are not altered, since, with the neglect of the term, $\frac{e^2\mathcal{H}^2}{8\mu^2c^2}\rho^2$, the radial wave equation is exactly the same as before.

One can interpret the change of energy levels in a fairly simple way. One can readily show* that an electron circulating in an orbit with angular momentum L has a magnetic moment, $\boldsymbol{M} = eL/2\mu c$. The energy of a magnetic moment in a magnetic field $\boldsymbol{\mathcal{H}}$ is

$$W = \boldsymbol{M}\cdot\boldsymbol{\mathcal{H}} = \frac{e\boldsymbol{L}\cdot\boldsymbol{\mathcal{H}}}{2\mu c} = \frac{e\mathcal{H}}{2\mu c}L_z \tag{52}$$

($\boldsymbol{\mathcal{H}}$ is assumed to be in the z direction.) Writing $L_z = m\hbar$, we obtain

$$W = \frac{e\mathcal{H}}{2\mu c}\hbar m \tag{53}$$

The magnetic moment of an electron with unit angular momentum $\hbar$ is called a *Bohr magneton*. It is equal to $e\hbar/2\mu c$. Thus, the magnetic moment of an electron in an atom is some integral multiple of a Bohr magneton.

The splitting of energy levels derived above gives rise to a change in the pattern of spectral lines emitted by an atom. This change is known as the *Zeeman effect*. The effects of electron spin must be considered before a complete theory of the Zeeman effect can be given.†

The quadratic term, $\frac{e^2\mathcal{H}^2}{8\mu c^2}\rho^2$, may become important for very strong magnetic fields, where it leads to a general shifting and reordering of energy levels, which is connected with the Paschen-Back effect.‡

* See, for example, Richtmeyer and Kennard, p. 384.

† We shall discuss the Zeeman effect without spin in Chap. 18. For a treatment including the effects of spin, see Richtmeyer and Kennard, p. 399. See also White, *Introduction to Atomic Spectra*. New York: McGraw-Hill Book Company, Inc., 1934.

‡ White, *ibid.*

CHAPTER 16

Matrix Formulation of Quantum Theory

THUS FAR, WE HAVE FORMULATED quantum theory in terms of a wave function, $\psi(x)$, and in terms of linear operators, which operate on this function, and which are, in general, combinations of functions of x and $p = \frac{\hbar}{i}\frac{\partial}{\partial x}$. In this chapter, we shall develop an alternative formulation, originated by Heisenberg, in which the operators are expressed in terms of certain arrays of numbers known as matrices. We shall also demonstrate the equivalence of the two formulations. The matrix formulation has the advantage of greater generality, but the disadvantage that it is very difficult to use in the solution of special problems of appreciable complexity, such as for example, the stationary states of atoms.

1. Matrix Representation of an Operator. In order to obtain the matrix representation of an operator A, let us begin with some wave function, $\psi_m(x)$, which is any member of a complete orthonormal set of wave functions, $\psi_n(x)$. For example, ψ_n might be $\exp(2\pi inx/L)$ if we consider the orthonormal set involved in a Fourier series, or it might be $\exp(-x^2/2)h_n(x)$ where h_n is a Hermite polynomial. We now consider the new wave function $\varphi_m(x)$, obtained by operating on $\psi_m(x)$ with the operator A; i.e.,

$$A\psi_m(x) = \varphi_m(x)$$

Since the ψ_n form a complete orthonormal set, it must be possible to expand φ_m as a series of ψ_n. Thus,

$$A\psi_m(x) = \varphi_m(x) = \sum_n a_{nm}\psi_n(x) \tag{1}$$

If the numbers a_{nm} are known for all m and n, then the effect of the operator A on any wave function, $\psi = \sum_m C_m\psi_m$ can be represented as follows:

$$A\psi = A\sum_m C_m\psi_m = \sum_m C_m A\psi_m = \sum_n\sum_m C_m A_{nm}\psi_n$$

Moreover, if the operator A is given, the numbers a_{nm} can always be found. To solve for the a_{nm}, we need merely multiply eq. (1) by $\psi_r^*(x)$ and integrate over all x. From the normalization and orthogonality of the ψ_n we obtain

$$a_{rm} = \int \psi_r^*(x) A \psi_m(x)\, dx \tag{2}$$

The numbers a_{nm} (which are generally complex) form a square array that can be written schematically as

$$\begin{bmatrix} a_{1,1} & a_{1,2} & a_{1,3} & a_{1,4} & \cdot & \cdot \\ a_{2,1} & a_{2,2} & a_{2,3} & \cdot & \cdot & \cdot \\ a_{3,1} & a_{3,2} & \cdot & \cdot & \cdot & \cdot \\ a_{4,1} & \cdot & \cdot & \cdot & & \\ \cdot & \cdot & \cdot & & \cdot & \\ \cdot & \cdot & \cdot & & & \cdot \end{bmatrix}$$

It can easily be shown that the a_{mn} have all of the properties of a set of quantities known in mathematics as matrices. Each number a_{mn} is called an *element* (or a *component*) of the matrix. The symbol A is often used to represent the totality of all matrix elements. This is also represented by (a_{mn}). The matrix elements may be represented either by (A_{mn}) or by a_{mn}.

2. Properties of Matrices. The significant properties of a matrix are the following:

(1) Two matrices a_{mn} and b_{mn} can be added to yield a new matrix, with components that are the sums of the corresponding components of the separate matrices.

$$(A + B)_{mn} = a_{mn} + b_{mn} \tag{3}$$

(2) A matrix (a_{mn}) can be multiplied by an arbitrary complex number to yield a new matrix, as follows:

$$(KA)_{mn} = Ka_{mn} \tag{4}$$

(3) Two matrices are equal only when each element of the first is equal to the corresponding element of the second. Furthermore, a matrix is zero only when all of its elements are zero.

(4) Two matrices can be multiplied together as follows:

$$(AB)_{mn} = \sum_r a_{mr} b_{rn} \tag{5}$$

Example: Consider the formula for the rotation of a vector of components x_1 and x_2 through an angle ϑ_1, about an axis which is normal to x_1 and x_2. We get

$$\left.\begin{aligned} x_1 &= x_1' a_{1,1} + x_2' a_{1,2} \\ x_2 &= x_1' a_{2,1} + x_2' a_{2,2} \end{aligned}\right\} \tag{6}$$

where

$$a_{1,1} = \cos\vartheta_1, \quad a_{1,2} = -\sin\vartheta_1, \quad a_{2,1} = \sin\vartheta_1, \quad \text{and} \quad a_{2,2} = \cos\vartheta_1$$

The coefficients a_{ij} form a square array

$$\begin{pmatrix} \cos\vartheta_1 & -\sin\vartheta_1 \\ \sin\vartheta_1 & \cos\vartheta_1 \end{pmatrix} \tag{7}$$

The transformation can conveniently be written

$$x_i = \sum_j a_{ij} x'_j \tag{7a}$$

Let us now consider a second rotation through an angle ϑ_2, which defines the transformation

$$x'_j = \sum_k b_{jk} x''_k \tag{7b}$$

where the b_{jk} form the square array

$$\begin{pmatrix} \cos\vartheta_2 & -\sin\vartheta_2 \\ \sin\vartheta_1 & \cos\vartheta_2 \end{pmatrix}$$

By replacing x'_k in eq. (7a) by its transform as given in eq. (7b), we obtain

$$x_i = \sum_{j,k} a_{ij} b_{jk} x''_k \tag{8}$$

It is readily verified that $\sum_j a_{ij} b_{jk} = (AB)_{ik} = ik$th element of the product matrix AB. Thus, the application of two rotations in succession produces a transformation matrix that can be represented as the product of the separate transformation matrices.

Problem 1: Prove that the matrix $(AB)_{ij}$ is equal to

$$\begin{pmatrix} \cos(\vartheta_1 + \vartheta_2) & -\sin(\vartheta_1 + \vartheta_2) \\ \sin(\vartheta_1 + \vartheta_2) & \cos(\vartheta_1 + \vartheta_2) \end{pmatrix}$$

and thus show that two rotations carried out successively about the same axis are equivalent to a single combined rotation, whose angle is the sum of the angles of the separate rotations.

Commutation of Matrices. It is clear that the commutator of two matrices

$$(ab - ba)_{ik} = \sum_j (a_{ij} b_{jk} - b_{ij} a_{jk})$$

is not in general zero.

Example: Consider the matrices

$$a = \begin{pmatrix} 1 & 0 \\ 0 & -1 \end{pmatrix} \quad \text{and} \quad b = \begin{pmatrix} 0 & 1 \\ 1 & 0 \end{pmatrix}$$

We get

$$ba = \begin{pmatrix} 0 & -1 \\ 1 & 0 \end{pmatrix} \quad ab = \begin{pmatrix} 0 & 1 \\ -1 & 0 \end{pmatrix}$$

so that

$$ba - ab = \begin{pmatrix} 0 & -2 \\ 2 & 0 \end{pmatrix}$$

Problem 2: Prove that the matrices a_{ij} and b_{ij}, which represent the rotation through angles ϑ_1 and ϑ_2, respectively, defined in eq. (7), commute. Show that this corresponds to the fact that the same result is obtained when two rotations *about the same axis* are carried out in either of their two possible orders.

We see from the above that although matrices do not commute in general, it is possible for them to commute in special cases.

Diagonal Matrices. A matrix (a_{ij}) for which all elements are zero except where $i = j$ is called a *diagonal matrix.* In a square array, it looks like this:

$$\begin{pmatrix} a_{1,1} & 0 & 0 & \cdots \\ 0 & a_{2,2} & 0 & \cdots \\ 0 & 0 & a_{3,3} & \cdots \\ \cdot & \cdot & \cdot & \cdots \end{pmatrix} \tag{9}$$

A diagonal matrix can always be written as

$$a_{ij} = a_{ii}\delta_{ij}$$

where δ_{ij} is a symbol which is zero when $i \neq j$ and unity when $i = j$. It is called the *Kronecker delta.*

The Unit Matrix. A special case of the diagonal matrix is the unit matrix obtained by putting 1's in all the diagonal elements. It therefore has the form $a_{ij} = \delta_{ij}$. The unit matrix is often denoted by the symbol $(1)_{ij}$. It is readily verified that the unit matrix when multiplied by an arbitrary matrix leads to the same matrix. Thus

$$(M, 1)_{ij} = (1, M)_{ij} = M_{ij} \tag{10}$$

Problem 3: Prove the preceding result, and thus show that the unit matrix commutes with an arbitrary matrix.

The Reciprocal of a Matrix. In many cases, one can define a reciprocal of a matrix, which is analogous to the reciprocal of a number. The reciprocal A^{-1} of a matrix A has the property that

$$A^{-1}A = AA^{-1} = 1 \tag{11}$$

Note that, by definition, every matrix commutes with its reciprocal (if the reciprocal exists).

To obtain the reciprocal to a given matrix, let us write

$$(A)_{ij} = a_{ij} \qquad \text{and} \qquad (A^{-1})_{ij} = b_{ij}$$

Then, we must have

$$\sum_k a_{ik}b_{kj} = \sum_k b_{ik}a_{kj} = \delta_{ij} \tag{12}$$

Equation (12) may be regarded as a set of inhomogeneous linear equations defining the b_{ij} in terms of the a_{ij}. To solve these equations, we write

$$b_{ij} = \frac{[a]_{ij}}{[a]} \tag{13}$$

where $[a]$ represents the determinant formed from the elements a_{ij}, and $[a]_{ij}$ represents the ij minor of this determinant.

A necessary and sufficient condition for the existence of the reciprocal is that the determinant $[a]$ shall not vanish.

Problem 4: If the equation $A^{-1}A = 1$ can be solved, then can $AA^{-1} = 1$ also be solved? Give a proof for your answer.

3. Proof that Quantum-mechanical Operators have a Matrix Representation. In order to prove that the quantities a_{ij} appearing in eqs. (1) and (2) are matrices, it is necessary only to show that they satisfy requirements of (3), (4), and (5). That they satisfy (3) and (4) is quite easily seen.

Problem 5: Prove that quantum-mechanical operators lead to a_{ij} that satisfies (3) and (4).

To prove requirement (5) is satisfied, we consider two operators A and B, with matrix elements a_{ij} and b_{ij}. The product of the operators (AB) has a matrix element given by

$$(AB)_{ij} = \int \psi_i^*(x) AB\psi_j(x)\, dx = \int \psi_i^*(x) A \sum_k b_{kj}\psi_k\, dx = \sum_k a_{ik}b_{kj} \tag{14}$$

This, however, is exactly what is obtained by multiplying the corresponding matrices according to eq. (5).

Problem 6: Prove that the matrix of the unit operator is the unit matrix, i.e., that

$$(1)_{mn} = \delta_{mn}$$

4. An Example: Harmonic-oscillator Wave Functions. Consider a representation in which the wave function is represented as a series of harmonic-oscillator wave functions [see eq. (22), Chap. 13]. The matrix element of $x + ip/\hbar$ is then easy to calculate. Thus

$$\left(x + \frac{ip}{\hbar}\right)_{mn} = \int \psi_n^*(x) \left(x + \frac{ip}{\hbar}\right) \psi_m(x)\, dx = \int \psi_n^*(x) \left(x + \frac{\partial}{\partial x}\right) \psi_m(x)\, dx$$

According to eq. (39), Chap. 13,

$$\left(x + \frac{\partial}{\partial x}\right) \psi_m(x) = \sqrt{2m}\, \psi_{m-1}(x)$$

We therefore obtain

$$\left(x + \frac{ip}{\hbar}\right)_{mn} = \sqrt{2m} \int \psi_n^*(x)\psi_{m-1}(x)\, dx = \sqrt{2m}\, \delta_{m-1,n}$$

The matrix therefore looks like this:

$$\begin{array}{c} \\ n \\ \downarrow \end{array} \sqrt{2}\begin{bmatrix} 0 & \sqrt{2} & 0 & \cdot & \cdot & \cdot & \cdot \\ 0 & 0 & \sqrt{3} & \cdot & \cdot & \cdot & \cdot \\ 0 & 0 & 0 & \sqrt{4} & \cdot & \cdot & \cdot \\ 0 & 0 & 0 & 0 & 0 & \sqrt{5} & \cdot \\ \cdot & \cdot & \cdot & \cdot & \cdot & \cdot & \cdot \end{bmatrix} \quad m\rightarrow \tag{15}$$

In other words, all elements are zero, except those in the column to the right of the diagonal.

Problem 7: Obtain the matrices for $(x - ip/\hbar)$, and x and p.

5. Hermitean Matrices and Hermitean Conjugate Matrices. From the definition of a Hermitean operator [see eq. (13), Chap. 9], it is readily proved that the matrix corresponding to such an operator has the following property:

$$(H)_{ij} = (H^*)_{ji} \tag{16}$$

In other words, each matrix element is equal to the complex conjugate of the element of the transposed matrix (the element obtained by interchanging rows and columns).

If the operator M is not Hermitean, then it can also be shown that the matrix elements of the Hermitean conjugate operators, $M^\dagger$, satisfy the relation

$$(M^\dagger)_{ij} = (M^*)_{ji} \tag{17}$$

In other words, the Hermitean conjugate matrix is obtained by interchanging rows with columns, and taking the complex conjugate of every element.

Problem 8: Prove eqs. (16) and (17).

6. Diagonal Representation of Operators. If we choose for our orthonormal set in eq. (1) the eigenfunctions of the Hermitean operator A, then we obtain‡

$$A\psi_i = a_i\psi_i = \sum_j a_i\delta_{ij}\psi_j \tag{18}$$

Thus, we see that each Hermitean operator has a representation as a diagonal matrix, provided that the wave function is expanded in terms of its eigenfunctions.

7. Commutation of Diagonal Matrices. It is readily shown that all diagonal matrices commute. Thus suppose

$$a_{ij} = a_j\delta_{ij} \qquad \text{and} \qquad b_{ij} = b_i\delta_{ij}$$

‡ A must be restricted to Hermitean operators here because only then will the expansion postulate be applicable.

Then

$$(ab - ba)_{ij} = \sum_k (a_i\delta_{ik}b_k\delta_{kj} - b_i\delta_{ik}a_k\delta_{kj}) = 0$$

8. Continuous Matrices. Thus far we have considered the expansion of an arbitrary wave function in only a discrete set of functions and have thus obtained discrete matrices. If ψ is expanded in terms of a continuous orthonormal set of functions, one obtains continuous matrices. As an example, consider a Fourier integral. We write

$$\psi = \frac{1}{\sqrt{2\pi}} \int \varphi(\boldsymbol{k})e^{i\boldsymbol{k}\cdot\boldsymbol{x}}\, d\boldsymbol{k}$$

where the orthonormal functions are now the continuous set, $e^{i\boldsymbol{k}\cdot\boldsymbol{x}}$, and the $\varphi(\boldsymbol{k})$ are the corresponding expansion coefficients. A matrix element can then be written in analogy with eq. (2)

$$a_{kk'} = \frac{1}{2\pi} \int e^{-i\boldsymbol{k}\boldsymbol{x}} A e^{i\boldsymbol{k}\boldsymbol{x}}\, d\boldsymbol{x}$$

$a_{kk'}$ is a continuous function of k and k', but it may be regarded as the limit of a discrete square array in which the elements are allowed to approach closer and closer to each other. Thus, we obtain the concept of the continuous matrix.

More generally, we may use any continuous set ψ_p. Thus

$$a_{pp'} = \int \psi_p^* A \psi_{p'}\, dx \tag{19}$$

Continuous matrices may be treated in essentially the same way as are discrete matrices. Thus, we may represent the operator A as follows:

$$A\psi(x) = \int C_p A_{pp'} \psi_{p'}(x)\, dp\, dp' \tag{20}$$

where

$$\psi(x) = \int C_p \psi_p(x)\, dp \tag{21}$$

It is readily shown that the unit matrix becomes the Dirac δ function, $\delta(p - p')$ and that a diagonal matrix takes the form $a(p)\delta(p - p')$. One can also show that the rule for taking products of continuous matrices is

$$(AB)_{pp'} = \int A_{pp''} B_{p''p'}\, dp'' \tag{22}$$

Problem 9: Prove the preceding rule for multiplying continuous matrices.

Examples: (a) In the momentum representation, p becomes the diagonal matrix

$$(p)_{kk'} = \hbar k \delta(k - k') \tag{23}$$

As shown in eq. (44a), Chap. 10, the δ function can be represented as the following Fourier integral:

$$\delta(k - k') = \lim_{K \to \infty} \int_{-K}^{K} e^{i(k-k')x} \frac{dx}{2\pi} = \int_{-\infty}^{\infty} e^{ix(k-k')} \frac{dx}{2\pi} \tag{24}$$

We shall often find this representation of the δ function very convenient.

(b) In the position representation, x becomes a diagonal matrix

$$(x)_{y,y'} = y\delta(y - y') \tag{25}$$

(c) In the position representation, p is an off-diagonal matrix. To obtain the matrix of p in this representation, we can use eq. (2) which defines the matrix element associated with any two wave functions. We wish the matrix element of the operator p associated with the eigenfunctions of the operator x, namely, $\delta(x - x_1)$ and $\delta(x - x_2)$. This matrix element is*

$$(p)_{x_1x_2} = \int_{-\infty}^{\infty} \delta(x - x_1) \frac{\hbar}{i} \frac{\partial}{\partial x} \delta(x - x_2)\, dx = \frac{\hbar}{i} \frac{\partial}{\partial x_1} \delta(x_1 - x_2) \tag{26}$$

At first sight, it would seem that p is a diagonal matrix, since it vanishes unless $x = x'$. But, according to Sec. 7, all diagonal matrices should commute. Yet we know that p and x do not commute. What is the source of this paradox?

The answer is that we must be more careful in discussing continuous matrices which are singular, i.e., which have infinite terms such as $\delta(x - x')$ and $\frac{\partial}{\partial x} \delta(x - x')$. To give these terms a meaning, we must regard them as the limits of finite, but sharply peaked, functions, as shown in Chap. 10, Sec. 14. Now $\frac{\partial}{\partial x} \delta(x - x')$ is actually the limit of a function which is zero when $x = x'$, but which consists of two adjacent and very sharp peaks of opposite sign, located on either side of $x = x'$. This function is not a diagonal matrix. In this way, we see that the failure of commutation of p and x is not contradicted.

Problem 10: By regarding the δ function and its derivative as the limit of suitable, sharply peaked functions, obtain with the use of continuous matrices the commutation relation $xp - px = i\hbar$.

Problem 11: Obtain the matrix representation of the operator D, where

$$D\psi(x) = \psi(x + C)$$

C being constant.

(a) In the position representation.

(b) In the momentum representation.

9. Column Representation of the Wave Function. Suppose that in a particular representation, we write

$$\psi(x) = \sum_n C_n \psi_n(x)$$

Now the wave function is specified by specifying all of the C_n. These may be written in a column, as below

* The δ functions are not normalized in the usual way, but are normalized so that their integral is unity (Chap. 10, Sec. 14). The normalization is more convenient for an operator with continuous eigenvalues. It is straightforward to show that the usual formulas for matrix multiplication apply in this case too.

$$\begin{bmatrix} C_1 \\ C_2 \\ C_3 \\ \cdot \\ \cdot \\ \cdot \\ C_n \\ \cdot \\ \cdot \\ \cdot \end{bmatrix} \tag{27}$$

This notation is equivalent to a generalization of the concept of a vector to an infinite dimensional space. If one imagines one axis for each function ψ_n, then each C_n corresponds to the (complex) component of a vector in the direction of this axis.

A matrix operating on the wave function can now be represented as a linear transformation. Thus, in three dimensions, the transformation $x'_i = \sum_j a_{ij}x_j$ represents, in general, some combination of rotation, shearing, and stretch. In quantum theory, we simply extend this notion to the infinite dimensional space defined by the C_n. The quantity

$$A\psi = \sum_{m,n} C_n a_{mn} \psi_m = \sum_m C'_m \psi_m$$

can be represented by a new vector $C'_m = \sum_n a_{mn}C_n$. Every linear operator therefore corresponds to a process which changes each vector into some other (specified) vector.

In a continuous representation, the column is replaced by $C(\alpha)$, a continuous function of the eigenvalues α by which the orthonormal set is labeled.

10. Normalization and Orthogonality of Wave Functions in Column Representation. To obtain the conditions for normalizing a wave function in the column representation, we write

$$\int \psi^*\psi \, dx = \int \sum_{m,n} C_m^* C_n \psi_m^*(x) \, \psi_n(x) \, dx = \sum_{m,n} C_m^* C_n \delta_{mn} = \sum_m C_m^* C_m$$

The above quantity may be regarded as the analogue of the "length" of a three-dimensional vector. Thus, a normalization wave function corresponds to a vector of unit "length."

If two wave functions $\psi_a(x)$ and $\psi_b(x)$ are orthogonal, then we have

$$0 = \int \psi_a^* \psi_b \, dx = \int \sum_{m,n} C_{am}^* C_{bn} \psi_m^*(x) \psi_n(x) \, dx = \sum_{m,n} C_{am}^* C_{bn} \delta_{mn} = \sum C_{am}^* C_{bn}$$

The condition for the orthogonality of two wave functions is then the analogue of the condition that two vectors be perpendicular in three-dimensional space. This is the origin of the term "orthogonality."

11. Average Value of an Operator. The average value of an operator A is

$$\bar{A} = \int \psi^* A \psi \, dx \tag{28}$$

It is often convenient to express this in terms of the matrix elements a_{ij} and the expansion coefficients C_i of the wave function. Thus, we write

$$\psi = \sum_i C_i \psi_i(x)$$

and obtain

$$\bar{A} = \int \sum_i \sum_j C_j^* C_i \psi_j^*(x) A \psi_i(x) \, dx = \sum_{i,j} C_j^* a_{ji} C_i \tag{29}$$

This means that the average value of any operator can always be calculated from its matrix elements in any representation, provided that we know the expansion coefficients of the wave function in that representation.

12. Eigenvalues and Eigenvectors of Matrices. In order to obtain the eigenvalues of an operator A from its matrix representation, we begin with $A\Psi_r = a_r\Psi_r$ where a_r is the rth eigenvalue of the operator A. We then expand Ψ_r in an orthonormal series $\bar{\Psi}_r = \sum_n C_n \psi_n$, and obtain $\sum_n C_n A \psi_n = a_r \sum_n C_n \psi_n$. Finally, we multiply by ψ_m^* and integrate over x, obtaining

$$\sum_n C_n a_{mn} = a_r C_m \tag{30}$$

The above provides a set of homogeneous linear equations defining the C_m in terms of the a_{mn} and the a_r. The condition for a solution is the vanishing of the determinant of the coefficients of the C's.

$$|a_{mn} - \delta_{mn} a_r| = 0 \tag{31}$$

This equation is often called the *secular equation*.

It is readily seen that (31) provides an equation defining a_r, which is of the same order as the number of rows and columns in the determinant. Each solution provides an eigenvalue a_r. Once we have chosen the a_r, then we can solve for the C_n, and thus obtain the eigenvector associated with this eigenvalue. The eigenvector is simply the column representation of the corresponding eigenfunction of the operator A, in terms of the coefficients of the orthonormal set $\psi_n(x)$.

We now consider as an example the matrix $\begin{pmatrix} 0 & 1 \\ 1 & 0 \end{pmatrix}$. To obtain its eigenvalues and eigenvectors, we write

$$\begin{pmatrix} 0 & 1 \\ 1 & 0 \end{pmatrix} \begin{pmatrix} C_1 \\ C_2 \end{pmatrix} = \lambda \begin{pmatrix} C_1 \\ C_2 \end{pmatrix}$$

where λ is the eigenvalue and $\begin{pmatrix} C_1 \\ C_2 \end{pmatrix}$ the eigenvector.

The above is equivalent to

$$\lambda C_1 - C_2 = 0 \qquad \text{and} \qquad C_1 - \lambda C_2 = 0$$

The condition for a solution is the vanishing of the determinant of the coefficient of the C's, and this reduces to

$$\lambda^2 = 1 \qquad \text{or} \qquad \lambda = \pm 1$$

Thus, the two eigenvalues are $+1$ and -1. For each eigenvalue, we obtain the corresponding eigenvector by substituting the actual value of λ in the equations for the C's. One obtains

$$\begin{aligned} C_2 &= C_1 \qquad &(\lambda = 1) \\ C_2 &= -C_1 \qquad &(\lambda = -1) \end{aligned}$$

The normalized eigenvectors are

$$\frac{1}{\sqrt{2}} \begin{pmatrix} 1 \\ 1 \end{pmatrix} \quad \text{for } \lambda = 1 \qquad \frac{1}{\sqrt{2}} \begin{pmatrix} 1 \\ -1 \end{pmatrix} \quad \text{for } \lambda = -1$$

13. Change of Representation. Suppose that we have the matrices of a given set of operators expressed in any one representation, and we wish to change to some other orthonormal set of functions. An example of such a change is from the "x" representation of the "p" representation, or from a "p" representation to the set of Hermite polynomials (see Chap. 13, Sec. 7). To deal with such a general change of representation, suppose that we begin by expanding the wave function in some complete set of functions, $\psi_n(x)$, which we take for convenience to be discrete, although essentially the same methods will apply to a continuous set. Thus, we begin with $\psi = \sum_n C_n \psi_n(x)$. In this representation, the matrix of an operator A takes the form

$$(A)_{mn} = \int \psi_m^* A \psi_n \, dx \tag{32a}$$

Let us now consider a new orthonormal set of functions, $\varphi_p(x)$, and the associated matrix elements $(A')_{pq} = \int \varphi_p^* A \varphi_q \, dx$. Our objective is to find the relation between the $(A)_{mn}$ and the $(A')_{pq}$.

In order to obtain this relation, we first note that because of the expansion postulate, the $\psi_n(x)$ can be expanded in terms of the $\varphi_p(x)$. Thus we write

$$\psi_m(x) = \sum_p \alpha_{pm}\varphi_p(x) \tag{32b}$$

α_{pm} is itself a matrix and is called the transformation matrix.

By expanding $\psi_m^*(x)$ and $\psi_n(x)$ in eq. (32a), we then obtain

$$(A)_{mn} = \int \sum_{p,q} \alpha_{pm}^*\alpha_{qn}\varphi_p^* A\varphi_q \, dx = \sum_{p,q} \alpha_{pm}^*(A')_{pq}\alpha_{qn}$$

In order to reduce this expression to a more convenient form, we note that $(\alpha^\dagger)_{mp} = \alpha_{pm}^*$, where $\alpha^\dagger$ is the Hermitean conjugate of α [see eq. (17)]. The above equation then yields $(A)_{mn} = \sum_{p,q} (\alpha^\dagger)_{mp}(A')_{pq}\alpha_{qn}$. But this is equivalent to

$$(A)_{mn} = (\alpha^\dagger A'\alpha)_{mn} \tag{33}$$

We conclude that the change of representation can be expressed as a linear transformation, which replaces the matrix element $(A)_{mn}$ by a linear combination of the transformed matrix elements.

14. An Important Property of the α Matrices. Their Unitary Character. We now obtain an important property of the α matrices. To do this, we consider the expression

$$\delta_{pq} = \int \varphi_p^*\varphi_q \, dx = \int \sum_{r,s} \alpha_{rp}^*\alpha_{sq}\psi_r^*\psi_s \, dx = \sum_{r,s} \alpha_{rp}^*\alpha_{sq}\delta_{rs} = \sum_r \alpha_{rp}^*\alpha_{rq}$$
$$= \sum_r (\alpha^\dagger)_{pr}\alpha_{rq} = (\alpha^\dagger\alpha)_{pq} \tag{34}$$

This shows that $\alpha^\dagger\alpha$ is equal to the unit matrix, so that $\alpha^\dagger = \alpha^{-1}$. A matrix having this property is called a unitary matrix, and a transformation carried out with such a matrix [as in eq. (33)] is called a unitary transformation.

By a similar argument, the old wave function can be written in terms of the new with the aid of eq. (34). Thus, we get

$$\psi = \sum_n C_n\psi_n = \sum_{p,n} \alpha_{pn}C_n\varphi_p = \sum_p C'_p\varphi_p$$

The new coefficients are given in terms of the old by

$$C'_p = \sum_n \alpha_{pn}C_n \tag{35a}$$

By multiplying by $(\alpha^\dagger)_{mp}$, summing over p, and using the unitary character of α, we obtain

$$\sum_p (\alpha^\dagger)_{mp}C'_p = \sum_{p,n} (\alpha^\dagger)_{mp}\alpha_{pn}C_n = \sum_n \delta_{mn}C_n = C_m \tag{35b}$$

15. Significance of the Unitary Transformation. The important properties of a unitary transformation are the following:

(1) The normalization of an arbitrary wave function is left unchanged. To prove this, suppose that we start with an arbitrary wave function

$$\psi = \sum_n C_n \psi_n(x)$$

The integrated probability (which must be set equal to unity) is

$$\int \psi^* \psi \, dx = \int \sum_{m,n} C_m^* C_n \psi_m^* \psi_n \, dx = \sum_{m,n} C_m^* C_n \delta_{mn} = \sum_n C_n^* C_n$$

We apply eq. (35) and obtain

$$\sum_n C_n^* C_n = \sum_{n,p,p'} (\alpha_{np}^\dagger)^* \alpha_{np'}^\dagger C_p^{*\prime} C_p'$$

Now,

$$(\alpha_{np}^\dagger)^* = \alpha_{pn}$$

We therefore find that

$$\sum_n C_n^* C_n = \sum_{n,p,p'} (\alpha_{pn} \alpha_{np'}^\dagger) C_p^{*\prime} C_{p'}' = \sum_{p,p'} \delta_{pp'} C_p'^* C_{p'}' = \sum_p C_p'^* C_p' \tag{36}$$

Thus, the normalization is left unaltered. In Sec. 10, we saw that $\sum_n C_n^* C_n$ corresponds to the square of the length of the column vector associated with the wave function. Since a unitary transformation leaves this quantity unchanged, we conclude that it corresponds to a generalization of a rotation in three-dimensional space, which also leaves all vectors unaltered in length. Non-unitary transformations would then correspond to shearing and stretching.

(2) A unitary transformation causes wave functions that were originally orthogonal to be transformed into wave functions which remain orthogonal. In this respect, they also resemble a three-dimensional rotation, which transforms any two mutually perpendicular vectors into a new set of mutually perpendicular vectors.

To prove this property, we consider the following integral, which is zero for two orthogonal functions ψ_1 and ψ_2:

$$\int \psi_1^*(x) \psi_2(x) \, dx = \int \psi_2^*(x) \psi_1(x) \, dx = 0$$

If the ψ are expanded in a series of ψ_m, we have

$$\psi_1 = \sum_m C_{1m} \psi_m(x) \qquad \text{and} \qquad \psi_2 = \sum_n C_{2n} \psi_n(x)$$

and

$$\int \psi_1^* \psi_2 \, dx = \int \sum_{m,n} C_{1m}^* C_{2n} \psi_m^*(x) \psi_n(x) \, dx = \sum_m C_{1m}^* C_{2m}$$

Under a unitary transformation, we have $C_{2m} = \sum_p (\alpha^\dagger)_{mp} C_{2p}'$

$$C_{1m}^* = \sum_p (\alpha^\dagger)_{mp}^* C_{1p}'^* = \sum_p (\alpha)_{pm} C_{1p}'^*$$

and

$$\int \psi_1^* \psi_2 \, dx = \sum_m C_{1m}^* C_{2m} = \sum_m \sum_{p,q} (\alpha)_{pm} (\alpha^\dagger)_{mq} C_{1p}'^* C_{2q}' = \sum_{m,p,q} \delta_{pq} C_{1p}'^* C_{2q}'$$
$$= \sum_p C_{1p}'^* C_{2p}'$$

We conclude that the expansion of $\int \psi_1^* \psi_2 \, dx$ takes the same form in all representations, so that if it is zero in any one representation, it is also zero after a unitary transformation has been carried out. Thus, the orthogonality properties of a set of wave functions are left unchanged by a unitary transformation.

(3) Relationships between transformed operators are the same as those between the corresponding untransformed operators.

Consider for example a matrix operator

$$O = AB$$

The transformed matrix becomes

$$O' = \alpha^\dagger (AB) \alpha = \alpha^\dagger A \alpha \cdot \alpha^\dagger B \alpha \qquad \text{(by virtue of } \alpha\alpha^\dagger = 1\text{)}$$

and, therefore, $$O' = A'B'$$

A similar proof can be carried out for any operator function that can be expressed as a series of products. For example, it is easily seen that the commutator of two operators goes over into the commutator of the transformed operators, i.e.,

$$(AB - BA)' = (A'B' - B'A')$$

(4) The eigenvalues of a matrix are not changed by a unitary transformation.

Problem 12: Prove the above statement.

We conclude that from a given representation, one can, by means of a unitary transformation obtain an equivalent representation of all quantum-mechanical relationships. It is often convenient to transform from one representation to another in this way, because it usually turns out that each problem has some representation in which it is most simply expressed. For example, the average momentum of a particle is most easily evaluated in the momentum representation, whereas the average position is most easily evaluated in the position representation.‡

Problem 13: Prove that the transformation from x to p is unitary, and evaluate the transformation matrix (which is in this case continuous).

It can be shown that, in the classical limit, a unitary transformation of the wave functions produces a canonical transformation of the classical

‡ The transformation given in Chap. 14, Sec. 19, between rotated systems of co-ordinates is an example of a unitary transformation.

variables p and q. Thus, a unitary transformation is the quantum generalization of the classical concept of a canonical transformation.* For this reason a unitary transformation is often called a canonical transformation (also a contact transformation).

16. The Trace of a Matrix. A quantity that is often very useful in calculations is the *trace* (or *spur*) of a matrix. This is defined as the sum of the diagonal elements.

$$TrA = \sum_i a_{ii} \tag{37}$$

An important theorem is that the trace of a matrix is not changed by unitary transformation. Thus, we write

$$TrA' = Tr(\alpha^\dagger A\alpha) = \sum_{i,j,k} \alpha_{ij}^\dagger A_{jk}\alpha_{ki}$$

We can rewrite the above as

$$\sum_{j,k}\sum_i \alpha_{ki}\alpha_{ij}^\dagger A_{jk}$$

Because α is a unitary matrix, this reduces to

$$\sum_{j,k} \delta_{kj}A_{jk} = \sum_k A_{kk}$$

Thus, we see that the trace is invariant. This means that it can be evaluated in whatever representation is most convenient.

If we choose that unitary transformation which diagonalizes the matrix, then we see that $TrA = \sum_i a_i$. Thus the trace of a matrix is also the sum of its eigenvalues.

17. Simultaneous Eigenfunctions of Commuting Operators. From the expansion postulate, we know that we can expand an arbitrary wave function as a series of eigenfunctions of a Hermitean operator A, i.e., $\psi = \sum_a C_a\psi_a(x)$. We shall now show that if two operators commute, it is possible to expand an arbitrary wave function as a series of simultaneous eigenfunctions of both operators.

To do this, we first note that if A and B commute, and that if ψ_a is an eigenfunction of A belonging to the eigenvalue a, one has

$$(AB)\psi_a - (BA)\psi_a = A(B\psi_a) - a(B\psi_a) = 0$$

Thus, $B\psi_a$ is also an eigenfunction of A belonging to the eigenvalue a This means that $B\psi_a$ must be a linear combination of eigenfunctions of A, belonging to the eigenvalue a. (If A is nondegenerate, only one such eigenfunction exists, otherwise more than one.) To denote this fact, we write

* See Dirac, 3d. ed., pp. 121–130.

$$B\psi_{am} = \sum_n b_{amn}\psi_{an}$$

where ψ_{am} represents the mth eigenfunction of A belonging to the eigenvalue a.

Now, any set of functions such as ψ_{an} can always be regrouped by a suitable linear combination into an orthonormal set.† Let us suppose that this has been done so that the ψ_{an} form an orthonormal set. One can then express an arbitrary wave function as a series

$$\psi = \sum_{a,m} C_{am}\psi_{am}(x)$$

The equation defining the eigenfunctions and eigenvalues of B is

$$B\psi = \sum_{a,m,n} C_{am}b_{amn}\psi_{an}(x) = \lambda \sum_{a,m} C_{am}\psi_{am}(x)$$

We now multiply by $\psi^*_{a'm'}(x)$ and integrate over x. Let us note that for $a \neq a'$, the ψ_{am} are orthogonal to $\psi^*_{a'm'}$ because they are two eigenfunctions of the Hermitean operator A, corresponding to different eigenvalues (see Chap. 10, Sec. 24). For $a = a'$ and $m \neq m'$, they are orthogonal by hypothesis. Thus, we obtain

$$\lambda C_{a'm'} = \sum_m C_{a'm}b_{a'mm'} \tag{38}$$

This is a set of linear equations for the $C_{a'm}$, and the condition for a solution is the vanishing of the determinant of the coefficients of the $C_{a'm}$,

$$|b_{a'mm'} - \lambda\delta_{mm'}| = 0 \tag{39}$$

The most important characteristic of eqs. (38) and (39) is that they refer to a given value of a only. This means that the eigenfunctions of B, as obtained in this way, are simultaneously eigenfunctions of A. Moreover, since the coefficients C_{am} permitted the expansion of an arbitrary function, we see that it is possible to obtain a complete set of eigenfunctions of B in this way. We conclude that an arbitrary function can be expanded in a series of simultaneous eigenfunctions of A and B. Thus,

$$\psi = \sum_{a,b} C_{ab}\psi_{ab}(x) \tag{40}$$

If there are more than two commuting operators, a corresponding theorem can be proved. When we have exhausted all the physically significant operators which commute with each other, we are said to have a "complete set of commuting observables," and the most detailed possible information is obtained by specifying all of the associated expansion

† This can be done by what is called the Schmidt orthogonalization process. See, for example, E. Wigner, *Gruppentheorie und ihre Anwendung auf die Quantenmechanik der Atomspektren*. Braunschweig: Friedr. Vieweg und Sohn, 1931, p. 31.

coefficients. It is only when we have a complete commuting set that the specification of the wave function is unambiguous.

In some cases, the complete commuting set consists of only one operator. Thus, in a one-dimensional problem for a single particle without spin, *either* the operator x or the operator p provides a complete set, but, of course, one cannot use both simultaneously. In a three-dimensional problem, the three co-ordinates or the three momenta will serve. If the potential is spherically symmetrical (see Chap. 14, Sec. 1), then one can choose H, L^2, and L_z as the complete commuting set. Three variables are needed here, because H, L^2, and L_z are degenerate, so that a specification of only one or two does not define the wave function completely. On the other hand, if the most general nonspherically symmetric potential is used (see Chap. 14, Sec. 6), H ceases to commute with L^2 and L_z. H also becomes nondegenerate, so that the wave function can be completely specified by specifying the energy alone. When a given operator, such as H, is degenerate, however, one needs one or more additional operators to form a complete commuting set, just because the specification of an eigenvalue of H is not sufficient to define the wave function completely.

18. The Specification of an Arbitrary Operator in Terms of Its Commutators with a Complete Commuting Set of Operators. It can be shown that an arbitrary operator can be defined in terms of its commutator with a complete set of commuting observables. We shall not prove the general theorem here, but shall only give as an example the case where a single operator serves as a complete commuting set. This operator will be taken as the Hamiltonian operator in a one-dimensional nondegenerate case, such as the harmonic oscillator.

Now, it is adequate to define an operator in any single representation, since its form in any other representation can then be obtained from a unitary transformation. Let us choose here the representation in which H is diagonal, so that $H_{ij} = \epsilon_i \delta_{ij}$. Suppose that the commutator of an arbitrary operator A with H is known, so that we can write

$$(AH - HA)_{ij} = C_{ij}$$

Because H is diagonal, we obtain

$$(\epsilon_j - \epsilon_i)A_{ij} = C_{ij} \qquad \text{or} \qquad A_{ij} = \frac{C_{ij}}{\epsilon_j - \epsilon_i} \tag{41}$$

Thus, provided that $\epsilon_j \neq \epsilon_i$, we can solve* for the matrix A_{ij}, once we know C_{ij}. If the operator H had been degenerate, however, then H would not have been a complete commuting set, and additional operators

* This procedure does not define the diagonal element A_{ii}. The reader will readily verify that the commutator $(AH-HA)$ will not be changed by the addition to A of a matrix of the form $A_{ii}\delta_{ij}$, where A_{ii} is arbitrary. However, we shall see that in practice this degree of arbitrariness is not very important.

would have been needed to define the wave function. As has been pointed out, one can give a more general proof that shows that for this case also, it would be possible to obtain the operator A, once C is known.*

The definition of the commutators of operators is therefore one of the most important steps in the formulation of the quantum theory. Thus far, we have achieved this definition by restricting ourselves to Hermitean functions of x and p, which can be expressed as a power series. Since the commutator of p and x has already been found in Chap. 9, Sec. 11, we therefore have an adequate definition for all operators of this general type. Some alternative rules for defining commutators will be given in Sec. 23.

19. Schrödinger's Equation in an Arbitrary Representation. We have now reached a point of view, in which the expression of wave functions and operators in terms of the position, x, or the momentum, p, must be regarded as special cases of the more general method, which involves the specification of the coefficients C_i in the column representation of the wave function. In fact, the wave function in the position representation may be regarded as just such a column representation. Thus, in terms of the eigenfunctions of the position operator, $\delta(x - x')$, we have

$$\psi(x) = \int \psi(x')\delta(x - x')\, dx'$$

Thus, we may regard the ψ's as if they were written in a column,

$$\begin{bmatrix} \psi(x_1) \\ \psi(x_2) \\ \psi(x_3) \\ \cdot \\ \cdot \\ \cdot \end{bmatrix}$$

where the points x_1, x_2, etc., are allowed in the limit to become infinitely close.

Schrödinger's equation may now be regarded as an equation for the expansion coefficients $\psi(x_i)$. If we shift our representation to the eigenfunctions of some other operator, A, which we take for the sake of convenience to be discrete, we write $\psi = \sum_a C_a \psi_a(x)$. The analogue of Schrödinger's equation should then be an equation specifying the time rate of change of the coefficients C_a.

To obtain this analogue, we begin with Schrödinger's equation in the position representation

$$i\hbar \frac{\partial \psi}{\partial t} = H\psi$$

* This proof holds only in a discrete representation. It can however be modified in such a way as to deal with a continuous representation also

We now express ψ as a series of the $\psi_a(x)$. Since ψ is a function of time, the C_a must also be functions of time. We then obtain

$$i\hbar \sum_{a'} \dot{C}_{a'}\psi_{a'}(x) = \sum_{a'} C_{a'}H\psi_{a'}(x)$$

Now multiply this equation by $\psi_a^*(x)$ and integrate over all x. We then obtain (using normalization and orthogonality of the ψ_a)

$$i\hbar\dot{C}_a = \sum_{a'} H_{aa'}C_{a'} \tag{42}$$

where
$$H_{aa'} = \int \psi_a^*(x)H\psi_{a'}(x)\,dx$$

This equation completely defines how the C_a change, whenever the C_a are known at any one time.

Problem 14: Show that eq. (42) can be obtained from a unitary transformation, starting from Schrödinger's equation in the position representation where the transformation matrix is $\psi_a(x)$ (the eigenfunctions of A in the x representation). We regard this function as a matrix in the variables a and x, although it is discrete in one and continuous in the other.

20. The Hamiltonian Representation. A particularly useful representation is the one in which H is diagonal, or $H_{ij} = E_i\delta_{ij}$. The transformation matrix to go from the x representation to the Hamiltonian representation is just $\psi_{E_i}(x)$, the eigenfunction of H belonging to the energy level E_i.

In this representation, Schrödinger's equation (42), becomes

$$i\hbar \frac{dC_i}{dt} = \sum_j H_{ij}C_j = E_iC_i \tag{43}$$

and the solution is

$$C_i = e^{-iE_it/\hbar}C_i^0 \tag{44}$$

where C_i^0 is a constant. Thus, in the Hamiltonian representation, the C_i's oscillate with simple harmonic motion. An example of the Hamiltonian representation is given in eq. (15), where the operator $x + ip/\hbar$ is given as a matrix in the representation in which the Hamiltonian of a harmonic oscillator is diagonal.

21. The Heisenberg Representation. Let us now consider a transformation in which we go from the C_i to the C_i^0 as basic variables. The transformation is defined by eq. (44). The transformation matrix is readily verified to be

$$\alpha_{ij} = \delta_{ij}\,e^{-iE_it/\hbar} \tag{45a}$$

That the transformation is unitary can be proved by evaluation:

$$(\alpha^\dagger\alpha)_{ij} = \sum_k (\alpha^\dagger)_{ik}\alpha_{kj} = \sum_k \delta_{ik}\delta_{kj}\,e^{-i(E_j-E_k)t/\hbar} = \delta_{ij} \tag{45b}$$

This transformation yields what is known as the Heisenberg representation. In this representation the wave function "vector," C_i^o, is a constant. The matrix element of an operator becomes, however,

$$A_{ij}^o = \sum_k \alpha_{ik}^\dagger A_{kl} \alpha_{lj} = e^{-i(E_j - E_i)t/\hbar} A_{ij} \tag{46}$$

From eq. (45a) we can easily verify that the above is equivalent to

$$A_{EE'}^o = \int \psi_E^*(x)\, e^{iEt/\hbar} A \psi_{E'}(x)\, e^{-iE't/\hbar}\, dx \tag{47}$$

The use of the Heisenberg representation is equivalent to expanding the wave function in a series of the functions $\psi_E(x)\, e^{-iEt/\hbar}$.

$$\psi = \sum_E \psi_E(x)\, e^{-iEt/\hbar} C_E^o \tag{48}$$

From eq. (46) we see that the matrix elements now oscillate harmonically with time. We started, however, in the Schrödinger representation, where most operators, such as x, p, H, are represented by constant matrices, whereas the wave function varies in the time. It can easily be seen that in computing averages, such as $\bar{A} = \sum_{i,j} C_i^* A_{ij} C_j$, it makes no difference whether we regard the C_i's as oscillating as $e^{-iE_it/\hbar}$, while the A_{ij}'s are constant, or whether we regard the C_i's as constant, while the A_{ij}'s oscillate as $e^{-i(E_j - E_i)t/\hbar}$. Thus, as is always the case in a unitary transformation, we simply describe the same phenomena in a different language.

22. Time Rate of Change of Operators in a Heisenberg Representation. It is clear that the Schrödinger wave function in the Hamiltonian representation, C_i, is obtained from the C_i^o in the Heisenberg representation by a unitary transformation which is the reciprocal of (45a). Since a unitary transformation is equivalent to a rotation in the "wave function space" (see Sec. 14), we conclude that the motion of the system is equivalent in effect to some (generally complicated) rotation in this space.‡ The transformation from Schrödinger to the Heisenberg representation is then analogous to a transformation from a stationary system of axes to a system that rotates with the wave-function vectors, so that in this latter system, the wave function appears to be a constant. On the other hand, the operators that were constant in the nonrotating frame now become functions of the time in the rotating system.

Whether we use a Schrödinger or Heisenberg representation depends entirely on which is more convenient in the problem with which we are dealing.

Let us now compute the rate of change of $A_{EE'}$ [$A_{EE'}$ is defined by eq. (2)]

‡ In the classical limit, this becomes equivalent to the well-known result that the motion can be represented as a series of infinitesimal canonical transformations.

$$\frac{dA_{EE'}}{dt} = \frac{i}{\hbar}(E - E') \int \psi_E^*(x) A \psi_{E'}(x) e^{i(E-E')t/\hbar}\, dx + \int \psi_E^*(x) \frac{\partial A}{\partial t} \psi_E(x) e^{i(E-E')t/\hbar}\, dx \tag{49}$$

But, by definition, this is just equal to

$$\frac{dA_{EE'}}{dt} = \frac{i}{\hbar}(E - E')A_{EE'} + \left(\frac{\partial A}{\partial t}\right)_{EE'} \tag{50a}$$

where the matrix elements are in the Heisenberg representation. If we note that the Hamiltonian matrix is $H_{EE'} = E\delta_{EE'}$, we can easily show that the above is equivalent to

$$\frac{dA_{EE'}}{dt} = \frac{i}{\hbar}(HA - AH)_{EE'} + \left(\frac{\partial A}{\partial t}\right)_{EE'} \tag{50b}$$

Problem 15: Prove that the above result is invariant to any unitary transformation which is not a function of the time.

Thus far, we have used the Heisenberg representation only under conditions in which the Hamiltonian is diagonal. One can, however, generalize the Heisenberg representation by expanding the wave function in terms of any complete set, $\phi_n(x, t)$, of solutions of Schrödinger's equation. The most general solution of Schrödinger's equation can be expanded as a series of eigenfunctions of H. Thus

$$\phi_n(x, t) = \sum_E \alpha_{En} \psi_E(x)\, e^{-iEt/\hbar} \tag{51}$$

where the α_{En} are constant expansion coefficients. If the $\phi_n(x, t)$ are [like the $\psi_E(x)$] a complete orthonormal set, then as shown in Sec. 14, α_{En} is a unitary matrix and the transformation from the $\psi_E(x)\, e^{-iEt/\hbar}$ to the $\phi_n(x, t)$ is a unitary transformation. In fact to obtain the transformation explicitly, we multiply (51) by $(\alpha^\dagger)_{nE'}$ and sum over n. Using the unitary character of $(\alpha^\dagger)_{nE'}$, we obtain

$$\sum_n (\alpha^\dagger)_{nE'} \phi_n(x, t) = \sum_n \sum_E (\alpha^\dagger_{nE'} \alpha_{En}) \psi_E(x) e^{-iEt/\hbar} = \psi_{E'}(x)\, e^{-iE't/\hbar}$$

From Problem 15, one readily deduces that eq. (50b) still gives the rate of change of a matrix element, even in the most general Heisenberg representation. [Note the analogy to eq. (37), Chap. 9.] It must be pointed out, however, that eq. (50b) applies *only* in the Heisenberg representation.

As a special case, one can show from eq. (50b) that the basic commutation relations between p and x are constants of the motion. Thus

$$\frac{d}{dt}(xp - px) = \frac{i}{\hbar}[H(px - xp) - (px - xp)H]$$

If we choose $px - xp = \hbar/i$ initially, then the above equation shows that $\frac{d}{dt}(px - xp) = 0$, because $\hbar/i$ commutes with any operator. Thus we prove that if the above commutation relation holds at $t = 0$, the commutator will remain constant for all time. This is an essential step in demonstrating the consistency of our choice of commutation relations; for in general, a given set of assumed commutation relations will not necessarily be propagated by the equations of motion.

Problem 16: Prove from eq. (50b) that eq. (37), Chap. 9 follows.

Problem 17: Starting from $H = \frac{p^2}{2m} + V(x)$, show that we obtain

$$\frac{d}{dt}(x_{ij}) = \frac{p_{ij}}{m} \qquad \frac{d}{dt}(p_{ij}) = -\left(\frac{\partial V}{\partial x}\right)_{ij}$$

From the above problem, we see that eq. (50b) contains the quantum equations that replace the classical equations of motion.

23. Poisson Brackets. Equation (50b) is sometimes called the quantum equation of motion for the operators p and x. It is analogous to the classical equation for a function

$$\frac{dA}{dt} = \frac{\partial A}{\partial p}\frac{dp}{dt} + \frac{\partial A}{\partial q}\frac{dq}{dt} + \frac{\partial A}{\partial t} = \left[\frac{\partial A}{\partial q}\frac{\partial H}{\partial p} - \frac{\partial A}{\partial p}\frac{\partial H}{\partial q}\right] + \frac{\partial A}{\partial t} \tag{52}$$

The expression in the brackets above is known as the "Poisson bracket" of A and H. More generally, for the case of one variable, the Poisson bracket of two functions A and B is*

$$[A, B] = \left[\frac{\partial A}{\partial q}\frac{\partial B}{\partial p} - \frac{\partial A}{\partial p}\frac{\partial B}{\partial q}\right] \tag{53}$$

Since $\frac{d}{dt}(\bar{A}) = \frac{i}{\hbar}(\overline{HA - AH}) + \overline{\frac{\partial A}{\partial t}}$, it is clear that in the classical limit, the commutator must approach $\hbar/i$ times the corresponding Poisson bracket. It can be shown more generally that this must hold for all operators, i.e.,

$$(AB - BA) \rightarrow \frac{\hbar}{i}[A, B] \tag{54}$$

in the classical limit. It turns out, in fact, that for most operators which occur in practice, the commutator is *equal* to $\hbar/i$ times the Poisson bracket *considered as an operator.*

Problem 18: Prove that $(xp - px) = \frac{\hbar}{i}[x, p]$.

Problem 19: Prove that $(x^2p^2 - p^2x^2) = \frac{\hbar}{i}[x^2, p^2]$, provided that the Poisson bracket is first symmetrized in the order in which x and p appear.

* For a fuller treatment of this subject, see Dirac or Rojansky. (See list of references on p. 2.)

Problem 20: Prove that $(Ox - xO) = \frac{\hbar}{i}[O, x]$ where O is an arbitrary Hermitean operator, which can be represented as a power series

$$O = \sum_{m,n} C_{mn} \frac{(x^n p^m + p^m x^n)}{2}$$

Since, according to Sec. 18, the definition of an arbitrary operator requires only the definition of its commutators with a complete commuting set of operators, and since (in one dimension) x is itself a complete set, we see from Problem 20 that an alternative method of formulating quantum theory can be obtained in terms of the assumption as a postulate that the Poisson bracket of any operator with x is equal to $\hbar/i$ times the corresponding commutator.* This is, in fact, a formulation which is very frequently adopted,† but in this book, we have tried to develop the theory from a somewhat less abstract point of view.

Problem 21: Consider the commutator of the Hermitean operators

$$A = \frac{x^n p^m + p^m x^n}{2} \qquad B = \frac{x^r p^s + p^s x^r}{2}$$

Is the Poisson bracket $[A, B]$ (symmetrized to be made Hermitean) identically equal to $\frac{i}{\hbar}(AB - BA)$?

24. Heisenberg's Formulation of Quantum Theory. In this book, we have derived the matrix formulation of quantum theory from the wave theory, following what is, in essence, the line of development initiated by de Broglie and Schrödinger. Actually, the matrix method was obtained independently by Heisenberg slightly before the wave theory was worked out. The equivalence of the two methods was then proved a few years later by means of the theory of unitary transformations.‡

* With this definition, the Poisson bracket $[A, B]$ is not necessarily identically equal to the commutator, $\frac{i}{\hbar}(AB - BA)$, but the two are, of course, always equal in the classical limit. See Problem 21.

† As shown in Sec. 18, the diagonal elements of operators are not defined by this procedure. The diagonal elements can be defined, however, from the requirement that the mean value of an operator in the position representation shall approach the correct classical value in the classical limit. This still leaves some ambiguity in the domain of small quantum numbers, but here one can be guided by the heuristic requirement that the theory is to be made as simple as possible, subject to general requirements of consistency. When this is done, the usual theory is obtained, in which we replace the classical number, p, by the operator $\frac{\hbar}{i}\frac{\partial}{\partial q}$ wherever it occurs in a function, and make the resulting operator Hermitean by a suitable symmetrization of order of factors. See Chap. 9, Sec. 13.

‡ For a fuller treatment see Heisenberg, *The Physical Principles of the Quantum Theory.*

25. Physical Interpretation of Matrix Representations and Transformation Theory. We are now in a position to suggest a physical interpretation of the matrix representations and transformation theory, an interpretation that has already been given in qualitative form in the discussion of complementarity appearing in Chap. 8, Sec. 15.

We begin by noting that associated with each observable, A, is a series of eigenfunctions, ψ_a, belonging respectively to the eigenvalues denoted by a. When the wave function is ψ_a, then our physical interpretation of this fact is that the observable, A, has the definite value, a. Moreover, we note that according to the expansion postulate, an arbitrary wave function can be written as a series of eigenfunctions of any observable A. Thus,

$$\psi = \sum_a C_a \psi_a$$

It is clear that when several of the ψ_a appear in the wave function, the value of A cannot be regarded as well defined. The quantities, $|C_a|^2$ then yield the probability that a measurement of A carried out on a system having the wave function ψ will yield the definite result a. However, A does not exist with a definite value before the measurement takes place, but only as an incompletely defined potentiality, which is realized in a more definite form as a result of interaction with the measuring apparatus.*

The full physical content of the C_a is not exhausted by the above interpretation of $|C_a|^2$, because, as we shall see, the phase relations among the C_a help determine the probability distribution for variables that do not commute with A. In order to show that this is so, we consider such an observable B (which can most generally be represented by a matrix, $B_{aa'}$ in the A representation). Let the eigenfunctions of B be denoted by ϕ_b. Then according to Sec. 14, there will be a unitary transformation matrix, β_{ab}, such that

$$\phi_b = \sum_a \beta_{ab} \psi_a$$

This means that each eigenfunction of B will in general be a linear combination of eigenfunctions of A. Thus, when B has a definite value, it will be necessary that A shall spread over a range of variables, determined by the range in which β_{ab} differs appreciably from zero.† Moreover, the observable, B, can have the definite value b only when the functions ψ_a are combined with both the amplitudes and the *phases* implied by the coefficients β_{ab}. This means that the phase relations with which the ψ_a are combined will, in general, have physical significance, since they will determine, for example, whether or not B has a definite value.

* Chapter 6, Secs. 9 and 13; Chap. 8, Secs. 14 and 15.

† It is readily shown that if B does not commute with A, then more than one eigenfunction, ψ_a, is required in the expansion of ϕ_b.

Let us now consider a case in which the wave function is ϕ_b, so that B has a definite value and A does not. Suppose now that the system interacts with a device that can be used to measure A. After the process of interaction is over, then according to Chap. 6, Sec. 3, each ψ_a is multiplied by an uncontrollable phase factor, $e^{i\alpha_a}$, so that ϕ_b becomes

$$\chi_b = \sum_a \beta_{ab}\, e^{i\alpha_a}\psi_a$$

Thus, the phase relations needed to produce a definite value of B have been destroyed in the process of measurement of A. (This is the meaning of the noncommutativity of A and B.)

Before interaction with the apparatus took place, the system had interference properties associated with many values of a at once and, therefore, literally covered these values simultaneously. After the measurement, the system has no definite phase relations between the ψ_a, so that its subsequent behavior can be understood in terms of the notion that it has a definite value of A at this time, with a probability $|\beta_{ab}|^2$, that this value is a (compare with Chap. 6, Sec. 4 and Chap. 22, Sec. 9). On the other hand, the wave function now spreads over a range of values of B. Thus, the system has undergone a transformation from a state in which B had a definite value, while A was an incompletely defined quantity which was potentially capable of taking on a more definite value, to a state in which A has a definite value, while B has become incompletely defined and potentially capable of obtaining a more definite value. Thus, each observable has two aspects, since it may exist either in a definite form or as an incompletely defined potentiality.*

The fact that an observable may be in part a potentiality suggests a really striking difference between the nature of matter, as implied by quantum theory and that implied by classical theory. For each observable corresponds to some physical property in terms of which the system can manifest itself. Such an observable can be said to categorize (or classify) the possible results of a measurement of this physical property, for if a measurement of a given observable is valid, the result must come out as some single one of a range of logically alternative possible results. When two observables do not commute, then the two systems of categorization associated with the corresponding experiments cannot both apply simultaneously. Thus, when one of the observables is measured, the system of categories associated with another observable not commuting with the measured observable is literally dissolved, since as we have seen, the measurement of any one observable causes the system to spread out over a range of values of a noncommuting observable. Such a behavior is in striking contrast to that described in classical theory. Thus, classically, every particle can have its physical state categorized in terms of the

* Compare with discussion of angular momentum variables in Chap. 14, Sec. 21.

values of its position and momentum. This system of categorization never changes; only the values of the quantities associated with these categories will change. But in quantum theory, the system can either be categorized in terms of a definite position or a definite momentum, but not in terms of both together. In a process of measurement, the system of categories is, in general, actually transformed, and this mathematical transformation is reflected physically in transformations between particle-like behavior (associated with the position categorization) to wavelike behavior (associated with the momentum categorization). But there are in principle an infinite number of systems of categorization cutting across both momentum and position. Thus, one can expand the wave function in terms of eigenfunctions of the harmonic oscillator (Chap. 13) or eigenfunctions of the hydrogen atom (Chap. 15), or in still other ways that will occur to the reader. In these intermediate systems of categorization, the system spreads over a range of positions and momenta.

Finally, we see that the concept of transformation of categories provides a natural interpretation of the representation of an observable in terms of a matrix. For if two operators, A and B, do not commute, so that the categories associated with them do not apply simultaneously, then the observable B will have to be associated with many values of A at once. This property is reflected in the representation of the observable B in terms of matrix elements, $B_{aa'}$ belonging symmetrically to two values of a. It is only in a representation in which the operator B is diagonal that it can be described completely in terms of elements B_{bb}, each of which is associated with only a single value of b.

CHAPTER 17

Spin and Angular Momentum

IN CHAP. 14 WE STUDIED the quantum properties of the angular momentum of single-particle systems. We wish now to extend this treatment to take into account the angular momentum of a system of particles. We shall also discuss the treatment of the additional angular momentum arising from the fact that the electron has an intrinsic spin.

1. Electron Spin. Although the Schrödinger wave equation gives excellent general agreement with experiment in predicting the frequencies of spectral lines, small discrepancies are found, which can be explained in terms of the postulate that the electron has, besides its usual orbital angular momentum, an additional intrinsic angular momentum that acts as if it came from a spinning solid body.* It was found that agreement with experiment could be obtained by means of the assumption that the magnitude of this additional angular momentum was $\hbar/2$. The magnetic moment needed to obtain agreement with the Zeeman effect was, however, $\mu = e\hbar/2mc$, which is exactly the same as that arising from an orbital angular moment of $\hbar$.† The gyromagnetic ratio, i.e., the ratio of magnetic moment to angular momentum is therefore twice as great for electron spin as it is for orbital motion.

Many efforts were made to connect this intrinsic angular momentum to an actual spin of the electron, considered as a rigid body. In fact, the gyromagnetic ratio needed is exactly that which would be obtained if the electron consisted of a uniform spherical shell spinning about a definite axis. The systematic development of such a theory met, however, with such great difficulties that no one was able to carry it through to a definite conclusion.‡ Somewhat later, Dirac derived a relativistic wave equation for the electron, in which the spin and charge were shown to be bound up in a way that can be understood only in connection with the requirements of relativistic invariance.§ In the nonrelativistic limit, however, the electron still acts as if it had an intrinsic angular momentum

* See Kramers, *Die Grundlagen der Quantentheorie*.

† It should be noted that because it is of the order of $\hbar$, spin is an essentially quantum-mechanical property. In the classical limit, its effects are too small to be seen. Thus, as pointed out in Chap. 9, Sec. 28, one cannot obtain it by requiring that the quantum theory approach the correct classical limit.

‡ *Ibid.*

§ See Dirac, *The Principles of Quantum Mechanics*.

. /2. In this chapter, we shall therefore treat the nonrelativistic ıeory of spin in the form originally developed by Pauli, and we shall merely accept the spin as an empirically required addition to the angular momentum, without attempting to understand its origin a deeper way.

2. Matrix Representation of Angular-momentum Operators. We shall find it convenient in this chapter to use the matrix representation for angular momentum operators. According to eq. (35), Chap. 14, the eigenfunctions for angular-momentum operators can be described in terms of two quantum numbers, l and m, where

$$L^2\psi_l^m = \hbar^2 l(l+1)\psi_l^m \qquad \text{and} \qquad L_z\psi_l^m = \hbar m\psi_l^m$$

In a representation in which L^2 and L_z are diagonal, one obtains for the matrix elements of the above operator

$$L^2_{l,l';m,m'} = \int \psi_l^{m*} L^2 \psi_{l'}^{m'}\, d\Omega = \hbar^2 l(l+1)\delta_{ll'}\delta_{mm'} \tag{1a}$$

$$(L_z)_{l,l';m,m'} = \int \psi_l^{m*} L_z \psi_{l'}^{m'}\, d\Omega = \hbar m \delta_{ll'}\delta_{mm'} \tag{1b}$$

An arbitrary wave function can be expanded as a series

$$\psi = \sum_{l,m} a_{lm}\psi_l^m \tag{2}$$

The a_{lm} are a generalization of the column representation of the wave function. (See Chap. 16, Sec. 9), in which the eigenvectors can be regarded as a rectangular array, instead of a single column. Matrix elements then involve the four subscripts l, l'; m, m' as shown above in eqs. (1a) and (1b). The generalization of matrix multiplication to this case is straightforward.

There remains the problem of obtaining the matrices for L_x and L_y. To do this, it is convenient to work in terms of $L_x + iL_y$ and $L_x - iL_y$. According to Chap. 14, eqs. (27) and (28).

$$(L_x + iL_y)\psi_l^m = C_l^m \psi_l^{m+1} \qquad (L_x - iL_y)\psi_l^m = C_l'^m \psi_l^{m-1} \tag{3}$$

where C_l^m and $C_l'^m$ are appropriate constants which will be determined later. To obtain the matrix elements of $(L_x + iL_y)$, we simply use the definition given in eq. (2), Chap. 16

$$\begin{aligned}(L_x + iL_y)_{l,l';m,m'} &= \int \psi_l^{m*}(L_x + iL_y)\psi_{l'}^{m'}\, d\Omega = C_{l'}^{m'} \int \psi_l^{m*}\psi_{l'}^{m'+1}\, d\Omega \\ &= C_{l'}^{m'}\delta_{ll'}\delta_{m,m'+1}\end{aligned} \tag{4a}$$

Similarly

$$(L_x - iL_y)_{l,l;m,m'} = C_{l'}'^{m'}\delta_{ll'}\delta_{m,m'-1} \tag{4b}$$

This means that $(L_x + iL_y)$ and $(L_x - iL_y)$ are represented by matrices which are diagonal in l, but in which all elements are one space off the diagonal in m.

3. The Allowed Values of l and m; Half-integral Angular-momentum Quantum Numbers. We shall now reinvestigate the question of what determines the allowed values of l and m. We shall see that on the basis of our more general matrix point of view, we can obtain half-integral as well as integral values for these quantities, and that the results of Chap. 14, which gave only integral values, follow from certain excessively restrictive conditions that are actually correct for orbital angular momenta, but not for spin.

To determine the allowed values of l and m, we begin with the fact that if we are given a wave function ψ_l^m, we can always generate a wave function ψ_l^{m+1} or ψ_l^{m-1} by operating respectively† with $(L_x + iL_y)$ or $(L_x - iL_y)$. Unless this procedure eventually leads to $(L_x + iL_y)\psi_l^{m_1} = 0$ and $(L_x - iL_y)\psi_l^{m_2} = 0$ we will obtain arbitrarily large values of $|m|$. But according to eq. (30), Chap. 14, $\hbar^2|m|^2 < L^2$. Thus, we know that there must be a maximum value, $L_z = m_1\hbar$, and a minimum, $L_z = m_2\hbar$. Because we took only integral steps in going from m_1 to m_2, $m_1 - m_2$ must be an integer. But in eqs. (33) and (34), Chap. 14, it was shown that $m_2 = -m_1$. Thus, we find that $m_1 - m_2 = 2m_1$ is an integer, so that $m_1 = l$ may be either an integer or a half-integer. In Chap. 14 we chose only integral values of l, but as far as the abstract definition of the operator in terms of its commutators is concerned, half-integral angular-momentum quantum numbers are also permitted.

With this result in mind, let us re-examine the requirement used in Chap. 14 that the wave function be a single-valued function of position. All that we can really require is that all physically observable quantities be single valued. This would be achieved by making the average value of an arbitrary observable, $\bar{A} = \int \psi^* A \psi \, d\Omega$, a single-valued function. This requirement is certainly satisfied by a choice of only integral values of l. It can also be satisfied, however, by choosing only half-integral values, for then we can expand an arbitrary wave function

$$\psi = \sum_m C_m \, e^{i(m+\frac{1}{2})\varphi}$$

When φ changes by 2π, ψ is multiplied by -1, but $\psi^* A \psi$ is left unchanged. On the other hand, if both integral and half-integral angular momenta were present simultaneously, then even the probabilities would not be single valued. Thus, with

$$\psi = C_1 \, e^{i\varphi/2} + C_2 \, e^{i\varphi}$$

we obtain $$\psi^*\psi = |C_1|^2 + |C_2|^2 + C_1^* C_2 \, e^{i\varphi/2} + C_2^* C_1 \, e^{-i\varphi/2}$$

When φ is changed by 2π, this changes to

$$\psi^*\psi = |C_1|^2 + |C_2|^2 - (C_1^* C_2 \, e^{i\varphi/2} + C_2^* C_1 \, e^{-i\varphi/2})$$

† See eq. (3).

We conclude that a sensible theory could be made for orbital angular momenta, if the angular momenta were either all integral, or all half-integral, but not if both were present together. Experiment shows that only integral orbital angular momenta are actually present. For example, the choice of half-integral l would give a hydrogen spectrum very different from what is observed. When we quantize the intrinsic angular momentum of the electron, however, there is no a priori reason for either integral or half-integral spins, and to obtain agreement with experiment, it turns out that we must choose half-integral spins.

4. Matrices for $(L_x + iL_y)$ and $(L_x - iL_y)$. We shall now evaluate the constants appearing in eq. (4). Since L_x and L_y are Hermitean, $(L_x + iL_y)$ and $(L_x - iL_y)$ are Hermitean conjugates. Using this fact, we write eq. (31), Chap. 14, in matrix notation, obtaining†

$$[(L_x - iL_y)(L_x + iL_y)]_{mm'} + \hbar^2 m(m+1)\delta_{mm'} = \hbar^2 l(l+1)\delta_{mm'} \quad (5)$$

Now

$$[(L_x - iL_y)(L_x + iL_y)]_{mm'} = \sum_n (L_x - iL_y)_{mn}(L_x + iL_y)_{nm'}$$
$$= \sum_n C^{m*}\delta_{m+1,n}C^{m'}\delta_{n,m'+1} = C^{m*}C^{m'}\delta_{mm'}$$

For the case $m = m'$, we obtain

$$(C)^{m*}C^m = \hbar^2[l(l+1) - m(m+1)] = \hbar^2(l-m)(l+m+1)$$
$$|C^m| = \hbar\sqrt{(l-m)(l+m+1)} \quad (6)$$

Note that the phases of the C_m have not been determined by this procedure, because any choice of phases will lead to the satisfaction of the commutation rules. This means that we can write

$$(L_x + iL_y)_{mm'} = \hbar\sqrt{(l-m')(l+m'+1)}\,\delta_{m,m'+1}\,e^{i\phi_{m'}} \quad (7)$$

where ϕ_m is an arbitrary real number.

Since $(L_x - iL_y)$ is the Hermitean conjugate of $(L_x + iL_y)$, we have

$$(L_x - iL_y)_{mm'} = \hbar\sqrt{(l-m)(l+m+1)}\,\delta_{m+1,m'}\,e^{-i\phi_m} \quad (8)$$

We shall now show that by including a suitable phase factor in the definition of the wave functions we can eliminate the phase factor in the matrix elements. To do this, we refer to our definition of the only non-vanishing matrix elements of $(L_x + iL_y)$, viz.,

$$(L_x + iL_y)_{m,m-1} = \int \psi_m^*(L_x + iL_y)\psi_{m-1}\,d\Omega \quad (9)$$

† Because all significant operators (i.e., L_x, L_y, L_z, L^2) are diagonal in l, we shall hereafter drop the subscript l, unless we are considering an application in which matrices occur that are not diagonal in l.

Problem 1: Prove with the aid of eq. (9) that by multiplying the wave function ψ_m by the phase factor $\exp\left(-i \sum_{n+1-l}^{n=m} \phi_n\right)$, the matrix element will be multiplied by $e^{-i\phi_m}$, so that the phase factors in eq. (8) will be cancelled out. Verify also that the matrix elements $(L_z)_{mm'}$ and $L^2{}_{mm'}$ are left unchanged by this transformation.

From the problem, we can show that the matrix elements obtained by multiplication of the ψ_m by suitable constant phase factors provide just as good a representation as does the old set. We can therefore always assume that we have transformed to such a representation, and choose all the phase factors equal to unity. We then obtain

$$(L_x + iL_y)_{mm'} = \hbar \sqrt{(l - m')(l + m' + 1)}\; \delta_{m,m'+1} \tag{10}$$

$$(L_x - iL_y)_{mm'} = (L_x + iL_y)^\dagger_{mm'} = \hbar \sqrt{(l - m)(l + m + 1)}\; \delta_{m',m+1} \tag{11}$$

As examples, we write down the matrix elements for the case of $l = \frac{1}{2}$

$$(L_x + iL_y) = \hbar \begin{pmatrix} 0 & 1 \\ 0 & 0 \end{pmatrix} \qquad (L_x - iL_y) = \hbar \begin{pmatrix} 0 & 0 \\ 1 & 0 \end{pmatrix}$$

$$L_z = \frac{\hbar}{2}\begin{pmatrix} 1 & 0 \\ 0 & -1 \end{pmatrix} \tag{12}$$

$$L_x = \frac{\hbar}{2}\begin{pmatrix} 0 & 1 \\ 1 & 0 \end{pmatrix} \qquad L_y = \frac{\hbar}{2}\begin{pmatrix} 0 & -i \\ i & 0 \end{pmatrix}$$

$$L_z = \frac{\hbar}{2}\begin{pmatrix} 1 & 0 \\ 0 & -1 \end{pmatrix} \tag{13}$$

where the rows and columns correspond to $m' = \pm\frac{1}{2}$ and $m = \pm\frac{1}{2}$, respectively.

Problem 2: Work out L_x, L_y, L_z for the case $l = 1$, and show from eqs. (15) and (61), Chap. 14, that the same result is obtained from the spherical harmonics for $l = 1$.

The angular-momentum matrices for spin, $\hbar/2$, were first worked out by Pauli. These three matrices, called the *Pauli matrices*, are written as

$$L_x = \frac{\hbar}{2}\sigma_x \qquad L_y = \frac{\hbar}{2}\sigma_y \qquad L_z = \frac{\hbar}{2}\sigma_z \tag{14}$$

where the σ are matrices defined in eq. (13).

Because the σ matrices are proportional to angular-momentum operators, they satisfy the following commutation rules

$$\sigma_x\sigma_y - \sigma_y\sigma_x = 2i\sigma_z \tag{15}$$

with the other rules obtained by cyclic exchange of x, y, z. The three commutation rules are contained in the vector equation, $\boldsymbol{\sigma} \times \boldsymbol{\sigma} = 2i\boldsymbol{\sigma}$.

It can readily be shown by direct computation that

$$\sigma_x\sigma_y + \sigma_y\sigma_x = 0 \tag{16}$$

From this and from eq. (15), we obtain

$$\sigma_x \sigma_y = i\sigma_z \tag{17}$$

or more generally

$$\boldsymbol{\sigma} \times \boldsymbol{\sigma} = 2i\boldsymbol{\sigma} \tag{18}$$

It can be shown by direct computation that $\sigma_x^2 = \sigma_y^2 = \sigma_z^2 = 1$, so that we obtain

$$\sigma^2 = 3 \tag{19}$$

and

$$L^2 = \frac{\hbar^2}{4}(\sigma_x^2 + \sigma_y^2 + \sigma_z^2) = \frac{3}{4}\hbar^2 \tag{20}$$

This is clearly in agreement with the result obtained from eq. (1a) with $l = \frac{1}{2}$.

5. The Eigenfunctions of the σ Operators. As in Chap. 16, Sec. 9, we can represent the wave function as a column matrix $\begin{pmatrix} C_1 \\ C_2 \end{pmatrix}$, where $|C_1|^2$ represents the probability that $L_z = \hbar/2$ and $|C_2|^2$ represents the probability that $L_z = -\hbar/2$. To normalize the wave function, we must have

$$|C_1|^2 + |C_2|^2 = 1 \tag{21}$$

If the wave function is a function of x, the distribution of spin directions may depend on the position. Thus, in the most general case, C_1 and C_2 will be different functions of x, and the wave function can be represented by the column matrix $\begin{pmatrix} \psi_1(x) \\ \psi_2(x) \end{pmatrix}$ with

$$\int_{-\infty}^{\infty} (|\psi_1(x)|^2 + |\psi_2(x)|^2)\, dx = 1 \tag{22}$$

This means that the existence of spin can be regarded as leading to the use of two wave functions rather than one.* If the spin is independent of position, both $\psi_1(x)$ and $\psi_2(x)$ will vary in the same way, so that the wave function can be factored as below

$$\psi(x) \begin{pmatrix} C_1 \\ C_2 \end{pmatrix} \tag{23}$$

The normalized wave functions corresponding to $L_z = \hbar/2$ and $L_z = -\hbar/2$ respectively are

$$\psi_1 = \begin{pmatrix} 1 \\ 0 \end{pmatrix} \quad \text{and} \quad \psi_2 = \begin{pmatrix} 0 \\ 1 \end{pmatrix} \tag{24}$$

To test for orthogonality of two wave functions, $\begin{pmatrix} a_1 \\ a_2 \end{pmatrix}$ and $\begin{pmatrix} b_1 \\ b_2 \end{pmatrix}$, we evaluate

* This behavior is somewhat analogous to the appearance of several components of the potential in the description of electromagnetic waves.

$$(a_1^* \; a_2^*) \begin{pmatrix} b_1 \\ b_2 \end{pmatrix} = a_1^* b_1 + a_2^* b_2$$

It is clear that ψ_1 and ψ_2 are orthogonal.

6. Eigenfunctions of σ_x and σ_y. To obtain the eigenfunctions of σ_x, we require that $\sigma_x \psi = \alpha \psi$ where α is the eigenvalue, or

$$\begin{pmatrix} 0 & 1 \\ 1 & 0 \end{pmatrix} \begin{pmatrix} C_1 \\ C_2 \end{pmatrix} = \alpha \begin{pmatrix} C_1 \\ C_2 \end{pmatrix}$$

This reduces to

$$C_2 = \alpha C_1$$
$$C_1 = \alpha C_2$$
$$\alpha^2 = 1 \qquad \text{or} \qquad \alpha = \pm 1$$

Thus, as we expected, the allowed values of σ_x are ± 1. The respective normalized wave functions are

$$(\psi_+)_x = \frac{1}{\sqrt{2}} \begin{pmatrix} 1 \\ 1 \end{pmatrix} \qquad \text{and} \qquad (\psi_-)_x = \frac{1}{\sqrt{2}} \begin{pmatrix} 1 \\ -1 \end{pmatrix} \tag{25}$$

In a similar way, we obtain the eigenvalues of σ_y from the equation

$$\begin{pmatrix} 0 & -i \\ i & 0 \end{pmatrix} \begin{pmatrix} C_1 \\ C_2 \end{pmatrix} = \alpha \begin{pmatrix} C_1 \\ C_2 \end{pmatrix}$$
$$-iC_2 = \alpha C_1$$
$$iC_1 = \alpha C_2$$
$$\alpha^2 = 1 \qquad \alpha = \pm 1$$

The normalized wave functions are

$$(\psi_+)_y = \frac{1}{\sqrt{2}} \begin{pmatrix} 1 \\ i \end{pmatrix} \qquad \text{and} \qquad (\psi_-)_y = \frac{1}{\sqrt{2}} \begin{pmatrix} 1 \\ -i \end{pmatrix} \tag{26}$$

Problem 3: Prove that $(\psi_+)_y$ and $(\psi_-)_y$ are orthogonal.

As in the case of integral angular momentum (see Chap. 14, Secs. 12 and 18) the system can obtain a definite angular momentum in the x or y direction only as a result of interference of the states with $\sigma_z = +1$ and $\sigma_z = -1$. This means that when the angular momentum in any one direction is definite, the system must be regarded as covering all possible values of the other two angular momenta simultaneously. In this connection, note that although $(L_z)^2 = \hbar^2/4$, $L^2 = \frac{3}{4}\hbar^2$. This means that even when the z component of the spin is well defined, the other two components are not zero, but must be regarded as fluctuating between $\hbar/2$ and $-\hbar/2$.

7. Spinor Transformations. If we wish to obtain the average value of the spin in an arbitrary direction, then we take advantage of the vector character of angular momentum, and write

$$\sigma_n = \sigma_x \cos\alpha + \sigma_y \cos\beta + \sigma_z \cos\gamma \tag{27}$$

where α, β, γ are the respective angles between this direction and the x, y, z axes.

As a matrix, σ_n takes the form

$$\sigma_n = \begin{pmatrix} \cos\gamma, & \cos\alpha - i\cos\beta \\ \cos\alpha + i\cos\beta, & -\cos\gamma \end{pmatrix} \tag{28}$$

Let us now solve for the eigenvalues and eigenfunctions of σ_n. We obtain

$$C_1 \cos\gamma + C_2(\cos\alpha - i\cos\beta) = SC_1$$
$$C_1(\cos\alpha + i\cos\beta) - C_2\cos\gamma = SC_2$$

where S is the eigenvalue at σ_n. The condition for a solution is

$$(S - \cos\gamma)(S + \cos\gamma) = \cos^2\alpha + \cos^2\beta = 1 - \cos^2\gamma$$
$$S^2 = 1 \qquad S = \pm 1$$

We see then that the possible eigenvalues of the component of the spin in an arbitrary direction are always ± 1. The eigenfunctions are

$$\left.\begin{aligned} \psi'_+ &= \frac{1}{\sqrt{\cos^2\alpha + \cos^2\beta + \cos^2\gamma + 1 + 2\cos\gamma}} \begin{pmatrix} 1 + \cos\gamma \\ \cos\alpha + i\cos\beta \end{pmatrix} \\ &= \frac{1}{\sqrt{2}} \begin{pmatrix} \sqrt{1 + \cos\gamma} \\ \dfrac{\cos\alpha + i\cos\beta}{\sqrt{1 + \cos\gamma}} \end{pmatrix} = \begin{vmatrix} \cos\dfrac{\gamma}{2} \\ \dfrac{\cos\alpha + i\cos\beta}{2\cos\dfrac{\gamma}{2}} \end{vmatrix} \end{aligned}\right\} \tag{29}$$

$$\psi'_- = \frac{1}{\sqrt{2}} \begin{pmatrix} -\sqrt{1 - \cos\gamma} \\ \dfrac{\cos\alpha + i\cos\beta}{\sqrt{1 - \cos\gamma}} \end{pmatrix} = \begin{vmatrix} -\sin\dfrac{\gamma}{2} \\ \dfrac{\cos\alpha + i\cos\beta}{2\sin\dfrac{\gamma}{2}} \end{vmatrix} \tag{29a}$$

Problem 4: Prove that ψ_+' and ψ_-' are normal and orthogonal.

It is now clear that the spin theory can be set up in an equivalent way, using an arbitrary direction as a z axis. Whenever we obtain a series of equivalent ways of formulating the theory, then, according to Chap. 16, Sec. 15, we know that these different formulations must be connected by unitary transformations. To obtain the unitary transformations connecting ψ'_+, ψ'_- with ψ_+, ψ_-, we note that the following identity is true:

$$\left.\begin{aligned} \psi'_+ &= \cos\frac{\gamma}{2}\psi_+ + \frac{(\cos\alpha + i\cos\beta)}{2\cos\frac{\gamma}{2}}\psi_- \\ \psi'_- &= -\sin\frac{\gamma}{2}\psi'_+ + \frac{\cos\alpha + i\cos\beta}{2\sin\frac{\gamma}{2}}\psi_- \end{aligned}\right\} \tag{30}$$

The transformation may be written $\psi'_i = \sum_j \alpha_{ij}\psi_j$

To prove that α_{ij} is a unitary matrix, we must show that $\sum_k \alpha_{ik}\alpha^*_{kj} = \delta_{ij}$.

Problem 5: Prove that the unitary character of α_{ij} follows from the fact that the pairs (ψ_+', ψ_-') and (ψ_+, ψ_-) are normalized and respectively orthogonal.

Since the above transformation rotates the z axis into a definite direction, it is equivalent to a rotation through an angle γ about some axis in the xy plane. By choosing $\cos\beta = 0$ and $\cos\alpha = \sin\gamma$, for example, we obtain a rotation about the y axis. The associated matrix is then easily shown to be

$$\begin{pmatrix} \cos\frac{\gamma}{2} & \sin\frac{\gamma}{2} \\ -\sin\frac{\gamma}{2} & \cos\frac{\gamma}{2} \end{pmatrix} \tag{31}$$

and the transformation becomes

$$\begin{aligned} \psi'_+ &= \cos\frac{\gamma}{2}\psi_+ + \sin\frac{\gamma}{2}\psi_- \\ \psi'_- &= -\sin\frac{\gamma}{2}\psi_+ + \cos\frac{\gamma}{2}\psi_- \end{aligned} \tag{31a}$$

We see that the transformation resembles that of a rotation of a vector [see eq. (6), Chap. 16], but that the *half*-angle of rotation is involved. We shall return to this point later.

Let us now extend our treatment to include rotation about the z axis. To deal with this problem, let us note that it should be possible to obtain the average value of the spin in a given direction by using the transformed operators σ'_x and the transformed wave functions $\begin{pmatrix} C'_1 \\ C'_2 \end{pmatrix}$. Suppose, for example, that we start in a given co-ordinate system, and evaluate

$$\overline{\sigma_x} = (C_1^* \, C_2^*)\begin{pmatrix} 0 & 1 \\ 1 & 0 \end{pmatrix}\begin{pmatrix} C_1 \\ C_2 \end{pmatrix} = C_1^*C_2 + C_2^*C_1 \tag{32}$$

We now rotate the co-ordinate system about the z axis, through an angle φ. We suppose that the wave function becomes $\begin{pmatrix} C'_1 \\ C'_2 \end{pmatrix}$, whereas the operator σ_x can be expressed as follows:

$$\sigma_x = \sigma'_x \cos\varphi + \sigma'_y \sin\varphi = \begin{pmatrix} 0 & e^{-i\varphi} \\ e^{i\varphi} & 0 \end{pmatrix} \tag{33}$$

The average of σ_x thus becomes

$$\overline{\sigma_x} = (C_1'^* \, C_2'^*) \begin{pmatrix} 0 & e^{-i\varphi} \\ e^{i\varphi} & 0 \end{pmatrix} \begin{pmatrix} C_1' \\ C_2' \end{pmatrix} = e^{-i\varphi} C_1'^* C_2' + e^{i\varphi} C_2'^* C_1' \tag{34}$$

Now, according to Chap. 16, Sec. 15, it should be possible by means of a unitary transformation to express C_1' and C_2' in terms of C_1 and C_2. This unitary transformation must have the property that the average value of a transformed operator can be obtained from the transformed wave functions in the same way that the average of the original operator is obtained from the original wave functions. Thus, we wish to obtain (for arbitrary C_1, C_2)

$$\overline{\sigma_x} = C_1^* C_2 + C_2^* C_1 = e^{-i\varphi} C_1'^* C_2' + e^{i\varphi} C_2'^* C_1' \tag{35}$$

It is evident from inspection of (35) that this requires that

$$C_1' = e^{-i\varphi/2} C_1 \qquad \text{and} \qquad C_2' = e^{i\varphi/2} C_2 \tag{36}$$

The reader will readily verify that the above is a special case of a unitary transformation. In the matrix notation, eq. (36) becomes

$$\begin{pmatrix} C_1 \\ C_2 \end{pmatrix} = \begin{pmatrix} e^{i\varphi/2} & 0 \\ 0 & e^{-i\varphi/2} \end{pmatrix} \begin{pmatrix} C_1' \\ C_2' \end{pmatrix} \tag{37}$$

The most general rotation can be compounded out of successive rotations about the x, y, and z axes. Thus, our treatment is easily generalized to enable one to calculate the unitary transformation corresponsing to an arbitrary rotation.

Note that in eq. (37), as in (31), the *half*-angle of rotation appears in the matrix elements. These equations should be compared with eq. (16), Chap. 16, where the matrices defining the rotation of a vector about the z axis are defined. Here we see that the *full* angles of rotation appear in the matrix elements. In other words, the complex column vectors, $\begin{pmatrix} C_1 \\ C_2 \end{pmatrix}$, undergo a transformation on rotation, reminiscent of that undergone by a vector, but differing in that only half-angles of rotation are involved. The column vectors, $\begin{pmatrix} C_1 \\ C_2 \end{pmatrix}$, are therefore a new kind of quantity, analogous to a vector, but not the same thing. They are often called "spinors," or "semivectors." It can be shown† that spinors provide the most fundamental representation of the rotation group,

† E. Wigner, *Gruppentheorie*. Braunschweig: Friedrich Vieweg und Sohn, 1931.

because out of them can be formed all of the usual vectors and tensors, and also new representations, which are not included in the usual theory of vectors and tensors. The spinors are also closely connected with quaternions, and also with the Cayley-Klein parameters.*

8. The Addition of Angular Momenta. We wish now to study the problem of how angular momenta are to be added in quantum theory. For example, we may wish to know the combined angular momentum of two particles, or else we may wish to know how spin and orbital angular momentum are to be combined.

To solve this problem, we begin by noting that the orbital angular momentum commutes with the spin. This is because the two types of operations do not affect each other. The combined angular momentum produced by the spin and orbital motion is

$$J = L + S \tag{38}$$

The total combined angular momentum is

$$J^2 = (L + S)^2 = L^2 + S^2 + 2L \cdot S = l(l + 1)\hbar^2 + \tfrac{3}{4}\hbar^2 + 2L \cdot S \tag{39}$$

If we have more than one particle, we note that operators belonging to separate particles also commute. The combined orbital angular momentum is

$$L = \sum_i L_i \qquad L^2 = \Big(\sum_i L_i\Big)^2 \tag{40}$$

The combined spin is

$$S = \sum_i S_i \qquad S^2 = \Big(\sum_i S_i\Big)^2 \tag{41}$$

The combined angular momentum from all sources is

$$J = \sum_i L_i + \sum_i S_i = L + S \tag{42}$$

$$J^2 = \Big(\sum_i L_i + \sum_i S_i\Big)^2 = (L + S)^2 \tag{43}$$

Now, it is readily verified that because the L_i and S_i commute, the respective components of L, S, and M all have the same commutation rules as do the components of orbital angular momentum of a single particle.

Problem 6: Prove the above statement.

From Sec. 3, it follows that the eigenvalues of L^2, J^2, and S^2 are, respectively,

$$L^2 = l(l + 1) \qquad J^2 = j(j + 1) \qquad S^2 = S(S + 1) \tag{44}$$

where l, j, and S are either half-integers or whole integers.

* E. T. Whittaker, *A Treatise on the Analytical Dynamics of Particles and Rigid Bodies.* London: Cambridge University Press, 1927, Chap. 1. See also H. Goldstein *Classical Mechanics.* Cambridge, Mass.: Addison-Wesley Press, 1950, Chap. 4.

We shall often be faced with the problem of defining the simultaneous eigenvalues and eigenfunctions of a system having two different contributions to its angular momentum. For example, we may have a system consisting of two particles, with respective orbital angular momenta l_1 and l_2, and we may wish to know the eigenvalues and eigenfunctions of the combined angular momentum. In the old quantum theory, a rule was obtained for doing this which later turned out to be justified by the exact treatment. This is the well-known vector addition rule.* The rule requires us to consider two vectors of length l_1 and l_2, respectively. Suppose, for example, that $l_1 \geq l_2$. Then we assert that l_2 can have only integral projections on the direction of l_1 as shown in Fig. 1. Let this projection be p. The allowed values of the combined angular momentum l are then equal to $l_1 + p$, where p runs from $-l_2$ to $+l_2$. If $l_2 > l_1$, then we project l_1 on l_2 instead and obtain the result that l lies between $l_2 + l_1$ and $l_2 - l_1$. If the angular momentum that is being projected is half-integral, the same rules apply, except that the projections now run over half-integers. We shall discuss the general proof of this rule later, but illustrate it first in a few special cases.

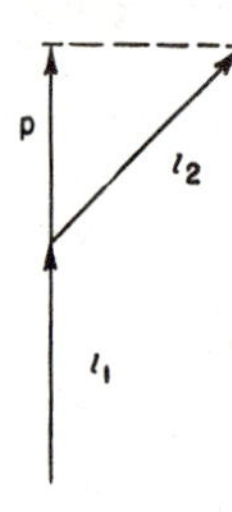

FIG. 1

9. Addition of Spin Angular Momenta of Two Separate Particles
Consider for example, two particles, each having a spin of $\hbar/2$. The relevant combined angular momenta are:

$$\begin{aligned} S &= S_1 + S_2 \\ S^2 &= (S_1 + S_2)^2 = S_1^2 + S_2^2 + 2S_1 \cdot S_2 = \tfrac{3}{2}\hbar^2 + 2S_1 \cdot S_2 \end{aligned} \tag{45}$$

In order to denote the wave functions of this system, we must, as shown in Chap. 10, Sec. 11, take the products of spin wave functions of the separate particles. Thus, from the column vectors, $\psi_+ = \begin{pmatrix}1\\0\end{pmatrix}$, $\psi_- = \begin{pmatrix}0\\1\end{pmatrix}$, which refer to single-particle wave functions, we can construct a total of four independent two-particle wave functions:

$$\begin{aligned} \psi_a &= \psi_+(1)\psi_+(2) \qquad & \psi_c &= \psi_-(1)\psi_+(2) \\ \psi_b &= \psi_+(1)\psi_-(2) \qquad & \psi_d &= \psi_-(1)\psi_-(2) \end{aligned} \tag{46}$$

$\psi_+(1)\psi_-(2)$ means, for example, that particle number 1 has a spin of $+\hbar/2$, whereas particle number 2 has $-\hbar/2$.

Since the above are the most general functions that can be constructed from the spin functions of two particles, we conclude, according to the expansion postulate, that an arbitrary function can be expanded as a series of these four functions, as shown below:

$$\psi = C_1\psi_a + C_2\psi_b + C_3\psi_c + C_4\psi_d \tag{47}$$

* See Ruark and Urey; also Richtmeyer and Kennard, p. 341 and 356.

The wave function can also be represented as a double-column vector, where the first column refers to the spin quantum number of the first particle, whereas the second column refers to that of the second particle. Thus, we obtain

$$\psi_a = \begin{pmatrix}1\\0\end{pmatrix}\begin{pmatrix}1\\0\end{pmatrix} \quad \psi_b = \begin{pmatrix}1\\0\end{pmatrix}\begin{pmatrix}0\\1\end{pmatrix} \quad \psi_c = \begin{pmatrix}0\\1\end{pmatrix}\begin{pmatrix}1\\0\end{pmatrix} \quad \psi_d = \begin{pmatrix}0\\1\end{pmatrix}\begin{pmatrix}0\\1\end{pmatrix} \tag{48}$$

To normalize a wave function, we sum over the spin quantum numbers of each particle separately and multiply the results of each summation together. Thus, to normalize ψ_a we consider $\begin{matrix}(1 & 0)\\(1 & 0)\end{matrix}\begin{pmatrix}1\\0\end{pmatrix}\begin{pmatrix}1\\0\end{pmatrix}$, where the upper row vector $(1 \quad 0)$ operates on the first-column vector, while the lower row vector operates on the second. It is clear from this definition that $\psi_a, \psi_b, \psi_c, \psi_d$ are already normalized. Orthogonality is tested for in a similar way. Thus, to see whether ψ_a and ψ_b are orthogonal, we consider

$$\begin{matrix}(1 & 0)\\(0 & 1)\end{matrix}\begin{pmatrix}1\\0\end{pmatrix}\begin{pmatrix}1\\0\end{pmatrix}$$

The summation over the spin of the second particle multiplies the above by a factor of zero, thus demonstrating the orthogonality of ψ_a and ψ_b. Similarly, it can be shown that all four ψ's are orthogonal.

Problem 7: Prove that the condition for normalization of the arbitrary wave function (47) is

$$|C_1|^2 + |C_2|^2 + |C_3|^2 + |C_4|^2 = 1 \tag{49}$$

In order to operate on the wave functions (46) and (48), we note that, for example, σ_{1z} operates only on the left-hand member of the product $\psi_+(1)\psi_+(2)$, while σ_{2z} operates on the right-hand member. Thus, we obtain, using $S_z = \frac{\hbar}{2}(\sigma_{1z} + \sigma_{2z})$,

$$\left.\begin{aligned} S_z\psi_a &= \hbar\psi_a \\ S_z\psi_b &= S_z\psi_c = 0 \\ S_z\psi_d &= -\hbar\psi_d \end{aligned}\right\} \tag{50}$$

This means that the ψ's are already eigenfunctions of S_z, and that ψ_a corresponds to a z component of the combined angular momentum of $\hbar$, ψ_d to $-\hbar$, while ψ_b and ψ_c correspond to a zero eigenvalue of S_z.

We wish now to construct simultaneous eigenfunctions of S^2 and of S_z. From eq. (45) we note that this is equivalent to the problem of obtaining simultaneous eigenfunctions of S_z and

$$\boldsymbol{\sigma}_1 \cdot \boldsymbol{\sigma}_2 = \sigma_{1x}\sigma_{2x} + \sigma_{1y}\sigma_{2y} + \sigma_{1z}\sigma_{2z}$$

The term $\sigma_{1x}\sigma_{2x}$, for example, means that σ_{1x} operates on the left-hand member of the double-column vector in eq. (48) while σ_{2x} operates on the right-hand member. The operation corresponding to σ_{2x} is to be followed by that corresponding to σ_{1x}, but since the two operators commute, the order of operation is immaterial.

It is quickly verified that

$$\left.\begin{array}{ll}(\boldsymbol{\sigma}_1 \cdot \boldsymbol{\sigma}_2)\psi_a = \psi_a & (\boldsymbol{\sigma}_1 \cdot \boldsymbol{\sigma}_2)\psi_d = \psi_d \\ (\boldsymbol{\sigma}_1 \cdot \boldsymbol{\sigma}_2)\psi_b = -\psi_b + 2\psi_c & (\boldsymbol{\sigma}_1 \cdot \boldsymbol{\sigma}_2)\psi_c = -\psi_c + 2\psi_b\end{array}\right\} \tag{51}$$

Problem 8: Verify eq. (51).

This shows that ψ_a and ψ_d are already eigenfunctions of $\boldsymbol{\sigma}_1 \cdot \boldsymbol{\sigma}_2$ corresponding to $\boldsymbol{\sigma}_1 \cdot \boldsymbol{\sigma}_2 = 1$, but that ψ_b and ψ_c are not. We seek therefore a wave function, $\psi = b\psi_b + c\psi_c$, which is an eigenfunction of $\boldsymbol{\sigma}_1 \cdot \boldsymbol{\sigma}_2$. This can be obtained by solving the equation

$$(\boldsymbol{\sigma}_1 \cdot \boldsymbol{\sigma}_2)(b\psi_b + c\psi_c) = \lambda(b\psi_b + c\psi_c) \tag{52}$$

where λ is an eigenvalue of $\boldsymbol{\sigma}_1 \cdot \boldsymbol{\sigma}_2$.

We now apply eq. (51) obtaining

$$\psi_b(-b + 2c) + \psi_c(-c + 2b) = \lambda(b\psi_b + c\psi_c) \tag{53}$$

By multiplying the above by ψ_b^* and summing over the spin indices of the separate particles using the orthogonality of ψ_b and ψ_c, we obtain

$$\begin{array}{ll}b(\lambda + 1) = 2c & c(\lambda + 1) = 2b \\ (\lambda + 1)^2 = 4 & \lambda = -1 \pm 2\end{array} \tag{54}$$

The eigenvalues of λ for this case are therefore 1 and -3. The corresponding normalized eigenfunctions are respectively:

$$\left.\begin{array}{l}\psi_1 = \dfrac{1}{\sqrt{2}}(\psi_b + \psi_c) = \dfrac{1}{\sqrt{2}}[\psi_+(1)\psi_-(2) + \psi_-(1)\psi_+(2)] \\ \psi_2 = \dfrac{1}{\sqrt{2}}(\psi_b - \psi_c) = \dfrac{1}{\sqrt{2}}[\psi_+(1)\psi_-(2) - \psi_-(1)\psi_+(2)]\end{array}\right\} \tag{55}$$

The three functions ψ_a, ψ_1, ψ_d all correspond to $\boldsymbol{\sigma}_1 \cdot \boldsymbol{\sigma}_2 = 1$ or, according to eq. (45), $S^2 = 2\hbar^2$. But this is just what is needed for an angular momentum of $\hbar$. These three functions therefore correspond to a total spin of $\hbar$ and to the three possible components in the z direction. The function ψ_2 corresponds to $\boldsymbol{\sigma}_1 \cdot \boldsymbol{\sigma}_2 = -3$, or $S^2 = 0$. This is just the case oi zero angular momentum. We see that the angular momenta that can be obtained from two particles of spin $\frac{1}{2}$ are just those predicted by the vector addition rule of Sec. (7). $S = \hbar$ corresponds to parallel spins in the vector model, $S = 0$ to antiparallel spins. The three states of parallel spin are sometimes called the triplet state, while the single state of antiparallel spin is called the "singlet" state.

10. Probability Distribution of Spin States in a Statistical Ensemble. Very often, electrons or other particles appear with a random statistical distribution of spin directions. Thus, if an electron boils out of a metal, it is equally likely that the spin in any given direction be positive or negative. A problem which often arises is that of finding the probability for a given combined spin of two such particles when the spin of each of them is random. Such a problem might arise, for example, in the treatment of the scattering of one electron on another electron when each electron comes from an independent source, so that there is no correlation between their spin directions.

We shall now show that under these conditions, it is equally likely that the combined system have a state corresponding to ψ_a, ψ_d, ψ_1, or ψ_2 so that each of the three triplet states is equally likely to occur, and each of these is just as likely as the singlet state. This means, however, that since there are three times as many triplet states as singlet states, the spins will turn out to be parallel $\frac{3}{4}$ of the time and antiparallel only $\frac{1}{4}$ of the time.

To treat this problem, we note that the correct single-particle wave functions representing a situation in which it is equally likely that σ_z is positive or negative are

$$\phi_1 = \frac{1}{\sqrt{2}}[(e^{i\alpha_{1,1}}\psi_+(1) + e^{i\alpha_{1,2}}\psi_-(1)] \tag{56a}$$

for the first particle, and

$$\phi_2 = \frac{1}{\sqrt{2}}[(e^{i\alpha_{2,1}}\psi_+(2) + e^{i\alpha_{2,2}}\psi_-(2)] \tag{56b}$$

for the second particle, where $\alpha_{1,1}$, $\alpha_{1,2}$, $\alpha_{2,1}$, and $\alpha_{2,2}$ are random uncontrollable phase factors (see Chap. 6, Sec. 4). The combined wave function for the two particles is

$$\psi = \phi_1\phi_2 = \frac{1}{2}\begin{bmatrix} e^{i(\alpha_{1,1}+\alpha_{2,1})}\psi_a + e^{i(\alpha_{1,2}+\alpha_{2,2})}\psi_d \\ +e^{i(\alpha_{1,1}+\alpha_{2,2})}\psi_b + e^{i(\alpha_{1,2}+\alpha_{2,1})}\psi_c \end{bmatrix} \tag{57}$$

or
$$\psi = \frac{1}{2}\left[\begin{matrix} e^{i(\alpha_{1,1}+\alpha_{2,1})}\psi_a + e^{i(\alpha_{1,2}+\alpha_{2,2})}\psi_d \\ + \dfrac{(e^{i(\alpha_{1,1}+\alpha_{2,2})} + e^{i(\alpha_{1,2}+\alpha_{2,1})})}{\sqrt{2}}\psi_1 + \dfrac{(e^{i(\alpha_{1,1}+\alpha_{2,2})} - e^{i(\alpha_{1,2}+\alpha_{2,1})})}{\sqrt{2}}\psi_2 \end{matrix}\right]$$

The probability function $\psi^*\psi$ is then equal to

$$\psi^*\psi = \tfrac{1}{4}(\psi_a^*\psi_a + \psi_d^*\psi_d + \psi_1^*\psi_1 + \psi_2^*\psi_2 + \text{terms involving random phase factors}) \tag{58}$$

All of the terms involving random phase factors will on the average cancel out. We conclude, then, that in a series of many experiments the four states, ψ_a, ψ_d, ψ_1, and ψ_2 will all occur with equal frequency.

11. Addition of Orbital and Spin Angular Momenta of a Given Particle. The next problem that we shall consider is how to add the orbital and spin angular momenta of a given particle. If we have a particle of a given L^2, L_z, and σ_z, the wave function takes the form

$$\psi = Y_l^m(\vartheta, \varphi)\psi_s \tag{59}$$

where ψ_s is one of the two spin functions given in eq. (24). It is clear that

$$J_z\psi = (L_z + S_z)\psi = \left(m + \frac{\sigma_z}{2}\right)\hbar\psi = k\hbar\psi \tag{60}$$

Hence, J_z is diagonal in this representation. We therefore represent the above wave function by the notation

$$\psi_{l,s}^k = Y_l^m(\vartheta, \varphi)\psi_s \qquad \text{where} \qquad k = m + S \tag{61}$$

To diagonalize J^2, we must have

$$J^2\psi_{l,s} = \left(L + \frac{\hbar\mathbf{\sigma}}{2}\right)\psi_{l,s} = \hbar^2\left[l(l+1) + \frac{3}{4} + \frac{L \cdot \mathbf{\sigma}}{\hbar}\right]\psi_{l,s} = \gamma\hbar^2\psi_{l,s} \tag{62}$$

It is necessary, therefore, to obtain eigenfunctions of the operator $L \cdot \mathbf{\sigma}$ made up of functions corresponding to a definite value of l.

The matrix $L \cdot \mathbf{\sigma}$ may be written as follows:

$$L \cdot \mathbf{\sigma} = \begin{pmatrix} L_z & L_x - iL_y \\ L_x + iL_y & -L_z \end{pmatrix} \tag{63}$$

We shall find it convenient to write our wave function as a column vector, where the components are functions of ϑ and φ:

$$\psi = \begin{pmatrix} f_1(\vartheta, \varphi) \\ f_2(\vartheta, \varphi) \end{pmatrix} \tag{64}$$

For an eigenfunction with eigenvalue λ, we obtain

$$L \cdot \mathbf{\sigma}\psi = \begin{pmatrix} L_zf_1 + (L_x - iL_y)f_2 \\ -L_zf_2 + (L_x + iL_y)f_1 \end{pmatrix} = \lambda\hbar\begin{pmatrix} f_1 \\ f_2 \end{pmatrix} \tag{65}$$

Our equations become

$$\left.\begin{aligned} L_zf_1 + (L_x - iL_y)f_2 &= \lambda\hbar f_1 \\ -L_zf_2 + (L_x + iL_y)f_1 &= \lambda\hbar f_2 \end{aligned}\right\} \tag{66}$$

We note from eq. (3) that $(L_x + iL_y)\psi_m \sim \psi_{m+1}$ and $(L_x - iL_y)\psi_m \sim \psi_{m-1}$. If we tentatively choose $f_2 = C_2Y_l^m(\vartheta, \varphi)$ and $f_1 = C_1Y_l^{m-1}(\vartheta, \varphi)$, we can then satisfy these two equations, for according to eqs. (10) and (11), we can write

$$(L_x - iL_y)Y_l^m = \hbar\sqrt{(l+m)(l-m+1)}\;Y_l^{m-1}$$
$$(L_x + iL_y)Y_l^{m-1} = \hbar\sqrt{(l+m)(l-m+1)}\;Y_l^m$$

We use the quantum number m here for convenience, since previous results of operations on spherical harmonics are given in terms of it. For this treatment it must be remembered, however, that m is defined in terms of the quantum number k of our observable J_z

$$m = k - \frac{S_z}{\hbar}$$

Our resulting functions will, as a rule, not be eigenfunctions of L_z.

Equation (66) then becomes (with $L_z Y_l^m = mY_l^m$)

$$\left.\begin{aligned} (m-1)C_1 + \sqrt{(l+m)(l-m+1)}\,C_2 &= \lambda C_1 \\ -mC_2 + \sqrt{(l+m)(l-m+1)}\,C_1 &= \lambda C_2 \end{aligned}\right\} \tag{67}$$

The equation defining λ is

$$(\lambda - m + 1)(\lambda + m) = (l+m)(l-m+1) = l(l+1) - m(m-1)$$

This reduces to

$$\lambda^2 + \lambda = l^2 + l \tag{68}$$

The solutions are $\lambda = l$, or $\lambda = -1 - l$. Insertion of these values of λ into eq. (62) yields

$$\left.\begin{aligned} \frac{J_a^2}{\hbar^2} &= l(l+1) + \frac{3}{4} + l = \left(l + \frac{1}{2}\right)\left(l + \frac{3}{2}\right) \\ \frac{J_b^2}{\hbar^2} &= l(l+1) - l - \frac{1}{4} = \left(l - \frac{1}{2}\right)\left(l + \frac{1}{2}\right) \end{aligned}\right\} \tag{69}$$

where J_a and J_b refer respectively to the eigenvalues l and $-1 - l$. Writing

$$j_a = l + \tfrac{1}{2} \qquad j_b = l - \tfrac{1}{2} \tag{70}$$

we obtain for both cases

$$\frac{J^2}{\hbar^2} = j(j+1) \tag{71}$$

Thus we obtain a result that is in agreement with that given by the vector rule, which says that the values of j are $l + \frac{1}{2}$ and $l - \frac{1}{2}$.

The eigenfunctions corresponding to a definite value of J^2 can be obtained by inserting the associated values of λ into eq. (67). In order to designate these functions, we note that they are simultaneous eigenfunctions, corresponding to

$$\begin{aligned} J_z &= k\hbar \\ J^2 &= j(j+1)\hbar^2 \\ L^2 &= l(l+1)\hbar^2 \end{aligned}$$

Thus, we write for the wave function

$$\psi = \varphi_{l,j}^k \tag{72}$$

The (normalized) eigenfunctions are then

$$\begin{aligned} \varphi_{l,l+1/2}^k &= \frac{1}{\sqrt{2l+1}}\begin{pmatrix} \sqrt{l+k}\, Y_l^{k-1/2}(\vartheta, \varphi) \\ \sqrt{l-k+1}\, Y_l^{k+1/2}(\vartheta, \varphi) \end{pmatrix} \\ \varphi_{l,l-1/2}^k &= \frac{1}{\sqrt{2l+1}}\begin{pmatrix} \sqrt{l-k+1}\, Y_l^{k-1/2}(\vartheta, \varphi) \\ -\sqrt{l+k}\, Y_l^{k+1/2}(\vartheta, \varphi) \end{pmatrix} \end{aligned} \tag{73}$$

Note that k is a half-integer here, so that $k + \frac{1}{2}$ and $k - \frac{1}{2}$ are integers. To justify the fact that we have designated the above as eigenfunctions of $J_z = L_z + \frac{\sigma_z \hbar}{2}$, we simply operate directly with this operator. The reader will quickly verify that one obtains $J_z = \hbar k$.

In the column representation, the original functions, $\psi_{l,s}^k$ take the form

$$\left. \begin{aligned} \psi_{l,1}^k &= \begin{pmatrix} Y_l^{k-\frac{1}{2}}(\vartheta, \varphi) \\ 0 \end{pmatrix} \\ \psi_{l,-1}^k &= \begin{pmatrix} 0 \\ Y_l^{k+\frac{1}{2}}(\vartheta, \varphi) \end{pmatrix} \end{aligned} \right\} \tag{74}$$

It is clear that the $\varphi_{l,j}^k$ are linear combinations of the ψ's listed above. Thus,

$$\left. \begin{aligned} \varphi_{l,l+\frac{1}{2}}^k &= \frac{1}{\sqrt{2l+1}} \left(\sqrt{l+k}\, \psi_{l,1}^k + \sqrt{l-k+1}\, \psi_{l,-1}^k\right) \\ \varphi_{l,l-\frac{1}{2}}^k &= \frac{1}{\sqrt{2l+1}} \left(\sqrt{l-k+1}\, \psi_{l,1}^k - \sqrt{l+k}\, \psi_{l,-1}^k\right) \end{aligned} \right\} \tag{75}$$

These equations can also be solved for the ψ's. The result is

$$\left. \begin{aligned} \psi_{l,1}^k &= \frac{1}{\sqrt{2l+1}} \left(\sqrt{l+k}\, \varphi_{l,l+\frac{1}{2}}^k + \sqrt{l-k+1}\, \varphi_{l,l-\frac{1}{2}}^k\right) \\ \psi_{l,-1}^k &= \frac{1}{\sqrt{2l+1}} \left(\sqrt{l-k+1}\, \varphi_{l,l+\frac{1}{2}}^k - \sqrt{l+k}\, \varphi_{l,l-\frac{1}{2}}^k\right) \end{aligned} \right\} \tag{76}$$

Thus, the φ's and the ψ's are related by a linear transformation. We conclude also that the allowed values of j for this case are precisely those given by the vector rule.

12. Discussion of General Problem of Adding Angular Momenta. We now proceed to a discussion of the more general problem of finding the wave function of a combined system containing two different sources of angular momentum. This problem can be treated by methods that are very similar to those used in the special cases considered here,* or else by the more powerful methods of group theory.† Although the details are rather complex, the same general result is obtained. That is, if $\psi_{l_1}^{m_1}$ and $\psi_{l_2}^{m_2}$ are wave functions of systems, having, respectively,

$$L^2 = l_1(l_1+1)\hbar^2 \qquad L_z = m_1\hbar$$

and

$$L^2 = l_2(l_2+1)\hbar^2 \qquad L_z = m_2\hbar \tag{77}$$

* E. U. Condon and G. H. Shortley, *The Theory of Atomic Spectra.* New York The Macmillan Company, 1935, Chap. 3.

† Wigner, *Gruppentheorie.* Braunschweig: Friedr. Vieweg und Sohn, 1931.

where $l_1 > l_2$, then the product can be expanded as a series, analogous to eq. (76).

$$\psi_{l_1}^{m_1}\psi_{l_2}^{m_2} = \sum_{x=-l_2}^{x=+l_2} \varphi_{l_1+x}^{m_1+m_2} C_{l_1,m_1,m_2,x} \tag{77a}$$

$\psi_{l+x}^{m_1+m_2}$ is a wave function with*

$$M^2 = (l + x)(l + x + 1)$$

and $M_z = m_1 + m_2$, and the C's are suitable constants.† This result means that the range of angular momenta that can appear in the combined system are precisely those given by the vector rule. Equation (77) will prove to be exceedingly useful in Chap. 18 for deriving selection rules.

13. Energy of a Spinning Electron.‡ As pointed out in Sec. 1, the electron spin has associated with it a magnetic moment of $-e\hbar/2mc$,§ where m is the electronic mass. This means that in a magnetic field $\mathcal{H}$ the spin makes the following contribution to the Hamiltonian operator:

$$W = -\frac{e\hbar}{2mc}\,\boldsymbol{\sigma}\cdot\mathcal{H} = -\frac{e\hbar}{2mc}\begin{pmatrix}\mathcal{H}_z & \mathcal{H}_x - i\mathcal{H}_y \\ \mathcal{H}_x + i\mathcal{H}_y & -\mathcal{H}_z\end{pmatrix} \tag{78}$$

The above is the nonrelativistic expression for the spin energy. A complete relativistic treatment can be given only by means of the Dirac equation. A treatment that is correct to order v/c, however, can be obtained by assuming that eq. (78) describes the energy in a Lorentz frame in which the electron is at rest. The relativistic generalization of eq. (78) to an arbitrary frame would then yield (to first order in v/c)

$$W = -\frac{e\hbar}{2mc}\left(\boldsymbol{\sigma}\cdot\mathcal{H} + \boldsymbol{\sigma}\cdot\frac{v\times\mathcal{E}}{c}\right) \tag{79}$$

The above expression, however, must be corrected further to take into account another relativistic effect known as the *Thomas precession*.‖ This reduces the contribution of the electric field by a factor of 2. We finally obtain

$$W_{sp} = -\frac{e\hbar}{2mc}\left[\boldsymbol{\sigma}\cdot\mathcal{H} + \frac{1}{2}\boldsymbol{\sigma}\cdot\left(\frac{v}{c}\times\mathcal{E}\right)\right] \tag{80}$$

* M^2 and M_z refer to the eigenvalues for total orbital angular momentum and for its z component.

† The constants are evaluated by Condon and Shortley and by Wigner.

‡ For a discussion of spin energy, see Schiff, p. 223 and 331. For a qualitative discussion, see White, *Introduction to Atomic Spectra*, Chap. 8.

§ e stands for the *absolute value* of electronic charge.

‖ Ruark and Urey, p. 162.

The term $\frac{\boldsymbol{\sigma}}{2}\cdot\left(\frac{v}{c}\times\mathbf{\mathcal{E}}\right)$ leads to what is known as *spin-orbit* interaction, in an atom where $\mathbf{\mathcal{E}}$ stands for the electric field of the nucleus (and the other electrons). Let us write $\mathbf{\mathcal{E}} = -\nabla\phi$. In a spherically symmetric atom, $\phi = \phi(r)$ and $\mathbf{\mathcal{E}} = \frac{-\mathbf{r}}{r}\,\phi'(r)$. The spin-orbit energy becomes

$$W_{so} = \frac{e\hbar}{4mc^2}\,\frac{\phi'(r)}{mr}\,\boldsymbol{\sigma}\cdot(\mathbf{p}\times\mathbf{r}) = \frac{-e\hbar}{4mc^2}\,\frac{\phi'(r)}{mr}\,(\mathbf{L}\cdot\boldsymbol{\sigma}) \tag{81}$$

PART IV

METHODS OF APPROXIMATE SOLUTION OF SCHRÖDINGER'S EQUATION

CHAPTER 18

Perturbation Theory. Time Dependent and Time Independent

1. Introduction to Part IV. In Part IV we shall develop a number of approximate methods of solving Schrödinger's equation. We shall begin with the method of variation of constants, which will be applied to the calculation of rates of transition, especially to those transitions involving emission and absorption of radiation. We shall then discuss small adiabatic perturbations, which lead to shifts in the energy levels and eigenfunctions. This will lead to the problem of large, but slowly varying, perturbations (general adiabatic approximation). Finally, we shall discuss a treatment that deals with the case of sudden changes in potential (impulsive approximation). This will complete our study of some of the common methods of approximation used in the solution of Schrödinger's equation.

2. Case of a Small Perturbation (Method of Variation of Constants). In this problem, we begin with a system for which the wave equation can be solved exactly, and then ask what will happen to this system under the action of a small external disturbance. For example, consider a hydrogen atom or a harmonic oscillator to which is applied a weak external electromagnetic field that could come from an incident light wave or from a constant externally impressed electric field. From experiments, we know that the atom can absorb a light quantum and go to a higher energy level; also, if the externally impressed electric field is constant with time, we obtain a shift in energy levels known as the *Stark effect.* Thus, an external disturbance certainly causes changes in the system with which we started.

In principle, the effect of the external disturbance could be obtained theoretically by solving Schrödinger's equation, if the impressed scalar potential ϕ and the vector potential A were included in the equation. In most cases, the resulting equation is, unfortunately, too complex to be solved exactly. An approximation technique can be developed, however, which is based on the reasonable assumption that a small change in the Hamiltonian produces a correspondingly small change in the wave function. With the aid of this assumption we can develop a method of successive approximations, in a manner somewhat analogous to the development of the series for S in the WKB approximation [see eq. (7) Chap. 12]. This method is also known as *perturbation theory.*

To apply this method, we start with the wave equation

$$i\hbar \frac{\partial \psi}{\partial t} = H\psi \tag{1}$$

Our perturbation theory will be valid only when the Hamiltonian operator can be written as a sum of two terms

$$H = H_0 + \lambda V(x, p, t) \tag{2}$$

where H_0 is the Hamiltonian operator of the unperturbed system, for which we assume the eigenvalues and eigenfunctions are known, while λV is the small perturbing term. The coefficient λ represents a constant, in terms of which the strength of the perturbation is measured. An example of such a problem arises when a hydrogen atom is placed in a uniform electric field that is weak in comparison with atomic electric fields. The Hamiltonian is then

$$H = -\frac{\hbar^2}{2m}\nabla^2 - \frac{e^2}{r} + e\mathcal{E}x \tag{3}$$

where $\mathcal{E}$ is the strength of the perturbing electric field, which we take to be in the x direction. In this case, $H_0 = -\frac{\hbar^2}{2m}\nabla^2 - \frac{e^2}{r}$ and the perturbing potential is $\lambda V = e\mathcal{E}x$. For this problem we may define the parameter λ to be just the electric field $\mathcal{E}$. More generally, the perturbing term λV may involve the momentum operators $\mathbf{p}$, as well as the co-ordinates $\mathbf{x}$. It may also involve the time. For example, the applied electric field in the above example might have been a function of the time.

If λ is small enough, i.e., if the perturbing forces are weak enough, the solution of the wave equation will not differ much from the solution that we get for $\lambda = 0$. But when $\lambda = 0$, the solution can be expanded as a series of eigenfunctions of H_0, which we shall denote by $U_n(x)e^{-iE_n^0t/\hbar}$ (E_n^0 represents the nth eigenvalue of H_0).

$$\psi_{\lambda=0} = \sum_n C_n U_n e^{-iE_n^0t/\hbar} \tag{4}$$

where C_n is an arbitrary constant.

The method used to obtain a solution when $\lambda \neq 0$ is to note that in general we can at any time t expand an arbitrary function $\psi(x)$ as a series of the $U_n(x)$. Since the function ψ is changing with time, the coefficients of the $U_n(x)$ must, in general, be functions of the time. If the coefficients take the special time variation $C_n\, e^{-iE_n{}^0t/\hbar}$, and C_n is a constant, then the series will be a solution of the unperturbed wave equation $\left(i\hbar \dfrac{\partial\psi}{\partial t} = H_0\psi\right)$. More generally, the coefficients vary with time in a more complex way, so that if we express the wave function as the series $\psi = \sum_n C_n\, e^{-iE_n{}^0t/\hbar} U_n(x)$, then the C_n will turn out to be functions of the time. This method is, for this reason, called the *method of variation of constants.*

To obtain a solution, let us insert the above series (with C_n a function of the time) into Schrödinger's equation [eq. (1)]. The result is

$$\sum_n (i\hbar\dot{C}_n + E_n^0 C_n) U_n(x) e^{-iE_n{}^0t/\hbar} = \sum_n E_n^0\, e^{-iE_n{}^0t/\hbar} U_n(x) C_n + \lambda \sum_n V C_n U_n(x)\, e^{-iE_n{}^0t/\hbar} \quad (5)$$

This reduces to

$$i\hbar \sum_n \dot{C}_n(t)\, U_n(x) e^{-iE_n{}^0t/\hbar} = \lambda \sum_n C_n V U_n(x) e^{-iE_n{}^0t/\hbar}$$

Let us now multiply this equation by $U_m^*(x)e^{iE_m{}^0t/\hbar}$, and then integrate over all x. Using the normalization and orthogonality of the U_n, we obtain

$$i\hbar\dot{C}_m = \lambda \sum_n C_n\, e^{i(E_m{}^0 - E_n{}^0)t/\hbar} V_{mn} \quad (6)$$

where

$$V_{mn} = \int U_m^*(x) V(x, p, t) U_n(x)\, dx \quad (7)$$

and dx represents the volume element $dxdydz$.

V_{mn} is simply the (m, n)th matrix element of V in the representation in which H_0 is diagonal [see eq. (2), Chap. 16]. Note that V_{mn} is, in general, a function of the time.

Equation (6) constitutes, in general, an infinite set of linear equations defining each C_m in terms of all the C_n. The exact form of the solution depends on the value of each of the V_{mn} and on the initial values of each of the C_n. The value of the V_{mn} is, in turn, determined by the form of the perturbing potential and by the eigenfunctions U_n of the unperturbed Hamiltonian. The time variation of the C_n therefore depends both on the form of the perturbing term and on the type of unperturbed system with which we started.

The procedure adopted here is essentially the equivalent of expanding the wave function in a series of solutions of Schrödinger's equation for the unperturbed system. If the interaction energy were zero, these would be exact solutions for the whole system, and we would have a Heisenberg representation of the wave function (see Chap. 16, Sec. 21). When the interaction energy does not vanish, however, we do not have a Heisenberg representation, because the U_n terms are no longer eigenfunctions of the energy operator, which is now $H_0 + \lambda V$.

3. Boundary Conditions. The boundary conditions on these equations are usually determined with the aid of the assumption that before some time t_0 the perturbing potential was not present. We may ask the physical meaning of the assumption that the perturbing potential was absent before $t = t_0$. With a light wave, for example, we can form a packet that first strikes the atom at $t = t_0$. With a constant electric field, t_0 would denote the time at which the field was first turned on. With other perturbations we can see that in a similar way there will usually be some time before which the strength of the perturbation was negligible. The most general possible state before the time $t = t_0$ is, according to the expansion theorem $\psi = \sum_n A_n \exp(-iE_n^0 t/\hbar) U_n(x)$, where the A_n are arbitrary constants except for requirements of normalization. A special possibility, very often realized in practice, is that the system is in a single stationary state, so that the wave function is

$$U_s\, e^{-iE_s^0 t/\hbar}\, e^{i\phi}$$

where ϕ is a constant phase factor of no physical significance (see Chap. 6, Sec. 3). Such a wave function might occur, for example, if we start with an atom in the ground state and then shine light on it, or else apply an electric or magnetic field.

We shall consider here only the boundary condition that the system starts out in one of its possible stationary states. The more general boundary conditions, which are of little physical interest, are readily treated by straightforward applications of the methods developed in this chapter.

4. Methods of Approximation. After the time $t = t_0$, the preceding wave function will no longer be a solution to the wave equation. Our problem will now be to find approximately how the C_n's change as a result of the appearance of the perturbing potential. Our method of approximation is based on the fact that, as can be seen from eq. (6), the changes of C_n with time are proportional to λ. Now, at $t = t_0$, we have assumed that all the C_n are zero except one, namely, C_s, which we may take to be unity except for an arbitrary phase factor of no significance. Because of the smallness of $\dot{C}_n$, we can say that at least for some period of time after $t = t_0$ (the length of which depends on λ) all of the C_n are small and, in fact, proportional to λ, while C_s remains close to unity. Thus, to a first approximation, we can solve for $\dot{C}_m$ when $m \neq s$, by inserting $C_n = 0$ and $C_s = 1$ on the right-hand side of eq. (6). We then obtain

$$i\hbar \dot{C}_m = \lambda e^{i(E_m^0 - E_s^0)t/\hbar} V_{ms}(t) \tag{8}$$

This equation will be a good approximation until the C_m terms, as calculated from it, become large. The conditions under which this happens will be discussed in Sec. 7.

Integration of eq. (8) yields

$$C_m = -\frac{i}{\hbar}\lambda \int_{t_0}^{t} e^{i(E_m{}^0 - E_s{}^0)t/\hbar} V_{ms}(t)\, dt \tag{9a}$$

It is also of interest to compute the first approximation to C_s; i.e., to the coefficient of the eigenfunction with which we started. Note that

$$i\hbar \dot{C}_s = \lambda V_{ss}(t) C_s + \lambda \sum_{n \neq s} e^{i(E_s{}^0 - E_n{}^0)t/\hbar} V_{sn}(t) \tag{9b}$$

Since C_n is proportional to λ when $n \neq s$, it follows that the summation on the right-hand side of the above equation is proportional to λ^2 and therefore can be neglected in a first-order treatment. We obtain

$$i\hbar C_s \cong \lambda V_{ss}(t) C_s \tag{10}$$

This can be integrated to yield

$$\dot{C}_s \cong e^{-i\lambda \int_t^t V_{ss} dt/\hbar} \tag{11}$$

When V_{ss} is not a function of the time, eq. (11) becomes

$$C_s \cong e^{-i\lambda V_{ss}(t - t_0)/\hbar} \tag{12a}$$

We shall have occasion to refer to the above equation later. There are two points connected with the above result which should be noted:

(1) The term involving V_{ss} comes in only in the exponential, so that it does not change the absolute value of C_s. Thus, no changes in probability and no transitions result from it.

(2) To a first approximation, the term in V_{ss} has only the effect of changing the angular frequency of oscillation of the wave function by the amount $V_{ss}/\hbar$. This is equivalent to changing the unperturbed energy by V_{ss}. To a first approximation, the energy therefore becomes

$$E = E_s + \lambda V_{ss} \tag{12b}$$

But $\lambda V_{ss} = \lambda \int U_s^* V U_s\, dx$, which is just the average value of the perturbing potential, taken with the unperturbed wave function. A similar result is obtained in classical perturbation theory, where the first approximation to the correction to the energy can be obtained from the time average of the perturbing potential taken over a period.†

5. Interpretation of the $|C_m|^2$ in Terms of Transition Probabilities. In Chap. 10, Sec. 29, it was shown that $|C_m|^2$ yields the probability that the system can be found in a state in which H_0, the unperturbed Hamiltonian, has the eigenvalue, E_m^0. Since this probability was taken to be

† See Born, *Mechanics of the Atom.*

zero at $t = t_0$, we conclude that $|C_m|^2$ yields the probability that a transition has taken place from the sth to the mth eigenstate of H_0, since the time $t = t_0$. Even though the C_m's change continuously at a rate determined by Schrödinger's equation and by the boundary conditions at $t = t_0$, the system actually undergoes a discontinuous and indivisible transition from one state to the other. The existence of this transition could be demonstrated, for example, if the perturbing potential were turned off a short time after $t = t_0$, while the C_m's were still very small. If this experiment were done many times in succession, it would be found that the system was always left in some eigenstate of H_0. In the overwhelming majority of cases, the system would be left in its original state, but in a number of cases, proportional to $|C_m|^2$, the system would be left in the mth state. Thus, the perturbing potential must be thought of as causing indivisible transitions to other eigenstates of H_0.

6. Evaluation of the C_m. The general expression for the C_m depends on exactly how V_{mn} varies with the time. There are three cases, however, which are easy to solve and occur very frequently in actual problems. These are:

(a) V_{mn} is turned on abruptly at the time $t = t_0$.
(b) V_{mn} oscillates trigonometrically with time.
(c) V_{mn} is turned on very slowly with time (adiabatic case).

7. Case a: V_{mn} Turned on Suddenly (Calculation to First Order in λ). For this case, C_m can be integrated directly from eq. (9a), when the system was originally in the sth eigenstate. The result is:

$$C_m = \frac{e^{i(E_m^0 - E_s^0)\frac{t_0}{\hbar}}}{E_m^0 - E_s^0}\left(1 - e^{i(E_m^0 - E_s^0)\frac{(t-t_0)}{\hbar}}\right)\lambda V_{ms} \tag{13}$$

Thus, we see that C_m is an oscillatory function of the time. The probability that the system is in the mth eigenstate of H_0 is

$$|C_m|^2 = \frac{\lambda^2|V_{ms}|^2|(1 - e^{i(E_m^0 - E_s^0)(t-t_0)/\hbar})|^2}{(E_m^0 - E_s^0)^2}$$
$$= \frac{4\lambda^2|V_{ms}|^2}{(E_m^0 - E_s^0)^2}\sin^2\left[\frac{(E_m^0 - E_s^0)(t - t_0)}{2\hbar}\right] \tag{14}$$

This probability oscillates with angular frequency, $\omega = (E_m^0 - E_s^0)/\hbar$ and it reaches a maximum every time $\frac{(E_m^0 - E_s^0)(t - t_0)}{2\hbar} = \left(N + \frac{1}{2}\right)\pi$. The maximum value is given by

$$|C_m|^2_{\max} = \frac{4\lambda^2|V_{ms}|^2}{(E_m^0 - E_s^0)^2} \tag{14a}$$

This behavior is reminiscent of the harmonic oscillator undergoing forced oscillations as a result of a periodic impressed force having a frequency ω

which is different from the natural frequency ω_0 of the oscillator [see eq. (37), Chap. 2]. In this case, the amplitude of oscillation increases and decreases with the beat frequency $\omega - \omega_0$.

The total probability that the system has made a transition away from the sth state is just the sum of all the $|C_m|^2$, for all m except $m = s$. This probability is

$$P = \sum_{s \neq m} |C_m|^2 = 4\lambda^2 \sum_{m \neq s} \frac{|V_{ms}|^2}{(E_m^0 - E_s^0)^2} \sin^2 \left[\frac{(E_m^0 - E_s^0)(t - t_0)}{2\hbar} \right] \tag{15}$$

In order that the perturbation theory be valid, in the approximation used thus far, it is necessary that P be small compared with unity. If this requirement is satisfied, then the C_m will all be small when $m \neq s$ and C_s will not change much compared with unity. Since

$$\sin^2 \frac{(E_m^0 - E_s^0)(t - t_0)}{2\hbar} \leq 1$$

we can write

$$P \leq 4\lambda^2 \sum_{m \neq s} \frac{|V_{ms}|^2}{(E_m^0 - E_s^0)^2} \tag{15a}$$

Thus a sufficient condition for the validity of the perturbation theory for all time is that

$$4\lambda^2 \sum_{m \neq s} \frac{|V_{ms}|^2}{(E_m^0 - E_s^0)^2} \ll 1 \tag{15b}$$

The preceding condition can always be satisfied by making λ very small, unless there are degenerate energy levels $E_m^0 = E_s^0$. We can therefore see that the question of degeneracy of energy levels may be important in a perturbation problem.

8. Degenerate Perturbations. If there are energy levels for which $E_m^0 = E_s^0$, eq. (14) becomes inapplicable. For this case, C_m can be obtained from eq. (8)

$$C_m = -\frac{i}{\hbar} \lambda V_{ms}(t - t_0) \tag{16}$$

In this case, therefore, C_m increases indefinitely with the time.* We conclude that if the system is degenerate, the perturbation theory must fail after a sufficient length of time. The method of treating the degeneracy problem over long periods of time will be discussed in Chap. 19.

9. A Description of Transitions in Terms of Quantum Fluctuations. We wish now to provide a way of picturing the transition processes described previously. Let us first consider the nondegenerate case

* Compare this with the harmonic oscillator, for the case that the forcing term is in resonance with the natural frequency [see eq. (35), Chap. 2].

where, as we have seen in eq. (14), the C_m's grow for a while, then become smaller, then larger, etc., but never exceed a certain bounded size. This means that the system starts to make transitions to other quantum states, but that these transitions are reversed in a time of the order of $\tau = h/(E_m^0 - E_s^0)$. Thus, the smaller $E_m^0 - E_s^0$, the longer the time available to make transitions to the mth level, and the larger will be the resulting maximum value of C_m.

In this connection, it must be remembered that the total energy of the system is not H_0, but $H_0 + \lambda V$, so that the description of transitions in terms of the eigenstates of H_0 is *not* a description in terms of transitions between definite energy levels. Nevertheless, because λ is small, the contribution of the perturbing potential to the total energy is also small, [see eq. (12b)]. This means that E_s^0 and E_m^0 can still be interpreted as approximate eigenvalues of the energy.

On the basis of the above remarks, we picture a system acted on by a perturbing potential as being in a state of continual fluctuation from one eigenstate of the unperturbed Hamiltonian H_0 to another and back again. In other words, when the perturbation is turned on, the system begins to make transitions toward all possible energy levels. Now, if the system remained permanently in an eigenstate of H_0 corresponding to an unperturbed energy E_m^0, which was very different from the initial value E_s^0 of the unperturbed energy, we would obtain a contradiction of the law of conservation of energy. This occurs, as we have seen, because the contribution of the perturbing potential to the energy is very small whereas the difference $E_m^0 - E_s^0$ can, in general, be fairly large. This contradiction is avoided, however, by the fact that the system stays in the new state for a period of time so short that, according to the uncertainty principle, the energy is not defined to within $E_m^0 - E_s^0$. Only if $E_m^0 = E_s^0$; i.e., if the system is degenerate, can the transition proceed indefinitely in the same direction without violating the law of conservation of energy.

The preceding description involves the replacement of the classical notion that a system moves along some definite path by the idea that under the influence of the perturbing potential, the system tends to make transitions in all directions at once. Only certain types of transitions can, however, proceed indefinitely in the same direction, namely, those which conserve energy. In many ways, the above concept resembles the idea of evolution in biology, which states that all kinds of species can appear as a result of mutations, but that only certain species can survive indefinitely, namely, those satisfying certain requirements for survival in the specific environment surrounding the species. Nevertheless, the analogy must not be carried too far, because a living system must belong either to one species or another and not to two at once. On the other hand, as we have seen in Chap. 6, when the wave function contains a sum of contributions from many quantum states, the system must be thought of as covering all these states at once, because important physical properties may depend on interference between the wave functions corresponding to these various states.†

Sometimes permanent (i.e., energy-conserving) transitions are called *real* transitions, to distinguish them from the so called *virtual* transitions, which do not conserve energy and which must therefore reverse before they have gone too far. This terminology is unfortunate, because it implies that virtual transitions have no real effects. On the contrary, they are often of the greatest importance, for a great many physical processes are the result of these so-called *virtual* transitions. For example, we shall see in Chap. 19, Sec. 13, the van der Waals attraction between molecules arises from virtual transitions.

10. Microscopic Reversibility of Transition Processes. From eq. (23), Chap. 9, we see that because V is Hermitean, $V_{mn} = V^*_{nm}$. But $|V_{mn}|^2$ is proportional to the probability that a system originally in the nth state makes a transition to the mth state, while $|V_{nm}|^2$ is proportional to the probability that a system, originally in the mth state, makes a transition to the nth state. Because of the above result, the two probabilities are equal. This property is often referred to as the *microscopic reversibility* of quantum processes. It is the quantum analogue of the microscopic reversibility of the classical equations of motion.‡ In fact, the microscopic reversibility of all quantum processes must lead, in the correspondence limit, to the microscopic reversibility of all classical motions.

† As shown in Chap. 6, Secs. 9, 10, and 13, and in Chap. 16, Sec. 25, a quantum system should be described in terms of incompletely defined potentialities, which are more definitely realized only in interaction with appropriate external systems. As long as definite phase relations between the C_n exist, however, the system cannot be regarded as having a single definite (but unknown) value of the unperturbed Hamiltonian. Only after it interacts with a suitable system (such as an apparatus that measures H_0) will it develop a definite value of H_0, and $|C_n(t)|^2$ represents the probability that this value will be E_n, provided that (as suggested in Sec. 5) the perturbing potential is turned off at the time t. Compare also with the description of the transition process in Chap. 22, Sec. 14.

‡ For any solution of the equations of motion, there is always another solution in which all particles have exactly the opposite velocities, so that they therefore execute the reverse motions. This property is called *microscopic reversibility* in order to distinguish it from the properties of macroscopic systems, which show, in general, an irreversible character in their motions. In the study of statistical mechanics, it is shown that the macroscopic irreversibility arises from the fact that there are so many microscopically different states of the system which are not distinguishable in the macroscopic (or thermodynamic) sense. As a result, when the particles move and scatter each other, the net effect is to produce a random shuffling, in which it becomes very unlikely that the original state will ever be reproduced. In quantum theory, as we have seen, the basic processes are also microscopically reversible, but in a macroscopic system, irreversibility is introduced by the same kind of random shuffling effects (see, for example, Tolman, *The Principles of Statistical Mechanics.* New York: Oxford University Press, 1938).

11. Conservation of Probability. From eq. (15), we have seen that the total probability that a transition away from the sth level has taken place is proportional to λ^2. Since probability is conserved, this probability of transition must be compensated by an equal decrease of probability that the system is in the initial state. But because the change of probability is of second order in λ, we must calculate C_s to second order also, in order to demonstrate this decrease in $|C_s|^2$.

C_s can be calculated to second order from eq. (6) by using the first-order approximation for C_m. Before doing this, however, it is convenient to make the substitution $C_n = e^{-i\lambda V_{nn}t/\hbar}A_n$. The eqs. (6) reduce to†

$$i\hbar\dot{A}_m = \lambda \sum_{n\neq m} e^{i(E_m{}^0-E_n{}^0)t/\hbar}V_{mn}A_n \tag{17}$$

Setting $m = s$ in the above equation, we obtain

$$i\hbar\dot{A}_s = \lambda \sum_{n\neq s} e^{i(E_s{}^0-E_n{}^0)t/\hbar}V_{sn}A_n \tag{18}$$

Taking A_n from eq. (13) and neglecting terms of second order or higher, we obtain

$$i\hbar\dot{A}_s = -\lambda^2 \sum_{n\neq s} e^{i(E_s{}^0-E_n{}^0)t/\hbar}V_{sn}V_{ns}\frac{[e^{\frac{i(E_n{}^0-E_s{}^0)t}{\hbar}} - e^{\frac{i(E_n{}^0-E_s{}^0)t_0}{\hbar}}]}{(E_n^0 - E_s^0)}$$

Integration yields [noting that $(A_s)_{t=t_0} = 1$]

$$A_s = 1 + \lambda^2 \sum_{n\neq s} \frac{V_{sn}V_{ns}}{(E_s^0 - E_n^0)^2}[e^{\frac{i(E_s{}^0-E_n{}^0)(t-t_0)}{\hbar}} - 1] - \frac{i\lambda^2}{\hbar}\sum_{n\neq s}\frac{V_{sn}V_{ns}(t-t_0)}{(E_s^0 - E_n^0)} \tag{19}$$

The square of the absolute value of the above expression is (noting from Sec. 10 that $V_{ns} = V_{sn}^*$ and retaining only terms up to λ^2)

$$|A_s|^2 = 1 - 2\lambda^2 \sum_{n\neq s}\left\{1 - \cos\left[\frac{(E_s^0 - E_n^0)}{\hbar}(t - t_0)\right]\right\}\frac{|V_{sn}|^2}{(E_s^0 - E_n^0)^2}$$
$$= 1 - 4\lambda^2 \sum_{n\neq s}\frac{|V_{sn}|^2}{(E_s^0 - E_n^0)^2}\sin^2\left[\frac{(E_s^0 - E_n^0)}{\hbar}(t - t_0)\right] \tag{20}$$

According to eq. (15), however, we see that the decrease of $|A_s|^2$ is just equal to the total probability that the system has made a transition away from the ground state. Probability is therefore conserved.

† Note that we have neglected λV_{nn} when it appears in the exponential, because it produces, at most, a correction proportional to λ^3 in the final answer.

Furthermore, we have seen that $|A_s|^2$ differs from unity only by a second-order term. The error in the calculation of A_m because of the assumption that $A_s = 1$ is therefore, at most, a third-order effect.

12. Case b: Trigonometric Variation of V_{mn} with Time, with Application to Absorption and Emission of Light. There are many important problems in which V_{mn} varies trigonometrically with the time. For example, an atom can be placed in a weak electric field that oscillates with some angular frequency ω. In this case, taking the field in the x direction, we must add to the Hamiltonian the perturbing term

$$e\mathcal{E}_0 x \cos \omega t \tag{21}$$

A more important example consists of the problem of finding what happens to an atom irradiated with light of a definite angular frequency ω. For an electromagnetic wave, one need consider only the vector potential, $\boldsymbol{a}$ (see Chap. 1, Sec. 3). The Hamiltonian can then be written† [see eq. (47) Chap. 15]

$$H = \frac{p^2}{2m} + V - \frac{e}{2mc}(\boldsymbol{a} \cdot \boldsymbol{p} + \boldsymbol{p} \cdot \boldsymbol{a}) + \frac{e^2}{2mc^2}a^2 \tag{22}$$

(V is the potential produced by all forces on the atom other than those coming from the incident electromagnetic radiation.)

Since we are restricting ourselves to the case in which the electromagnetic field is a small disturbance, we can neglect the term‡ involving a^2, which is of second order. The term V_{mn} is then given by

$$\lambda V_{mn} = -\frac{e}{2mc}\int U_m^*(x)(\boldsymbol{a} \cdot \boldsymbol{p} + \boldsymbol{p} \cdot \boldsymbol{a})U_n(x)\,dx \tag{23}$$

For a light wave of definite angular frequency, we can write

$$\boldsymbol{a} = \boldsymbol{G}(x)e^{-i\omega t} + \boldsymbol{G}^*(x)e^{i\omega t}$$

Note that the complex conjugate term must be added to keep $\boldsymbol{a}$ real. We then obtain

$$\lambda V_{mn} = \frac{-e}{2mc}\left[e^{-i\omega t}\int U_m^*(x)(\boldsymbol{G} \cdot \boldsymbol{p} + \boldsymbol{p} \cdot \boldsymbol{G})U_n(x)\,dx + e^{i\omega t}\int U_m^*(x)(\boldsymbol{G}^* \cdot \boldsymbol{p} + \boldsymbol{p} \cdot \boldsymbol{G}^*)U_n\,dx\right]$$

Let us define

$$G_{mn} = -\frac{e}{mc}\int U_m^*(x)\,\frac{\mathbf{G} \cdot \boldsymbol{p} + \boldsymbol{p} \cdot \mathbf{G}}{2}\,U_n(x)\,dx \tag{24}$$

† We are neglecting spin in this treatment. The effects of spin will be discussed in Sec. 50.

‡ This neglect is valid when, as is usually the case, the matrix element V_{mn} does not vanish. If V_{mn} vanishes, then the A^2 term must be retained, because it is then the main term responsible for transitions.

then the complex conjugate of G_{mn} is obtained by taking the complex conjugate of all parts of the integral above

$$G^*_{mn} = -\frac{e}{mc}\int U_m(x)\,\frac{\boldsymbol{G}^*\cdot\boldsymbol{p}^* + \boldsymbol{p}^*\cdot\boldsymbol{G}^*}{2}\,U^*_n(x)\,dx \tag{25}$$

If we write $\boldsymbol{p} = \frac{\hbar}{i}\nabla$, and note that $\boldsymbol{G} = \boldsymbol{G}(\boldsymbol{x})$, we can easily show by integration by parts (noting that the integrated part vanishes) that

$$G^*_{mn} = -\frac{e}{mc}\int U^*_n(x)\,\frac{\boldsymbol{G}^*\cdot\boldsymbol{p} + \boldsymbol{p}\cdot\boldsymbol{G}^*}{2}\,U_m(x)\,dx \tag{26}$$

(In other words, the $\boldsymbol{p}$ can operate on U_m instead of on U^*_n.) With these definitions, we obtain the result that

$$\lambda V_{mn} = G_{mn}\,e^{-i\omega t} + G^*_{nm}\,e^{i\omega t} \tag{27}$$

We can now calculate C_m from eq. (9a), obtaining

$$C_m = -\frac{i\lambda}{\hbar}\int_{t_0}^{t} e^{i(E_m{}^0 - E_s{}^0)t/\hbar}V_{ms}(t)\,dt = -[G_{ms}F(\omega) + G^*_{sm}F(-\omega)] \tag{28}$$

where $$F(\omega) = e^{i(E_m{}^0 - E_s{}^0 - \omega\hbar)t/\hbar}\,\frac{[1 - e^{-i(E_m{}^0 - E_s{}^0 - \omega\hbar)(t-t_0)/\hbar}]}{(E^0_m - E^0_s - \omega\hbar)} \tag{29}$$

13. Application to Plane Wave. Usually G is chosen such that the light wave is a plane wave, although this is not necessary. We can, for example, choose a wave traveling in the x direction, with $\boldsymbol{a}$ in the z direction.† (Note that the condition for transversality of the wave is satisfied by having $\boldsymbol{a}$ normal to the direction in which the wave travels.) Thus, one writes

$$a_z = a_0\,e^{i(kx-\omega t)} + a_0^*\,e^{-i(kx-\omega t)} \tag{30}$$

To calculate C_m, we merely evaluate G_{mn} with the aid of the above choice of $\boldsymbol{a}$. We obtain (noting that $p_z\,e^{ikx} + e^{ikx}p_z = 2p_z\,e^{ikx}$)

$$G_{ms} = -\frac{ea_0}{mc}\int U^*_m(x)p_z\,e^{ikx}U_s(x)\,dx = \frac{e}{mc}\,a_0\alpha_{ms} \tag{31a}$$

where $$\alpha_{ms} = \int U^*_m(x)p_z\,e^{ikx}U_s(x)\,dx \tag{31b}$$

We then obtain

$$C_m = -\frac{e}{mc}\,a_0[\alpha_{ms}F(\omega) + \alpha^*_{sm}F(-\omega)] \tag{32}$$

14. Interpretation of Results. The preceding result resembles that obtained with a constant potential, except that $E^0_m - E^0_s$ is replaced by $E^0_m - E^0_s \pm \hbar\omega$. The general result will be that $|C_m|^2$ fluctuates, just as when V_{ms} was constant, except when $E_m - E_s = \pm\hbar\omega$. In the latter

† See Chap. 1, eq. (21).

cases, one of the terms [either $F(\omega)$ or $F(-\omega)$] provides a contribution that increases indefinitely with time. This shows that when the perturbing term in the Hamiltonian oscillates with angular frequency $\hbar\omega = |E_m^0 - E_s^0|$, it can cause nonreversing transitions from the mth to the sth level, and also from the sth to the mth level. This means that light with angular frequency ω can permanently exchange energy with an electron only when the condition $E_m^0 - E_s^0 = \pm\hbar\omega$ is satisfied. Because of the appearance of the $\pm$ sign, we conclude that an electron can either emit or absorb a quantum of energy, $E_m^0 - E_s^0 = \hbar\omega$. This process is, of course, in agreement with experiment.

These results can easily be described in terms of the transitions developed in Sec. 9. As in Sec. 9 we say that, in response to the perturbing potential, the system begins to fluctuate in all possible directions. In a periodically varying perturbation, however, the condition for a nonreversing transition is not the conservation of energy but the Einstein condition $E_m^0 - E_s^0 = \pm\hbar\omega$. If we wished to pursue our biological analogy, we could say that the replacement of a time constant perturbation by a time varying one corresponds to a change of environment for an organism, favoring the survival of a different kind of species. Once again, however, we caution that the analogy has a limited validity. It is important to note that, in both cases, the systems which do not survive indefinitely can still produce physically significant effects. In other words, "virtual" transitions must not be considered as "unreal."

15. Present Treatment Does Not Quantize Radiation Field. In the approximate treatment we are now using, it is not immediately evident why the light beam of angular frequency ω should be able to supply an energy of just $\hbar\omega$. This point becomes clear only when the electromagnetic field is quantized,* because one then describes the process of absorption of a photon not only as a transition of the electron from the mth to the nth level, but also as a transition of the radiation oscillators to an energy state that is lower by $\hbar\omega$ than the original state. Thus, the combined energies of electron and radiation oscillators are conserved in all processes which survive for a long time.

In the present treatment, we have taken the vector potential to be a number that can be specified with arbitrarily high precision at each point in space and time, by means of the classical Maxwell equations,† whereas the behavior of the electron has been taken to be quantized. In a more complete treatment it is necessary also to quantize the electromagnetic field. Just as the momentum of the particle $\boldsymbol{p}$ is replaced by an operator, the vector potential $\boldsymbol{a}$ must also be replaced by an operator. We must also add the Hamiltonian of the radiation field to the total Hamiltonian for the system. This program can be realized by regarding the electro-

* See, for example, Schiff, *Quantum Mechanics*, Chap. 14.

† Maxwell's equations are given in Chap. 1, Sec. 3.

magnetic fields as a collection of harmonic oscillators, one for each wave vector $\mathbf{p}$ and direction of polarization $\boldsymbol{\mu}$ in the manner discussed in Chap. 1. Each oscillator must be quantized in the same way as was done with material harmonic oscillators. We then say that if an oscillator is in the Nth excited state, there are N of the corresponding photons in the electromagnetic field. According to the correspondence principle, if many photons are present, we can describe the electromagnetic field approximately as a classical system. This is exactly what we have been doing thus far in our theory. The present treatment is, therefore, rigorous only when the radiation field is highly excited (i.e., many photons in it). Yet, it turns out that the results derived in this way are substantially correct, even when only one photon is present. This is a fortunate accident, resulting from the fact that the correspondence treatment of a harmonic oscillator is good down to small quantum numbers. In this book, we shall restrict ourselves to a classical treatment of the radiation field, remembering that this is completely rigorous only in a very intense beam of radiation.

16. Calculation of Rate of Transition. We are now ready to calculate the probability $|C_m|^2$ that the system is in the mth state. According to eq. (28), this is

$$|C_m|^2 = |G_{ms}F(\omega) + G^*_{sm}F(-\omega)|^2 \tag{33}$$

Now, we shall be interested only in the long-range transition processes which alone can lead to large $|C_m|^2$ when G_{ms} is small. This means that we must either have $E^0_m - E^0_s = \hbar\omega$, or $E^0_m - E^0_s = -\hbar\omega$. The former case corresponds to absorption of energy from the perturbing field, and the latter to emission of energy into the perturbing field. Let us first consider the former case. Then only the term involving $[F(\omega)F^*(\omega)]$ will become really large after the passage of a long time. Terms involving $F(-\omega)$ will tend to oscillate and produce small corrections that we shall neglect. Thus, we write approximately

$$|C_m|^2 \cong |G_{ms}|^2|F(\omega)|^2$$

Evaluation of $F(\omega)^2$ from eq. (29) yields

$$|C_m|^2 = 4\,\frac{|G_{ms}|^2 \sin^2\left[(E^0_m - E^0_s - \omega\hbar)\,\dfrac{(t-t_0)}{2\hbar}\right]}{(E^0_m - E^0_s - \hbar\omega)^2} \tag{34}$$

17. Relation between Vector Potential and Intensity. At this point it will be convenient to express eq. (34) in terms of the intensity of light. From eq. (31), we obtain

$$|G_{ms}|^2 = \frac{e^2}{m^2c^2}\,|a_0|^2|\alpha_{ms}|^2$$

To obtain $|a_0|^2$, we use the fact that the intensity of radiation, i.e., the rate of transport of energy per unit area per unit time, is given by Poynting's vector

$$I = \frac{c(\mathcal{E} \times \mathcal{H})}{4\pi}$$

where $\mathcal{E} = -\frac{1}{c}\frac{\partial \boldsymbol{a}}{\partial t}$ and $\mathcal{H} = \nabla \times \boldsymbol{a}$.† Since

$$\boldsymbol{a} = \boldsymbol{a}_0\, e^{i(\boldsymbol{k}\cdot\boldsymbol{x}-\omega t)} + \boldsymbol{a}_0^*\, e^{-i(\boldsymbol{k}\cdot\boldsymbol{x}-\omega t)}$$

we obtain

$$\mathcal{E} = \frac{i\omega}{c}\,[\boldsymbol{a}_0\, e^{i(\boldsymbol{k}\cdot\boldsymbol{x}-\omega t)} - \boldsymbol{a}_0^*\, e^{-i(\boldsymbol{k}\cdot\boldsymbol{x}-\omega t)}]$$

In free space, $|\mathcal{E}| = |\mathcal{H}|$, and $\mathcal{E}$ is normal to $\mathcal{H}$. Thus $\mathcal{E} \times \mathcal{H}$ is a vector in the direction of propagation $\boldsymbol{k}$ and has a magnitude of $|\mathcal{E}|^2$. We obtain for the intensity

$$\frac{2\omega^2}{c}\frac{|\boldsymbol{a}_0|^2}{4\pi} - \frac{\omega^2}{4\pi c}\,[\boldsymbol{a}_0^2\, e^{2i(\boldsymbol{k}\cdot\boldsymbol{x}-\omega t)} - (\boldsymbol{a}_0^*)^2\, e^{-2i(\boldsymbol{k}\cdot\boldsymbol{x}-\omega t)}]$$

The latter terms are oscillatory and, therefore, average out to zero. The time average intensity is then equal to

$$I = \frac{\omega^2}{2\pi c}\,|\boldsymbol{a}_0|^2$$

so that we obtain

$$|\boldsymbol{a}_0|^2 = \frac{2\pi c}{\omega^2}\, I \tag{35}$$

The probability of transition [eq. (34)] then becomes

$$P = \frac{8\pi e^2}{m^2 c\omega^2}\frac{I|\alpha_{ms}|^2}{(E_m^0 - E_s^0 - \omega\hbar)^2}\sin^2\left[\left(\frac{E_m^0 - E_s^0 - \omega\hbar}{2\hbar}\right)(t - t_0)\right] \tag{36}$$

18. Effect of Distribution of Frequencies of the Incident Light Wave. Equation (36) yields the probability that a light wave of a given angular frequency ω will produce a transition during the time $t - t_0$. Note that for times so short that $(E_m - E_n - \hbar\omega)(t - t_0)/2\hbar \ll 1$, this probability increases with $(t - t_0)^2$, as can be seen by expanding the sine function and retaining only the first term. Thus, if ω is perfectly definite, the probability does not, as one would at first expect, increase linearly with the time but, instead, it increases quadratically. Furthermore, just as in the case of a time constant perturbation, the probability eventually oscillates between zero and some maximum value. A similar result has, however, already been obtained in Chap. 2, Sec. 16, in connection with the classical theory of absorption of radiant energy by an atom. In this

† We take $\phi = 0$ for empty space. This is justified in Chap. 1, Sec. 4.

case, it was shown that when one takes into account the fact that the incident radiation is actually distributed over a range of frequencies, the energy absorbed comes out proportional to the time of irradiation. Similarly, we must, in the present quantum treatment, integrate the transition probability over a range of frequencies.

Now eq. (34) is correct for light of a definite frequency. If a beam of light contains many different frequencies, eq. (31) implies that all the different contributions must first be added together to give the complete vector potential, $\boldsymbol{a}(t) = \int_{-\infty}^{\infty} \boldsymbol{a}(\omega)e^{-i\omega t}\,d\omega$, and α_{ms} must then be calculated with the aid of this potential. In general, the different frequencies will interfere, and the effects of this interference will be to produce pulses of radiation (see, for example, Chap. 3, Sec. 16). On the other hand, if there are no simple phase relations between adjacent $\boldsymbol{a}(\omega)$, then we obtain not a pulse but instead something analogous to random noise in radio waves. If this is the case, the interference terms between different $\boldsymbol{a}(\omega)$ will average out to zero in the expression for $|G_{ms}|^2$. One will then be able to calculate $|C_m|^2$ by summing over the separate contributions of each frequency, using eq. (36), and thus a great simplification will be made possible.

Now, in a real light source, the radiation comes from atoms that are on the average widely spaced in comparison with a wavelength, but which frequently collide with other atoms. Thus, the vibrations of each atom tend to have a fairly random phase relative to those of other atoms. Furthermore the atoms move at different speeds and thus have different Doppler shifts. This means that each different frequency tends to come in with a phase essentially unrelated to that of other frequencies. As a result, one concludes that in a typical light beam, we can ignore phase relations between contributions of different frequencies and, instead, simply sum up the probabilities resulting from each frequency separately, as suggested in the preceding paragraph.

To carry out this program, we first replace the intensity I in eq. (36), which refers to a definite frequency, by $I(\nu)\,d\nu$, the intensity lying between ν and $\nu + d\nu$, and we then integrate over all ν. Setting

$$d\nu = \frac{d\omega}{2\pi}$$

we obtain for the total probability of transition

$$|C_m|^2 = \frac{4e^2}{m^2c}\int_0^{\infty} \frac{|\alpha_{ms}|^2 I(\nu)}{\omega^2(E_m^0 - E_s^0 - \hbar\omega)^2}\sin^2\left[\frac{(E_m^0 - E_s^0 - \hbar\omega)}{2\hbar}(t - t_0)\right]d\omega \tag{37}$$

Now, as shown in Chap. 2, Sec. 16, when $(t - t_0)$ is large, the integrand is appreciable only in a narrow region near $\hbar\omega = E_m^0 - E_s^0$. We can therefore take $|\alpha_{ms}|^2/\omega^2$ outside the integral sign and evaluate it at $\omega_0 = (E_m^0 - E_s^0)/\hbar$. The remaining integral can then be calculated by methods given in Chap. 2, Sec. 16, and one finally obtains

$$|C_m|^2 = \frac{2\pi e^2}{m^2 c\hbar} I(\nu_0)|\alpha_{ms}|^2 \frac{(t - t_0)}{\omega_0^2} \tag{38}$$

19. Discussion of Results. (1) The transition probability is now proportional to the time but, as in the classical theory (see Chap. 2, Sec. 16), the result is large only for a narrow band of frequencies near $\omega = \omega_0$. The width of this band is $\Delta\omega \cong 1/(t - t_0)$. Thus, the longer one waits, the closer must ω be to ω_0 in order to obtain a large probability of transition.

In practice, one seldom measures absorption in less than 10^{-7} sec, and normally much longer times than this are involved in such measurements. Since the light has an angular frequency of the order of 10^{16} sec^{-1}, it is clear that the band of frequencies contributing for a sharply defined level involves a percentage change of 1 part in 10^9. This is an effect which is too small to be measured, since the natural width of the energy level itself is greater. (See Chap. 10, Sec. 34, for a discussion of natural width.)

(2) Two approximations are required to make eq. (38) valid. One is that $(t - t_0)$ be short enough so that $|C_m|^2$ does not approach unity; otherwise the perturbation theory would no longer be valid. The second is that $(t - t_0)$ be long enough so that

$$\frac{\sin^2\left[\dfrac{(E_m^0 - E_s^0 - \hbar\omega)}{2\hbar}(t - t_0)\right]}{(E_m^0 - E_s^0 - \hbar\omega)^2}$$

regarded as a function of ω be small except within a very narrow range of frequencies, near $\omega = (E_m^0 - E_s^0)/\hbar$. This condition is used in eq. (37) in order to justify taking part of the integrand out from under the integral sign. The only way that we can satisfy both requirements simultaneously is to make $|G_{ms}|^2$ small; in other words, to have a weak perturbation.

(3) The rate of transition will depend on G_{ms}. The calculation of G_{ms} will therefore be a key problem in solving for transition probabilities.

20. Induced Emission of Quanta. In eq. (33), it was shown that permanent transitions could occur not only when $E_m^0 - E_s^0 = \hbar\omega$, but also when $E_m^0 - E_s^0 = -\hbar\omega$. The former correspond, as we have seen, to absorption of quanta, whereas the latter must correspond to emission. From eq. (31), it follows that $|\alpha_{sm}|^2 = |\alpha_{ms}|^2$. Thus we conclude from eq. (33) that the probability of absorption is the same as that of emission.* We therefore conclude that if an atom is in a state in which it

* This is an example of microscopic reversibility. See Sec. 10.

can lose energy $\hbar\omega$ by going from an excited level to a lower energy level, it will do so at a rate proportional to the intensity of the radiation that is already present. This effect is known as "induced emission" of radiation.

21. Classical Analogue of Induced Emission. Induced emission of radiation appears even in the classical theory. In an electrical field $\mathcal{E} = \mathcal{E}_0 \cos(\omega t - \phi)$, where ϕ is an arbitrary phase angle, the rate at which a moving charge absorbs energy is

$$\frac{dW}{dt} = e\mathcal{E} \cdot \boldsymbol{v} = e(\mathcal{E}_0 \cdot \boldsymbol{v}) \cos(\omega t - \phi)$$

where $\boldsymbol{v}$ is the velocity of the charge. In order that there be a great deal of absorption of energy, it is necessary that the electromagnetic field oscillate at a frequency that is in resonance with that of the motion of the charge. Thus, we write $\boldsymbol{v} = \boldsymbol{v}_0 \cos \omega t$. This gives

$$\frac{dW}{dt} = (e\boldsymbol{v}_0 \cdot \mathcal{E}_0) \cos(\omega t - \phi) \cos \omega t$$
$$= \frac{e\mathcal{E}_0 \cdot \boldsymbol{v}_0}{2} [\cos \phi(1 + \cos 2\omega t) + \sin \phi \sin 2\omega t]$$

The latter terms average out to zero over a period. The average rate of absorption of energy therefore depends only on ϕ, the angle between the phase of the oscillating radiation field and that of the oscillating electron. If this is zero, then dW/dt is a maximum. For phases between $-\pi/2$ and $\pi/2$, dW/dt is positive, and for others, it is negative. In other words, if the light wave happens to be 180° out of phase with the motion of the electron, the electron loses energy to the electromagnetic field at a rate proportional to $e\mathcal{E}_0 \cdot \boldsymbol{v}$. Thus, we have induced emission. Because we are assuming various contributions to the incident light wave which have more or less random phases, there will be some instances in which the phase is such that energy is absorbed by the electron, and some in which it is emitted. Thus, both absorption and induced emission will take place.

22. Spontaneous Emission. The theory given above does not predict the well-known result that an accelerated electron should radiate, even when no light is incident. This process is called "spontaneous emission." The reason that it is not predicted is that the theory does not take into account the fact that the radiation field has a Hamiltonian function, and can absorb energy, just as does a material particle. When this is taken into account in the classical theory, the correct spontaneous emission is predicted. The same also occurs in the quantum theory.*

23. Einstein's Treatment of Spontaneous Emission. Although the rate of spontaneous emission of quanta by excited atoms can be calculated

* See, for example, Schiff, Chap. 14.

rigorously by quantizing the electromagnetic field, we shall give here an earlier treatment, developed by Einstein, who obtained this quantity from considerations on thermodynamic equilibrium with the surrounding atoms, which are at some temperature T. The argument is that when the system is in thermodynamic equilibrium, the atoms should be in a steady state in which the probability, P_{mn}, of a transition from the mth state to the nth state, with the emission of a quantum, is balanced by the probability, P_{nm}, of a transition from the nth to the mth state, with the absorption of a quantum. Now, the probability of absorption is just (according to eq. 38),

$$P_{nm} = 2\pi|\alpha_{nm}|^2 \left(\frac{e}{mc}\right)^2 \frac{c}{\omega^2\hbar^2} I(\nu)p_n = A_{nm}I(\nu)p_n$$

where p_n is the probability that the atom is in the nth state. (This equation defines A_{nm}. Note that $A_{nm} = A_{mn}$.)

The probability of emission is compounded of two parts, both of which are proportional to p_m, the probability that the atom is in the mth state. The first of these is just the probability of induced emission, which is

$$P_{mn} = A_{nm}I(\nu)p_m$$

The second is the probability of spontaneous emission, which we denote as $B_{mn}p_m$. The latter does not depend on $I(\nu)$.

The condition for equilibrium is

$$p_nA_{mn}I(\nu) = p_m[B_{mn} + A_{mn}I(\nu)]$$

or

$$\frac{p_m}{p_n} = \frac{A_{mn}I(\nu)}{B_{mn} + A_{mn}I(\nu)} \tag{39}$$

Now, we also know that in statistical equilibrium, the different quantum states of the atoms must have a Maxwellian distribution,* or

$$\frac{p_m}{p_n} = \frac{e^{-E_m/\kappa T}}{e^{-E_n/\kappa T}} = e^{(E_n-E_m)\kappa T} = e^{-h\nu/\kappa T} \tag{40}$$

Thus, we obtain

$$e^{-h\nu/\kappa T} = \frac{A_{mn}I(\nu)}{B_{mn} + A_{mn}I(\nu)} \quad \text{or} \quad \frac{B_{mn}}{A_{mn}} = I(\nu)(e^{h\nu/\kappa T} - 1) \tag{41}$$

Another expression for $I(\nu)$ can be obtained from Chap. 1, eqs. (31) and (32), which give the density of radiation in a black box in thermal equilibrium;

$$\rho(\nu) = \frac{8\pi h\nu^3}{c^3}\frac{1}{e^{h\nu/\kappa T} - 1}$$

* It is shown that atoms obey the Maxwell Boltzman statistics. See, for example, Tolman, *The Principles of Statistical Mechanics.*

If the radiation is isotropic, it can easily be shown that the intensity of radiation per unit solid angle, having any given polarization, is

$$I(\nu) = \frac{c\rho(\nu)}{8\pi} = \frac{h\nu^3}{c^2}\frac{1}{e^{h\nu/\kappa T} - 1} \tag{42}$$

Problem 1: Prove the above statement.

Insertion of this value of $I(\nu)$ into eq. (41) yields

$$\frac{B_{mn}}{A_{mn}} = \frac{h\nu^3}{c^2}$$

We see, therefore, that the rate of spontaneous emission is proportional to the rate of absorption. More specifically, we obtain

$$B_{mn} = 2\pi|\alpha_{mn}|^2\left(\frac{e}{mc}\right)^2\frac{h\nu^3}{c\hbar^2\omega^2} \tag{43}$$

24. Applications of Transition Theory. We are now ready to apply the theory which has been developed thus far to various systems. Before doing this, it is necessary to calculate the matrix elements, α_{mn} (see eq. 31). Note that the α_{mn} depend on the form of the perturbing potential (which was in this case assumed to be a plane wave), and symmetrically on the wave function of both the initial and the final state. This is a specifically quantum-mechanical feature. In classical theory, we should not expect the rates of transition processes to depend on what the particle is going to do after it has made its transition from one state to the other. In quantum theory, however, the process of transition is indivisible, and the electron in transition between two states must be thought of as covering both states at once, with a probability, however, that changes with time in such a way that the probability of being in the final state is steadily increasing. The appearance of the final-state wave function in the formula for the transition probability therefore reflects the indivisibility of the transition process.

The integral that must be evaluated is

$$\alpha_{mn} = \int U_m^*(p_z\, e^{ikx})U_n\, dx = \frac{\hbar}{i}\int U_m^*\, e^{ikx}\frac{\partial U_n}{\partial z}\, dx \tag{44}$$

This integral can be approximated, if one notes that $kx = 2\pi/\lambda$, where λ is the length of the light wave. Now, the factors $U_m(x)$ and $U_n(x)$ are wave functions that are large only in the region of the order of the size of an atom, which is about 3×10^{-8} cm, at most. On the other hand, the lengths of light waves are of the order of 6×10^{-5} cm. The exponential may therefore be expanded as a power series.

$$e^{ikx} = e^{ikx_0}\left[1 + ik(x - x_0) - \frac{k^2}{2}(x - x_0)^2 + \cdots\right]$$

where x_0 is the co-ordinate of the center of the atom. Unless the integral over the first term vanishes, the exponential may, with very small error, be replaced by e^{ikx_0}, because $k(x - x_0)$ is very small in the region in which U_m and U_n are large.† The effect of the higher terms in the series will be discussed later in connection with forbidden transitions.

The matrix element then reduces to

$$\alpha_{nm} \cong \frac{\hbar}{i} e^{ikx_0} \int U_m^* \frac{\partial U_n}{\partial z} \, dx \tag{45}$$

25. Electric Dipole Approximation. We shall now show that the approximation of neglecting $k(x - x_0)$ in the exponential is equivalent to replacing the atom by an electric dipole of moment equal to that of the actual charge, taken about the center of the atom.

To do this, we first take advantage of an important property of the U_m's. Suppose that $\psi_m(x, t) = U_m(x)e^{-iE_m{}^0t/\hbar}$ is a solution of Schrödinger's equation for the unperturbed system, i.e., when no light wave is present. One can then show that

$$\frac{d}{dt} \int \psi_m^* z \psi_n \, dx = \frac{\hbar}{im} \int \psi_m^* \frac{\partial}{\partial z} \psi_n \, dx \tag{46}$$

This follows from the relations

$$\frac{d}{dt} \int \psi_m^* z \psi_n \, dx = \int \left(\frac{\partial \psi_m^*}{\partial t} z \psi_n + \psi_m^* z \frac{\partial \psi_n}{\partial t} \right) dx$$

and

$$i\hbar \frac{\partial \psi_n}{\partial t} = H_0 \psi_n \qquad -i\hbar \frac{\partial \psi_n^*}{\partial t} = H_0^* \psi_n^*$$

with

$$H_0 = -\frac{\hbar^2}{2m} \nabla^2 + V(x)$$

From the Hermiticity of H_0, the reader will then readily verify eq. (46).

Problem 1: Prove eq. (46) in the way outlined above. Show also that it follows from Chap. 16, eq. (49), by noting that $\int \psi_m{}^* z \psi_n \, dx$ is the matrix element of z in a Heisenberg representation.

We now insert into eq. (46) the following solution of Schrödinger's equation for the unperturbed system, $\psi_m = U_m \, e^{-iE_m{}^0t/\hbar}$. The result is

$$\frac{d}{dt} [e^{i(E_m{}^0 - E_n{}^0)t/\hbar}] \int U_m^* z U_n \, dx = e^{i(E_m{}^0 - E_n{}^0)t/\hbar} \int U_m^* \frac{\hbar}{im} \frac{\partial}{\partial z} U_n \, dx$$

Writing $(E_m^0 - E_n^0) = \hbar\omega_{mn}$, we finally obtain

$$\frac{\hbar}{im} \int U_m^* \frac{\partial}{\partial z} U_n \, dx = i\omega_{mn} \int U_m^* z U_n \, dx$$

† Compare this treatment with Chap. 2, Sec. 16.

The integral on the right is a sort of average of the co-ordinate z, taken with a weighting function, $U_m^*U_n$, which involves both the initial and the final state. It is something analogous to a dipole moment,† except that it involves two states at once. It is sometimes called the "dipole moment between the mth and the nth states." It is denoted by

$$z_{mn} = \int U_m^* z U_n \, dx \tag{47}$$

We then obtain

$$\alpha_{mn} = i\omega_{mn} m z_{mn} \tag{48}$$

The probability of absorption of radiation which is incident in the x direction and polarized in the z direction is then given by eq. (38)

$$P = 2\pi \left(\frac{e}{c\hbar}\right)^2 cI(\nu)|z_{nm}|^2 \tag{49}$$

The probability of spontaneous emission of radiation into the element of solid angle, $d\Omega$, in the x direction, and polarized in the z direction, is given by eq. (43)

$$R \, d\Omega = 8\pi^3 \left(\frac{e}{c}\right)^2 \frac{\nu^3}{ch} |z_{nm}|^2 \, d\Omega$$

If we wish to obtain the probability of radiation into some other direction, and with some other polarization, then it is merely necessary to note that the same development would have gone through if we had originally chosen in eq. (30) a plane wave going in an arbitrary direction with arbitrary polarization. In the final result, the only change would be that $|z_{nm}|^2$ would become $|\xi_{nm}|^2$, where ξ is the value of the co-ordinate of the particle taken in the direction of polarization of the wave. Thus, we write

$$R = 8\pi^3 \left(\frac{e}{c}\right)^2 \frac{\nu^3}{ch} |\xi_{nm}|^2 \tag{50}$$

If α, β, and γ are the angles of the direction of polarization relative to the x, y, and z axes respectively, then we can write

$$\xi = x \cos \alpha + y \cos \beta + z \cos \gamma$$

and

$$R = 8\pi^3 \left(\frac{e}{c}\right)^2 \frac{\nu^3}{ch} |(x \cos \alpha + y \cos \beta + z \cos \gamma)_{mn}|^2 \tag{51}$$

In order to illustrate the angles involved, let us refer to Fig. 1. The direction of propagation is taken to be that of the line OP which makes an angle A with the z axis, and the projection of which on the xy plane

† Because an electric dipole of moment $M = ez$ would lead to the same matrix elements, we conclude that the approximation 45 which neglects $k(x - x_0)$ is equivalent to the replacement of the actual charge by such a dipole.

makes an angle B with the x axis. Such a wave can be analyzed into waves having one of two directions of polarization. It is therefore sufficient to consider separately waves which are polarized in the plane of OPZ and normal to this plane. If they are polarized in the plane of OPZ, then ξ is given by

$$\xi = -x \cos B \cos A - y \sin B \cos A + Z \sin A \quad (52)$$

If they are polarized normal to the plane OPZ, then

$$\xi = x \sin B - y \cos B \quad (53)$$

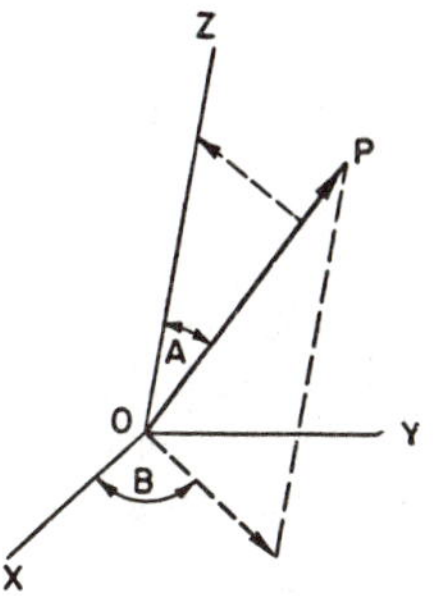

FIG. 1

26. Evaluation of α_{nm} for Isotropic Harmonic Oscillator. Let us consider an isotropic three-dimensional harmonic oscillator. For convenience, let us consider a wave moving in the x direction and polarized in the z direction. We must now evaluate z_{nm}. According to eq. (47), this is just

$$z_{mn} = \int U_m^* z U_n \, dx$$

where U_m, U_n are normalized eigenfunctions belonging to different eigenstates. In this case, they are eigenfunctions for the three-dimensional harmonic oscillator [see Chap. 15, eq. (40)].

The eigenfunctions may be written out more fully as

$$U_m = \psi_{m_x}\psi_{m_y}\psi_{m_z} \qquad \text{and} \qquad U_n = \psi_{n_x}\psi_{n_y}\psi_{n_z}$$

where ψ_{m_x} is the eigenfunction for a harmonic oscillator in the x direction, in the m_x state, etc. The value of this integral is most easily evaluated with the aid of Chap. 13 eqs. (38) and (39). We first set $z = \sqrt{\hbar/m\omega}\, q$ [see eqs. (2) and (3), Chap. 13] and then write

$$z = \frac{1}{2}\sqrt{\frac{\hbar}{m\omega}}\left[\left(\frac{\partial}{\partial q} + q\right) - \left(\frac{\partial}{\partial q} - q\right)\right] \quad (54)$$

According to eqs. (38) and (39), Chap. 13,

$$\left(\frac{\partial}{\partial q} - q\right)\psi_{n_z} = \sqrt{2(n_z + 1)}\,\psi_{(n_z+1)} \qquad \text{and} \qquad \left(\frac{\partial}{\partial q} + q\right)\psi_{n_z} = \sqrt{2n_z}\,\psi_{(n_z-1)}$$

Thus,

$$z_{mn} = -\sqrt{\frac{\hbar}{2m\omega}}\int\int\int dx\, dy\, dz \psi_{m_x}^*\psi_{n_x}\psi_{m_y}^*\psi_{n_y}\psi_{m_z}^*\left(\sqrt{n_z + 1}\,\psi_{(n_z+1)} - \sqrt{n_z}\,\psi_{(n_z-1)}\right) \quad (55)$$

Because of the orthogonality of the ψ's, z_{mn} vanishes unless $m_x = n_x$ and $m_y = n_y$. There is therefore no change in the state of oscillation in the x and y directions in any transition involving light that is polarized only in the z direction. If $m_x = n_x$ and $m_y = n_y$, then the integrals over x and y are unity because the ψ's are assumed to be normalized.

As for the integrals over z, these will vanish unless either

$$m_z = n_z + 1 \qquad \text{or} \qquad m_z = n_z - 1 \tag{56}$$

In other words, transitions can occur only to states in which the z component of the oscillation changes either to the next higher state or to the next lower state. In the latter two cases, the integral over z is unity. We can therefore write

$$z_{mn} = \begin{cases} -\sqrt{\dfrac{\hbar}{2m\omega}}\sqrt{n_z + 1} & \text{if } m_z = n_z + 1 \quad \text{(Absorption)} \\ +\sqrt{\dfrac{\hbar}{2m\omega}}\sqrt{n_z} & \text{if } m_z = n_z - 1 \quad \text{(Emission)} \end{cases} \tag{57}$$

The first of the above cases corresponds to absorption of a photon since the energy of the atom increases as a result of the transition, while the second corresponds to emission since the energy of the atom decreases.

The mean rates of spontaneous emission of quanta into a unit solid angle in a direction normal to z is given by inserting the above result into eq. (51)

$$R = \frac{\pi}{m}\left(\frac{e}{c}\right)^2 \frac{\nu^2}{c} n_z \tag{58}$$

Note that the rate of absorption and emission both increase with increasing excitation of the oscillator. In the ground state ($n_z = 0$) there is, of course, no emission, but there is still a definite possibility of absorption.

27. Selection Rules for Harmonic Oscillator. We saw above that z_{mn} vanishes for most transitions (for example, when $m_z = n_z + 2$). This means that, at least in the dipole approximation which we have been using, these other transitions cannot occur. In the older terminology, they are said to be "forbidden." We shall see in Secs. 34 and 35 that when the higher terms in the expansion* of $e^{ik(z-z_0)}$ are taken into account, then they can occur, but with a probability that is much less than that of the "allowed" transitions. They should therefore really be called "improbable" transitions, since they are not totally forbidden.

The rule, $m_z = n_z + 1$ or $m_z = n_z - 1$, which specifies the allowed transitions, is commonly called a *selection rule*. We shall obtain many examples of such selection rules.

* See eq. (44).

28. Connection of Selection Rules with Correspondence Principle. In the classical dipole approximation, we assume that the orbit of the electron is so small in comparison with a wavelength that the oscillating electron acts like an oscillating dipole of negligible size, as far as producing radiation is concerned. If this dipole undergoes a simple harmonic oscillation with angular frequency ω, it should radiate and absorb only light of the same frequency. The quantum-mechanical probabilities of transition must yield the same result in the classical limit. But quantum-mechanically, $\pm\hbar\omega = E_m - E_n$. (The $\pm$ sign indicates the possibility of emission or absorption.) Since $E_{n_z} = (n_z + \frac{1}{2})\hbar$, we will obtain the correct classical frequency in the classical limit only if all transitions are subject to the restriction $m_z - n_z = \pm 1$. But this is just the restriction obtained quantum-mechanically in eq. (57). Thus, our selection rule guarantees the correct classical frequency in the classical limit. If it were not present, then any multiple of this frequency could be radiated by having transitions in which m_z changed by more than unity. Before the modern quantum theory was available, the existence of such selection rules was, in fact, guessed by the requirement that the frequencies emitted in the correspondence limit of large quantum numbers agree with the classical frequency. We shall return to this point later.

29. Introduction of Parity. It is convenient especially for complex systems containing many particles to introduce a classification of wave functions according to a property called "parity." This property of parity depends on whether or not the wave function changes sign when the value of each co-ordinate of every particle is replaced by the negative of that co-ordinate. To investigate the parity, we must therefore consider

$$\psi(-x_1, -y_1, -z_1; \quad -x_2, -y_2, -z_2; \quad -x_3, -y_3, -z_3; \quad \ldots)$$

where x_1, y_1, z_1 are the co-ordinates of the first particle, etc. The above may be abbreviated to $\psi(-\boldsymbol{x}_i)$, where $\boldsymbol{x}_i$ is the position vector of the ith particle. In general, there is no particular relation between $\psi(\boldsymbol{x}_i)$ and $\psi(-\boldsymbol{x}_i)$. For a system in which the potential function $V(\boldsymbol{x}_i)$ does not change when $\boldsymbol{x}_i$ is replaced by $-\boldsymbol{x}_i$ [i.e., where $V(\boldsymbol{x}_i) = V(-\boldsymbol{x}_i)$], it may be shown, however, that all eigenstates can be grouped according to whether they have one of the two following properties

$$\psi(\boldsymbol{x}_i) = \psi(-\boldsymbol{x}_i) \qquad \text{or} \qquad \psi(\boldsymbol{x}_i) = -\psi(-\boldsymbol{x}_i) \tag{59}$$

The first type of state is said to have even parity; the second, odd parity.

To prove the possibility of this classification, let us consider a non-degenerate eigenfunction of the Hamiltonian $H\psi_E = E\psi_E$. Now, the kinetic energy is not changed when $\boldsymbol{x}_i$ is replaced by $-\boldsymbol{x}_i$. Since we are assuming that V also is not changed by this operation, it is clear that H

is not changed either. Replacing x_i by $-x_i$ in the above expression, we obtain

$$H\psi_E(-x_i) = E\psi_E(-x_i) \tag{60}$$

Hence if $\psi_E(x_i)$ is an eigenfunction belonging to E, so also is $\psi(-x_i)$. But if the energy level is nondegenerate, there is only one such eigenfunction. This means that $\psi_E(-x_i) = C\psi_E(x_i)$, where C is some constant. To evaluate the constant, we again replace x_i by $-x_i$ in the equation $\psi_E(-x_i) = C\psi_E(x_i)$. We obtain $\psi_E(x_i) = C\psi_E(-x_i) = C^2\psi_E(x_i)$. Therefore $C^2 = 1$, and $C = \pm 1$. We conclude that every nondegenerate eigenfunction of H must either have the property that $\psi_E(-x_i) = \psi_E(x_i)$, or that $\psi_E(-x_i) = -\psi_E(-x_i)$. It should be noted, however, that this property is required only when $H(x_i) = H(-x_i)$.

When the energy levels are degenerate, the above reasoning does not necessarily follow. It may be shown, however, that it is still always possible to classify states according to their parity, but we shall not do so here.

Examples: In a one-particle problem with a radially symmetrical potential, we have $V(x_i) = V(-x_i)$. Each eigenstate should therefore have a definite parity. According to eq. (1), Chap. 15, the eigenfunctions are $f_{l,n}(r)Y_l^m(\vartheta, \varphi)$, where l is the total angular-momentum quantum number, and m is the azimuthal quantum number. Since r is by definition positive, the term $f_{l,n}(r)$ does not change when x is replaced by $-x$. The term $Y_l^m(\vartheta, \varphi)$, however, may or may not change sign. For example, $P_1(\cos\vartheta) = P_1(z/r) = z/r$ changes its sign when z is replaced by $-z$. On the other hand, $P_2(\cos\vartheta) = \left(\frac{3z^2}{r^2} - 1\right)$ does not change sign. It may be shown that $P_l^m(\cos\vartheta)e^{im\varphi}$ has an odd parity when l is odd, and an even parity when l is even.

Problem 3: Prove the above statement.

The usefulness of parity as a means of classification of states is that it can still be applied even when there are many particles, and when the potential is not spherically symmetrical. For example, in a symmetrical diatomic molecule the potential is unchanged when x_i is replaced by $-x_i$, provided that x_i is measured from the center of the molecule. Parity will still be a good quantum number, but angular momentum will not be because the Hamiltonian is not spherically symmetrical.

30. Selection Rules on Parity. Let us consider a matrix element of the perturbing term $V(x_i)$,

$$\int \psi_m^*(x_i)V(x_i)\psi_n(x_i)\,dx_i$$

Since this is an integration over all x_i, it must not be changed if all x_i are replaced by $-x_i$. Thus we can write

$$\int \psi_m^*(x_i)V(x_i)\psi_n(x_i)\,dx_i = \int \psi_m^*(-x_i)V(-x_i)\psi_n(-x_i)\,dx_i \tag{61}$$

Now, suppose that V has the property that $V(x_i) = V(-x_i)$, and suppose further that ψ_m and ψ_n have definite parity. If ψ_m and ψ_n have the same parity, the above relation becomes an identity. But if they have opposite parity, we obtain

$$\int \psi_m^*(x_i) V(x_i) \psi_n(x_i)\, dx_i = -\int \psi_m^*(x_i) V(x_i) \psi_n(x_i)\, dx_i \tag{62}$$

This can only be true if the integral vanishes. We therefore obtain the selection rule that a perturbing term of even parity can cause transitions only between wave functions of like parity.

In a similar way, it may be shown that if V has odd parity, $V(-x_i) = -V(x_i)$ then it can cause transitions only between states of unlike parity.

Problem 4: Prove the above statement.

Example: In the one-particle problem, the perturbing term for dipole radiation is $x \cos \alpha + y \cos \beta + z \cos \gamma$ [see eq. (51)]. This obviously has odd parity. Thus, in any dipole transition, the parity must change. Parity-selection rules are especially significant in complex systems, where the angular-momentum selection rules (to be obtained in the next few sections) are no longer useful. Furthermore, there exist many nonspherically symmetrical systems, such as crystal lattices, for which parity selection rules remain valid, even though there are no angular-momentum selection rules at all.

31. Selection Rules for Spherically Symmetric Potential, with the Neglect of Spin. We now study the selection rules for a single particle moving in a spherically symmetrical potential, but neglecting the effects of spin. The eigenfunctions of such a system are given in eq. (1), Chap. 15.

We begin with the case of a light wave moving in the x direction and polarized in the z direction. For this case we must evaluate the matrix element of $z = r \cos \vartheta$,

$$z_{n',l',m';n,l,m} = \int_0^\infty dr \int_0^\pi d\vartheta \int_0^{2\pi} d\varphi\, r^2 \sin \vartheta f_{l',n'}(r) Y_{l'}^{*m'}(\vartheta, \varphi) r \cos \vartheta f_{l,n}(r) Y_l^m(\vartheta, \varphi) \tag{63}$$

Since $Y_l^m(\vartheta, \varphi) \sim P_l^m(\vartheta) e^{im\varphi}$, it is clear that the integration over φ involves the factor $\int_0^{2\pi} e^{i(m-m')\phi}\, d\phi$. This integral vanishes unless $m' = m$.

Let us now consider the integration over ϑ. We must evaluate

$$\int_0^\pi Y_{l'}^{m'}(\cos \vartheta) \cos \vartheta Y_l^m(\cos \vartheta)\, d(\cos \vartheta) \tag{64}$$

We first restrict ourselves to the special case $m = 0$. Here we have $Y_l^0 = \sqrt{\frac{2l+1}{2}} P_l(\cos \vartheta)$ [see eq. (52a), Chap. 14]. From eq. (54a), Chap 14,

$$\cos\vartheta P_l(\cos\vartheta) = \frac{(l+1)}{2l+1}P_{l+1}(\cos\vartheta) + \frac{l}{2l+1}P_{l-1}(\cos\vartheta) \tag{65}$$

The integral to be evaluated becomes

$$\sqrt{\frac{(2l'+1)}{2}\cdot\frac{(2l+1)}{2}}\int_0^\pi P_{l'}(\cos\vartheta) \left[\frac{(l+1)}{2l+1}P_{l+1}(\cos\vartheta) + \frac{l}{2l+1}P_{l-1}(\cos\vartheta)\right]d(\cos\vartheta) \tag{66}$$

Because the Legendre polynomials are orthogonal, this vanishes unless

$$l' = l+1 \qquad \text{or} \qquad l' = l-1 \tag{67}$$

We then obtain

$$z_{n',l',m';n,l,m} = \frac{1}{2}\int_0^\infty f_{l',n'}(r)f_{l,n}(r)r^3\,dr \begin{Bmatrix} (l+1)\sqrt{\dfrac{2l+3}{2l+1}} & \text{when } l' = l+1 \\ l\sqrt{\dfrac{2l-1}{2l+1}} & \text{when } l' = l-1 \end{Bmatrix} \tag{68}$$

The integration over r does not in general vanish for any particular choices of n or l.

Selection Rules. For the case in which $m' = m = 0$, we conclude that light polarized in the z direction can be emitted or absorbed only if

$$m' = m$$
$$l' = l+1 \qquad \text{or} \qquad l' = l-1 \tag{69}$$

These are the selection rules for this case.

Generalization to Arbitrary m and m'. Similar methods could be used to generalize these results to arbitrary m and m', but it is easier to use the general result quoted in eq. (77b), Chap. 17, expressing the product of two angular-momentum wave functions as a sum of eigenfunctions of the angular momentum. In our case, we write

$$P_1(\cos\vartheta)Y_l^m(\vartheta,\varphi) = C_1Y_{l-1}^m(\vartheta,\varphi) + C_2Y_l^m(\vartheta,\varphi) + C_3Y_{l+1}^m(\vartheta,\varphi)$$

The matrix element becomes

$$\int Y_{l'}^{m'*}P_1(\cos\vartheta)Y_l^m\,d\Omega = \delta_{m',m}(C_1\delta_{l',l-1} + C_2\delta_{l',l} + C_3\delta_{l',l+1}) \tag{70}$$

Now the function $P_1(\cos\vartheta) = z/r$ has odd parity. This means, according to Sec. 29, that only states of differing parity can fail to vanish in the matrix elements of $P_1(\cos\vartheta)$. We conclude that C_2 must vanish, and therefore obtain once again the selection rules $m' = m$, $l' = l \pm 1$.

Extension to Arbitrary Directions of Propagation and Polarization. We can also extend the selection rules to the case of a plane wave polarized in an arbitrary direction, given by cosines which are respectively $\cos\alpha$, $\cos\beta$, and $\cos\gamma$. To do this, we must use eq. (51). We see that the matrix element is

$$\int f_{l',n'}(r) Y_{l'}^{m'*}(\vartheta, \varphi)[r(\cos\gamma\cos\vartheta + \cos\alpha\sin\vartheta\cos\varphi + \cos\beta\sin\vartheta\sin\varphi)] Y_l^m(\vartheta, \varphi) f_{l,n}(r) r^2\, dr \quad (71)$$

It is readily seen that the terms involving $\sin\vartheta\cos\varphi$ and $\sin\vartheta\sin\varphi$ will vanish after integration over ϑ and φ, unless $l' = l \pm 1$. To show this, it is necessary merely to rotate the co-ordinate axis into such a direction that the new z axis points in the direction of the old x axis. The term, $x = r\cos\varphi\sin\vartheta$ then becomes $z' = r'\cos\vartheta'$. We have already seen that for this case the matrix element vanishes unless

$$l' = l \pm 1$$

But the value of l is not changed by a rotation since $l(l+1)\hbar^2$ is just the square of the absolute value of the angular momentum, which is a scalar. The same selection rules on l must therefore prevail in the original co-ordinate system.

An additional selection rule can be obtained by noting that

$$\int_0^{2\pi} e^{-im'\varphi}\cos\varphi e^{im\varphi}\, d\varphi \qquad \text{and} \qquad \int_0^{2\pi} e^{im'\varphi}\sin\varphi e^{im\varphi}\, d\varphi$$

both vanish unless $m' = m \pm 1$. If the wave has a component of polarization in the x or y directions, m can change only by ± 1.

Summary of Selection Rules for Dipole Transitions.

(1) $\Delta l = \pm 1$.
(2) $\Delta m = 0$ for waves polarized in z direction.
$\Delta m = \pm 1$ for waves polarized in x or y direction.

If the waves are not polarized in any particular direction, then $\Delta m = 0$ or ± 1.

32. Forbidden Transitions, Electric Quadripole Radiation. When the dipole matrix element, ξ_{mn} vanishes, this does not mean that no transition can take place. What happens is that the higher terms in the expansion of eq. (44) can no longer be neglected. We shall see that these can still cause transitions, but with a considerably reduced probability.

The first term that we have neglected in α_{nm} is (for a wave polarized in the z direction)

$$\hbar k \int U_m^*(\mathbf{x}) x \frac{\partial}{\partial z} U_n(\mathbf{x})\, d\mathbf{x} = \frac{\hbar k}{2} \int U_m^*(\mathbf{x}) \left[\left(x\frac{\partial}{\partial z} + z\frac{\partial}{\partial x}\right) + \left(x\frac{\partial}{\partial z} - z\frac{\partial}{\partial x}\right)\right] U_n(\mathbf{x})\, d\mathbf{x} \quad (72)$$

Let us first consider $\hbar \frac{k}{2} \int U_m^*(x) \left(x \frac{\partial}{\partial z} + z \frac{\partial}{\partial x} \right) U_n(x)\, dx$. By methods similar to those used to obtain eq. (48) for the dipole transitions, we can show that

$$\frac{\hbar k}{2} \int U_m^*(x) \left(x \frac{\partial}{\partial z} + z \frac{\partial}{\partial x} \right) U_n(x)\, dx = im\omega_{mn} \int U_m^*(x) xz U_n(x)\, dx \quad (73)$$

Problem 5: Prove the above statement.

The above integral is a sort of average of one component of the quadripole moment† of the charge, taken with a weighting function $U_m^*(x) U_n(x)$. Writing $k = 2\pi/\lambda$, we see that this integral involves in the integrand the same factors as were in the dipole moment, and in addition the factor $2\pi x/\lambda$. Since the atom is much smaller than a wavelength, this integral will tend to be smaller than typical dipole moments by the factor $2\pi a/\lambda$, where a is the atomic radius. For typical wavelengths λ and atomic radii a, this ratio is about $\frac{1}{100}$. Since the probability of transition is proportional to $|\alpha_{nm}|^2$, this means that when dipole transitions are forbidden, quadripole transitions can still, in general, occur, but with a probability reduced by a factor of about 10,000.

33. Magnetic Dipole Radiation. The term in eq. (72) involving $x \frac{\partial}{\partial z} - z \frac{\partial}{\partial x}$ is (except for a factor of i) just the angular-momentum operator, L_y. In fact, this term resembles an average of the magnetic moment, taken with the weighting factor $U_m^*(x) U_n(y)$. For this reason, transitions resulting from this term are called magnetic dipole radiation. It can be shown that these terms lead to the same distribution of radiation in angle as would be produced by an oscillating magnetic dipole. One can also show that the probability of magnetic dipole radiation is of the same order as that of electric quadripole radiation. Physically, the reason for the smallness of magnetic dipole radiation is that the magnetic dipole moment of the moving electron is smaller than the electric dipole moment by a factor of v/c, which is of the order of $\frac{1}{100}$ for electrons in most atoms.

34. Selection Rules for Electric Quadripole Radiation.

(1) *Parity:* The matrix element has even parity, so that only transitions with no change of parity will occur as a result of this type of perturbation. Since we have shown that functions have even or odd parity accordingly as they have even or odd l terms, it is clear that in quadripole transitions we must have Δl as an even number.

† The quadripole moment of a charge distribution is a tensor with two indices, defined as follows:

$\varphi_{ij} = \int \rho(x) x_i x_j \, dx_j$ where x_i represents the co-ordinate (i.e., $x_1 = x$, $x_2 = y$, $x_3 = z$). It can be shown that these terms lead to the same distribution of radiation as would an oscillating electric quadripole. See, for example, Stratton, *Electromagnetic Theory*, p. 177.

(2) *Angular Momentum:* It can be shown that the elements of the quadripole moment tensor $x_i x_j$ can be expressed as linear combinations of the spherical harmonics Y_2^m.

Problem 6: Prove the preceding statement.

The selection rules are therefore the same as those obtained from the matrix elements

$$\int Y_{l'}^{m'} Y_2^{m''} Y_l^m \, d\Omega$$

By eq. (77b), Chap. 17, we write

$$Y_2^{m''} Y_l^m = C_2 Y_{l-2}^{m+m''} + C_{-1} Y_{l-1}^{m+m''} + C_0 Y_l^{m+m''} + C_1 Y_{l+1}^{m+m''} + C_2 Y_{l+2}^{m+m''}$$

Since the parity does not change, however, we know that C_{-1} and C_1 must vanish. This leaves us the selection rules

$$\Delta l = 0, \pm 2 \qquad \Delta m = 0, \pm 1, \pm 2 \tag{74}$$

35. Selection Rules for Magnetic Dipole Radiation.

(1) *Parity:* $x \dfrac{\partial}{\partial z} - z \dfrac{\partial}{\partial x}$ has even parity; so that this type of transition also leaves the parity unchanged.

(2) *Angular Momentum:* The magnetic dipole matrix element is proportional to

$$\int Y_{l'}^{m'}(\vartheta, \varphi) L_y Y_l^m(\vartheta, \varphi) \, d\Omega$$

Now in eq. (27), Chap. 14, we showed that

$$(L_x + iL_y)\psi_l^m(\vartheta, \varphi) \sim \psi_l^{m-1}(\vartheta, \varphi)$$

Similarly,

$$(L_x - iL_y)\psi_l^m(\vartheta, \varphi) \sim \psi_l^{m+1}$$

The above matrix element will therefore vanish unless $l = l'$ and

$$m' = m \pm 1$$

The above, however, is not the most general possible matrix element for magnetic dipole radiation. For example, we could have chosen radiation moving in the x direction and polarized in the y direction. We should then have obtained $\left(x \dfrac{\partial}{\partial y} - y \dfrac{\partial}{\partial x}\right) = \dfrac{iL_z}{\hbar}$ in the integral. But $L_z \psi_l^m(\vartheta, \varphi) = m\psi_l^m$. In the emission or absorption of such a wave, we should have $m' = m$; thus, if all possible polarizations are included, the selection rules are

$$\Delta l = 0 \qquad \Delta m = 0, \pm 1 \tag{75}$$

36. Higher Order Transitions. In those transitions in which electric dipole, magnetic dipole, and the electric quadripole transitions are all forbidden (for example, $\Delta l > 2$) it is necessary to go to still higher terms in the expansion of the exponential in eq. (31b). It can be shown that such terms will, in the classical limit, produce radiation patterns that are the same as those of higher order electric and magnetic multipoles. For example, the k^2x^2 term in expansion leads to magnetic quadripole and electric octopole radiation. Each time one goes to a higher order term in the expansion, one obtains an integral reduced by a factor of the order of $2\pi a/\lambda$, where a is the atomic radius, and therefore a transition probability reduced by a factor of the order of $(2\pi a/\lambda)^2$, which is of the order of 1/10,000.

37. $l = 0$ to $l = 0$ Transitions Totally Forbidden. One can obtain a rather important selection rule, namely, that transitions from $l = 0$ to $l = 0$ are totally forbidden for *all* orders of multipole radiation. To prove this, consider the general matrix element, noting that U_m and U_n are both functions only of r because they represent s states:

$$\frac{\hbar}{i}\int f_m(r)\,\frac{\partial}{\partial x_j}\,e^{i\boldsymbol{k}\cdot\boldsymbol{x}}f_n(r)\,dx$$

$\partial/\partial x_j$ refers to differentiation in the direction in which the wave is polarized. Note that transversality requires that $\boldsymbol{k}$ be normal to x_j. The above integral may be written as

$$\frac{\hbar}{i}\int f_m(r)e^{i\boldsymbol{k}\cdot\boldsymbol{x}}x_j\,\frac{f_n'(r)}{r}\,dx$$

Since the integrand is an odd function of x_j, it must vanish when integrated over x_j. Thus, the matrix element vanishes for all multipole orders.

The above selection rule may result in a very long-lived metastable state. For example, if it turns out, as it does for many atoms, that both the ground state and the next lowest state have $l = 0$, then an atom that gets into the state just above the ground state cannot get rid of its energy by radiation, so that it may remain excited for a long time. Hydrogen, and the noble gas atoms, such as helium, neon, and argon, are among those which have metastable levels of this kind.

38. Total Rate of Radiation. Thus far, we have calculated only the rate of emission of radiation into a given direction and with a given direction of polarization. To obtain the total rate of radiation one must integrate over all directions of emission and sum over both directions of polarization. Let us carry out this procedure here for the special case of a dipole transition in which $\Delta m = 0$. The only matrix element that does not vanish for this case will be z_{nm}.

Consider a light wave that is emitted in a direction with a latitude angle A and azimuthal angle B, as shown in Fig. 1. Such a wave may have two directions of polarization: one in the plane that includes the direction of the ray and the z axis, and the other in a perpendicular direction. The wave that is polarized in the perpendicular direction has an electric vector normal to the z axis; hence the matrix element ξ_{nm} must vanish for this wave, since in this transition only z_{nm} fails to vanish. For the other polarization, we have [see eq. (52)]

$$\xi_{nm} = z_{nm} \sin A - x_{nm} \cos A \cos B - x_{nm} \cos A \sin B$$

The only nonvanishing part is

$$\xi_{nm} = z_{nm} \sin A \tag{76}$$

Thus, according to eq. (51), the probability of spontaneous emission of a wave into the direction given by A and B (per unit solid angle) is

$$dR = 8\pi^3 \left(\frac{e}{c}\right)^2 \frac{\nu^3}{hc} |z_{nm}|^2 \sin^2 A \, d\Omega \tag{77}$$

Note that in this transition the intensity of radiation is proportional to $\sin^2 A$. This is a typical dipole pattern. Each type of multipole has its own characteristic intensity pattern, which can be calculated from the matrix elements. The total rate of radiation is given by integrating the above over all solid angles of emission with the weighting factor $\sin A \, dA \, dB$. This procedure yields

$$R = \frac{64\pi^4}{3} \left(\frac{e}{c}\right)^2 \frac{\nu^3}{hc} |z_{nm}|^2 \tag{78}$$

The above is the rate at which quanta are emitted. To obtain the net rate of emission of energy, one should multiply it by the energy per quantum, $h\nu$. This gives

$$\frac{dW}{dt} = Rh\nu = \frac{64\pi^4}{3} \frac{e^2}{c^3} \nu^4 |z_{nm}|^2 \tag{79}$$

39. Comparison with Classical Theory. According to classical electrodynamics, the mean rate of radiation of energy by a moving charge is*

$$\frac{dW}{dt} = \frac{2}{3} \frac{e^2}{c^3} |\ddot{\mathbf{x}}|^2 \qquad \text{where } \ddot{\mathbf{x}} \text{ is the acceleration.} \tag{80}$$

Let us, for simplicity, consider the case of a harmonic oscillator which is excited in the z direction. Thus, its motion is given by $z = z_0 \cos \omega t$, so that

$$\ddot{z} = -\omega^2 z_0 \cos \omega t \qquad \text{and} \qquad (\ddot{z})^2 = \omega^4 z_0^2 \cos^2 \omega t = \frac{\omega^4 z_0^2}{2} (1 + \cos 2\omega t)$$

When this is averaged over a period of oscillation, the $\cos 2\omega t$ term drops out, and we find

$$\overline{(\ddot{z})^2} = \frac{\omega^4 z_0^2}{2} \tag{81}$$

* Chap. 2, eq. (45).

Now, the energy of a harmonic oscillator is given by $W = \frac{m\dot{z}^2}{2} + m\omega^2\frac{z^2}{2}$. When z reaches its maximum ($z = z_0$), $\dot{z} = 0$, so that we obtain

$$W = \frac{m\omega^2 z_0^2}{2} \quad \text{or} \quad z_0^2 = \frac{2W}{m\omega^2}$$

and

$$\overline{(\ddot{z})^2} = \frac{\omega^2 W}{m} \tag{82}$$

The mean classical rate of radiation is then given by

$$\frac{dW}{dt} = \frac{2}{3}\frac{e^2}{c^3}|\ddot{x}|^2 = \frac{2}{3}\frac{e^2}{mc^3}\omega^2 W \tag{83}$$

Let us compare this with the quantum-mechanical rate of radiation by a harmonic oscillator. One obtains z_{nm} from eq. (57), and using eq. (79) for dW/dt, one is led to

$$\frac{dW}{dt} = \frac{\hbar}{2m\omega} n_z \nu^4 \frac{64\pi^4}{3}\frac{e^2}{c^3} \tag{84}$$

To obtain a comparison with the classical rate, we write $\hbar\omega n_z \cong W$ (in the classical limit)

$$\frac{dW}{dt} = \frac{32\pi^4}{3}\frac{W}{m}\frac{e^2}{c^3}\frac{\nu^4}{\omega^2} = \frac{2}{3}\frac{e^2}{mc^3}\omega^2 W \tag{85}$$

This result is the same as the classical result [eq. (83)]. Thus, we conclude that for a harmonic oscillator the quantum theory yields the same rate of radiation as does classical theory.

In order to make a comparison between quantum and classical rates of radiation for a general system, we use the result of eq. (48), Chap. 2, that the time average rate of radiation of energy for the nth harmonic in a classical system of angular frequency ω_0 is

$$\bar{R} = \frac{1}{3}\frac{e^2}{c^3}\omega_0^4 |a_n|^2 n^4 \tag{86}$$

As shown in eq. (53), Chap. 2, the nth harmonic corresponds to a jump over n quantum states at once, with an energy charge of $\Delta E = n\omega_0\hbar$. Now, we have seen that the mean rate of radiation of quanta is proportional to the square of the absolute value of the matrix element corresponding to the transition. Now it can be shown* that for the limit of high quantum numbers the matrix element $a_{m,m+n}$ approaches the classical Fourier component, a_n. Thus quantum theory predicts in the classical limit a rate of radiation of the nth harmonic proportional to $|a_n|^2$, in agreement with the results of classical theory. It can be shown by means of a fairly straightforward treatment that the constant of proportionality is such as to lead exactly to eq. (86).

* See W. Heisenberg.

40. Sum Rules for Evaluating Matrix Elements. A number of useful rules for evaluating the sums of the matrix elements of transitions from or to a given quantum state can be obtained by making use of a few of the mathematical properties of matrices. In order to illustrate these rules, let us begin with the operator relation

$$px - xp = i\hbar \tag{87}$$

Now, in the Heisenberg representation, we have, according to Problem 17, Chap. 16, the following matrix relation

$$p = m\dot{x} \tag{88}$$

If we make the Hamiltonian diagonal, then as we saw in eq. (46), Chap. 16, each matrix element oscillates with the exponential $e^{i(E_m - E_n)t/\hbar}$. Thus, we obtain

$$(\dot{x})_{mn} = \frac{i}{\hbar}(E_m - E_n)x_{mn} \tag{89}$$

Equation (87) then becomes

$$\frac{mi}{\hbar}\sum_n [(E_m - E_n)x_{mn}x_{nr} - x_{mn}x_{nr}(E_n - E_r)] = i\hbar\delta_{mr} \tag{90}$$

(Note that the unit matrix is represented by δ_{mr}.)

If we set $m = r$, we obtain

$$\sum_n (E_m - E_n)x_{mn}x_{nm} = \frac{\hbar^2}{2m}$$

Because x_{mn} is Hermitean, we write $x_{nm} = x^*_{mn}$, obtaining

$$\sum_n (E_m - E_n)|x_{mn}|^2 = \frac{\hbar^2}{2m} \tag{91}$$

This expression is often useful because it provides a relation between the quantities $|x_{mn}|^2$ entering into the transition probabilities to and from the nth level. Such a rule is called a *sum rule*. In practice, $|x_{mn}|^2$ becomes small for large n, so that this relation can be checked experimentally by observing the transitions for a few values of n near m.

Another example of a sum rule is obtained from the matrix relation

$$\sum_n |x_{mn}|^2 = \sum_n x_{mn}x_{nm} = (x^2)_{mm} = \int \psi^*_m x^2 \psi_m \, dx \tag{92}$$

This means that the sum of the squares of the matrix elements involving the mth level can be obtained by simply finding the mean value of x^2 in the mth state.

41. Circular Polarization. Thus far, we have discussed plane polarized light waves. Let us now see what happens with circularly polarized light.

A left-hand circularly polarized light beam moving in the z direction can be described as follows:

$$\left.\begin{aligned} a_x &= a_0 \cos(\omega t - kz) \\ a_y &= a_0 \sin(\omega t - kz) \end{aligned}\right\} \tag{93}$$

(Note that a_y is 90° out of phase with a_x.)

In such a wave, the direction of a rotates with angular frequency, ω, in a counterclockwise direction. In order to obtain a wave that rotates in a clockwise direction, we can write

$$\left.\begin{aligned} a_x &= a_0 \cos(\omega t - kz) \\ a_y &= -a_0 \sin(\omega t - kz) \end{aligned}\right\} \tag{94}$$

The above results may be written more conveniently for our purposes as follows:

$$\left.\begin{aligned} &\text{Right Hand:} && a_x = [a_0\, e^{i(kz-\omega t)} + a_0^*\, e^{-i(kz-\omega t)}] \\ & && a_y = i[a_0\, e^{i(kz-\omega t)} - a_0^*\, e^{-i(kz-\omega t)}] \\ &\text{Left Hand:} && a_x = [a_0\, e^{i(kz-\omega t)} + a_0^*\, e^{-i(kz-\omega t)}] \\ & && a_y = -i[a_0\, e^{i(kz-\omega t)} - a_0^*\, e^{-i(kz-\omega t)}] \end{aligned}\right\} \tag{95}$$

One can make a plane polarized wave by interference of two oppositely circularly polarized waves. For example, let a_+ represent the vector potential of a right-hand wave, a_- that of a left-hand wave. Then $a_+ + a_-$ is a plane polarized wave, polarized in the x direction. This follows from the fact that the y components of the two waves cancel out. Similarly, $a_+ - a_-$ is a plane polarized wave, polarized in the y direction.

42. Elliptical Polarization. If we take two plane waves at right angles that are not 90° out of phase, or which do not have equal intensities, we obtain elliptically polarized waves. We shall not discuss what happens in these cases in any detail, but shall merely point out that the elliptically polarized waves can be treated by fairly simple generalizations of the methods used for circularly polarized waves.

43. Quantum Treatment. In order to compute the rate of transition resulting from a circularly polarized wave, one need merely insert the complete vector potential for this wave into eq. (23), in the calculation of V_{nm}. For the sake of simplicity, let us consider a wave moving in the z direction which is circularly polarized in the xy plane. Using eq. (46), we see that the matrix element will involve

$$a_0 \int U_m^*(x) \left[\frac{\partial}{\partial x} \pm i \frac{\partial}{\partial y}\right] U_n(x)\, dx + a_0^* \int U_m^*(x) \left[\frac{\partial}{\partial x} \mp i \frac{\partial}{\partial y}\right] U_n(x)\, dx \tag{96}$$

(The + sign refers to right-hand polarization, the − sign to left-hand polarization.) By a treatment similar to that which led to eq. (48), one can show that the above matrix element is proportional to

$$i\omega \int U_m^*(x)[a_0(x \pm iy) + a_0^*(x \mp iy)]U_n(x)\,dx \tag{97}$$

44. Selection Rules. Let us consider a transition in which

$$U_m = f_{l',n'}(r)P_{l'}^{m'}(\vartheta)e^{im'\varphi} \quad \text{and} \quad U_n = f_{l,n}(r)P_l^m(\vartheta)e^{im\varphi} \tag{98}$$

Writing $x \pm iy = re^{\pm i\varphi}$, we note that for left-hand circularly polarized light the matrix element vanishes unless $m' = m - 1$, while for right-hand polarization it fails to vanish only when $m' = m + 1$. If these conditions are not satisfied, the integral over φ will be zero. A transition in which $m' = m + 1$ will therefore result in right-hand circularly polarized light, at least for a wave going in the z direction, whereas $m' = m - 1$ yields a wave of opposite polarization.

For light which is emitted normal to the z axis, however, the polarization will be linear in any transition in which the change of m is defined. For example, consider light going in the x direction. It can have two directions of polarization, either in the z or in the y direction. As has already been shown, a transition in which $\Delta m = 0$ can produce only light which is polarized in the z direction, while if $\Delta m = \pm 1$, it must be linearly polarized normal to the z direction.

This does not mean, however, that circularly polarized light cannot be emitted in the x direction. It only means that in a transition in which the change of z component of the angular momentum (Δm) is well-defined, the light going normal to z is linearly polarized. In a transition in which the change of the x component of $\boldsymbol{L}$ was well-defined, one could obtain circularly polarized light moving in the x direction.

If the change of L_z is well-defined, then one can show that light moving in a direction that is neither parallel to z nor normal to z will be elliptically polarized.

45. Application to Normal Zeeman Effect.

Classical Treatment. Larmor Precession. When an atom is placed in a weak magnetic field, then with the aid of classical theory, one can show that the orbit precesses about an axis parallel to the magnetic field with angular frequency given by $\Omega = e\mathcal{H}/2mc$, where m is the electronic mass. The above precession is called the Larmor precession, and the frequency is called the Larmor frequency.† (Note that this frequency is only one-half the "cyclotron frequency," i.e., the frequency with which a *free* electron goes around in a circle in a magnetic field.)

Let us first consider a particular orbit, in which, for example, the particle is rotating about the z axis with angular frequency $\pm\omega$. $+\omega$

† G. Herzberg, *Atomic Spectra*, New York: Prentice-Hall, Inc., 1937, p. 103; also, White, *Introduction to Atomic Spectra.*

indicates counterclockwise rotation, $-\omega$ indicates clockwise rotation. If we view the atom in the z direction, we will obtain light that is circularly polarized with the electric vector rotating in the same direction as the electron. If we view it normal to the z axis, the electric field that reaches us will depend only on the projection of the electronic motion on the axis normal to the direction of viewing. The projection of circular motion on such an axis is simple harmonic motion, so that the light viewed in a

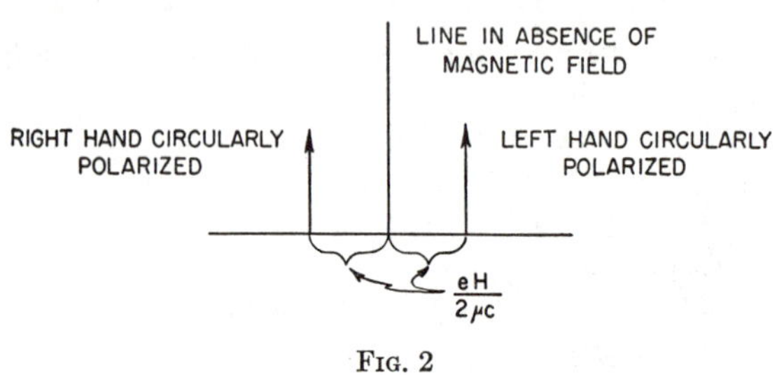

FIG. 2

direction parallel to the plane of motion will be polarized in a direction normal to the direction of viewing and parallel to the plane of electronic motion.

If a magnetic field is turned on in the z direction, then the components of motion in the z direction are left unchanged, whereas the components of motion in the xy plane are altered. Those atoms with counterclockwise orbits will have their frequencies of rotation in the xy

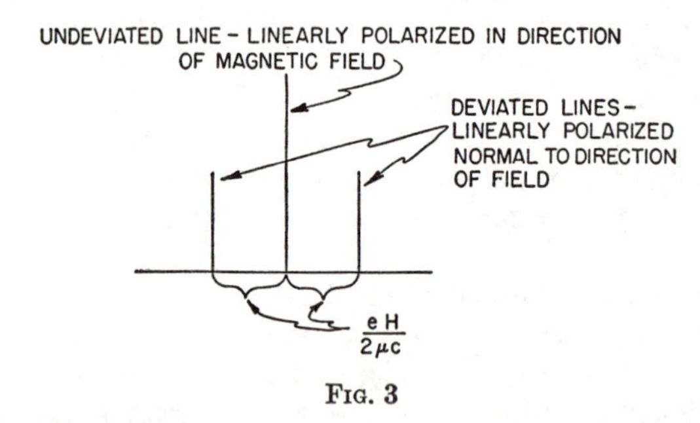

FIG. 3

plane increased by $e\mathcal{H}/2mc$, while those with clockwise orbits will have their frequencies decreased by $e\mathcal{H}/2mc$. When viewed along the direction of the magnetic field, the radiated line will therefore split into two lines, each with opposite circular polarization as shown in Fig. 2. The electrons moving in the z direction cannot radiate in the z direction; as a result, there will be no undeviated lines in the light emitted in the z direction. If the atom is viewed normal to the z direction, then there will be an undeviated line, produced by electrons which move in the z direction. This will be polarized in the z direction. The components of electronic motion in the xy plane will produce two deviated lines, each linearly polarized in a direction normal to z. The effect is illustrated in Fig. 3. If the radiation is viewed in an intermediate direction, one obtains elliptically polarized light.

46. Quantum Description of Normal Zeeman Effect. As shown in eq. (53), Chap. 15, the energy levels of an atom in a magnetic field are displaced by the amount

$$\Delta E = -\frac{e\mathcal{H}}{2\mu c} m\hbar \tag{99}$$

where m is the z component of the angular momentum. In order to see how this displacement affects the radiated frequencies, we apply the selection rules for a dipole transition, $\Delta m = 0$ or ± 1. In a transition in

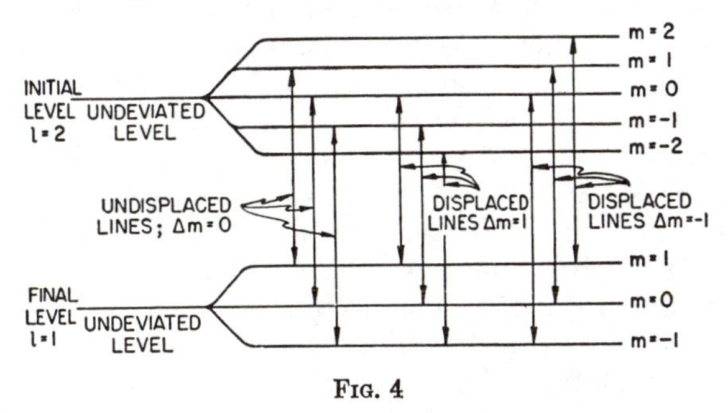

FIG. 4

which $\Delta m = 0$, the initial level is displaced just as much as the final level so that the radiated frequency is not changed. If $\Delta m = +1$, the final level is raised more than the initial level and the radiated angular frequency is reduced by

$$\Delta\omega = \frac{\Delta E}{\hbar} = -\frac{e\mathcal{H}}{2\mu c} \tag{100}$$

Conversely, when $\Delta m = -1$, the radiated angular frequency is increased by

$$\Delta\omega = \frac{e\mathcal{H}}{2\mu c}$$

Thus the spectral line will, in general, be split into three components, exactly as predicted by the classical theory.

The selection rules are illustrated in the transition scheme, shown in Fig. 4, for the case in which transitions occur from a level with $l = 2$ to a level with $l = 1$. (Note that $\Delta l = \pm 1$ for dipole transitions.)

If the radiation is viewed along the direction of the magnetic field, then only x and y can appear in the matrix element, so that the transition with $\Delta m = 0$ does not contribute to this line. The transition with $\Delta m = +1$ leads to right-hand circularly polarized light, while $\Delta m = -1$

yields left-hand circularly polarized light. As a result, only two lines appear along the direction of the magnetic field, each displaced away from the original line equally and in opposite directions, and each polarized circularly in opposite directions.

If the light is viewed normal to the magnetic field, say in the x direction, then it can be polarized either in the z or in the y direction. In a transition in which $\Delta m = 0$, we have already seen that the only matrix element which does not vanish is z_{mn}. This means that transitions with $\Delta m = 0$ lead to light polarized in the z direction. The net result is that the line is split into three parts; first an undeviated part polarized in the z direction, and second, two parts displaced respectively by $\pm e\mathcal{H}/2\mu c$ and polarized in the direction normal to z.

47. The Anomalous Zeeman Effect. We see that the quantum predictions for the Zeeman effect are exactly the same as those predicted classically. On the other hand, it is found that with most atoms, the observed Zeeman pattern is considerably more complex than is the one outlined above. Such atoms are said to exhibit the "anomalous Zeeman effect," as contrasted with the above pattern, which is called "the normal Zeeman effect." When the electron spin has been taken into account, however, the anomalous Zeeman patterns can be predicted exactly. Only atoms for which the total electron spin is zero will exhibit the simple or "normal" pattern described above.*

48. General Methods in Calculating Transition Probabilities. In this section we have calculated only transition probabilities which are produced by radiation. The same methods, however, can clearly be used with any perturbing potential, for example, one that would exist if we had a time-varying perturbing force of any origin whatever. The essential problem will always be to calculate the matrix element, V_{mn}. Selection rules are always obtained by finding those transitions in which V_{mn} vanishes.

Problem 7: Calculate the mean lifetime for emission of photons by a hydrogen atom in the state $n = 2$, $l = 1$, $m = 0$. Use eqs. (63) and (78). An atom in this state can, according to the selection rules, make transitions to the ground state ($n = 1$, $l = 0$, $m = 0$).

Problem 8: What kind of transition is needed for a hydrogen atom in the state $l = 2$, $n = 3$, to go to the ground state? Roughly, what will be the comparative rates of this transition and the one specified in the previous problem?

49. Effects of Electron Spin on Transition Probabilities. Let us now consider how the interaction between electron spin and the radiation field modifies the Hamiltonian function. In addition to the usual term [see Eq. (22)] involving the vector potential, we shall also have to include the term given in eq. (79), Chap. 17:

* For a treatment of the anomalous Zeeman effect, see Herzberg, *Atomic Spectra*, or White, *Introduction to Atomic Spectra*

$$W_{sp} = \frac{e\hbar}{2mc}\left(\boldsymbol{\sigma} \cdot \mathfrak{H} + \boldsymbol{\sigma} \cdot \frac{v}{c} \times \boldsymbol{\varepsilon}\right) \tag{101}$$

Because $|\boldsymbol{\varepsilon}| = |\mathfrak{H}|$ for an electromagnetic wave, and because $v/c \ll 1$ in a typical atom, we can neglect the second term in the right-hand side of eq. (101). For a plane wave $\boldsymbol{a} = \boldsymbol{a}_0\, e^{i(\boldsymbol{k}\cdot\boldsymbol{x}-\omega t)}$ we can write

$$\mathfrak{H} = \nabla \times \boldsymbol{a} = i(\boldsymbol{k} \times \boldsymbol{a}_0)e^{i(\boldsymbol{k}\cdot\boldsymbol{x}-\omega t)}$$

With these simplifications, we obtain†

$$W_{sp} \cong -\frac{e\hbar}{2mc} Rl[e^{i(\boldsymbol{k}\cdot\boldsymbol{x}_0-\omega t)}\, e^{i\boldsymbol{k}\cdot(\boldsymbol{x}-\boldsymbol{x}_0)}\boldsymbol{\sigma} \cdot (\boldsymbol{k} \times \boldsymbol{a})_0] \tag{102}$$

where $\boldsymbol{x}_0$ is the position of the center of the atom.

As was done in obtaining eq. (45), we expand the exponential, $e^{i\boldsymbol{k}\cdot(\boldsymbol{x}-\boldsymbol{x}_0)}$, retaining, however, the first power of $(\boldsymbol{x} - \boldsymbol{x}_0)$ this time, because the zeroth power, which does not involve $\boldsymbol{x}$, will yield vanishing matrix elements between any two orthogonal wave functions. Thus, we can leave out the first term in the expansion of $e^{i\boldsymbol{k}\cdot(\boldsymbol{x}-\boldsymbol{x}_0)}$ and obtain

$$W_{sp} \cong -\frac{e\hbar}{2mc} Rl[ie^{i(\boldsymbol{k}\cdot\boldsymbol{x}_0-\omega t)}\, \boldsymbol{k} \cdot (\boldsymbol{x} - \boldsymbol{x}_0)\boldsymbol{\sigma} \cdot (\boldsymbol{k} \times \boldsymbol{a}_0)] \tag{103}$$

The spin matrix element between two states will then be

$$(W_{sp})_{ab} = \int \psi_a^* W_{sp} \psi_b\, d\boldsymbol{x} \tag{104}$$

ψ_a and ψ_b contain both space and spin variables of the electron. In order to exhibit the above quantities more explicitly, we now express ψ_a and ψ_b in terms of the column vector representation (see Chap. 17, Sec. 5).

$$\psi_a = \begin{pmatrix} \psi_{a1} \\ \psi_{a2} \end{pmatrix} \qquad \psi_b = \begin{pmatrix} \psi_{b1} \\ \psi_{b2} \end{pmatrix} \tag{105}$$

$(\boldsymbol{\sigma} \cdot \mathfrak{H})$ is equal to

$$\begin{pmatrix} \mathfrak{H}_z & \mathfrak{H}_x - i\mathfrak{H}_y \\ \mathfrak{H}_x + i\mathfrak{H}_y & -\mathfrak{H}_z \end{pmatrix}$$

It is useful to estimate the order of magnitude of this matrix element. Now the integration of $(\boldsymbol{x} - \boldsymbol{x}_0)$ will yield terms of the same order of magnitude as those appearing in the dipole moment, z_{mn}, of eq. (47). The spin operator, $\boldsymbol{\sigma}$, contributes matrix elements of the order of unity. (This is because the nonvanishing matrix elements of σ_x, σ_y, σ_z all have an absolute value of unity.) The ratio of spin matrix elements to orbital matrix elements will then be of the order of $\hbar k^2/2m\omega$. Writing $\omega = ck$, we obtain for this ratio

$$\frac{\hbar k}{2mc} = \frac{h}{2\lambda mc} \tag{107}$$

† Rl means "the real part of."

For a typical case ($\lambda \cong 5 \times 10^{-5}$ cm) this ratio is about 10^{-6}. Since the transition probability varies as the square of the matrix element, we conclude that transitions arising from the spin terms are only 10^{-12} as fast as those arising from electric dipole terms. This means that the spin terms will ordinarily be very unimportant in causing transitions, unless the vector potential terms are highly forbidden.

50. Case c: V_{mn} Varies Slowly with the Time (Adiabatic Case). It often happens that a perturbation is turned on very slowly; for example, in the Zeeman effect the magnetic field is turned on in a time which is long compared with atomic periods. After the perturbation has been built up to its full value, it remains constant. Yet the theory developed thus far for time constant V_{mn} cannot be applied because, in deriving our results, we have assumed that the potential was turned on suddenly at the time $t = t_0$. Let us now see what happens when the perturbation is turned on gradually.

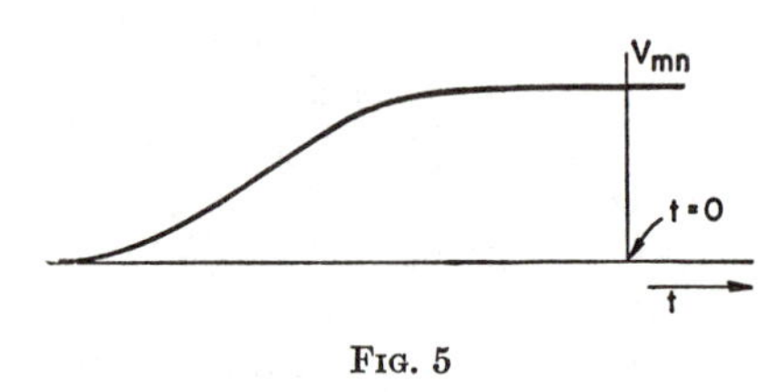

FIG. 5

A typical behavior of V_{mn} with time is shown in Fig. 5. As $t \to -\infty$, $V_{mn} \to 0$ asymptotically. For positive values of t, V_{mn} approaches a constant asymptotically. In between, the variation of V_{mn} with time is smooth and slow.

We must begin with eq. (9a) for C_m. Since $V_{mn} \to 0$ in an integrable way as $t \to -\infty$, we can, with negligible error, replace t_0 by $-\infty$. This gives

$$C_m = -\frac{i}{\hbar}\int_{-\infty}^{t} \lambda V_{ms}(t)e^{i(E_m{}^0-E_s{}^0)t/\hbar}\,dt$$

(C_m has to remain small for this to be valid.)

Let us integrate the above equation by parts, noting that

$$V_{ms}(-\infty) = 0$$

We obtain

$$C_m = -\frac{\lambda V_{ms}(t)e^{i(E_m{}^0-E_s{}^0)t/\hbar}}{(E_m^0 - E_s^0)} + \int_{-\infty}^{t} \frac{e^{i(E_m{}^0-E_s{}^0)t/\hbar}}{E_m^0 - E_s^0}\frac{d(\lambda V_{ms})}{dt}\,dt \qquad (108)$$

We now note that since V_{ms} approaches a constant for large positive t, $dV_{ms}/dt \to 0$. As a result, the limits of integration may, with negligible error, be taken as $\pm\infty$, if we wish to consider times greater than $t = 0$. This gives

$$C_m = -\frac{\lambda V_{ms}\,e^{i(E_m{}^0-E_s{}^0)t/\hbar}}{E_m^0 - E_s^0} + \int_{-\infty}^{\infty} \frac{e^{i(E_m{}^0-E_s{}^0)t/\hbar}}{E_m^0 - E_s^0}\frac{d}{dt}(\lambda V_{ms})\,dt \qquad (109)$$

The integral on the right is just proportional to the Fourier component of dV_{ms}/dt corresponding to the frequency $\omega_{ms} = (E_m^0 - E_s^0)/\hbar$. Now dV_{ms}/dt looks more or less as shown in Fig. 6, starting out at zero as $t \to -\infty$, increasing to a maximum, and then decreasing to zero at $t \to +\infty$. There will be some mean interval of time Δt during which dV_{ms}/dt is large. From our work on wave packets, we know that the Fourier components of dV_{ms}/dt will be large only in a region $\Delta\omega \sim 1/\Delta t$. The integral on the right-hand side of eq. (109) will therefore be negligible if

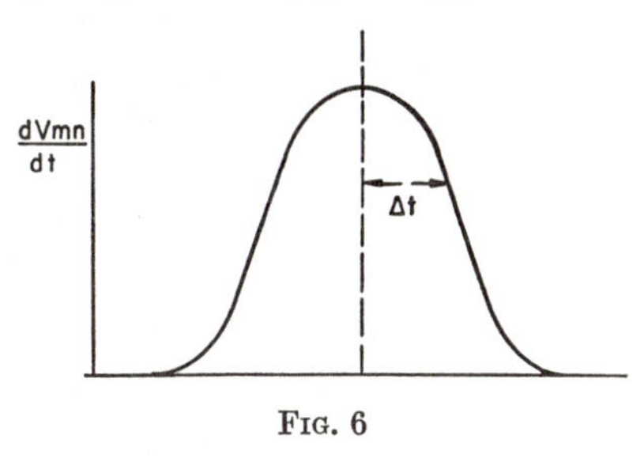

FIG. 6

$$\Delta\omega = \frac{(E_m^0 - E_s^0)}{\hbar} > \frac{1}{\Delta t} \qquad \text{or} \qquad E_m^0 - E_s^0 > \frac{\hbar}{\Delta t}$$

Problem 9: Suppose that $dV_{ms}/dt = \lambda \exp[-t^2/2(\Delta t)^2]$. Show that when $\Delta t \gg \hbar/E_m{}^0 - E_s{}^0$, the Fourier component in eq. (109) becomes negligible.

Thus, when $(E_m^0 - E_s^0)/\hbar > 1/\Delta t$, we can write

$$C_m \cong -\frac{\lambda V_{ms}}{E_m^0 - E_s^0} e^{i(E_m{}^0 - E_s{}^0)t/\hbar} \tag{109a}$$

We therefore conclude that if the potential is turned on infinitely slowly, only the term which oscillates with the angular frequency $(E_m^0 - E_s^0)/\hbar$ will appear in C_m. Comparing this with the result for the sudden turning on of the perturbation (eq. 13), we see that, in the latter case, there is an additional term, which does not oscillate with time.

The result is rather analogous to what happens to a harmonic oscillator of natural angular frequency ω_0 subjected to an external harmonically varying force of angular frequency ω. The equation of motion of such a system is

$$m(\ddot{x} + \omega_0^2 x) = Fe^{i\omega t}$$

The general solution is

$$x = Ae^{i\omega_0 t} + Be^{-i\omega_0 t} + \frac{F}{m} \frac{e^{i\omega t}}{(\omega_0^2 - \omega^2)}$$

One can satisfy any particular boundary conditions, for example,

$$x = \dot{x} = 0 \quad \text{at } t = 0$$

by proper choice of A and B. In general, A and B will not vanish, and therefore, there will be so-called "free oscillation" with the angular frequency ω_0. If it turns out, however, that A and B vanish, one has only "forced" oscillation with angular frequency ω equal to that of the forcing term.

How can such pure "forced" oscillations be excited? One way is to turn on the forcing term very slowly (or adiabatically, as one may say) in comparison with the period of an oscillation. If one increases the amplitude of the forcing term very slowly, then one can show that in the limit of an infinitely slow process of building F up to its final value, one obtains only the "forced" oscillations, and $A = B = 0$.

Problem 10: Prove the above statement.

In a similar way, one can regard the eqs. (109a) for C_m as determining the rate of oscillation of the C_m's. The term in eq. (108) involving $V_{ms}\, e^{i(E_m{}^0 - E_s{}^0)t/\hbar}$ acts as a "forcing term" tending to make C_m vibrate with the angular frequency $(E_m^0 - E_s^0)/\hbar$. If V_{ms} is built up very slowly (or adiabatically) from zero, then the C_m's respond by oscillating only with the impressed frequency. If V_{ms} changes appreciably, however, in a time comparable with $\hbar/(E_m^0 - E_s^0)$, then some "free" oscillation of C_m is produced. In this case, the frequency of free oscillation happens to be zero, so that this type of term is just a constant. The zero value can be seen by noting that when the forcing term in eq. (8) is absent, the equation becomes $i\hbar\dot{C}_m = 0$; thus, the natural frequency must, in this case, be regarded as zero.

Equation (108) is a general equation telling how the way in which the potential is turned on controls the value of C_m. If $(dV/dt)_{mn}$ has Fourier components corresponding to an angular frequency of $(E_m - E_n)/\hbar$, then the C_m will have a large constant term added to it.

One can even describe the sudden turning on of a perturbation with this equation. To do this, we say that V_{ms} is zero until $t = t_0$, after which it is a constant. Thus V_{ms} is a constant times the "step function," $S(t - t_0)$. [The step function $S(t - t_0)$ is zero for $t < t_0$ and unity for $t > t_0$.] Now, it can be shown that the derivative of a step function is a δ function.

Problem 11: Prove that $\dfrac{dS(t - t_0)}{dt} = \delta(t - t_0)$.

HINT: Consider the integral of a δ function.

Thus, our integral becomes

$$\frac{\lambda V_{ms}\, e^{i(E_m{}^0 - E_s{}^0)t_0/\hbar}}{E_m^0 - E_s^0}$$

and

$$C_m = \frac{\lambda V_{ms}}{E_m^0 - E_s^0}\left[e^{i(E_m{}^0 - E_s{}^0)t_0/\hbar} - e^{i(E_m{}^0 - E_s{}^0)t/\hbar}\right]$$

Comparison with eq. (13) shows that the two are the same.

51. Adiabatic Turning on of Potential Results in Perturbed Stationary State. Equation (109a) has two important consequences:

First, the complete wave function can be written [from eq. (4)]

$$\psi = C_s U_s e^{-iE_s^0 t/\hbar} + \sum_{n \neq s} C_n U_n e^{-iE_n t/\hbar} \tag{110}$$

Now, according to eq. (11), $C_s \cong e^{-i\int_{-\infty}^{t} \lambda V_{ss}\, dt/\hbar}$, but since V_{ss} approaches a constant for $t > 0$, we have, for $t > 0$,

$$\int_{-\infty}^{t} \lambda V_{ss} \frac{dt}{\hbar} = \int_{-\infty}^{0} \lambda V_{ss} \frac{dt}{\hbar} + \int_{0}^{t} \lambda V_{ss} \frac{dt}{\hbar} = \text{constant} + \frac{\lambda V_{ss} t}{\hbar}$$

thus $C_s = C_{s0}\, e^{-i\lambda V_{ss} t/\hbar}$ and $C_{s0} = e^{i\phi}$, where ϕ is a constant phase factor. The constant phase factor has no physical significance; it can be absorbed into the definition of U_s.

For the adiabatic case, we evaluate C_n from eq. (109a), obtaining

$$\psi = U_s\, e^{-i(E_s^0 + \lambda V_{ss})t/\hbar} + \sum_{n \neq s} \frac{\lambda V_{ns} U_n}{E_s^0 - E_n^0}\, e^{-i(E_n^0 + E_s^0 - E_n^0)t/\hbar}$$

Since the sum $n \neq s$ is proportional to λ, we can multiply it by $e^{-i\lambda V_{ss} t/\hbar}$, making an error at most of order λ^2. Thus, to first order, we obtain

$$\psi = \left[U_s + \lambda \sum_{n \neq s} \frac{V_{ns} U_n(x)}{E_s^0 - E_n^0} \right] e^{-i(E_s^0 + \lambda V_{ss})t/\hbar} = f(x) e^{-i(E_s^0 + \lambda V_{ss})t/\hbar} \tag{111a}$$

The second important result which we obtain from eq. (109a) is that

$$|C_m|^2 = \frac{\lambda^2 |V_{ms}|^2}{(E_m^0 - E_s^0)^2} \tag{111b}$$

These two results are of considerable interest. The first of them (eq. 111a) shows that the whole wave function oscillates (to first order in λ) with the angular frequency $(E_s^0 + \lambda V_{ss})/\hbar$. The system is therefore in a stationary state, and all probabilities remain constant with time. For example, the probability that the system can be found in the mth state is given by eq. (111b).

This result is valid only when the perturbation is turned on very slowly; if it had been turned on rapidly, ψ would not have taken the form $f(x)e^{-i(E_s^0 + \lambda V_{ss})t/\hbar}$; instead, various other frequencies of oscillation would have been present. The system would therefore not have been in a stationary state, and probabilities would have fluctuated with time, as shown in eq. (14), which describes the case in which the perturbation is turned on suddenly at $t = t_0$.

How can one picture the origin of such a stationary state? One can still use the picture developed in Sec. 9, where perturbations are regarded as creating a continual tendency for the system to make transitions to other eigenstates of the unperturbed energy, H_0. If the perturbation is turned on slowly, however, the system remains in a state in which transitions to other states are balanced by transitions back to the original state. As a result, the probability of finding an electron in some other state remains constant, except for the slow increase of probability which occurs while the perturbation is being turned on. If the perturbation had been turned on rapidly, there would have been an unbalanced, and therefore rapidly fluctuating, probability that the system had made a transition to some other state, which would have been analogous to the appearance of free oscillations in a suddenly excited harmonic oscillation. It should be pointed out again, however, that this fluctuation picture has only limited validity, because of the possibility of interference between the different functions $U_n(x)$. To take this effect into account, it is necessary to imagine that the system fluctuates simultaneously into all possible states, so that it covers all states simultaneously.*

One may ask why there is no contradiction with the law of conservation of energy, even though there is a constant probability that the particle can be found in the mth state, with an energy different from its original value of $E_m^0 - E_s^0$. The reason is that the energy is now the sum $(H_0 + \lambda V)$. There is a slight uncertainty in H_0 resulting from the presence of other eigenfunctions of H_0 in the wave function with small coefficients, C_n. But the total energy is just h times the frequency of oscillation of the wave function, and this is

$$E = E_s^0 + \lambda V_{ss} \tag{112}$$

The system has the approximate spatial wave function

$$f(x) = U_s(x) + \lambda \sum_{n \neq s} \frac{U_n(x) V_{ns}}{E_s^0 - E_n^0}$$

and the net energy, $H_0 + \lambda V$, has a definite value even though H_0 does not have a definite value. (Of course, all of this is accurate only to first order in λ.) The function $f(x)$ is therefore just the first approximation for the eigenfunction of the operator $H_0 + \lambda V$, corresponding to the first approximate eigenvalue, $E_s^0 + \lambda V_{ss}$.

Importance of Degeneracy. If any levels are degenerate, it is clear that no matter how slowly one turns on the perturbation, it will be impossible to satisfy the condition $t > \hbar/E_s^0 - E_n^0$. Thus, for degenerate levels, this treatment breaks down, not only because the perturbation theory is, as we have seen, not valid for an indefinite length of time, but

* As pointed out in Sec. 9, as long as definite phase relations exist between the $U_n(x)$, the system cannot correctly be regarded as being in a definite but unknown eigenstate of H_0. Instead, it should be regarded as something that is potentially capable of developing a definite value of H_0 in interaction with an apparatus that can be used to provide a measurement of H_0. The probability that the nth eigenstate will be obtained in such a measurement is equal to $C_n{}^2$.

also because the adiabatic condition cannot be applied. The treatment of the case of degeneracy will be discussed in Chap. 19.

52. Perturbation of Stationary-state Wave Functions. We have seen that if a perturbation is turned on very slowly, a perturbed stationary state will result. Although the method that we have used in solving for the wave function by means of time-dependent perturbation theory is perfectly valid, it is somewhat unwieldy. In this section, we shall treat the same problem by solving directly for the time-independent eigenfunctions of the Hamiltonian, $H_0 + \lambda V$, using perturbation theory. This method is better for systematically carrying through the perturbation theory for stationary states than is the time-dependent method, but the physical meaning of the results is less evident.

We begin by writing down Schrödinger's equation for the sth eigenfunction,

$$(H_0 + \lambda V)\psi_s = E_s\psi_s$$

The complete time-dependent wave function is

$$\psi = \psi_s\, e^{-iE_st/\hbar} \tag{113a}$$

Just as in the method of variation of constants, we express ψ as a series of eigenfunctions, $U_n(x)$, of the unperturbed Hamiltonian. Since ψ_s now represents a stationary state, the coefficients of the $U_n(x)$ will all be constants,

$$\psi_s = \sum_n C_{ns}U_n(x) \tag{113b}$$

Insertion of this value of ψ_s into the above equation yields

$$(H_0 + \lambda V)\sum_n C_{ns}U_n(x) = E_s\sum_n C_{ns}U_n(x)$$

With the aid of the relation, $H_0U_n(x) = E_n^0U_n(x)$, we obtain

$$\sum_n C_{ns}(E_s - E_n^0)U_n(x) = \sum_n \lambda V C_{ns}U_n(x)$$

Let us now multiply the above equation by $U_m^*(x)$, and integrate over all x. Because of normalization and orthogonality of the U_n, we find

$$C_{ms}(E_s - E_m^0) = \lambda\sum_n V_{mn}C_{ns} \tag{114}$$

where $V_{mn} = \int U_m^*(x)VU_n(x)\,dx$.

The above equation is analagous to eq. (6), except that the C_m's are constant here. We have, in general, an infinite number of linear equations in an infinite number of variables. These must be solved for the C_{ms} and for the allowed values of the energy, E_s. We shall solve them here with the assumption that λ is small, so that neither the wave functions nor the energy levels are changed very much from what they would be if λ were zero. If λ were zero, the eigenfunctions would be just U_s and the eigenvalues would be E_s^0. Thus, we would have

$$C_{ms} = \delta_{ms} = \begin{cases} 0 & m \neq s \\ 1 & m = s \end{cases}$$
$$E_s = E_s^0$$

We now suppose that the changes of C_{ms} and E_s can be expressed as a series of powers of λ:

$$\begin{aligned} C_{ms} &= \delta_{ms} + \lambda C_{ms}^{(1)} + \lambda^2 C_{ms}^{(2)} + \cdots \\ E_s &= E_s^0 + \lambda E_s^{(1)} + \lambda^2 E_s^{(2)} + \cdots \end{aligned} \tag{115}$$

We see that all the C_{ms} are of first order in λ except C_{ss}.

Inserting this equation into eq. (114) and collecting coefficients of equal powers of λ, we obtain

$$\begin{aligned} 0 = \delta_{ms}[E_s^0 - E_m^0] &+ \lambda[\delta_{ms}E_s^{(1)} + C_{ms}^{(1)}(E_s^0 - E_m^0) - V_{ms}] \\ &+ \lambda^2\left[\delta_{ms}E_s^{(2)} + C_{ms}^{(1)}E_s^{(1)} + C_{ms}^{(2)}(E_s^0 - E_m^0) - \sum_n V_{mn}C_{ns}^{(1)}\right] \end{aligned} \tag{116}$$

The coefficients of each power of λ must separately be zero. It is clear that the coefficient of (λ^0) is zero, since δ_{ms} is zero except when $m = s$, and, in the latter case, $E_s^0 - E_m^0$ is zero. This merely reflects the fact that the $U_s(x)$ are solutions when $\lambda = 0$. The coefficients of λ yield the following relations:

First-order Theory:

$$C_{ms}^{(1)} = \frac{V_{ms}}{E_s^0 - E_m^0} \qquad m \neq s \tag{117}$$

$$E_s^{(1)} = V_{ss} \tag{118}$$

We note that the first-order wave function is then

$$\psi_s = U_s(x) + \lambda C_{ss}^{(1)}U_s(x) + \sum_{n \neq s} \frac{\lambda U_n(x)V_{ns}}{E_s^0 - E_n^0} \tag{119}$$

[see eq. (113a) for time variation of ψ]. This is exactly what we derived from the assumption that the potential was turned on slowly. [See eq. (111).] The first-order energy is

$$E_s = E_s^0 + \lambda V_{ss}$$

Again, this is exactly what we obtained from the time-dependent method eq. (112).

Let us note that the first-order equations do not define the coefficient $C_{ss}^{(1)}$, since the relation obtained by setting $m = s$ (eq. 118) gave us not $C_{ss}^{(1)}$, but $E_s^{(1)}$ instead. This means that $C_{ss}^{(1)}$ is arbitrary, as far as the present approximations are concerned. One can easily show that the appearance of $C_{ss}^{(1)}$ is equivalent to multiplying the unperturbed wave function by a constant. In fact, in each order of approximation, C_{ss} is defined such that the entire wave function is normalized. We have

$$1 = \int \psi_s^* \psi_s \, dx \cong \int dx \left[(1 + \lambda C_{ss}^{(1)*}) U_s^*(x) + \lambda \sum_{n \neq s} C_{ns}^{(1)*} U_n^*(x) \right] \left[(1 + \lambda C_{ss}^{(1)}) U_s(x) + \lambda \sum_{n \neq s} C_{ns}^{(1)} U_n(x) \right] = 1 + \lambda (C_{ss}^{(1)*} + C_{ss}^{(1)}) + \text{terms in } \lambda^2 \quad (120)$$

Thus, we obtain $C_{ss}^{(1)*} + C_{ss}^{(1)} = 0$, so that to order λ, $C_{ss}^{(1)}$ must be a pure imaginary, which we can therefore denote by $i\xi$. Then

$$C_{ss} = 1 + i\lambda\xi + \text{terms of order } \lambda^2$$

But we can also write $e^{i\lambda\xi} = 1 + i\lambda\xi + \text{terms of order } \lambda^2$, and, in eq. (113b), we obtain

$$\psi_s = e^{i\lambda\xi} U_s(x) + \sum_{n \neq s} C_{ns} U_n(x)$$

This means that the choice of $C_{ss}^{(1)}$ is equivalent (to order λ) to multiplying the unperturbed wave function, $U_s(x)$, by an arbitrary phase factor. This procedure clearly can produce no physically observable changes.

Second-order Theory. When $m \neq s$, the vanishing of the coefficient of λ^2 yields the following relation:†

$$C_{ms}^{(2)} = \frac{1}{E_s^0 - E_m^0} \left(\sum_n V_{mn} C_{ns}^{(1)} - C_{ms}^{(1)} E_s^{(1)} \right)$$

Using eqs. (117) and (118), and $C_{ss}^{(1)} = 0$, we obtain

$$C_{ms}^{(2)} = \frac{1}{E_m^0 - E_s^0} \left(\sum_{n \neq s} \frac{V_{mn} V_{ns}}{E_n^0 - E_s^0} - \frac{V_{ms} V_{ss}}{E_m^0 - E_s^0} \right) \quad (121)$$

When $m = s$, we obtain from eq. (116)

$$E_s^{(2)} = \sum_{n \neq s} V_{sn} C_{ns}^{(1)} = \sum_{n \neq s} \frac{V_{sn} V_{ns}}{E_s^0 - E_n^0} = \sum_{n \neq s} \frac{|V_{ns}|^2}{E_s^0 - E_n^0} \quad (122)$$

(We assume $V_{ns} = V_{sn}^*$, which follows from the Hermiticity of V.)

As in the case of $C_{ss}^{(1)}$, we find that our equations do not define $C_{ss}^{(2)}$. We shall, as before, define it from the requirement that the wave function be normalized (this time to second order). This means that

† We could have carried the time-dependent adiabatic perturbation theory of Sec. 51 up to second order and obtained the same results, but the procedure would have been very unwieldy.

$$1 = \int dx \left[U_s^*(1 + \lambda^2 C_{ss}^{(2)*}) + \sum_{n \neq s} (\lambda C_{ns}^{(1)*} + \lambda^2 C_{ns}^{(2)*}) U_n^* \right]$$
$$\left[U_s(1 + \lambda^2 C_{ss}^{(2)}) + \sum_{n \neq s} (\lambda C_{ns}^{(1)} + \lambda^2 C_{ns}^{(2)}) U_n \right]$$

Using normalization and orthogonality of the U's and retaining only terms up to order λ^2, we get

$$1 + \lambda^2(C_{ss}^{(2)*} + C_{ss}^{(2)}) + \lambda^2 \sum_{n \neq s} |C_{ns}^{(1)}|^2 = 1$$

or

$$C_{ss}^{(2)*} + C_{ss}^{(2)} = -\sum_{n \neq s} \frac{|V_{ns}|^2}{(E_s^0 - E_n^0)^2} \tag{122a}$$

One possible choice of $C_{ss}^{(2)}$ is

$$C_{ss}^{(2)} = -\frac{1}{2} \sum_{n \neq s} \frac{|V_{ns}|^2}{(E_s^0 - E_n^0)^2} \tag{122b}$$

One could add an arbitrary imaginary constant to $C_{ss}^{(2)}$, but this would merely correspond to a shift in phase of the unperturbed wave function, which is of no physical significance. We shall therefore retain the above value for $C_{ss}^{(2)}$.

Higher approximations can be carried out in a systematic way by evaluating coefficients of higher powers of λ in eq. (116).

53. Interpretation of Second-order Formulas for Energy. We have seen that the first-order value of the energy, $(E_s^0 + \lambda V_{ss})$, is just the mean value of the Hamiltonian, $H_0 + \lambda V$, taken with the zeroth-order wave function, $U_s(x)$. We shall now show that the second-order value of the energy is equal to the mean value of the Hamiltonian, taken with the normalized first-order wave functions. This is part of a general rule, which is, that the $(n + 1)$th approximation to the energy can be obtained by averaging the Hamiltonian with the nth normalized approximate wave functions.

The first step is to write out the normalized first-order wave function, which we obtain from eq. (119). Since eq. (119) is normalized only to first order, we must multiply the entire wave function by a suitable factor, which we denote by A. Thus we obtain

$$\psi_s = A \left[U_s(x) + \lambda \sum_{n \neq s} \frac{V_{ns} U_n(x)}{E_s^0 - E_n^0} \right] \tag{123}$$

where A is the normalizing coefficient, defined by the relation

$$\int |\psi_s|^2 \, dx = 1 = |A|^2 \int dx \left[U_s^*(x) + \lambda \sum_{n \neq s} \frac{V_{ns}^* U_n^*(x)}{E_s^0 - E_n^0} \right]$$
$$\left[U_s(x) + \lambda \sum_{n \neq s} \frac{V_{ns} U_n(x)}{E_s^0 - E_n^0} \right] \tag{123a}$$

One can simplify the above integral by taking advantage of the normalization and orthogonality of the $U_n(x)$. One obtains

$$|A|^2 = \frac{1}{1 + \lambda^2 \sum_{n \neq s} \frac{|V_{ns}|^2}{(E_s^0 - E_n^0)^2}} \tag{123b}$$

We note that $|A|^2$ differs from unity only by a second-order term. When $|A|^2$ is expanded as a series in λ^2, one obtains (to second-order terms)

$$|A|^2 \cong 1 - \lambda^2 \sum_{n \neq s} \frac{|V_{ns}|^2}{(E_s^0 - E_n^0)^2} \tag{123c}$$

The next step is to obtain the mean value of the Hamiltonian. This is

$$\bar{H} = \int \psi_s^*(H_0 + \lambda V)\psi_s\, dx = \int \psi_s^* E_s^0 \psi_s\, dx + \int \psi_s^*(H_0 - E_s^0)\psi_s\, dx + \lambda \int \psi_s^* V \psi_s\, dx \tag{124}$$

Because ψ_s is normalized, the first integral on the right side of the above equation is just E_s^0. The second integral is [using $H_0 U_n(x) = E_n^0 U_n(x)$]

$$|A|^2 \int \left[U_s^* + \lambda \sum_{n \neq s} \frac{V_{ns}^* U_n^*}{E_s^0 - E_n^0}\right] \left[(E_s^0 - E_s^0)U_s + \lambda \sum_{n \neq s} \frac{V_{ns}(E_n^0 - E_s^0)}{E_s^0 - E_n^0} U_n\right] dx \tag{125}$$

Because of the normalization and orthogonality of the $U_n(x)$, the above integral becomes

$$-|A|^2\lambda^2 \sum_{n \neq s} \frac{|V_{ns}|^2}{E_s^0 - E_n^0} \cong -\lambda^2 \sum_{n \neq s} \frac{|V_{ns}|^2}{E_s^0 - E_n^0} \tag{125a}$$

(neglecting terms of order higher than λ^2). The third integral on the right is

$$\lambda \int \psi_s^* V \psi_s\, dx$$
$$= |A|^2 \lambda \int dx \left(U_s^* + \lambda \sum_{n \neq s} \frac{V_{ns}^* U_n^*}{E_s^0 - E_n^0}\right) V(x) \left(U_s + \lambda \sum_{n \neq s} \frac{V_{ns} U_n}{E_s^0 - E_n^0}\right)$$

With the neglect of terms of third order or higher, the above becomes

$$|A|^2 \left\{\lambda \int U_s^* V U_s\, dx + \lambda^2 \int \sum_{n \neq s} \frac{1}{E_s^0 - E_n^0} [V_{ns}^* U_n^*(x) V(x) U_s(x) + U_s^*(x) V(x) U_n(x) V_{ns}]\right\}$$
$$= \left[\lambda V_{ss} + 2\lambda^2 \sum_{n \neq s} \frac{|V_{sn}|^2}{E_s^0 - E_n^0}\right] |A|^2 = \lambda V_{ss} + 2\lambda^2 \sum_{n \neq s} \frac{|V_{sn}|^2}{E_s^0 - E_n^0}$$

Note that $A = 1$ to second order. The combined result for all three integrals is

$$\bar{H} = E^0 + \lambda V_{ss} + \lambda^2 \sum_{n \neq s} \frac{|V_{sn}|^2}{E_s^0 - E_n^0}$$

But the above is just the same as the second approximation to the energy (eq. 122). This proves our theorem, up to the second approximation. A similar proof may be given in the higher approximations.

54. Application of Perturbation Theory. We shall now treat a few applications of steady-state perturbation theory to the calculation of the energy levels of atoms. In Sec. 51 we showed that this theory applies when a perturbation is turned on very slowly. The same theory applies also, however, whenever any system has been standing around long enough to come to a steady state. In other words, the energy levels of a system that has reached a stationary state must be the same as those obtained by turning on a perturbation very slowly. The result is analogous to a similar one in thermodynamics, viz., that the state of thermodynamic equilibrium reached by any system after a long time is the same as that obtained in a quasi-static process.

(1) *First Approximation to Energy in Atoms Other than Hydrogen.* The first application that we shall make is to the calculation of the displacement of energy levels in atoms other than hydrogen. Let us recall that in a Coulomb force field, the energy levels possess a special degeneracy, namely that levels of the same n and different l quantum numbers have the same energy.* Now, the outermost electron of the alkali metals moves in a force field which is the same as that of hydrogen as long as the electron remains outside the inner shells of the electrons; these shells screen the outer electron from most of the nuclear charge. But if the electron enters the inner shells, it is no longer screened so effectively, so that the potential is deeper than it would be if it were a pure Coulomb force. The behavior of the potential is illustrated in Fig. 7. For the lighter alkali atoms, the difference between the actual potential and the Coulomb potential may be regarded as a small perturbation. As we go to atoms of higher atomic number, this difference becomes so large that it can no longer be treated as a small correction.

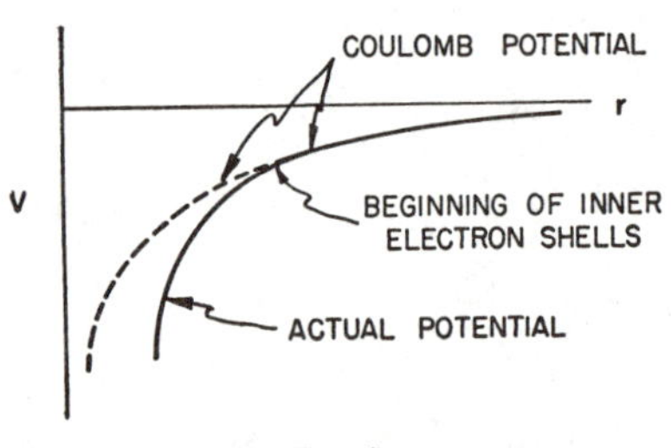

FIG. 7

For the alkali atoms, then, one can get some idea of the change in energy levels by writing $V = -\frac{e^2}{r} - \delta V$, where $-\delta V$ is the correction

* Chap. 15, Sec. 13.

to the Coulomb potential. Note that this correction is taken to be negative because the effect of incomplete screening in the inner shells is always to increase the force of binding of the electron to the nucleus. According to eq. (118), the first-order correction to the sth energy level will be

$$\lambda V_{ss} = -\int U_s^* \,\delta V U_s \, dx$$

where $U_s(x)$ is the sth eigenfunction of the unperturbed Hamiltonian.

We note that in order that $E_s^{(1)}$ be large, it is necessary that $U_s(x)$ be large in the same place that δV is large, namely, at small radii. Now, in Chap. 15, Sec. 8, it was shown that wave functions of the same n and different l differ in that the higher l becomes, the smaller will be the wave function near the origin. This effect is, as we have seen, produced by the centrifugal potential, which keeps the electron away from the nucleus when l is large. We therefore conclude that the above integral will be largest for the lowest value of l. The s state is thus depressed the most, the p state the next, and so on. This means that levels of the same n and different l are split by an amount that increases with increasing deviation from a Coulomb force, or, in other words, with increasing atomic number. (In classical theory, the levels of smallest l correspond to the most penetrating orbits.)

(2) *Stark Effect in Atoms Other than Hydrogen.* An application of the second-order calculation of the energy occurs in the Stark effect of atoms other than hydrogen. (Hydrogen must be given a special treatment because of the degeneracy of levels of the same n and different l.) The Stark effect consists of a shift of energy levels in an external electric field. Suppose that this field is taken to be in the z direction. Then the perturbing potential on an electron is

$$\lambda V = e\mathcal{E}z$$

where $\mathcal{E}$ is the electric field strength.

It is easily shown that the first-order correction to the energy vanishes for any eigenstate of the unperturbed energy. To show this, we write

$$\lambda V_{ss} = e\mathcal{E}\int U_s^* z U_s \, dx \tag{126}$$

Now $U_s^* U_s$ will always be an even function of z, for any spherical harmonic, $Y_l^m(\vartheta, \varphi)$. To prove this, we write

$$U_s = f_l(r) P_l^m(\cos\vartheta) e^{im\varphi}$$

and

$$|U_s|^2 = |f_l(r)|^2 [P_l^m(\zeta)]^2$$

Now, from the definition of P_l^m in eq. (60), Chap. 14, we see that it is either an even or odd function of $\zeta = \cos\vartheta = z/r$, according to whether $l - m$ is even or odd. Thus $[P_l^m(\zeta)]^2$ is always an even function of z.

This means that the integral in eq. (126) vanishes, since the integrand is odd because of the multiplication of z by an even function.

We must now calculate the second-order contribution to the energy. To do this, we use eq. (122). We must first calculate V_{ns}. We note from eq. (70) that Z_{ns} vanishes except between states for which $m' = m$ and $\Delta l = \pm 1$. As a result, only these states can contribute to the energy, and we need sum only over these states. If l and m are the quantum numbers of the state in question, we obtain

$$E_s^{(2)} = e^2 \mathcal{E}^2 \sum_{\substack{l' = l \pm 1 \\ m' = m \\ n'}} \frac{|z|^2_{n',l',m';n,l,m}}{(E^0_{l,m,n} - E^0_{l',m',n'})} \tag{127}$$

This equation has several interesting consequences:

(a) The shift in energy is proportional to $\mathcal{E}^2$; this effect is therefore called the "quadratic" Stark effect, to distinguish it from the much larger shift in hydrogen, which, as we shall see in Chap. 19, is proportional to $\mathcal{E}$ because of the degeneracy.

(b) The closer are the levels, $E^0_{s,l}$, the larger will be the energy shift. Thus, in the lighter alkali atoms, for which the degeneracy has hardly been removed, a rather large quadratic Stark effect is to be expected.

(c) $E_s^{(2)}$ is proportional to $|z|^2_{n,l';s,l}$

where
$$z_{n,l';s,l} = \int U^*_{n,l'} z U_{s,l}\, dx$$

The above integral is roughly proportional to the size of the region in which $U_{s,l}$ and $U_{n,l'}$ are large, so that the quadratic Stark effect increases roughly as the square of the size of the orbit. Thus, it tends to be larger in orbits of large n than in orbits of small n. This dependence on n means, in general, that the two levels involved in a transition will have different shifts, and that the spectral line will therefore be displaced. It is this displacement which offers a means of observing the quadratic Stark effect.

(3) *Polarizability of Atoms.* When an atom is placed in an external electral field, the classical orbit is disturbed. If the field is turned on rapidly in comparison to atomic periods, this disturbance will cause the orbit to wobble and precess in a way that is, in general, very complicated.† If, however, it is turned on adiabatically, the orbit will be modified, but will retain a steady shape. The main effect will be a displacement of the orbit as a whole in the direction of the electric field. This displacement is resisted by the electric field of the nucleus, which tends to pull the orbit back to its original position, centering on the nucleus. To a first approximation, the restoring force is proportional to the displacement of the orbit. The polarizability is defined as the mean displacement of the orbit per

† See Chap. 2, Sec. 14.

unit electric field. Thus we write for the polarizability

$$P = \frac{d}{\mathcal{E}} \tag{128}$$

where d is the mean displacement of the orbit in the direction of the electric field $\mathcal{E}$.

The displacement of the orbit in an electric field is illustrated in Fig. 8.

In quantum theory, one must consider the mean value of the displacement of the orbit also, but this time the mean value is given by

$$d = \int \psi^* z \psi \, dx = \bar{z} \tag{129}$$

where z is the value of the coordinate in the direction of the electric field. We have already seen that $\bar{z}$ vanishes in the absence of an electric field (i.e., for an eigenfunction of the unperturbed energy). Hence, we obtain the reasonable result that an isolated atom is unpolarized. Our next problem is to find the value of d when an electric field is present. To do this, we must use the perturbed wave functions given by eq. (119). We obtain

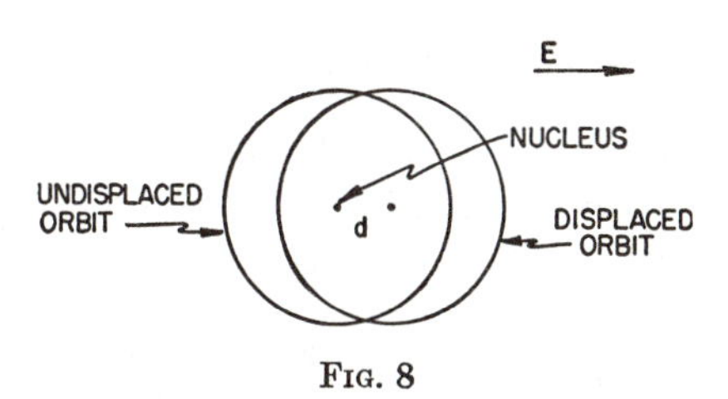

Fig. 8

$$d_s = \int dx \left[U_s^* + \lambda \sum_{n \neq s} \frac{V_{ns}^* U_n^*}{E_s^0 - E_n^0} \right] z \left[U_s + \lambda \sum_{n \neq s} \frac{V_{ns} U_n}{E_s^0 - E_n^0} \right] \tag{130a}$$

Noting that $\int U_s^* z U_s \, dx = 0$, and retaining only first order terms, we get

$$d_s = \lambda \sum_{n \neq s} \frac{1}{E_s^0 - E_n^0} (V_{ns}^* z_{ns} + V_{ns} z_{sn}) \tag{130b}$$

Now, the perturbing potential is $\lambda V = e\mathcal{E}z$. Thus, the above becomes

$$d_s = 2e\mathcal{E} \sum_{n \neq s} \frac{|z_{sn}|^2}{E_s^0 - E_n^0} \tag{130c}$$

and the polarizability is

$$P = \frac{d_s}{\mathcal{E}} = 2e \sum_{n \neq s} \frac{|z_{sn}|^2}{E_s^0 - E_n^0} \tag{130d}$$

We note that the polarizability is very closely related to the shift in energy levels [eq. (122)]. In fact, we obtain

$$E_s^{(2)} = \frac{e\mathcal{E}^2 P}{2} \tag{131}$$

This is a well-known result, which is true for any system for which the polarization is proportional to the electric field.

CHAPTER 19

Degenerate Perturbations

1. Introduction. It is clear that if any energy levels are degenerate, the perturbation theory that we have developed thus far cannot be applied to perturbations that last a long time. As we have seen, non-reversing transitions occur between degenerate energy levels, so that eventually the different degenerate eigenstates are all mixed up with each other, and the assumption that the wave function is close to its initial form breaks down. Similarly, in the stationary-state method, the energy difference in the denominator of eq. (117), Chap. 18 vanishes, and again we conclude that the effect of the perturbation may be large, even though λ is small.

The problem of degeneracy is of considerable importance, because it arises in so many different systems. A brief summary of some of the common kinds of degeneracy that we have met thus far is given below:

(1) For a free particle, the energy depends only on the absolute magnitude of the momentum ($E = p^2/2m$) and not on the direction. This degeneracy is removed (at least, in part) whenever a potential is present.

(2) In a spherically symmetrical potential, the energy is degenerate with regard to changes of the z component (or any other component) of the angular momentum, provided that the total angular momentum $\sqrt{L^2}$ is kept constant. This degeneracy is removed whenever the Hamiltonian is made to depend on the angle, e.g., in the presence of an external electric field, another atom, or a magnetic field.

(3) In a Coulomb field, the energy is also degenerate with regard to changes of the total angular momentum quantum number l when the principal quantum number n is kept fixed. This degeneracy is removed by an external electric field (Stark effect) or by a modification of the spherical potential away from a Coulomb form. (See Chap. 18, Sec. 54.)

(4) In a three-dimensional harmonic oscillator, the energy will be degenerate if the frequencies of oscillation in the directions of the three principal axes are not linearly independent. (See Chap. 15, Sec. 16.)

How can we deal with the problem of degeneracy? The first step is to neglect all transitions to nondegenerate states, and to solve the resulting equations exactly, taking into account only the transitions between states of the same energy. This is called the "zeroth-order" solution.

The next step will be to take these "zeroth-order" solutions and to apply the usual perturbation theory to them instead of to the original eigenfunctions. The justification for this procedure is that transitions to levels of energy different from the original energy will produce, as we have seen, comparatively small changes in the wave function. It is therefore reasonable to solve first for the large effects resulting from degenerate perturbations and then later to include the comparatively small effects of nondegenerate perturbations.

2. Example: Doubly Degenerate Level. To illustrate the method, let us suppose that there are two degenerate energy levels for which the unperturbed eigenfunctions are U_1 and U_2, and for which the common energy is E_0. We now go back to eq. (114), Chap. 18, and consider only those terms involving U_1 and U_2, temporarily neglecting all other terms. We obtain the following equations:

$$\left.\begin{aligned} C_1(E_s - E_0) &= \lambda(V_{1,1}C_1 + V_{1,2}C_2) \\ \text{or} \quad C_1(E_s - E_0 - \lambda V_{1,1}) &= \lambda C_2 V_{1,2} \end{aligned}\right\} \tag{1a}$$

$$\left.\begin{aligned} C_2(E_s - E_0) &= \lambda(V_{2,1}C_1 + V_{2,2}C_2) \\ \text{or} \quad C_2(E_s - E_0 - \lambda V_{2,2}) &= \lambda C_1 V_{2,1} \end{aligned}\right\} \tag{1b}$$

These constitute two homogeneous equations for two unknowns. In order that there exist a solution, the determinant of the coefficients of the C's must vanish. Noting that $V_{2,1} = V_{1,2}^*$, we then obtain

$$(E_s - E_0 - \lambda V_{1,1})(E_s - E_0 - \lambda V_{2,2}) = \lambda^2 |V_{1,2}|^2 \tag{2}$$

The above is a quadratic equation for E_s; there are two solutions. One must, of course, choose only one of these at a time. In order to simplify the discussion, let us suppose that $V_{1,1}$ and $V_{2,2}$ are equal. No essential generality is lost as a result of this simplification. The result is

$$E_s - E_0 - \lambda V_{1,1} = \pm\lambda\sqrt{|V_{1,2}|^2} = \pm\lambda|V_{1,2}|$$

where

$$V_{1,2} = |V_{1,2}|e^{-i\phi} \tag{3}$$

From eqs. (1a) and (1b) we can write

$$\frac{C_2}{C_1} = \pm e^{i\phi} \tag{4}$$

The zeroth approximate wave functions will be given by

$$v_\pm = \frac{1}{\sqrt{2}}[u_1(x) \pm e^{i\phi}u_2(x)]; \qquad \left(\frac{1}{\sqrt{2}} \text{ is a normalizing factor}\right) \tag{5}$$

Problem 1: Prove that the above functions are normalized if u_1 and u_2 are normalized and orthogonal.

The approximate energies associated with the above functions are, from eq. (3),

$$E_\pm = E_0 + \lambda V_{1\,1} \pm \lambda|V_{1,2}| \tag{6}$$

Problem 2: Solve for the values of E and v when $V_{1,1}$ and $V_{2,2}$ are not equal.

It usually turns out that $\phi = 0$; hence, unless otherwise specified, we shall use $\phi = 0$ in subsequent work.

3. Interpretation of Results. The first important point is that as a result of the long-time action of the perturbation, the approximate wave functions undergo a large change. In fact, the new stationary states have an equal mixture of u_1 and u_2; so that there is an equal probability of finding the system in either of the unperturbed eigenstates.

The second point is that the two different stationary states will have different energies (unless $V_{1,2}$ vanishes in which case a special treatment is required). *The effect of a perturbing potential will therefore usually be to remove the degeneracy.*

4. Important Properties of Approximate Solutions. There are two important properties of the approximate solutions.

(1) The solutions are orthogonal. To prove it for this case, we can write

$$\int v_+^*(x)v_-(x)\,dx = \tfrac{1}{2}\int dx[u_1^*(x) + u_2^*(x)][u_1(x) - u_2(x)]$$

Using orthogonality and normalization of u_1 and u_2, we obtain

$$\tfrac{1}{2}\int(|u_1(x)|^2 - |u_2(x)|^2)\,dx = 0$$

(2) The matrix elements of λV between v_+ and v_- vanish. To prove this, we write

$$V_{+,-} = \tfrac{1}{2}\int dx(u_1^* + u_2^*)V(u_1 - u_2) = \tfrac{1}{2}(V_{1,1} - V_{2,2} - V_{1,2} + V_{2,1})$$

In our case, we assumed that $V_{1,1} = V_{2,2}$. Also, $V_{1,2} = V_{2,1}^*$, but since we have assumed $\phi = 0$, $V_{1,2}$ is a real number and $V_{1,2} = V_{2,1}$. We therefore conclude that $V_{+,-} = 0$.

This result is also obtained in the more general case, when $V_{1,1}$ and $V_{2,2}$ are not equal and also when ϕ is not zero.

Problem 3: Prove the above statement.

5. Higher Approximations. The higher approximations can now be carried out directly. Instead of expanding an arbitrary wave function as a series of the $u_n(x)$, one uses the $v_\pm$ functions which are obtained by solving eqs. (1a) and (1b) for each set of degenerate levels. Because the v's are normal and orthogonal, the same treatment goes through as for the u's. But now the matrix elements between v's corresponding to the same unperturbed energy vanish. Thus, when we solve for the first approximate wave functions [eq. (117), Chap. 18], only transitions to nondegenerate levels will occur, and the perturbation theory will thereafter be valid. Furthermore, in the higher approximations, the removal of the degeneracy in the zeroth approximation will mean that no more nonreversing (energy-conserving) transitions can occur; this guarantees that a small perturbation will produce a correspondingly small change in the wave function for arbitrarily long times, and thus justifies the use of perturbation theory.†

6. More Than Two Degenerate Levels. If there are more than two degenerate levels, we proceed in the same way. In eq. (114), Chap. 18, we begin by considering only the C's corresponding to a given set of degenerate levels. Let the C's be denoted by C_i and let E_0 be the common unperturbed energy. The equations become

$$C_i(E_s - E_0) = \lambda \sum_{j=1}^{N} V_{ij}C_j \quad \text{or} \quad \sum_{j=1}^{N} [\lambda V_{ij} - \delta_{ij}(E_s - E_0)]C_j = 0 \tag{7}$$

(N is the total number of degenerate levels). The above equations are similar to the set in eq. (114), Chap. 18, except that here we consider only a finite number of equations and a finite number of unknowns. The condition for a solution of these equations is

$$\text{Determinant } |\lambda V_{ij} - \delta_{ij}(E_s - E_0)| = 0 \tag{8}$$

The above is an Nth order equation; so that there are, in general, N roots. These roots correspond to the N possible zeroth order energies. Each root leads to a different solution. Thus, there will be N solutions. The sth solution can be written as follows:

$$v_s = \sum_j C_{js}v_j(x) \tag{9}$$

We shall now show that the matrix of λV between any two of the v_s vanishes, i.e.,

$$\lambda \int v_s^* V v_r \, dx = 0 \tag{10a}$$

To evaluate the above integral, we insert the values of v_s and v_r given by eq. (9). The result is

$$\lambda \int \sum_m \sum_n C_{ms}^* u_m^* V C_{nr} u_n \, dx = \lambda \sum_m \sum_n C_{ms}^* V_{mn} C_{nr} \tag{10b}$$

We now use eq. (7), which says that for the rth solution,

$$C_{mr}(E_r - E_0) = \lambda \sum_{n=1}^{N} V_{mn}C_{nr}$$

The integral in eq. (10a) then becomes

$$(E_r - E_0) \sum_m C_{ms}^* C_{mr} = \lambda \int v_s^* V v_r \, dx \tag{12}$$

† This is because the energy denominators, $E_n{}^0 - E_s{}^0$, in eqs. 112 and 119, Chap. 18 will no longer vanish if we use for $E_n{}^0$ and $E_s{}^0$ the values obtained by removing the degeneracy.

But in equation (10b) we could have summed over m first instead of n, using the relation $V_{mn} = V^*_{nm}$. We would then have

$$\sum_m C^*_{ms} V_{mn} = \sum_m C^*_{ms} V^*_{nm} = \frac{(E_s - E_0)}{\lambda} C^*_{ns}$$

Equating the two sums, we obtain

$$(E_s - E_r) \sum_n C^*_{nr} C_{ns} = 0$$

Now, in general $E_s \neq E_r$, that is, the degeneracy has been removed; if this is the case we conclude that

$$\sum_n C^*_{nr} C_{ns} = 0 \tag{13}$$

and from eqs. (10) and (12)

$$\lambda \int v^*_r V v_s \, dx = 0 \tag{14}$$

This is what we wished to prove.

Orthogonality of the v_i. We also wish to show that the different v_i are orthogonal. Let us consider the integral

$$\int v^*_s v_r \, dx = \int \sum_{m,n} C^*_{ms} C_{nr} u^*_m u_n \, dx = \sum_m C^*_{ms} C_{mr}$$

(By virtue of orthogonality and normalization of the u_m.)

But by eq. (13), the above is zero if $r \neq s$. Thus, we conclude that the v's are orthogonal.

7. General Solution to Higher Orders. Because the v's are linear combinations of the u's and because there are just as many independent v's as there are u's, one can expand an arbitrary function in terms of the v's. Since the v's are also orthogonal, the procedure of expansion is the same as for the u's. Finally, because there are no matrix elements of V between the v_i belonging to the same unperturbed energy, the perturbation theory can be used in the same manner as that described for the case where only two degenerate levels are present.

8. Time-dependent Solution for Special Case of Two Degenerate Levels. It is very instructive to show how the wave function changes with time when there is degeneracy. We shall consider here the special case of two degenerate levels. (Because the transitions to nondegenerate levels do not go very far, for a small perturbation, it will be an adequate approximation to consider only the transitions between the degenerate levels and back again.)

Let us also consider the case given in eqs. (1a) and (1b), putting

$$V_{1,1} = V_{2,2} = 0$$

Let us also suppose, for simplicity, that $\phi = 0$. The two eigenfunctions are then

$$v_1 = \frac{1}{\sqrt{2}}(u_1 + u_2) \qquad \text{and} \qquad v_2 = \frac{1}{\sqrt{2}}(u_1 - u_2) \tag{14a}$$

Since v_1 and v_2 are approximate stationary states, their time variation can be found by multiplying each respectively by $e^{-i(E_0+\lambda V_{1,2})t/\hbar}$ and $e^{-i(E_0-\lambda V_{1,2})t/\hbar}$. The time-dependent wave functions then become

$$\psi_1(x, t) = \frac{1}{\sqrt{2}}(u_1 + u_2)e^{-i(E_0+\lambda V_{1,2})t/\hbar}$$

and

$$\psi_2(x, t) = \frac{(u_1 - u_2)}{\sqrt{2}} e^{-i(E_0-\lambda V_{1,2})t/\hbar} \tag{14b}$$

Let us suppose that at the time $t = 0$, the wave function was given by $\psi = u_1(x)$. This is just the problem discussed by perturbation theory in Chap. 18, Sec. 3, for nondegenerate levels. One can write

$$\psi_{t=0} = u_1 = \frac{1}{\sqrt{2}}(v_1 + v_2) \tag{15}$$

In order to find the time variation of ψ, we must multiply v_1 and v_2 separately by the frequencies with which each oscillates, as given in (14b). We then obtain

$$\psi(x, t) = \frac{e^{-iE_0t/\hbar}}{\sqrt{2}}(v_1 e^{-i\lambda V_{1,2}t/\hbar} + v_2 e^{i\lambda V_{1,2}t/\hbar}) \tag{16}$$

Let us now eliminate v_1 and v_2 in terms of u_1 and u_2 with the aid of eqs. (14a)

$$\psi = e^{-iE_0t/\hbar}\left(u_1 \cos\frac{\lambda V_{1,2}t}{\hbar} - iu_2 \sin\frac{\lambda V_{1,2}t}{\hbar}\right) \tag{17}$$

9. Quantum-mechanical "Resonance." It is clear that at $t = 0$ the above solution is equal to u_1, and that, at later times, the function u_2 enters. This means that transitions from u_1 to u_2 are taking place. At first, the wave function u_2 grows linearly with the time, at the same rate as predicted by perturbation theory [see eq. (9a), Chap. 18]. Eventually, however, the rate deviates from linearity. This is because perturbation theory has broken down. Meanwhile, u_1 decreases. When

$$\frac{\lambda V_{1,2}t}{\hbar} = \frac{\pi}{2}$$

the system will be entirely in the state u_2. Then it goes back to u_1, etc. This process is very similar formally to that of two resonant harmonic oscillators that are weakly coupled. If one of the oscillators is initially excited, the energy is transferred back and forth between the two oscillators at a rate proportional to the strength of the coupling force between the two. In the quantum problem, the wave amplitude, and therefore the probability, goes back and forth between the two degenerate states at a rate proportional to $\lambda V_{1,2}$, which may be regarded as a sort of coupling term. Thus, whenever there are degenerate eigenstates, there is always the possibility of this type of "resonance," as it is called. If there are more than two eigenstates, the resonance is more complex, just as in the analogous case where there are more than two coupled harmonic oscillators. In both cases, the "excitation" is transferred between the resonant systems in a more or less complex way, depending on the nature of the coupling terms.

As has been shown in Sec. 3, when the system is in a stationary state, the wave function is $v = (u_1 \pm u_2)/\sqrt{2}$, in which case there is an equal probability that the system can be found in state 1 or 2. In order to

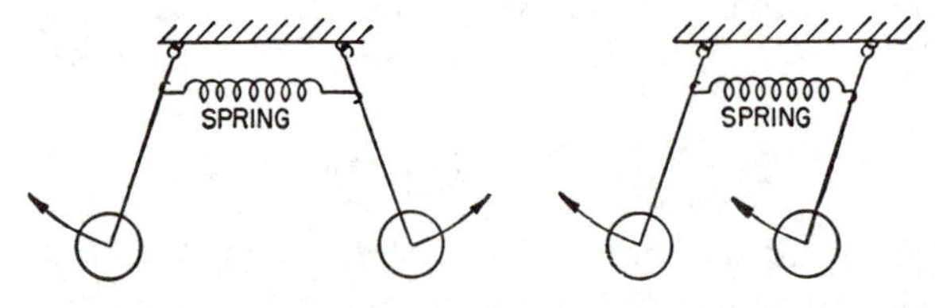

FIG. 1

put the system into the state corresponding to the wave function u_1, it was necessary to include the two wave functions v_1 and v_2 of different energy, so that the system was no longer in a stationary state. [See eq. (17).] Instead, it continually made transitions between u_1 and u_2. The analogy to harmonic oscillators is, as we have already pointed out, to have two equal pendulums coupled by a weak spring (see Fig. 1). If the pendulums oscillate in phase or out of phase as shown above, the system is in a state of stationary oscillation, in the sense that each pendulum retains a constant energy. But if one of the pendulums is started out while the other is at rest, a continual transfer of energy takes place between the pendulums.

If we took pendulums of very different period, such a resonant transfer of energy would not occur. This is because the successive impulses delivered by the one pendulum through the loose spring would be far from being in the right phase to build up the amplitude of the other. Only if the two pendulums have almost equal periods will successive impulses from the one be delivered to the other in such a phase as to result in a cumulative transfer of energy after many oscillations. The same happens in quantum theory, where, as we have seen, transitions to states of different energy are reversed before they can get very far (see Chap. 18, Sec. 9), but transitions to states of the same energy result in resonant transfer of probability from one state to the other.

The reason that resonance results from equal frequencies in classical theory and equal energies in quantum theory is simply because of the de Broglie relation, $E = h\nu$. Thus the wave function oscillates with the frequency $\nu = E/h$, and if two wave functions have the same energy, they have the same frequency. Resonance is therefore really a characteristic of oscillatory phenomena, both classically and in quantum theory.

Because of the definite phase relations between u_1 and u_2 appearing in eq. (17), the picture of quantum-mechanical resonance as a transfer of probability between degenerate eigenstates is incomplete; for the system cannot correctly be regarded as being in a definite but unknown eigenstate of H_0. (This follows from the fact that important physical properties depend on interference between u_1 and u_2. See Chap. 6 and Chap. 16, Sec. 25.) Instead it is better to think of the system as having potentialities for development of a definite value of H_0, in interaction with an apparatus that can be used to provide a measurement of H_0. The changing coefficients of u_1 and u_2 in eq. (17) then imply changing probabilities for realization of these potentialities in such a measurement process. (See also Chap. 22, Sec. 14.)

10. Analogy of Degeneracy Problem to Principal Axis Transformation. The transformation from the u's to the v's is formally very similar to a transformation to a set of principal axes. Consider, for example, a non isotropic classical three-dimensional harmonic oscillator. In terms of an arbitrary set of axes, the equations of motion are

$$m\ddot{x}_i = \sum_j a_{ij}x_j \tag{18}$$

x_i represents the co-ordinates, x, y, z, as i runs from 1 to 3, respectively.

The transformation to principal axes consists of a linear transformation to a new set of co-ordinates, ξ_k,

$$x_i = \sum_k \alpha_{ik}\xi_k \tag{19a}$$

such that in the new set of axes, the equations of motion take the form

$$m\ddot{\xi}_k = b_k\xi_k \tag{19b}$$

In other words, the ξ_k all undergo simple harmonic motion, each in general, with its own period. The ξ_k are the co-ordinates along the principal axes of the system. The principal axes have the property that an oscillation in their direction is not coupled to those in other directions.

The equations satisfied by the u_i are formally very similar to those satisfied by the x's. One may regard the u_i as components of a vector in a space having as many dimensions as there are energy levels. Then the transformation from the u's to the v's is analogous to a transformation to principal axes in u_i space.† The v_i have the property that they can oscillate independently of each other, i.e., there is no coupling between them.

Applications of Degenerate Perturbations

11. First-order Stark Effect. In the Stark effect, as we have seen [eq. (70), Chap. 18], the matrix elements will vanish unless $\Delta m = 0$ and $\Delta l = \pm 1$. In hydrogen, states of the same n and different l have the same energy so that the theory of degenerate perturbations must be used. Only for the ground state ($n = 1$, $l = m = 0$) is this degeneracy absent.

Let us investigate this problem for the next simplest case, namely, $n = 2$. We shall have transitions between $l = 0$ and $l = 1$ for the case $m = 0$. For $m = \pm 1$, no such transitions occur, and these levels may be treated by nondegenerate perturbation theory.

Let us denote the state $l = 0$ by $u_1(x)$ and $l = 1$, $m = 0$, by $u_2(x)$. Then as shown in Chapter 18, eq. (70), $V_{1,1}$ and $V_{2,2}$ vanish, for a uniform electric field. We may therefore use the development leading to eqs. (5) and (6). The energy level will then split into two levels, given by

$$E = E_0 \pm \lambda |V_{1,2}| = E_0 \pm e\mathcal{E}|z_{1,2}| \tag{20}$$

where

$$z_{1,2} = \int u_1^* z u_2 \, dx$$

There are several important consequences of this result.

(1) The shift of energy is *linear* in $\mathcal{E}$, as contrasted with the quadratic shift obtained in the nondegenerate case.‡ The linear shift is normally much larger than the quadratic shift obtained with most atoms. The reason for the linear shift in energy lies in the degeneracy, which causes the wave function to undergo its full change even when there is only a small perturbing force. Now, the change in energy is proportional to the product of the effect of the change in wave function and the change in Hamiltonian. In the nondegenerate case, both are proportional to $\mathcal{E}$ [see eqs. (2) and (114), Chap. 18], so that the energy shift is proportional to $\mathcal{E}^2$. In the degenerate case, the full change of wave function occurs even for the weakest perturbation [see eq. (5)], so that the change of wave function is independent of $\mathcal{E}$, and the net effect is linear in $\mathcal{E}$.

(2) We now obtain some conclusions about the effect of the electric field on the emitted spectral line. First, we note that except for a minute quadratic Stark effect, the energy levels, $n = 1$, and $n = 2$, $l = 1$, $m = \pm 1$, are unshifted. This means that those dipole transitions in which $\Delta l = -1$ and $\Delta m = \pm 1$ will have essentially unaltered frequencies. On the other hand, the line $\Delta l = -1$, $\Delta m = 0$ will split into two parts, one of slightly higher and one of slightly lower frequency. Thus, in

† The transformation to principal axes and the transformation from the u's to the v's are special cases of a canonical transformation (see Chap. 16, Sec. 15).

‡ Chap. 18, eq. (127).

general, the line will split into 3 parts. Now $\Delta m = \pm 1$ leads to polarization normal to the z axis, and $\Delta m = 0$ leads to polarization along the z axis. (See Chap. 18, Sec. 45.) If the light is viewed along the direction of the electric field, only the polarization normal to z will be observed, and, as we have seen, this corresponds to $\Delta m = \pm 1$, which is unshifted. If the light is viewed normal to the z direction, we shall obtain the unshifted component polarized normal to the electric field, and also the two shifted components polarized along the direction of the electric field (see Fig. 2).

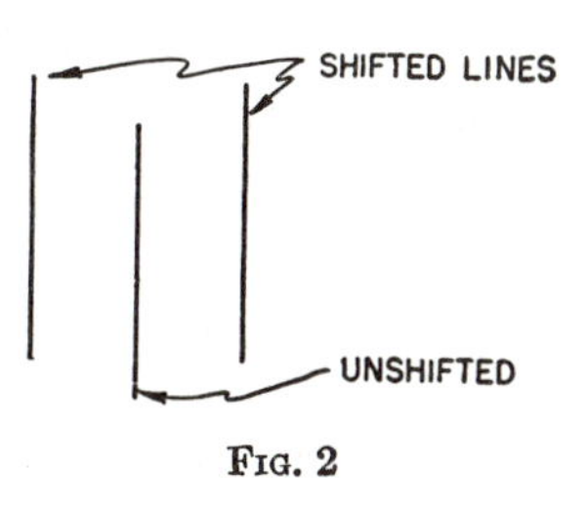

FIG. 2

The higher levels will exhibit more complex patterns for the Stark shift, because more degenerate levels will, in general, be involved.

(3) The value of the shift depends on $z_{1,2}$. This is a quantity of the order of some mean between the sizes of the atom in the states 1 and 2. In fact, it may be evaluated by fairly straightforward methods, and one obtains for the Stark shift for the transition $l = 1$, $m = 0$, $n = 2$, to $l = 0$, $m = 0$, $n = 1$

$$\lambda V_{1,2} = 3e\mathcal{E}a_0 \tag{21}$$

where a_0 is a Bohr radius.

Problem 4: Obtain the result given in eq. (21).

The Stark effect in hydrogen atoms can be treated exactly by means of transformation to parabolic co-ordinates.*

12. Classical Interpretation of Linear Stark Effect. The quantum degeneracy of levels of the same n and different l is reflected in the classical degeneracy between frequencies of rotation and of radial oscillation. With a general law of force, these two frequencies differ;† this means that there will be no closed orbit; instead, a noncircular orbit will precess at a rate determined by the difference of radial and angular frequencies. Only for the Coulomb force do these two frequencies become equal so that one obtains closed, nonprecessing elliptical orbits. The system is degenerate with respect to the direction of the major axis of the ellipse, i.e., no energy is required to rotate this major axis so long as the focus remains fixed. Since the average position of the electron in its orbit is not at the focus of the ellipse, such a rotation will shift this average position, as shown in Fig. 3. If a very weak electrical field is applied, the orbit will tend to line up along the direction of this field; in so doing, the average co-ordinate of the electron will shift by a quan-

* See, for example, Schiff, *Quantum Mechanics*, and A. Sommerfeld, *Atombau und Spektralinien*. Braunschweig, Friedr. Vieweg und Sohn, 1939. See also, eq. (94), Chap. 21 and associated footnote.

† See Chap. 2, Sec. 14.

tity d of the order of the mean radius of the orbit. The energy given off will be

$$W = e\mathcal{E}d$$

Thus, one obtains a first-order shift in energy, just because the displacement d is independent of the electric field. The weakest electric field is able to produce the full displacement. In a sense, a degenerate system like this is infinitely polarizable.

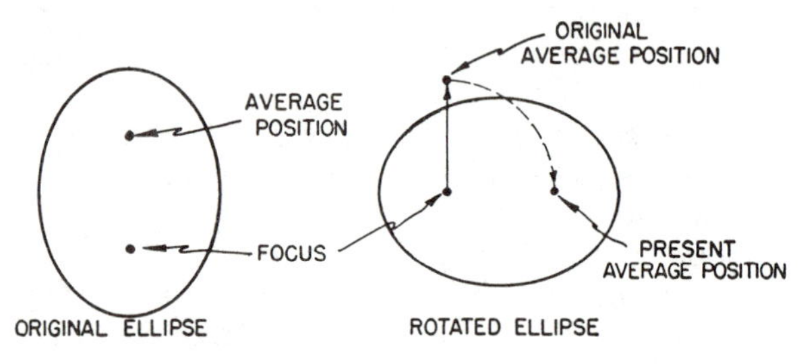

FIG. 3

Problem 5: By taking a classical elliptic orbit corresponding to $L_z = \hbar$ and with an energy equal to that of a quantum state with $n = 3$, show that the shift obtained by the model suggested above is of the right order of magnitude.

In a nondegenerate system, there are no such elliptical orbits, but, instead, the orbit precesses around rapidly as shown in Chap. 2, Fig. 5. In such a precessing orbit, the time average position of the electron is at the center of the atom. When a weak electric field is turned on slowly, then, as shown in Chap. 18, Sec. 54, the orbit shifts slightly in the direction of the electric field. In terms of the present description, the reason for the difference in behavior is that a permanent alignment of the orbit is prevented by the precession, which tends to make the average position go back to zero. The net polarization is the result of a balance between the two processes, which results in a mean displacement proportional to the electric field, $\mathcal{E}$, and therefore a mean energy proportional to $\mathcal{E}^2$. As one approaches degeneracy, i.e., as the radial and angular frequencies approach equality, the rate of precession approaches zero. There is then more time for alignment of the orbit along the electric field before this alignment is destroyed by precession so that as the atom approaches degeneracy, its polarizability becomes large.

13. Van der Waals Forces between Atoms. Consider two atoms that are gradually brought closer and closer to each other. As long as they remain more than an atomic diameter distant from each other, the electronic charge of each atom will tend to shield its own nucleus, so that, in the zeroth approximation, there will be no net force between the atoms. If we consider what happens more carefully, however, we can see that there should be a small residual force. This arises from the fact that the electrons move around in orbits. As a result, the potential produced by any given atom will undergo small fluctuations. These fluctuations will produce small electric fields that polarize the other atom, and create a dipole moment, which we denote by $\boldsymbol{M}$. Since the force on $\boldsymbol{M}$ is $\boldsymbol{F} = (\boldsymbol{M} \cdot \nabla)\boldsymbol{\varepsilon}$, where $\boldsymbol{\varepsilon}$ is the electric field and, since $\boldsymbol{M} = P\boldsymbol{\varepsilon}$, where P is the polarizability, one obtains

$$\boldsymbol{F} = P(\boldsymbol{\varepsilon} \cdot \nabla)\boldsymbol{\varepsilon}$$

This small residual force is called the *van der Waals force.*

Although the fluctuating electric field $\boldsymbol{\varepsilon}$ averages out to zero each time that the electron goes around in its orbit, we observe that the force given above depends quadratically on $\boldsymbol{\varepsilon}$, so that its time average is not zero. This is because of the fact that when $\boldsymbol{\varepsilon}$ changes its sign, the induced dipole moment, $\boldsymbol{M} = P\boldsymbol{\varepsilon}$, makes a compensating change.

Let us now consider how to treat this problem quantum-mechanically. We shall consider a case in which there is only one "valence electron"; this means that the rest of the electrons are bound so tightly to the nucleus that they shield it very effectively and therefore need not be considered in this problem. The effective nuclear charge is then unity.

The co-ordinates that are significant in the problem are illustrated in Fig. 4. P_1 and P_2 are the locations of the nuclei of the two atoms, e_1 and e_2 those of the two electrons. $\boldsymbol{R}$ is the distance between the two nuclei, $\boldsymbol{r}_1$ is the distance from the first electron to the first nucleus, $\boldsymbol{r}_2$ is the distance from the second electron to the second nucleus, $\boldsymbol{R}_{1,2}$ is the distance from the first nucleus to the second electron, and $\boldsymbol{R}_{2,1}$ from the second nucleus to the first electron. $\boldsymbol{r}_{1,2}$ is the distance between electrons.

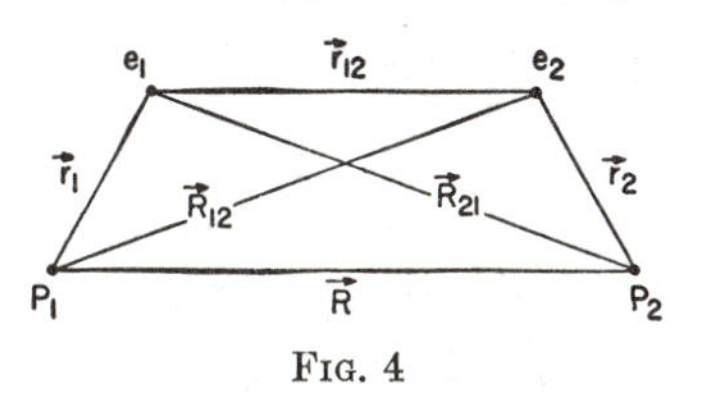

FIG. 4

Note that we can neglect the kinetic energies of the nuclei because they are so heavy that they can be localized very accurately with very little kinetic energy. This follows from the uncertainty principle ($\delta P \cong \hbar/\delta x$), plus the fact that the kinetic energy is $T = p^2/2m$. Thus, $\delta T \cong \hbar^2/2m(\delta x)^2$. If m is large, very little kinetic energy is required to fix the position accurately, and one can then neglect the nuclear kinetic energy as a first approximation.* Another way of seeing the meaning of the neglect of the kinetic energy of the nuclei is to go to the classical limit and consider the orbits. The electrons move so much more rapidly than the nuclei that they execute many revolutions in their orbits before the nuclei move appreciably. To a first approximation,

* For a fuller discussion, see H. Margenau, *Rev. Mod. Phys.*, **11**, 1 (1939); also L. Pauling, *The Nature of the Chemical Bond.* Ithaca, N. Y., Cornell University Press, 1940.

one can therefore solve for the electronic motion under the assumption that the nuclei are fixed. This is essentially the adiabatic approximation, which will be discussed in Chap. 20.

The Hamiltonian can then be written as follows:

$$H = H_1(\boldsymbol{r}_1) + H_2(\boldsymbol{r}_2) + V$$

where

$$V = e^2\left[\frac{1}{R} + \frac{1}{r_{1,2}} - \frac{1}{R_{1,2}} - \frac{1}{R_{2,1}}\right] \tag{22}$$

$H_1(\boldsymbol{r}_1)$ is the Hamiltonian of the first electron in the absence of the second atom and $H_2(\boldsymbol{r}_2)$ is the Hamiltonian of the second electron in the absence of the first atom.

When the atoms are far apart, the interaction energy between them, which appears in the brackets above, can be neglected. A solution to Schrödinger's equation is then

$$\psi_{n_1,n_2} = u_{n_1}(\boldsymbol{r}_1)v_{n_2}(\boldsymbol{r}_2) \tag{23}$$

$u_{n_1}(\boldsymbol{r}_1)$ represents the n_1th state of the first atom; $v_{n_2}(\boldsymbol{r}_2)$ represents the n_2th state of the second atom. Note that the two atoms need not be the same; we designate this possibility by distinguishing between the u's and the v's. If the two atoms are the same, the u's and the v's will be the same functions. The u's and v's satisfy the following equations:

$$H_1u_{n_1} = E^0_{n_1}u_{n_1} \qquad \text{and} \qquad H_2v_{n_2} = E^0_{n_2}v_{n_2} \tag{24}$$

As a result of the interaction energy V this wave function will be changed when the atoms are brought together. One can treat this problem by perturbation theory as long as V is much smaller than the potential of the electron in the field of its own nucleus. This will be possible if the interatomic separation R is appreciably larger than the mean atomic diameter. This means that

$$r_1 \ll R \qquad \text{and} \qquad r_2 \ll R$$

If this is true, however, the potential can be approximated by a simpler expression. In doing this, we shall find it convenient to note that V is just the expression for the energy of the second atom resulting from the potential produced by the first atom. When r_1/R and r_2/R are each much less than unity, the electrostatic potential arising from the first atom is approximately equal to that of a dipole of moment $M = -e\boldsymbol{r}_1$. The potential produced by such a dipole at the point $\boldsymbol{R}$ is

$$\phi = -\frac{e\boldsymbol{r}_1 \cdot \boldsymbol{R}}{R^3} = -\frac{e(x_1X + y_1Y + z_1Z)}{R^3}$$

The electric field is found by differentiating the above potential, i.e.,

$$\boldsymbol{\mathcal{E}} = -\boldsymbol{\nabla}\phi$$

To find the energy of the second atom in the field of the first, we regard it as equivalent to a dipole of moment $M = -e\mathbf{r}_2$. The energy is then equal to

$$W = e^2(\mathbf{r}_2 \cdot \nabla)\left(\frac{\mathbf{r}_1 \cdot \mathbf{R}}{R^3}\right)$$

where the differentiation is carried out with respect to R. By a little algebra, the above reduces to

$$W = \frac{e^2}{R^3}\left[\mathbf{r}_1 \cdot \mathbf{r}_2 - 3\,\frac{(\mathbf{r}_1 \cdot \mathbf{R})(\mathbf{r}_2 \cdot \mathbf{R})}{R^2}\right] \tag{25}$$

This is the first approximation to the expression for V, good when $r_1/R \ll 1$ and $r_2/R \ll 1$. It should be recognized as the expression for the mutual energy of two dipoles of moments $\mathbf{M}_1 = -e\mathbf{r}_1$ and $\mathbf{M}_2 = -e\mathbf{r}_2$, separated by a distance R.

(1) *Nondegenerate Case.* If there is no degeneracy, the problem can be treated by the ordinary nondegenerate perturbation theory. As the atoms are brought together, the wave functions are distorted because of the interaction between the atoms, and the perturbed states will contain in them a small amount of the higher unperturbed energy wave functions. The perturbation matrix element will be

$$V_{n_1',n_2';n_1,n_2} = \frac{e^2}{R^3}\int\int u^*_{n_1'}(\mathbf{x}_1)x_1u_{n_1}(\mathbf{x}_1)v^*_{n_2'}(\mathbf{x}_2)x_2v_{n_2}(\mathbf{x}_2)\,d\mathbf{x}_1\,d\mathbf{x}_2 \tag{26}$$

$$+ \text{ other similar terms involving } y_1y_2 \text{ and } z_1z_2$$

But the above is just the product of matrix elements which appear in the theory of radiative dipole transitions (see Chap. 18, Sec. 25) and also in the Stark effect and polarizability of atoms (see Chap. 18, Sec. 54). To obtain the complete matrix element, it is convenient to choose the z axis along the line between centers of atoms. From eq. (25), we then obtain

$$V_{n_1',n_2';n_1,n_2} = \frac{e^2}{R^3}\,(x_1x_2 + y_1y_2 - 2z_1z_2)_{n_1',n_2';n_1,n_2} \tag{27}$$

According to the selection rules [Chap. 18, eq. (70)], these can cause transitions only when $\Delta l = \pm 1$, $\Delta m = 0, \pm 1$. Thus, only levels connected with the unperturbed state in this way will contribute to the energy. We note also that $V_{n_1n_2;n_1n_2} = 0$. This is because the mean value of the coordinate, $\mathbf{x}$, has been shown to be zero for any stationary state. The correction to the energy will therefore come only from the second-order term [see eq. (122), Chap. 18].

$$\Delta E = \frac{e^4}{R^6}\sum_{\substack{n_1' \neq n_1 \\ n_2' \neq n_2}} \frac{|(x_1x_2 + y_1y_2 - 2z_1z_2)_{n_1',n_2';n_1,n_2}|^2}{E^0_{n_1} + E^0_{n_2} - E^0_{n_1'} - E^0_{n_2'}} \tag{28}$$

The change in energy is proportional to R^{-6}, which is in agreement with what is required to explain van de Waals forces. One can easily see the reason for this dependence. The electric field caused by a dipole is proportional to R^{-3}, the dipole moment induced in the other atom is proportional to this field, so that the energy, which is proportional to the product of the two, involves R^{-6}.

The sum in eq. (28) is obviously closely related to the atomic polarizability. The sum will be large only when the matrix elements, x_1x_2, are large, or when the energy differences in the denominator are small. The former case will occur when the atoms are large, the latter when the atomic states are nearly degenerate. The alkaline metals like sodium have nearly degenerate levels, since their wave functions are nearly like those of hydrogen, and the van der Waals forces will be large. Since the levels are far from degenerate in noble gases, the result that van der Waals forces are observed to be very small for such gases is reasonable on the basis of the above picture.

(2) *Degenerate Case. Resonant Transfer of Excitation Energy between Atoms.* If the two atoms are different, or if both are in the ground state, there will usually be no degeneracy. But an important degeneracy arises if an atom in an excited state comes near a like atom in the ground state. Such a case arises, for example, if some atoms of sodium or mercury vapor in a discharge tube are excited either electrically or by incident radiation. As a result of random molecular motion, such an excited atom is bound to come near an unexcited atom. Now, because the two atoms are alike, another state of the same unperturbed energy will be brought about by transfer of the energy from the one atom to the other. As we shall see, this degeneracy has many important consequences.

To treat this case, we once again neglect all nondegenerate transitions, which will produce only comparatively small effects. Let us take the z axis to be on the line of centers of the two atoms. For the sake of definiteness, let us suppose that the ground state is given by $n = 1, l = 0, m = 0$, and let us consider only the excited state $n = 2, l = 1, m = 0$. In a complete treatment, however, all excited states must be taken into account. Then, according to our selection rules, the matrix elements of x_1x_2 and y_1y_2 will vanish, since the matrix elements of x and y correspond to radiation polarized in the x and y directions, and these vanish when $\Delta m = 0$ (see Chap. 18, Sec. 45). Only z_1z_2 will survive.

Let us denote by $\psi_1(r_1, r_2)$ the wave function for the state in which the first atom is excited, and by $\psi_2(r_1, r_2)$ the wave function for the state in which the second atom is excited. These wave functions are

$$\psi_1 = u_{1,1,0}(r_1)u_{0,0,0}(r_2) = u_1(r_1)u_0(r_2) \tag{29a}$$

$$\psi_2 = u_{1,1,0}(r_2)u_{0,0,0}(r_1) = u_1(r_2)u_0(r_1) \tag{29b}$$

The only significant matrix elements that do not vanish are then

$$V_{1,0;0,1} = -\frac{2e^2}{R^3} z_{1,0}z_{0,1} = -\frac{2e^2}{R^3} |z_{1,0}|^2 \qquad V_{0,1;1,0} = -\frac{2e^2}{R^3} |z_{1,0}|^2 \tag{30}$$

Applying the degenerate perturbation theory,* we obtain for the wave functions

$$\left.\begin{aligned} \psi_a &= \frac{1}{\sqrt{2}}[u_1(r_1)u_0(r_2) + u_1(r_2)u_0(r_1)] \\ \psi_b &= \frac{1}{\sqrt{2}}[u_1(r_1)u_0(r_2) - u_1(r_2)u_0(r_1)] \end{aligned}\right\} \tag{31}$$

The shift in energy is

$$E_a = -2e^2\frac{|z_{1,0}|^2}{R^3} \qquad \text{and} \qquad E_b = 2e^2\frac{|z_{1,0}|^2}{R^3} \tag{32}$$

Interpretation of Results. We notice that for the stationary-state wave function ψ_a, the energy of interaction of the atoms, E_a, is negative, while for ψ_b it is positive. In both cases, it is proportional to R^{-3} as contrasted to the R^{-6} obtained in the nondegenerate case. Furthermore, in each stationary state the excitation covers both atoms simultaneously, so that there is equal probability that if an observation of the energy is made, the excitation will be found on either atom.† On the other hand, if the excitation energy had been definitely on any one of the atoms at a given time, then according to the discussion in Sec. 9, we would have had a nonstationary state. For such a state, the excitation is transferred back and forth between the atoms with a frequency

$$\nu = \frac{2e^2|z_{1,0}|^2}{hR^3} \tag{33}$$

Thus, the closer the atoms get together, the more rapid is this "resonant" transfer of energy between them.

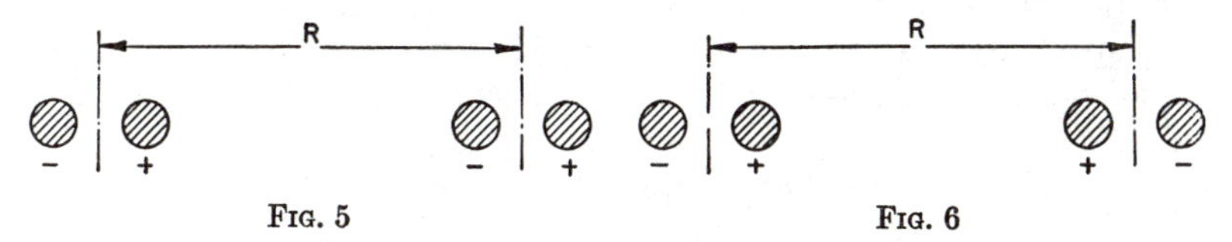

FIG. 5 FIG. 6

One can understand these results by considering the analogous classical problem in which two oscillating electric dipoles of the same natural frequency are brought near each other. If they oscillate in phase as shown in Fig. 5, they attract each other, but if they oscillate out of phase, as shown in Fig. 6, they repel each other. The force between them is

* See eqs. (5) and (6).

† In a stationary state, where $\psi = \frac{1}{\sqrt{2}}(\psi_1 \pm \psi_2)$, each atom must, because of interference effects, be regarded as covering both states simultaneously. In other words, it is incorrect to say that each atom is always in a definite but unknown state. Instead, one should say that the state of each atom is incompletely defined, but potentially capable of becoming better defined in interaction with a device that can be used to provide a measurement of the energy of that atom. (See Chaps. 6 and 8, and Chap. 16, Sec. 25.)

proportional to R^{-3}. This is because the electric field caused by any one of them is given by

$$\mathcal{E} = \cos \omega t \nabla \frac{\mathbf{M}_1 \cdot \mathbf{R}}{R^3} \tag{34}$$

where $\mathbf{M}_1$ is the maximum moment of the first dipole. The energy of interaction is $W = \mathbf{M}_2 \cdot \nabla \left(\frac{\mathbf{M}_1 \cdot \mathbf{R}}{R^3} \right) \cos \omega t \cos (\omega t + \phi)$, where ϕ is the phase of the second oscillator relative to that of the first.

If $\phi = 0$, the average value of W over a period is negative; if $\phi = \pi$ it is positive. This shows more precisely that oscillations in phase attract, and those out of phase repel. The R^{-3} law follows from the fact that as long as the oscillators can stay in phase, the moment of the second one is independent of the field of the first so that only the R^{-3} term resulting from the field comes in. (If the oscillator frequencies were different they would rapidly get out of phase, and the mean value of the contribution of this term to the energy would vanish. We should then have only the term coming from polarization of the second oscillator by the first, which is, as we have seen, proportional to R^{-6}.) Choosing the z axis along the direction of the separation, $\mathbf{R}$, between the two dipoles, we can, by an argument similar to that leading to eq. (25), obtain for dipoles oscillating only in the z direction

$$W = -\frac{2M_1M_2}{R^3} \cos \omega t \cos (\omega t + \phi)$$

14. Quantum-mechanical Analogue of Oscillator Phase. Is there a quantum-mechanical analogue of the correlation of oscillator phases, which explained the R^{-3} variation of the energy in the classical problem? To see that there is, let us look at the wave function, ψ_a. The probability function is

$$P_a(\mathbf{r}_1, \mathbf{r}_2) = |\psi_a|^2 = \tfrac{1}{2}|u_1(\mathbf{r}_1)u_0(\mathbf{r}_2) + u_1(\mathbf{r}_2)u_0(\mathbf{r}_1)|^2 \tag{35}$$

When $\mathbf{r}_1$ and $\mathbf{r}_2$ are the same, then the two terms contributing to ψ_a are equal; they add up, and produce a large probability. Now, suppose that $u_0(\mathbf{r})$ represents an s wave, and $u_1(\mathbf{r})$ is a p wave. The s wave has even parity, the p wave odd parity. Hence, if $\mathbf{r}_1$ and $\mathbf{r}_2$ are given opposite signs, then $u_1(\mathbf{r}_1)u_0(\mathbf{r}_2)$ has a sign opposite to that of $u_1(\mathbf{r}_2)u_0(\mathbf{r}_1)$. This means that for $\mathbf{r}_1 = \mathbf{r}_2$, the sum vanishes, so that there is no probability that $\mathbf{r}_1$ and $\mathbf{r}_2$ are the negatives of each other, and only a small probability that $\mathbf{r}_1$ is close to $-\mathbf{r}_2$. Thus we see that the exact correlation of phases in the classical problem has been replaced by a statistical correlation in the quantum problem. For the wave function

$$\psi_b = \frac{1}{\sqrt{2}}[u_1(\mathbf{r}_1)u_0(\mathbf{r}_2) - u_1(\mathbf{r}_2)u_0(\mathbf{r}_1)]$$

the correlation is obviously reversed, since the wave function vanishes when $r_1 = r_2$, and is large when $r_1 = -r_2$. We see then that ψ_a corresponds to a statistical tendency to oscillate in phase, for which both classically and quantum-mechanically the systems are found to attract each other while ψ_b corresponds to a similar statistical tendency to be out of phase and to repel.

One can prove the correlation more quantitatively by evaluating the correlation function for z_1 and z_2, given by $C_{1,1}$ in eq. (4), Chap. 10.

$$C_{1,1} = \overline{z_1 z_2} - \bar{z}_1 \bar{z}_2 = \overline{z_1 z_2} \tag{36a}$$

(since $\bar{z} = \bar{z}_2 = 0$, for this case). We get

$$C_{1,1} = \tfrac{1}{2}\int(\psi_1^* \pm \psi_2^*) z_1 z_2 (\psi_1 \pm \psi_2)\, dr_1\, dr_2$$

Now $\int \psi_1^* z_1 z_2 \psi_1 \, dr_1\, dr_2 = \int u_1^*(r_1) z_1 u_1(r_1)\, dr_1 \int u_0^*(r_2) z_2 u_0(r_2)\, dr_2 = \bar{z}_1 \bar{z}_2 = 0$

Similarly, $\int \psi_2^* z_1 z_2 \psi_2 \, dr_1\, dr_2 = 0$

We are left with

$$\pm\tfrac{1}{2}\int\int(\psi_1^*\psi_2 + \psi_2^*\psi_1) z_1 z_2 \, dr_1\, dr_2$$
$$= \pm\tfrac{1}{2}\left[\int u_1^*(r_1) z_1 u_0(r_1)\, dr_1 \int u_0^*(r_2) z_2 u_1(r_2)\, dr_2 + \text{complex conjugate}\right]$$

We obtain

$$\bar{C}_{1,1} = \pm |z_{0,1}|^2 \tag{36b}$$

We see that for ψ_a, which corresponds to the plus sign, there is positive correlation between z_1 and z_2, while for ψ_b, which corresponds to the negative sign, there is negative correlation.

15. Experimental Consequences of Degeneracy.

(1) The force between an excited and an unexcited atom should vary as R^{-3}, and should therefore have a much longer range than the usual van der Waals forces. It may be either attractive or repulsive, depending on whether the wave function is ψ_a or ψ_b. There is an equal probability of either in a random distribution. Some of the atoms of a gas will therefore attract and some will repel.

(2) The transfer of excitation has the effect of producing a very large broadening of spectral lines. This is because the excitation is carried to another atom, which, in general, does not radiate in phase with the original atom. The lifetime of the excited state of the first atom is therefore shortened, so that by the uncertainty principle, the line should be broadened. The broadening of a line by such resonant transfers to atoms of the *same* kind is thus much more important than that resulting from the second-order perturbing effects of other atoms.†

† A. C. Mitchell and M. W. Zemansky, *Resonance Radiation and Excited Atoms*. New York: The Macmillan Company, 1934; also Mott and Massey, *Theory of Atomic Collisions*, Chap. 13; L. Pauling, *The Nature of the Chemical Bond;* and P. M. Morse, *Rev. Mod. Phys.*, **4**, 577 (1932).

16. Exchange Degeneracy. An important source of degeneracy arises whenever there is more than one particle of a given kind. For example, in a helium atom, there are two electrons. If these two electrons are interchanged, we obtain a wave function which is, in general, not the same as the one with which we started. But because all electrons are equivalent, the exchange of any two cannot change the energy of the system.* The two wave functions must therefore correspond to degenerate energy levels.

The Hamiltonian for the electrons in a helium atom is

$$H = \frac{p_1^2 + p_2^2}{2m} - \frac{2e^2}{r_1} - \frac{2e^2}{r_2} + \frac{e^2}{r_{1,2}} \tag{37}$$

p_1 and r_1 are, respectively, the momentum of the first electron and its distance from the center of the nucleus; p_2 and r_2 refer to the same quantities for the second electron. $r_{1,2}$ is the distance between the two electrons

$$r_{1,2}^2 = (x_1 - x_2)^2 + (y_1 - y_2)^2 + (z_1 - z_2)^2$$

Now, to a first approximation, one can neglect the term $e^2/r_{1,2}$, which represents the interaction between the electrons. This term is perhaps about $\frac{1}{5}$ as large on the average as the potential energy terms, $2e^2\left(\frac{1}{r_1} + \frac{1}{r_2}\right)$, which have been taken into account in the zeroth approximation. It is therefore not particularly small, but it turns out to be just barely small enough so that a fair degree of approximation is provided by the perturbation theory.

A solution of the zeroth-order wave equation is then given by

$$\psi_{a,b} = u_a(\mathbf{r}_1)u_b(\mathbf{r}_2) \qquad \text{with} \qquad E = E_a + E_b$$

$u_a(\mathbf{r}_1)$ represents a solution of the equation $\left(\frac{p_1^2}{2m} - \frac{2e^2}{r_1}\right)u_a(\mathbf{r}_1) = E_a u_a(\mathbf{r}_1)$, whereas $u_b(\mathbf{r}_2)$ is a solution of the corresponding equation for particle number 2.

When the term $e^2/r_{1,2}$ is taken into account, the above ψ_{ab} is no longer a solution of Schrödinger's equation. We could, however, try to obtain a solution by means of perturbation theory. To do this, we must expand ψ as a series of the ψ_{ab} with arbitrary coefficients C_{ab}

$$\psi = \sum_{a,b} C_{ab}\psi_{ab} = \sum_{a,b} C_{ab}u_a(\mathbf{r}_1)u_b(\mathbf{r}_2)$$

*The consequences of this fact will be discussed in more detail in Secs. 20 to 29.

Because the two electrons are identical, we know that corresponding to each wave function ψ_{ab}, there will be another function ψ_{ba} that has the same unperturbed energy, so that the level is degenerate.

Any pair of functions belonging to the two degenerate levels can then be expressed as follows:

$$\psi_{ab} = u_a(\mathbf{r}_1)u_b(\mathbf{r}_2) \qquad \psi_{ba} = u_a(\mathbf{r}_2)u_b(\mathbf{r}_1)$$

Clearly, the two functions differ only in that the electrons have been interchanged. This degeneracy is therefore called "exchange degeneracy." Obviously, it can occur only when the two particles are equivalent, because otherwise the energy would, in general, be changed by exchanging particles.

17. Solution of Problem. It is necessary first to remove the degeneracy, i.e., to solve the zeroth-order equations connected with degenerate perturbation theory. Otherwise, the energy denominators appearing in eq. (117), Chap. 18, will be infinite, and the perturbation theory will be inapplicable.

Let us denote $u_a(\mathbf{r}_1)u_b(\mathbf{r}_2)$ by ψ_1, and $u_a(\mathbf{r}_2)u_b(\mathbf{r}_1)$ by ψ_2. The perturbing term is, in this case, $\lambda V = e^2/r_{1,2}$. Let us now write down the significant matrix elements.

$$\left.\begin{aligned} \lambda V_{1,1} &= e^2 \int \frac{\psi_1^*\psi_1}{r_{1,2}}\, d\mathbf{r}_1\, d\mathbf{r}_2 \qquad & \lambda V_{2,2} &= e^2 \int \frac{\psi_2^*\psi_2}{r_{1,2}}\, d\mathbf{r}_1\, d\mathbf{r}_2 \\ \lambda V_{1,2} &= e^2 \int \frac{\psi_1^*\psi_2}{r_{1,2}}\, d\mathbf{r}_1\, d\mathbf{r}_2 \qquad & \lambda V_{2,1} &= e^2 \int \frac{\psi_2^*\psi_1}{r_{1,2}}\, d\mathbf{r}_1\, d\mathbf{r}_2 \end{aligned}\right\} \tag{38}$$

It is clear from the symmetry of the problem that $V_{1,1} = V_{2,2}$. Moreover, we can also prove that $V_{1,2} = V_{2,1}$, for

$$\lambda V_{1,2} = e^2 \int \frac{u_a^*(\mathbf{r}_1)u_b^*(\mathbf{r}_2)u_b(\mathbf{r}_1)u_a(\mathbf{r}_2)}{r_{1,2}}\, d\mathbf{r}_1\, d\mathbf{r}_2$$

We can interchange the labels of $\mathbf{r}_1$ and $\mathbf{r}_2$ without changing the value of the integral, obtaining

$$\lambda V_{1,2} = e^2 \int \frac{u_a^*(\mathbf{r}_2)u_b^*(\mathbf{r}_1)u_b(\mathbf{r}_2)u_a(\mathbf{r}_1)}{r_{1,2}}\, d\mathbf{r}_1\, d\mathbf{r}_2 = \lambda V_{2,1}$$

Now since $V_{2,1} = V_{1,2}^*$, we conclude that $V_{1,2} = V_{1,2}^*$, so that $V_{1,2}$ and $V_{2,1}$ are both real.

We are now ready to evaluate the correct zeroth-order wave functions. From eq. (5), we now obtain†

$$\psi_\pm = \frac{1}{\sqrt{2}}(\psi_1 \pm \psi_2) = \frac{1}{\sqrt{2}}[u_a(\mathbf{r}_1)u_b(\mathbf{r}_2) \pm u_b(\mathbf{r}_1)u_a(\mathbf{r}_2)] \tag{38a}$$

† Note that if $V_{1,2}$ is real, the quantity $e^{i\phi}$ occurring in eq. (5) is either $+1$ or -1. In either case, the solutions for ψ will take the form given in eq. (38a). We shall see in Sec. 19, however, that $V_{1,2}$ is always positive for a perturbation consisting of a Coulomb potential, so that for this case, we have $\phi = 0$.

18. Symmetric and Antisymmetric Functions. The functions in eq. (38a) have the property that they are multiplied respectively either by $+1$ or by -1, whenever the two particles are interchanged, i.e., when r_1 and r_2 are interchanged. The first type of function is called "symmetric," the second type "antisymmetric." It can be seen that in either case, the probability function is left unchanged by interchanging the particles. This means that the following situations are equally likely:

Electron (1) is in state a, electron (2) is in state b.

Electron (2) is in state a, electron (1) is in state b.

Actually, however, it is not completely accurate to say that either electron (1) is in state a and electron (2) is in state b, or vice versa. This is because, as we shall see, important physical properties depend on interference between ψ_1 and ψ_2. Thus, it is better to regard each electron as covering, in some sense, both states at once.†

Special Case: Both Particles in Same State

A special treatment must be given when both particles are in the same state. In this case, $u_a = u_b$. The function $u_a(r_1)u_b(r_2)$ is therefore automatically symmetric. On the other hand, the antisymmetric function vanishes identically. This means that there is only one state, and that there is no degeneracy. For example, in the ground state of the helium atom, both electrons are in the same state, and there is only one energy level.‡ If one of the electrons is excited, however, while the other is left in the ground state, then there will be degeneracy and the energy level will, in general, split into two, as we shall see in Sec. 19.

19. Evaluation of the Energy. Let us now evaluate the energy in the zeroth approximation. There are two ways of doing this. First, we may use eq. (6) writing

$$V_{1,1} = e^2 \int \frac{\psi_1^*\psi_1}{r_{1,2}}\, dr_1\, dr_2 = e^2 \int \frac{|u_a(r_1)|^2|u_b(r_2)|^2}{r_{1,2}}\, dr_1\, dr_2 \tag{39a}$$

$$V_{1,2} = e^2 \int \frac{\psi_1^*\psi_2}{r_{1,2}}\, dr_1\, dr_2 = e^2 \int \frac{u_a^*(r_1)u_b^*(r_2)u_a(r_2)u_b(r_1)}{r_{1,2}}\, dr_1\, dr_2 \tag{39b}$$

where ψ_1 and ψ_2 are defined in connection with eq. (38). Thus, according to (6), the energy becomes§

$$E_\pm = E_a + E_b + V_{1,1} \pm V_{1,2} \tag{40}$$

† We shall see in Sec. 29 that neither electron can correctly be ascribed to a definite state, but that instead, the state occupied by each electron should be regarded as somewhat indefinite, and potentially capable of becoming more definite in interaction with a suitable system.

‡ This is a special case of the Pauli exclusion principle. See Sec. 26.

§ We use the fact that $V_{1,1} = V_{2,2}$ and that $V_{1,2}$ and $V_{2,1}$ are real. Note that whether the value of $e^{i\phi}$ occurring in eq. (5) is $+1$ or -1, the result obtained in eq. (40) is correct. Actually, we shall see presently that $V_{1,2}$ is always positive, so that we must take $\phi = 0$. In this connection, see also Sec. 17.

The second way of obtaining the above result is simply to evaluate the mean energy using the correct zeroth-order wave function

$$\psi_{\pm} = \frac{1}{\sqrt{2}}(\psi_1 \pm \psi_2)$$

given in (38a). The result is

$$E_{\pm} = \frac{e^2}{2}\int \frac{(\psi_1^* \pm \psi_2^*)(\psi_1 \pm \psi_2)}{r_{1,2}}\,d\mathbf{r}_1\,d\mathbf{r}_2 + E_a + E_b$$

$$= \frac{e^2}{2}\int \frac{(\psi_1^*\psi_1 + \psi_2^*\psi_2)}{r_{1,2}}\,d\mathbf{r}_1\,d\mathbf{r}_2 \pm \frac{e^2}{2}\int \frac{(\psi_1^*\psi_2 + \psi_2^*\psi_1)}{r_{1,2}}\,d\mathbf{r}_1\,d\mathbf{r}_2 + E_a + E_b$$

A simple calculation shows that the above is equivalent to eqs. (39a) and (39b), where the first term is equal to $V_{1,1}$, and the second to $\pm V_{1,2}$.

The term $V_{1,1}$ is formally the equivalent of the potential energy of interaction between the two continuously smeared out charge distributions below:

$$\rho_1 = |u_a(\mathbf{r}_1)|^2$$
$$\rho_2 = |u_b(\mathbf{r}_2)|^2$$

That is,

$$V_{1.1} = \int \frac{\rho_1(\mathbf{r}_1)\rho_2(\mathbf{r}_2)}{r_{1,2}}\,d\mathbf{r}_1\,d\mathbf{r}_2 \tag{41}$$

This is the value we should expect for the mean Coulomb energy, if the wave function were unchanged by the perturbation.

The quantity $V_{1,2}$ is called the "exchange integral" because from it one can calculate the change in energy, $\pm V_{1,2}$, resulting from the exchange degeneracy.† In the most general case, the sign and magnitude of $V_{1,2}$ will depend both on the form of the perturbing potential and on the unperturbed functions, u_a and u_b. For the special case that we are now considering, however (i.e., where the perturbing potential arises from a Coulomb interaction between two particles) we shall presently see that $V_{1,2}$ is always positive.

The physical meaning of the exchange energy can be understood in terms of correlations between electronic positions that are inevitably present whenever the wave function is symmetric or antisymmetric. To demonstrate the existence of such correlations, we note that for the symmetric function, ψ_+, the wave function is a maximum when $\mathbf{r}_1 = \mathbf{r}_2$, while for the antisymmetric function, ψ_-, the wave function is zero for $\mathbf{r}_1 = \mathbf{r}_2$ and very small when $\mathbf{r}_1$ is close to $\mathbf{r}_2$. Thus, a symmetric wave function implies an unusually *large* probability that the two electrons will be close together, and an antisymmetric function implies an unusually *small* probability that they will be close together.

† The part of the energy, $\pm V_{1,2}$, which depends on the sign with which ψ_1 and ψ_2 are combined is often called the "exchange energy."

Although the relative positions of the two electrons are thus correlated, we must be careful to point out that the position of each electron relative to the *nucleus* is not affected by symmetrization or antisymmetrization of the wave function. To prove this, we evaluate the average of an arbitrary function, $f(\mathbf{r}_1)$, of the position of one of the particles, which we take to be the first. This is

$$\begin{aligned}\bar{f}(\mathbf{r}_1) &= \tfrac{1}{2}\int(\psi_1^* \pm \psi_2^*)f(\mathbf{r}_1)(\psi_1 \pm \psi_2)\,d\mathbf{r}_1\,d\mathbf{r}_2 \\ &= \tfrac{1}{2}\int[|u_a(\mathbf{r}_1)|^2|u_b(\mathbf{r}_2)|^2 + |u_a(\mathbf{r}_2)|^2|u_b(\mathbf{r}_1)|^2]f(\mathbf{r}_1)\,d\mathbf{r}_1\,d\mathbf{r}_2 \\ &\quad \pm \tfrac{1}{2}\int[u_a^*(\mathbf{r}_1)u_b(\mathbf{r}_1)u_b^*(\mathbf{r}_2)u_a(\mathbf{r}_2) + u_b^*(\mathbf{r}_1)u_a(\mathbf{r}_1)u_a^*(\mathbf{r}_2)u_b(\mathbf{r}_2)]f(\mathbf{r}_1)\,d\mathbf{r}_1\,d\mathbf{r}_2\end{aligned}$$

Because of orthogonality of u_a and u_b, the second integral vanishes. Because of the normalization of these functions, we then obtain

$$\bar{f}(\mathbf{r}_1) = \tfrac{1}{2}\int[|u_a(\mathbf{r}_1)|^2 + |u_b(\mathbf{r}_1)|^2]f(\mathbf{r}_1)\,d\mathbf{r}_1\,d\mathbf{r}_2 \tag{42}$$

This is just the mean value of f, averaged between the states corresponding respectively to u_a and to u_b. We see then that $\bar{f}(\mathbf{r}_1)$ (and therefore the mean position of each electron) does not depend on whether the wave function is symmetric or antisymmetric.

How then are we to interpret the correlations in electronic position which are, as we have seen, associated with the symmetry of the wave function? The interpretation is that for an antisymmetric wave function, the two electrons tend to be on opposite sides of the nucleus with a higher probability than would be present in a random distribution, whereas for a symmetric wave function, there is a statistical tendency to favor their being on the same side of the nucleus. Since the Coulomb energy of interaction between electrons, $e^2/r_{1,2}$, depends on the interelectronic distance, we see that for a symmetric wave function, this energy must be larger than for an antisymmetric wave function. Because the energy difference between symmetric and antisymmetric wave functions is $2V_{1,2}$ [see eq. (40)], we conclude that the exchange integral, $V_{1,2}$, is positive for a Coulomb potential. Moreover, we see also that the so-called "exchange energy" is merely a part of the usual Coulomb energy, resulting from the quantum-mechanical correlations of the relative positions of the two electrons.

To obtain a further qualitative picture of the effects of exchange degeneracy, let us suppose, for example, that we were able to put one of the electrons into an excited state of the helium atom, while the other was in the ground state, so that the combined wave function was initially $\psi_1 = u_1(\mathbf{r}_1)u_0(\mathbf{r}_2)$.† Because of the perturbation arising from the

† In this connection, see Sec. 29, where it will be shown that this state can never actually be realized because of the requirement of antisymmetry of all electronic wave functions.

Coulomb interaction between electrons, there would be a tendency to make a transition to the state $\psi_2 = u_1(r_2)u_0(r_1)$. In this state, the excitation energy has been exchanged between the electrons. After a long time, the new wave function would be very different from the original one, since because of the existence of degeneracy, the process of transfer of energy can go a long way (see Sec. 1). The original unperturbed wave function would therefore be a very poor one to use as a starting basis in perturbation theory. There do exist, however, two wave functions, ψ_+ and ψ_-, in which the flow of probability of excitation from one electron to the other is balanced by an equal flow back. These wave functions are stationary states in the zeroth approximation, and may therefore serve as a good basis for higher order perturbation theory.

A further discussion of the significance of these wave functions will be given in Sec. 29. (In this connection, see also Sec. 18.)

20. Higher Approximations. Thus far, we have discussed only the removal of the degeneracy in the zeroth approximation. The use of perturbation theory in the higher approximations will naturally further modify the wave functions and energy levels. Yet one can draw some conclusions about the nature of these modifications without actually solving the problem completely. These conclusions are based on the fact that if two particles are identical, then the complete Hamiltonian operator must be a symmetrical function of the co-ordinates of each particle. If it were not, then the two particles could not be identical because we mean by identity that under all possible perturbations, the two particles must act in the same way. For example, in our special case of the helium atom, we see that the complete Hamiltonian is indeed symmetric in the two particles. [See eq. (37).]

Now, in general, the perturbed wave function will depend on the matrix elements between the zeroth approximate states and the other states. One can easily show that if the Hamiltonian is symmetric, then V_{mn} vanishes for any transition between a symmetric and an antisymmetric function. To prove this, consider such a matrix element

$$\int \psi_+^* V \psi_- \, d\mathbf{r}_1 \, d\mathbf{r}_2 = I$$

where ψ_+ is a symmetric function, ψ_- is antisymmetric. Now the above integral is an integral over $\mathbf{r}_1$ and $\mathbf{r}_2$, hence it should not be changed by interchanging $\mathbf{r}_1$ and $\mathbf{r}_2$, since this is just a relabeling of the variable of integration. Such an interchange leaves $V(\mathbf{r}_1, \mathbf{r}_2)$ unchanged, and also $\psi_+(\mathbf{r}_1, \mathbf{r}_2)$ since these are symmetric. $\psi_-(\mathbf{r}_1, \mathbf{r}_2)$, however, is reversed in sign. An interchange of the particles therefore reverses the sign of the integrand. Thus, we obtain $I = -I$, which can be satisfied only if $I = 0$.

The above result means that if we start out with a zeroth-order function of a definite symmetry, the subsequent wave functions obtained in the higher approximations must have the same symmetry. All stationary states of a system containing two identical particles must therefore have either symmetric or antisymmetric wave functions. This means that the classification of levels into symmetric and antisymmetric, obtained first in the zeroth approximation, will continue to be valid in all approximations.

Another way of looking at the problem is by means of the time-dependent theory. If a function starts out with a definite symmetry, and if it is subjected to symmetrical perturbations, as will happen if the two particles are identical, then, because the associated matrix elements vanish, no transitions to functions of any other symmetry can ever take place. Thus, the symmetry class is retained for all time; it is a constant of the motion.

21. Effects of Spin. Thus far, we have neglected the spin-dependent terms in the Hamiltonian [see Chap. 17, eq. (80)]. To the extent that this is a good approximation, the complete wave function can be written as a product of space and spin functions. Thus, we begin with the unperturbed functions for the problem in which neither spin nor interaction between electrons has been taken into account.

$$\psi_0 = u_a(r_1)u_b(r_2)v_m(1)v_n(2) \tag{43}$$

where $v_m(1)$ is the spin function for the first particle and $v_n(2)$ is that of the second particle; m and n are either $+1$ or -1.

The removal of the degeneracy brought about by Coulomb interaction produces zeroth-order wave functions, which are either symmetric or antisymmetric in the exchange of two space co-ordinates, but the spin wave functions are not affected. We therefore obtain for our wave functions

$$\psi = \psi_\pm(r_1, r_2)v_m(1)v_n(2) \tag{44}$$

The Coulomb energy does not, however, include all of the interaction energy between electrons, because there is still a spin-dependent term. This term arises from the spin energy [Chap. 17, eq. (80)].

$$W_{sp} = \frac{-e\hbar}{2mc}\left\{\boldsymbol{\sigma}_1 \cdot \left[\mathcal{H}(r_1) + \frac{p_1}{2mc} \times \mathcal{E}(r_1)\right] + \boldsymbol{\sigma}_2 \cdot \left[\mathcal{H}(r_2) + \frac{p_2}{2mc} \times \mathcal{E}(r_2)\right]\right\} \tag{45}$$

p_1 and p_2 represent the momenta of the first and second particles, respectively. $\mathcal{H}(r_1)$ represents the magnetic field at the position of the first particle, $\mathcal{H}(r_2)$ represents that at the second particle. Similarly $\mathcal{E}(r_1)$ and $\mathcal{E}(r_2)$ are the corresponding electric fields.

The magnetic field at the first particle produced by the orbital motion of the second particle is given by the Biot-Savart law.*

* In the above equations, e stands for the *absolute value* of electronic charge.

$$\mathcal{H}_{1,o}(r_1) = -\frac{e}{mc} p_2 \times \frac{(r_1 - r_2)}{|r_1 - r_2|^3} \tag{46a}$$

The corresponding magnetic field produced by the spin of the second particle is

$$\mathcal{H}_{1,s}(r_1) = \frac{e\hbar}{2mc} \nabla_1 \left(\frac{(r_1 - r_2) \cdot \sigma_2}{|r_1 - r_2|^2} \right) \tag{46b}$$

The total electric field acting on the first particle is

$$\mathcal{E}_1(r_1) = \frac{Zer_1}{r_1^3} - e\frac{(r_1 - r_2)}{|r_1 - r_2|^3} \tag{46c}$$

where Z is the atomic number of the nucleus. We obtain for the total spin energy

$$\begin{aligned} W_{sp} = \frac{e^2\hbar}{2m^2c^2} &\left[\sigma_1 \cdot p_2 \times \frac{(r_1 - r_2)}{|r_1 - r_2|^3} + \sigma_2 \cdot p_1 \times \frac{(r_2 - r_1)}{|r_2 - r_1|^3} \right] \\ - \frac{e^2\hbar^2}{4m^2c^2} &\left\{ \sigma_1 \cdot \nabla_1 \left[\frac{\sigma_2 \cdot (r_1 - r_2)}{|r_1 - r_2|^3} \right] + \sigma_2 \cdot \nabla_2 \left[\frac{\sigma_1 \cdot (r_2 - r_1)}{|r_2 - r_1|^3} \right] \right\} \\ - \frac{Ze^2\hbar}{4m^2c^2} &\left(\frac{\sigma_1 \cdot p_1 \times r_1}{r_1^3} + \frac{\sigma_2 \cdot p_2 \times r_2}{r_2^3} \right) \\ &+ \frac{e^2\hbar}{4m^2c^2} \left[\sigma_1 \cdot p_1 \times \frac{(r_1 - r_2)}{|r_1 - r_2|^3} + \sigma_2 \cdot p_2 \times \frac{(r_2 - r_1)}{|r_2 - r_1|^3} \right] \end{aligned} \tag{47}$$

The above term in the Hamiltonian tends to produce transitions between different spin states. Among the possible transitions are those in which the two particles exchange their spins. Since the unperturbed energy does not contain the spin, we conclude that these two levels are degenerate. The removal of degeneracy is carried out in the same way as was done for the case of exchange of space co-ordinates of the electrons [eq. (38)] and, in a similar way, one finds that the correct zeroth-order spin wave functions are

$$\left. \begin{aligned} \phi_+ &= \frac{1}{\sqrt{2}} [v_m(1)v_n(2) + v_m(2)v_n(1)] \\ \phi_- &= \frac{1}{\sqrt{2}} [v_m(1)v_n(2) - v_m(2)v_n(1)] \end{aligned} \right\} \tag{48}$$

ϕ_+ is symmetric in the exchange of the two spins, ϕ_- is antisymmetric.

The complete zeroth-order wave functions that remove both space exchange and spin exchange degeneracy are then

$$\psi_{\pm,\pm} = \psi_\pm \phi_\pm \tag{49}$$

It is of interest to consider the combined symmetry properties for simultaneous exchange of space and spin co-ordinates of the two electrons. The wave functions $\psi_{+,+}$ and $\psi_{-,-}$ are symmetric for such an exchange, while the other two, $\psi_{+,-}$ and $\psi_{-,+}$, are antisymmetric.

22. The Antisymmetry of Electron Wave Functions. From the above discussion we should expect that the excited states of helium would have, in general, four levels. Instead, there are only two. This fact and similar considerations obtained from a study of other atoms show that not all of the wave functions that are solutions of Schrödinger's equation actually appear in nature. It is found instead that all observed spectral lines can be explained correctly by assuming that the only wave functions actually appearing are those which are antisymmetric with respect to simultaneous exchange of both space and spin co-ordinates of any two electrons. This rule leads, as we shall see in Sec. 26, to the Pauli exclusion principle. It has been found that the rule of antisymmetry is obeyed, not only by electrons, but also by many other elementary particles, including neutrons, protons, and neutrinos. In fact, each type of elementary particle is characterized either by a wave function that is always antisymmetric in the exchange of two particles or else by one which is always symmetric.* No particular symmetry relations exist as far as exchanges of different types of particles are concerned.

23. Correlation between Exchange Energy and Electron Spin Brought about by Antisymmetry of Wave Functions. In order to satisfy the requirement of complete antisymmetry, it is necessary to choose either symmetric spin wave functions and antisymmetric space functions, or antisymmetric spin and symmetric space functions. Now, according to Chap. 17, Sec. 9, symmetric spin functions represent parallel spins and antisymmetric functions represent antiparallel spins. We therefore conclude that when the spins are parallel, the negative sign must be taken in eq. (39) for the exchange energy, whereas when the spins are antiparallel, the positive sign should be taken. In this way, one obtains an energy that is apparently produced by spin interactions, but that is actually the result of the fact that the mean Coulomb energy happens to be correlated with the spin. There is actually another term in the energy arising from the genuine magnetic interaction between spins [see eq. (47)], but this term is much smaller than the apparent interaction energy between spins introduced by the correlation between spin directions and the spatial symmetry of the wave function.

This apparent interaction between spins has important consequences, especially in spectroscopy and in the theory of ferromagnetism. Thus, in helium it creates a fairly large separation between singlet and triplet states. Because the exchange integral [see eq. (39b)] is positive in helium,

* There is some evidence that certain kinds of particles called mesons may have symmetrical wave functions (see Wentzel, *Quantum Theory of Fields*). If one wishes to regard photons as equivalent particles, then one can show that they should also have symmetrical wave functions (see Dirac, *Principles of Quantum Mechanics*). The reason for this general restriction to symmetric or antisymmetric wave functions is not known at present, but it is suspected that it is connected with requirements of relativistic invariance [see W. Pauli, *Rev. Mod. Phys.* **13**, 203 (1941)].

the triplet state which has an antisymmetric space wave function has a lower energy than does the singlet state.

In the problem of ferromagnetism,* we are concerned with a tendency (present in ferromagnetic materials only) for the spins of electrons in neighboring atoms to become parallel to each other, and thus to create a strong magnetization arising from the cumulative contributions of the magnetic moments of all the electrons. According to statistical mechanics,† the only reason why a state in which all spins are parallel might be thermodynamically stable is that the energy of such a system is lower than it is for a system in which spins are oriented at random. The first attempts at a theory of ferromagnetism were based on the assumption that neighboring molecular magnets tended to line up because of the magnetic energy given off when such a line-up occurred. It had long been known, however, that magnetic energies were several hundred times too small to account for the persistence of ferromagnetism up to temperatures of several hundreds of degrees centigrade. To understand why this is so, we must note that the tendency for spins to become parallel is resisted by thermal agitation, which tends to cause spins to point in more or less random directions. At very high temperatures, the effect of this agitation is, in fact, so great that the magnetization averages out to zero. As the temperature is lowered, however, a critical point, known as the Curie point, is reached, below which the forces tending to align the spins of neighboring atoms become great enough to overcome the effects of thermal agitation, with the result that the average magnetization is no longer zero. The Curie point is determined (very roughly) as that point at which the mean energy of thermal agitation κT becomes equal to the energy given off when neighboring dipoles line up. From the Curie temperature one can therefore roughly estimate the energy of interaction between dipoles and prove that it is much too large to be explained by the assumption of a purely magnetic interaction.

The cause of the line-up of neighboring spins was explained by Heisenberg, who first called attention to the fact that if the exchange integral $V_{1,2}$ is positive, the exchange energy will produce a tendency for neighboring electron spins to become parallel. This is because the antisymmetry of the complete electronic wave function requires that the negative sign be taken in eq. (39b) when the spins of two electrons are parallel, the positive sign when they are antiparallel. In this way, one obtains an energy that is apparently a result of spin interactions, but is actually a result of the correlation between mean Coulomb energy and spin. This term is several hundred times as large as the magnetic energy of interaction between spins and is therefore large enough to account for the

* See N. F. Mott and H. Jones, *Theory of Properties of Metals and Alloys*. London, Clarendon Press, 1936.

† Tolman, *Statistical Mechanics*.

observed temperatures at which ferromagnetism* ceases to exist. Moreover, the fact that only certain materials are ferromagnetic is now understood because these are the materials for which the exchange integral $V_{1,2}$ is positive. (If $V_{1,2}$ is negative, then the energy of the system will be increased when neighboring spins are parallel.)

24. Formal Expression of Exchange Energy in Terms of Spin Operators. The apparent spin interactions can be expressed formally with the aid of the spin operators. From eq. (54), Chap. 17, we note that the operator, $\boldsymbol{\sigma}_1 \cdot \boldsymbol{\sigma}_2$, is $+1$ when the spins are parallel and -3 when they are antiparallel. The operator

$$P_{1,2} = \frac{(1 + \boldsymbol{\sigma}_1 \cdot \boldsymbol{\sigma}_2)}{2} \tag{50}$$

therefore has the property that it is $+1$ when the spins are parallel, -1 when they are antiparallel. The exchange energy [eq. (40)] can then be written as

$$J_{1,2} = \pm V_{1,2} = (1 + \boldsymbol{\sigma}_1 \cdot \boldsymbol{\sigma}_2) V_{1,2} \tag{51}$$

Thus the exchange energy depends on the angle between the two spins in a way that is formally somewhat analogous to the energy of magnetic interaction of a pair of dipoles, even though it is actually part of the electrostatic interaction.†

25. A System of Many Electrons. We now proceed to extend the theory to a system having an arbitrary number of electrons. If we denote the normalized unperturbed wave functions of the individual particles (combining space and spin) by $w_1(x_1)$, $w_2(x_2) \ldots w_N(x_N)$, then a typical unperturbed wave function for the combined system is the product

$$\psi = w_1(x_1) w_2(x_2) \ldots w_N(x_N) \tag{52}$$

This wave function is degenerate, in the sense that the same energy is obtained whenever any two particles are exchanged. In general, a total of $N!$ different wave functions can be obtained by exchanging particles, one for each permutation of the particles among the wave functions. The correct zeroth-order wave functions removing the degeneracy must be some linear combination of these $N!$ unperturbed functions.

The general problem of removal of degeneracy is fairly complicated. If, however, we restrict ourselves to the case of the electron, then the problem is greatly simplified, because the wave function must then be

* The large magnetizability associated with ferromagnetism is made possible because of the large electrostatic forces favoring a line-up of all spin directions. In paramagnetic substances, the line-up of spins is merely the result of the comparatively weak effects of external fields.

† According to eq. (51), two electrons attract each other when their spins are parallel; this is analogous to the behavior of two magnetic dipoles placed end to end. On the other hand, dipoles placed side by side show the opposite behavior.

antisymmetric in the exchange of any two particles. A function of this kind is called a totally antisymmetric function. Slater has shown that such a function is given by the determinant*

$$\psi = \begin{vmatrix} w_1(x_1) & w_2(x_1) & \cdots\ \cdots & w_N(x_1) \\ w_1(x_2) & w_2(x_2) & \cdots\ \cdots & w_N(x_2) \\ \cdots & \cdots & \cdots\ \cdots & \cdots \\ \cdots & \cdots & \cdots\ \cdots & \cdots \\ w_1(x_N) & w_2(x_N) & \cdots\ \cdots & w_N(x_N) \end{vmatrix} \tag{53}$$

It is readily verified that this is the function that we want. First, we note that the determinant is equal to $\sum_P (-1)^P P w_1(x_1) w_2(x_2) \ldots w_N(x_N)$ where the symbol P stands for some permutation of the particles among the wave functions. The sign of each term is either + or −, depending on whether the permutation is even or odd. The sum is taken over all possible permutations. The above determinant is therefore a linear combination of degenerate eigenfunctions. To prove that it is antisymmetric in the exchange of any two particles, we note that such an exchange produces an interchange of the two rows of the determinant. It is well known that a determinant changes sign when two rows are interchanged.

Problem 6: Prove from Schrödinger's equation that if the Hamiltonian is a symmetric function of the space and spin co-ordinates of all the particles and if the wave function is initially proportional to the Slater determinant [eq. (53)], then, provided that we neglect transitions to states of a different energy, it remains proportional to the same Slater determinant for all time.

It now follows immediately from Problem 6 that the exchange degeneracy is removed by a choice of the antisymmetric function; for we have obtained a linear combination of degenerate eigenfunctions which is not changed in the zeroth order with the passage of time.

26. Pauli Exclusion Principle. We shall now prove that the antisymmetry of the wave functions has as a consequence that no two electrons can have the same quantum state. This result, which is known as the Pauli exclusion principle, is found by experiment to be true in all cases that have ever been investigated. To derive it, we note that if two of the separate electronic wave functions in the determinant (53) are the same, then the determinant vanishes, because it has two identical columns. Thus, in all nonvanishing completely antisymmetric wave functions each electron must be in a different quantum state.

The Pauli exclusion principle is of the greatest importance in predicting atomic energy levels. To applv it, we note that since there are only two quantum states of the spin, no more than two electrons can have a

* F. Seitz, *Modern Theory of Solids*, p. 237.

given set of orbital quantum numbers, and these two must have opposite values of the spin. This result has as a consequence, for example, that in the ground state of helium, in which both electrons are in s states, the two electrons have opposite spins so that the total spin is zero. It also leads to the well-known shell structure of atoms. Thus, in an atom of higher atomic number, one first fills the state $n = 1$, $l = 0$, and stops when one has a full shell, which is, in this case, two electrons. The next electrons can be either $n = 2$ and $l = 0$ or $n = 2$ and $l = 1$, and both of these have a considerably higher unperturbed energy than does the $n = 0$ level. According to Chap. 18, Sec. 54, the state with $l = 0$ will usually be the one of lower energy because its orbits are the more penetrating. Two more electrons can go into this level. Since there are three levels with $l = 1$, ($m = 0, \pm 1$) a total of six electrons can go into the level $l = 1$, $n = 2$. At this point, one obtains a full shell and the total spin must again be zero. By considerations of this kind, one is able to account qualitatively for the general electronic structure of the elements. Of course, this is still the zeroth approximation, and, in order to obtain more precise predictions about the energy levels, one needs a more accurate solution.*

27. The Solution to Higher Degrees of Approximations. The Slater determinants provide the correct zeroth approximate wave functions. Since the exact wave function must also be antisymmetric, it too can be expanded in a series of Slater determinants corresponding to all possible levels of the unperturbed system.

28. Totally Symmetric Wave Functions. As pointed out in Sec. 22, all elementary particles which do not have completely antisymmetric wave functions have completely symmetric wave functions. Such a wave function is given by

$$\psi = \sum_P P w_1(x_1) w_2(x_2) \ . \ . \ . \ w_N(x_N) \tag{54}$$

Problem 7: Prove that the above wave function is completely symmetric.

Problem 8: Prove that any completely symmetric wave function is orthogonal to any completely antisymmetric wave function.

Problem 9: Prove that the matrix element of a symmetrical Hamiltonian between a symmetric and an antisymmetric wave function is zero.

When there are only two particles, the most general possible wave function must be some linear combination of symmetric and antisymmetric functions. When there are more than two particles, however, then there exist functions of intermediate symmetry that also remove the degeneracy. Such functions are neither completely symmetric nor completely antisymmetric, but are symmetric with respect to some

* See, for example, Condon and Shortley, *The Theory of Atomic Spectra*. London: Cambridge University Press, 1935.

exchanges and antisymmetric with respect to others. Such functions however, do not actually appear, because they do not satisfy the exclusion principle (see, for example, Problem 10).

Problem 10: Consider a system of three identical particles for which the unperturbed wave functions are u_1, u_2, and u_3, and for which the perturbing terms in the Hamiltonian is $V(r_{1,2}) + V(r_{2,3}) + V(r_{1,3})$ where $r_{1,2}$ is the distance between particle 1 and particle 2. Carry out the procedure removing the degeneracy, and show that one obtains three energy levels, one corresponding to a completely symmetric wave function, one to a completely antisymmetric wave function, and one to a set of wave functions of intermediate symmetry.

29. Indistinguishability of Equivalent Particles. In classical physics there are two ways of identifying particles. The first takes advantage of the fact that different particles act differently. Thus, they may reflect or scatter light differently, or react to electric or magnetic forces in a different way. In order to "label" a particle in this way, we need to use at least one property that is unique for that particle. Since all actions of the particle are determined by the Hamiltonian, we therefore require that in some circumstances, at least, the Hamiltonian for each different particle be different.

If a pair of particles is completely equivalent, i.e., if each has the same form of Hamiltonian under all conditions, then the labeling method of identification does not work. It is still possible, however, in classical physics to identify the particles by the continuity of their trajectories, because this property enables an observer to follow each particle.

In the quantum theory, the problem of identifying equivalent objects, such as electrons, is more difficult, mainly because of the wave properties of matter. Even if electrons were not restricted to antisymmetric wave functions, for example, one would not always be able to identify an electron by following its trajectory, simply because each electron has a wave packet of finite width. If these wave packets overlap, then it becomes impossible to identify a given electron by tracing its trajectory. Nevertheless, as long as the electrons did not have the same wave functions it might be possible to continue to identify them by some property other than the position; for example, the momentum or the angular momentum, or perhaps some other observable.

We shall now see, however, that when the wave functions are restricted to being completely antisymmetric or completely symmetric one cannot even give a meaning to the notion of identifying separate electrons. Suppose, for example, that we consider a situation in which the first electron occupies a region of space near $x = x_a$ with a wave packet $f(x_1 - x_a)$, while the second electron occupies a region of space near $x = x_b$, with a wave packet $g(x_2 - x_b)$. The combined wave function for this system is

$$\psi_1 = f(x_1 - x_a)g(x_2 - x_b) \tag{55}$$

On the other hand, the corresponding antisymmetric wave function is

$$\psi = \frac{1}{\sqrt{2}} [f(x_1 - x_a)g(x_2 - x_b) - f(x_2 - x_a)g(x_1 - x_b)] \tag{56}$$

When the wave function is antisymmetric, each electron should not, however, be ascribed to a definite (but unknown) packet opposite to that of the other,† because important physical properties may depend on interference between the functions representing states in which the two electrons have respectively been interchanged. Thus, the probability function is

$$P(x_1, x_2) = \tfrac{1}{2}\{|f(x_1 - x_a)g(x_2 - x_b)|^2 + |f(x_2 - x_a)g(x_1 - x_b)|^2 + [f^*(x_1 - x_a)g(x_1 - x_b)g^*(x_2 - x_b)f(x_2 - x_a) + \text{complex conjugate}]\} \tag{57}$$

The first two terms represent the "classical probability" terms, whereas the remaining terms are characteristic quantum-mechanical interference effects. To the extent that such interference effects are important (for example, in the determination of "exchange energy" treated in Sec. 19), we cannot correctly regard each electron as having a definite identity. Because interference properties are characteristic of the wave-like aspects of matter (see Chap. 6), they are better understood in terms of a description of the two-electron system as something showing, in these applications, a greater resemblance to a six-dimensional wave than to a pair of distinct particles. Of course, if the system were ever to interact with an apparatus that treated the two particles differently, the wave function would cease to be completely antisymmetric (see Sec. 20), and we could then distinguish the two electrons.‡ Thus, such an apparatus would tend to bring about the realization of the system's potentialities§ for developing into something showing less resemblance to a six-dimensional wave and more resemblance to a pair of distinct particles. But because the electrons are identical in *all* interactions, these potentialities can *never* be realized, since the requisite apparatus cannot actually be constructed.

It follows then that for totally symmetric or antisymmetric wave functions, different electrons do not have an identity, since they do not even act like separate and distinct objects, which are capable, in principle, of being identified. An example of this property is that when two sets of electronic variables are interchanged, one obtains the same wave function, except for a minus sign, so that the quantum state of the system is not altered by such an exchange. On the other hand, if the electrons

† See Sec. 18.

‡ A two-electron system can act completely like a pair of distinct objects only to the extent that its wave function is separable into a product of independent wave functions, such as (55). In this connection, see also Chap. 22, Sec. 17.

§ See Chap. 6, Chap. 8, and Chap. 16, Sec. 25.

were distinct and identifiable objects, one would expect a new quantum state to result from such an exchange. In fact, for the wave function (55), an exchange of electronic variables does bring about a new quantum state. The failure of an exchange of particle variables to produce a new quantum state has important consequences in statistical mechanics,* and leads to "Bose-Einstein" statistics for particles with completely symmetric wave functions, and "Fermi-Dirac" statistics for particles with completely antisymmetric wave functions.

* Tolman, *The Principles of Statistical Mechanics.*

CHAPTER 20

Sudden and Adiabatic Perturbations

1. General Adiabatic Perturbations. Thus far, we have treated the case of a slowly varying potential only when the potential is so small that perturbation theory can be used.* It is possible, however, to extend this treatment to a more general problem in which the perturbing potential undergoes large changes, but over such a long period of time that the change in potential during the period of the light that is emitted in transition to the nearest neighboring state is small compared with the change of energy involved in this transition. More precisely, this requirement is that

$$\frac{\tau}{(E_s^0 - E_n^0)} \frac{\partial V}{\partial t} \ll 1$$

where E_s^0 is the initial energy, E_n^0 the energy of the nearest neighboring state, and τ the period in question. Since the period of the light emitted in transition is $\tau = h/(E_s^0 - E_n)$, our requirement is

$$\frac{h}{(E_s^0 - E_n^0)^2} \frac{\partial V}{\partial t} \ll 1 \tag{1}$$

The basic idea of this approximation is that if $\partial V/\partial t$ is small enough to satisfy the above condition, then the wave function is, at any instant of time, very nearly equal to that which would be obtained if $\partial V/\partial t$ were zero, and V were equal to its instantaneous value.

To illustrate the method, let us suppose that the Hamiltonian operator is given by

$$H(t) = -\frac{\hbar^2}{2m} \nabla^2 + V(x, p, t) \tag{2}$$

Schrödinger's equation becomes

$$i\hbar \frac{\partial \psi}{\partial t} = H(t)\psi$$

Thus if $H(t)$ varies slowly enough, we may expect that a good approximate solution should be given by solving Schrödinger's equation at each instant of time under the assumption that H is a constant and equal to

* Chap. 18, Sec. 51

its instantaneous value, $H(\theta)$, where θ is the value of t at which we wish to evaluate H. The stationary-state wave functions, obtained by setting $t = \theta =$ constant, would satisfy the equation

$$H(\theta)u_n(x, \theta) = E_n(\theta)u_n(x, \theta) \tag{3}$$

One may now expect that if H is a slowly varying function of θ, a good approximate solution is

$$\psi_n = u_n(x, t)e^{-i\int_0^t \frac{E_n(\theta)\,d\theta}{\hbar}} \tag{4}$$

This means simply that the space variation of the wave function at the time, t, is that of the "instantaneous" eigenfunction of $H(t)$, whereas the angular frequency is given by the instantaneous value of $E_n(t)/\hbar$. We shall discuss the meaning of this equation later.

To prove that the above is a good approximate solution when $\partial H/\partial t$ is small, we note that the ψ_n form a complete orthonormal set; hence the correct wave function can be expanded as a series of the ψ_n, with coefficients C_n, which are in general functions of the time:

$$\psi = \sum_n C_n(t)u_n(x, t)e^{-i\int_0^t \frac{E_n(\theta)\,d\theta}{\hbar}}$$

Inserting this into Schrödinger's equation and using eq. (3), we obtain

$$i\hbar \sum_n \left(\dot{C}_n u_n + \frac{\partial u_n}{\partial t} C_n\right) e^{-i\int_0^t \frac{E_n(\theta)\,d\theta}{\hbar}} + \sum_n C_n u_n E_n\, e^{-i\int_0^t \frac{E_n(\theta)\,d\theta}{\hbar}} = \sum_n C_n u_n E_n\, e^{-i\int_0^t \frac{E_n(\theta)\,d\theta}{\hbar}}$$

We now multiply by $u_m^* e^{i\int_0^t \frac{E_m(\theta)\,d\theta}{\hbar}}$ and integrate over all space. Using normalization and orthogonality of the U_m, we obtain

$$\dot{C}_m + \sum_n C_n \int u_m^* \frac{\partial u_n}{\partial t} e^{-i\int \frac{(E_n - E_m)\,d\theta}{\hbar}}\, dx = 0 \tag{5}$$

We wish now to simplify these equations somewhat by transforming away the term in the sum for which $m = n$. To do this, we first show that the coefficient of this term, namely $\gamma_m(t) = \int u_m^* \frac{\partial u_m}{\partial t}\, dx$, is a pure imaginary. To prove this, we begin with the normalization condition, $\int u_m^* u_m\, dx = 1$. Differentiation of this equation yields

$$\int \left(u_m^* \frac{\partial u_m}{\partial t} + \frac{\partial u_m^*}{\partial t} u_m\right) dx = \gamma_m(t) + \gamma_m^*(t) = 0$$

The above shows that the real part of γ_m vanishes, so that we can write $\gamma_m = i\beta_m$, where β_m is real. We now make the substitution

$$v_m = u_m e^{-i\int_0^t \beta_m(\theta)\, d\theta} \qquad E_{n'} = E_n + \beta_n \tag{6}$$

This substitution leads to the equations

$$\left.\begin{aligned} &\dot{C}_m + \sum_{n \neq m} C_n \left(\int v_m^* \frac{\partial v_n}{\partial t}\, dx \right) e^{-i\int_0^t (E_n' - E_m')\, d\theta} = 0 \\ \text{or} \quad & \\ &\dot{C}_m + \sum_{n \neq m} C_n \alpha_{mn}\, e^{-i\int_0^t (E_n' - E_m')\, d\theta} = 0 \end{aligned}\right\} \tag{7}$$

where

$$\alpha_{mn} = \int v_m^* \frac{\partial v_n}{\partial t}\, dx \tag{8}$$

We note that the above substitution leads to a set of functions, v_m, which are still orthonormal, so that it amounts only to a trivial phase change. This substitution also produces a change in the energy, which is small whenever H changes slowly with the time.

Our next problem is to show that if we start with an approximate solution v_s, then the coefficient of other states C_m will remain small for all time. To do this, we begin with eq. (3)

$$H(t)v_n(t) = E_n(t)v_n(t)$$

Differentiation with respect to t yields

$$\frac{\partial H}{\partial t} v_n + H \frac{\partial v_n}{\partial t} = \frac{\partial E_n}{\partial t} v_n + E_n \frac{\partial v_n}{\partial t}$$

Multiplying by v_m^* (where $m \neq n$) and integrating over all space, we obtain

$$\int v_m^* \frac{\partial H}{\partial t} v_n\, dx + \int v_m^* H \frac{\partial v_n}{\partial t}\, dx = \frac{\partial E_n}{\partial t} \int v_m^* v_n\, dx + E_n \int v_m^* \frac{\partial v_n}{\partial t}\, dx$$

We now use the fact that H is Hermitean, so that in the second term on the left, we can operate on v_m^* instead of on $\partial v_n/\partial t$. We also note that the first term on the right disappears because of orthogonality of v_m and v_n. The result is

$$\int v_m^* \frac{\partial H}{\partial t} v_n\, dx = (E_n - E_m) \int v_m^* \frac{\partial v_n}{\partial t}\, dx$$

and from eq. (8)

$$\alpha_{mn} = \frac{1}{E_n - E_m} \int v_m^* \frac{\partial H}{\partial t} v_n\, dx \tag{9}$$

We finally obtain

$$C_m + \sum_{n \neq m} \frac{C_n \int v_m^* \frac{\partial H}{\partial t} v_n \, dx \, e^{-i\int_0^t (E_n' - E_m') \, d\theta}}{E_n - E_m} = 0 \tag{10}$$

We are now ready to proceed more or less as in the method of variation of constants. Suppose that the system starts with $C_s = 1$ and $C_n = 0$ for $n \neq s$. Then one can solve for C_m by successive approximations. For the first approximation, we obtain

$$\dot{C}_{ms} + \frac{(\partial H/\partial t)_{ms}}{E_s - E_m} e^{-i\int_0^t \frac{(E_s' - E_m') \, d\theta}{\hbar}} = 0 \tag{11}$$

where
$$\left(\frac{\partial H}{\partial t}\right)_{ms} = \int v_m^* \frac{\partial H}{\partial t} v_s \, dx$$

If E_s' and E_m' are slowly varying functions of θ, then in any particular time interval, the integral in the exponential may be replaced approximately by $(E_s' - E_m')t/\hbar$. Furthermore, β is usually small, so that E_s' and E_m' can be replaced by E_s and E_m. The result is

$$\dot{C}_{ms} + \frac{1}{(E_s - E_m)} \left(\frac{\partial H}{\partial t}\right)_{ms} e^{-i(E_s - E_m)t/\hbar} = 0 \tag{12}$$

To estimate the matrix element, we can neglect the slow change of $(\partial H/\partial t)_{ms}$. We then obtain

$$C_{ms} \cong \frac{\hbar}{i(E_s - E_m)^2} \left(\frac{\partial H}{\partial t}\right)_{ms} [e^{-i(E_s - E_m)t/\hbar} - e^{-i(E_s - E_m)t_0/\hbar}] \tag{13}$$

The exponential factor in eq. (13) is at most of the order of unity. Hence, the total probability of transition to the mth level is less than

$$|C_{ms}|^2 \cong \frac{4\hbar^2}{(E_m - E_s)^4} \left| \left(\frac{\partial H}{\partial t}\right)_{ms} \right|^2 \tag{14}$$

We thus argue that if $(\partial H/\partial t)_{ms}$ is small enough [i.e., if condition (1) is satisfied], a negligible error is made by neglecting $|C_{ms}|^2$ and saying that the system remains in the state $v_s(x, t)$, even though v_s itself is changing with time. This is known as the "adiabatic approximation." This result is formally very similar to that obtained from the variation of constants [eq. (14a), Chap. 18], except that λv_{ms} has been replaced by $\frac{\hbar}{E_m - E_s}\left(\frac{\partial H}{\partial t}\right)_{ms}$. Now, $\frac{h}{E_s - E_m}$ is just the period τ of the light emitted in the transition from s to m; hence

$$\frac{\hbar}{E_s - E_m}\left(\frac{\partial H}{\partial t}\right)_{ms} = \frac{\tau}{2\pi}\left(\frac{\partial H}{\partial t}\right)_{ms}$$

is just the matrix element of the change of H during the time $\tau/2\pi$. The condition for validity of the adiabatic approximation is then equivalent to the requirement that the change of H during a period is small compared with the energy difference, $E_s - E_m$.

If this condition is met, eq. (4) is a good approximate solution to Schrödinger's equation.

The above criterion can often be written more conveniently in terms of the angular frequency, $\omega_{ms} = (E_m - E_s)/\hbar$. To make the adiabatic approximation valid, we must have

$$\frac{\hbar}{(E_m - E_s)^2}\left(\frac{\partial H}{\partial t}\right)_{ms} = \frac{1}{\hbar\omega_{ms}^2}\left(\frac{\partial H}{\partial t}\right)_{ms} = \frac{(\partial H/\partial t)_{ms}}{\omega_{ms}(E_m - E_s)} \ll 1 \qquad (15)$$

2. Interpretation of Results. Let us imagine that we slowly changed the shape of the potential energy which binds the electron to the nucleus. This could be done, for example, with an intense external field or perhaps by slowly bringing another charged particle near the atom in question. The wave function would become slowly distorted, but, as we have seen, the quantum number n remains constant. This is because to form a new quantum state, we must allow the wave function to make another oscillation in the region of positive kinetic energy, and this requires a large change of energy.* Since the potential is changing very slowly with time, it seems reasonable that this large change of wave function does not occur, but that instead, there is a gradual change of shape of the wave function, such that it accommodates itself to the changing potential by retaining a constant number of nodes. Only if the potential were changed rapidly in comparison to $\hbar/(E_s - E_n)^2$ would there occur transitions to other quantum states, i.e., to states containing other numbers of nodes. The same result has been obtained for the case of small slowly varying perturbations (Chap. 18, Sec. 51), for which we have seen that at each instant of time the wave function is equal to the stationary-state wave function appropriate to the value of the Hamiltonian at that instant of time.

In the classical limit, we have shown in the discussion of the WKB approximation (Chap. 12, Sec. 13) that the number of nodes in the wave function is equal to $J/\hbar$, where J is the action variable. We therefore conclude that in an adiabatic change of the Hamiltonian, the action J remains constant. It is, in fact, a well-known theorem of classical mechanics that J does indeed remain constant in an adiabatic process.† In fact, Ehrenfest originally argued from the adiabatic invariance of J that this was the only classical quantity that could sensibly be quantized. The reason is that one can always produce in any system an arbitrary

* See Chap. 11, Sec. 12.

† See Born, *Atomic Physics*.

slowly varying change of the Hamiltonian, for example, by applying an external field. If any quantity is quantized, it can change only by a minimum discrete amount. On the other hand, the energy of the system is observed classically to change continuously. The only way to make the quantum theory approach classical theory properly for this problem is to have the quantity which is quantized be a classical constant under adiabatic changes, and to have the relation of this quantity to the energy change in a continuous way.

An example of a classical adiabatic change is the slow shortening of the length of an oscillating pendulum. For a simple harmonic motion $J = E/\nu$ (Chap. 2, Sec. 11), where ν is the frequency. According to the theorem that J is an adiabatic invariant, we conclude that E is proportional to ν. If the string is shortened, the energy in the pendulum therefore increases. This increase can easily be verified by directly calculating the work necessary to shorten the string against the centrifugal force of the oscillating pendulum.*

3. Applications. (*a*) *Stern-Gerlach Experiment. Deflection of Atoms in an Inhomogeneous Magnetic Field.* In Chap. 14, Sec. 16, we have discussed the Stern-Gerlach experiment, in which a beam of atoms is sent through an inhomogeneous magnetic field and suffers a deflecting force. In discussing this problem we neglected the possibility that the magnetic field could cause transitions in which the angular momentum was changed. We must remember, however, that the magnetic field changes in intensity and direction where the atom enters or leaves the region of the field, as shown in Fig. 1.

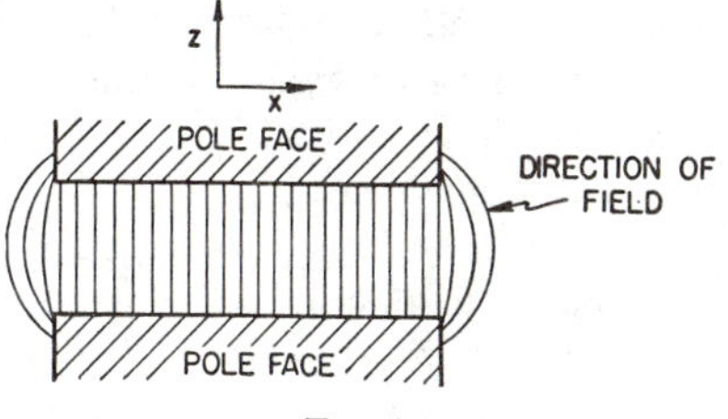

FIG. 1

When the atom is passing through the edge of the magnet, it experiences a time-varying field, which, as we have seen in the section on the method of variation of constants,† can cause transitions in which the component of the angular momentum is changed. If this were to happen to any appreciable extent, the conclusions drawn from the experiment would be invalidated.

What, then, are the conditions that no appreciable number of transitions occur? These obviously are just those of adiabatic invariance. In other words, the time spent by the atom in the region of varying field must be long compared to the period, $\tau = h/(E_n - E_s)$, of whatever transitions may take place. If the particle enters the magnetic field slowly enough, it will then have exactly the same value of L_z inside the field

* M. Born, *Mechanics of the Atom.*
† Chap. 18, Secs. 7 and 12.

as it had when it was outside. Furthermore, it will also leave the region of the field without changing its value of L_z if the magnetic field decreases at a correspondingly slow rate.

To treat this problem quantitatively, we begin with the additional perturbing term on an atom in a magnetic field [see Chap. 15, eq. (52)]

$$\lambda V = \frac{e\mathcal{H}}{2\mu c} \cdot L$$

where μ is the electron mass, $\mathcal{H}$ is the magnetic field, and L is the angular momentum. (The validity of this formula requires that the fractional change of the magnetic field $\mathcal{H}$ across the atom be small.)

We shall make the assumption that the atom moves in the x direction and that the field has a line of symmetry $y = 0$ and $z = 0$. On this line of symmetry, $\mathcal{H}$ is in the z direction. We shall further assume that the motion of the atom as a whole is classically describable; this is permissible because it is so heavy that the quantum effects of the uncertainty principle produce negligible changes of velocity, as was shown in Chap. 19, Sec. 13. We therefore assume that the atom moves on a track that is essentially in the x direction and that we can write $x = vt$, $y =$ constant, and $z =$ constant. Of course, the particle experiences a small deflecting force in the z direction, but this does not make an appreciable change in the z co-ordinate until long after the particle has left the magnetic field; we therefore neglect it.

The magnetic field is a function of position and, since the atom's position is changing with time, $\mathcal{H}$ becomes a function of time. Thus, we write

$$\left.\begin{aligned} \mathcal{H} = \mathcal{H}(x, y, z) &= \mathcal{H}(vt, y, z) \\ \frac{\partial \mathcal{H}}{\partial t} &= v\frac{\partial \mathcal{H}}{\partial x} \end{aligned}\right\} \qquad (16)$$

What we must now do is to investigate whether this changing potential can cause transitions in which the value of L_z is changed, i.e., will the direction of angular momentum be flipped? Let us begin with the case of a particle which moves along the line of symmetry. For this particle, $\mathcal{H}$ remains always in the z direction, and we obtain

$$\lambda V = \frac{e\mathcal{H}}{2\mu c} L_z$$

$$\frac{\partial}{\partial t}(\lambda V) = \frac{ev}{2\mu c}\frac{\partial \mathcal{H}}{\partial x} L_z$$

Now the operator L_z has the property that the wave functions, $e^{im\phi}$, are its eigenfunctions. This means that matrix elements corresponding to a change of m will vanish. Let us recall that $\partial\mathcal{H}/\partial x$ is just the numerical value of the gradient of the field at the center of the atom. Thus, we conclude that along the central line of symmetry, no transitions will take place.

If, however, the particle has some other value of z, the field will not be entirely in the z direction as it enters the magnet. To obtain an estimate of the x component of $\mathcal{H}$ as a function of z, for example, one can expand H_x as a series of powers of z,

$$H_x \cong (H_x)_{z=0} + z\left(\frac{\partial H_x}{\partial z}\right)_{z=0} + \cdots$$

where $(H_x)_{z=0} = 0$ by hypothesis. Because $\nabla \times \mathcal{H} = 0$ in free space, one obtains $\partial \mathcal{H}_x/\partial z = \partial \mathcal{H}_z/\partial x$. Thus we obtain

$$\mathcal{H}_x \cong z\left(\frac{\partial \mathcal{H}_z}{\partial x}\right)_{z=0}$$

The perturbing term in the Hamiltonian becomes

$$\lambda V = \frac{e}{2\mu c}(H_z L_z + \mathcal{H}_x L_x)$$

Now, we saw that H_z causes no transitions to other values of m. On the other hand, the term L_x does cause such transitions, as one can see, for example, by noting from Chap. 14, Sec. 10 that $(L_x + iL_y)\psi_m \sim \psi_{m+1}$ and $(L_x - iL_y)\psi_m \sim \psi_{m-1}$. The matrix elements of L_x between different values of m therefore do not vanish, but are of the order of $\hbar$. In classical physics, this corresponds to the fact that the z component of H exerts no torque on the z component of the magnetic moment, but a field with an x component definitely exerts a torque on a moment which is in the z direction.

In order to apply the criterion for adiabatic motion, we must evaluate

$$\left[\frac{\partial}{\partial t}(\lambda V)\right]_{ns} \cong \frac{ez}{2\mu c}\frac{\partial^2 \mathcal{H}_z}{\partial t\, \partial x}\int \psi_n^* L_x \psi_s \, dx$$

(The integration is carried out over the co-ordinates of the electron in the atom under the assumption that $\mathcal{H}_x$ is approximately constant over the space where the electron's wave function is large, i.e., over the size of the atom.) As we have seen, the integral is of the order of $\hbar$. We also have from eq. (16)

$$\frac{\partial \mathcal{H}_z}{\partial t} = v\frac{\partial \mathcal{H}_z}{\partial x}$$

Our matrix element is therefore of the order of

$$\frac{\hbar evz}{2\mu c}\frac{\partial^2 \mathcal{H}_z}{\partial x^2}$$

To test adiabatic approximation, we must evaluate the expression in eq. (15)

$$\frac{1}{\hbar}\frac{(\partial\mathcal{H}/\partial t)_{ns}}{\omega_{ns}^2} \cong \frac{1}{\omega_{ns}^2}\frac{ezv}{2\mu c}\frac{\partial^2\mathcal{H}_z}{\partial x^2}$$

where $\omega_{ns} = \dfrac{E_n - E_s}{\hbar} = \dfrac{e\mathcal{H}_z}{2\mu c}(m - m') = \dfrac{e\mathcal{H}_z}{2\mu c}$ (Larmor frequency)

The requirement (15) then becomes

$$\frac{vz}{\omega_{ns}}\left(\frac{\partial^2\mathcal{H}/\partial z^2}{\mathcal{H}_z}\right) \ll 1$$

If Δx is the distance in which the magnetic field builds up from zero to its full value $\mathcal{H}_z$, then one can write roughly

$$\frac{\partial^2\mathcal{H}_z}{\partial x^2} \cong \frac{\mathcal{H}_z}{(\Delta x)^2}$$

The criterion for adiabatic invariance becomes

$$\frac{v}{\omega_{ns}\,\Delta x}\frac{z}{\Delta x} \ll 1 \tag{17}$$

The significance of the above terms is as follows:

$v/\omega_{ns}\,\Delta x$ is the ratio of the distance moved by the particle during the period of a Larmor precession to the distance Δx, in which the field undergoes its major variation. $z/\Delta x$ is the ratio of the mean distance of the particles in the beam from the line of symmetry to Δx. The ratio $z/\Delta x$ is usually not very small, since the distance Δx is of the order of the distance between pole pieces, and the height of the beam is usually a fair fraction of this latter ratio. The validity of the adiabatic approximation therefore usually requires in practice that $v/\omega\,\Delta x$ be small.

For magnetic moments arising from atomic electrons,

$$\omega = e\mathcal{H}/2\mu c \cong 10^7\mathcal{H}$$

where $\mathcal{H}$ is measured in gauss. Thermal velocities are of the order of 10^4 cm/sec, while Δx is of the order of 1 mm. Thus, $v/(\omega\,\Delta x) \cong 10^{-2}/\mathcal{H}$. The above ratio is easily made small, even with very weak fields.

In the study of nuclear magnetic moments, however, the Larmor frequency is determined by a quantity of the order of a proton mass. In this case, the ratio in question will be of the order of $20/\mathcal{H}$. It is clear that one must go to moderately large magnetic fields to make this ratio small.

Relation to Curvature of Magnetic Field. If a particle is moving through a magnetic field that changes its direction from point to point, and if the adiabatic condition is satisfied, i.e., if the direction of the field does not curve too rapidly, the component of L in the direction of the *field* will remain constant, despite the change in direction of the field.

Resonant Flipping of Angular Momentum With Radio Frequency Fields. If the adiabatic condition is not satisfied, one obtains transitions between different components of L_z. In some experiments, one tries on purpose to obtain such transitions by using a rapidly oscillating magnetic field* (of radio frequency). We shall not, however, discuss this point in detail here.

(*b*) *Collisions of Gas Molecules.* If two gas atoms approach each other in a collision, an important question is whether the forces resulting from their interaction can cause transitions among the electronic states. In other words, can molecular kinetic energy be transferred to electronic excitation, or vice versa, can excited electrons make transitions to the ground state, giving up their energies to the kinetic energy of the molecules? Now, molecular velocities are usually fairly low (about 10^4 cm/sec), whereas velocities of electrons in atoms are much higher ($\cong 10^8$ cm/sec). During an electronic period of rotation in the atom, the molecule therefore does not move very far, so that the interaction energy does not change much. This means that the collision may usually be regarded as an adiabatic process, in which the electron remains in its original quantum state. As a result, the collision will be elastic, in the sense that after it is all over, the electrons will neither have gained energy from nor lost energy to the motion of the molecules. [This property has already been used, for example, in describing the van der Waals forces (Chap. 19, Sec. 13), where we neglected the effects of motion of the molecules.]

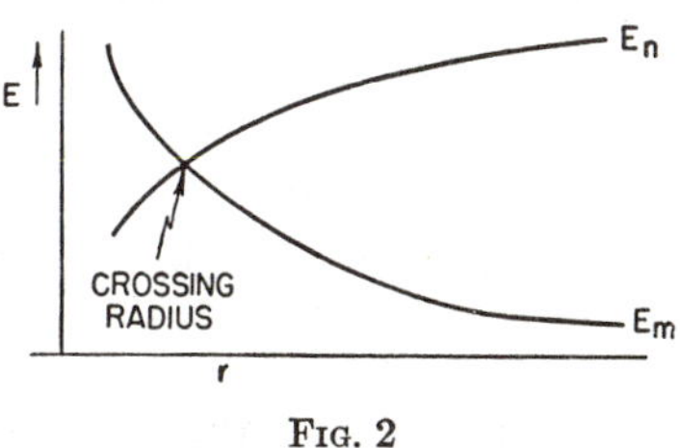

FIG. 2

There are, however, many cases in which this adiabatic property breaks down. Let us recall that the validity of the adiabatic approximation requires not only the smallness of $\partial H/\partial t$, but also that $E_n - E_s$ shall not become too small. When the atoms are far apart, the electronic states are usually fairly widely separated. But as the atoms approach each other, the energies of the electronic states are changed, because each electron is in the combined field of force of both atoms. It may happen that this change is in such a direction as to make electronic levels cross at some radius, as shown in Fig. 2. If the particles get as close as the crossing radius, there is a large probability of a change of quantum state. Then, when they recede, they may be in another quantum state. For example, suppose the atom is originally in the excited state. If it collides with another atom and if the atoms get as close as the crossing radius,

* J. B. M. Kellogg and S. Millman, *Rev. Mod. Phys.*, **18**, 323 (1946).

there may be a transition to the ground state. Then when the atoms recede, the electron may be left in the ground state, and the electronic energy will have gone into molecular kinetic energy. This process is known as a "collision of the second kind."*

(*c*) *Energy Loss of Fast Charged Particles to Atoms.* When a heavy charged particle, such as a proton, or an α-particle, moves past an atom, the force between it and the electrons may result in the transfer of energy to the electrons, thus causing the excitation or ionization of the atom. The resulting energy loss slows down the fast particle. Eventually, after enough of such transfers, the fast particle will be brought to rest. The probability of such energy transfers will then determine the mean range of the fast particle in the material in question.

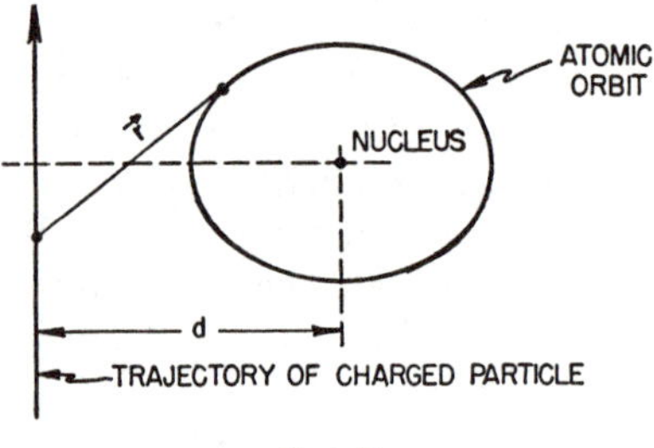

FIG. 3

It is clear that the energy transfer in any particular collision will depend on how close the charged particle comes to the atom. We wish to obtain a more precise idea of how the energy transfer depends on the distance of closest approach, d. The nature of the problem is illustrated in Fig. 3.

Let us first give this process a classical description. The force between the charged particle and an atomic electron is $Ze^2\mathbf{r}/r^3$, where $\mathbf{r}$ is the vector representing the distance between the electron and the particle. This force will be large only for a time during which the charged particle remains within a distance of the order of d from the point of closest approach; thereafter, it decreases very rapidly with increasing distance. Thus, if v is the velocity of the particle, the time during which energy transfer takes place is of the order of $d/v = \tau$. Now, if this time is very short compared with the period of rotation of the electron in its orbit, τ_e, the collision will be over before the atomic electron can move very far. Such a collision is called an "impulsive" collision, and it will normally result in an appreciable transfer of energy to the electron. On the other hand, if the distance d is so large that $\tau \gg \tau_e$, the electron will make many revolutions while the collision is taking place. In the limit of very long time of collision, the electronic orbit will adjust adiabatically to the change of potential resulting from the presence of the heavy charged particle. In other words, the electron will move approximately in the orbit in which it would move if the heavy charged particle were fixed at

* For a more complete discussion, see L. Pauling, *The Nature of the Chemical Bond;* Ruark and Urey, *Atoms, Molecules and Quanta*, pp. 386–403; Mott and Massey, *The Theory of Atomic Collisions*, pp. 243–250; C. Zener, *Phys. Rev.* **37**, 556 (1931); E. Stueckelberg, *Helv. Physica Acta*, **5**, 6, 369 (1932).

its instantaneous position. As the heavy particle moves, the electronic orbit changes slowly and reversibly (i.e., adiabatically), so that after the collision is over, the electron is left in the same orbit as before the collision. As a result, no energy will be transferred in an adiabatic collision. Since the collision becomes adiabatic when $d/v > \tau_e$, one concludes that collisions in which $d > \tau_e v$ will not transfer appreciable energies. This result is important in computing the penetrating power of charged particles in matter. Bohr has worked out the theory of this problem in detail [*Phil. Mag.*, **25,** 10 (1913)].

The variation of energy transfer with the velocity of the incident particle can also be understood as follows: At a given distance, d, of closest approach, the collision will be impulsive when the particle moves fast enough. The faster the particle, however, the smaller will be the momentum transfer. Thus, the energy loss tends to increase as the particle slows down, until the particle becomes so slow that the adiabatic approximation applies, and then the energy transfer decreases. As a result, there will be some velocity (of the order of the velocities of electrons in atoms) at which the energy transfer will be a maximum.

In quantum theory, the problem is more or less the same, except that one must compare d/v with $\hbar/(E_n - E_0)$, where E_n and E_0 are the atomic energy levels. But since $\hbar/(E_n - E_0)$ is usually of the same order as the classically calculated periods, the classical and quantum-mechanical adiabatic conditions are essentially the same.

4. The Approximation of Sudden Change of Potential. In many cases, one has a large disturbance, which is, however, turned on very rapidly, in comparison with the period involved in a transition ($\tau \cong \hbar/E_n - E_0$). We have already treated the case of a small disturbance of this kind by the method of variation of constants.* It is easy, however, to generalize this treatment to the case of a large disturbance. To do this, let us suppose that at the time $t = 0$, the Hamiltonian suddenly changes from H_0 to H_1 and remains constant thereafter. Up until $t = 0$, the eigenfunctions are given by $u_n\, e^{-iE_n{}^0t/\hbar}$, where $H_0 u_n = E_n^0 u_n$. After the time $t = 0$, the eigenfunctions of the Hamiltonian operator will be denoted by v_m. They satisfy the equation

$$H_1 v_m = E_m v_m \tag{18}$$

and they will have the time variation, $v_m\, e^{-iE_m t/\hbar}$.

If the system has been left to itself for a long time before $t = 0$, it will settle down to some stationary state, which is, in this case, an eigenstate of H_0. Let us suppose that it is in the nth eigenstate. The wave function at $t = 0$ will then be

$$\psi = u_n(x) \tag{19}$$

* See Chap. 18, Sec. 7.

After $t = 0$, $u_n e^{-iE_n^0 t/\hbar}$ will no longer be a solution of Schrödinger's equation, because the Hamiltonian suddenly changes to H_1. The wave function must remain continuous at this time, but according to the equation $i\hbar \frac{\partial \psi}{\partial t} = H\psi$, its rate of change will be altered abruptly when H_0 goes over into H_1. To find how ψ changes after $t = 0$, we adopt the usual procedure [see Chap. 10, eq. (73)] of expanding it in a series of solutions of Schrödinger's equation, which are in this case, $v_m e^{-iE_m t/\hbar}$. Thus, at $t = 0$, we write

$$\psi = u_n(x) = \sum_m C_{mn} v_m(x) \tag{20}$$

The coefficients C_{mn} are obtained by multiplying by $v_m^*(x)$ and integrating over x. Using orthogonality and normalization of the v's, we obtain

$$C_{mn} = \int v_m^*(x) u_n(x)\, dx \tag{21}$$

After $t = 0$, the wave function becomes

$$\psi_n(y) = \sum_m C_{mn} v_m(y) e^{-iE_m t/\hbar} = \sum_m \int v_m^*(x) u_n(x)\, dx\, v_m(y) e^{-iE_m t/\hbar} \tag{22}$$

The above derivation certainly holds for instantaneous changes of the Hamiltonian and must therefore also hold for sufficiently rapid, but not instantaneous, changes. The essential simplification resulting from the instantaneous change was that the wave function did not change, while the Hamiltonian was changing. The change in the wave function during the time, τ, during which the Hamiltonian is changing, is determined by an exponential factor of the order of $e^{\frac{i(\epsilon_m - \epsilon_n)t}{\hbar}}$ where ϵ_m and ϵ_n are the instantaneous values of the eigenfunctions of the Hamiltonian during this time (ϵ_m will usually be somewhere between the initial energy level, E_m^0, and the final level, E_m). In order that this change of wave function be small, it is necessary that $(\epsilon_m - \epsilon_n)\tau/\hbar \ll 1$, where ϵ_m and ϵ_n are the energy levels that are involved in the transitions under investigation.

5. Application. Emission of Electron from Nucleus in β-Decay. In the process of β decay, an electron is emitted from the nucleus with a speed that is, in most cases, close to that of light.* The electron leaves the atom in a time of the order of r/c, where r is the atomic radius. On the other hand, the periods of electrons in the atoms are of the order of $2\pi r/v$, which is usually at least 100 times as great (v is the speed of atomic electrons). This means that for all practical purposes one can say that the nuclear charge is suddenly increased from Z to $Z + 1$. At the moment that this change has occurred, the electronic wave function, $u_n(x)$, is that appropriate to a stationary state for an atom of charge Z.

* For a discussion of β-decay, see H. Bethe, *Elementary Nuclear Physics.*

In the new atom of charge $Z + 1$, this wave function no longer corresponds to a stationary state, but must be expanded in terms of the stationary-state wave functions for the new charge, $Z + 1$, as shown in eq. (20). This means that there will be a certain probability that the atom will be left in an excited state of the new atom as a result of the suddenness of the process of β-decay. This excitation can be detected by the subsequent emission of radiation, which is usually in the x ray region.

Actually, there is a whole spectrum of electron energies emitted in β-decay. A small fraction of the electrons are emitted at very low velocities. Those electrons with velocities well below those of the mean atomic electronic velocities will tend to produce adiabatic perturbations of the atomic electrons, and these will not leave the atom excited. There are so few of these low-velocity electrons, however, that their effects are hard to detect.

Problem 1: A harmonic oscillator of angular frequency ω and mass m is in its ground state. A constant force is applied in the direction of its oscillation for a time τ, which is short compared with the period of oscillation. Calculate the probability that the atom is found in its first excited state after the force has been turned off. HINT: If the constant force is a, the addition to the potential is ax, the Hamiltonian is

$$H = \frac{p^2}{2m} + \frac{m\omega^2}{2}\left(x - \frac{a}{m\omega^2}\right)^2 - \frac{a^2}{2m\omega^2}$$

The above represents simply an oscillator with a new equilibrium point. One must then use the wave functions of the oscillator with the new equilibrium point, carrying the expansion in eq. (20) out to the first-order Hermite polynomials.

6. Relation between Perturbation Theory and Theory of Sudden Transitions. In Chap. 18, Sec. 7, we treated the case of a small perturbation, which was turned on suddenly at $t = t_0$ and saw that this perturbation could be regarded as causing transitions to other levels of the unperturbed Hamiltonian H_0. An exact treatment, however, would be to use the method developed in this section for dealing with a sudden change in the Hamiltonian. As soon as the perturbation is turned on, the eigenfunctions of H_0 cease to be stationary states. One can, however, expand these eigenfunctions as a series of the true eigenfunctions of the Hamiltonian. Thus,

$$u_n = \sum_m C_{mn} v_m(x) e^{-iE_m t/\hbar}$$

In this description, the wave function changes because there is a linear combination of true stationary states, each oscillating with its own phase factor. The changes are, however, completely equivalent to those obtained when one regards the perturbation as the cause of transition to other eigenstates of the unperturbed Hamiltonian.

Problem 2: Prove that for a small perturbation the method of variation of constants leads to the same results as does the "sudden approximation."

PART V
THEORY OF SCATTERING

CHAPTER 21

1. Introduction. Whenever a beam of particles of any kind is directed at matter, the particles will be deflected out of their original paths as a result of collision with the particles of matter which they encounter. The problem of studying this scattering process is important for two reasons: First, a great many interesting effects, such as the stopping of electrons in gaseous discharges, the collisions of gas molecules, and the stopping of radioactive and cosmic ray particles, are all determined, at least in part, by the probability of scattering. Second, and perhaps even more important, is the fact that from a detailed study of the results of scattering, much can be learned about the nature of the particles that are being scattered, and as well as of those that are doing the scattering. A large part of our knowledge of atomic and nuclear physics has come from studies of just such measurements.

2. Classical Theory of Scattering. The early idea of an atom was of a perfectly elastic object, more or less spherical in shape. Since the atoms of a gas are moving in random directions, they must occasionally collide with each other and thus suffer deflections in their directions of motion. The probability of collision depends on three factors: the density of molecules, their sizes, and their mean velocities.

If the molecules are spherical in shape, with a radius a, a collision will occur whenever the centers of two molecules come closer than $d = 2a$. To compute the probability that in the short time dt a given particle collides with another, consider a cylinder with a base of area of πd^2 and height equal to the distance $dx = v\,dt$ traveled by the particle during this time. The probability of collision is then just equal to the probability that the center of another particle lies in this cylindrical region. In terms of the particle density ρ, this probability is just

$$dP = \rho \pi d^2 v\,dt \tag{1a}$$

Strictly speaking, this probability is correct only for times so short that dP is small. This is because if we wait a longer time, the cylinder discussed above may contain so many molecules that some will get in the way of others, i.e., some may be in the shadows of others. This situation is illustrated in Fig. 1. The molecule A is the one that we are following. Because it can strike B, the probability of striking C is reduced since it may be deflected or brought to rest before it strikes C. When the path is so long that the probability of collision is large, one must discuss the possibility of more than one impact; this requires a theory of multiple scattering.* We shall not treat the case of multiple scattering, however, and shall restrict ourselves to thicknesses of matter so small that multiple

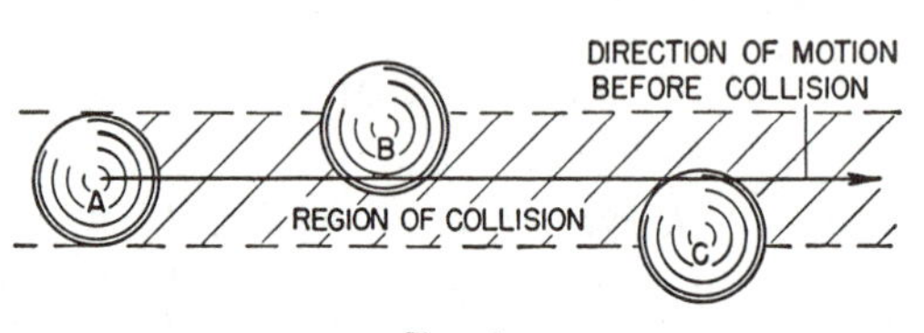

FIG. 1

scattering may be neglected. This restriction means a "thin" target, as opposed to a "thick" target.

3. Definition of Cross Section. The probability that a particle will be scattered as it traverses a given thickness of matter dx can be expressed in terms of a quantity called the "scattering cross section." To do this, let us note that each molecule presents to the oncoming particles a target area equal to $\sigma = \pi d^2$. This target area is just a cross section of the region within which a collision can take place, as viewed along the direction of motion of the beam. This is where the name "scattering cross section" comes from.

If, as is usually the case, we are dealing with a specimen containing many molecules, then the total target area is just the sum of the cross sections of the individual molecules. Actually, the above statement is true only as long as the specimen is so thin that it is unlikely that any one molecule will block the path of another; if this condition is not satisfied, then the total target area will be less than the sum of the cross sections of the separate molecules. If the target is thin enough, however, a sheet of material of area A and thickness dx (containing $\rho A\ dx$ molecules) will present an effective target area equal to $\rho A\sigma\ dx$. The fraction of the total area, A, which is "blocked" by molecules is then $\rho A\sigma\ dx/A = \rho\sigma\ dx$. The probability that an oncoming particle makes a collision is just equal to this fraction. Thus, we obtain

$$dP = \rho\sigma\ dx \tag{1b}$$

Writing $\sigma = \pi d^2$, we see that the above is the same as eq. (1). Equation (1b) yields the basic relation, connecting the probability of collision with the scattering cross section.

* See Richtmeyer and Kennard, p. 221.

4. Distribution of Free Paths. The free path is defined as the distance moved by a particle before it makes a collision. This free path will obviously vary in a more or less random fashion, depending on where it happens that a scattering molecule gets into the way of the impinging particle. Just because the scatterers are distributed at random, it occasionally will happen that a particle will have a very long free path. Yet, on the average, there will be a statistical distribution of free paths, such that most of them are close to a mean.

In order to obtain the mean free path, we begin by calculating the probability, $Q(x)$, that a particle does not make a collision in the distance x. This gives the probability that the free path is equal to x or longer. To compute this probability, we note that in the distance dx, Q is decreased by an amount equal to the probability that a collision occurs within this distance. This, however, is equal to the probability that the particle reaches the point x without collision, times the probability that if it is in this region, a collision will occur. According to eq. (1b) the latter is just $\rho\sigma\, dx$. Thus we obtain $dQ = -Q\rho\sigma\, dx$, or

$$Q = e^{-\rho\sigma x} \qquad (\text{Note that } Q = 1 \text{ at } x = 0)$$

The probability that the free path lies between x and $x + dx$ is obtained by differentiating the above,

$$R(x) = \left|\frac{dQ}{dx}\right| = \rho\sigma e^{-\rho\sigma x}$$

The mean free path is just

$$l = \int_0^\infty xR(x)\, dx = \int_0^\infty \rho\sigma e^{-\rho\sigma}x\, dx = \frac{1}{\rho\sigma} \tag{2}$$

5. Cross Section as a Function of Scattering Angle. So far, we have not considered the distribution of scattering angles which may occur as a result of collision. To study this problem, let us begin with the special case in which the scattered molecule is very light compared with the scattering molecule; the latter may then be assumed to remain essentially at rest in the process of a collision. The general case will be discussed in Sec. 12. Let us also begin by assuming that each molecule is a hard elastic sphere of radius a. The angular deflection θ of the particle is then defined as the angle between the directions of motion before and after collision. The angle of collision will depend on how directly the two particles collide. For example, we may have the two extremes of a "head-on" collision, which results in a deflection close to π, and a "glancing" collision, which results in a comparatively small deflection.

To treat this problem, consider the diagram given in Fig. 2. The angular deflection θ will clearly depend on the distance b between the original line of approach and the center O of the scattering particle. This distance is called the "collision parameter." If the spheres are perfectly elastic, the angle of deflection will be just twice the angle ψ between the original direction of motion, and the tangent to the two spheres at their point of contact. Thus, we obtain

$$\theta = 2\psi$$

Furthermore, a little geometry shows that

$$\cos \psi = \frac{b}{2a} \qquad \text{or} \qquad \theta = 2 \cos^{-1} \left(\frac{b}{2a}\right)$$

All particles striking with b smaller than $2a \cos \psi$ will receive deflections larger than $\theta = 2\psi$. Thus, if we define a cross section $S(\theta)$ equal

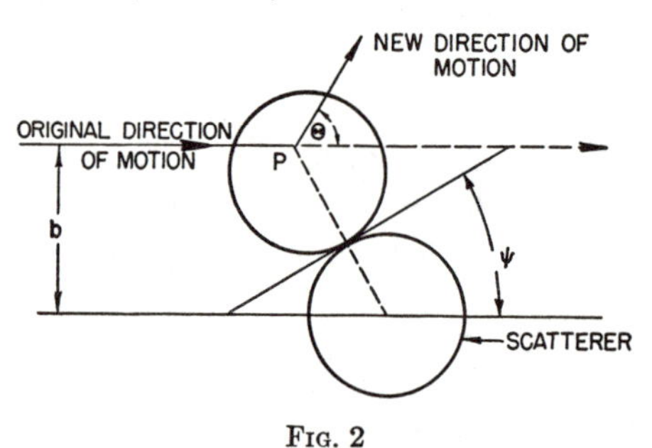

FIG. 2

to the effective area for producing collisions with deflections larger than θ, we obtain (for the elastic-sphere model)

$$S(\theta) = \pi b^2 = 4\pi a^2 \cos^2 \psi = 4\pi a^2 \cos^2 \frac{\theta}{2} \tag{3}$$

$S(\theta)$ is called the total cross section for scattering through an angle θ or greater. It is clear that only part of the scattering sphere is effective in producing large deflections; hence $S(\theta)$ decreases with increasing θ.

6. Differential Cross Sections. Another cross section that is important is the differential cross section, $q(\theta)$, which is defined such that $q(\theta)\, d\theta$ is the cross section for producing deflections that lie between θ and $\theta + d\theta$. This is obtained by differentiating $S(\theta)$.

$$q(\theta) = \left|\frac{dS}{d\theta}\right| \tag{4}$$

In the case of hard elastic spheres, $q(\theta)$ becomes

$$q(\theta) = 4\pi a^2 \sin \frac{\theta}{2} \cos \frac{\theta}{2} = 2\pi a^2 \sin \theta \tag{5}$$

Still another cross section of importance is the differential cross section per unit of solid angle. To illustrate the angles involved, consider the diagram given in Fig. 3. θ is the angle of deflection; and ϕ is the azimuthal angle made by the motion of the deflected particle relative to some standard direction. The cross section per unit solid angle, $\sigma(\theta, \phi)$, is then defined such that the effective area for deflection into the element of solid angle, $d\Omega = \sin\theta\, d\theta\, d\phi$ is equal to*

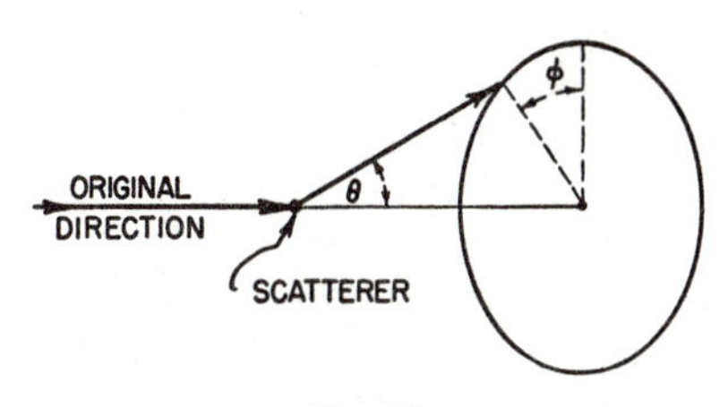

FIG. 3

$$\sigma(\theta, \phi) \sin\theta\, d\theta\, d\phi$$

For the case of hard spheres, $\sigma(\theta, \phi)$ is not a function of ϕ. If the particle were nonspherical in shape, then it is clear that the probability of deflection into different elements of $d\phi$ would be different. The relation between $\sigma(\theta, \phi)$ and $q(\theta)$ is, in general,

$$q(\theta) = \sin\theta \int_0^{2\pi} \sigma(\theta, \phi)\, d\phi \tag{6}$$

For the case where σ is not a function of ϕ, one obtains

$$q(\theta) = 2\pi \sin\theta\, \sigma(\theta) \tag{6a}$$

It is clear from the above definition that, for hard spheres,

$$\sigma = a^2 \tag{6b}$$

This means that for hard spheres there is a uniform probability of scattering into any element of solid angle.

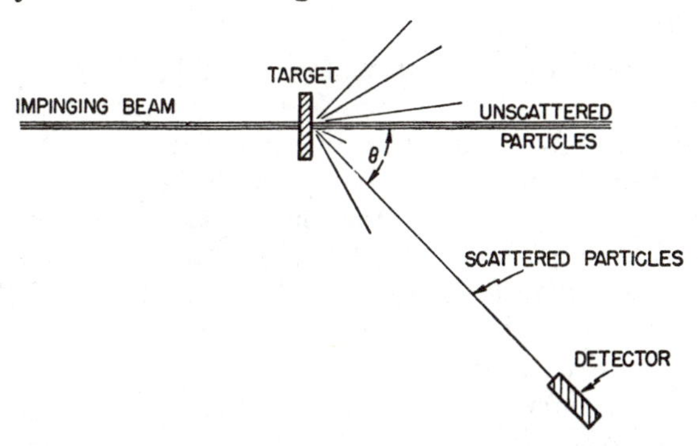

FIG. 4

A typical experimental arrangement in a scattering problem is shown in Fig. 4. The scattered particles are counted with the aid of the detector. The number of particles scattered into the detector per unit

*The σ appearing below is not to be confused with the σ introduced in Sec. 3, where it represents the total cross-section. Hereafter, unless otherwise specified, σ will refer to cross-section per unit solid angle.

time is $j\rho\sigma\, dx\, d\Omega$, where j is the incident current per unit area, and $d\Omega$ is the solid angle subtended by the detector at the target. From the measured value of this number, one can calculate σ, if ρ and j are known.

For gaseous targets, the experimental problem is usually more difficult, but these difficulties are often overcome in various ways.

7. More General Theory of Scattering. So far, we have discussed the scattering process under the assumption that the particles behave as if they are hard elastic spheres. Now, we know that this assumption is not entirely true. For example, the forces between atoms can be described by means of a potential curve as shown in Fig. 5. The atoms actually attract each other at long distances and repel each other at short distances. Because the repulsive force rises rather sharply as the atoms approach very close to each other, there is some radius, r_0, which may be defined as a rough value of the effective atomic radius, closer than which it is very difficult to bring atoms together. A hard sphere would have a

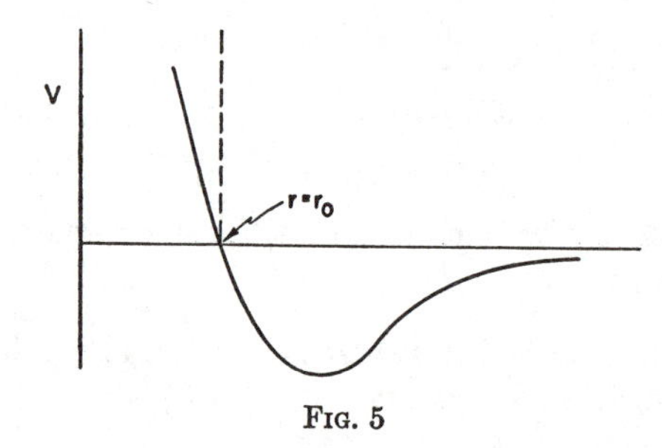

FIG. 5

potential that was zero everywhere for $r > r_0$ but infinite for $r < r_0$. Some systems approach hard spheres more closely than do others. For example, with noble gas atoms the attractive forces are very small, while the repulsive forces rise very steeply. As a result, they act very nearly like "hard spheres." On the other hand, sodium atoms are much "softer" in the sense that the force does not appear so abruptly. The potential between charged particles ($V = e^2/r$) provides a still "softer" force, so "soft" in fact, that the concept of hard spheres is very far from being a good approximation.

We must now extend our treatment so that we can calculate cross sections for an arbitrary spherically symmetrical law of force. To do this, we note that the orbit for a spherically symmetrical force will always lie in a plane. Such an orbit is illustrated in Fig. 6. We first define the collision parameter b, which is, as in the case of hard spheres, just the distance between the original line of approach and the center of the scattering force. The particle will follow some orbit as shown. (The above case is for a repulsive law of force; for an attractive law, the orbit would curve the other way.) The net deflection is denoted by θ and the distance of closest approach by a. At any instant, the position of the particle is described by the polar angle ϕ and the radius r.

In general, if the equations of motion are solved, it will turn out that the deflection, θ, will always be some function of the collision parameter b. Thus, we can write

$$\theta = \theta(b)$$

or alternatively

$$b = b(\theta)$$

The cross section for scattering into an angle between θ and $\theta + d\theta$ will

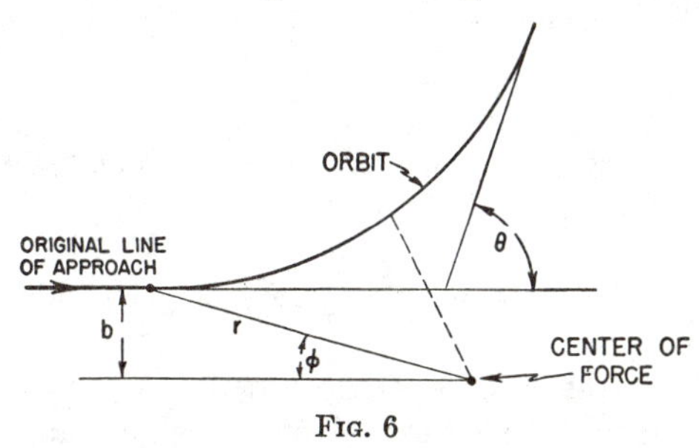

FIG. 6

just be the area of the ring ($2\pi b\, db$) within which particles must be if they are to be scattered into the above range of angles. Thus

$$q(\theta)\, d\theta = 2\pi b \frac{db}{d\theta}\, d\theta \tag{7}$$

The total cross section for scattering into angles of θ and greater is obtained by integrating the above from $b = 0$ to $b = b(\theta)$.

$$S(\theta) = \int_0^{b(\theta)} 2\pi b\, db = \pi b^2(\theta) \tag{7a}$$

This is just the area inside a circle of radius $b(\theta)$.

The total cross section for scattering through all possible angles (0 or greater) is found by setting $\theta = 0$ in the above expression:

$$S(0) = \int_0^{\pi} q(\theta)\, d\theta = \pi b^2(0) \tag{7b}$$

In order to compute the various cross sections, it is necessary, in principle, at least, to obtain the orbit of the particle, and to use the equations for the orbit to solve for θ as a function of b.

8. The Approximation of Small Deflections. Classical Perturbation Theory. We shall present here an approximate method of solving for $\theta(b)$, which is good whenever θ is a small angle. Small deflections will, in general, be the results of weak forces, and the forces will usually be weakest when the particle remains farthest away from the center, i.e., when b is large.

We begin by obtaining an expression for the angle of deflection, θ. Let us choose the x axis along the initial direction of motion, and the y axis normal to it. Let $p =$ the initial momentum, which is of course all in the x direction. As a result of the force, the particle obtains a y component of the momentum, which we shall denote by p_y. The angle of deflection is then given by

$$\sin\theta = \frac{p_y}{p}$$

The next step is to solve for p_y. Since p_y is initially zero, we can write from Newton's laws of motion

$$p_y = \int_{-\infty}^{\infty} F_y\,dt$$

If the force is spherically symmetrical, then F_y will be equal to $\frac{y}{r}F$ where F is the total force. Thus

$$p_y = \int_{-\infty}^{\infty} y\frac{F(r)}{r}\,dt$$

In order to evaluate this integral exactly, we must know r and y as functions of t. This amounts to having solved the equations of motion. Our method of approximation, however, is based on the fact that if the deflecting force is small, the particle travels a path which is almost the same as the original straight line, at a velocity which is almost constant. Since p_y is already a small quantity, the differences resulting from evaluating r as equal to what it would have been in the absence of the force will be of second order. A good approximation will therefore be to evaluate r along the "unperturbed orbit," i.e., along the straight line that the particle would have followed if the force had been zero. Thus, we can write

$$y \cong b, \qquad x \cong vt, \qquad r \cong \sqrt{b^2 + v^2t^2}$$

and

$$\theta \cong \sin\theta = \frac{p_y}{p} \cong \int_{-\infty}^{\infty} \frac{bF(\sqrt{b^2+v^2t^2})\,dt}{p\sqrt{b^2+v^2t^2}}$$

It is convenient to adopt the new variable $t = bu/v$. We obtain (with $p = mv$ and $E = mv^2/2$)

$$\theta \cong \frac{b}{2E}\int_{-\infty}^{\infty} \frac{F(b\sqrt{1+u^2})\,du}{\sqrt{1+u^2}} \tag{8}$$

The above is the result that we are seeking. We shall now apply it to several examples:

(*a*) *Coulomb Force.* For this case, $F = Z_1Z_2e^2/r^2$ where Z_1 is the charge of the scattering particles in units of the electronic charge and Z_2 that of the scattered particle. This force leads to

$$\theta \cong \frac{Z_1Z_2e^2}{2bE}\int_{-\infty}^{\infty} \frac{du}{(1+u^2)^{3/2}} = \frac{Z_1Z_2e^2}{Eb} \tag{9a}$$

The above relation shows that the angle of deflection is inversely proportional to the collision parameter b. This is an important result. The cross section is

$$q(\theta) = 2\pi b \left|\frac{db}{d\theta}\right| = \frac{2\pi(Z_1Z_2e^2)^2}{E^2\theta^3} \tag{9b}$$

This result has several significant properties:

(1) The cross section for a given angle θ is a rapidly decreasing function of energy. Physically, this is because it takes more force to deflect a faster particle, and this additional force can be obtained only with smaller collision parameters; hence the rapid decrease of θ with E.

(2) The cross section approaches ∞ as θ approaches zero. In fact, the integrated cross section $S(\theta)$ also approaches infinity. The reason is that the Coulomb force has such a long range. If one is willing to consider smaller and smaller deflections, one can always obtain them at larger and larger collision parameters; as a result, the cross section becomes larger and larger.

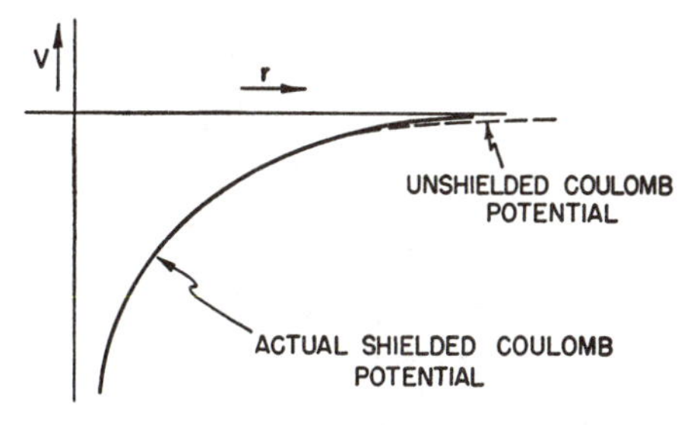

FIG. 7

(3) Actually, it is always an abstraction to assume that the Coulomb force continues unmodified out to arbitrarily large radii. For example, the Coulomb forces resulting from atomic nuclei are screened (or shielded) by the atomic electrons beyond distances of the order of a few atomic radii. The resulting shape of the potential is shown in Fig. 7. Similarly, in an ion gas or in an electrolyte, ions of a given sign are always surrounded by a cloud of charge consisting of ions of opposite sign that eventually shield out the Coulomb potential at large enough radii.* In general, such shielding will always exist in any real problem.

A good approximation to a shielded Coulomb potential is

$$V = \frac{Ze^2}{r}\exp\left(-\frac{r}{r_0}\right) \tag{10}$$

The exponential factor causes the force to become negligible when r/r_0 is much greater than unity.

* White, *Introduction to Atomic Spectra*, p. 314.

(4) With a shielded Coulomb potential, θ will approach zero with increasing b much more rapidly than $1/b$, as soon as b goes beyond the shielding radius. In fact, shortly beyond the shielding radius, the entire scattering effect can be neglected. The minimum angle, below which the cross section ceases to increase, is given by setting $b = r_0$ in eq. (9a); i.e.,

$$\theta_{\min} = \frac{Z_1 Z_2 e^2}{E r_0} \tag{11}$$

As a function of angle, the unshielded Coulomb cross section is shown in Fig. 8. The shielded Coulomb cross section is shown in Fig. 9.

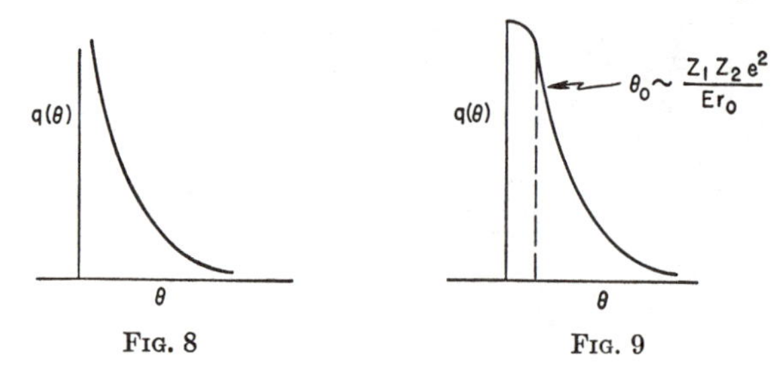

FIG. 8 FIG. 9

(5) It should be recalled that the perturbation theory breaks down if θ is large ($\cong \frac{1}{2}$). We shall, however, obtain the exact result for all θ in Section 10.

(*b*) $1/r^3$ *Law of Force.* For this case, we write $F = K/r^3$, and

$$\theta = \frac{K}{2b^2E}\int_{-\infty}^{\infty}\frac{du}{(1+u^2)^2} = \frac{\pi K}{4b^2E} \tag{12a}$$

$$b^2 = \frac{\pi K}{4E\theta} \tag{12b}$$

The differential cross section is

$$q(\theta) = \pi\left|\frac{d}{d\theta}[b^2(\theta)]\right| = \frac{\pi^2 K}{4E\theta^2} \tag{12c}$$

Problem 1: Obtain the cross section for $F = r^{-n}$.

We see that the energy and angular dependence of the cross section depend on the law of force. Thus, an experimental study of these quantities provides information about the law of force.* We shall return to this point in Sec. 11.

9. Cross Section for Energy and Momentum Transfer. In several cases [eqs. (9b) and (12c)] we have obtained cross sections, which are not only infinite at $\theta = 0$, but which also yield infinite results when inte-

* The formulas obtained in this section apply only in the classical limit. For an extension of these results to quantum theory, see Secs. 23 and 47.

grated over θ. As shown in Sec. 8, such infinite cross sections indicate merely that if one is willing to consider a small enough deflection, one can obtain it at a very large collision parameter. Such minute deflections, however, usually produce only correspondingly minute physical effects. For example, the stopping power of matter for charged particles depends on the mean transfer of energy from the direction of original motion into a direction at right angles to this. Now, the loss of energy in the original direction of motion in a collision is

$$\Delta E = \frac{(\Delta p)^2}{2m} = \frac{p^2 \sin^2 \theta}{2m} \cong \frac{p^2\theta^2}{2m}$$

Thus, the mean energy transfer is

$$\overline{\Delta E} = \int_0^\pi q(\theta)\, \Delta E(\theta)\, d\theta \cong \frac{p^2}{2m} \int_0^\pi q(\theta)\theta^2\, d\theta$$

As a rule, the mean energy transfer remains finite, even when the cross section itself is infinite. For example, for the force law, $F = K/r^3$, one obtains [see eq. (12c)]

$$\overline{\Delta E} \cong \frac{p^2}{2m} \frac{\pi^2 K}{4E} \int_0^\pi \frac{\theta^2\, d\theta}{\theta^2} = \frac{\pi^3 p^2 K}{8mE} \tag{13a}$$

For the Coulomb cross section, one obtains

$$\overline{\Delta E} \cong \frac{p^2}{2m}\, 2\pi \frac{(Z_1 Z_2 e^2)^2}{E^2} \int_{\theta_{\min}}^\pi \frac{\theta^2\, d\theta}{\theta^3} = \frac{\pi p^2}{m} \frac{(Z_1 Z_2 e^2)^2}{E^2} \ln\left(\frac{\pi}{\theta_{\min}}\right) \tag{13b}$$

where $\theta_{\min}$ is the minimum angle for Coulomb scattering (as determined by the shielding radius).

Although this result becomes infinite as $\theta_{\min} \to 0$, the logarithm changes so slowly with $\theta_{\min}$ that, in practice, the result is not very sensitive to the actual value of $\theta_{\min}$ within very wide limits.* Thus, a crude estimate of $\theta_{\min}$ will usually give an adequate approximation to $\overline{\Delta E}$.

10. Exact Solution for Scattering. In order to obtain a theory of large-angle scattering, it is necessary to solve exactly for the motion of the particle. We shall use the following two equations:

$$mr^2 \frac{d\phi}{dt} = mvb \qquad \text{(Conservation of Angular Momentum)} \tag{14a}$$

$$\frac{m}{2}\left[\left(\frac{dr}{dt}\right)^2 + r^2\left(\frac{d\phi}{dt}\right)^2\right] + V(r) = \frac{mv^2}{2} \qquad \text{(Conservation of Energy)} \tag{14b}$$

* The limitations on minimum angle of scattering given above apply only to the extent that classical theory is applicable. Analogous limitations are obtained in the quantum domain which, however, are not precisely the same as those that are valid in the classical limit (see Secs. 21, 32, 35, and 38).

(Note that we assume that $V(r) \to 0$ as $r \to \infty$.) Insertion of (14a) into (14b) yields

$$\frac{dr}{dt} = \pm \sqrt{v^2 - \frac{b^2v^2}{r^2} - \frac{2}{m} V(r)}$$

Division of (14a) by the above yields

$$\frac{d\phi}{dr} = \pm \frac{vb}{r^2 \sqrt{v^2 - \frac{b^2v^2}{r^2} - \frac{2}{m} V(r)}}$$

One can now solve for the deflection, θ, by integrating the above from $r = \infty$ down to $r = a$ = distance of closest approach, and back out to ∞ again. Consultation of Fig. 6 shows that if we start with $\phi = 0$, we obtain, after integration,

$$\Delta\phi = \pi - \theta \qquad \text{or} \qquad \theta = \pi - \Delta\phi$$

But since the integrand runs through the same series of values on the inward integration as it does on the outward, one can simply double the result of integration over r from a to ∞. One obtains

$$\Delta\phi = 2vb \int_a^\infty \frac{dr}{r^2 \sqrt{v^2 - \frac{2V(r)}{m} - \frac{b^2v^2}{r^2}}} \tag{15}$$

Insertion of any particular value of $V(r)$ into the above equation will now enable one, in principle, to calculate $\theta(b)$, and from this $b(\theta)$ and then $q(\theta)$.

Example: Coulomb Scattering. Rutherford Cross Section.
Let us set $V = Z_1Z_2e^2/r$. We obtain

$$\Delta\phi = 2vb \int_a^\infty \frac{dr}{r^2 \sqrt{v^2 - \frac{2Z_1Z_2e^2}{mr} - \frac{b^2v^2}{r^2}}}$$

It is convenient to make the substitution $r = 1/u$, $du = -dr/r^2$. This yields

$$\Delta\phi = 2vb \int_0^{1/a} \frac{du}{\sqrt{v^2 - \frac{2Z_1Z_2e^2}{m}u - b^2v^2u^2}} = 2 \int_0^{1/a} \frac{du}{\left(\frac{1}{b^2} - \frac{2Z_1Z_2e^2}{mb^2v^2}u - u^2\right)^{1/2}}$$

Now the distance of closest approach is defined to be the place where $dr/dt = 0$; this is, however, exactly the place where the denominator in the integrand vanishes. When we carry out the above integral, we then obtain

$$\Delta\phi = \pi - \theta = 2\cos^{-1}\frac{Z_1Z_2e^2}{mbv^2} \qquad \text{or} \qquad \frac{Z_1Z_2e^2}{mbv^2} = \sin\frac{\theta}{2}$$

and

$$b = \frac{Z_1Z_2e^2}{mv^2 \sin\frac{\theta}{2}} = \frac{Z_1Z_2e^2}{2E\sin\frac{\theta}{2}} \tag{16a}$$

The differential cross section is

$$q(\theta) = 2\pi b \left|\frac{db}{d\theta}\right| = \frac{\pi(Z_1Z_2e^2)^2}{4E^2} \frac{\cos\frac{\theta}{2}}{\sin^3\frac{\theta}{2}}$$

For small θ, one readily verifies that, in agreement with the approximate equation (9b), we obtain

$$q(\theta) = 2\pi \frac{(Z_1Z_2e^2)^2}{E^2\theta^3} \tag{16b}$$

The cross section per unit solid angle is

$$\sigma(\theta) = \frac{1}{2\pi}\frac{q(\theta)}{\sin\theta} = \frac{(Z_1Z_2e^2)^2}{16E^2\sin^4\frac{\theta}{2}} \tag{16c}$$

This is the well-known Rutherford cross section.

11. Use of Cross Sections to Investigate Law of Force. So far, we have assumed that the law of force is known, and tried to investigate the cross section. Very often, however, one tries to use cross-sectional data to investigate the unknown law of force. There are several ways to do this. The most common way is to assume that the potential takes some simple shape, such as $Ke^{-r/r_0}/r^n$, and to see whether K, r_0, and n can be chosen so as to obtain a fit to the data. In doing this, one should have a clear idea of what range of radii are being probed by particles of a given range energy and angular deflections. For example, if the force is a Coulomb force, eq. (16a) shows that for small deflections

$$b = \frac{Z_1Z_2e^2}{2E\sin\frac{\theta}{2}}$$

In general, particles with a given collision parameter will obtain a deflection that depends most strongly on the intensity of the force at radii of the order of the collision parameter. This is because the force usually decreases fairly rapidly with distance, so that the largest deflecting force is experienced when the particle is within a region of width of the order of the collision parameter. Furthermore, since dr/dt approaches zero in this region, the particle also spends more time there. Thus, according to the above formula, to probe the nature of the force at small radii, we need large E or large θ, or both. There is a limit to θ, namely π. As a result, there is a minimum particle energy that will probe the force at a given distance. This is

$$E = \frac{Z_1Z_2e^2}{2b} \tag{17}$$

Problem 2: For α particles scattering off Beryllium nuclei, what energy is needed to obtain $\theta = 10°$, with a collision parameter of 10^{-12} cm?

If one can obtain the integrated cross section $S(\theta)$ for a particular value of θ, then one has even more valuable information than that yielded by the differential cross section alone. This is because a measurement of $S(\theta)$ is equivalent to measuring the collision parameter $b = [S(\theta)/\pi]^{1/2}$, so that one then knows what collision parameter is needed to yield a given deflection θ. Since the momentum transfer is $\Delta p = p \sin \theta$, we are provided with information on the strength of the force in the general region of radii of the order of $b = (S/\pi)^{1/2}$.

A careful investigation of the scattering of α-particles from various nuclei was made by Rutherford, who showed that the scattering predicted on the assumption of a Coulomb force was obeyed remarkably well down to very small radii. It was on the basis of these results that the current atomic theory was justified, i.e., the picture of a highly localized charged nucleus surrounded by planetary electrons was demanded in order to agree with these scattering experiments. As such experiments were done to higher and higher energies, however, deviations from the scattering predicted by Coulomb theory were obtained. These deviations were obtained at energies of a few hundred kev, from which one was able to conclude that at radii of the order of 10^{-12} cm or smaller (see Problem 2) new forces of a non-Coulombic nature were coming into play. By a careful investigation of how the cross sections varied with energy and angle, many properties of these so-called "nuclear forces" were deduced. We shall discuss these forces later, in connection with the quantum theory of scattering, because quantum effects are important in describing them. At present, we shall merely note in a qualitative way that the dependence of scattering cross section on angle and energy provides us with a measure of the "softness" of the law of force. For a hard sphere, for example, the cross section is always $4\pi a^2$, independent of angle, regardless of how high the energy is. But if the force is "soft," like, for example, a Coulomb force, a high-energy particle must come very close to the nucleus before it can suffer an appreciable deflection, so that the cross section for a given angle of scattering decreases rapidly with increasing energy. Also, because of the long range of the unshielded Coulomb force, there is an enormous target area within which very small deflections can be obtained; hence the infinite cross section as θ approaches zero.

12. Transformation from Center-of-Mass System to Laboratory System of Co-ordinates. Our results thus far have been discussed with the assumption that the scatterer remains at rest during the collision, because it is so much heavier than the scattered particle. In order to deal with the more general case, we start with the well-known classical mechanical result* that in a co-ordinate system which moves with the

* See Richtmeyer and Kennard, p. 120.

center of mass of the two particles, the equations of motion for the relative co-ordinates, $\xi = r_1 - r_2$, are the same as those for a single particle under the same potential, $V(\xi)$, but with a reduced mass

$$\mu = \frac{m_1 m_2}{m_1 + m_2}$$

and a reduced energy $E = \frac{\mu}{2}\left(\frac{d\xi}{dt}\right)^2$. The same general result is true in quantum theory. The equations for scattering may therefore be solved in exactly the same way as we have been doing, provided that we are careful to identify the constants correctly.

In the center-of-mass system of co-ordinates, each particle begins by approaching the center of mass in opposite directions, at such velocities that the total momentum is zero. We therefore have

$$m_1 v_1 = m_2 v_2$$

The particles must scatter in opposite directions in order that the total momentum remain zero after collision. The two orbits therefore resemble the figure shown in Fig. 10. Our problem is now to transform the cross section $q(\theta')$, calculated in the center-of-mass system, back into the laboratory system, in which cross sections are always observed.

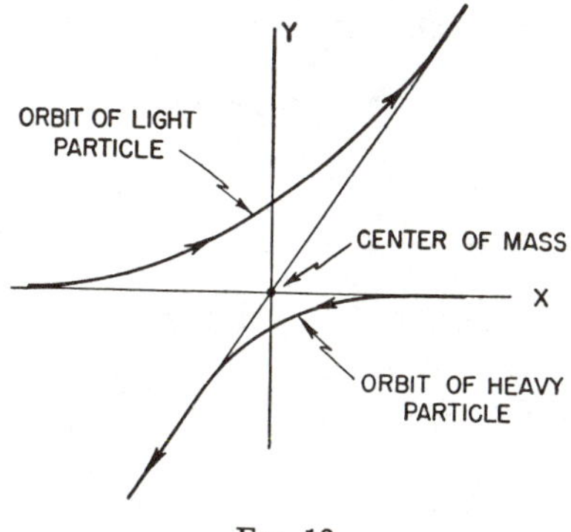

FIG. 10

To do this, it is necessary, first, to transform the angles θ', measured in the center-of-mass system, back into the laboratory system. Collisions usually involve firing particles at other particles that are at rest in the laboratory system. Let the mass of the latter particles be m_1, and let the mass of the moving particles be m_2. Let the moving particles be moving initially (before collision) with a velocity v, which is taken to be in the x direction. The velocity of the center of mass of the system is then in the x direction and it is equal to $w = \dfrac{m_2 v}{m_1 + m_2}$.

In the center-of-mass system, the relative speed, $|d\xi/dt|$, is still v. But each particle now has a velocity inversely proportional to its mass. Thus, before collision, we have for the first particle

$$(U_{10})_x = -\frac{m_2}{m_1 + m_2} v \qquad (U_{10})_y = 0$$

For the second particle

$$(U_{20})_x = \frac{m_1 v}{m_1 + m_2} \qquad (U_{20})_y = 0$$

After a collision which results in scattering through an angle θ' in the center-of-mass system, one obtains

$$(U_1)_x = -\left(\frac{m_2}{m_1 + m_2}\right) v \cos \theta' \quad \text{and} \quad (U_1)_y = -\left(\frac{m_2}{m_1 + m_2}\right) v \sin \theta'$$

$$(U_2)_x = \left(\frac{m_1}{m_1 + m_2}\right) v \cos \theta' \quad \text{and} \quad (U_2)_y = \left(\frac{m_1}{m_1 + m_2}\right) v \sin \theta'$$

Note that the relative velocity v is left unchanged by the collision.

To obtain the velocities in the laboratory system after scattering, one adds the velocity of motion of the center of mass to the x components of the above velocities. This gives

$$(U_1)_x = \frac{(m_2 - m_2 \cos \theta')}{m_1 + m_2} v \quad \text{and} \quad (U_1)_y = -\left(\frac{m_2}{m_1 + m_2}\right) v \sin \theta'$$

$$(U_2)_x = \frac{(m_2 + m_1 \cos \theta')}{(m_1 + m_2)} v \quad \text{and} \quad (U_2)_y = \left(\frac{m_1}{m_1 + m_2}\right) v \sin \theta'$$

The angles of motion in the laboratory system are then given by

$$\tan \theta_1 = \frac{(U_1)_y}{(U_1)_x} = -\frac{\sin \theta'}{1 - \cos \theta'} = -\cot \frac{\theta'}{2}$$

$$\text{and} \qquad \tan \theta_2 = \frac{(U_2)_y}{(U_2)_x} = \frac{m_1 \sin \theta'}{m_2 + m_1 \cos \theta'} \tag{18}$$

These equations completely define the angles at which each of the two particles comes off as a function of the angle of scattering in the center-of-mass system. In order to obtain the cross section in the laboratory system, one uses the fact that $q(\theta')\, d\theta'$ is proportional to the number of particles scattered into angles lying between θ' and $\theta' + d\theta'$, whereas $q(\theta)\, d\theta$ is the number scattered into angles lying between θ and $\theta + d\theta$. If we choose θ such that it is related to θ' by this relationship, then the number of particles in corresponding ranges of $d\theta$ and $d\theta'$ must, by definition, be equal. Thus we obtain

$$q(\theta)\, d\theta = q(\theta')\, d\theta'$$

or

$$q(\theta) = q(\theta') \frac{d\theta'}{d\theta}$$

Now for the scattered particle, θ is given by θ_2 of eq. (18). By differentiating the above relation, we obtain

$$\sec^2 \theta_2 \frac{d\theta_2}{d\theta'} = m_1 \frac{(m_1 + m_2 \cos \theta')}{(m_2 + m_1 \cos \theta')^2}$$

As a result,

$$q(\theta) = q(\theta') \frac{\sec^2 \theta (m_2 + m_1 \cos \theta')^2}{m_1(m_1 + m_2 \cos \theta')} \tag{19}$$

To obtain the cross section as a function of θ, it is necessary to eliminate θ' in terms of θ through eq. (18).

13. Discussion of Results.

Case A: $m_2 < m_1$

This is the case when the bombarding particle is lighter than the particle that is being struck. It is clear from eq. (18) that for small θ,

$$\theta \cong \frac{m_1}{m_1 + m_2} \theta' \tag{20}$$

The relation between θ and θ' is rather complex for large θ. For example, one obtains $\theta = \pi/2$ when $\cos \theta' = -m_2/m$. This always happens for $\theta' > \pi/2$. The maximum angle, θ, of scattering is always π.

Case B: $m_2 > m_1$

In this case, one can easily see that the maximum of θ is less than $\pi/2$. Equation (20) still holds for small θ.

Case C: $m_1 = m_2$

Here, we obtain $\theta = \theta'/2$. The maximum of θ is then $\pi/2$. The angle in the laboratory system is just half the angle in the center-of-mass system.

Problem 3: There appears to be an abrupt discontinuity in the form of the cross section, since for $m_1 = m_2$ the maximum value of θ is $\pi/2$, while for m_2 very slightly less than m_1, the maximum value of θ suddenly jumps to π. Show that there is no real physical discontinuity, because the cross section, $q(\theta)$, approaches zero for angles greater than $\pi/2$, as m_2 approaches m_1.

Very often, one can measure the angle θ_2 with which the struck particle is ejected. From eq. (18), we can obtain the result that

$$\theta_1 = -\left(\frac{\pi}{2} - \frac{\theta'}{2}\right)$$

14. Identical Particles. If both particles are the same, then one cannot distinguish the struck particle from the one that was originally moving. One must therefore add to $q(\theta')$ the cross section for the process in which, in the center-of-mass system, the bombarded particle is scattered through an angle of $\pi - \theta'$. Reference to Fig. 9 will show that such a particle will contribute to the stream of scattered particles in exactly the same way as would the original particle scattered through an angle of θ. Thus, the cross section $q(\theta')$ should be replaced by $q(\theta') + q(\pi - \theta')$.

15. Quantum Theory of Scattering. In order to treat the scattering problem quantum mechanically, we must take into account the fact that the motion of particles cannot be described with complete accuracy by the classical orbits, but that one must, instead, use wave packets whose average co-ordinates give the classical orbits. The scattering process must therefore be described by wave functions that are solutions of Schrödinger's equations, rather than by particle trajectories that are solutions of the classical equations of motion.

16. Condition for Validity of Classical Theory of Scattering. The conditions under which classical theory becomes inadequate and quantum theory becomes necessary can easily be obtained. If a classical description is to be applicable, one must be able, without seriously altering any significant results, to obtain this classical description by forming a wave packet. Since the angle of scattering for a particular trajectory is determined mostly by the magnitude of the forces in the neighborhood of the distance of closest approach, the wave packet must be narrower than this distance; otherwise there is no way of being sure that the particle experiences a definitely predictable force from which the deflection can be calculated in the classical way.

To obtain a rough estimate of the validity of the classical description, we can safely assume that the distance of closest approach is of the same order of magnitude as the collision parameter b. In order to form a wave packet that is smaller than b, it is, of course, necessary that one use a range of wavelengths of the order of b or smaller. Thus, the first requirement is that the momentum of the incident particles be considerably larger than $p \cong \hbar/2b$. In defining the position of this packet, moreover, we will make the momentum of the particle uncertain by a quantity much greater than $\delta p \cong \hbar/2b$. This uncertainty will cause the angle of deflection to be made uncertain by a quantity much greater than $\delta\theta \cong \delta p/p$. In order that the classical description be applicable, the above uncertainty ought to be a great deal smaller than the deflection itself; otherwise the entire calculation of the deflection by classical methods will be meaningless. This requirement, however, is equivalent to the requirement that the uncertainty in momentum be much smaller than the net momentum, Δp, transferred during the collision, or that

$$\frac{\delta p}{\Delta p} \cong \frac{\hbar}{2b\,\Delta p} \ll 1 \tag{21}$$

Now Δp must be obtained by the use of classical orbit theory. A general discussion for scattering through arbitrarily large angles is rather complicated, but for the case in which the angle of deflection is small, one can use classical perturbation theory. This theory is applicable only when the scattering angle is large compared with the quantum fluctuations, but small compared with π. One can then compute Δp from Sec. 8, obtaining

$$\Delta p \cong b \int_{-\infty}^{\infty} \frac{F(r)\,dt}{r} \cong \frac{b}{v} \int_{-\infty}^{\infty} \frac{F(r)\,dx}{r}$$

where, as in Sec. 8, we write

$$r = \sqrt{b^2 + x^2}$$

Thus, a classical description will be valid whenever*

$$\frac{2b\,\Delta p}{\hbar} = \frac{2b^2}{\hbar v}\int_{-\infty}^{\infty}\frac{F(r)\,dx}{r} \gg 1 \qquad \text{(for small deflections)} \qquad (21a)$$

We shall discuss some applications of this criterion in Secs. 35 and 38.

17. Quantum Description of Scattering. We have seen that the classical theory requires that one specify an orbit, along which one can calculate the detailed transfers of momentum to the particle at every point. In the quantum theory, however, the particle cannot have a definite momentum when its position is well defined, so that the scattering process cannot be analyzed in the classical way. Instead, one has a choice of specifying either the momentum or the position, but not both simultaneously.

The former choice involves the use of the momentum representation of the wave function (Chap. 9, Sec. 8), and corresponds to the causal description† of the scattering process. In other words, one discusses the deflection as something that is caused by the deflecting force, but one cannot specify exactly where the momentum was transferred within the region of a wave packet. Instead, one must imagine that all parts of the potential covered by a wave packet contribute simultaneously to the scattering process. This is similar to what happens in the electron diffraction experiments of Davisson and Germer (Chap. 3, Sec. 11), where all parts of the crystal must be assumed to co-operate simultaneously.

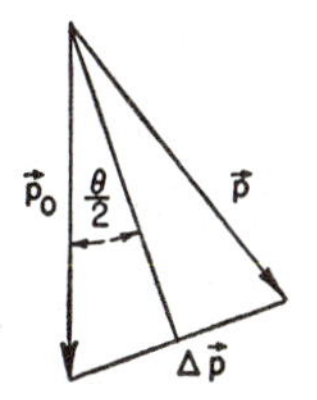

FIG. 11

The mathematical expression of this point of view requires, as we have already stated, the use of the momentum representation of the wave function. One starts with a particle with an initial momentum, $\boldsymbol{p}_0$, and as a result of the scattering potential the particle obtains a new momentum, $\boldsymbol{p}$. In elastic collisions,‡ the absolute value of $\boldsymbol{p}$ is the same as that of $\boldsymbol{p}_0$, i.e., energy is conserved, but, more generally, this need not be so. The momentum transferred during the collision is $\Delta\boldsymbol{p} = \boldsymbol{p} - \boldsymbol{p}_0$. For elastic collisions, one obtains (see Fig. 11)

$$\Delta p = 2p \sin\frac{\theta}{2} \qquad (22)$$

where θ is the angle of deflection. The probability of scattering through a given angle θ is then obtained by first finding the probability of the corresponding momentum transfer $\Delta\boldsymbol{p}$.

* Note that a large force favors the validity of the classical approximation.

† Chap. 8, Secs. 13 and 14.

‡ This statement applies either for an infinitely heavy scattering center, or more generally, in the center-of-mass coordinate system for an arbitrary scattering center. In the latter case, we must use reduced mass and reduced energy. See Secs. 12 and 13.

The alternative procedure is to describe the scattering process by means of the wave picture, which gives a space-time description.* In this description, one begins with an incident wave packet that is very large in comparison with the scattering system. This packet is produced by sending the incident particles through a collimating slit. The time during which particles pass through the slit is under normal experimental conditions rather poorly defined in comparison with the time necessary for the particle to pass through the region in which the scattering potential is appreciable. It is therefore a good approximation to replace the actual wave packet by an incident plane wave of infinite extent. This situation is illustrated in Fig. 12.

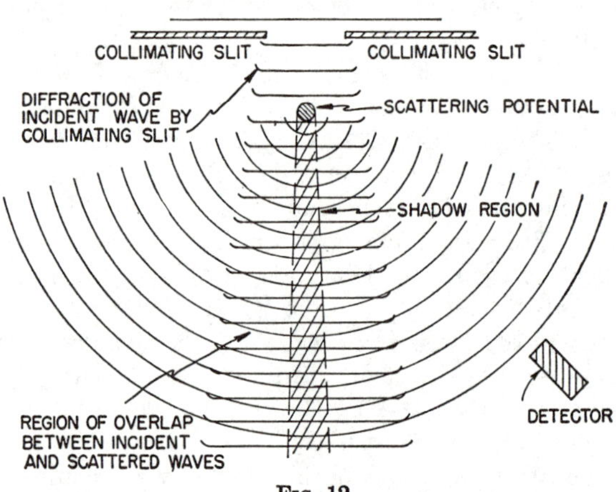

FIG. 12

As the wave packet comes through the collimating slit, it is diffracted slightly near the edges, but because the slit is usually much larger than an electronic wavelength, this diffraction can be neglected. When the wave enters the scattering potential, it enters a region of changing index of refraction (see Chap. 11), Sec. 2, where it is both refracted and diffracted. If the index changes slowly in comparison with a wavelength (i.e., the potential is smooth and slowly varying), the WKB approximation† holds, and the diffraction can be neglected. In other words, the bending of the wave can be described by the bending of rays that are normal to the wave front and that are, of course, just the classical particle trajectories. In this case, the classical approximation can be used. But if the potential changes rapidly within a wavelength, the WKB approximation breaks down and characteristic diffraction effects take place. If the potential has sharp edges, there may also be reflection (Chap. 11, Sec. 4). In any case, the wave description becomes essential.

* See Chap. 8, Sec. 14.
† See Chap. 12.

Whether the correct description is classical or quantum mechanical, a scattered wave appears. The intensity of the scattered wave yields the probability that the particle has been scattered through a given angle. Where the incident and scattered waves overlap, they may interfere. This region of interference includes, first of all, the "shadow" of the scattering potential, i.e., a region in which the incident wave is weakened as a result of loss to the scattered wave, and second, a region in which incident and scattered waves overlap. Since the scatterer usually contains many atoms, the shadow region will be roughly the region immediately behind the target. In order that the scattered wave be clearly distinguishable from the incident wave, it is necessary that the detector be placed somewhere outside the region accessible to the incident wave, as shown in Fig. 12.

The above yields a space-time description of the scattering process. This however, has been achieved at the expense of giving up a detailed causal description of how the electron obtains its momentum. As far as the probability of scattering through a given angle is concerned, the space-time description must, of course, yield the same results as the causal description,* but each method gives a considerably different description of the intermediate mechanism. It is only to the extent that one can form a wave packet appreciably smaller than the distance of closest approach without introducing significant uncertainties in the angle of scattering that one can give simultaneously detailed space-time and causal descriptions. Otherwise, the scattering process must be regarded as made up of indivisible elements. This indivisibility is reflected in the fact that any attempt to follow the scattering process in detail by means of observations must use quanta that impart enough momentum to change the angle of scattering to a significant, but incompletely predictable and controllable, extent.†

18. Scattering Considered as a Transition between Different States in Momentum Space. We shall begin by treating the scattering problem in the momentum representation. In this procedure the scattering potential is regarded as something which causes transitions from one state in momentum space to another. In order to represent the state of the system before the particle is scattered, one must form an incident wave packet that has a small range of momenta centering around some value p_0. This range is normally so small that one can, in practice, replace the wave function in momentum space by a δ function, i.e.,

$$\phi_0(p) = \delta(p - p_0)$$

* As shown in Chap. 8, Secs. 13 and 14, the causal description is the same as a description in momentum space.

† See Chap. 5, Sec. 14, for a description of similar effects which occur when one tries to follow an electron in its orbit.

This means that in configuration space the initial wave function is being represented as a plane wave,

$$\psi = L^{-3/2} e^{i\mathbf{p}_0\cdot\mathbf{r}/\hbar}$$

(The $L^{-3/2}$ factor is necessary for normalization.) As time goes on, other momenta appear, and the wave function must, in general, be represented as a Fourier integral covering all possible momenta. We shall find it convenient, however, to use a Fourier series in which ψ is developed as a function that is periodic within a large box of side L, where L is so big that the walls will have a negligible effect on the scattering process. Just as in the electromagnetic problem (see Chap. 1, Sec. 4), we use periodic boundary conditions at the walls of the box. An arbitrary function can now be expanded in a Fourier series as follows:

$$\psi = L^{-3/2} \sum_{\mathbf{p}} a_{\mathbf{p}}\, e^{i\mathbf{p}\cdot\mathbf{x}/\hbar} \tag{23}$$

The permissible values of $\mathbf{p}$ are those which make the function spatially periodic with a period equal to L; i.e., $p_x = 2\pi\hbar l/L$, $p_y = 2\pi\hbar m/L$, $p_z = 2\pi\hbar n/L$. (l, m, and n are arbitrary integers.)

In order to solve for the probability that the momentum has changed, one must use Schrödinger's equation,

$$i\hbar\frac{\partial\psi}{\partial t} = \left(-\frac{\hbar^2}{2m}\nabla^2 + V\right)\psi$$

Insertion of the above series into Schrödinger's equation yields

$$\sum_{\mathbf{p}} i\hbar\frac{\partial a_{\mathbf{p}}}{\partial t} e^{i\mathbf{p}\cdot\mathbf{r}/\hbar} = \sum_{\mathbf{p}}\left[\frac{p^2}{2m} + V(\mathbf{r})\right] a_{\mathbf{p}}\, e^{i\mathbf{p}\cdot\mathbf{r}} \tag{23a}$$

Let us now multiply the above equation by $L^{-3/2} e^{-i\mathbf{p}'\cdot\mathbf{r}/\hbar}$ and integrate over the entire box. Using normality and orthogonality of the exponential functions, we obtain

$$i\hbar\frac{\partial a_{\mathbf{p}}}{\partial t} = \frac{p^2}{2m} a_{\mathbf{p}} + \sum_{\mathbf{p}'} L^{-3} V(\mathbf{p} - \mathbf{p}') a_{\mathbf{p}'} \tag{24}$$

where

$$V(\mathbf{p} - \mathbf{p}') = \int e^{i(\mathbf{p}-\mathbf{p}')\cdot\mathbf{r}/\hbar}\, V(\mathbf{r})\, d\mathbf{r} \tag{25}$$

The above is just Schrödinger's equation in the momentum representation.* Since the sum over $\mathbf{p}'$ is replaced by an integral as the walls recede to infinity, it is essentially an integro-differential equation. Thus we see that the form of Schrödinger's equation depends strongly on the representation that we use.

* This equation could have been obtained directly by going to the momentum representation [see Chap. 16, eq. (42)].

It is now convenient to make the substitution

$$a_{p'} = C_{p'} e^{-iE_{p'}t/\hbar}$$

where
$$E_{p'} = \frac{(p'^2)}{2m}$$

Equation (24) then becomes

$$i\hbar \dot{C}_p = L^{-3} \sum_{p'} V(p - p') C_{p'} e^{i(E_p - E_{p'})t/\hbar} \tag{26}$$

Equation (26) is exactly the same as that used in the method of variation of constants. [See Chap. 18, eq. (6).] In fact, we could have obtained it directly from this equation, but it is more instructive in some ways to obtain it from the momentum representation.

19. Born Approximation. Perturbation Theory. If the scattering potential is not large, one can solve this problem by perturbation theory, setting $C_{p_0} = 1$ at some initial time, which we call $t = 0$, while all the other C_p are zero at this time. To the first order, C_p can then be obtained by putting these values into the right side of eq. (26). This procedure yields

$$i\hbar \dot{C}_p = L^{-3} V(p - p_0) e^{i(E_p - E_{p_0})t/\hbar} \tag{27}$$

The above approximation is equivalent to the assumption that the incident wave is not seriously distorted by the scattering potential. When this assumption, which is essentially just the use of perturbation theory, is made in a scattering problem, it is called the Born approximation.

To find the probability of scattering, we integrate the above equation. Setting $C_p = 0$ at $t = 0$, we obtain

$$C_p = V(p - p_0) L^{-3} \frac{[e^{i(E_p - E_{p_0})t/\hbar} - 1]}{(E_p - E_{p_0})} \tag{28a}$$

$$|C_p|^2 = 4L^{-6} |V(p - p_0)|^2 \frac{\sin^2 \left[(E_p - E_{p_0}) \frac{t}{2\hbar} \right]}{(E_p - E_{p_0})^2} \tag{28b}$$

The above yields the probability that at the time t a transition has taken place from an initial momentum state of p_0 to a final momentum state of p, so that $|C_p|^2$ is equal to the probability that the corresponding deflection has occurred.

There are several points in connection with the above equation that must be discussed. First, our boundary conditions are excessively abstract, in that we have chosen a plane wave of infinite extent. This means that at $t = 0$, the incident plane wave is assumed to cover all space, *including the scatterer itself*. Actually, it is necessary to form a packet, which initially has not yet struck the scatterer. We shall see in Sec. 29, however, that the error resulting from the use of the wrong boundary condition is negligible.

The second important point is that for a given p the probability of scattering is, for short times, proportional to t^2. The problem is very similar to that appearing in radiation theory (see Chap. 2, Sec. 16, and Chap. 18, Sec. 18), and its solution is very much the same. Actually, one must integrate over a range of incident particle energies, and this integration yields a probability of transition that is proportional to the time. In this case, however, one can obtain the same result in another way, which is rather instructive, namely, by assuming a definite initial momentum p_0, but summing over a range of final energies, $E_p = p^2/2m$.

In discussing the range of final energies, one must use the fact that the box is very large. Successive values of p are therefore so close together that none of the quantities that are being summed will change appreciably as p goes from one value to the next. The sum may therefore be replaced by an integral. To do this, we note that in summing over a small range of states, $dp_x\, dp_y\, dp_z$, we include a number of states

$$\delta N = \rho(p)\, dp_x\, dp_y\, dp_z$$

where $\rho(p)$ is the density of states in momentum space. The total probability that the particle makes a transition into the range $dp_x\, dp_y\, dp_z$ is therefore equal to

$$dP = \rho(p)|C_p|^2\, dp_x\, dp_y\, dp_z \tag{29}$$

where $|C_p|^2$ can be obtained from eq. (28a).

The density of state is given in eq. (26), Chap. 1, in terms of k space, as

$$dN = \left(\frac{L}{2\pi}\right)^3 dk_x\, dk_y\, dk_z$$

Writing $k = p/\hbar$, we obtain

$$dN = \left(\frac{L}{h}\right)^3 dp_x\, dp_y\, dp_z$$

and

$$\rho(p) = \left(\frac{L}{h}\right)^3$$

It is now convenient to transform to polar co-ordinates in momentum space. Equation (29) then becomes

$$dP = \rho(p)|C_p|^2p^2\, dp\, d\Omega = \left(\frac{L}{h}\right)^3 |C_p|^2p^2\, dp\, d\Omega \tag{29a}$$

This equation yields the probability that the particle makes a transition to a momentum p, directed within the range of solid angles $d\Omega$. It will also be convenient, however, to replace the momentum by the energy, according to the relations

$$E = \frac{p^2}{2m} \qquad \text{and} \qquad dP = \sqrt{\frac{m}{2E}}\, dE$$

We obtain

$$dP = m\sqrt{2Em}\,\rho(p)|C_p|^2\,dE\,d\Omega = |C_p|^2\rho(E)\,dE\,d\Omega \tag{29b}$$

where

$$\rho(E) = m\sqrt{2mE}\,\rho(p) = m^2v\rho(p) \tag{29c}$$

and v is the particle velocity.

Obtaining $|C_p|$ from eq. (28), and dP from eq. (29b), we get

$$dP = 4L^{-6}\rho(E)|V(p - p_0)|^2 \frac{\sin^2\left[(E_p - E_{p_0})\dfrac{t}{2\hbar}\right]}{(E_p - E_{p_0})^2}\,dE_p\,d\Omega \tag{29d}$$

We observe that the above expression contains the factor

$$\frac{\sin^2 (E_p - E_{p_0})\dfrac{t}{2\hbar}}{(E_p - E_{p_0})^2}$$

As shown in Chap. 2, Sec. 16, this factor becomes a sharply peaked function of $E_p - E_{p_0}$ when t is large. This means that although there is always some probability of transition to any energy E_p, the overwhelming probability after a long time is that the transition takes place to a level that conserves energy, within the limits, $\Delta E \cong \hbar/t$, within which the energy is definable. Thus, most of the transitions will take place to a small range of energies near $E_p = E_{p_0}$. This range grows narrower and narrower with the passage of time.*

Because $V(p - p_0)$ and $\rho(E)$ are smoothly varying functions, they remain practically constant inside the narrow range within which the integrand is large, so that they may be taken out of the integral, and evaluated at $E_p = E_{p_0}$. We then obtain for the probability of scattering into the range of solid angles $d\Omega$

$$\delta P = 4L^{-6}\rho(E_{p_0})|V(p - p_0)|^2\,d\Omega \int_0^\infty \frac{\sin^2\left[(E_p - E_{p_0})\dfrac{t}{2\hbar}\right]}{(E_p - E_{p_0})^2}\,dE_p \tag{30}$$

In accordance with conservation of energy, as discussed in the previous paragraph, $|V(p - p_0)|$ should be evaluated* at $E_p = E_{p_0}$ or at $|p| = |p_0|$; the direction of p may of course be different from that of p_0.

Since the integrand will usually be negligible for negative values of E_p, we can, as was done in Chap. 2, Sec. 16, simplify the result by allowing the limits of integration to run from $-\infty$ to ∞. We then make the substitution $(E_p - E_{p_0})\dfrac{t}{2\hbar} = x$, and obtain

* This means, of course, that the collision is elastic.

$$\delta P = 4L^{-6}\rho(E_{p_0})|V(p - p_0)|^2 \frac{d\Omega\, t}{2\hbar} \int_{-\infty}^{\infty} \frac{\sin^2 x\, dx}{x^2}$$
$$= \frac{2\pi\rho(E_{p_0})}{\hbar} L^{-6}|V(p - p_0)|^2 t\, d\Omega \quad (30a)$$

The above is a result of very general applicability. In any problem in which transitions occur to a continuous range of final states with density $\rho(E)$, one obtains the result that

$$\delta P = \frac{2\pi}{\hbar} \rho(E)|W_{1,2}|^2 t\, d\Omega \quad (30b)$$

where $W_{1,2}$ is the matrix element between the two states, defined by the relation $W_{1,2} = \int \psi_1^* V \psi_2\, dr$, and where ψ_1 and ψ_2, respectively, are the normalized wave functions of the initial and final states. We see from eq. (25) that in our case, $W_{1,2} = L^{-3}V(p - p_0)$.

Obtaining $\rho(E)$ from (29c), we get for the probability of transition per unit time

$$\frac{\delta P}{t} = \frac{4\pi^2 m^2 v}{h^4} L^{-3}|V(p - p_0)|^2\, d\Omega \quad (31)$$

20. Evaluation of Cross Section. To evaluate the cross section, we note that the latter can also be expressed in terms of the probability of scattering per unit time. According to Sec. 6, the probability of scattering into the element of solid angle, $d\Omega$ in the distance, dx, is

$$\delta P = \rho\sigma\, d\Omega\, dx$$

where ρ is the density of scatterers.

Writing $dx = v\, dt$ we obtain (setting $dt = t =$ a small interval of time)

$$\frac{\delta P}{t} = \rho v \sigma\, d\Omega \quad (32)$$

In our case, we have been studying a problem in which there is one particle within a box volume, L^3. Hence $\rho = L^{-3}$. This yields

$$\frac{\delta P}{t} = v\sigma L^{-3}\, d\Omega$$

Equating the above with (31), we obtain

$$\sigma = |V(p - p_0)|^2 \frac{4\pi^2 m^2}{h^4} \quad (33)$$

The above is independent of the size of the box; this is, of course, quite reasonable.

If $V(r)$ is spherically symmetric, one can show that $V(p - p_0)$ is a function only of $|p - p_0|$ and not of the direction of $p - p_0$.

Problem 4: Prove the above statement.

$|p - p_0|$ can be expressed in terms of the angle of scattering by means of eq. (22). We obtain

$$\sigma = \left|V\left(2p \sin \frac{\theta}{2}\right)\right|^2 \frac{4\pi^2 m^2}{h^4} \tag{33a}$$

Note that the cross section is determined by the Fourier components of the potential. This exemplifies the fact that the quantum description of the scattering process involves the entire potential acting as a whole, rather than just the parts covered by a particle in an orbit. This is because the particle is described by a wave packet, rather than by a trajectory. In the quantum domain, we must choose a wave packet covering the entire atom; otherwise, according to Sec. 16, the uncertainty in momentum resulting from defining the orbit to a higher accuracy will destroy the scattering pattern. Only if condition (21a) is satisfied can the classical orbit description be used.

21. Example of Application: The Shielded Coulomb Force. As an illustrative example, let us apply the above to the case of the shielded Coulomb potential

$$V = \frac{Z_1 Z_2 e^2}{r} \exp\left(-\frac{r}{r_0}\right) \tag{34}$$

We must evaluate the quantity

$$V(\boldsymbol{k} - \boldsymbol{k}_0) = Z_1 Z_2 e^2 \int \exp\left(-\frac{r}{r_0}\right) \exp \frac{[i(\boldsymbol{k} - \boldsymbol{k}_0) \cdot \boldsymbol{r}]\, d\boldsymbol{r}}{r}$$

An evaluation of this integral yields

$$V(\boldsymbol{k} - \boldsymbol{k}_0) = \frac{4\pi Z_1 Z_2 e^2}{|\boldsymbol{k} - \boldsymbol{k}_0|^2 + \left(\frac{1}{r_0}\right)^2} \tag{34a}$$

Problem 5: Obtain the above result.

The net result for the scattering cross section is

$$\sigma = \frac{4m^2 (Z_1 Z_2)^2 e^4}{\left(4p^2 \sin^2 \frac{\theta}{2} + \frac{\hbar^2}{r_0^2}\right)^2} \tag{35}$$

As r_0 approaches ∞ (i.e., when there is no shielding) we obtain

$$\sigma = \frac{m^2 (Z_1 Z_2 e^2)^2}{4p^4 \sin^4 \frac{\theta}{2}} = \frac{(Z_1 Z_2 e^2)^2}{16E^2 \sin^4 \frac{\theta}{2}} \tag{36}$$

This is the same as the exact Rutherford law, obtained classically [see eq. (16c)]. The complete agreement between the two for all angles of scattering is a special property of the Coulomb law of force; for, as we shall see in Sec. 38, the classical and quantum results do not agree for an arbitrary law of force.

The general appearance of the cross section for a shielded Coulomb force as a function of angle is shown in Fig. 13. The curve rises steeply with decreasing θ, as is characteristic of the Rutherford cross section, until

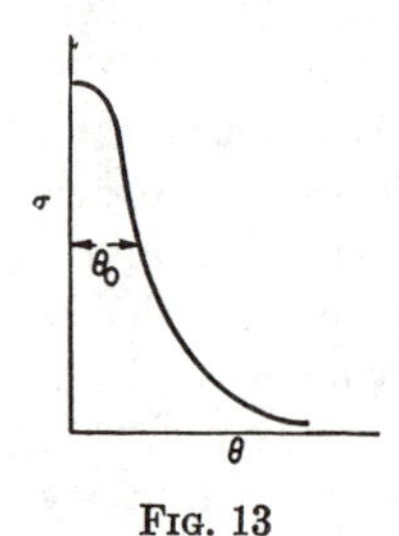

FIG. 13

$$\sin\frac{\theta_0}{2} \cong \frac{\hbar}{2pr_0} \tag{36a}$$

For angles smaller than θ_0, the rise of σ is comparatively small. Thus, θ_0 may be regarded as a sort of minimum angle, below which Rutherford scattering ceases, as a result of the effects of shielding.

22. Relation between Born Approximation and Fourier Analysis of the Potential. Equations (25 and 33) show that the cross section depends only on the absolute value of the Fourier component of the potential corresponding to

$$|\boldsymbol{k} - \boldsymbol{k}_0| = \frac{2p}{\hbar}\sin\frac{\theta}{2} = \frac{2\sqrt{2mE}}{\hbar}\sin\frac{\theta}{2} \tag{37}$$

This means that a detailed study of the dependence of scattering on energy and angle will enable one to Fourier analyze the potential, and thus to obtain a good idea of the range and shape of the potential, *provided that the Born approximation is valid.* As we shall see in Sec. 39, when the Born approximation fails, the same information can still be obtained, but a more complex procedure is required.

A very important property of the deflections is that a given momentum change, $\Delta\boldsymbol{p} = \boldsymbol{p} - \boldsymbol{p}_0$ can be produced only if the potential has such a shape that this Fourier component is present. A very large deflection can be produced in this way by a very small force, provided that the force varies rapidly enough in space. This will happen if, for example, the region in which the potential is large is very narrow. A small force will then mean only a small *probability* of a deflection. This is in contrast to classical theory, which says that large deflections produced in small distances always require large forces.

How can these two results be made consistent? Let us remember that in the Born approximation the deflection process is described as a single indivisible transition from one momentum state to another. The fact that there is only one transition was contained in eq. (27), which said that a given $\dot{C}_p$ always came from C_{p_0}, and not from any other C_p. In higher approximations, however, one would have processes in which the particle suffered many successive elementary deflections by the *same* atom. This would be described in perturbation theory by going to the second, or to higher approximations, since the latter would show how a given value of $\dot{C}_p$ could be contributed to not only by p_0, but also by other first-order C_p's, which arise from the first deflection process. In general, a large potential tends to favor the breakdown of perturbation theory, and thus to produce many successive deflections in the same scattering process. If there are enough successive deflections, the scattering process will begin to seem continuous, and it will approach a classical behavior. Thus, we see in another way* why a strong force tends to produce a classical behavior; also we see how the apparently continuous classical deflection arises, despite the indivisible nature of the elementary processes of deflection.

23. Illustration: Comparison of Cross Sections for Gaussian Potential and Square Well. To illustrate how one can investigate the shape of the potential curve, let us consider the differences in cross section resulting from the following two potentials,

$$V = V_0 \exp\left[-\frac{1}{2}\left(\frac{r}{r_0}\right)^2\right] \qquad \text{(Gaussian potential)}$$

$$V = V_0 \quad \text{when } r < r_0 \quad \text{and} \quad V = 0 \quad \text{when } r > r_0 \qquad \text{(square well)}$$

We shall begin by obtaining a general expression for the Fourier component of V, whenever V is a spherically symmetrical potential. Let the spherical polar co-ordinates be designated by r, α, β. Then the Fourier component that we wish to evaluate is

$$V(\boldsymbol{k} - \boldsymbol{k}_0) = \int_0^\infty \int_0^\pi \int_0^{2\pi} V(r) e^{i(\boldsymbol{k}-\boldsymbol{k}_0)\cdot \boldsymbol{r}}\, r^2 \sin\alpha \, dr\, d\alpha\, d\beta \qquad (38)$$

Let us choose our z axis in the direction of $\boldsymbol{k} - \boldsymbol{k}_0$. Then

$$(\boldsymbol{k} - \boldsymbol{k}_0) \cdot \boldsymbol{r} = |\boldsymbol{k} - \boldsymbol{k}_0| r \cos\alpha$$

The integral over β can be carried out directly, and it yields 2π. The integral over α then yields

$$V(\boldsymbol{k} - \boldsymbol{k}_0) = \frac{2\pi}{i|\boldsymbol{k} - \boldsymbol{k}_0|} \int_0^\infty V(r)\left(e^{i|\boldsymbol{k}-\boldsymbol{k}|r} - e^{-i|\boldsymbol{k}-\boldsymbol{k}|r}\right) r\, dr \qquad (38a)$$

If $V(r)$ is an even function of r, we can write further,

$$V(\boldsymbol{k} - \boldsymbol{k}_0) = \frac{2\pi}{i|\boldsymbol{k} - \boldsymbol{k}_0|} \int_{-\infty}^\infty V(r) e^{i|\boldsymbol{k}-\boldsymbol{k}_0|r}\, r\, dr \qquad (38b)$$

For the Gaussian potential, we obtain

$$V(\boldsymbol{k} - \boldsymbol{k}_0) = \frac{2\pi V_0}{i|\boldsymbol{k} - \boldsymbol{k}_0|} \int_{-\infty}^\infty \exp\left[-\frac{1}{2}\frac{r^2}{r_0^2} + i|\boldsymbol{k} - \boldsymbol{k}_0| r\right] r\, dr$$
$$= (2\pi)^{3/2} r_0^3 V_0\, e^{-\frac{1}{2}|\boldsymbol{k}-\boldsymbol{k}_0|^2 r_0^2} \qquad (39a)$$

* It was shown in Sec. 16 that a strong force favors the validity of the classical approximation.

For the square well potential, we obtain

$$V(k - k_0) = \frac{2\pi V_0}{i|k - k_0|}\int_{-r_0}^{r_0} e^{i|k-k_0|r}r\,dr$$
$$= -\frac{4\pi V_0}{|k - k_0|^2}\left(r_0 \cos|k - k_0|r_0 - \frac{\sin|k - k_0|r_0}{|k - k_0|}\right) \quad (39b)$$

Problem 6: Obtain the preceding results.

Interpretation of Results. From eq. (37) we obtain

$$|k - k_0| = \frac{2p}{\hbar}\sin\frac{\theta}{2}$$

where θ is the angle of deflection. Small $|k - k_0|$ therefore corresponds either to small deflections or small momentum (slow particles). We observe that for small $|k - k_0|$, both cases above, as well as that of shielded Coulomb scattering (eq. 35), all have the property that

$$V(k - k_0) \cong V(0) + \frac{V''(0)}{2}|k - k_0|^2 = V(0) + \frac{V''(0)}{2\hbar^2}\left(2p\sin\frac{\theta}{2}\right)^2$$

In other words, the term linear in $|k - k_0|$ is absent. (This may be verified directly by expansion in each case.) Thus, for very small momenta, the cross section does not depend much on angle. The momentum at which the cross section begins to depend appreciably on angle will be given by

$$\left|\frac{V''(0)}{V(0)}\right|\frac{|k - k_0|^2}{2} \cong 1 \quad (40)$$

In most cases, it turns out that this happens where $|k - k_0|r_0 \cong 1$ or $p\sin\frac{\theta}{2} \cong \frac{\hbar}{2r_0}$. Thus appreciable angular dependence will begin to occur only for momenta above $2\hbar/r_0$. This latter, however, is of the order of the momentum demanded by the uncertainty principle to localize a particle within a radius of the order of r_0.

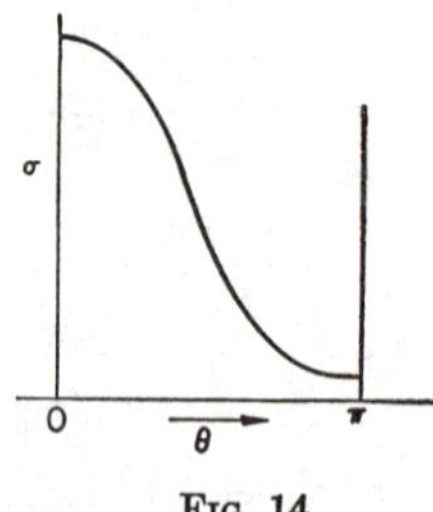

FIG. 14

When $|k_0|$ becomes large enough so that $|k_0|r_0 \cong 1$, i.e., when the incident wave is short enough so that it oscillates appreciably as it crosses the region in which the potential is large, the scattered wave begins to depend on angle. Until $|k_0|r_0 \cong 1$, there is no way to tell one shape of potential well from any other one; at large $|k|$, however, the scattering cross section depends strongly on the shape. For example, in the case of the Gaussian well, we obtain for large $|k_0|$, a rapid but smooth fall off of cross section with increasing angle, as shown in Fig. 14. For the square well, the cross section tends to oscillate, although it does fall off with increasing θ. A typical behavior at large $|\boldsymbol{k}|$ appears in Fig. 15. The oscillatory nature of σ arises from the sharp "edge" in the square potential. If the edge were smoothed out, the Fourier components would vary in a more regular fashion, and something like the Gaussian distribution would result.

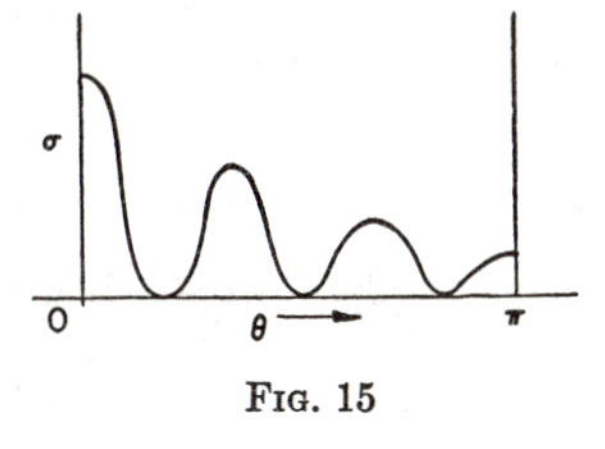

Fig. 15

In any case, it is clear that if one investigates the angular dependence of σ at high $|\boldsymbol{k}|$, one can obtain much information about the shape of the cross section by this method, provided, of course, that the Born approximation is good. (The conditions for validity of the Born approximation will be discussed in Sec. 31.) The finer the details of the shape that one wishes to investigate, the higher the values of $|\boldsymbol{k} - \boldsymbol{k}_0|$ that are needed in the Fourier analysis; hence the higher are the momenta of incident particles to which one must go.

24. The Space-time Representation of Scattering. We shall now go on to the second method of treating the scattering problem, namely, the space-time description of the scattering process by means of the wave model. In the momentum representation, we found it convenient to solve for the time dependent transition probabilities. In discussing the space-time representation, however, we shall find it convenient to start with the stationary-state wave functions and later to get the time dependence by forming wave packets, more or less as has already been done in the case of the free particle (Chap. 3, Sec. 2) and in the resonant trapping of particles in a potential well (Chap. 11, Sec. 17). In this description, one begins with a steady incident wave. Part of this wave is deflected, in the manner described in Sec. 17, and from the intensity of the scattered wave, one computes the probability of scattering. Of course, the actual incident wave is a packet in the sense that it is collimated and has a finite duration in time. Relative to atomic dimensions, however, the size of the packet is so great that a negligible error is made by assuming an incident plane wave of infinite extent. The incident beam is then described by the wave function

$$\psi_0 = e^{i\boldsymbol{k}_0\cdot\boldsymbol{r}}$$

As the incident wave enters the region of the scattering potential, a scattered wave is produced, which we denote by $g(\boldsymbol{r})$. The complete wave function is then

$$\psi = e^{i\boldsymbol{k}_0\cdot\boldsymbol{r}} + g(\boldsymbol{r}) \tag{41}$$

Since the incident beam and the scattered beam remain steady, all probabilities become independent of time and one can write for the complete time-dependent wave function

$$\Psi = \psi e^{-iEt/\hbar} = [e^{i\mathbf{k}_0\cdot\mathbf{r}} + g(\mathbf{r})]e^{-iEt/\hbar} \tag{41a}$$

If the potential vanishes as $r \to \infty$, as it usually does, E is just the value of the kinetic energy of the incident beam, which is

$$E = \frac{p^2}{2m} = \frac{\hbar^2 k_0^2}{2m}$$

Schrödinger's equation becomes

$$H\psi = \left(\frac{p^2}{2m} + V\right)\psi = E\psi$$

We note that $\left(\frac{p^2}{2m} - E\right) e^{i\mathbf{k}_0\cdot\mathbf{r}} = 0$. Thus, we obtain

$$\left(\frac{p^2}{2m} - E\right) g(\mathbf{r}) = -V(\mathbf{r})[e^{i\mathbf{k}_0\cdot\mathbf{r}} + g(\mathbf{r})] = -V(\mathbf{r})\psi(\mathbf{r}) \tag{42}$$

Since the potential energy is assumed to vanish as $r \to \infty$, the particles must approach free motion at large values of r. Because the incident wave is already represented by $e^{i\mathbf{k}_0\cdot\mathbf{r}}$, the function $g(\mathbf{r})$ represents only an outgoing stream. The function $g(\mathbf{r})$ must therefore approach asymptotically,

$$g(\mathbf{r}) \xrightarrow[r\to\infty]{} f(\theta, \phi)\,\frac{e^{ikr}}{r} \tag{43}$$

The above corresponds to the most general possible outgoing wave. The amplitude is a function of θ and ϕ; this indicates that the strength of the scattered wave depends on the angle of scattering. The nature of the wave involved is illustrated in Fig. 12.

The most general asymptotic solution to Schrödinger's equation [when $V(r) \to 0$ as $r \to \infty$] is

$$f(\theta, \phi)\,\frac{e^{ikr}}{r} + h(\theta, \phi)\,\frac{e^{-ikr}}{r}$$

The latter term, however, corresponds to an ingoing wave. Although such a wave is conceivable, it is never realized in practice, and, instead, one has just an incident plus an outgoing wave. We therefore conclude that $h(\theta, \phi) = 0$.

To obtain the scattering cross section, we first evaluate the incident current of particles. In the incident beam, the probability density is $P = |\psi_0|^2 = 1$ (since $\psi_0 = e^{i\mathbf{k}_0\cdot\mathbf{r}}$). The incident current per unit area is then $Pv = \hbar k_0/m$. In the scattered beam, the density is $|f(\theta, \phi)|^2/r^2$; the outgoing current per unit area is then $\frac{\hbar k}{mr^2}\,|f(\theta, \phi)|^2$. On a sphere (which is assumed to be very large compared with the size of the atom) the element of area is $r^2\,d\Omega$, where $d\Omega$ is the element of solid angle. The current into the element solid angle $d\Omega$ is then

$$j = \frac{\hbar k}{m}\,|f(\theta, \phi)|^2\,d\Omega \tag{44}$$

By definition (Secs. 3 and 6), however, the cross section $\sigma\,d\Omega$ is numerically equal to the probability that a particle within a beam of unit area will be deflected into the element of solid angle $d\Omega$. In terms of an incident current, I, the current of deflected particles is just $I\sigma\,d\Omega$. The cross section, $\sigma\,d\Omega$ is therefore just the ratio of the current of deflected particles to that of incident particles. We then obtain from eq. (44)

$$\sigma\,d\Omega = \frac{|k|}{|k_0|}\,|f(\theta, \phi)|^2\,d\Omega \tag{45a}$$

Since $|k| = |k_0|$ in our present problem, the above reduces to

$$\sigma\,d\Omega = |f(\theta, \phi)|^2\,d\Omega \tag{45b}$$

25. New Form for Schrödinger's Equation. The problem of evaluating σ is thus reduced to the problem of obtaining the strength of the outgoing wave. This requires, strictly speaking, a solution of Schrödinger's equation. We wish, however, to develop methods of approximation, and this can most conveniently be done by replacing Schrödinger's differential equation by an equivalent integral equation.*

To do this, we begin with eq. (42), which may be written

$$\left(\frac{\hbar^2}{2m}\nabla^2 + \frac{\hbar^2 k_0^2}{2m}\right) g = V\psi$$

Setting $\frac{2m}{\hbar^2} V\psi = U(\boldsymbol{r})$, one obtains

$$(\nabla^2 + k_0^2) g = U(\boldsymbol{r}) \tag{46}$$

Our objective is now to express g as a function of U. To do this, we make use of the following theorem

$$(\nabla^2 + k_0^2)\left(\frac{e^{ik_0|\boldsymbol{r} - \boldsymbol{r}_1|}}{|\boldsymbol{r} - \boldsymbol{r}'|}\right) = 4\pi\delta(x - x')\delta(y - y')\delta(z - z') \tag{47}$$

To prove this, we notice by direct differentiation that when $r \neq r'$,

$$(\nabla^2 + k_0^2)\frac{e^{ik_0|\boldsymbol{r} - \boldsymbol{r}'|}}{|\boldsymbol{r} - \boldsymbol{r}'|} = \lambda(\boldsymbol{r} - \boldsymbol{r}') = 0$$

Problem 7: Verify the preceding equation.

Thus, λ satisfies the first requisite of a δ function, namely, that it is zero everywhere, except at $\boldsymbol{r} = \boldsymbol{r}'$. In order to prove that it is a δ func-

* The method adopted in this section is also discussed in Mott and Massey, *Theory of Atomic Collisions*, Chap. 7.

tion, it will suffice to show that the integral of the above function taken over an arbitrary region surrounding the origin is a finite constant, independent of the size or shape of the region. (See definition of δ-function, Chap. 10, Sec. 15).

Since λ is zero when $r \neq r'$, the value of $\int \lambda(r - r')\, dr'$ is obviously the same for all regions of integration that include the point r; so that this integral may be evaluated by finding its limit as $|r - r'|$ approaches zero. Let us therefore choose for our range of integration a sphere of radius $|r - r'| = \epsilon$. As $\epsilon \to 0$, one can easily show that $\int k_0^2 \frac{e^{ik_0|r-r'|}}{|r - r'|}\, dr'$ approaches zero.

Problem 8: Prove the preceding statement.

There remains only the problem of evaluating

$$\int \nabla^2 \left(\frac{e^{ik_0|r-r'|}}{|r - r'|} \right) dr' = \int \operatorname{div} \left[\nabla \left(\frac{e^{ik_0|r-r'|}}{|r - r'|} \right) \right] dr'$$

By Green's theorem, we obtain for the above the following surface integral

$$\int \nabla \left(\frac{e^{ik_0|r-r'|}}{|r - r'|} \right) \cdot dS = \int \frac{\partial}{\partial r'} \left(\frac{e^{ik_0|r'-r|}}{|r' - r|} \right) dS$$

Writing $dS = |r - r'|^2\, d\Omega$, where $d\Omega$ is the element of solid angle, integrating over $d\Omega$, and setting $|r' - r| = \epsilon$, we finally obtain for the limit as $r \to 0$,

$$\int \lambda\, dr = -4\pi$$

Thus, we have proved that the integral of λ is a finite constant, independent of the region of integration as long as the latter includes the point $r = r'$. This completes the proof that λ is a δ function.*

* $\frac{e^{ik_0|r-r'|}}{|r - r'|}$ is a special case of a general class of functions, which are called *Green's functions* in mathematics. Green's functions may be used to solve linear differential equations of the form $\alpha(D)\Psi = U(r)$, where $\alpha(D)$ represents an arbitrary linear function of differentiation operators,

$$\alpha(D) = A(r) + B(r)D + C(r)D^2 + \cdots$$

A Green's function $G(r - r')$ has the following properties:

(a) $\alpha(D)G(r - r') = 0$ if $r \neq r'$ (i.e., it is a solution of the homogeneous equation, obtained by setting $U = 0$, when $r \neq r'$.

(b) $\alpha(D)G(r - r')$ is singular at $r = r'$.

(c) The approach to ∞ is such that the integral over an element containing the origin is finite. Thus, $\alpha(D)G(r - r')$ satisfies all prerequisites of a δ function.

The general problem of obtaining a solution of the equation

$$\alpha(D)\Psi = U(r)$$

can be solved once G is known. The solution is

$$\Psi = \int G(r - r')U(r')\, dr'$$

Let us now apply this theorem to our problem by constructing the function

$$g(\mathbf{r}) = -\frac{1}{4\pi}\int U(\mathbf{r}')\frac{e^{ik_0|\mathbf{r}-\mathbf{r}'|}}{|\mathbf{r}-\mathbf{r}'|}\,d\mathbf{r}' \tag{48}$$

Applying eq. (47), we see that $g(\mathbf{r})$ satisfies eq. (46). The only additional requirement to prove that it is the desired solution is to show that $g(\mathbf{r})$ contains only outgoing waves, i.e., that it takes the form $f(\theta, \phi)e^{ikr}/r$ as $r \to \infty$. To prove this, we note that $U(\mathbf{r}')$ must approach zero as $r' \to \infty$, just because $V(\mathbf{r}') \to 0$ and ψ remains finite. All contributions to the above integral therefore come from limited values of $\mathbf{r}'$. When $\mathbf{r}$ is very large, $|\mathbf{r} - \mathbf{r}'|$ may therefore be expanded as a series of powers of $|\mathbf{r}'|/|\mathbf{r}|$. A little geometry shows that for large r, we obtain

$$|\mathbf{r} - \mathbf{r}'| \cong r - \mathbf{r}'\cdot\mathbf{n}$$

where $\mathbf{n}$ is a unit vector in the direction of $\mathbf{r}$. We are then led to the following expansion,

$$\frac{e^{ik|\mathbf{r}-\mathbf{r}'|}}{|\mathbf{r}-\mathbf{r}'|} \cong \frac{e^{ik_0(r-\mathbf{r}'\cdot\mathbf{n})}}{r}\left(1 - \frac{\mathbf{r}'\cdot\mathbf{n}}{r} + \cdots\right)$$

Thus, as $r \to \infty$, we do indeed get an outgoing wave; if we had chosen $\frac{e^{-ik|\mathbf{r}-\mathbf{r}'|}}{|\mathbf{r}-\mathbf{r}'|}$, we would have obtained instead an ingoing wave.

This completes our reformulation of Schrödinger's equation. Setting $U = \frac{2m}{\hbar^2}V\psi$, we obtain

$$g(\mathbf{r}) = -\frac{m}{2\pi\hbar^2}\int V(\mathbf{r}')\psi(\mathbf{r}')\frac{e^{ik|\mathbf{r}-\mathbf{r}'|}}{|\mathbf{r}-\mathbf{r}'|}\,d\mathbf{r}' \tag{49}$$

As $r \to \infty$, the above becomes

$$g(\mathbf{r}) \to -\frac{m}{2\pi\hbar^2}\int e^{-ik(\mathbf{n}\cdot\mathbf{r}')}V(\mathbf{r}')\psi(\mathbf{r}')\,d\mathbf{r}' \tag{50}$$

The above may be simplified by noting that the vector, $\mathbf{k}'$, which goes in the direction of the outgoing wave, is equal to $k\mathbf{n}$. Thus, we obtain

$$g(\mathbf{r}) \xrightarrow[r\to\infty]{} -\frac{m}{2\pi\hbar^2}\frac{e^{ikr}}{r}\int e^{-i\mathbf{k}'\cdot\mathbf{r}'}V(\mathbf{r}')\psi(\mathbf{r}')\,d\mathbf{r}' \tag{51}$$

We can also write [see eqs. (43) and (45)]

$$f(\theta, \phi) = -\frac{m}{2\pi\hbar^2}\int e^{-i\mathbf{k}'\cdot\mathbf{r}'}V(\mathbf{r}')\psi(\mathbf{r}')\,d\mathbf{r}' \tag{51a}$$

and

$$\sigma = |f(\theta, \phi)|^2 = \left(\frac{m}{2\pi\hbar^2}\right)^2\left|\int e^{-i\mathbf{k}'\cdot\mathbf{r}'}V(\mathbf{r}')\psi(\mathbf{r}')\,d\mathbf{r}'\right|^2 \tag{51b}$$

26. Interpretation of Results. Equation (49) is an integral equation whose solution satisfies Schrödinger's equation with the correct boundary conditions. If we write $\psi = g(r) + e^{ik_0\cdot r}$, we obtain

$$g(r) = -\frac{m}{2\pi\hbar^2}\int g(r')\,\frac{e^{ik|\,-r'|}}{|r-r'|}\,V(r')\,dr' - \frac{m}{2\pi\hbar^2}\int e^{ik_0\cdot r'}\,\frac{e^{ik|r-r'|}}{|r-r'|}\,V(r')\,dr' \tag{52}$$

The above is a standard integral equation defining $g(r)$. It can be solved approximately by standard methods, which will be discussed in Sec. 27. As we shall see, it is in a form that makes the application of perturbation theory very easy.

One can obtain a simple physical picture of eq. (49). To do this, we note that $g(r)$ may be regarded as that part of the wave function produced by the scattering potential. $g(r)$ is an integral involving the function $e^{ik|r-r'|}/|r-r'|$. This is just, however, a spherical wave that spreads out from the point r', with a wavelength $\lambda = 2\pi/k$. Each spherical wave is weighted with the amplitude factor $V(r')\psi(r')$. In other words, each point contributes according to the product of the potential at that point, and the wave function, $\psi(r')$. Note that $\psi(r')$ is the *total* wave function, *including* all contributions from the scattered waves, as well as that of the incident wave. This picture corresponds exactly to Fresnel diffraction in optics.*

The asymptotic form of the wave (eq. 51) states that the outgoing wave, moving in the direction n is the sum of a set of wavelets originating at the points r'. Each point contributes an amplitude $V(r')\psi(r')$ and the phase is changed by the factor $e^{-ik(n\cdot r')}$. This picture corresponds exactly to the Fraunhofer* diffraction in optics, i.e., to the diffraction pattern at infinite distance. To illustrate this point, we show in Fig. 16 how the diffraction pattern of a grating is calculated. This is done by taking a wave proportional to the wave amplitude existing at the grating, and adding up the contributions of each part with a phase of

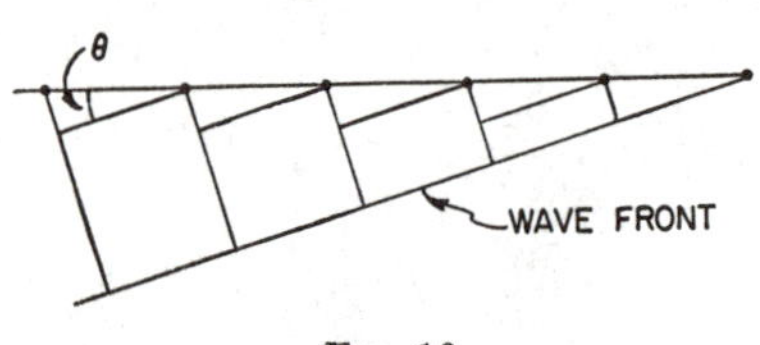

FIG. 16

$$e^{-ikn\cdot r'} = e^{-ikx\sin\theta}$$

where n is the direction of viewing, x is the co-ordinate measured along the direction of the grating, and θ is the angle shown in Fig. 16. Equations (50) and (51) may be regarded as a rigorous expression of Huyghens' principle for electron waves.†

* See Jenkins and White, *Fundamentals of Physical Optics.*

† See Chap. 6, Sec. 3; also R. P. Feynman, *Rev. Mod. Phys.*, **20**, 377 (1948), Sec. 7.

27. The Born Approximation. The Born approximation is applicable whenever V is fairly small. The idea is simply that of successive approximations. If V is small enough, then terms on the right-hand side of eq. (52) involving Vg will be of second order, because g is already of first order. This amounts to replacing ψ by the incident wave, $e^{i\mathbf{k}_0\cdot\mathbf{r}}$, a procedure that is valid when the scattered wave is small compared with the incident wave. As a result, we are neglecting the rescattering of the scattered wave. When this approximation is inserted into eq. (52), one obtains

$$g \cong -\frac{m}{2\pi\hbar^2}\int V(\mathbf{r}')e^{i\mathbf{k}_0\cdot\mathbf{r}'}\frac{e^{ik|\mathbf{r}-\mathbf{r}'|}}{|\mathbf{r}-\mathbf{r}'|}\,d\mathbf{r}' \tag{53}$$

and for $r'/r \ll 1$

$$f(\theta, \phi) \cong -\frac{m}{2\pi\hbar^2}\int e^{i(\mathbf{k}_0-\mathbf{k}')\cdot\mathbf{r}'}\,V(\mathbf{r}')\,d\mathbf{r}' \tag{54}$$

and

$$\sigma(\theta, \phi) = \left(\frac{m}{2\pi\hbar^2}\right)^2\left|\int e^{i(\mathbf{k}_0-\mathbf{k}')\cdot\mathbf{r}'}V(\mathbf{r}')\,d\mathbf{r}'\right|^2 \tag{55}$$

The expression for the cross section is exactly what was obtained from the theory which regarded scattering as a transition [see eq. (33)].

The Born approximation is commonly used in optics, but its use is not, as a rule, explicitly stated. In calculating the diffraction from a slit, for example, one assumes that the wave amplitude in the slit is equal to that of the incident wave only. A complete treatment, however, would require that one also add the amplitude of the diffracted wave at the slit, in computing the net intensity by a Huyghens' construction. Since the diffracted wave, in turn, contributes to the net diffraction pattern, one can see that this is a complex problem. It can be treated rigorously only by a complete solution of the wave equation, which takes into account the change of wave amplitude in the slit resulting from electric currents that are induced in the slit by the total wave, *including* the part produced by the currents themselves. For a slit that is wide in comparison to a wavelength, the wave amplitude inside the slit is not very different from the incident amplitude, so that the Born approximation can be used in computing the diffraction pattern. For a narrow slit, however, the modification of the wave by the slit is so great that one needs a much better solution of the wave equation.

28. Relation of Space-time and Causal Descriptions. The treatment in terms of a position representation, which we have just developed, constitutes a point of view that is complementary to a treatment in terms of the momentum representation. In the position representation we predict the probability of scattering into a definite range of angles by finding the contributions of all the different wavelets scattered from the various parts of the potential. With this method it is easy to see why the shape of the potential is so important, since interference from different parts of the potential will determine the net scattering pattern. It is also easy to see why slow particles cannot very effectively probe the shape of the potential [see eq. (53)], for if the wavelength is too long, the phase of the wave changes so little over the entire region of the potential that it hardly matters at all how the potential is distributed. This method fails only in that it does not give us a clear picture of why the scattered entity turns up as a concentrated particle, and not as a scattered wave* (see Sec. 17). The momentum representation pictures the scattering process as an indivisible transition from one momentum state to the other. It does not permit one to analyze in detail how the transition takes place. The entire process is lumped into a single act, which *must not* be analyzed into smaller processes. In this sense, the description is incomplete. Yet, it gives a very fine account of how the particle comes off in some definite direction.

In different circumstances, one of the methods may be more appropriate than the other. We shall therefore use either, as the occasion demands. It should be observed that both methods always leads to the same final result. Since each involves merely the solution of Schrödinger's equation in a different way, it is only in making a picture that one is more convenient than the other.

29. Relation of Stationary-state Method to Time Dependent Descriptions. Let us recall that, as pointed out in Sec. 17, the use of a stationary incident plane wave does not correspond to the actual boundary conditions. Instead, there is actually a wave packet, which is incident on the scattering potential as t approaches $-\infty$. After the packet strikes the potential, a scattered packet appears, and recedes from the scattering center as t approaches $+\infty$.

To form a packet, we must multiply the stationary state function by its appropriate time-dependent factor $e^{-i\hbar k^2 t/2m}$ and integrate over a small range of incident momenta† k_z, with a weighting factor, $f(k_z - k_{0z})$. Using the asymptotic form of the wave function given in eqs. (41) and (43), we obtain for the time-dependent wave function at large radii,

$$\psi(x, t) = \int f(k_z - k_{0z})\, dk_z \left(e^{ik_{0z}} + f\frac{(\theta, \phi)e^{ikr}}{r}\right) e^{-\frac{i\hbar k^2}{2m}t} \tag{56}$$

The first term on the right-hand side yields the incident wave. The center of this packet occurs where the phase has an extremum, or where

* This is, of course, an example of the wave-particle duality of the properties of matter, discussed in more detail in Chaps. 6 and 8. Thus, when an electron scatters from a potential existing in a region small compared with the electronic wave length, the wave-like potentialities of the electron are emphasized; but when it interacts with a position-measuring device, its particle-like potentialities are emphasized.

† One also forms a packet in the x and y directions, but this is usually so broad that the spread can be neglected in these directions, even in comparison with that in the z direction.

$z = \hbar k_0 t/m = p_0 t/m$. This packet represents an electron moving in the z direction, starting at large negative values of z, as t approaches $-\infty$, passing the scattering potential near $t = 0$, and going on to positive values of z as t approaches $+\infty$.

The scattered wave is represented by the second term on the right side of eq. (56). Writing $f(\theta, \phi) = |f(\theta, \phi)|e^{i\alpha}$, we obtain for the center of this packet,

$$r = \frac{\hbar k t}{m} - \frac{\partial \alpha}{\partial k}$$

Since, by definition, only positive values of r have significance, it is clear that for large negative values of t, there is no scattered packet, and that for large positive values of t, there is a scattered packet receding from the origin. The term $\partial\alpha/\partial k$ represents a time delay (or advance) which is brought about by the action of the potential. (Compare with Chap. 11, Sec. 19.)

If the packet is large in comparison with the range of the scattering potential, then the exact size of the packet will not affect the cross section in any critical manner. This is because the range of wave numbers in the packet will then be small compared with those present in the potential; thus, the uncertainty in momentum arising from the width of the packet will produce deflections that are negligible compared with those resulting from the scattering potential itself. In this way, we justify the use of the infinite plane wave to compute the cross section.

We shall now consider the question of why cross sections obtained from the time-dependent perturbation theory* with the momentum representation are the same as those obtained with the use of wave packets made up of eigenfunctions of the Hamiltonian. Let us recall that the time-dependent theory involved the assumption that at $t = 0$ the potential was suddenly turned on, and also that the wave function was a plane wave, $e^{ik_0 z}$. Such a boundary condition is certainly not a very accurate representation of what actually happens. In the derivation of eq. (31), however, the assumption was made that the perturbation lasts for so long a time that the function $\dfrac{\sin^2\left[(E - E_{p_0})\dfrac{t}{2\hbar}\right]}{(E_p - E_{p_0})^2}$ appearing in eq. (30) becomes very sharply peaked, and essentially equivalent to a δ function. This means that the scattering process is assumed to last so long that the contribution of "edge" effects introduced by the sudden turning on of a potential at $t = 0$ are negligible. Thus, although the boundary conditions are unrealistic, they will lead to the right answer. It is of interest that the replacement of a wave packet by an infinite plane

* See Sec. 20.

wave in the stationary state method is also justified on similar grounds, i.e., the wave packet is so broad that "edge" effects are negligible.

30. Another Application of the Born Approximation: Scattering from a Crystal Lattice. At this point, it is instructive to apply the Born approximation to the problem of scattering from a crystal lattice. In a lattice, the atoms will be spaced regularly at position vectors that are given by

$$\boldsymbol{r}_j = l_j\boldsymbol{a} + m_j\boldsymbol{b} + n_j\boldsymbol{c}$$

l_j, m_j, n_j are integers, and $\boldsymbol{a}$, $\boldsymbol{b}$, $\boldsymbol{c}$ are the three basic lattice vectors. If the three vectors are equal and orthogonal, for example, one has a simple cubic lattice.

The total electrostatic potential resulting from all the atoms in the lattice is just the sum of the potentials caused by each atom. Each atom contributes a potential $V_j = V_j(\boldsymbol{r} - \boldsymbol{r}_j)$, which is, to a first approximation, spherically symmetrical about the center of the atom, $\boldsymbol{r} = \boldsymbol{r}_j$. The potential resembles roughly a shielded Coulomb potential (see Fig. 7). More accurately, however, the potential deviates somewhat from spherical symmetry, especially at large radii, and takes the crystal symmetry instead. The total potential is then

$$V = \sum_j V(\boldsymbol{r} - \boldsymbol{r}_j)$$

Note that V is periodic, in the sense that it is unchanged if $\boldsymbol{r}$ is displaced by any integral multiple of $\boldsymbol{a}$, $\boldsymbol{b}$, or $\boldsymbol{c}$.

Let us now solve for the probability of scattering of an electron in this potential. We obtain

$$f(\theta, \phi) = -\frac{m}{2\pi\hbar^2}\int e^{i(\boldsymbol{k}-\boldsymbol{k}_0)\cdot\boldsymbol{r}'} \sum_i V(\boldsymbol{r}' - \boldsymbol{r}_j)\, d\boldsymbol{r}'$$

$$= -\frac{m}{2\pi\hbar^2}\sum_j e^{i(\boldsymbol{k}-\boldsymbol{k}_0)\cdot\boldsymbol{r}_j}\int V(\boldsymbol{r}' - \boldsymbol{r}_j)e^{i(\boldsymbol{k}-\boldsymbol{k}_0)\cdot(\boldsymbol{r}'-\boldsymbol{r}_1)}\, d\boldsymbol{r}'$$

We note that the integral in the above equation is independent of j; it may therefore be denoted by $g(\boldsymbol{k} - \boldsymbol{k}_0)$. This is just the Fourier coefficient of the potential of any *one* of the atoms. We obtain

$$f(\theta, \phi) = g(\boldsymbol{k} - \boldsymbol{k}_0)\sum_j e^{i(\boldsymbol{k}-\boldsymbol{k}_0)\cdot\boldsymbol{r}_j} \tag{57}$$

Writing $\boldsymbol{r}_j = l_j\boldsymbol{a} + m_j\boldsymbol{b} + n_j\boldsymbol{c}$ we see that the above sum vanishes unless

$$(\boldsymbol{k} - \boldsymbol{k}_0)\cdot\boldsymbol{a} = 2\pi\alpha$$
$$(\boldsymbol{k} - \boldsymbol{k}_0)\cdot\boldsymbol{b} = 2\pi\beta$$
$$(\boldsymbol{k} - \boldsymbol{k}_0)\cdot\boldsymbol{c} = 2\pi\gamma$$

where α, β, and γ are integers. This, however, is just the well known condition for Bragg reflection from crystals.*

The complete periodicity of V implies a crystal of infinite extent. Real crystals, of course, have only a finite number of atoms, although this number may be very large. When the sum (57) is carried out over a finite number of values of j, one obtains a function that is sharply peaked near the Bragg angles. The width of the peak is inversely proportional to the size of the crystal. The problem is very similar to that of the resolving power of a finite diffraction grating in optics.

The function $|g(\boldsymbol{k} - \boldsymbol{k}_0)|^2$ is called the "atomic form factor." It determines the strength of reflection of the electrons in any allowed direction. It is clear that to obtain large changes of momentum one needs atoms with potentials that change sharply as a function of position; otherwise one would not obtain high Fourier components.

Electron diffraction is an important tool in investigating crystal structure. It can also be used to investigate the "atomic form factor," and thus provide information about the distribution of potential inside the atom. Finally, it can be used to investigate molecular structure.

Problem 9: Assuming that the potential resulting from an atom is $e^{-r/r_0}/r$, calculate the diffraction pattern to be expected from a diatomic molecule, with atomic separation, a. Assuming $r_0 = 10^{-8}$ cm, and $a = 3 \times 10^{-8}$ cm, estimate the minimum electronic energy needed to give a clear indication of the separation of the two atoms in the molecule. Note that one must average over all possible orientations of the molecule.

31. Conditions for Validity of Born Approximation. The Born approximation involved replacing the total wave function, ψ, by the incident wave function $e^{i\boldsymbol{k}_0\cdot\boldsymbol{r}}$ in eq. (50). It will therefore be valid whenever the scattered wave, $g(\boldsymbol{r})$ is small compared to $e^{i\boldsymbol{k}_0\cdot\boldsymbol{r}}$ in the region where $V(\boldsymbol{r})$ is large. In most cases, both $V(\boldsymbol{r})$ and $g(\boldsymbol{r})$ are largest near the origin, so that a rough criterion for the validity of the Born approximation is

$$|g(\boldsymbol{r})|^2 \ll 1$$

for small values of r. It may sometimes happen, however, that $|g(\boldsymbol{r})|$ is small when r is small, but large for intermediate values of r, such that $V(\boldsymbol{r})$ is still appreciable. One must therefore use some care in applying this criterion. Furthermore, it can also happen that the Born approximation still gives the right answer when the criterion is not satisfied. Having $|g(\boldsymbol{r})|$ small everywhere provides a sufficient condition for the validity of the approximation, but not a necessary condition.

With the aid of eq. (53) our criterion becomes

$$|g(0)|^2 = \left(\frac{m}{2\pi\hbar^2}\right)^2 \left| \int V(\boldsymbol{r}') \frac{e^{i(\boldsymbol{k}_0'\cdot\boldsymbol{r}'+kr')}}{r'} d\boldsymbol{r}' \right|^2 \ll 1 \qquad (58a)$$

* Richtmeyer and Kennard, 3d. ed., pp. 486–495.

(It normally suffices to evaluate $|g(r)|$ at $r = 0$, since the wave function is usually largest at the centers.)

If the potential is spherically symmetric, the above can be integrated over θ (choosing the z axis in the direction of $\boldsymbol{k}_0$) to yield

$$4\pi^2 \left(\frac{m}{2\pi\hbar^2 k}\right)^2 \left| \int_0^\infty V(r') e^{ikr'} (e^{ik_0 r'} - e^{-ik_0 r'})\, dr' \right|^2 \ll 1$$

Setting $k = k_0$, we obtain

$$\left(\frac{m}{\hbar^2 k}\right)^2 \left| \int_0^\infty V(r')(e^{2ikr'} - 1)\, dr' \right|^2 \ll 1 \tag{58b}$$

32. Application to Screened Coulomb Scattering. Let us test the validity of the Born approximation for the screened Coulomb potential

$$V = \frac{Z_1 Z_2 e^2}{r} e^{-ar}$$

where $a = 1/r_0$. We must evaluate the integral

$$I = \int_0^\infty e^{-ar'}(e^{2ikr'} - 1) \frac{dr'}{r'}$$

To compute I, we first differentiate with respect to a,

$$\frac{\partial I}{\partial a} = \int_0^\infty -r'\, e^{-ar'}(e^{2ikr'} - 1)\, dr' = \frac{1}{a} - \frac{1}{a - 2ik}$$

We now integrate with respect to a, and find

$$I = \ln (a) - \ln (a - 2ik) + C$$

where C is a constant of integration. It can be shown that for $a = \infty$, $I = 0$. Thus, we must choose $C = 0$. We obtain

$$I = -\ln\left(1 - \frac{2ik}{a}\right) = -\ln (1 - 2ikr_0)$$

Let us write $(1 - 2ikr_0) = \sqrt{1 + 4k^2 r_0^2}\, e^{i\phi}$ where $\phi = -\tan^{-1} 2kr_0$. Then

$$I = -\ln \sqrt{1 + 4k^2 r_0^2} - i\phi = -\ln \sqrt{1 + 4k^2 r_0^2} + i \tan^{-1} 2kr_0$$

The condition for validity of the Born approximation is then

$$\left(\frac{m}{\hbar^2 k}\right)^2 (Z_1 Z_2 e^2)^2 [(\ln \sqrt{1 + 4k^2 r_0^2})^2 + (\tan^{-1} 2kr_0)^2] \ll 1$$

Writing $k\hbar/m = v$ = particle velocity at infinity, we obtain

$$\left(\frac{Z_1 Z_2 e^2}{\hbar v}\right)^2 [(\ln \sqrt{1 + 4k^2 r_0^2})^2 + (\tan^{-1} 2kr_0)^2] \ll 1$$

The factors in the brackets will not normally become very large. $\mathrm{Tan}^{-1} 2kr_0$ is limited to $\pi/2$, while the logarithmic factor grows very slowly with increasing r_0. For most problems, it is not a great deal larger than unity. The main requirement for the validity of the Born approximation is that

$$\left(\frac{Z_1 Z_2 e^2}{\hbar v}\right) \ll 1 \tag{59}$$

This criterion is practically independent of the shielding radius, except for the very weak dependence because of $\ln \sqrt{1 + 4k^2 r_0^2}$ appearing in eq. (59).

How well is this criterion satisfied? For a typical case of electrons of 10 kev, we have $v = 6 \times 10^9$ cm/sec. Suppose we choose $Z = 10$. Then we obtain $(Z_1 Z_2 e^2/\hbar v) \cong 0.4$. The Born approximation is barely satisfied for this case. If we choose $Z = 1$, however, the Born approximation will be satisfied fairly well. In order to satisfy the Born approximation, we need, therefore, incident particles of high velocity and scatterers of low atomic number. We shall see, however, that for the special case of Coulomb scattering, there are certain reasons why the Born approximation still yields a good approximation to the correct results, even when this criterion is not satisfied.

33. Another Criterion for Validity of Born Approximation. If we recall that a change of potential acts like a change of index of refraction in optics, we can derive another criterion for the validity of the Born approximation. Inside the region of the potential, the wave vector is given roughly by $k = \sqrt{2m(E - V)}/\hbar$. The phase of the wave at the edge of the atom will then differ from the phase that it would have in the absence of a potential by the quantity

$$\Delta\phi = \int_0^\infty \sqrt{\frac{2m}{\hbar}} (\sqrt{E - V} - \sqrt{E})\, dr \tag{60}$$

If this difference is small compared with unity, one may take it as an indication that the wave function is not very different from what it would have been in the absence of the potential.

If $V/E \ll 1$, this criterion may be simplified by expanding the square root. We obtain

$$\frac{1}{\hbar} \int_0^\infty \sqrt{\frac{m}{2E}}\, V\, dr \ll 1$$

If we define $\bar{V}$ as the average potential, and $\bar{r}$ as the mean range over which the potential spreads, then we obtain

$$\frac{1}{\hbar} \sqrt{\frac{m}{2E}}\, \bar{V} \bar{r} \ll 1$$

For a square well of radius, a, and a depth, $V_0 \ll E$, we obtain as a criterion for the validity of the Born approximation

$$E \gg \frac{m}{2}\left(\frac{V_0 a}{\hbar}\right)^2$$

34. Relation of Validity of Classical Approximation to Breakdown of Born Approximation. From the point of view in which scattering is regarded as a transition from one momentum state to another, the breakdown of the Born approximation means, as has already been stated, that one must go to a higher approximation, in which the system may make many successive transitions. In other words, it is scattered, and rescattered by the same atom, the number of rescatterings depending on the extent to which the Born approximation has broken down. This, however, is exactly what will lead to the possibility of describing the scattering process classically. In other words, under conditions of extreme breakdown of the Born approximation, the deflection of a particle by a given scatterer results from so many successive quantum processes that we may hope to be able to regard the entire process as approximately continuous and classically describable. We have already derived a criterion for the validity of the classical approximation [eq. (21a)], which shows that, in general, the classical approximation does indeed improve as the potential (and therefore the force) grows larger, so that if the Born approximation is bad enough, we may, in general, hope to be able to use the classical approximation.

One must, however, use some caution in applying classical theory when the Born approximation fails badly. This is because in the derivation of eq. (21a) the momentum delivered to the particle was calculated under the assumption that the particle is not deflected through a large angle; for large deflections, one ought to evaluate this quantity by a more accurate method. In most cases, however, the small deflection approximation will still yield results of the right order of magnitude even when the deflection is large.

35. Classical vs. Born Approximations for a Coulomb Force. In this problem, it is most convenient to start with eq. (21a) which is the condition for validity of the classical theory. Insertion of $F = Z_1 Z_2 e^2/r^2$ yields

$$\frac{4Z_1 Z_2 e^2}{\hbar v} \gg 1 \tag{61}$$

Comparison of the above result with eq. (59) verifies that, for Coulomb scattering, the classical theory becomes applicable just when the Born approximation breaks down very badly.

36. Unusual Properties of Coulomb Force. The Coulomb force has the unusual property that the classical approximation [eq. (61)] and the

Born approximation [eq. (59)] both lead to the same scattering cross section. Furthermore, as we shall see in Sec. 59, the exact quantum-mechanical cross section is also the same as the Rutherford cross section, even in the intermediate region, where neither the classical nor the Born approximations hold. This result is true of no other law of force; in fact, as we shall see, even the shielded Coulomb force shows differences between classical and quantum scattering in the region of angles where shielding is important.

This unusual coincidence produces the important result that in the scattering of electrons from atoms, the Born approximation often yields surprisingly good results, even though none of the general criteria for its validity are satisfied. The reason is as follows: Near the edge of the atom, the force is far from Coulombic, but it is weak because most of the nuclear charge is shielded out by the electrons. The Born approximation therefore applies here simply because V is so small. Inside the atom, V is so large that one might expect the Born approximation to break down, but here the shielding is absent and the force is Coulombic, so that by accident, the exact result is close to that given by the Born approximation. Thus, over the whole atom, the Born approximation gives a fairly good result, even though it cannot be justified on general grounds.

37. Lack of Applicability of Born Approximation to Nuclei. It is fortunate for the development of atomic theory that the Born approximation was so good, because otherwise a complete theory of atomic structure might not have been developed for a long time, simply because of mathematical complications. In the nucleus, however, one can easily show that the Born approximation is no good, except at very high bombarding energies ($\cong$ 100 mev or higher). To see that this is so, let us use the square well as a model of nuclear forces, as discussed in Chap. 11, Sec. 3. One represents the potential between a neutron and a proton by a well of radius $a = 2.8 \times 10^{-13}$ cm and with a depth of $V_0 \cong 20$ mev. According to eq. (58b) the validity of the Born approximation requires that

$$\left(\frac{m}{k\hbar^2}\right)^2 \left|\int_0^a V_0(e^{2ikr} - 1)\, dr\right|^2 = \left(\frac{m}{k\hbar^2}\right) V_0^2 \left|\frac{e^{2ika}}{2k} - a\right|^2 \ll 1$$

We shall say rather arbitrarily that the Born approximation begins to become reliable when the above number is $\frac{1}{2}$. In neutron-proton scattering, one uses the reduced mass, $\mu = m/2$ with $m = 1.6 \times 10^{-24}$ gram. A simple calculation shows that the approximation becomes valid when the relative energy is of the order of 50 mev, or when the bombarding energy (see Chap. 15, Sec. 5) is 100 mev, or higher. Since most nuclear experiments are done at lower energies, we shall have to use more accurate methods to obtain a prediction of nuclear scattering. It unfortunately turns out that whereas the Born approximation fails in nuclei, it does not fail badly enough to make the classical treatment valid, so that one must deal with the complex intermediate region by means of more accurate methods that will be described in Sec. 39.

Problem 10: Check the preceding result by the criterion derived from the smallness of phase shift (eq. 60).

38. Application to Shielded Coulomb Force. With an unshielded Coulomb force, we have seen that the scattering cross section turns out to be the same, regardless of whether or not the Born approximation is valid. With a shielded Coulomb force, however, we shall now show that the minimum angle, below which Rutherford scattering fails, is different, according to which approximation is valid. If the Born approximation can be used, we obtain this angle from eq. (36a)

$$(\theta_0)_{\text{Born}} \cong \sin \theta_0 \cong \frac{\hbar}{pr_0} \tag{62a}$$

If classical theory is valid, the angle is, as given in eq. (11),

$$(\theta_0)_{\text{classical}} \cong \frac{Z_1 Z_2 e^2}{E r_0} \tag{62b}$$

The ratio of the two is

$$\frac{(\theta_0)_{\text{classical}}}{(\theta_0)_{\text{Born}}} = \frac{Z_1 Z_2 e^2}{E\hbar} p = \frac{2Z_1 Z_2 e^2}{v\hbar} \tag{62c}$$

When $(Z_1 Z_2 e^2/v\hbar) \gg 1$, i.e., when the classical approximation applies, the classical result is much bigger than that given by the Born approximation. When $(Z_1 Z_2 e^2/v\hbar) \ll 1$, i.e., when the Born approximation applies, the classical result is then the smaller of the two. We conclude that, in general, the true minimum angle is always equal to the *larger* of the two possibilities.

One can interpret the above rule of always taking the larger of the two minimum angles in the following way. In all cases, the fundamental scattering process is, of course, quantum mechanical, and the classical theory becomes valid only when the momentum transfer involves many successive indivisible processes. In each basic quantum-mechanical process of momentum transfer, the minimum angle below which Rutherford scattering cannot apply is determined in the following way: If the particle is going to be scattered at all, it must enter the region of the potential, which covers a radius of the order of r_0. But while it is in this region, it cannot have a completely definite momentum, simply because, as we have seen in Chap. 8, the very structure of a localized particle requires that its possible momenta be distributed more or less uniformly over a range of the order of at least $\Delta p \cong \hbar/2r_0$. This range of momenta results in a range of scattering angles of the order of $\theta_0 \cong \Delta p/p \cong \hbar/2r_0 p$, which is the same as that given in eq. (62a). Since the angles must be distributed more or less uniformly over this range, it is not possible that there be a peak in the distribution, as predicted by unshielded Coulomb scattering.

This means that if the classically predicted minimum angle for Rutherford scattering turns out to be smaller than $\hbar/2pr_0$, the classical theory is wrong, since its results neglect the effects of the uncertainty principle. The minimum angle can, therefore, never be smaller than $\hbar/2pr_0$.

The above discussion explains the quantum-mechanical minimum deflection. It can also be understood on the basis of the picture of diffraction of electron waves, for it is a well-known result that a wave that is diffracted by a region of the order of r_0 in size will have a diffraction pattern with a minimum angular width of the order of $\lambda/r_0 \cong \hbar/2pr_0$.

Problem 11: Prove the above result.

When the Born approximation fails, the particle receives many successive deflections, each at least as large as the above minimum. Thus, where the classical approximation is valid, the resulting deflection will always be larger than that predicted by the Born approximation. This explains the rule that the classical result is to be taken only when the minimum deflection that it predicts is larger than that given by the Born approximation, whereas the quantum result must be taken when the classical result is the smaller.

39. Method of Partial Waves. (Rayleigh, Faxen and Holtsmark). When the Born approximation fails, a more accurate method is needed to solve the scattering problem. One such method, applicable to a spherically symmetrical potential, is to expand the wave function as a series of spherical harmonics multiplied by radial wave functions, just as was done, for example, with the hydrogen atom. This method was originally applied by Rayleigh* to the scattering of sound waves, and later by Faxen and Holtsmark† to the scattering of Schrödinger waves.

We begin by noting that the wave function possesses cylindrical symmetry about a line in the direction of the incident wave, which we shall call the z direction. The wave function can now be expanded as a series of Legendre polynomials; let us note that the associated Legendre functions are not needed because ψ possesses cylindrical symmetry and is therefore not a function of ϕ. Thus, just as in Chap. 15, Sec. 1, we obtain

$$\psi = \sum_l f_l(r) P_l(\cos\theta) \tag{63}$$

Each term in the above expansion is called a "partial wave," corresponding to a particular value of l. The $f_l(r)$ satisfy differential equations

* Rayleigh, *Theory of Sound*, 2d ed., London: The Macmillan Company, 1894–96. p. 323.

† Mott and Massey, *Theory of Atomic Collisions*, Chap. 2.

given in eq. (1), Chap. 15. It is convenient to deal with

$$g_l = rf_l(r)$$

The equations are

$$-\frac{d^2g_l}{dr^2} + \left\{\frac{l(l+1)}{r^2} - \frac{2m}{\hbar^2}[E - V(r)]\right\} g_l = 0 \tag{64}$$

In order to solve the scattering problem, we must first solve the above set of equations, subject to the boundary condition that (see Chap. 15, Sec. 3)

$$g_l \to r^{l+1} \qquad \text{as} \qquad r \to 0$$

40. General Nature of Solutions. We always assume that $V(r) \to 0$ as $r \to \infty$. At large r, the wave functions therefore approach asymptotically those functions obtained by neglecting $V(r)$ and $l(l+1)/r^2$ in eq. (64). Since E is always positive in a scattering process, these functions take the following asymptotic form:

$$g_l \cong A_l \sin (kr + \Delta_l) \tag{65}$$

where $k = \sqrt{2mE}/\hbar$. A_l and Δ_l are constants, which we must determine by solving the differential equation.

We can see what determines the phase Δ_l by looking at the general nature of the solutions, using the methods developed in connection with the hydrogen atom (Chap. 15, Sec. 12). For s waves, for example, the solution (see Fig. 17) starts out with $g \sim r$. Then, because the potential is large near the origin, the wave function curves rapidly there, and the wavelength is short. As r increases, the potential decreases, and eventually the wavelength becomes equal to that corresponding to a free particle. The phase of the wave, however, depends on the cumulative effects of the potential on the curvature of the wave function at smaller radii. Thus, Δ_0 will depend, in general, on $V(r)$ and on the incident energy E.

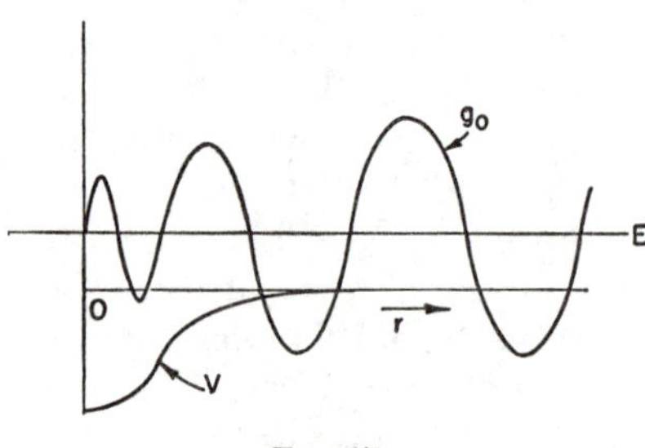

FIG. 17

If $l \neq 0$, the general nature of the process by which Δ_l is determined will be very similar to that for $l = 0$. $g_l(r)$, however, starts out as r^{l+1} and does not begin to curve back downward until the "effective kinetic energy," $\left[E - V(r) - \frac{l(l+1)\hbar^2}{2mr^2}\right]$, is positive. The wave function will look like the graph shown in Fig. 18.

41. Special Case: Coulomb Potential. The assumption that Δ_l approaches a constant as $r \to \infty$ requires not only that $V(r) \to 0$ as

$r \to \infty$ but also that $V(r) \to 0$ faster than $1/r$. To prove this, we use the WKB approximation, restricting ourselves to the case of s waves, but noting that the same results hold for all values of l. The WKB approximate wave function is*

$$g \cong \frac{\sin\left\{\int_0^r \sqrt{2m[E - V(r)]}\,\frac{dr}{\hbar}\right\}}{p^{1/2}}$$

Although this may not be a good approximation at small r, it is always a good approximation at large r just because V is very small, so that the

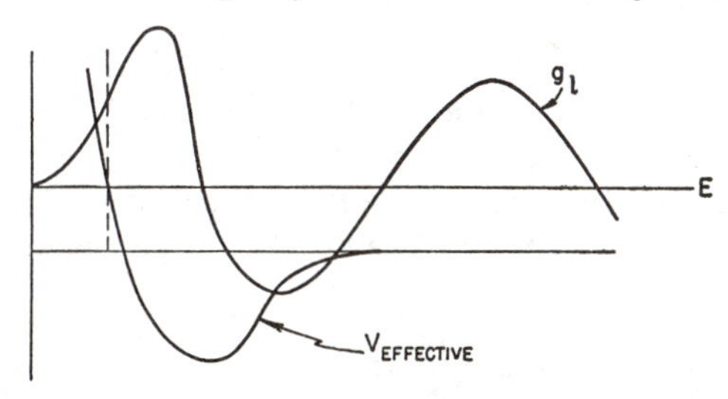

FIG. 18

fractional change of wavelength occurring within a wavelength is also small. At large r, we can also expand

$$\sqrt{E - V} \cong \sqrt{E} - \frac{V}{2\sqrt{E}}$$

Thus

$$g \sim p^{-1/2} \sin\left[\int_a^r \sqrt{2mE}\,\frac{dr}{\hbar} - \int_a^r \sqrt{\frac{m}{2E}}\,V\,\frac{dr}{\hbar} + \int_0^a \sqrt{2m(E - V)}\,\frac{dr}{\hbar}\right]$$

a is an arbitrary radius, beyond which the expansion is good.

If we set $V = Ze/r$, we obtain (with $\sqrt{2mE}/\hbar = k$)

$$g \sim p^{-1/2} \sin\left[kr + \int_0^a \sqrt{2m(E - V)}\,\frac{dr}{\hbar} - Ze\sqrt{\frac{m}{2E}}\ln\frac{r}{a}\right]$$

We see that as $r \to \infty$, the phase does not approach a constant, but instead varies as $\ln r$. If we had chosen $V = Ze^2/r^{1+n}$, where $n > 0$, then we should have obtained a constant phase as $r \to \infty$. Thus, the assumption that g approaches the form given in eq. (65) is good only if $V(r)$ falls off with increasing r more rapidly than does the Coulomb potential. Since this property will prove to be very important for the validity of the method of partial waves, the Coulomb potential has to be given a special treatment, which we shall discuss in Sec. 58. For the

* Chap. 15, eq. (14a).

present, we restrict ourselves to forces that fall off more rapidly than $1/r$.

42. Partial Waves for Free Particle. In order to illustrate the method, and to obtain some results that will be useful later, we shall solve the problem of the free particle by the method of partial waves. The differential equation (64) is

$$\frac{d^2g_l}{dr^2} - \frac{l(l+1)}{r^2}g_l = -k^2g_l \tag{66}$$

The most general solution is*

$$g_l = A\sqrt{kr}\,J_{l+\frac{1}{2}}(kr) + B\sqrt{kr}\,J_{-l-\frac{1}{2}}(kr) \tag{67}$$

Because $J_{-(l+\frac{1}{2})}(kr)$ starts out as $(kr)^{-(l+\frac{1}{2})}$ at small r, it is not an admissible solution; hence we must choose $B = 0$.

It turns out that for Bessel's functions of half-integral order, one can find an expression involving a finite series of trigonometric terms. For example,

$$\left.\begin{aligned} g_0 &= A\sqrt{\frac{2}{\pi}}\sin kr \\ g_1 &= A\sqrt{\frac{2}{\pi}}\left(\cos kr - \frac{\sin kr}{kr}\right) \end{aligned}\right\} \tag{68}$$

The reader will easily verify that the above are solutions of the differential equation (66). For the higher order Bessel functions of this type, the expressions are somewhat unwieldly, but they can readily be obtained.†

43. Asymptotic Form of Bessel's Function. It is a well-known mathematical theorem† that for large x,

$$J_n(x) \xrightarrow[x\to\infty]{} \sqrt{\frac{2}{\pi x}}\cos\left(x - \frac{\pi}{4} - n\frac{\pi}{2}\right) \tag{69}$$

Thus, in our case, we obtain

$$g_l \xrightarrow[r\to\infty]{} A\sqrt{\frac{2}{\pi}}\cos\left[kr - \frac{\pi}{4} - \left(l + \frac{1}{2}\right)\frac{\pi}{2}\right] = A\sqrt{\frac{2}{\pi}}\sin\left(kr - \frac{l\pi}{2}\right) \tag{70}$$

We obtain

$$\Delta_l = -\frac{l\pi}{2} \tag{71}$$

44. Interpretation of Partial Waves.

Case A: $l = 0$; s waves

We see from eq. (68) that the wave function is just

$$\psi_0 = \frac{g_0}{r} \sim \sqrt{\frac{2}{\pi}}\frac{\sin kr}{r} = \sqrt{\frac{2}{\pi}}\left(\frac{e^{ikr}}{r} - \frac{e^{-ikr}}{r}\right) \tag{72}$$

* Watson, *Bessel Functions.* London, Cambridge University Press, 1922.

† *Ibid.*

The wave function is spherical. It is a sum of ingoing and outgoing waves, each moving in the radial direction. This wave function corresponds to a condition in which waves are made to converge on the origin (e^{-ikr}/r) after which they diverge away (e^{ikr}/r). To obtain conservation of probability, one needs both ingoing and outgoing waves. Furthermore, to avoid an infinite value of ψ at the origin, one needs to subtract the ingoing from the outgoing wave, just as is done above. (A sum of ingoing and outgoing waves would yield $\psi = \infty$ at $r = 0$.)

Case B: $l = 1$ (p waves)

The complete wave function is

$$\psi \sim P_1(\cos\theta)\,\frac{g_1(r)}{r} \sim \sqrt{\frac{2}{\pi}}\left(\frac{\cos kr}{r} - \frac{\sin kr}{kr^2}\right)\cos\theta \tag{73}$$

It is easily seen that $|\psi|$ is proportional to r for small r, and that it reaches a maximum somewhere near $kr \cong 1$, after which it decreases. It is therefore improbable that the particle gets much closer to the origin than

$$r_0 \cong \frac{1}{k} = \frac{\lambda}{2\pi}$$

This can be interpreted by saying that particles of unit angular momentum are not likely to be nearer to the origin than the distance r_0, at which their angular momentum measured classically, $pr_0 = \hbar k r_0$, would be of the order of $\hbar$. For higher angular momentum, one can show that in a similar way the minimum distance at which $|\psi|^2$ is large is given by $pr_0 \cong \hbar l$.

The above result may have to be modified somewhat in the presence of a strong attractive potential. In this case, the minimum distance is obtained by evaluating the momentum from the total kinetic energy $p^2 = 2m(E - V)$. Thus, the criterion for the minimum probable radius is

$$\sqrt{2m[E - V(r)]}\, r_0 \cong \hbar l \tag{74}$$

If V is large and negative, the particle may be pulled fairly close to the origin despite the repulsive effects of the "centrifugal potential."

The complete wave function for $l = 1$ is, of course, proportional to $\cos\theta$. Asymptotically, these waves are just the sum of an ingoing and an outgoing component [see eq. (73)], but, near the origin, the behavior is more complex because the wave does not exactly hit the origin, as does the s wave, but instead, tends to avoid the origin because of the angular momentum.

45. Boundary Conditions on Partial Waves for Free Particle. Thus far, we have studied separately the various partial waves representing possible wave functions for a free particle. All of these waves correspond to situations in which waves are made, more or less, to approach the origin, after which they recede again. None of them corresponds to the usual boundary conditions at infinity for a free particle, namely, that there is an incident plane wave.

Since a plane wave is a solution of Schrödinger's equation for a free particle and since each partial wave is also a solution of this equation, it should be possible to expand the plane wave as a series of partial waves, because from such a series an arbitrary solution can be obtained according to the expansion theorem. Thus, we ought to be able to write

$$e^{ikz} = e^{ikr\cos\theta} = \sum_l C_l \frac{g_l(r)}{r} P_l(\cos\theta) \tag{75}$$

We can solve for $g_n(r)$ by multiplying by $P_n(\cos\theta)$ and integrating over all θ. Writing $\cos\theta = x$ and using normalization and orthogonality of the $P_l(x)$, we obtain

$$\frac{C_n g_n(r)}{r} = \frac{2n+1}{2}\int_{-1}^{1} e^{ikrx}P_n(x)\,dx$$

Now, it can be shown mathematically that the above integral is indeed just the right Bessel function to give us the correct result for $g_n(r)$. In fact, one obtains*

$$\int_{-1}^{1} e^{ikrx}P_n(x)\,dx = \frac{\sqrt{2\pi}\,i^n}{\sqrt{kr}} J_{n+\frac{1}{2}}(kr) \tag{76a}$$

Comparison with eq. (67) shows that this is the right function, and that

$$C_n = \frac{i^n(2n+1)}{k}$$

The expansion of a plane wave in Legendre polynomials is therefore

$$e^{ikr\cos\theta} = \frac{1}{k}\sum_l \frac{g_l(kr)}{r}(i)^l(2l+1)P_l(\cos\theta)$$

$$= \frac{1}{k}\sum_l \frac{g_l(kr)}{r}(i)^l(2l+1)P_l(\cos\theta) \tag{76b}$$

This means that in order to describe a plane wave, we must make up a sum of spherical waves. Such a plane wave has all possible angular momenta in it. It is clear that these angular momenta are necessary, for example, in the classical limit. If a beam of particles is directed at an atom, then all possible angular momenta are present, because the angular momentum is pr_0, where r_0 is the collision parameter, which takes on all possible values. In the quantum theory, however, the possible

* *Ibid.*

angular momenta are quantized and each one is associated with a wave function of the appropriate angular dependence $P_l(\cos \vartheta)$.

46. Imposition of Boundary Conditions When a Potential is Present. In order to impose boundary conditions when a potential is present, we first observe that according to eqs. (70) and (76b), the asymptotic expansion for a plane wave is

$$e^{ikr\cos\vartheta} \sim \sum_l \frac{(i)^l}{kr}(2l+1)P_l(\cos\vartheta)\sin\left(kr - \frac{l\pi}{2}\right)$$

Writing $(i)^l = e^{\pi il/2}$, we obtain

$$e^{ikr\cos\vartheta} = \sum_l \frac{(2l+1)}{2ikr}P_l(\cos\vartheta)(e^{ikr} - e^{il\pi}e^{-ikr}) \tag{77}$$

When a potential is present, then, according to eq. (65), the asymptotic form of the wave function is

$$r\psi = \sum_l P_l(\cos\vartheta)g_l(r) \sim \sum_l A_l P_l(\cos\vartheta)\sin(kr + \Delta_l)$$

At this point, we shall find it convenient to introduce

$$\Delta_l = \delta_l - \frac{l\pi}{2} \qquad A_l = e^{il\pi/2}(2l+1)\frac{e^{i\delta_l}B_l}{k}$$

We then obtain

$$\psi = \sum_l \frac{B_l(2l+1)}{2ikr}(e^{ikr+2i\delta_l} - e^{il\pi}e^{-ikr})P_l(\cos\theta) \tag{78}$$

The coefficients B_l can now most conveniently be determined by constructing a wave packet. It is clear that before the wave packet strikes the potential, its form must be in the same as that of a plane wave, and that modifications in its form can occur only after the wave has actually struck the potential. This means that the ingoing part of the actual wave packet must be identical with the ingoing part of a packet of plane waves. To satisfy these boundary conditions, one must choose $B_l = 1$. To prove that this is the correct choice, we multiply ψ by $e^{-i\hbar k^2t/2m}$ and integrate over a small range of k. The center of the packet will then be at the point where the phase of the wave function is an extremum. With the choice $B_l = 1$, we obtain the center of the ingoing packet from the equation

$$\frac{\partial\phi_l}{\partial k} = \frac{\partial}{\partial k}\left(-kr - \frac{\hbar k^2}{2m}t + l\pi\right) = 0$$

or

$$r = -\frac{\hbar kt}{m} = -vt$$

Thus, for large negative times, we obtain an ingoing packet. Since only positive values of r exist, the ingoing packet disappears at $t = 0$, after which it is replaced by an outgoing packet. The center of the outgoing packet corresponding to the lth wave occurs where

$$\frac{\partial \phi_l'}{\partial k} = 0 \qquad \text{where} \qquad \phi_l' = kr - \frac{\hbar k^2}{2m} t + 2\delta_l$$

or at

$$r = vt - 2\frac{\partial \delta_l}{\partial k}$$

The outgoing packet thus appears with a time delay (or advance) of $2\frac{\partial \delta_l}{\partial k}$, which is the result of the action of the potential (see, for example, Sec. 29, and Chap. 11, Sec. 19).

We conclude from the above that the incident packet is identical with the incident part of a packet of plane waves, but that the outgoing packet will be modified by the actions of the potential.

47. Formula for Scattering Cross Section. To obtain the strength of the scattered wave, we note that even if there were no potential, there would still be an outgoing wave, which is just the outgoing part of a plane wave. The test for a scattered wave is to see whether the outgoing packet has been modified. We therefore obtain the asymptotic form of the scattered wave by subtracting from the actual outgoing wave the outgoing wave that would be present if there were no potential. That is, according to (77) and (78),

$$F_{\text{scatt}} = \sum \frac{e^{ikr}}{r} \frac{(e^{2i\delta} - 1)P_l(\cos\theta)(2l+1)}{2ik} = \frac{e^{ikr}}{r} f(\theta) \qquad (79)$$

where F_{scatt} is the asymptotic form of the scattered wave. The complete asymptotic wave function is now

$$e^{ikz} + \frac{f(\theta)e^{ikr}}{r}$$

Comparing with eq. (45), we see that the cross section is

$$\sigma = |f(\theta)|^2 = \frac{1}{k^2}\left|\sum_l \frac{(2l+1)}{2} P_l(\cos\theta)(e^{2i\delta_l} - 1)\right|^2 \qquad (80)$$

The above formula yields the angular-dependent cross section, once we know δ_l. (The latter must be obtained by solving Schrödinger's equation.) This angular dependence arises, in part, from the interference of waves of different l. For example, suppose we have scattered waves with $l = 0$ alone. Then there is no angular dependence, i.e., the cross section is spherically symmetric. With $l = 1$ alone, the cross section is proportional to $\cos^2\theta$. If both are present, as in $f(\theta) = a + b\cos\theta$, then

$$\sigma = |a|^2 + |b|^2 \cos^2 \theta + (ab^* + ba^*) \cos \theta$$

A few typical curves are shown in Fig. 19. Thus, the angular dependence of the cross section involves interference between different l terms. If higher angular momenta are included, the pattern may grow still more complex. In the classical limit ($l \to \infty$) one can form a packet of waves of different l in such a way that they build up to a maximum at a definite value of θ. This corresponds to a classical orbit in which particles come in with a definite collision parameter and scatter through a definite angle.

48. Total Cross Section. To find the total cross section, we integrate σ over all solid angle, using the orthogonality of the $P_l(\cos \theta)$ and the

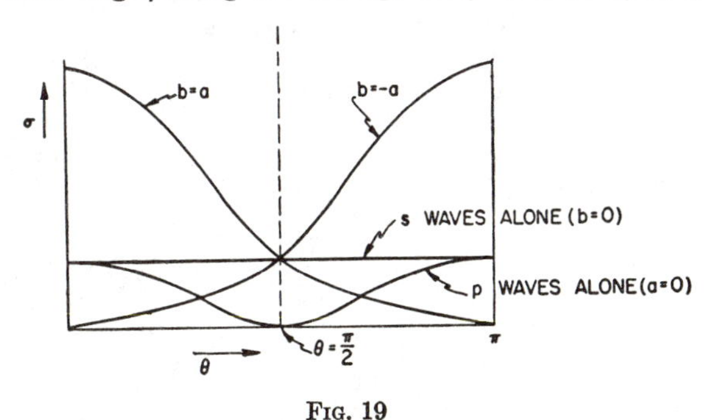

FIG. 19

normalization conditions [Chap. 14, eq. (52a)]. We obtain for the total cross section,

$$S = \sum_l \frac{4\pi}{k^2} (2l + 1) \sin^2 \delta_l \tag{81a}$$

The above result means that in the total cross section, the various partial waves do not interfere. It is only in determining the angular distribution that they interfere.

The maximum cross section corresponding to a given value of l is

$$(S_l)_{\max} = \frac{4\pi(2l + 1)}{k^2} \tag{81b}$$

This will occur if $\delta_l = \pi/2$. Writing $k = 2\pi/\lambda$, we obtain

$$(S_l)_{\max} = \frac{(2l + 1)\lambda^2}{\pi} \tag{81c}$$

For s waves, for example, the maximum cross section corresponds to a circle of radius λ/π, and for higher l, it is still higher. This cross section can actually be produced by a scatterer that is much smaller than λ, provided that conditions are such as to make $\delta_l = \pi/2$.

49. Calculation of Phase for Impenetrable Sphere. Because the sphere is impenetrable, we must have $\psi = 0$ at the edge of the sphere, which we assume has a radius a. For s waves, the differential equation outside the sphere is just $-d^2g/dr^2 = k^2 g$. The solution is

$$g_0 = A \sin (kr + \delta)$$

To have $g = 0$ at $r = a$, we must have $\delta_0 = -ka$. The partial scattering cross section for s waves is therefore

$$S = \frac{4\pi}{k^2} \sin^2 ka \tag{82a}$$

For higher angular momenta, the solutions are

$$g = \sqrt{kr}\,[AJ_{l+\frac{1}{2}}(kr) + BJ_{l-\frac{1}{2}}(kr)]$$

[Note that since the origin is now excluded, $J_{-l-\frac{1}{2}}(kr)$ must now be retained.] The boundary condition at $r = a$ yields

$$g(a) = 0 \qquad \text{or} \qquad \frac{B_l}{A_l} = -\frac{J_{l+\frac{1}{2}}(ka)}{J_{-l-\frac{1}{2}}(ka)}$$

The phase can be calculated from the asymptotic form of the wave function [see eq. (70)]. For large r,

$$\begin{aligned} g \sim A \cos\left[kr - \left(l + \frac{1}{2}\right)\frac{\pi}{2} - \frac{\pi}{4}\right] &+ B \cos\left[kr + \left(l + \frac{1}{2}\right)\frac{\pi}{2} - \frac{\pi}{4}\right] \\ = A \sin\left(kr - \frac{l\pi}{2}\right) &+ (-1)^l B \cos\left(kr - \frac{l\pi}{2}\right) \\ &= \sqrt{A^2 + B^2}\sin\left(kr - \frac{l\pi}{2} + \delta_l\right) \end{aligned}$$

where

$$\tan \delta_l = (-1)^l \frac{B_l}{A_l}$$

Special Case: $ka \ll 1$. If the wave length is so large that $ka \ll 1$, it is readily seen that the values of δ_l for successive l rapidly become very small. This is because particles of a given angular momentum, l, will scatter heavily only if it is likely that they strike the potential, or only if $pa \gtrsim \hbar l$ or $ka \gtrsim l$ (see Sec. 44). This can also be shown by evaluating δ_l from the formula given above.

Problem 2: Using the series expansion of Bessel's functions,* evaluate $\tan \delta_l$ for small ka, and show that

$$\frac{\delta_1}{\delta_0} \ll 1, \qquad \frac{\delta_l + 1}{\delta_l} \ll 1$$

* *Ibid.*

For small ka, the cross section is therefore given almost entirely by the s waves. Thus, we can use eq. (82a). With the expansion of $\sin^2 ka$, this equation yields

$$S \cong 4\pi a^2 \tag{82b}$$

Note that this result is four times the classical result for a hard sphere* (eq. 3). The increase is the result of quantum-mechanical diffraction effects.

It is of some interest to follow the transition from quantum to classical scattering, since the cross section must drop from $4\pi a^2$ to πa^2 as this transition takes place. Quantum scattering occurs when $ka \ll 1$, i.e., when $\lambda \gg 2\pi a$. As the wavelength goes below the size of the sphere, the first effect will be to introduce waves of higher angular momentum, so that the cross section becomes angular dependent. As the wavelength is made still shorter, however, and the classical region is approached, the cross section once again becomes spherically symmetrical, with a value reduced to πa^2. except for a region near $\theta = 0$ with an angular width of the order of $\Delta\theta \cong \lambda/2\pi a$. The polar intensity pattern is shown for large λ in Fig. 20. The large projection in the forward direction is essentially a diffraction effect, containing a total cross section of πa^2. Thus, for very short wavelengths, the total cross section is $2\pi a^2$, in contrast to the value of $4\pi a^2$, obtained with very long wavelengths. In the classical limit, however, the wavelength becomes so short that the large projection near the forward direction corresponds to deflections too small to produce significant results. Thus, for all practical purposes, the effective classical cross section is only πa^2.

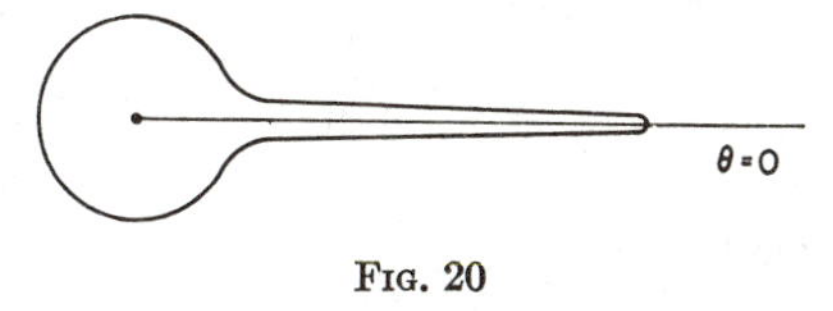

FIG. 20

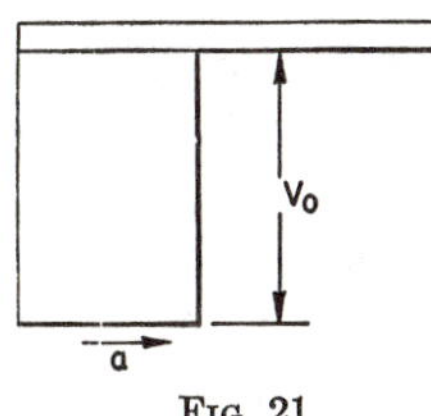

FIG. 21

Problem 3: With a sphere of radius 1 cm, and electrons of energy 1 ev, compute the angular width of the diffraction pattern in the forward direction and show that it is too small to be important in practice. (Use Huyghens' principle, as in optics.)

50. Application of Exact Method to Scattering from Square Well for *s* Waves. Consider a square well of radius a, depth V_0, as shown in Fig. 21. Suppose that particles are incident with energy E. We wish to compute the cross section, restricting ourselves to s waves only. This restriction will be valid only if $ka \ll 1$. Inside the well the radial equation is

* In eq. (3), a is by definition equal to only half the distance between centers of the spheres, whereas in (82b), it is equal to the full distance between centers. (We are considering only identical particles here.)

$$-\frac{d^2g}{dr^2} = k_1^2 g \tag{83}$$

where
$$k_1^2 = (E + V_0)\frac{2m}{\hbar^2}$$

Since $g(r)$ must vanish at the origin,* the most general admissible solution is

$$g = A \sin k_1 r \tag{83a}$$

where A is an arbitrary constant. Outside the well, the most general solution is

$$g = B \sin (kr + \delta_0) \tag{83b}$$

where
$$k^2 = \frac{2m}{\hbar^2} E$$

B and δ must be obtained by requiring that $g(r)$ and $g'(r)$ be continuous at $r = a$. To solve for δ alone, however, it is sufficient to make g'/g continuous. Setting

$$\frac{g'}{g} = \alpha \tag{84}$$

we obtain

$$k \cot (ka + \delta_0) = \alpha \tag{84a}$$

where
$$\alpha = k_1 \cot k_1 a$$

or
$$\tan (ka + \delta_0) = \frac{\tan ka + \tan \delta_0}{1 - \tan ka \tan \delta_0} = \alpha \tag{84b}$$

Solving for $\tan \delta_0$, we obtain

$$\tan \delta_0 = \frac{\dfrac{k}{\alpha} - \tan ka}{1 + \dfrac{k}{\alpha}\tan ka} = \frac{k\left(\dfrac{1}{\alpha} - \dfrac{\tan ka}{k}\right)}{1 + \dfrac{k}{\alpha}\tan ka} \tag{84c}$$

The total cross section is [see eq. (81a)]

$$S = \frac{4\pi \sin^2 \delta_0}{k^2} = \frac{4\pi}{k^2(1 + \cot^2 \delta_0)} = \frac{4\pi}{k^2 + \dfrac{(\alpha + k \tan ka)^2}{\left(1 - \alpha\dfrac{\tan ka}{k}\right)^2}} \tag{85}$$

By evaluating α from eq. (84a), one obtains the cross section.

51. Ramsauer Effect. We observe from eq. (85) that if the scattering phase is equal to some integral multiple of π for nonzero k, the cross section vanishes. If δ is an integral multiple of π, then $\tan \delta_0 = 0$. For

* See Chap. 15, Sec. 3.

a square well, we obtain the condition for the vanishing of tan δ_0 from eq. (84c):

$$\frac{1}{\alpha} = \frac{\tan ka}{k} \tag{86}$$

Obtaining α from eq. (84a) one finds

$$\frac{\tan k_1 a}{k_1} = \frac{\tan ka}{k}$$

For small k, $ka \ll 1$. Replacement of tan ka by ka then yields

$$\tan (k_1 a) \cong k_1 a$$

For small k, k_1 is given approximately by $\sqrt{2mV_0}/\hbar$. If V_0 and a are such that the eq. (86) is satisfied, the scattering cross section will be zero, and if it is nearly satisfied, the cross section will be very small. This vanishing of the scattering cross section for a non-zero potential is peculiar to the wave properties of matter. It would occur, for example, with light waves which were being scattered from small transparent spheres with a high index of refraction, so chosen that the sin δ_0 corresponding to the scattered wave vanished. This means, essentially, that the contributions of the various parts of the potential to the scattered wave [see Sec. 26] interfere destructively, leaving only an unscattered wave. Although this result was derived for a square well, it can easily be extended to any well that has the property that it is fairly localized in space. This is because the vanishing of the phase is determined by the cumulative phase shifts suffered by the wave throughout the entire well, so that it is always possible to obtain a phase shift of $n\pi$ by properly choosing the magnitude and range of the potential.

For slow electrons scattered from noble gas atoms, it turns out that the sin δ_0 is very small and the cross section for electron-atom scattering is therefore much smaller than the gas-kinetic cross section. This effect is known as the Ramsauer effect. As the electron energy is increased, the phase of the scattered wave changes, and, eventually, at higher energies above 25 ev the usual gas-kinetic cross section is approached.

The Ramsauer effect is somewhat analogous to the transmission resonances obtained in the one dimensional potential (see Chap. 11, Sec. 9). The analogy, however, is not complete, because the condition for the Ramsauer effect [eq. (86)] is not exactly the same as that for a transmission resonance in a one-dimensional well [eq. (50), Chap. 11]. The reason for the difference is that in the one-dimensional case we define the transmitted wave as the total wave that comes through the well. In the scattering problems, we have an incident wave that converges on the well. Some of it enters the well and some of it is reflected at the edge of the well. The net effect is to produce an outgoing wave, whose phase depends on what happens to the wave at the well. The question of how much of this outgoing wave corresponds to a scattered wave depends on how large a phase shift it has suffered relative to the outgoing wave which would have been present in the absence of a potential. Thus we see that the intensity of the scattered wave depends on properties of the potential that are somewhat different from those determining the intensity of that part of the wave that is transmitted through the potential and out again on the other side. The vanishing of the cross section in the Ramsauer effect is, as we have already seen, a result of the fact that the contributions of different parts of the potential all add up in such a way as to produce a wave that cannot be distinguished from one which has not been inside a potential at all.

52. Approximation for Small k. For small k, we can expand the expression for $\tan \delta_0$, retaining only terms up to order k^3. We obtain

$$\tan \delta_0 \cong k \left[\frac{\left(\frac{1}{\alpha} - a\right) - \frac{k^2 a^3}{3}}{1 + k^2 \frac{a}{\alpha}} \right] \tag{87}$$

We see that as $k \to 0$, the phase also approaches zero. The sign of the phase at small k depends on the sign of $\frac{1}{\alpha} - a$.

If k is so small that $k^2 a^2 \ll 1$ and $k^2 a/\alpha \ll 1$, then the above expression simplifies to

$$\tan \delta_0 \cong k \left(\frac{1}{\alpha} - a\right) \tag{87a}$$

The cross section is (in this approximation)

$$S \cong \frac{4\pi}{k^2 + \frac{\alpha^2}{(1 - \alpha a)^2}} = \left(\frac{2\pi \hbar^2}{m}\right) \frac{1}{E + \frac{\hbar^2 \alpha^2}{2m} \frac{1}{(1 - \alpha a)^2}} \tag{88}$$

To obtain a good idea of the low-energy cross section, we need only obtain α, which is defined in eq. (84a).

53. Application to Nuclear Scattering. We shall now make some applications in the field of nuclear scattering. Before doing this, however, we wish to point out that very little certain knowledge exists concerning nuclear forces. The main reason for studying the problem in this book is to illustrate how one uses the quantum theory to try to make advances in new fields, where the fundamentals are still uncertain. In this way, we hope to show that the application of the theory is not necessarily always restricted to the mere calculation of various kinds of numerical results, on the basis of a known and defined theory.

As has been stated in Chap. 11, Sec. 3, evidence exists indicating that the potential energy of a neutron in the field of a proton can be represented by a well that is about 20 mev deep and has a radius of the order of 2.8×10^{-13} cm. This well is almost certainly not exactly square, but many of its main features can be represented roughly with the aid of a square well.

We can obtain, however, many important results without making any specific assumptions about the shape of the well, other than that beyond some radius, which is of the order of 3×10^{-3} cm, the potential is small enough to be neglected. For this reason, it is convenient to separate the problem of solving Schrödinger's equation into two parts, namely, that of solving the problem inside the well and that of solving it outside. Since there is no appreciable potential outside, the solution is just that for a free particle [see eq. (83b)]. Inside the well, the general problem of solving the wave equation is complicated, but the result of this procedure, starting with $g = 0$ at $r = 0$, will always be to determine the ratio $g'/g = \alpha$ at the point $r = a$. All that remains to be done is to make g'/g continuous at the point a by proper choice of the phase δ.

For s waves, the procedure is exactly the same as that leading to eq. (84), so that the same equations hold, provided that we interpret α as the ratio, $(g'/g)_{r=a}$ obtained by solving Schrödinger's equation with the actual potential, whatever it may be.

54. Approximate Expression of Low-energy Cross Section in Terms of Binding Energy of Deuteron. Although we cannot solve for α directly unless we know the details of the shape of the potential, we shall nevertheless be able to obtain a good deal of information about α by comparing the observed cross sections with those which are predicted as a function of α. In this work, we shall use an approximate value of α, obtained with the aid of the observed result that there is a bound state of the deuteron at $E = -2.23$ mev. The value of α for a bound state is easily calculated from the fact that outside the potential the wave function is just a decaying exponential (for s waves), $g = A \exp(-\sqrt{2mB}\, r/\hbar)$ where B is the binding energy. Thus, we obtain for the value of α_0 in the bound state

$$\alpha_0 = -\frac{\sqrt{2mB}}{\hbar}$$

We observe that α_0 must be *negative* for a bound state; this is because the wave function inside the well has gone past a maximum and is decreasing with radius to meet a decaying exponential at $r = a$.

Now, the potential is of the order of 20 mev deep; hence α_0 undergoes only a small change as E is increased from -2.23 mev to a value of zero or slightly above, simply because the wavelength at any particular point is not changed much by this small fractional increase in kinetic energy.

For example, with a square well $\alpha_0 = k_1 \cot k_1 a$ is changed by about 20 per cent as E is increased from -2.16 mev to zero.*

Approximating α_0 by its value in the bound state, we obtain for the cross section (from eq. 88)

$$S \cong \left(\frac{2\pi\hbar^2}{m}\right) \frac{1}{E + \dfrac{B}{(1 - \alpha_0 a)^2}} \tag{89}$$

In terms of the actual proton mass and the energy in the laboratory system of co-ordinates, which is twice the relative energy, we obtain

$$S = \frac{4\pi\hbar^2}{m_L} \frac{1}{\dfrac{E_L}{2} + \dfrac{B}{(1 - \alpha_0 a)^2}} \tag{90}$$

where E_L is the energy in the laboratory system, and m_L is the actual proton mass.

Problem 14: Evaluate the above cross section at $E = 0$. (The result is of the order of 3×10^{-24} cm².)

The above cross section has a maximum at $E = 0$, and it decreases more or less uniformly thereafter. The approximation is fairly good (to about 25 per cent) up to $E \cong 5$ mev. At higher energies, more accurate formulae must be used. Furthermore, the p waves begin to come in, and these will affect both the total cross section and the angular dependence.

55. Spin-dependent Forces. We have obtained, in the previous section, a general approximate expression for the low-energy neutron-proton scattering cross section expressed as a function of the binding energy of the deuteron only; it is independent of the details of the shape of the potential function. Comparison with experiment should therefore provide a good check on the validity of our basic ideas of nuclear forces. Experiment shows that the low energy cross section is of the order of 20×10^{-24} cm², whereas our predictions are of the order of only 3×10^{-24} cm².

This discrepancy was explained by Wigner, who observed that to obtain a larger zero-energy cross section, one must, according to eq. (88), have a well for which the value of α is far below that which is obtained from the deuteron binding energy. To obtain such a low value of α, one needs a well which is shallower than that needed to yield the correct deuteron binding energy. This will cause the wave function to curve less within the region of the potential, and therefore to reach $r = a$ with a smaller slope. A comparison between the properties of a well deep

* Our procedure is therefore to evaluate α_0 empirically for a slightly negative value of the energy, and to use this value as an approximation to α_0 for slightly positive values of the energy.

enough to explain the deuteron binding energy and a well that explains scattering by yielding a small value of α is shown in Fig. 22.

In order to reconcile the different potential depths demanded by scattering data and by deuteron binding energy, he suggested that the nuclear forces were spin dependent, in such a way that when the spins of the particles are parallel the well is deeper than when they are antiparallel. Now it is known on independent grounds* that in the deuteron, the neutron and proton spins are parallel. On the other hand, in a beam of incident neutrons, for example, the relative orientations of spin to that of any proton in the target are random, so that both possibilities occur. This means that the large scattering cross section is produced in those cases of antiparallel orientation, while the parallel orientation has a deep enough well to explain the binding energy of the deuteron.

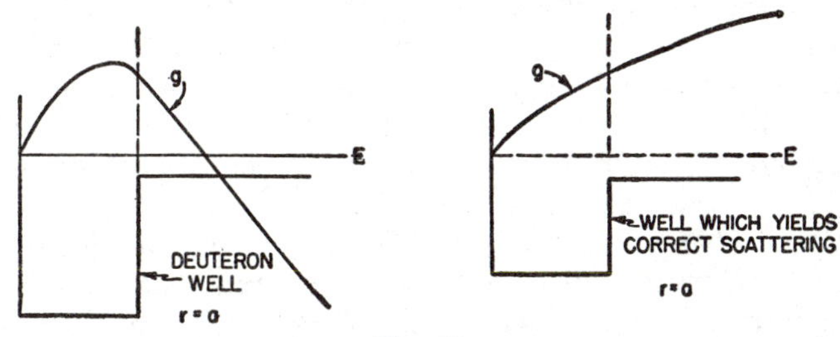

Fig. 22

In a beam in which neutron spins are oriented at random, the neutron spin will be parallel on the average, to that of an arbitrary proton for $\frac{3}{4}$ of the incident particles, and antiparallel for $\frac{1}{4}$ of them. This result follows from a study of the properties of the spin variables, which shows that there are three times as many ways to make the spins parallel as there are to make the spins antiparallel.† This means that the total cross section is

$$S = \tfrac{3}{4}S_p + \tfrac{1}{4}S_a$$

where S_p and S_a are respectively the cross sections for parallel and antiparallel spins.

Setting S equal to its observed value of 21×10^{-24} cm², and

$$S_p = 3 \times 10^{-24} \text{ cm}^2$$

we obtain

$$S_a \cong 75 \times 10^{-24} \text{ cm}^2$$

This is indeed a rather large cross-section. The cross-sectional area of the potential well is only about 0.3×10^{-24} cm². The possibility of so large a cross section comes entirely from the wave properties of matter and is, as we shall see, connected with the existence of a resonance near

* H. A. Bethe, *Elementary Nuclear Physics.*
† See, for example, Chap. 17, Sec. 10.

$E = 0$, resulting from the small slope of the wave function at $r = a$. Note that for a square well, the condition for a resonance is that the slope of the wave function vanish at $r = a$ [see Chap. 11, eq. (50)].

56. Solution for Depth of Single Well. One can obtain agreement with experimental values of low-energy scattering by choosing α^2 for the singlet (antiparallel) well to be about $\frac{1}{50}$ of the value for the triplet (parallel) well [see eq. (90)]. This result can be achieved in either of two ways:

(1) There could be a real bound state, very close to $E = 0$. The binding energy would be $B \cong 2.23/50$ mev $\cong 40$ kev.

(2) There could be a virtual level* at $E \cong +40$ kev. This would correspond to a positive value of α at $E = 0$, but for $E \cong 50$ kev, the curvature of the wave function would increase sufficiently to bring $\alpha = g'/g$ down to zero at $r = a$, thus yielding a virtual level at this point.

There is no way from neutron-proton scattering alone to distinguish between these two possibilities. Let us observe, however, from eq. (87a) that for positive α the phase at small k is positive, while for negative α, it is negative. Now, it is possible to obtain the relative signs of the phases of two scattered waves by looking for interference. To do this, one scatters neutrons separately from hydrogen molecules in which the spins of the two protons are parallel (ortho-hydrogen) and atoms in which they are antiparallel (para-hydrogen). If neutrons of wavelength much longer than a molecular diameter are incident on these molecules, the scattered waves from the two nuclei should interfere to a significant extent. In ortho-hydrogen, both scattered waves will certainly have the same sign of phase. In para-hydrogen, however, the interference between the waves scattered from different atoms will be destructive, if the respective values of α for singlet and triplet states have opposite signs, constructive if they have the same sign. Hence, if the scattering from para-hydrogen is less than that from ortho-hydrogen, one can conclude that α is positive and that the singlet level is virtual. Experiment shows that this is indeed the case.†

Knowing the value of α, we can now solve for the depth of the singlet potential provided that we assume a square shape. From eqs. (83a), (83b), and (84), we obtain $\alpha = k_1 \cot k_1 a$. Since α is evaluated at $E = 0$, we have $k_1 \cong \sqrt{2mV_s}/\hbar$ where V_s is the depth of the singlet well. V_s can be found by solving this equation. Since $\alpha \cong 0$, an approximate solution is $k_1 a \cong \pi/2$, and

$$V_s \cong \left(\frac{\pi}{2}\frac{\hbar}{a}\right)^2 \frac{1}{2m}$$

* See Chap. 11, Sec. 20 and Chap. 12, Sec. 14, for a definition of a virtual level.

† For a discussion of this problem, see H. A. Bethe, *Rev. Mod. Phys.*, **8**, 117–118 (1936).

The depth turns out to be about 12 mev. The approximate formula for the singlet cross-section (good at low energies) is then [according to eq. (88)]

$$S_a \cong \frac{2\pi\hbar^2}{m} \frac{1}{E + \dfrac{\hbar^2\alpha^2}{2m}}$$

[Note that because α is small for the singlet state, we have neglected the factor $1/(1 - \alpha a)^2$ appearing in eq. (88).] It is convenient to set $\hbar^2\alpha^2/2m = W$. This quantity has the dimensions of energy, and numerically, it is roughly equal to the energy of the "virtual," or resonance, level, which exists near $E = 0$ for antiparallel spins. The complete cross section is then

$$S = \frac{3}{4} S_p + \frac{1}{4} S_a \cong \frac{2\pi\hbar^2}{m} \left[\frac{3}{4} \frac{1}{E + \dfrac{B}{(1 - \alpha_0 a)^2}} + \frac{1}{4} \frac{1}{E + W} \right] \tag{91}$$

The general shape of the cross section as a function of energy is shown in Fig. 23. The sharp rise at low energies results of course, from the resonance in the singlet state. The expression (91) is good only at low energies. To obtain more accurate results, or to go to higher energies, one must either use more accurate expansions for α,* or solve Schrödinger's equation rigorously.

57. Comparison with Experiment; Measurements of Radius of Potential. In the approximations that we have used so far, the predicted results for the cross section as a function of energy do not depend significantly on the assumed radius of the potential at all, but are determined mainly by the binding energy of the deuteron and the low-energy scattering cross sections [see eq. (91)]. When a more accurate approximation is used, it is found that the predicted variation of cross section with energy does depend somewhat on the assumed radius of the potential. Up to 5 mev, however, any variations of the radius that are reasonable on the basis of our general knowledge of nuclear physics would result in the prediction of about a 25 per cent variation, at most, in the cross section. Since

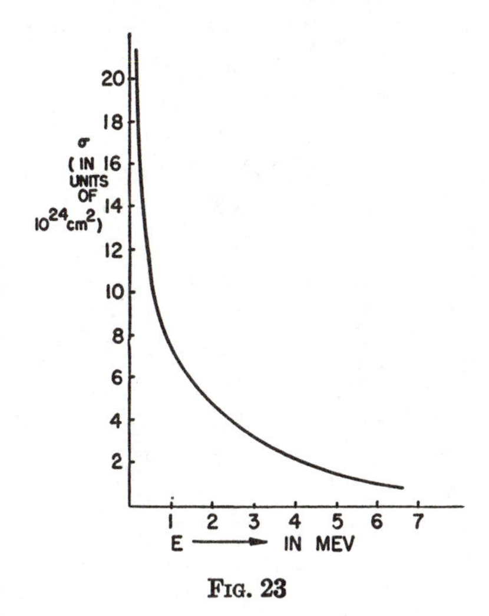

FIG. 23

* See H. A. Bethe, *Phys. Rev.*, **76**, 38 (1949).

the experiments on neutron-proton scattering are not more accurate than about 10 per cent, it is difficult to use these results to fix the radius very precisely. One can, however, fix it within rough limits. The radius obtained in this way depends considerably, however, on the exact shape of the potential, and there is much doubt that a square well is the right shape.

To sum up, we note that the triplet scattering up to about 5 mev. is determined approximately by the deuteron binding energy. The singlet scattering in this range is determined approximately from the requirement of fitting neutron-proton scattering cross sections at low energies (close to zero). (See, for example, John M. Blatt and J. David Jackson, *Phys. Rev.*, **76,** 18, 1949, and H. A. Bethe, *Phys. Rev.*, **76,** 38, 1949.) In order to obtain a more precise fit to scattering data, as well as to other data discussed in the above references, we must set certain limitations on the radius of the potential (i.e., the range in which it is appreciable). It is found that triplet and singlet wells should have different ranges, the best values being of the order of $(1.5 \pm .5) \times 10^{-13}$ cm. for the triplet range and $(2.6 \pm .5) \times 10^{-13}$ cm. for the singlet range. Thus, the range and depth of the potential are now moderately well known, but further details (such as shape) can be obtained only from more accurate scattering data, or from data obtained at a higher energy (where, as we recall, the cross section depends more critically on the shape of the well). It should be repeated, however, that our knowledge of nuclear forces is still tentative, and it is by no means certain that the concept of a nuclear potential will necessarily be adequate to describe what happens at higher energies. This concept does, however, seem to be adequate at least up to about 25 mev.

In this connection, it is worth-while to mention that accurate proton-proton scattering experiments are much easier to do than are neutron-proton scattering experiments. This is because the bombarding energy of the protons can be very accurately controlled with the aid of the electrical and magnetic properties of the proton, which latter also make the detection of the particles much easier. It is necessary, however, to study both systems, since, in the absence of any reason to the contrary, there is no cause for us to suppose that neutron-proton forces are exactly the same as proton-proton forces. Present experimental evidence indicates, however, that they are very similar except for the absence of a Coulomb force between neutron and proton. Unfortunately, this Coulomb force greatly increases the difficulty of treating proton-proton scattering theoretically.*

* For a discussion of Couloumb scattering, see Sec. 58. For a discussion of proton-proton scattering, see H. A. Bethe, and R. F. Bacher, *Rev. Mod. Phys.*, **8,** 82 (1936). For a more general discussion of the state of the problem of the scattering of elementary nuclear particles, see H. A. Bethe, *Phys. Rev.*, **76,** 38 (1949).

58. Coulomb Scattering. As shown in Sec. 41, the method of Rayleigh, Faxen, and Holtsmark breaks down for a Coulomb potential because the wave function does not approach sin $(kr + \delta)$ as r approaches infinity, where δ is a definite phase factor. Instead, the phase factor δ is proportional to ln r.

There are two ways to deal with this problem. First, one can use the fact that all Coulomb potentials are actually screened at some distance, r_0, and thus return to the method of Rayleigh, Faxen, and Holtsmark. This method, however, is likely to be clumsy because the phase shifts will be large, even for partial waves corresponding to very high angular momenta. Thus, many terms will have to be carried in the expansion (63) in partial waves. We shall give here a better method,* which does not require expansion in partial waves but treats the entire wave function as a unit. This method involves writing the wave function in parabolic co-ordinates. We first note that if we choose the z direction as that of the incident wave, then the entire system is cylindrically symmetrical, and the wave function is not a function of ϕ, but is instead a function only of r and z. Thus we write $\psi = \psi(r, z)$. The transformation to parabolic co-ordinates is the following:

$$\begin{aligned} \xi &= r - z \qquad & r &= \frac{\eta + \xi}{2} \\ \eta &= r + z \qquad & z &= \frac{\eta - \xi}{2} \end{aligned} \tag{92}$$

Problem 16: Prove that the lines ξ = constant and ν = constant are orthogonal parabolas (in the plane ϕ = constant).

Problem 17: Show that

$$\nabla^2\psi = \frac{4}{\xi + \eta}\left[\frac{\partial}{\partial\xi}\left(\xi\frac{\partial\psi}{\partial\xi}\right) + \frac{\partial}{\partial\eta}\left(\eta\frac{\partial\psi}{\partial\eta}\right) + \frac{1}{\xi\eta}\frac{\partial^2\psi}{\partial\phi^2}\right] \tag{93}$$

From the previous problem, we obtain the following wave equation expressed in parabolic co-ordinates, using the fact that ψ is not a function of ϕ:

$$-\left(\frac{\hbar^2}{2m}\right)\left(\frac{4}{\xi + \eta}\right)\left[\frac{\partial}{\partial\xi}\left(\xi\frac{\partial\psi}{\partial\xi}\right) + \frac{\partial}{\partial\eta}\left(\eta\frac{\partial\psi}{\partial\eta}\right)\right] + \frac{2Z_1Z_2e^2}{\xi + \eta}\psi = E\psi \qquad (94)\dagger$$

where Z_1 is the atomic number of the scattering nucleus, Z_2 that of the scattered particle, E is the reduced energy of the scattered particle, and m its reduced mass. (If the two particles have the same sign of charge, Z_1Z_2 is positive; otherwise it is negative.)

* For a fuller treatment of Coulomb scattering, see, Mott and Massey, *Theory of Atomic Collisions*, or L. Schiff, *Quantum Mechanics*.

† Note that if a uniform electric field is applied in the z direction, producing a potential energy of $e\mathcal{E}Z_1 = e\mathcal{E}(\eta - \xi)/2$, the wave function is still separable in parabolic co-ordinates. Thus, the Stark effect in hydrogen can be treated rigorously in this way. (See Chap. 19, Sec. 11.)

We now assert that the solution that we want can be written in the form

$$\psi = e^{ikz}f(\xi) = e^{ik\frac{(\eta-\xi)}{2}}f(\xi) \tag{95}$$

where

$$E = \frac{\hbar^2k^2}{2m} \tag{95a}$$

We shall prove that this is the right solution by showing that $f(\xi)$ can be chosen in such a way that ψ satisfies the differential equation (94) and the proper boundary conditions.

We first insert ψ into (94), obtaining

$$\xi\frac{d^2f}{d\xi^2} + (1 - ik\xi)\frac{df}{d\xi} - nkf = 0 \tag{96}$$

where

$$n = \frac{Z_1Z_2\,e^2m}{\hbar^2k} = \frac{Z_1Z_2\,e^2}{\hbar v} \tag{96a}$$

where v is the velocity of the incoming particle. This equation is the same one satisfied by the confluent hypergeometric function

$$Z\frac{d^2F}{dZ^2} + (b - z)\frac{dF}{dZ} - aF = 0 \tag{97}$$

$$F = F(a, b, Z) = \sum_{s=0}^{\infty}\frac{\Gamma(a+s)\Gamma(b)}{\Gamma(b+s)\Gamma(a)}\frac{Z^s}{s!} \tag{98}$$

is the solution of this equation* expanded in a power series which is good near the origin. We therefore obtain

$$f(\xi) = CF(-in, 1, ik\xi) \tag{99}$$

where C is a constant, to be determined.

To fit the boundary conditions, we require the asymptotic form of $\psi(\xi)$. The first two terms of this expansion are*

$$\psi \xrightarrow[r\to\infty]{} \frac{Ce^{\frac{1}{2}n\pi}}{\Gamma(1+in)}\left\{e^{i[kz-n\ln k(r-z)]}\left[1 - \frac{n^2}{2k(r-z)}\right] + \frac{f_C(\theta)}{r}e^{i(kr-n\ln 2kr)}\right\} \tag{100}$$

where

$$f_C(\theta) = \frac{\Gamma(1+in)}{i\Gamma(1-in)}\frac{e^{-in\ln\sin(\theta/2)}}{2k\sin^2(\theta/2)} = \frac{n}{2k\sin^2(\theta/2)}e^{-in\ln[\sin^2(\theta/2)]+i\pi+2i\alpha_0} \tag{101}$$

with

$$\alpha_0 = arg\Gamma(1+in) \tag{101a}$$

59. Interpretation of Above Result. We first note that the Coulomb wave function does not approach asymptotically the form (43),

$$\psi = e^{ikz} + f(\theta)\frac{e^{ikr}}{r}$$

* Whittaker and Watson, *Modern Analysis*, 3d ed., Chap. 14.

The first term on the right-hand side of eq. (100) does contain a factor, e^{ikz}, but this is modified by another factor, $e^{-in \ln k(r-z)} \left[1 - \frac{n^2}{ik(r-z)}\right]$. This factor indicates that the incident plane wave is slightly distorted, no matter how far away from the origin one chooses to go. This distortion is, of course, a consequence of the long range of the force. Similarly, we see that the outgoing wave contains the factor, $e^{-in \ln 2kr}$, so that it does not approach a definite phase. Despite these long-range effects, however, it is still possible to define a scattering cross section, because the distorting factors alter physically observable quantities, such as the mean current, in a way that goes to zero as r approaches infinity. To prove this, we evaluate the incident current obtained from the first term on the right-hand side of (100). We obtain

$$j = \frac{\hbar}{2mi}(\psi^*\nabla\psi - \psi\nabla\psi^*) \tag{102}$$

We note that when we differentiate the logarithm term, we bring down a factor proportional to $1/r$. For large r, this becomes negligible in comparison with the result of differentiating e^{ikz}. Thus, at large r, the incident current is very nearly in the z direction, and it has approximately the value

$$j_i = \frac{\hbar k}{m} \frac{|C|^2}{|\Gamma(1+in)|^2} \tag{103}$$

Similarly, the outgoing current per unit solid angle has the approximate value

$$j_o = \frac{\hbar k}{m} \frac{|C|^2 |f_C(\theta)|^2}{|\Gamma(1+in)|^2} \tag{104}$$

According to eq. (45a), the cross-section per unit solid angle is then

$$\sigma = |f_C(\theta)|^2 = \frac{n^2}{\left(2k \sin^2 \frac{\theta}{2}\right)^2} = \left(\frac{Z_1 Z_2 e^2}{2mv^2}\right)^2 \frac{1}{\sin^4 \frac{\theta}{2}} \tag{105}$$

The above is just the Rutherford cross section [see eq. (16c)]. As has already been pointed out, a Coulomb force has the unique property that the exact classical theory, the exact quantum theory, and the Born approximation in the quantum theory all yield the same scattering cross sections.

60. Exchange Effects in Coulomb Scattering. Whenever two equivalent charged particles are scattered from each other, the Coulomb cross section will be modified by exchange effects. Consider, for example, the scattering of α particles on each other. α particles have a total spin of zero, and have symmetrical wave functions. (The symmetry of the wave function follows from the fact that each α particle is made up of two neutrons and two protons. Thus, when two α-particles are exchanged, one exchanges four elementary particles at a time. The wave function is multiplied by $(-1)^4 = 1$, so that it is symmetric in such exchanges.)

To construct a suitable symmetric wave function, we note that from any solution of Schrödinger's equation for two equivalent particles $F(\mathbf{r}_1, \mathbf{r}_2)$ we obtain another one by interchanging $\mathbf{r}_1$ and $\mathbf{r}_2$. The desired symmetric function is then $F(\mathbf{r}_1, \mathbf{r}_2) + F(\mathbf{r}_2, \mathbf{r}_1)$. Now, in a two particle collision, the wave function takes the form $e^{i(\mathbf{k}_1+\mathbf{k}_2)\cdot(\mathbf{r}_1+\mathbf{r}_2)/m_1+m_2}f(\mathbf{r}_1 - \mathbf{r}_2)$. (See Chap. 15, Sec. 5.) The symmetric wave function is then

$$\psi = e^{i(\mathbf{k}_1+\mathbf{k}_2)\cdot(\mathbf{r}_1+\mathbf{r}_2)/m_1+m_2}[f(\mathbf{r}_1 - \mathbf{r}_2) + f(\mathbf{r}_2 - \mathbf{r}_1)] \tag{106}$$

(The exponential factor refers to the uniform motion of the center of mass, whereas the other factor is a function only of the relative co-ordinate.) Thus, the wave function of the relative co-ordinates can be found by taking $f(\mathbf{r}) + f(-\mathbf{r})$ where $\mathbf{r} = \mathbf{r}_1 - \mathbf{r}_2$.

From eq. (126), we obtain (noting that when we interchange $\mathbf{r}$ with $-\mathbf{r}$, we replace θ by $\pi - \theta$)

$$\psi \xrightarrow[r\to\infty]{} \frac{Ce^{\frac{1}{2}n\pi}}{\Gamma(1+in)} \left\{ e^{ikz-n\ln k(r-z)}\left[1 - \frac{n^2}{ik(r-z)}\right] \right.$$
$$\left. + e^{i[-kz-n\ln k(r+z)]}\left[1 - \frac{n^2}{ik(r+z)}\right] + \frac{e^{ikr-n\ln 2kr}}{r}[f_C(\theta) + f_C(\pi - \theta)] \right\}$$

This wave function corresponds to the fact that in the center of mass system one of the particles is incident from the right, whereas the other is incident from the left. We obtain the cross section, as with eq. (45a) by finding the ratio of the scattered current per unit solid angle to the incident current per unit area. Note that this time we do not divide the wave function by $\sqrt{2}$ when we symmetrize it, because we want to normalize the incident current for *each* particle per unit area to $\hbar k/m$. The cross section is

$$\sigma = |f_C(\theta) + f_C(\pi - \theta)|^2 \tag{107}$$

where m is the reduced mass $k = \sqrt{2mE}/\hbar$, and E is the reduced energy. From eq. (101), we finally obtain

$$\sigma = \left(\frac{Z_1Z_2e^2}{2mv^2}\right)^2\left[\frac{1}{\sin^4\theta/2} + \frac{1}{\cos^4\theta/2} + \frac{2\cos n(\ln\tan^2\theta/2)}{\sin^2\theta/2\cos^2\theta/2}\right] \tag{108}$$

The first two terms in the above are the same as would be obtained classically for equivalent particles.* The third term is the result of interference and is completely nonclassical. This expression is originally due to Mott and is called Mott scattering.

* The first two terms are obtained, for example, from Sec. 14.

For particles, such as electrons or protons, which have antisymmetric combined space and spin wave functions, one takes advantage of the fact that the space wave function is symmetric one-quarter of the time and antisymmetric three-quarters of the time.* One then obtains

$$\sigma = \left(\frac{Z_1 Z_2 e^2}{2mv^2}\right)^2 \left[\frac{1}{\sin^4 \theta/2} + \frac{1}{\cos^4 \theta/2} - \frac{\cos n(\ln \tan^2 \theta/2)}{\sin^2 \theta/2 \cos^2 \theta/2}\right] \quad (109)$$

The characteristic "exchange" effects are largest for $\theta = 90°$ (45° in laboratory system). At this angle, the exchange effects for α particles cause the cross section to be twice as large as that obtained with the neglect of exchange.

The additional interference terms characteristic of Mott scattering are in agreement with experiment.† Note however that in the classical limit, ($\hbar \to 0$ and $n \to \infty$), the "exchange" term becomes a very rapidly oscillating function of θ which averages out to zero. Thus, if the measurements have a given error $\Delta\theta$, then as $\hbar$ becomes smaller, the effects of the exchange terms will eventually be too small to be observed.‡

* See Chap. 17, Sec. 10.

† See Mott and Massey, 1st ed., p. 73.

‡ Thus, exchange effects are essentially quantum mechanical and disappear in the classical limit (see Chap. 19, Sec. 29).

PART VI

QUANTUM THEORY OF THE MEASUREMENT PROCESS

CHAPTER 22

1. Introduction. The quantum theory as developed thus far provides, in principle, a way to calculate the probable results of any measurement that one wishes to carry out. To calculate the average value of any observable A, we simply write $\bar{A} = \int \psi^* A \psi \, dx$, where ψ is the wave function of the system under investigation. If the quantum theory is to be able to provide a complete description of everything that can happen in the world, however, it should also be able to describe the process of observation itself in terms of the wave functions of the observing apparatus and those of the system under observation. Furthermore, in principle, it ought to be able to describe the human investigator as he looks at the observing apparatus and learns what the results of the experiment are, this time in terms of the wave functions of the various atoms that make up the investigator, as well as those of the observing apparatus and the system under observation. In other words, the quantum theory could not be regarded as a complete logical system unless it contained within it a prescription in principle for how all of these problems were to be dealt with.

In this chapter, we shall show how one can treat these problems within the framework of the quantum theory.†

2. The Nature of the Observing Apparatus. We shall begin by describing the general nature of an observing apparatus. In all cases, one obtains information by studying the interaction of the system of interest, which we denote hereafter by S, with the observing apparatus, which we denote by A. Any objects whose properties are understood, even if only in part, can in principle be utilized in the construction of the observing apparatus. For example, one frequently studies the forces between neutrons and protons by finding out how they scatter each other. In this case, the inadequately understood forces between particles are investigated by observing the effects that these forces have on the better

† For another treatment of this problem, see J. von Neumann, *Mathematische Grundlagen der Quantenmechanik*. Berlin: Julius Springer, 1932.

understood long-range motion of the same particles. Finally, it should be pointed out that not all of the observing apparatus has to be constructed by man, nor does it have to be located in a laboratory. Thus, the magnetic field of the earth can be regarded as part of a mass spectrograph that separates cosmic-ray particles according to their energies and charges.

Although every observation must be carried out by means of an interaction, the mere fact of interaction is not, by itself, sufficient to make possible a significant observation. The further requirement is that, after interaction has taken place, the state of the apparatus A must be correlated to the state of the system S in a reproducible and reliable way. This correlation is in general statistical, but in limiting cases it may approach any conceivable degree of exactness. Thus, one can measure the position of a star by observing the position of a spot on a photographic plate, which is produced by the light of the star after it travels through a telescope. The interaction between star and photographic plate is here brought about by the electromagnetic forces produced by light waves, which are able to change the chemical condition of the atoms of silver in the sensitive emulsion. An ideal telescope and camera system would produce a unique correlation between the position of the star and the position of the spot on the plate. No real telescope and camera system can do this with complete precision, first, because in practice there are unavoidable errors in its functioning and, second, because even in principle the wave nature of light produces a finite resolving power. Thus, in a typical observing apparatus we obtain a correlation such that each clearly distinguishable state of the apparatus corresponds to a range of possible states of the system under observation. This range may be called the *uncertainty*, or the *error*, in the measurement. The possibility of error usually arises from defects or inadequacies in design of the apparatus that are, in principle, avoidable. In extremely accurate measurements, however, it may arise from the quantum nature of matter, in which case a more accurate measurement cannot be made without changing what is observed in a fundamental way (see Chap. 5).

3. The Classical Stages of an Observing Apparatus. Let us now recall the main result of Chap. 8, that at the quantum level of accuracy the entire universe (including, of course, all observers of it), must be regarded as forming a single indivisible unit with every object linked to its surroundings by indivisible and incompletely controllable quanta.* If it were necessary to give *all* parts of the world a completely quantum-mechanical description, a person trying to apply quantum theory to the process of observation would be faced with an insoluble paradox. This would be so because he would then have to regard himself as something connected inseparably with the rest of the world. On the other hand,

* See Chap. 8, Secs. 23 and 24; also Chap. 6, Sec. 13.

the very idea of making an observation implies that what is observed is totally distinct from the person observing it.

This paradox is avoided by taking note of the fact that all real observations are, in their last stages, classically describable.* The observer can therefore ignore the indivisible quantum links between himself and the classically describable part of the observing apparatus from which he obtains his information, because these links produce effects that are too small to alter in any essential way the significance of what he sees.† In other words, the interaction between the observer and his apparatus is such that statistical fluctuations arising from the quantum nature of the interaction are negligible in comparison with experimental error. It is therefore correct for us to approximate the relation between the investigator and his observing apparatus, in terms of the simplified notion that these are two separate and distinct systems interacting only according to the laws of classical physics. Furthermore, any number of observers can interact with the same apparatus, without changing any essential property of the apparatus. The various possible configurations or states of the measuring apparatus, corresponding to the different possible results of a measurement, can therefore correctly be regarded as existing completely separately from and independently of all human observers. The quantum theory of the measurement process can in this way be reduced to a description of the relation between the state of the system under investigation and the state of some classically describable part of the observing apparatus. (In this sense, it is, of course, the same as the classical theory.)

The preceding discussion is, however, somewhat imprecise, and is only intended for the purpose of indicating the general approach by which one can justify the application of the customary classical procedure of regarding the observer and his apparatus as separate systems, even though they are actually linked by indivisible quanta. Throughout the rest of this chapter, we shall give a more precise, but comparatively simple, mathematical discussion of what happens in the process of observation, and show that essentially the same result is obtained.

If a sharp distinction could not be made between the observer and the systems observed, scientific research as we know it could not be carried out, because the observer would not know which aspects of an observation

* We may give as an example the usual practice in science, whereby one obtains data from meter readings, spots on a photographic plate, clicks of a Geiger counter, etc. All these objects and phenomena have the common property of being classically describable. A little reflection will convince the reader that all observations ever made in science have employed at least one such classically describable stage.

† If the investigator wishes to study the quantum properties of matter, he requires apparatus that amplifies the effects of individual quanta to a classically describable level.

originated in himself, and which originated in the outside systems of interest. We do not wish to imply, however, that scientific research is necessarily impossible whenever an observer interacts significantly with the things that he observes; for as long as the observer can correct for the effects of his interactions, on the basis of known causal laws, he can still distinguish between effects originating in him and those originating outside.* But if, for example, the interaction occurred through a single indivisible and uncontrollable quantum, this kind of correction could not be carried out. The observer would not then be able to tell whether what he saw related to him or to an outside object since the quantum connecting both belongs mutually and indivisibly to both. Because the interaction between human observer and measuring apparatus is in all real observations classically describable this difficulty will, of course, never actually arise.

4. Extent of Arbitrariness in Distinction Between the Observer and What He Sees. In the event that there is more than one classically describable stage of the apparatus, any one stage may be chosen as the point of separation between the observer and what he sees. Consider, for example, an experiment in which a person obtains information for a photograph. One possible description of this experiment is as follows: The system observed consists of the objects photographed, plus the camera, plus the light that connected object with image. The observer is then said to obtain his information by looking at the plate. Because this process is classically describable, there is a sharp distinction between the observer and the plate that he is looking at. An equally good description, however, involves regarding the system under investigation as the object itself. The camera and the plate can then be considered as part of the observer. A third description of the same procedure would be to say that the investigator observes the image on the retina of his eye, so that the retina of the eye, plus the rest of the world, including of course the photographic plate, are to be regarded as the system under observation.

To summarize the results of the preceding discussion, we say that the processes by which an observer obtains his information usually involve a series of classically describable stages. If the connection between the observer and what he sees are to permit him to obtain reliable information, these stages must function causally, and in such a way that a definite state of one stage is reflected in a one-to-one way in a corresponding definite state of the next stage. Thus, a definite spot on an object should produce a corresponding spot on the photographic plate, and this should produce a corresponding spot on the retina of the eye. To the extent that this correspondence exists, the point of division

* See, for example, Chap. 5, Sec. 9, where a case is discussed in which the effects of the apparatus are corrected for on the basis of known causal laws.

between the observer and what he sees can correctly be made at *any* classically describable stage.

We may now ask how far this point of distinction can be carried in either direction, i.e., into the object under investigation or into the brain of the investigator himself. Now, the criterion for a well-defined piece of apparatus is that it faithfully transmits information about the nature of the object in a one-to-one way. Thus, the only remaining restriction on how far into the object the point of distinction can be pushed is that the distinction must not be drawn at an essentially quantum-mechanical stage. If we wish to observe the position and momentum of an electron to a quantum level of accuracy, we must regard the electron, plus the light quanta used in making the observation, as part of an indivisible combined system. Eventually, however, these light quanta are able to activate classically describable processes, e.g., the production of spots on a photographic plate, and from there on the distinction may be drawn.

Let us now consider the problem of how far into the brain the point of distinction between the observer and what is observed can be pushed. Before doing this, however, we wish to stress that the question is completely irrelevant as far as the theory of measurements is concerned since, as we have already seen, it is necessary only to carry the analysis to some classically describable stage of the apparatus. Nevertheless, it is perhaps of some interest to indulge in a few speculations on this fascinating general problem, concerning which very little specific information is at present available.

If, for example, as suggested in Chap. 8, Sec. 28, the brain contains essentially quantum-mechanical elements, then the point of distinction cannot be pushed as far as these elements. Even if the brain functions in a classically describable way, however, the point of distinction may cease to be arbitrary, because the response of the brain may not be in a simple one-to-one correspondence with the behavior of the object under investigation. To illustrate the problems involved, we can begin with the optic nerve, which is almost certainly classically describable. This nerve seems to function solely as a signalling device, so that it responds in a one-to-one way to the image on the retina. Thus, the observer can be said to obtain visual information by observing the signals coming in along the optic nerve. Signals similar to those caused by light can be obtained by electrical or mechanical stimulation of this nerve. If we try to carry this type of description much farther into the brain, then we begin to reach more speculative grounds. It seems to be fairly certain, however, that before the observer can become conscious of these signals, they must go through several additional complex systems of nervous tissue that carry out actions essential for the recognition of the objects seen. The loss of certain parts of the brain, for example, is known to prevent recognition of objects even when the eye and optic nerve are in good condition. Thus, it seems likely that a person can be said to observe the signals after they have come through the part of the brain involved in the recognition of the object.

Practically nothing at all is known as yet about the details of what happens to the signal in the next stage. There is, however, a good reason to expect that the description in terms of the propagation of a signal which is in one-to-one correspondence with the behavior of the object eventually becomes inadequate. The reason is that nervous circuits in the brain frequently permit the feeding of impulses reaching a later point back into an earlier point. When this happens, it is no longer correct to say that the role of a given nerve is only to carry signals from outside, because each nerve may then be mixing in an inextricable (and nonlinear) way the effects of signals coming from other parts of the brain as well as from outside. When this stage is reached, the analysis in terms of a division between two distinct systems, i.e., the observer and the rest of the world, becomes inappropriate and, instead, it is probably better to say that all parts of the brain significantly coupled by feedback respond as a unit. It is this response as a unit that should probably be regarded as the process by which the observer becomes aware of the incoming signal. It therefore seems likely that the division between the observer and the rest of the world cannot be pushed arbitrarily far into the brain.

In view of the fact that we know so little about the details of the functioning of the brain, it seems fortunate that the analysis has to be carried only to a classically describable part of the apparatus.

5. Mathematical Treatment of Process of Observation. To describe the process of observation quantum mechanically, we must begin by solving Schrödinger's equation, taking into account the effects of the observing apparatus. Now, before the experiment begins, the observing apparatus A and the system S under observation are, in general, not coupled. For example, when we take a photograph, we couple the plate to the object under investigation by opening a shutter at some fairly definite time, before which there is no significant interaction between plate and object. The reader will readily convince himself that this requirement of lack of coupling before the experiment begins is satisfied in all actual measurement processes. At this time, the Hamiltonian operator can therefore be written

$$H = H_S + H_A = H_S(x) + H_A(y) \tag{1}$$

where H_S is the Hamiltonian of the system alone, and H_A is the Hamiltonian of the apparatus alone. The fact that there is no coupling between the two is taken into account by making H_S a function only of the system variables x, while H_A is a function only of the apparatus variables y.

In this elementary treatment, we assume, for the sake of simplicity, that the apparatus is in a fairly definite state before the measurement is begun. To denote the state of the apparatus, we assume an apparatus wave function of the form $f(y, t)$. The time dependence of $f(y, t)$ indicates that the apparatus may itself be in a changing state. At this stage of the discussion, it is not necessary to be very specific about the nature of $f(y, t)$. One should note, however, that because the measuring apparatus is to function in a classically describable way, the function $f(y, t)$ will generally take the form of a wave packet whose definition is much poorer than the limits of precision set by the uncertainty principle (Chap. 10, Sec. 9). To consider an example, we may allow y to represent the position of an ammeter needle. The wave packet $f(y, t)$ will then represent the extent to which the position of this needle is defined. Suppose that the natural period of oscillation of this needle is 0.1 sec. The distance between adjacent quantum states will then be

$$\Delta E = h\nu \cong 6.6 \times 10^{-26} \text{ erg}$$

To obtain an estimate of the maximum displacement x of the needle corresponding to a single quantum jump, we use the formula for the energy

$$E = \frac{m}{2} \omega^2 x^2, \qquad \Delta E = m\omega^2 x \, \Delta x, \qquad \text{or} \qquad \Delta x = \frac{\Delta E}{m\omega^2 \, \Delta x}$$

Taking $x \cong 1$ mm for the displacement of the needle, and $m \cong 1$ mg, we obtain $\Delta x \cong 10^{-26}$ cm. It is clear that under no circumstances will one ever use such an ammeter with a precision approaching a quantum level of accuracy.

As for the system under observation, its state is not known, but it is the object of the measurement to provide some information about it. Let us suppose that the wave function of the system S is expanded in terms of some orthonormal series $v_m(x, t)$ of solutions of Schrödinger's equation for the system alone, i.e.,

$$\psi_S = \sum_m C_m v_m(x, t) \tag{2}$$

where the C_m are unknown coefficients.

After the apparatus begins to interact with the system S, a third term will appear in the Hamiltonian, which we denote by $H_I(x, y)$. Thus we obtain

$$H = H_S(x) + H_A(y) + H_I(x, y) \tag{3}$$

It is the term $H_I(x, y)$ that introduces the correlation between the state of the system S and the apparatus A and thus makes possible a measurement. We shall see that before this correlation is strong enough to make possible a definite measurement, the interaction must be on for a certain minimum length of time Δt, which is inversely proportional to the strength of the interaction. If the interaction is turned off too soon, we shall therefore not have a measurement. The interaction can be turned off, however, at any time after Δt, and in most types of apparatus it must be turned off within some definite period of time in order to prevent the destruction of the record of the experiment. In a camera, for example, one controls the time of interaction by means of a shutter. If this time is too short, then the film will be underexposed, and one will not obtain an accurate picture. The minimum exposure time is inversely proportional to the intensity of light, i.e., to the strength of interaction. If the shutter is left open too long, however, the film will be overexposed, and the record of the observation will be destroyed.

Whenever the results of an observation are recorded (e.g., on a photograph, a wire tape, a punch card, by pencil marks in a notebook, or simply the change in position or momentum of some convenient object), one has a situation in which the record is decoupled from the system under observation, so that any number of observers can consult the record without affecting the system S. This property is, in fact, included in the very definition of what one means by making a record. Although it is not necessary that the results of all observations shall actually be recorded somewhere, it is certainly true that all observational data have the property that they can, in principle, be recorded. For our present purpose of showing that the quantum theory is able to give a consistent account of the process of measurement, it will be adequate to confine ourselves to the cases in which the data are actually recorded. In other words, we assume that, after some time Δt, the final stage of the apparatus (i.e., the part on which the results are recorded) is decoupled from the system under observation. Since the process of recording can always, in principle, be made completely automatic, it is clear that when the human observer obtains his information by looking at the record he need produce absolutely no changes in the system S under investigation. This means that all changes in the system S (which latter may be essentially quantum mechanical) are produced only by the actions of the apparatus, whereas the effects of the human observer as he obtains his information are confined to the classically describable parts of the apparatus where, as we have seen, they produce no significant changes.

After the interaction has taken place, the state of the system S may have changed for two reasons. First, the variables under observation may not even be constants of the motion of the undisturbed system S. Thus, the position of a freely moving particle changes continuously with the passage of time. Second, the interaction with the observing apparatus may introduce further changes in the variable under observation. Thus, in the measurement of the momentum of a charged particle by means of the track that is left in a cloud chamber, the apparatus changes this momentum, first, because the magnetic field deflects the particle in a systematic way, and second, because the gas molecules with which the particle collides give it small random deflections.

As far as the theory of measurements is concerned, changes of the variable under observation that occur during the course of a measurement introduce irrelevant complications. We shall see presently that it is, in principle, always possible to design an apparatus that measures any given variable without changing that variable during the course of the measurement.* (In accordance with the uncertainty principle, the complementary variables must, of course, change in an incompletely controllable way.) One method of accomplishing this result is as follows: First, we make what is called an *impulsive* measurement. This means that the interaction lasts for so short a time that the changes of the variable that would occur in the absence of a measurement are negligible. Thus, we may measure the position of a particle with the aid of a pulse of light of so short a duration that the particle does not move appreciably, while it scatters the pulse. In order to photograph a particle in a very short time, however, one needs an intense source of light. More generally, an impulsive measurement requires a strong interaction of very short duration between apparatus and system under observation. While the impulsive measurement is taking place, the terms $H_S(x)$ and $H_A(y)$ produce changes in the wave function which are negligible in comparison to those produced by the interaction term $H_I(x, y)$. Thus, during this time (and during this time only) Schrödinger's equation can be written as

$$i\hbar \frac{\partial \psi}{\partial t} = H_I \psi \tag{4}$$

Before and after the time of interaction, however, one has

$$i\hbar \frac{\partial \psi}{\partial t} = (H_A + H_S)\psi \tag{5}$$

During the time of interaction, the functions $u_n(y, t)$ and $v_m(x, t)$, which are solutions of Schrödinger's equation when $H_I = 0$, can therefore be regarded as constant in the time. We shall hereafter designate them as $u_x(y)$ and $v_m(x)$. The impulsive measurement therefore avoids all changes in the variable under observation that occur independently of the process of interaction.

In order to obtain an impulsive measurement, we have seen that a large interaction energy is needed. If the interaction energy is so large, how are we to avoid changes in the variable under observation that are brought about by the process of interaction itself? To answer this question, we denote the observable under consideration by the operator M, having eigenvalues m and eigenfunctions $v_m(x)$. If the state of the

* In this connection, see also Chap. 6, Sec. 3.

apparatus is to be correlated to that of the variable M, it is necessary that H_I depend at least on M, as well as on y. Let us now observe that if H_I is chosen in such a way that it is diagonal in the same representation in which M is diagonal, then the matrix elements corresponding to transitions from one value of m to another will vanish. This means that no matter how strong the interaction is, there will be no change of M, (although complementary observables may change a great deal). In order to have H_I diagonal when M is diagonal, we choose

$$H_I = H_I(M, y) \tag{6}$$

That is, H_I is a function only of M and y. In this way, we have accomplished our objective of designing an apparatus with which a given variable M can be measured without itself suffering any changes.

When the variable under observation does change in the course of the measurement, one can, in general, correct for these changes. For example, in the measurement of the momentum by means of the Doppler shift (Chap. 5, Sec. 9) we saw that momentum was changed in the measurement by a definite amount, so that a correction for this change could be made. We shall not prove here that such a correction is, in general, possible, but the reader can convince himself of the truth of this statement after some reflection.† In this work, we shall hereafter confine ourselves to the discussions of the simpler case in which the variable M does not change, while it is being measured. Since, as we have seen, it is always possible, in principle, to carry out the measurement in this way, we conclude that a treatment of this type of measurement alone is adequate to show that the quantum theory is able to give a consistent account of the process of measurement.

We are now ready to put Schrödinger's equation for the case of an impulsive measurement into a particularly convenient form. To do this, let us consider the combined wave function, $\psi(x, y, t)$, representing the state of the system S and the apparatus A during the time while the measurement is taking place. When ψ is regarded as a function of x, it is clear that the expansion postulate permits us to write

$$\psi(x, y, t) = \sum_m f_m v_m(x) \qquad \text{where} \qquad f_m = \int v_m^*(x)\psi(x, y, t)\, dx \tag{7a}$$

† In this connection, we should note that in some cases, the measurement may even be said to destroy the object under observation. Consider, for example, the absorption of a photon. Whether or not we regard the object under investigation as destroyed is, however, largely a matter of choice of terminology. For example, another description of the absorption of a photon is to say that a certain radiation oscillator went to a lower quantum state, thus giving up a quantum of energy. In this description, the variable under observation (in this case, the energy of a certain radiation oscillator) is not destroyed, but only changed. Thus, the general argument in terms of the notion that the variable may change while it is being measured also applies to this case.

The v_m are eigenfunctions of the operator M. We see from the above that the f_m will, in general, be functions of y and t. Thus, we obtain

$$\psi(x, y, t) = \sum_m f_m(y, t)v_m(x) \tag{7b}$$

Schrödinger's equation then becomes

$$i\hbar \sum_m \frac{\partial f_m(y, t)}{\partial t} v_m(x) = \sum_m H_I f_m(y, t)v_m(x) \tag{8a}$$

Because H_I is a function of M, the above reduces to

$$i\hbar \sum_m \frac{\partial f_m(y, t)}{\partial t} v_m(x) = \sum_m H_I(m, y)v_m(x)f_m(y, t) \tag{8b}$$

where we have replaced the operator M by the number m corresponding to the eigenfunction on which H_I is operating. Multiplication by $v_r^*(x)$ and integration over x yields

$$i\hbar \frac{\partial f_r(y, t)}{\partial t} = H_I(r, y)f_r(y, t) \tag{9}$$

This means that the apparatus will undergo a change of state that is different for each eigenvalue r of the system variable M. If the interaction is strong enough and if it is allowed to continue for a long enough time, the changes of the variables describing the apparatus will be so great that the state of the latter will depend primarily on the value of the observable M. It is this correlation between the two sets of observables that is needed before an interaction can be utilized for the purpose of making a measurement.

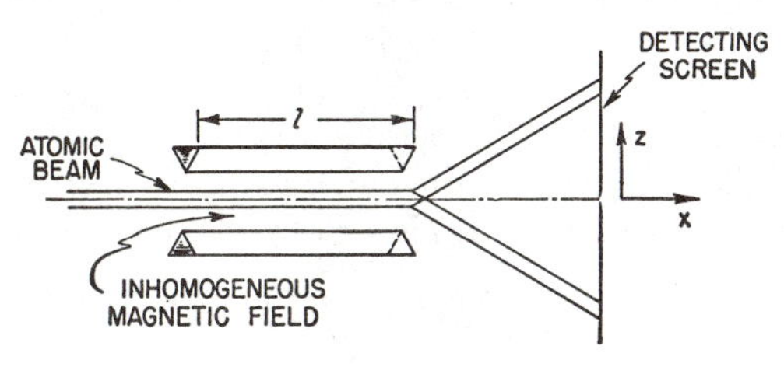

Fig. 1

6. An Example: The Measurement of the Spin of an Atom. As an example, we shall first consider a special case in which we measure the z component of the angular momentum of an atom whose angular momentum is $\hbar/2$. The system S under observation has only two possible states, which we denote respectively by $s = 1$ and $s = -1$. The spin will be measured by means of a Stern-Gerlach experiment, illustrated in Fig. 1 (see also Chap. 14, Sec. 16).

Let us suppose that the deflection of the atom by the inhomogeneous magnetic field is impulsive. This means that the z motion of the atom occurring while it is in the region of the inhomogeneous field can be neglected. The magnetic force then gives the particle a momentum that is directed up or down, according to whether the spin is up or down. The resulting z motion of the particle after it leaves the field carries it to a height that depends on the spin. In this way, a rather rough observation of the position enables us to tell whether the spin was up or down.

The interaction energy for this problem is* [see eq. (78), Chap. 17]

$$H_I = \mu(\boldsymbol{\sigma} \cdot \boldsymbol{\mathcal{H}}) \tag{10a}$$

where

$$\mu = -\frac{e\hbar}{2mc}$$

Now, in the median plane, near which the beam goes, the field is in the z direction. To a first approximation, we can write $\mathcal{H}_z \cong \mathcal{H}_0 + z\mathcal{H}_0'$; where $\mathcal{H}_0 = (\mathcal{H}_z)_{z=0}$ and $\mathcal{H}_0' = \left(\frac{\partial \mathcal{H}_z}{\partial z}\right)_{z=0}$. With these approximations, we obtain†

$$H_I \cong \mu(\mathcal{H}_0 + z\mathcal{H}_0')\sigma_z \tag{10b}$$

In this case, the position of the atom, z, is the apparatus co-ordinate, because by observing z we can find the value of the spin. The observing apparatus may be regarded as the combination of the inhomogeneous magnetic field, the co-ordinate of the atom, and the detecting screen. The function of the magnetic field is, of course, to introduce a correlation between the atomic spin and the apparatus co-ordinate.

Before the atom enters the magnetic field, it is necessary that it shall be in a fairly definite state; otherwise no conclusions about the value of σ_z can be drawn from this experiment. The z dependence of the atomic wave function will take the form of a packet, which we denote by $f_0(z)$. According to Sec. 3, the apparatus co-ordinate should be classically describable. This means that the definition of the state of the atom is much less precise than the limits allowed by the uncertainty principle, so that

$$\Delta p \, \Delta z \gg \hbar \tag{11}$$

We note that H_I and σ_z are diagonal in the same representation. This means that the Stern-Gerlach experiment measures σ_z without changing it. (σ_x and σ_y are, however, changed in an uncontrollable way.)

* We are assuming that the spin variable refers to a neutral atom.

† Strictly speaking, since $\frac{\partial \mathcal{H}_y}{\partial y} + \frac{\partial \mathcal{H}_z}{\partial z} = 0$, there is always an inhomogeneous component of the field, which should produce deflections in the y direction. Since these are of no interest to us here, we shall not include the y component of the field hereafter.

The initial wave function for this system is then

$$\psi_0 = f_0(z)(c_+v_+ + c_-v_-) \tag{12a}$$

where v_+ and v_- are the spin functions belonging respectively to $\sigma_z = 1$ and $\sigma_z = -1$, whereas c_+ and c_- are the unknown coefficients of these spin wave functions.

While the interaction is taking place, the wave function can still be expanded in terms of the two possible spin wave functions, but the coefficients will, as in eq. (7b'), become functions of z and t. Thus, we write

$$\psi = f_+(z, t)v_+ + f_-(z, t)v_- \tag{12b}$$

Schrödinger's equation becomes*

$$i\hbar \frac{\partial \psi}{\partial t} = H_I\psi$$

or

$$i\hbar \left(\frac{\partial f_+}{\partial t} v_+ + \frac{\partial f_-}{\partial t} v_-\right) = \mu(\mathcal{H}_0 + z\mathcal{H}_0')(f_+v_+ - f_-v_-) \tag{13a}$$

Since the coefficients of v_+ and v_- must be separately equal, we obtain

$$i\hbar \frac{\partial f_+(z, t)}{\partial t} = \mu(\mathcal{H}_0 + z\mathcal{H}_0')f_+(z, t)$$

$$i\hbar \frac{\partial f_-(z, t)}{\partial t} = -\mu(\mathcal{H}_0 + z\mathcal{H}_0')f_-(z, t) \tag{13b}$$

The boundary conditions are that at $t = 0$,

$$f_+ = f_0(z)c_+ \qquad \text{and} \qquad f_- = f_0(z)c_- \tag{14}$$

The above equations can easily be integrated. The solutions satisfying the correct boundary conditions are

$$f_+ = c_+f_0(z)e^{-i\mu(\mathcal{H}_0+z\mathcal{H}_0')t/\hbar} \qquad f_- = c_-f_0(z)e^{+i\mu(\mathcal{H}_0+z\mathcal{H}_0')t/\hbar} \tag{15a}$$

and

$$\psi = f_0(z)[c_+ e^{-i\mu(\mathcal{H}_0+z\mathcal{H}_0')t/\hbar}v_+ + c_- e^{+i\mu(\mathcal{H}_0+z\mathcal{H}_0')t/\hbar}v_-] \tag{15b}$$

We now come to the problem of determining the time, Δt, during which the interaction takes place. This is clearly the time that the particle spends in the magnetic field. Strictly speaking, this time is not perfectly definable, because one must set up a wave packet in the x direction, and the time at which the packet passes a given point is indefinite to within $\delta t \cong \hbar/\Delta E$. If one makes the length l of the field region large in comparison with the separation of the pole faces, however, the error in the time resulting from the width of the wave packet is made negligible. One can then treat the x motion as classical and say that the field acts for a time, $\Delta t = l/v$, where v is the velocity in the x direction. Because the

* Because the interaction is impulsive, we neglect all energies other than H_I [see eq. (4)]. In this case, the neglected term is the kinetic energy of mass motion of the atom ($p_z^2/2m$). The spin itself (in a nonrelativistic theory) does not have any energy associated with it, other than its energy of interaction with an electromagnetic field.

determination of the spin does not depend in a sensitive way on the precise evaluation of Δt, this procedure is adequate for our purposes. On the other hand, the z motion, which is clearly coupled to the spin through the term $H_I = \mu\sigma_z(\mathfrak{H}_0 + z\mathfrak{H}_0')$, must be treated quantum mechanically, because we are seeking a quantum description of what happens in the course of the measurement.

After the particle passes through the magnetic field, the wave function is therefore to be obtained from eq. (15b) by setting $t = \Delta t = l/v$. We note that v_+ and v_- are multiplied by phase factors of opposite sign. The phase factor $e^{-i\mu\mathfrak{H}_0'\Delta tz/\hbar}$, multiplying v_+ signifies that if the spin is positive, the momentum change is $\delta p_z = -\mathfrak{H}_0'\mu\,\Delta t$, while the factor, $\exp\,(i\mu\mathfrak{H}_0'\,\Delta tz/\hbar)$, multiplying v_-, shows that if the spin is negative, the particle obtains exactly the opposite momentum.* Thus, it would be possible in principle to measure the spin by measuring the momentum transmitted to the particle by the magnetic field. In this experiment, however, it turns out to be more convenient to measure the momentum indirectly by measuring the distance traveled by the particle by the time it reaches a distant screen. We must therefore follow the motion of the wave packet after the particle leaves the field. To do this, we Fourier analyze the initial wave packet, writing

$$\psi = \int g(k)(v_+c_+e^{ikz} + v_-c_-e^{ikz})\,dk \tag{16}$$

where $g(k)$ is a packet centering around $k = 0$. Immediately after the particle leaves the field, we then obtain

$$\psi = \int g(k)\left\{c_+v_+\exp\left[i\left(k - \frac{\mu\mathfrak{H}_0'\,\Delta t}{\hbar}\right)z - i\frac{\mu\mathfrak{H}_0\,\Delta t}{\hbar}\right]\right. \\ \left. + c_-v_-\exp\left[i\left(k + \frac{\mu\mathfrak{H}_0'\,\Delta t}{\hbar}\right)z + i\frac{\mu\mathfrak{H}_0\,\Delta t}{\hbar}\right]\right\}dk \tag{17a}$$

The kth Fourier component of the part of the wave function with positive spin now oscillates with angular frequency $\omega = \dfrac{p^2}{2m\hbar} = \dfrac{\hbar}{2m}\left(k - \dfrac{\mu\mathfrak{H}_0'\,\Delta t}{\hbar}\right)^2$, whereas the part with negative spin has $\omega = \dfrac{\hbar}{2m}\left(k + \dfrac{\mu\mathfrak{H}_0'\,\Delta t}{\hbar}\right)^2$. The wave function then becomes

$$\psi = \int dkg(k) \\ \left\{c_+v_+\exp\left[i\left(k - \frac{\mu\mathfrak{H}_0'\,\Delta t}{\hbar}\right)z - i\frac{\mu\mathfrak{H}_0\,\Delta t}{\hbar} - \frac{i\hbar t}{2m}\left(k - \frac{\mu\mathfrak{H}_0'\,\Delta t}{\hbar}\right)^2\right]\right. \\ \left. + c_-v_-\exp\left[i\left(k + \frac{\mu\mathfrak{H}_0'\,\Delta t}{\hbar}\right)z + i\frac{\mu\mathfrak{H}_0\,\Delta t}{\hbar} - \frac{i\hbar t}{2m}\left(k + \frac{\mu\mathfrak{H}_0'\,\Delta t}{\hbar}\right)^2\right]\right\} \tag{17b}$$

* Note that the wave function [eq. (15b)] takes the form of a packet, because it is multiplied by the function $f_0(z)$. The mean momentum of the packet, however, is changed according to the argument of the exponential.

The center of the wave packet occurs where the phase has an extremum, or where

$$z = -\frac{\mathcal{H}_0'\mu\,\Delta t}{\hbar}t \qquad \text{for positive spin}$$
$$z = \frac{\mathcal{H}_0'\mu\,\Delta t}{\hbar}t \qquad \text{for negative spin} \tag{18}$$

Thus, the wave function breaks up into two packets that move in different directions, according to whether the spin is positive or negative.

In order that the Stern-Gerlach experiment shall be able to make possible a measurement of the spin, it is necessary that the momentum gained by the particle from the magnetic field, $\delta p = \pm\mu\mathcal{H}_0'\,\Delta t$, shall be much greater than the initial uncertainty Δp_0 in the momentum of the beam. If this requirement is not satisfied, then the natural spread of the wave packet of the particle will be great enough to mask the deflections that depend on the spin. We therefore require that

$$\mu\mathcal{H}_0'\,\Delta t \gg \Delta p_0 \qquad \text{or} \qquad \mathcal{H}_0'\,\Delta t \gg \frac{\Delta p_0}{\mu} \tag{19}$$

In this way, we compute the minimum product of $\mathcal{H}_0'$ and Δt needed before the correlations introduced by the measuring apparatus are strong enough to make a good measurement possible. Under ideal circumstances, $\Delta p_0 \cong \hbar/\Delta x$, where Δx is the width of the packet. We have seen, however, that in practice the packet is always much wider (in momentum space) than the minimum width permitted for a given Δx by the uncertainty principle.

After the packet leaves the magnetic field, it will start to spread. The minimum width of the packet is, of course, limited by the uncertainty principle, but in any case, the packet will spread out to at least

$$\Delta z = \frac{\Delta p_0 t}{m}$$

by the time it reaches the detecting screen.* This spread will be in addition to the original width of the packet while it passed through the magnet. But if we satisfy condition (19), then the mean distance traveled by each packet (see eq. 18) will be much greater than the fluctuation in this distance, so that the original lack of complete definition of the beam will not be able to prevent us from obtaining a good measurement of the spin.

We give in Fig. 2 a graph showing the general shape of the packets in momentum space and in position space, when the particle reaches the detecting screen.

* For a discussion of the spread of wave packets, see Chap. 3, Sec. 5, and Chap. 10, Sec. 8.

It is now clear that even when the position and momentum of the beam are defined only to a classical level of accuracy, we can always make the product $\mathcal{H}_0' \Delta t$ so large that we obtain a classically describable

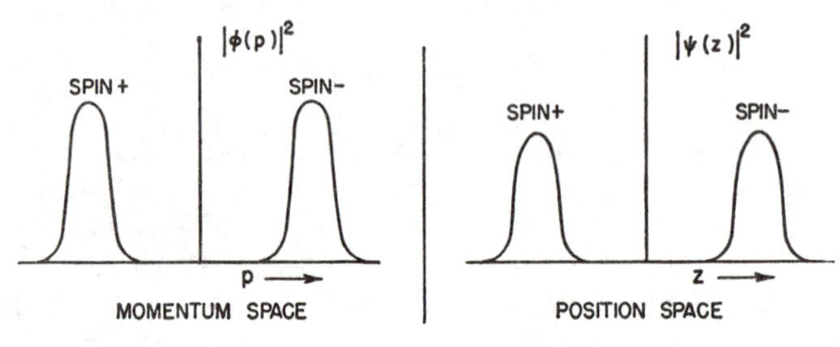

FIG. 2

separation between the beams corresponding respectively to positive and negative spins. We have thus attained our objective of reducing the theory of measurement of the spin to a description of the relation between the state of the quantum system under investigation and the state of a classically describable part of the observing apparatus.

7. Generalization to a Variable with an Arbitrarily Large Number of Eigenvalues. The generalization of these results to a variable having an arbitrarily large number of eigenvalues is straightforward. We see from eq. (9) that the wave function of the apparatus undergoes a change that depends on the quantum state, r, of the system under observation. If a good observation is to be made, the interaction between the apparatus and the system under observation must be so strong that adjacent quantum numbers, r, of the system under observation lead to classically distinguishable states of the apparatus, i.e., to wave functions of the apparatus separated by a great many quantum states. We shall see that it is always possible, in principle, to achieve this result by making the product of the strength of interaction and the time in which it acts large enough.

We begin with eq. (9). Let us first note that in the special case in which $H_I(r, y)$ is diagonal in the same representation in which y is diagonal (i.e., H_I is a function of y alone and contains no operators such as $\partial/\partial y$), eq. (9) is easily integrated. We use the boundary condition that at $t = 0$, the wave function of the system S is $\sum_r c_r v_r(x)$ [see eq. (2)], and that of the apparatus is $f_0(y)$,* so that the combined wave function is then $\psi_0 = f_0(y) \sum_r c_r v_r(x)$. Equation (9) then yields

$$f_r(y, t) = c_r f_0(y) e^{-iH_I(r,y)t/\hbar} \tag{20}$$

* We deal here only with the case of an impulsive measurement, so that the apparatus wave function, $f_0(y)$, is effectively a constant.

[This is the generalization of eq. (15), where we note that H_I was also a function only of z.]

If H_I is not diagonal in the same representation in which y is diagonal, we can always make a unitary transformation to a representation in which H_I is diagonal. (This is possible whenever H_I is a Hermitean operator.) After making the transformation, we can integrate Schrödinger's equation as was done above, and then carry out the inverse unitary transformation back to the original variables. In the subsequent work, we shall, however, restrict ourselves to the case in which H_I is diagonal in the same representation in which y is diagonal, noting that the treatment is readily generalized by the methods outlined above.

Let us recall that the quantum number of a given quantum state is equal to the number of nodes in the wave function for this state. Since the factor $f_0(y)$ is common to all the wave functions $f_r(y, t)$ in eq. (20), it is clear that the difference in the number of nodes for different values of r depends only on the factor

$$e^{-iH_I(r,y)t/\hbar} = \cos\left[H_I(r, y)\frac{t}{\hbar}\right] - i\sin\left[H_I(r, y)\frac{t}{\hbar}\right]$$

The real part of $e^{-iH_I(r,y)t/\hbar}$ will have a node* every time that

$$H_I(r, y)\frac{t}{\hbar} = \left(n + \frac{1}{2}\right)\pi$$

where n is an integer. The wave function is appreciable, however, only in the limited region Δy in which $f_0(y)$ is large. Within this region, one can usually approximate H_I by the first two terms of a power series. Thus

$$H_I(r, y) = H_0(r) + yH_0'(r) + \cdots \tag{21a}$$

where $\quad H_0(r) = H_I(0, r) \quad$ and $\quad H_0'(r) = \left[\frac{\partial}{\partial y}H_I(y, r)\right]_{y=0}$

In the region where the wave function is appreciable, the difference in the number of nodes for adjacent values of r is then of the order of

$$\Delta n = \frac{\Delta y}{\pi}\frac{t}{\hbar}[H_0'(r + 1) - H_0'(r)] \tag{21b}$$

It is clear that if $[H_0'(r + 1) - H_0'(r)]t/\hbar$ is made large enough, Δn can be made as large as we please. In particular, Δn can be made so large that adjacent quantum states r of the system S lead to classically distinguishable states of the apparatus.

It is of interest to note that when a good measurement has been made, the apparatus wave functions corresponding to different values of r will

* See Chap. 11, Sec. 12. The quantum number of a complex function is equal to the number of nodes in the real part.

be approximately orthogonal. To show this, we consider the integral

$$\int f_0^*(y) e^{iH_I(r,y)t/\hbar} f_0(y) e^{-iH_I(r-1,y)t/\hbar}\, dy \tag{22a}$$

which is the integral involved in testing for the orthogonality of apparatus wave functions corresponding to adjacent quantum states in the system S. The above integral is approximately equal to

$$\int dy\, f_0^*(y) f_0(y) e^{i\frac{yt}{\hbar}[H_0'(r) - H_0'(r-1)]}\, e^{-\frac{it}{\hbar}[H_0(r) - H_0(r-1)]} \tag{22b}$$

Now $f_0^*(y)f_0(y)$ is a function which resembles a wave packet. The orthogonality integral is just the Fourier component of $|f_0(y)|^2$ corresponding to $k = [H_0'(r) - H_0'(r-1)]t/\hbar$. Now this Fourier component will be large only in a limited region, $\Delta k \cong 1/\Delta y$. But according to eq. (21b), whenever one has a good measurement, one obtains

$$\Delta n = [H_0'(r) - H_0'(r-1)] \frac{t}{\hbar} \Delta y \gg 1 \tag{22c}$$

We conclude that if the interaction is strong enough to provide a good measurement, then $k \gg \Delta k$, so that the Fourier component of $|f_0(y)|^2$ corresponding to k will be very small. This means that the apparatus wave functions corresponding to adjacent values of r are very nearly orthogonal.†

As an example, we consider the spin wave function. In Sec. 6, it was shown that when a good measurement of the spin has been made, the wave packet corresponding to $s = +1$ is so far from the one corresponding to $s = -1$ that the two do not overlap to any appreciable extent. As a result, the integrand in the integral $\int f_+^*(z) f_-(z)\, dz$ is very small everywhere, so that the integral is very small.

8. Destruction of Interference in the Process of Measurement. We now come to a crucial problem that arises in the demonstration of the logical self-consistency of the quantum theory of measurements, namely, the destruction of interference that takes place in the course of a measurement. In Chap. 6, Sec. 3 and Chap. 10, Sec. 36, we stated that whenever a measurement of any variable is carried out, the interaction between the system under observation and the observing apparatus always multiplies each part of the wave function corresponding to a definite value of A by a random phase factor, $e^{i\alpha_a}$. Thus, if the wave function is $\sum_a c_a \psi_a(x)$ before the measurement, it is changed into $\Sigma c_a\, e^{i\alpha_a} \psi_a(x)$. The random phase factors cause interference between difference $\psi_a(x)$ to be destroyed. As shown in Chap. 6, Sec. 4, if interference were not destroyed under these circumstances, the quantum theory could be shown to lead to

† If adjacent wave functions are nearly orthogonal, then wave functions having very different values of r will clearly be even more nearly orthogonal.

absurd results. The proof that interference is, in fact, destroyed is therefore essential for the consistency of the theory.

In treating this problem, we shall restrict ourselves to the special case of the measurement of the z component of the spin, but the method of extension to a general case is fairly straightforward. We begin by noting that after the measurement has been carried out, the spin wave function and the apparatus wave function (i.e., the co-ordinate of the particle) are very closely correlated in a manner shown in eq. (15). Very often, however, one may wish to know the average of some function of the spin, without specifying the state of the apparatus. To obtain this quantity one must average over all possible states of the apparatus variable. To obtain the average of an arbtrary function of the spin, $g(\boldsymbol{\sigma})$, we therefore write

$$\overline{g(\boldsymbol{\sigma})} = \int [f_+^*(z)v_+^* + f_-^*(z)v_-^*]g(\boldsymbol{\sigma})[f_+(z)v_+ + f_-(z)v_-]\,dz \tag{23a}$$

where f_+ and f_- are defined in eq. (15). This is equal to

$$\overline{g(\boldsymbol{\sigma})} = \int |f_+(z)|^2 v_+^*\,g(\boldsymbol{\sigma})v_+\,dz + \int |f_-(z)|^2 v_-^* g(\boldsymbol{\sigma})v_-\,dz \\ + \int f_+^*(z)f_-(z)v_+^* g(\boldsymbol{\sigma})v_-\,dz + \int f_-^*(z)f_+(z)v_-^* g(\boldsymbol{\sigma})v_+\,dz \tag{23b}$$

Now, $\int |f_+(z)|^2\,dz$ is just the total probability that the particle is in the wave packet corresponding to spin $\hbar/2$ while $\int_{-\infty}^{\infty} |f_-(r)|^2\,dz$ is the probability that it is in the packet corresponding to a spin of $-\hbar/2$. The sum of the first two terms is then just

$$P_+\overline{g_+(\boldsymbol{\sigma})} + P_-\overline{g_-(\boldsymbol{\sigma})} \tag{23c}$$

where P_+ and P_- are respectively the probabilities that the spin is $\hbar/2$ and $-\hbar/2$, whereas $\overline{g_+(\boldsymbol{\sigma})}$ and $\overline{g_-(\boldsymbol{\sigma})}$ are respectively the mean values of $g(\boldsymbol{\sigma})$ when the spin is $\hbar/2$ and $-\hbar/2$. This expression may be called the "classical" contribution to the average, because it is just the value that would be obtained in a classical system for which the probabilities of positive and negative spin were respectively P_+ and P_-.

The third and fourth terms in eq. (23b) are the characteristic interference terms of quantum theory. As shown in Sec. 6, whenever a good measurement has been made, the separation between the centers of the packets, $f_+(z)$ and $f_-(z)$, is much greater than the width of these packets. This means that the product $f_+(z)f_-(z)$ is always very small, so that practically the entire contribution to $g(\boldsymbol{\sigma})$ comes from the "classical" terms in eq. (23b). As far as the average of any function of the spin is concerned, the characteristic quantum-mechanical interference terms between v_+ and v_- will no longer be present after a measurement which is good enough to define the value of the z component of the spin has taken place.

9. The Appearance of Random Phase Factors. Another way of representing the destruction of interference is with the aid of the concept of random phase factors, already described in Sec. 8 of this chapter, and in Chap. 6, Sec. 3. In order to obtain this formulation, we note that in eq. (15b) v_+ and v_- are multiplied by factors which depend on z. Now in the region, Δz, in which the wave function is appreciable, each of these factors varies by as much as $e^{i\alpha}$, where $\alpha \cong \frac{\mu \mathcal{H}_0' \,\Delta t}{\hbar} \Delta z$. But from eq. (19), we see that whenever $\mathcal{H}_0' \,\Delta t$ is large enough to provide a good measurement, $\mu \mathcal{H}_0' \,\Delta t / \Delta p \gg 1$. Since for a classical measurement, $\Delta z \gg \hbar / \Delta p$, we conclude that

$$\frac{\mu \mathcal{H}_0' \,\Delta t \,\Delta z}{\hbar} \gg \frac{\mu \mathcal{H}_0' \,\Delta t}{\Delta p} \gg 1 \tag{24}$$

The phase of each wave function therefore varies by a number much greater than 2π, in the region in which $f_0(z)$ is appreciable. We now note that *as far as the problem of evaluating averages of functions of the spin alone is concerned*, we can regard z, the apparatus co-ordinate, as a parameter on which the coefficients of the spin wave function depend. From one measurement to the next, the classically describable position of the apparatus co-ordinate will fluctuate over the whole range of values accessible to it. We conclude, therefore, that the phase factors, $e^{i\alpha_+}$ and $e^{i\alpha_-}$, respectively multiplying v_+ and v_-, will fluctuate at random over all possible values. Furthermore, the ratio of these phase factors is

$$\frac{c_+}{c_-} e^{-2i\mu \mathcal{H}_0 \,\Delta t/\hbar} \, e^{-2i\mu \mathcal{H}_0' \,\Delta t z/\hbar} \tag{25}$$

The phase factors $e^{i\alpha_+}$ and $e^{i\alpha_-}$ are therefore completely uncorrelated to each other (and also to the value of the spin). In this way, we justify the formulation of the destruction of interference given in Chap. 6, Sec. 3 and Chap. 10, Sec. 36.

A few numbers at this point will perhaps bring out more sharply how large a phase shift will occur in a typical experiment. Suppose that we consider a characteristic magnetic field of 1000 gauss, a magnetic field gradient of 10,000 gauss/cm, and a magnet of length 10 cm. A typical velocity for the atoms is 10^4 cm/sec. Then after the time $\Delta t = l/v = 10^{-3}$ sec has elapsed, we obtain for the ratio of phase factors from eq. (25) (using $c_+/c_- \cong 1$ and $e/mc \cong 10^7$)

$$e^{-10^{5} i} \, e^{-10^{6} iz}$$

We see then that the phase does indeed change by a very large number.

10. Interpretation of Combined Wave Function in Terms of a Statistical Ensemble of Wave Functions for the Spin Alone. After the spin has interacted with the apparatus that measures its value, there is clearly no single wave function belonging to the spin alone, but, instead, there is only a combined wave function in which spin and apparatus co-ordinates are inextricably bound up. Nevertheless, there is a procedure by means of which one can correctly interpret the wave function for the combined system in terms of a statistical ensemble of wave functions for the spin alone.*

To obtain this interpretation, we make use of the result of Sec. 8, that after σ_z has been measured, the interference terms between the spin wave functions v_+ and v_- can no longer contribute to the average of any function of the spin. From this result we conclude that in obtaining averages of functions of the spin alone, one can ignore the apparatus co-ordinates and assume instead that the spin wave function is either entirely v_+ or entirely v_-, but that the probabilities that each of these functions is actually the correct one are, respectively, $|a_+|^2$ and $|a_-|^2$, where the spin wave function before the measurement took place was $a_+v_+ + a_-v_-$. Thus, we have replaced the actual wave function for the combined system by a statistical ensemble of separate wave functions representing situations in which the spin wave function alone is either v_+ or v_-.

It should be noted, however, that the statistical ensemble of wave functions of the spin alone is an idealization which gives correct averages of functions of $\boldsymbol{\sigma}$ only when interaction between spin and observing apparatus has prepared the combined wave function by destroying interference between different eigenfunctions of the spin. If, for example, the product $\mathcal{H}_0' \Delta t$ occurring in eq. (24), were too small to provide a good measurement, the wave packets corresponding to positive and negative spin would overlap, so that the products $f_+(z)f_-(z)$ would not vanish, and interference terms would therefore be able to contribute to averages of functions of the spin. It would then no longer be correct to obtain such averages by assuming that the wave function was either entirely v_+ or entirely v_-.

The procedure of replacing the wave function of the combined system by a statistical ensemble of wave functions of the spin alone now makes it possible for us to interpret the experiment in the customary way as something that yields a single definite result out of all of the various logically alternative possible results. Thus, we say that after the apparatus has functioned, but before any observer has found out what the results of its functioning are, the system has the same average of an arbitrary function of the spin as it would have if it occupied some single one of the two spin states, with the appropriate probability, $|a_+|^2$ or $|a_-|^2$. When the observer looks at the apparatus, he then discovers in which

* This procedure is essentially the same as that given in Chap. 6, Sec. 4, but we shall now give another treatment here which repeats part of the earlier treatment, but is somewhat more general.

state the system actually is, by finding out in which of the two possible classically distinguishable states the observing apparatus is. At this point, he finds it appropriate to replace the statistical ensemble of wave functions by the single wave function corresponding to the actually observed value of the spin. The sudden replacement of the statistical ensemble of wave functions by a single wave function represents absolutely no change in the state of the spin, but is analogous to the sudden changes in classical probability functions which accompany an improvement of the observer's information (see Chap. 6, Sec. 4). The reason that such a sudden change of wave function has no physical significance is that the different members of the statistical ensemble of wave functions cannot interfere with each other. (If such a sudden change of wave function occurred while definite phase relations still existed, then, as shown in Chap. 6, Sec. 4, the quantum theory would make no sense at all.)

The statistical ensemble of wave functions of the spin alone, which replaces the combined spin and apparatus wave function is sometimes said to define a "mixed state" of the spin in contradistinction to a "pure state," in which the spin wave function is definite. This terminology is somewhat misleading because the term "quantum state" has already come to represent a situation in which many different parts of the wave functions all interfere in a definite way such that some aspects of the system are mutual or "interference" properties of the various component parts. This means that if one wishes to understand *all* of the properties of such a system, one cannot regard it as analyzable into more detailed "sub-states." On the other hand, with a statistical ensemble of wave functions, no such interference between different component wave functions can occur. Furthermore, as soon as the observer looks at the apparatus the spin goes from a "mixed" state to a "pure" state. It seems unwise to adopt a terminology that suggests that the spin changes its state (from mixed to pure) under circumstances in which nothing changes except the observer's information about the spin. The phrase "statistical ensemble of states" provides a more accurate description.

11. Inclusion of Apparatus Co-ordinates. The preceding discussion shows that in the evaluation of any function of the spin alone, there is no interference between different eigenfunctions of σ_z after the electron has interacted with an apparatus that measures the z component of its spin. But it is by no means obvious that the same conclusion holds for an arbitrary function of the spin and the apparatus co-ordinate together $f(z, s)$. In fact, the combined wave function of spin and apparatus is a pure wave function, so that one might, at first sight, expect that interference might be important in evaluating averages of functions like $f(z, s)$.

Consider, for example, a Stern-Gerlach experiment. One way of demonstrating interference effects between the two beams for the combined system would be to have some arrangement of magnetic fields that brings the two packets together after they have been separated. A schematic diagram of such an arrangement is shown in Fig. 3. If the uniform magnetic fields shown in the diagram are set up in exactly the right way, and if the second inhomogeneous field is an exact duplicate of the first one, the two wave packets can be brought together into a single coherent packet. Although the precision required to achieve this result would be fantastic, it is, in principle, attainable. In this way, by using an apparatus which acts in a way that depends simultaneously on both z and σ, we would be able to take advantage of the interference existing between the two packets. Whether the final beam had a definite spin in the x, y, or z direction would depend entirely on the relative phases with which the two beams were brought together. (For example, if the spin wave function became $(v_+ + v_-)/\sqrt{2}$, the resulting spin would be definite in the $+x$ direction.*)

If it were possible to use the apparatus in such a way that one could simultaneously measure the value of the spin in the z direction and allow the beams to come together again with coherent interference, an absurd result would follow. This is because each time the z component of the spin was measured some definite result would be obtained, i.e., either $\hbar/2$

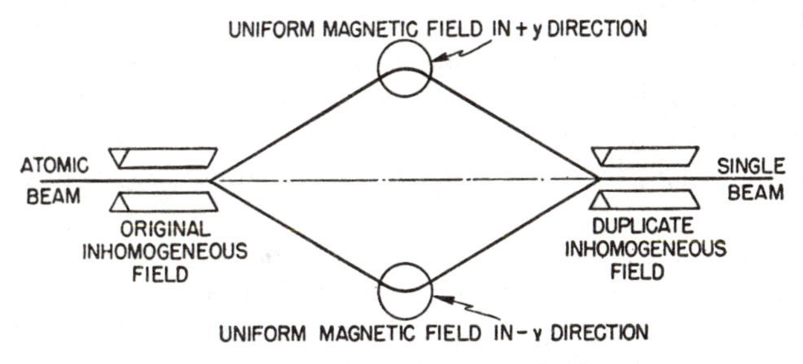

FIG. 3

or $-\hbar/2$. If, for example, $\hbar/2$ were obtained, one would immediately conclude that this particular atom was in the upper beam, so that the wave function for the lower beam would thereafter, have to be zero. In the next experiment, a definite result of $-\hbar/2$ might be obtained, leading us to a zero value for the wave function in the upper beam. In all cases, if there were any interference between the spin wave functions, v_+ and v_-, after the apparatus had functioned in such a way as to permit a measurement of the spin, this interference would therefore be destroyed at the moment that the observer became aware of the results of the functioning of his apparatus. But many aspects of the actual behavior of matter depend on the interference properties of various parts of the wave function. Thus, if interference between v_+ and v_- were not destroyed by the actions of the observing apparatus, no objective description of the world would be possible at all, since so much of the behavior of matter would then depend on whether or not observers were aware of what the electron had been doing. (In this connection see Chap. 6, Sec. 4.)

To see that interference is actually destroyed in the case that we have studied in detail here, i.e., that of the measurement of the z component

* See eq. (25), Chap. 17.

of the spin, one has only to note that, in order to measure this quantity, one must measure the co-ordinate of the atom before the two beams have been brought together again. But it is readily shown with the aid of the uncertainty principle (see Chap. 6, Sec. 2, for example) that the disturbance resulting from this measurement will actually destroy interference. Thus, we conclude that interference in the combined wave function of the spin and observing apparatus is, in principle, possible only when the apparatus itself is not observed by means of some other apparatus.

We may include the general ideas just developed in the notion of mixed wave functions by saying that after the z co-ordinate of the atom is observed, the combined wave function now includes the co-ordinates of three systems, viz., atomic spin, co-ordinate of the atom, and the co-ordinates of the device which measures the z co-ordinate of the atom. After the second measuring apparatus has functioned, the system consisting of spin and the z co-ordinate of the atom can be idealized in terms of an ensemble of wave functions because interference between different values of both σ and z has been destroyed. Thus, even functions like $f(z, \boldsymbol{\sigma})$ can now be evaluated in terms of the assumption that σ_z is either $+1$ or -1.

It may now be argued that one has only pushed the difficulties back another stage, because the three-fold system still has a pure wave function and can therefore, in principle, show interference effects. We may say that when the third system is observed, either by still another type of apparatus, or else by a human observer, then once again one obtains ensembles of wave functions for the three-fold system, but the combined wave function of the larger system, including the human observer if he has interacted with the system at any point, is still a pure wave function.

Can this difficulty ever finally be overcome, or is it necessary to assume that the analysis will always be incomplete? We shall see that this problem can be solved without carrying the analysis as far as the stage in which the apparatus interacts with a human observer. To do this, let us suppose that the experiment is set up in such a way that the apparatus functions completely automatically, so that the results of the experiment are recorded on some convenient device, such as a photographic plate. The entire system, consisting of spin, z co-ordinate of the atom, apparatus which measures the z co-ordinate, and apparatus which records the results of this measurement, is assumed to have some pure wave function when the experiment starts. (It is not necessary that any human observer know exactly what this wave function is.) After the interaction is over, the combined wave function will go over into some other pure wave function. We wish to show here that although the final wave function is indeed a pure one, the phase relations between parts of the wave function corresponding to different values of σ_z are so complicated that it is unlikely to the highest degree that any physical process, either now or in the future, will depend appreciably on such interference. Thus, if the system is allowed to function by itself without the intervention of any human observers, it goes over into a state in which, with overwhelming probability, physical results will be the same as if the spin were in one of a statistical ensemble of states. When a human observer interacts with the apparatus, the system is put into a genuine statistical ensemble of states. The destruction of definite phase relations involved in this process will, however, make no significant difference in the behavior of the system, because the effects of interference between wave functions corresponding to different values of the observable are already negligible. Thus, we are able to obtain a completely objective description of the process of measurement, which does not involve human observers in any way at all.*

We shall now demonstrate in terms of the Stern-Gerlach experiment that interference is effectively destroyed after the apparatus has functioned in such a way as to make a measurement possible. According to eqs. (18) and (19), a good measurement requires that the product of the strength of the interaction $\mathcal{H}_0$ and the time of interaction Δt be so great that the two beams obtain a classically distinguishable separation between them. As shown in Sec. 9, under these circumstances, the relative phase shift of the wave functions multiplying v_+ and v_- will indeed be very large. Although it is, in principle, possible by means of the apparatus shown in Fig. 3 to bring the two beams together, they will come together with relative phases that depend very sensitively on exactly how the apparatus is constructed. The slightest error or lack of reproducibility in the functioning of the apparatus would change the relative phase by a great deal, and thus change the resulting spin direction of the atom after the beams have come together. As soon as the position of the particle is observed by means of some other apparatus, then these phase relations would depend on the state of this additional apparatus. Now, before a piece of apparatus can be suitable to be used to make an observation, it is necessary that it produce in the last stage results that

* The treatment given in this chapter demonstrates the objectivity of the process of measurement with the aid of what is essentially a space-time description, carried out in terms of the wave function (see Chap. 8, Secs. 14 and 15). The same result was obtained in Sec. 3, however, with the aid of a causal description, i.e., a description in terms of the uncontrollable and indivisible quantum transfers from the observer to the measuring apparatus. In this case, it was pointed out that the classically describable stages of the measuring apparatus could adequately be regarded as having a separate existence, because the uncontrollable quantum transfers were too small to be significant. In the wave-function description, we are led to the same conclusion with the aid of the idea that the phase shifts occurring when a human observer looks at the classically describable part of the apparatus likewise produce no significant changes. Finally, we note that the causal factors (i.e., the momenta) always appear in the space-time description in terms of phases of the wave function. Thus, the two methods yield complementary descriptions of the same process.

are of macroscopic order of magnitude. An object which is so large has many complex degrees of freedom and it is, in practice, impossible to have motion in any of these degrees without coupling in all of the other degrees. Such coupling phenomena occur, among other places, in friction, where the mass motion of the axle of an ammeter needle, for example, may excite very complicated internal thermal motions of the molecules of the shaft and the bearings. It can safely be said that no macroscopic objects exist which do not do this to some degree. Every one of these new degrees of freedom would create new complicated phase shifts in the combined wave function for the entire system. Because the system is operating at a classical level, the phase shifts would all be large (as they were in the first stages of the Stern-Gerlach apparatus) and would depend very sensitively on exactly what all of these co-ordinates were doing.

In order to bring the beams into interference, it would then be necessary to adjust carefully the contributions of each of these degrees of freedom to the phase. Eventually the requirements for a definite interference pattern would become so complicated and so difficult to control that it is overwhelmingly improbable that such interference will ever be important in any physical process, either by accident or by design. When this stage is reached, we can say that to all intents and purposes, the system acts as if all interference between eigenfunctions of the measured variable has been destroyed.

12. Irreversibility of Process of Measurement and Its Fundamental Role in Quantum Theory. From the previous work it follows that a measurement process is irreversible in the sense that, after it has occurred, re-establishment of definite phase relations between the eigenfunctions of the measured variable is overwhelmingly unlikely. This irreversibility greatly resembles that which appears in thermodynamic processes, where a decrease of entropy is also an overwhelmingly unlikely possibility.*

* There is, in fact, a close connection between entropy and the process of measurement. See L. Szilard, *Zeits. f. Physik*, **53**, 840, 1929. The necessity for such a connection can be seen by considering a box divided by a partition into two equal parts, containing an equal number of gas molecules in each part. Suppose that in this box is placed a device that can provide a rough measurement of the position of each atom as it approaches the partition. This device is coupled automatically to a gate in the partition in such a way that the gate will be opened if a molecule approaches the gate from the right, but closed if it approaches from the left. Thus, in time, all the molecules can be made to accumulate on the left-hand side. In this way, the entropy of the gas decreases. If there were no compensating increase of entropy of the mechanism, then the second law of thermodynamics would be violated. We have seen, however, that in practice, every process which can provide a definite measurement disclosing in which side of the box the molecule actually is, must also be attended by irreversible changes in the measuring apparatus. In fact, it can be shown that these changes must be at least large enough to compensate for the decrease in entropy of the gas. Thus, the second law of thermodynamics cannot actually be violated in this way. This means, of course, that Maxwell's famous "sorting demon" cannot operate, if he is made of matter obeying all of the laws of physics. (See L. Brillouin, *American Scientist*, **38**, 594, 1950.)

Because the irreversible behavior of the measuring apparatus is essential for the destruction of definite phase relations and because, in turn, the destruction of definite phase relations is essential for the consistency of the quantum theory as a whole, it follows that thermodynamic irreversibility enters into the quantum theory in an integral way. This is in remarkable contrast to classical theory, where the concept of thermodynamic irreversibility plays no fundamental role in the basic sciences of mechanics and electrodynamics. Thus, whereas in classical theory fundamental variables (such as position or momentum of an elementary particle) are regarded as having definite values independently of whether the measuring apparatus is reversible or not, in quantum theory we find that such a quantity can take on a well defined value only when the system is coupled indivisibly to a classically describable system undergoing irreversible processes. The very definition of the state of any one system at the microscopic level therefore requires that matter in the large shall undergo irreversible processes. There is a strong analogy here to the behavior of biological systems, where, likewise, the very existence of the fundamental elements (for example, the cells) depends on the maintenance of irreversible processes involving the oxidation of food throughout an organism as a whole.* (A stoppage of these procesess would result in the dissolution of the cell.)

13. Wave vs. Particle Properties of Matter as Potentialities. In Chap. 6, Sec. 13 and Chap. 8, Sec. 24, it was shown that matter behaves like something that has properties that depend in part on the indivisible quantum links with its surroundings. The question of whether a given object, such as an electron, acts more like a wave or more like a particle is therefore not determined entirely by the electron itself but depends partly on the environment of the electron.

We have seen, for example, that when an electron interacts with an apparatus that measures its position, it produces in the apparatus a classically definable state that is equivalent to what would be produced by a classical particle localized in a small region. On the other hand, when it interacts with an apparatus (such as a grating) that measures its momentum, it comes off with a classically definable angle in much the same way as it would have done if it had been a classical wave. Thus, the electron may be regarded as an entity that has potentialities for developing either its particlelike or its wavelike aspects, depending on the type of matter with which it interacts.

Now, a quantum-mechanical system can produce classically describable effects, not only in measuring apparatus, but also in all kinds of

* In this connection, see the last paragraph in Chap. 23 and the footnote connected with it. Compare also the general concept of the relation between small-scale and large-scale properties of matter developed throughout Chap. 23, with the ideas suggested here.

systems that are not actually being used for the purpose of making measurements.* Thus, under all circumstances, we picture the electron as something that is itself not very definite in nature but that is continually producing effects which, whether they are actually observed by any human observers or not, call for the interpretation that the electron has a nature that varies in response to the environment. Only in so far as it is capable of producing classically describable results can we say that it has any definable model at all, and since the types of results that can be produced are so different, we need at different times the complementary models of wave and particle.

14. On the Relation between Continuity and Discontinuity in Quantum Transfers. We are now in a position to provide a qualitative picture for one of the most puzzling features of quantum processes; viz., the transition of a system from one discrete energy level to another. In such a transition, the energy changes discontinuously, and yet, the wave function moves continuously from the region of space associated with the one orbit into the region of space associated with the other orbit. To understand such a duality of properties, we refer again to Chap. 6, Secs. 9 and 13, where it is shown that at the quantum level of accuracy, the properties of a given object do not exist separately in that object alone, but are potentialities, which are brought out in a way that depends on the systems with which the objects interact. In particular, the energy of an electron and its position are opposing potentialities, each of which can be developed into a definite value only at the expense of the definiteness of the other. Suppose then that the electron starts out with a definite energy E. Its position is indefinite and spreads over the whole wave packet, but it has the latent possibility of developing a more definite position. In fact, if it interacts with an incident light quantum, then, for a short time, it realizes its potentiality for obtaining a more definite position, while simultaneously it spreads over a range of energies. During this time, it moves outward to another region of space, associated with another orbit of higher energy. Meanwhile it begins to lose its definite position and to realize again its potentialties for developing a definite energy. This will happen as the phase relations between states of definite energy are destroyed, so that the system ultimately acts to all intents and purposes as if it were in some single one of the energy states. Which of these states it will go into is not, in general, completely predictable from the state of the system before interaction, although the general statistical trend of the transition is predictable.†

* See Chap. 6, Sec. 10.

† The transition process described above is treated mathematically in Chap. 18, Secs. 1 to 10. The reader is referred particularly to Sec. 9, where it is pointed out that while the transition is taking place, the system is not in a definite (but unknown) eigenstate of the unperturbed Hamiltonian, H_0, but instead, covers a range of these states simultaneously. During this time, however, there is a continuously changing

We conclude that throughout the process of transition, the potentialities associated with the electron change in a continuous way, but the forms (i.e., the definite eigenvalues of the energy) in which these potentialities can be realized are discrete. The description of a quantum process as a discontinuous transition is therefore partly a consequence of the inadequacy of our customary language, which does not make clear the fact that the properties of the electron are always in part potential and incompletely defined. Thus, when we say that an electron has a definite energy *at a given time*, the customary usage of the language implies that an electron is *at all times* an object that has a definite energy, so that a continuous transition could only consist of a gradual change in the value of this energy. Yet, because the electron has the latent possibility of being transformed into something having a more definite position and a less definite energy, the transition is still, in a sense, continuous, even though the electron does not pass through the intermediate energies.*

The continuously changing potentialities and the discontinuous forms in which these potentialities may be realized are, in fact, opposing, but complementary, properties of the electron, each of which expresses an equally important aspect of the electron's behavior. (See Chap. 8, Sec. 15, for a discussion of the principle of complementarity.)

15. The Paradox of Einstein, Rosen, and Podolsky. In an article in the Physical Review,† Einstein, Rosen, and Podolsky raise a serious criticism of the validity of the generally accepted interpretation of quantum theory. This objection is raised in the form of a paradox to which they are led on the basis of their analysis of a certain hypothetical experiment, which we shall discuss in detail later. Their criticism has, in fact, been shown to be unjustified,‡ and based on assumptions concerning the nature of matter which implicitly contradict the quantum theory at the outset. Nevertheless, these implicit assumptions seem, at first sight, so natural and inevitable that a careful study of the points which the authors raised affords deep and penetrating insight into the difference between classical and quantum concepts of the nature of matter.

The authors first undertook to define criteria for a complete physical

---

probability for the development of any particular eigenstate of H_0 in a process of interaction with a system that can provide a measurement of H_0.

* See Chap. 8, Sec. 27, where analogies with indivisible transitions are discussed in connection with thought processes. Here we were led to describe certain aspects of thought processes in terms of discontinuous changes, because the definite logical forms to which such processes can lead are discretely different (for example, each logical category is conceived of as completely separate from all others). On the other hand, the intermediate stages of the thought process connecting these logically expressed definite concepts are continuous, and to some extent, resemble the incompletely defined potentialities of quantum theory.

† *Phys. Rev.*, **47,** 777 (1935).

‡ N. Bohr, *Phys. Rev.* **48,** 696 (1935); W. H. Furry, *Phys. Rev.* **49,** 393, 476 (1936).

theory. It seemed to them that a necessary requirement for a *complete* physical theory was the following:

(1) Every element of physical reality must have a counterpart in a *complete* physical theory.

As to what actually constituted the correct elements in terms of which physical theory should be expressed, they felt that this question can be decided finally only by recourse to experiments and observations. They nevertheless suggested the following criterion for recognizing an element of reality, which seemed to them a *sufficient* criterion:

(2) If, without in any way disturbing the system, we can predict with certainty (i.e., with probability equal to unity) the value of a physical quantity, then there exists an element of reality corresponding to this physical quantity.

The authors agreed that elements of physical reality might well be recognized in other ways also, but they intended to show that even if one restricted oneself to elements that could be recognized by means of *this* criterion alone, quantum theory as now interpreted led to contradictory results.

The use of the above explicit criteria rests, however, on certain implicit assumptions, which are an integral part of the treatment given by the authors, but which are never explicitly stated. These assumptions are:

(3) The world can correctly be analyzed in terms of distinct and separately existing "elements of reality,"

(4) Every one of these elements must be a counterpart of a *precisely* defined mathematical quantity appearing in a *complete* theory.*

We shall temporarily accept the above criteria and assumptions, in order to permit the further development of the arguments given by the authors, but in Sec. 18 we shall show that these criteria should not be applied at the quantum level of accuracy.

Now, let us recall that in the present quantum theory, one assumes that all relevant physical information about a system is contained in its wave function, so that when two systems have wave functions which differ by at most a constant phase factor, they are said to be in the same quantum state.† What the authors wished to do with their criteria for reality was to show that the above interpretation of the present quantum theory is untenable and that the wave function cannot possibly contain a complete description of all physically significant factors (or "elements of reality") existing within a system. If their contention could be proved, then one would be led to search for a more complete theory,

* This criterion is essentially a strengthened form of (1). Einstein, Rosen, and Podolsky do not restrict themselves to the assumption (1), that every element of reality always has a counterpart in a complete theory, but they also assume implicitly that this counterpart must always be precisely definable.

† See Chap. 9, Sec. 4.

perhaps containing something like hidden variables,* in terms of which the present quantum theory would be a limiting case.

Let us now consider an arbitrary observable A having a set of eigenfunctions, ψ_a, belonging to a series of eigenvalues which are denoted by a. When the wave function is ψ_a, then the system is said to be in a quantum state in which the observable A has the definite value a. In this situation, ERP would say that there is in the system an element of reality corresponding to the observable, A. But now let us consider another observable B which does not commute with A, so that there exists no wave function for which A and B have simultaneously definite values. Now if we adopt the implicit assumption (4) that every element of reality must be a counterpart of a *precisely* defined mathematical quantity appearing in a *complete* theory, then the usual assumption that the wave function provides a *complete* description of reality leads to the conclusion that A and B cannot exist simultaneously.† This follows from the fact that the supposedly complete wave theory contains no *precisely* defined mathematical elements corresponding to the simultaneous existence of A and B. From this point of view, we must also assume, however, that when B is measured and obtains a definite value, the elements corresponding to A are destroyed (since we have assumed that they cannot exist together with those corresponding to B). It seems natural to suppose that this destruction is brought about by the quanta that are transferred from the measuring apparatus to the system under observation. It is clear, however, that in such an interpretation of the noncommutativity of two observables, it is essential that in every measurement there shall actually be a disturbance arising from the apparatus that destroys all elements of reality corresponding to observables that do not commute with the measured variable. For if there were no such disturbance, then one could take a system initially having a definite value of A and then measure B without in any way altering the elements corresponding to A, thus obtaining a system in which the elements of reality corresponding to A and B exist together at the same time. Now, in the next section, we shall discuss a type of hypothetical experiment suggested by ERP that actually permits us to measure a given observable without in any way disturbing the associated system. With the aid of this type of hypothetical experiment, they are then able to obtain a contradiction between the assumption that the quantum theory provides a complete description of reality and the assumption that their criteria for reality must necessarily apply in any complete theory. If one accepts their

* Chap. 2, Sec. 5; Chap. 5, Sec. 3.

† In Sec. 18, we shall make the alternative assumption that elements of reality exist in a roughly defined form and do not necessarily have to be counterparts of precisely defined mathematical quantities appearing in a complete theory. Thus, we shall give up the implicit assumptions (3) and (4).

criteria, one is left with a single remaining alternative, viz., that quantum theory does not provide a complete description of reality. This is the conclusion that they originally set out to obtain.

16. The Hypothetical Experiment of Einstein, Rosen, and Podolsky. We shall now describe the hypothetical experiment of Einstein, Rosen, and Podolsky. We have modified the experiment somewhat, but the form is conceptually equivalent to that suggested by them, and considerably easier to treat mathematically.

Suppose that we have a molecule containing two atoms in a state in which the total spin is zero and that the spin of each atom is $\hbar/2$. Roughly speaking, this means that the spin of each particle points in a direction exactly opposite to that of the other, insofar as the spin may be said to have any definite direction at all. Now suppose that the molecule is disintegrated by some process that does not change the total angular momentum. The two atoms will begin to separate and will soon cease to interact appreciably. Their combined spin angular momentum, however, remains equal to zero, because by hypothesis, no torques have acted on the system.

Now, if the spin were a classical angular momentum variable, the interpretation of this process would be as follows: While the two atoms were together in the form of a molecule, each component of the angular momentum of each atom would have a definite value that was always opposite to that of the other, thus making the total angular momentum equal to zero. When the atoms separated, each atom would continue to have every component of its spin angular momentum opposite to that of the other. The two spin-angular-momentum vectors would therefore be correlated. These correlations were originally produced when the atoms interacted in such a way as to form a molecule of zero total spin, but after the atoms separate, the correlations are maintained by the deterministic equations of motion of each spin vector separately, which bring about conservation of each component of the separate spin-angular-momentum vectors.

Suppose now that one measures the spin angular momentum of any one of the particles, say No. 1. Because of the existence of correlations, one can immediately conclude that the angular-momentum vector of the other particle (No. 2) is equal and opposite to that of No. 1. In this way, one can measure the angular momentum of particle No. 2 indirectly by measuring the corresponding vector of particle No. 1.

Let us now consider how this experiment is to be described in the quantum theory. Here, the investigator can measure either the x, y, or z component of the spin of particle No. 1, but not more than one of these components, in any one experiment. Nevertheless, it still turns out as we shall see that whichever component is measured, the results are correlated, so that if the same component of the spin of atom No. 2 is measured, it will always turn out to have the opposite value. This means that a measurement of any component of the spin of atom No. 1 provides, as in classical theory, an indirect measurement of the same component of the spin of atom No. 2. Since, by hypothesis, the two particles no longer interact, we have obtained a way of measuring an arbitrary component of the spin of particle No. 2 without in any way disturbing that particle. If we accept the definition of an element of reality (2) suggested by ERP, it is clear that after we have measured σ_z for particle 1, then σ_z for particle 2 must be regarded as an element of reality, existing separately in particle No. 2 alone. If this is true, however, this element of reality must have existed in particle No. 2 even before the measurement of σ_z for particle No. 1 took place. For since there is no interaction with particle No. 2, the process of measurement cannot have affected this particle in any way. But now let us remember that, in each case, the observer is always free to reorient the apparatus in an arbitrary direction while the atoms are still in flight, and thus to obtain a definite (but unpredictable) value of the spin component in any direction that he chooses. Since this can be accomplished without in any way disturbing the second atom, we conclude that if criterion (2) of ERP is applicable, precisely defined elements of reality must exist in the second atom, corresponding to the simultaneous definition of all three components of its spin. Because the wave function can specify, at most, only one of these components at a time with complete precision, we are then led to the conclusion that the wave function does not provide a complete description of all elements of reality existing in the second atom.

If this conclusion were valid, then we should have to look for a new theory in terms of which a more nearly complete description was possible. We shall see, however, in Sec. 18, that the analysis given by ERP involves in an integral way the implicit assumptions (3) and (4) that the world is actually made up of separately existing and precisely defined "elements of reality." Quantum theory, however, implies a quite different picture of the structure of the world at the microscopic level. This picture leads, as we shall see, to a perfectly rational interpretation of the hypothetical experiment of ERP within the present framework of the theory.

17. Mathematical Analysis of Experiment According to Quantum Theory. Before discussing the physical interpretation that the present quantum theory gives to the hypothetical experiment of Einstein, Rosen and Podolsky, we shall first show how this experiment is to be described in mathematical terms.

The system containing the spin of two atoms has four basic wave functions, from which an arbitrary wave function can be constructed.*

* The complete wave function for the system is then obtained by multiplying the spin wave functions by appropriate space wave functions, which depend on the space co-ordinates of both particles.

These are

$$\psi_a = u_+(1)u_+(2) \qquad \psi_c = u_+(1)u_-(2)$$
$$\psi_b = u_-(1)u_-(2) \qquad \psi_d = u_-(1)u_+(2)$$

where u_+ and u_- are the one-particle spin wave functions representing, respectively, a spin $\hbar/2$ and $-\hbar/2$, and the argument (1) or (2) refers, respectively, to the particle which has this spin. Now ψ_c and ψ_d represent the two possible situations in which each particle has a definite z component of the spin in a direction which is opposite to that of the other. The wave function for a system of total spin zero is the following linear combination of ψ_c and ψ_d (see Chap. 17, Sec. 9):

$$\psi_0 = \frac{1}{\sqrt{2}}(\psi_c - \psi_d) \tag{26}$$

The particular sign with which ψ_c and ψ_d are combined is of crucial importance in determining the combined spin, for if they are combined with a $+$ sign, one obtains an angular momentum of $\hbar$ (but with a zero value of the z component of the angular momentum). We denote this result below:

$$\psi_1 = \frac{1}{\sqrt{2}}(\psi_c + \psi_d) \tag{27}$$

It is clear, then, that the total angular momentum is an interference property of ψ_c and ψ_d. On the other hand, the only states in which each particle has a definite spin opposite to that of the other are represented either by ψ_c or by ψ_d separately. Thus, in any state in which the value of σ_z for each particle is definite, the total angular momentum must be indefinite. Vice versa, whenever the total angular momentum is definite, then neither atom can correctly be regarded as having a definite value of its own spin, for if it did, there could be no interference between ψ_c and ψ_d, and it is just this interference which is required to produce a definite total angular momentum.

Besides leading to a definite value of the combined spin, however, definite phase relations between ψ_c and ψ_d have additional physical meaning, for they also imply that if the same component of the spin of each atom is measured, the results will be correlated. Such correlations can be demonstrated, for example, in a process in which the z component of the spin of each atom is measured by allowing each atom to pass through a separate Stern-Gerlach apparatus (see Fig. 1). For the sake of simplicity, we can suppose that both spins are measured at the same time, although no results will depend significantly on this assumption. The Hamiltonian at the time of measurement is then [see eqs. (10a) and (10b)]:

$$W = \mu(\mathcal{H}_0 + z\mathcal{H}_0')\sigma_{1,z} + \mu(\mathcal{H}_0 + z_2\mathcal{H}_0')\sigma_{2,z}$$

where z_1 is the z co-ordinate of the first atom and z_2 is the z co-ordinate of the second atom. (We assume that both pieces of apparatus are identical in construction.)

We now expand the spin wave function during the course of the measurement in terms of the four basic functions, ψ_a, ψ_b, ψ_c, and ψ_d. Since this measurement does not change σ_z, it will remain true that only ψ_c and ψ_d are needed during the course of the measurement.* Thus, we write

$$\psi = f_c\psi_c + f_d\psi_d$$

In our case, the initial value of f_c is $1/\sqrt{2}$, and the initial value of f_d is $-1/\sqrt{2}$. By methods similar to those leading to equation (13b), one derives

$$i\hbar \frac{\partial f_c}{\partial t} = \mu f_c[(\mathcal{H}_0 + \mathcal{H}_0' z_1) - (\mathcal{H}_0 + \mathcal{H}_0' z_2)]$$

$$i\hbar \frac{\partial f_d}{\partial t} = -\mu f_d[(\mathcal{H}_0 + \mathcal{H}_0' z_1) - (\mathcal{H}_0 + \mathcal{H}_0' z_2)]$$

The solution for f_c and f_d with the proper boundary conditions yields for the wave function just after the particles leave the magnetic field

$$f_c = \frac{1}{\sqrt{2}} e^{-i\frac{\mu\mathcal{H}_0'}{\hbar}(z_1 - z_2)\,\Delta t} \qquad f_d = -\frac{1}{\sqrt{2}} e^{i\frac{\mu\mathcal{H}_0'}{\hbar}(z_1 - z_2)\,\Delta t}$$

where we have inserted $t = \Delta t$ = time of interaction between atoms and the inhomogeneous magnetic field.

This wave function implies that the two results represented, respectively, by ψ_c and by ψ_d are equally probable. In the first possible result, atom No. 1 has a positive value of σ_z, while atom number 2 has a negative value. The factor $e^{-i\mu\mathcal{H}_0'\Delta t(z_1 - z_2)/\hbar}$ represents the fact that in the Stern-Gerlach experiment, each atom obtains an opposite momentum corresponding to its opposite spin. Similarly, in the second possible result, atom No. 2 has a negative value of σ_z, whereas atom No. 1 has a positive value. As in Secs. 9 and 11, we can show that because the apparatus is classically describable, the apparatus wave functions (which depend on z_1 and z_2), multiply the spin wave function by uncontrollable phase factors, so that we finally obtain

$$\psi = \frac{1}{\sqrt{2}} (\psi_c\, e^{i\alpha_c} + \psi_d\, e^{i\alpha_d})$$

where α_c and α_d are separate and uncontrollable phase factors.

This result shows that if the value of σ_z is measured for each atom, the result will come out a definite number for each, which is always

* Note that these are the only terms present initially.

opposite to that of the other. In this way, we prove that correlations resembling those of classical theory will also be obtained in the quantum theory. After the measurement is over, however, the system has been transformed from one that had a definite combined angular momentum and an indefinite value of σ_z for each particle to one which has a definite value of σ_z for each particle, but an indefinite combined angular momentum. Moreover, the precise value of σ_z which will be obtained for each particle is not related deterministically to the state of the system before the measurement, but only statistically.

Let us now describe the process of measurement of σ_x. The results are very similar, because the wave function for a system of zero total spin is the same when expressed in terms of v_+, v_- (the eigenfunctions of σ_x) as in terms of u_+, u_-. Thus, we obtain

$$\psi_0 = \frac{1}{\sqrt{2}}\,[v_+(1)v_-(2) - v_-(1)v_+(2)]$$

One can now describe the measurement of σ_x for each particle in exactly the same way as was done with σ_z, and after the interaction with the measuring apparatus, one obtains

$$\psi = \frac{1}{\sqrt{2}}\,[v_+(1)v_-(2)\,e^{i\alpha_1} + v_-(1)v_+(2)\,e^{i\alpha_2}]$$

where α_1 and α_2 are separate uncontrollable phase factors.

We conclude that the value of σ_x for each particle is also correlated to that of the other in such a way that the sum of the two is zero. Moreover, it is readily verified that if one had taken the function

$$\psi_1 = \frac{1}{\sqrt{2}}\,(\psi_a + \psi_b)$$

then with the substitution, $v_+ = \frac{1}{\sqrt{2}}(u_+ + u_-)$ and $v_- = \frac{1}{\sqrt{2}}(u_+ - u_-)$, one would have the wave function

$$\psi_1 = \frac{1}{\sqrt{2}}\,[v_+(1)v_+(2) + v_-(1)v_-(2)]$$

This represents a situation in which measurement of σ_x will disclose that both particles have a positive value together, or that both particles have a negative value together. We see therefore that the type of correlation of σ_x which can develop depends on the sign with which ψ_c and ψ_d are added, and therefore also on the combined angular momentum.

One more significant point arises in connection with this experiment; namely, that the existence of correlations does not imply that the behavior of either atom is affected in any way at all by what happens to the other, after the two have ceased to interact. To prove this statement, we first evaluate the mean value of any function $g(\boldsymbol{\sigma}_2)$ of the spin variables of particle No. 2 alone. With the wave function before a measurement took place, we obtain

$$\bar{g}_0(\boldsymbol{\sigma}_2) = \tfrac{1}{2}(\psi_c^* - \psi_d^*)g(\boldsymbol{\sigma}_2)(\psi_c - \psi_d) = \tfrac{1}{2}[\psi_c^* g(\boldsymbol{\sigma}_2)\psi_c + \psi_d^* g(\boldsymbol{\sigma}_2)\psi_d]$$

(By virtue of the orthogonality of ψ_c and $g(\boldsymbol{\sigma}_2)\psi_d$.) After the spin of the first particle is measured, the average of $g(\boldsymbol{\sigma}_2)$ becomes

$$\bar{g}_f(\boldsymbol{\sigma}_2) = \tfrac{1}{2}(\psi_c^* e^{-i\alpha_c} - \psi_d^* e^{-i\alpha_d})g(\boldsymbol{\sigma}_2)(\psi_c e^{i\alpha_c} - \psi_d e^{i\alpha_d})$$
$$= \tfrac{1}{2}[\psi_c^* g(\boldsymbol{\sigma}_2)\psi_c + \psi_d^* g(\boldsymbol{\sigma}_2)\psi_d]$$

This is the same as what was obtained without a measurement of the spin variables of particle No. 1. The behavior of the two spins is, however, correlated despite the fact that each behaves in a way that does not depend on what actually happens to the other after interaction has ceased.

18. Physical Description of Origin of Correlations. We have deduced mathematically that in a system of two atoms having a total spin of zero, the spin components of each atom in an arbitrary direction will be correlated, despite the fact that according to our present interpretation of quantum theory these spin components cannot all exist simultaneously in precisely defined forms. We wish to show now that the paradoxical results obtained by ERP in the interpretation of this fact will not be obtained if one avoids making their implicit assumptions (3) and (4); viz., that the world can correctly be analyzed into elements of reality, each of which is a counterpart of a precisely defined mathematical quantity appearing in a complete theory. These assumptions, which are at the root of all classical theory, might perhaps be called the hypothesis that reality is built upon a mathematical plan, for it is required that every element appearing in the real world shall correspond *precisely* to some term appearing in a complete set of mathematical equations. Although such a hypothesis seems quite natural to us at this time, it is by no means inescapable.† In fact, in quantum theory, one makes a quite different, but equally plausible, hypothesis concerning the fundamental nature of matter. Here, we assume that the one-to-one correspondence between mathematical theory and well-defined "elements of reality" exists only at the classical level of accuracy. For at the quantum level, the mathematical description provided by the wave function is certainly not in a one-to-one correspondence with the actual behavior of

† Historically speaking, it is a comparatively new idea, having arisen in connection with the great success of mathematical analysis in mechanics and electrodynamics during the period between the sixteenth and early twentieth centuries (see Chap. 8, Secs. 2 to 10).

the system under description, but only in a statistical correspondence.* Yet, we assert that the wave function (in principle) can provide the most complete possible description of the system that is consistent with the actual structure of matter. How can we reconcile these two aspects of the wave function? We do so in terms of the assumption that the properties of a given system exist, in general, only in an imprecisely defined form, and that on a more accurate level, they are not really well-defined properties at all, but instead only potentialities,† which are more definitely realized in interaction with an appropriate classical system, such as a measuring apparatus. For example, consider two noncommuting observables, such as momentum and position of an electron. We say that, in general, neither exists in a given system in a *precisely* defined form, but that both exist together in a roughly defined form, such that the uncertainty principle is not violated.‡ Either variable is potentially capable of becoming better defined at the expense of the degree of definition of the other, in interaction with a suitable measuring apparatus. We see then that the properties of position and momentum are not only incompletely defined and opposing potentialities, but also that in a very accurate description, they cannot be regarded as belonging to the electron alone; for the realization of these potentialities depends just as much on the systems with which it interacts as on the electron itself.§ This means that there are actually no precisely defined "elements of reality" belonging to the electron. Thus, we contradict the assumptions (3) and (4) of Einstein, Rosen, and Podolsky.

Quantum-mechanical spin variables must be interpreted in a similar way. Whereas ERP would say that the only existing component of the spin is the one which may happen to be defined precisely by the wave function, we say that, in general, all three components exist simultaneously in roughly defined forms, and that any one component has the potentiality for becoming better defined at the expense of the others if the associated atom interacts with a suitable measuring apparatus. The probability for the development of a definite value of any spin component in a suitable process of measurement is proportional to the square of the amplitude of the coefficient of the part of the wave function corresponding to this component. We must, however, recall that the complete spin wave function for a given atom can be expanded in terms of the eigenfunctions, u_+ and u_-, of the spin variables in *any* direction. Thus $\psi = a_+u_+ + a_-u_-$. In such an expansion, the phase relations between u_+ and u_- help determine the distribution over spin components in other directions.|| (Thus, if u_+, u_- represent eigenfunctions of σ_z, then

* See Chap. 6, Sec. 4.

† See Chap. 6, Secs. 9 and 13, Chap. 8, Secs. 14 and 15, Chap. 22, Sec. 13.

‡ See discussion of complementarity in Chap. 8, Sec. 15.

§ Chap. 6, Sec. 13, Chap. 8, Sec. 16.

|| See Chap. 17, Secs. 6 and 7.

an eigenfunction of σ_x is obtained when $\psi = \frac{1}{\sqrt{2}}(u_+ \pm u_-)$.) This means that as long as definite phase relations exist between u_+ and u_-, one cannot categorize (or classify) the system as having a spin which corresponds either *entirely* to u_+ or *entirely* to u_-, with respective probabilities,* $|a_+|^2$ and $|a_-|^2$. Instead, we must say that the system cuts across this method of classification, and in some sense, covers both states at once in a poorly defined way.† Thus, we must give up the classical picture of a precisely defined spin variable associated with each atom, and replace it by our quantum concept of a potentiality, the probability of whose development is given by the wave function. It is only when the wave function is an eigenfunction of a given spin component that the system is certain (in interaction with a suitable apparatus) to develop a predictable value of that spin component.

Now, when we come to our system of two atoms having a total spin of zero, we see from eq. (26) that because the wave function

$$\psi_0 = \frac{1}{\sqrt{2}}(\psi_c - \psi_d)$$

has definite phase relations between ψ_c and ψ_d, the system must cover the states corresponding to ψ_c and ψ_d simultaneously. Thus, for a given atom, *no* component of the spin of a given variable exists with a precisely defined value, until interaction with a suitable system, such as a measuring apparatus, has taken place. But as soon as either atom (say, No. 1) interacts with an apparatus measuring a given component of the spin, definite phase relations between ψ_c and ψ_d are destroyed. This means that the system then acts as if it is either in the state ψ_c or ψ_d. Thus, in every instance in which particle No. 1 develops a definite spin component in, for example, the z direction, the wave function of particle No. 2 will automatically take such a form that it guarantees the development of the opposite value of σ_z if this particle also interacts with an apparatus which measures the same component of the spin. The wave function therefore describes the propagation of correlated potentialities. Because the expansion of the wave function ψ_0 takes the same form when expanded in terms of the eigenfunctions of an arbitrary component of the spin, we conclude that similar correlations will be obtained if the same component of the spin of each atom in any direction is measured. Moreover, because the potentialities for development of a definite spin component are not realized irrevocably until interaction with the apparatus actually takes place, there is no inconsistency in the statement that while the atoms are still in flight, one can rotate the apparatus into an arbitrary

* Chap. 6, Sec. 4, Chap. 22, Sec. 10.
† Chap. 16, Sec. 25, and Chap. 8, Sec. 15.

direction, and thus choose to develop definite and correlated values for any desired spin component of each atom.

Finally, it is perhaps interesting to consider in a new light the fact that the mathematical description provided by the wave function is not in a one-to-one correspondence with the actual behavior of matter. From this fact, we are led to conclude that, contrary to general opinion, quantum theory is less mathematical in its philosophical basis than is classical theory, for, as we have seen, it does not assume that the world is constructed according to a precisely defined mathematical plan. Instead, we have come to the point of view that the wave function is an abstraction, providing a mathematical reflection of certain aspects of reality, but not a one-to-one mapping. To obtain a description of *all* aspects of the world, one must, in fact, supplement the mathematical description with a physical interpretation in terms of incompletely defined potentialities.* Moreover, the present form of quantum theory implies that the world cannot be put into a one-to-one correspondence with any conceivable kind of precisely defined mathematical quantities, and that a complete theory will always require concepts that are more general than that of analysis into precisely defined elements. We may probably expect that even the more general types of concepts provided by the present quantum theory will also ultimately be found to provide only a partial reflection of the infinitely complex and subtle structure of the world. As science develops, we may therefore look forward to the appearance of still newer concepts, which are only faintly foreshadowed at present, but there is no strong reason to suppose that these new concepts are likely to lead to a return to the comparatively simple idea of a one-to-one correspondence between the real world and precisely defined mathematical abstractions.

19. Proof that Quantum Theory Is Inconsistent with Hidden Variables. We can now use some of the results of the analysis of the paradox of Einstein, Rosen, and Podolsky to help prove that quantum theory is inconsistent with the assumption of hidden causal variables. (See Chap. 2, Sec. 5 and Chap. 5, Sec. 3.) We first note that the assumption that there are separately existing and precisely defined elements of reality would be at the base of any precise causal description in terms of hidden variables; for without such elements there would be nothing to which a precise causal description could apply. Similarly, as we saw in Chap. 8, Sec. 20, the existence of separate elements requires a precise causal theory of the relationships between these elements for its consistent application. Thus, the analysis of the world into pre-

* See Chap. 23 for a fuller discussion of how the wave function must be supplemented with its interpretation in terms of potentialities for the production of various classically describable results.

cisely defined elements and the synthesis of these elements according to precise causal laws must stand or fall together.

Now, from the reasoning of ERP we conclude that if the world can be explained in terms of such precisely defined elements, then the correct interpretation of two noncommuting variables, such as momentum and position, would be that they correspond to simultaneously existing elements of reality. To interpret the uncertainty principle, we would then have to assume that we are simply unable to measure the values of the two simultaneously with complete precision. But we saw in Chap. 6, Sec. 11, that any such assumption would lead to a contradiction with the uncertainty principle, which is one of the most fundamental deductions of the quantum theory. We conclude then that no theory of mechanically determined hidden variables can lead to *all* of the results of the quantum theory. Such a mechanical theory might conceivably be so ingeniously framed that it would agree with quantum theory for a wide range of predicted experimental results.* But the hypothetical experiment suggested in Chap. 6, Sec. 11 would then be an example of a crucial test of the theory. If, in this experiment, we were able to violate the uncertainty principle, then the theory of mechanically determined underlying variables would be strongly indicated, whereas if we were not able to violate the uncertainty principle, we should obtain a fairly convincing proof that no correct mechanical theory could ever be found. Unfortunately, such an experiment is still far beyond present techniques, but it is quite possible that it could some day be carried out. Until and unless some such disagreement between quantum theory and experiment is found, however, it seems wisest to assume that quantum theory is substantially correct, because it is a self-consistent theory yielding agreement with such a wide range of experiments not correctly treated by any other known theory.

* We do not wish to imply here that anyone has ever produced a concrete and successful example of such a theory, but only state that such a theory is, as far as we know, conceivable.

CHAPTER 23

Relationship between Quantum and Classical Concepts

THROUGHOUT THIS BOOK, we have tried to develop a qualitative description of the properties of matter implied by the quantum theory. In doing this, we were led to the conclusion that quantum concepts concerning the nature of matter are radically different from those associated with the previously existing classical theory. Nevertheless, despite this extreme difference, it was possible with the aid of the correspondence principle* to construct the quantum theory in such a way that it approached the classical theory in the classical limit. At first sight, one would then be temped to conclude that classical theory is merely a limiting form of the quantum theory, or in other words, that classical theory is logically a special case of the quantum theory. In this chapter, we wish to investigate the relationship between classical and quantum concepts more thoroughly, in order to show that quantum theory in its present form actually presupposes the correctness of classical concepts. We shall then be led to the conclusion that classical concepts cannot be regarded as limiting forms of quantum concepts, but must instead be combined with quantum concepts in such a way that, in a complete description, each complements the other.

We begin with a brief summary contrasting classical and quantum concepts. Classical concepts are characterized by three assumptions concerning the properties of matter:

(1) The world can be analyzed into distinct elements.

(2) The state of each element can be described in terms of dynamical variables that are specifiable with arbitrarily high precision.

(3) The interrelationship between parts of a system can be described with the aid of exact causal laws that define the changes of the above dynamical variables with time in terms of their initial values. The behavior of the system as a whole can be regarded as the result of the interaction of all of its parts.

It is characteristic of the classical domain that within it exist objects, phenomena, and events that are distinct and well-defined and that exhibit reliable and reproducible properties with the aid of which they can be identified and compared (see, for example, Chap. 8, Secs. 17 to 22). It is this aspect of the world that is most readily described in

* Chap. 2, Sec. 5, Chap. 3, Sec. 8, Chap. 9, Sec. 24.

terms of our customary scientific language, in which the ideal is to express every concept in terms of well-defined elements with well-defined logical relationships between them.

When we come to describe quantum concepts, however, we find that just because our customary scientific language aims for such precision, it leads to difficult and unwieldy modes of expression. For as we have seen,* the quantum properties of matter are to be associated with incompletely defined potentialities, which can be more definitely realized only in interaction with a classically describable system (a special case of which is a measuring apparatus.)† Because even the so-called "intrinsic" properties of a system (e.g., wave or particle) are brought out only in interactions with other systems, it is clear that the quantum properties of matter imply the indivisible unity of all interacting systems. Thus, we have contradicted assumptions (1) and (2) of the classical theory, since there exist at the quantum level neither well-defined elements nor well-defined dynamical variables, which describe the behavior of these elements. It is not surprising, then, that assumption (3) is also not satisfied in the quantum theory, since exact causal laws would be meaningless in a context in which there were no precisely defined variables to which they could apply. In fact, instead of having well-defined variables that are in a one-to-one correspondence with the actual behavior of matter, we have at the quantum level a wave function that is only in statistical correspondence with this behavior.‡

It is in connection with the interpretation of the wave function that classical and quantum theories meet. For the physical interpretation of the wave function is always in terms of the probability that when a system interacts with a suitable measuring apparatus, it will develop a definite value of the variable that is being measured. But, as we have seen, the last stages of a measuring apparatus are always classically describable.§ In fact, it is only at the classical level that definite results for an experiment can be obtained, in the form of distinct events which are associated in a one-to-one correspondence with the various possible values of the physical quantity that is being measured.|| This means that without an appeal to a classical level, quantum theory would have no meaning. *We conclude then that quantum theory presupposes the classical level and the general correctness of classical concepts in describing this level; it does not deduce classical concepts as limiting cases of quantum concepts.*¶

* Chap. 6, Secs. 9 and 13; Chap. 8, Secs. 14 and 15.
† Chap. 22.
‡ See Chap. 6, Sec. 4.
§ See Chap. 22, Sec. 3.
|| See Chap. 22, Secs. 3, 4, 11, and 13.
¶ As, for example, one deduces Newtonian mechanics as a limiting case of special relativity.

At first sight, one might object to the above conclusion by suggesting that one could eliminate the need for presupposing a classical level with the aid of the usual procedure of approaching the classical limit with the aid of the WKB approximation.* To show that this objection is not valid, let us recall that even when a wave packet is defined to only a classical order of accuracy, it will eventually spread over tremendous distances.† Yet the object in question (for example, an electron) can always be found within an arbitrarily small region of space when its position is measured. We conclude that a description at the quantum level (i.e., in terms of the wave function alone) does not, in general, adequately represent the definiteness of physical properties that the electron is capable of manifesting when it interacts with suitable measuring devices. In order to obtain a means of interpreting the wave function, we must therefore *at the outset* postulate a classical level in terms of which the definite results of a measurement can be realized. Thus, the correspondence principle is simply a consistency condition which requires that when the quantum theory plus its classical interpretation is carried to the limit of high quantum numbers, the simple classical theory will be obtained.

The necessity for presupposing a classical level and the appropriate classical concepts implies that the large scale behavior of a system is not completely expressible in terms of concepts that are appropriate at the small scale level. Thus, as we have seen, the concepts appropriate at the quantum level are those of incompletely defined potentialities. As we go from small scale to large scale level, new (classical) properties then appear which cannot be deduced from the quantum description in terms of the wave function alone, but which must nevertheless be consistent with this quantum description. These new properties manifest themselves, as we have seen, in the appearance of definite objects and events,‡ which cannot exist at the quantum level.

Large-scale and small-scale properties are not independent, but are actually in the closest inter-relationship. For, as we have seen, it is only in terms of well-defined classical events that quantum-mechanical potentialities can be realized. Moreover, this interdependence is reciprocal, for it is only in terms of a quantum theory of its component molecules that the large-scale behavior of a system can be fully understood.

* Chap. 12.

† Chap. 3, Sec. 5.

‡ In this connection, we point out that to the extent that a system is classically describable, its properties must be regarded as definite, whether they are known in detail by any observers or not. Thus, in a box containing gas molecules, each molecule is assumed classically to occupy a definite position at each instant of time, even though it is in practice impossible for observers to measure the positions of every molecule. It is only when quantum phenomena are important (for example, in a degenerate gas) that this assumption ceases to be valid.

Thus, large-scale and small-scale properties are both needed to describe complementary aspects of a more fundamental indivisible unit, namely, the system as a whole.

In order to express in more detail the actual relationships of large-scale and small-scale properties of matter, we can describe these two kinds of properties in terms of the interplay of two opposing trends. From the quantum level, one obtains a continual tendency for a system to cover the whole range of its potentialities; i.e., to escape the bounds of any system of categories that would, according to classical lines of reasoning, limit its behavior in any specific way.* On the other hand, at the classical level, one obtains, as we have seen, a continual tendency for things to become definite, i.e., for a specific potentiality to be realized irrevocably at the expense of all other potentialities. For example, in a process of measurement, the system settles down to a particular value of the measured variable and all other possibilities are discarded. (In this connection, see the discussion of "collapse" of the wave function in Chap. 6, Sec. 4, and Chap. 22, Sec. 10.) The appearance of a definite result of a measurement at the classical level is reflected back into the microscopic level in two ways: First, the system takes on a range of values of the measured property corresponding to that range that is consistent with the range of indeterminacy in the measurement. Thus, a narrowing down of potentialities at the classical level is accompanied by a similar narrowing of potentialities at the quantum level. But, in the very same process in which a quantum system obtains a more definite value of the measured variable, it suffers a corresponding decrease in definiteness of the complementary variable† (or variables). Thus, associated with the narrowing down of a given range of potentialities is always a compensating process of widening the range of new kinds of potentialities. The appearance of new potentialities will be reflected in further changes at the classical level, etc. This means that, in the continual interplay between the quantum potentialities and their classical realizations, the system is subject to an endless series of transformations.

To sum up, we state that quantum theory has actually evolved in such a way that it implies the need for a new concept or the relation between large scale and small scale properties of a given system. In this chapter, we have discussed two aspects of this new concept:

1. Quantum theory presupposes a classical level and the correctness of classical concepts in describing this level.

2. The classically definite aspects of large-scale systems cannot be deduced from the quantum-mechanical relationships of assumed small-scale elements. Instead, classical definiteness and quantum potentiali-

* Chap. 8, Sec. 15; Chap. 16, Sec. 25.

† See, for example, the discussion of the uncertainty principle in Chap. 5; also Chap. 6, Sec. 7.

ties complement each other in providing a complete description of the system as a whole.

Although these ideas are only implicit in the present form of the quantum theory, we wish to suggest here in a speculative way that the successful extension of quantum theory to the domain of nuclear dimensions may perhaps introduce more explicitly the idea that the nature of what can exist at the nuclear level depends to some extent on the macroscopic environment.*

* In this connection, see Chap. 22, Sec. 12, where it was shown that the definition of small scale properties of a system is possible only as a result of interaction with large scale systems undergoing irreversible processes. In line with the above suggestion, we propose also that irreversible processes taking place in the large scale environment may also have to appear explicitly in the fundamental equations describing phenomena at the nuclear level.

INDEX

A CATALOG OF SELECTED

DOVER BOOKS

IN SCIENCE AND MATHEMATICS

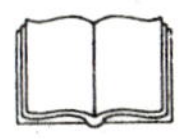

A CATALOG OF SELECTED

DOVER BOOKS

IN SCIENCE AND MATHEMATICS

Astronomy

BURNHAM'S CELESTIAL HANDBOOK, Robert Burnham, Jr. Thorough guide to the stars beyond our solar system. Exhaustive treatment. Alphabetical by constellation: Andromeda to Cetus in Vol. 1; Chamaeleon to Orion in Vol. 2; and Pavo to Vulpecula in Vol. 3. Hundreds of illustrations. Index in Vol. 3. 2,000pp. 6⅛ x 9¼.
23567-X, 23568-8, 23673-0 Three-vol. set

THE EXTRATERRESTRIAL LIFE DEBATE, 1750–1900, Michael J. Crowe. First detailed, scholarly study in English of the many ideas that developed from 1750 to 1900 regarding the existence of intelligent extraterrestrial life. Examines ideas of Kant, Herschel, Voltaire, Percival Lowell, many other scientists and thinkers. 16 illustrations. 704pp. 5⅜ x 8½. 40675-X

A HISTORY OF ASTRONOMY, A. Pannekoek. Well-balanced, carefully reasoned study covers such topics as Ptolemaic theory, work of Copernicus, Kepler, Newton, Eddington's work on stars, much more. Illustrated. References. 521pp. 5⅜ x 8½.
65994-1

AMATEUR ASTRONOMER'S HANDBOOK, J. B. Sidgwick. Timeless, comprehensive coverage of telescopes, mirrors, lenses, mountings, telescope drives, micrometers, spectroscopes, more. 189 illustrations. 576pp. 5⅝ x 8¼. (Available in U.S. only.)
24034-7

STARS AND RELATIVITY, Ya. B. Zel'dovich and I. D. Novikov. Vol. 1 of *Relativistic Astrophysics* by famed Russian scientists. General relativity, properties of matter under astrophysical conditions, stars, and stellar systems. Deep physical insights, clear presentation. 1971 edition. References. 544pp. 5⅝ x 8¼. 69424-0

Chemistry

CHEMICAL MAGIC, Leonard A. Ford. Second Edition, Revised by E. Winston Grundmeier. Over 100 unusual stunts demonstrating cold fire, dust explosions, much more. Text explains scientific principles and stresses safety precautions. 128pp. 5⅜ x 8½. 67628-5

THE DEVELOPMENT OF MODERN CHEMISTRY, Aaron J. Ihde. Authoritative history of chemistry from ancient Greek theory to 20th-century innovation. Covers major chemists and their discoveries. 209 illustrations. 14 tables. Bibliographies. Indices. Appendices. 851pp. 5⅜ x 8½. 64235-6

CATALYSIS IN CHEMISTRY AND ENZYMOLOGY, William P. Jencks. Exceptionally clear coverage of mechanisms for catalysis, forces in aqueous solution, carbonyl- and acyl-group reactions, practical kinetics, more. 864pp. 5⅜ x 8½.
65460-5

Math–Geometry and Topology

ELEMENTARY CONCEPTS OF TOPOLOGY, Paul Alexandroff. Elegant, intuitive approach to topology from set-theoretic topology to Betti groups; how concepts of topology are useful in math and physics. 25 figures. 57pp. 5⅜ x 8½. 60747-X

COMBINATORIAL TOPOLOGY, P. S. Alexandrov. Clearly written, well-organized, three-part text begins by dealing with certain classic problems without using the formal techniques of homology theory and advances to the central concept, the Betti groups. Numerous detailed examples. 654pp. 5⅜ x 8½. 40179-0

EXPERIMENTS IN TOPOLOGY, Stephen Barr. Classic, lively explanation of one of the byways of mathematics. Klein bottles, Moebius strips, projective planes, map coloring, problem of the Koenigsberg bridges, much more, described with clarity and wit. 43 figures. 210pp. 5⅜ x 8½. 25933-1

CONFORMAL MAPPING ON RIEMANN SURFACES, Harvey Cohn. Lucid, insightful book presents ideal coverage of subject. 334 exercises make book perfect for self-study. 55 figures. 352pp. 5⅝ x 8¼. 64025-6

THE GEOMETRY OF RENÉ DESCARTES, René Descartes. The great work founded analytical geometry. Original French text, Descartes's own diagrams, together with definitive Smith-Latham translation. 244pp. 5⅜ x 8½. 60068-8

THE THIRTEEN BOOKS OF EUCLID'S ELEMENTS, translated with introduction and commentary by Sir Thomas L. Heath. Definitive edition. Textual and linguistic notes, mathematical analysis. 2,500 years of critical commentary. Unabridged. 1,414pp. 5⅜ x 8½. Three-vol. set.
Vol. I: 60088-2 Vol. II: 60089-0 Vol. III: 60090-4

GEOMETRY OF COMPLEX NUMBERS, Hans Schwerdtfeger. Illuminating, widely praised book on analytic geometry of circles, the Moebius transformation, and two-dimensional non-Euclidean geometries. 200pp. 5⅝ x 8¼. 63830-8

DIFFERENTIAL GEOMETRY, Heinrich W. Guggenheimer. Local differential geometry as an application of advanced calculus and linear algebra. Curvature, transformation groups, surfaces, more. Exercises. 62 figures. 378pp. 5⅜ x 8½. 63433-7

CURVATURE AND HOMOLOGY: Enlarged Edition, Samuel I. Goldberg. Revised edition examines topology of differentiable manifolds; curvature, homology of Riemannian manifolds; compact Lie groups; complex manifolds; curvature, homology of Kaehler manifolds. New Preface. Four new appendixes. 416pp. 5⅜ x 8½. 40207-X

TOPOLOGY, John G. Hocking and Gail S. Young. Superb one-year course in classical topology. Topological spaces and functions, point-set topology, much more. Examples and problems. Bibliography. Index. 384pp. 5⅝ x 8¼. 65676-4

Physics

OPTICAL RESONANCE AND TWO-LEVEL ATOMS, L. Allen and J. H. Eberly. Clear, comprehensive introduction to basic principles behind all quantum optical resonance phenomena. 53 illustrations. Preface. Index. 256pp. 5⅜ x 8½. 65533-4

ULTRASONIC ABSORPTION: An Introduction to the Theory of Sound Absorption and Dispersion in Gases, Liquids and Solids, A. B. Bhatia. Standard reference in the field provides a clear, systematically organized introductory review of fundamental concepts for advanced graduate students, research workers. Numerous diagrams. Bibliography. 440pp. 5⅜ x 8½. 64917-2

QUANTUM THEORY, David Bohm. This advanced undergraduate-level text presents the quantum theory in terms of qualitative and imaginative concepts, followed by specific applications worked out in mathematical detail. Preface. Index. 655pp. 5⅜ x 8½. 65969-0

ATOMIC PHYSICS (8th edition), Max Born. Nobel laureate's lucid treatment of kinetic theory of gases, elementary particles, nuclear atom, wave-corpuscles, atomic structure and spectral lines, much more. Over 40 appendices, bibliography. 495pp. 5⅜ x 8½. 65984-4

AN INTRODUCTION TO HAMILTONIAN OPTICS, H. A. Buchdahl. Detailed account of the Hamiltonian treatment of aberration theory in geometrical optics. Many classes of optical systems defined in terms of the symmetries they possess. Problems with detailed solutions. 1970 edition. xv + 360pp. 5⅜ x 8½. 67597-1

THIRTY YEARS THAT SHOOK PHYSICS: The Story of Quantum Theory, George Gamow. Lucid, accessible introduction to influential theory of energy and matter. Careful explanations of Dirac's anti-particles, Bohr's model of the atom, much more. 12 plates. Numerous drawings. 240pp. 5⅜ x 8½. 24895-X

ELECTRONIC STRUCTURE AND THE PROPERTIES OF SOLIDS: The Physics of the Chemical Bond, Walter A. Harrison. Innovative text offers basic understanding of the electronic structure of covalent and ionic solids, simple metals, transition metals and their compounds. Problems. 1980 edition. 582pp. 6⅛ x 9¼. 66021-4

HYDRODYNAMIC AND HYDROMAGNETIC STABILITY, S. Chandrasekhar. Lucid examination of the Rayleigh-Benard problem; clear coverage of the theory of instabilities causing convection. 704pp. 5⅜ x 8¼. 64071-X

INVESTIGATIONS ON THE THEORY OF THE BROWNIAN MOVEMENT, Albert Einstein. Five papers (1905–8) investigating dynamics of Brownian motion and evolving elementary theory. Notes by R. Fürth. 122pp. 5⅜ x 8½. 60304-0

THE PHYSICS OF WAVES, William C. Elmore and Mark A. Heald. Unique overview of classical wave theory. Acoustics, optics, electromagnetic radiation, more. Ideal as classroom text or for self-study. Problems. 477pp. 5⅜ x 8½. 64926-1

PHYSICAL PRINCIPLES OF THE QUANTUM THEORY, Werner Heisenberg. Nobel Laureate discusses quantum theory, uncertainty, wave mechanics, work of Dirac, Schroedinger, Compton, Wilson, Einstein, etc. 184pp. 5⅜ x 8½. 60113-7

ATOMIC SPECTRA AND ATOMIC STRUCTURE, Gerhard Herzberg. One of best introductions; especially for specialist in other fields. Treatment is physical rather than mathematical. 80 illustrations. 257pp. 5⅜ x 8½. 60115-3

AN INTRODUCTION TO STATISTICAL THERMODYNAMICS, Terrell L. Hill. Excellent basic text offers wide-ranging coverage of quantum statistical mechanics, systems of interacting molecules, quantum statistics, more. 523pp. 5⅜ x 8½. 65242-4

THEORETICAL PHYSICS, Georg Joos, with Ira M. Freeman. Classic overview covers essential math, mechanics, electromagnetic theory, thermodynamics, quantum mechanics, nuclear physics, other topics. First paperback edition. xxiii + 885pp. 5⅜ x 8½. 65227-0

PROBLEMS AND SOLUTIONS IN QUANTUM CHEMISTRY AND PHYSICS, Charles S. Johnson, Jr. and Lee G. Pedersen. Unusually varied problems, detailed solutions in coverage of quantum mechanics, wave mechanics, angular momentum, molecular spectroscopy, more. 280 problems plus 139 supplementary exercises. 430pp. 6½ x 9¼. 65236-X

THEORETICAL SOLID STATE PHYSICS, Vol. 1: Perfect Lattices in Equilibrium; Vol. II: Non-Equilibrium and Disorder, William Jones and Norman H. March. Monumental reference work covers fundamental theory of equilibrium properties of perfect crystalline solids, non-equilibrium properties, defects and disordered systems. Appendices. Problems. Preface. Diagrams. Index. Bibliography. Total of 1,301pp. 5⅜ x 8½. Two volumes. Vol. I: 65015-4 Vol. II: 65016-2

A TREATISE ON ELECTRICITY AND MAGNETISM, James Clerk Maxwell. Important foundation work of modern physics. Brings to final form Maxwell's theory of electromagnetism and rigorously derives his general equations of field theory. 1,084pp. 5⅜ x 8½. Two-vol. set. Vol. I: 60636-8 Vol. II: 60637-6

OPTICKS, Sir Isaac Newton. Newton's own experiments with spectroscopy, colors, lenses, reflection, refraction, etc., in language the layman can follow. Foreword by Albert Einstein. 532pp. 5⅜ x 8½. 60205-2

THEORY OF ELECTROMAGNETIC WAVE PROPAGATION, Charles Herach Papas. Graduate-level study discusses the Maxwell field equations, radiation from wire antennas, the Doppler effect and more. xiii + 244pp. 5⅜ x 8½. 65678-5

INTRODUCTION TO QUANTUM MECHANICS With Applications to Chemistry, Linus Pauling & E. Bright Wilson, Jr. Classic undergraduate text by Nobel Prize winner applies quantum mechanics to chemical and physical problems. Numerous tables and figures enhance the text. Chapter bibliographies. Appendices. Index. 468pp. 5⅜ x 8½. 64871-0